SPECTRAL ELEMENT METHOD IN STRUCTURAL DYNAMICS

SPECTRAL ELEMENT METHOD IN STRUCTURAL DYNAMICS

Usik Lee

Inha University, Republic of Korea

John Wiley & Sons (Asia) Pte Ltd

Visit our Home Page on www.wiley.com

Other Wiley Editorial Offices

John Wiley & Sons, Ltd, The Atrium, Southern Gate, Chichester, West Sussex, PO19 8SQ, UK

John Wiley & Sons Inc., 111 River Street, Hoboken, NJ 07030, USA

Jossey-Bass, 989 Market Street, San Francisco, CA 94103-1741, USA

Wiley-VCH Verlag GmbH, Boschstrasse 12, D-69469 Weinheim, Germany

John Wiley & Sons Australia Ltd, 42 McDougall Street, Milton, Queensland 4064, Australia

John Wiley & Sons Canada Ltd, 5353 Dundas Street West, Suite 400, Toronto, ONT, M9B 6H8, Canada

Wiley also publishes its books in a variety of electronic formats. Some content that appears in print may not be available in electronic books.

Library of Congress Cataloging-in-Publication Data

Lee, Usik.
Spectral element method in structural dynamics / Usik Lee.
p. cm.
Includes index.
ISBN 978-0-470-82374-3 (cloth)
1. Structural dynamics–Mathematics. 2. Structural frams–Mathematical models.
3. Spectral theory (Mathematics) I. Title.
TA654.15.L44 2009
624.1'71–dc22
2009015525

ISBN 978-0-470- 82374-3(HB)

Typeset in 10/12pt Times by Thomson Digital, Noida, India.
Printed and bound in Singapore by Markono Print Media Pte Ltd, Singapore.
This book is printed on acid-free paper responsibly manufactured from sustainable forestry in which at least two trees are planted for each one used for paper production.

Contents

Preface

Owing to the rapid developments in computer technology, impressive progress in the computational methods used in engineering and science has been made over recent decades. The classical finite element method (FEM) has probably been the most popular in many areas of engineering and science, being one of the most convenient and easy-to-use computational methods. Though the FEM is applicable to most geometries, boundary conditions and material variations, it can be extremely expensive and it is often impossible to work out solutions to the large scale finite element models using a desktop computer. Thus, an alternative method that can provide accurate solutions while reducing the computational burden, but retaining the key advantage features of FEM, is mandatory, even today.

The FEM is a time-domain solution method in which the finite element equation is formulated in the time domain and solved by using a numerical integration method. On the other hand, the spectral element method (SEM) is a frequency-domain solution method in which the spectral element equation is formulated in the frequency domain and solved by using the fast Fourier transform (FFT) based spectral analysis method. In SEM, the exact dynamic stiffness matrix, known as the spectral element matrix, is formulated in the frequency domain by using exact wave solutions for the governing differential equations. Accordingly, in theory, the SEM will provide exact frequency-domain solutions while using only a minimum number of degrees-of-freedom.

Although there have been a huge number of journal publications since the basic concept of SEM was initially introduced more than two decades ago, very few books exist on the subject. Thus, the present book presents a concise introduction to the theory of SEM and its applications to various problems in structural dynamics and other related areas. It is intended as a valuable reference book for graduate students, professors, and professional researchers in the areas of mechanical engineering, civil engineering, aerospace engineering, naval architecture, structural engineering, applied mechanics, biomechanical engineering, and other related areas including computational methods. This book could be also used as a textbook for graduate students.

There are three parts to the book. The first part (Chapters 1 and 2) addresses the background and history of the SEM and the fundamentals of the spectral analysis of signals. In the second part (Chapters 3 and 4), the methods of spectral element formulation and the spectral element analysis method are addressed. The last part, from Chapter 5 to Chapter 16, presents the applications of SEM to various engineering problems in structural dynamics and related areas. The following overview summarizes the key features of each chapter.

Chapter 1 addresses key features of the SEM, together with a brief historical perspective on the development of SEM and its applications.

Chapter 2 introduces the fundamentals of the DFT (discrete Fourier transform) and FFT theories necessary for the spectral analysis of signals.

Chapter 3 is devoted to various methods of spectral element formulation: the force-displacement relation method, the variational method, and the state-vector equation method.

Chapter 4 addresses the general procedure of spectral element analysis: formulation and assembly of spectral elements; imposition of boundary conditions; and the computation of eigensolutions and time responses.

Chapter 5 deals with the dynamics of beams and Levy-type plates. For distributed dynamic forces, the two-element method is introduced as an approximation approach.

Chapter 6 is devoted to the spectral element modeling and analysis of the flow-induced vibrations of pipelines conveying two types of internal fluids: steady fluid and unsteady fluid.

Chapter 7 addresses the spectral element modeling and analysis of the dynamics and stability of axially moving structures, such as the string, the Bernoulli–Euler beam, the Timoshenko beam, and the thin plates.

Chapter 8 is devoted to the spectral element modeling and analysis of the dynamics of rotor systems, which consist of the flexible shafts, thin and thick rigid disks and bearing supports.

Chapter 9 discusses the spectral element modeling and analysis of the dynamics of multi-layered beam structures, such as the elastic–elastic two-layer beams and the elastic–viscoelastic–elastic passive constrained layer damping (PCLD) beams.

Chapter 10 discusses the spectral element modeling and analysis of the dynamics of adaptive structures, such as the elastic–piezoelectric two-layer beams and the active constrained layered damping (ACLD) beams with and without active control.

Chapter 11 is devoted to the spectral element modeling and analysis of the dynamics of composite laminate beams subjected to axial-bending-shear coupled vibration and to bending-torsion-shear coupled vibration.

Chapter 12 addresses the SEM-based continuum modeling method and the spectral transfer matrix method for the efficient dynamic analysis of various types of periodic lattice structures.

Chapter 13 extends the SEM application to a biomechanics problem: blood flow analysis. The pseudo-force based non-linear spectral element modeling is introduced.

Chapter 14 discusses the SEM-based methods to identify non-ideal structural boundaries and the joints in a beam structure from experimentally measured frequency response function (FRF) data.

Chapter 15 addresses the SEM-based linear and non-linear structural damage identification methods to identify multiple local damage generated in a structure from experimentally measured FRF data.

Chapter 16 discusses other promising SEM applications, such as the SEM–FEM hybrid method and the impact force identification, while providing a brief list of applications from the existing literature.

In most chapters the derivations of the associated governing equations are provided. Furthermore, the conventional finite element models are also provided in the appendices of each chapter for the use in validating or comparing with the corresponding spectral element models. Though no source codes are included in this book, some code examples (M files) are available for download from the book's companion website http://www.wiley.com/go/ulee.

This book is the product not only of my knowledge, research and teaching experience, but of numerous discussions with my past and present graduate students over a period of more than 15 years. I wish to acknowledge the contributions of my graduate students during the various stages of the manuscript, without which this book would not have been completed. All support and cooperation from the staff of John Wiley & Sons, senior commissioning editor Mr. James Murphy, project editor Mr. Roger Bullen, and production editor Ms. Sarah Karim are accordingly acknowledged. The author also acknowledges the financial support from Inha University (Inha University Research Grant) during the course of manuscript preparation. Finally, I dedicate this book to my mother Chungkyung Koh and to the memory of my father Seokbong Lee.

The author will be pleased to hear from readers who find misprints and errors, or who can provide hints to other ways of improving the book in any future editions.

Usik Lee
Inha University, Incheon, Republic of Korea

Part One

Introduction to the Spectral Element Method and Spectral Analysis of Signals

1

Introduction

1.1 Theoretical Background

1.1.1 Finite Element Method

As the dynamic behavior and characteristics of a structure are of great importance in engineering, it is necessary to predict them accurately in an efficient and economic manner. The finite element method (FEM) is probably one of the powerful and popular computation methods to have been used in many areas of engineering and science.

As an illustrative example, the vibration pattern of a structure certainly varies depending on the vibration frequency, and its wavelength at high frequency is very low. Because a sufficiently accurate dynamic response can only be obtained by capturing all necessary high frequency wave modes, the mesh (finite element) size used in the finite element modeling must be sufficiently small, comparable to the lowest wavelength of the vibrating structure.

However, as the conventional finite element models are formulated by using frequency-independent (static or fixed) polynomial shape (interpolation) functions, the FEM cannot capture all necessary high frequency wave modes of interest. Thus the FEM solutions become significantly inaccurate, especially at high frequencies, where associated wavelengths are very short. The so-called *h*-method is one of well-known approaches to improving the FEM accuracy by refining the meshes. Unfortunately this approach will make the size of the system extremely large, and hence from the computational aspect, the conventional FEM often becomes prohibitive for most complex, large flexible structures. It is known that, as a rough guide, the mesh size must be 10–20 times smaller than the wavelength of the highest frequency wave mode of interest [6].

An alternative approach to improve the solution accuracy is to use the shape functions, which can vary depending on vibration frequency. Accordingly, the candidate shape functions will be frequency dependent and they are known as dynamic shape functions in the literature. As the dynamic shape functions can readily capture all necessary high frequency wave modes of interest, extremely highly accurate solutions can be obtained, and the need to refine the meshes is no longer necessary. This elegant concept has led to the so-called dynamic stiffness method (DSM) [11, 108].

Spectral Element Method in Structural Dynamics Usik Lee

1.1.2 Dynamic Stiffness Method

The exact dynamic stiffness matrix is used in the DSM. The exact dynamic stiffness matrix is formulated in the frequency domain by using exact dynamic shape functions that are derived from exact wave solutions to the governing differential equations. To obtain the exact wave solutions in the frequency domain, the time-domain governing differential equations are transformed into the frequency domain by assuming harmonic solutions of a single frequency. Accordingly the exact dynamic stiffness matrix is also frequency dependent and it can be considered as a mixture of the inertia, stiffness and damping properties of a structure element. As the exact dynamic stiffness matrix is formulated by using exact frequency dependent dynamic shape functions derived from the exact wave solutions, it automatically deals with the continuous mass distribution in a structure element exactly. Thus the structure element represented by an exact dynamic stiffness matrix is often called a continuum element. Consequently, the DSM guarantees exact frequency-domain solutions to the governing differential equations (or the mathematical model) adopted for the problem under consideration: this is why the DSM is referred to as an exact solution method in the literature. Of course the absolute accuracy of the DSM will be limited to the accuracy level of the governing differential equations adopted to formulate the exact dynamic stiffness matrix. For instance, the DSM based on the Timoshenko-beam model will provide more accurate frequency-domain solutions when compared with that based on the Bernoulli–Euler beam model. However, as the assumptions made for the DSM will be less severe than those made for the conventional FEM and other approximate solution methods, the DSM will still provide better solutions.

As the exact dynamic stiffness matrix is formulated by using exact dynamic shape functions, it treats the mass distribution in a structure member exactly. Thus only one single element is sufficient for a regular part of a structure (without any structural or material discontinuities inside), regardless of its length between any two successive structural or material discontinuities, to acquire exact solutions. That is, we no longer need to refine a regular part of a structure into multiple fine meshes. This will significantly reduce the size of the problem, in other words, the total number of meshes and degrees-of-freedom (DOFs). In due course, this will significantly reduce the computation cost and time, together with improving the solution accuracy by reducing the computer round-off errors or numerical errors that are inevitable for most large size problems. In addition, the DSM provides an infinite number of eigensolutions from the exact dynamic stiffness matrix represented in terms of a minimum number of DOFs.

Because the exact dynamic stiffness matrices are stiffness formulated (as the conventional finite element stiffness matrices are), they can be assembled in a completely analogous way to that used in the FEM. Thus, the meshing and assembly features of the FEM can be applied equally to the DSM. The only difference from FEM is that the assembly must be repeated at all discrete frequencies by using the Do-Loop routine. By combining the advantage features of the DSM with those of the spectral analysis method (SAM), Beskos [124] introduced the fundamental concept of the spectral element method (SEM) for the first time in his 1978 work.

1.1.3 Spectral Analysis Method

The solution methods for the governing differential equations formulated in the time domain can be categorized into two major groups. The first group consists of the time-domain methods, such as the numerical integration methods and the modal analysis method, which is commonly used for the vibration analysis. The second group consists of the frequency-domain methods. The spectral analysis method (SAM) is one of the frequency-domain methods most popularly reported in the literature. It is worth remembering that throughout this book, the word "frequency" is used to mean the Fourier transform of the "time" (that is, "time frequency") rather than that of the "spatial" coordinate (that is, "space frequency").

In SAM, the solutions to the governing differential equations are represented by the superposition of an infinite number of wave modes of different frequencies (or periods). This corresponds to the continuous Fourier transform of the solutions. This approach involves determining an infinite set of spectral components (or Fourier coefficients) in the frequency domain and performing the inverse Fourier transform to reconstruct the time histories of the solutions. The continuous Fourier transform is feasible only when the function to be transformed is mathematically simple and the inverse transform is the biggest impediment to most practical cases, especially when dealing with digitized experimental data measured through a modern data acquisition system. Thus, instead of using the continuous Fourier transform, the discrete Fourier transform (DFT) is widely used in practice.

The DFT is an approximation of the continuous Fourier transform. In contrast to the continuous Fourier transform, the solutions are represented by a finite number of wave modes of discrete frequencies and thus, as an enormous advantage, one can use the fast Fourier transform (FFT) algorithm to compute the DFT and its inversion economically and quickly. Thus, the use of the FFT algorithm makes it possible to efficiently take into account as many spectral components as are needed up to the highest frequency of interest. Accordingly, the DFT/FFT-based SAM can provide very accurate solutions, while reducing the computation cost and time significantly.

It is worth mentioning that firstly, although the DFT is the approximation of the continuous Fourier transform, the DFT processing itself is exact in the sense that it does allow the time signal to be regained exactly at discrete times [127], and secondly, although one uses a computer to accomplish the DFT/FFT-based spectral analysis, it is certainly not a numerical method in the sense that the analytical descriptions of the Fourier transforms are used in the computation.

1.1.4 Spectral Element Method

As shown in Figure 1.1, the spectral element method (SEM) can be considered as the combination of the key features of the conventional FEM, DSM and SAM. The key features

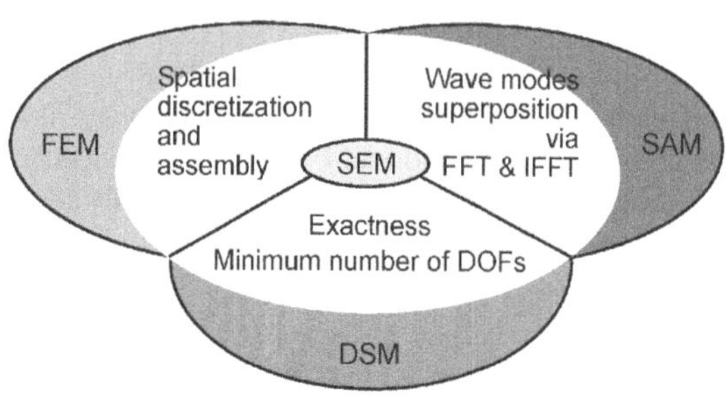

Figure 1.1 Key features of the spectral element method

of each method can be summarized as follows:

1. **Key features of FEM.** Meshing (spatial discretization) and the assembly of finite elements.
2. **Key features of DSM.** Exactness of the dynamic stiffness matrix formulated in terms of a minimum number of DOFs.
3. **Key features of SAM.** Superposition of wave modes via DFT theory and FFT algorithm.

It is worth mentioning that the above key features of the both FEM and SAM have been of little interest in most of the existing literature associated with the classical DSM.

In SEM, exact dynamic stiffness matrices are used as the element stiffness matrices for the finite elements in a structure. To formulate an exact dynamic stiffness matrix for the classical DSM, the dynamic responses of a structure are usually assumed to be the harmonic solutions of a single frequency. However, for the SEM, the dynamic responses are assumed to be the superposition of a finite number of wave modes of different discrete frequencies based on the DFT theory. Accordingly, the computation of the exact dynamic stiffness matrix must be repeated at all discrete frequencies up to the highest frequency of interest. By virtue of the exactness of the spectrally formulated dynamic stiffness matrix, one-element modeling will suffice for a regular structure member. The SEM is an element method, just like the conventional FEM. Thus, the mesh refining can also be applied in the SEM when any geometric or material discontinuities in the spatial domain of concern exist, and also when any externally applied forces exist. The details of the spectral element formulation procedure and some examples are provided in Chapter 3.

The spectrally formulated exact dynamic stiffness matrix to be used in the SEM in conjunction with the SAM is known as the "spectral element matrix" in this book and in much of the existing literature. In some of the literature, it is often called a "spectral finite element matrix" or "spectral dynamic stiffness matrix." Accordingly, the spectral element method is also often called the "spectral finite element method" and, sometimes the "continuum method." The finite length of the structure element represented by the spectral element matrix is called the "spectral element," as the terminology "finite element" is commonly used in the FEM. The terminologies "spectral element" and finite element" will be used interchangeably throughout this book without introducing any unnecessary confusion. The frequency-domain nodal DOFs specified on a spectral element are known as "spectral nodal DOFs."

It should be pointed out that unfortunately the same terminology "spectral element method" or "SEM" has also been used for a different class of finite element methods developed in 1984 by Anthony Patera at MIT [135]. In fact, the word "spectral" for Patera's SEM is not time-wise, but space-wise.

The SEM is stiffness formulated. Thus, the spectral elements can be assembled to form a global system matrix equation for the whole problem domain by using exactly the same assembly techniques as used in the conventional FEM. The global system matrix equation is then solved for the global spectral nodal DOFs, of course, repeatedly at all discrete frequencies. Finally, we use the inverse-FFT (IFFT) algorithm to compute the time histories of dynamic responses (time-domain solutions).

1.1.5 Advantages and Disadvantages of SEM

By virtue of the apparent advantages of the exact dynamic stiffness matrix used in conjunction with the DFT/FFT-based SAM, the major advantages of the SEM (although not all can be listed here) may be summarized as follow:

1. **Extremely high accuracy.** In theory, the SEM provides exact frequency-domain solutions, such as the eigensolutions (natural frequencies and modes) and the frequency response functions (system transfer functions). The SEM may also provide extremely accurate time-domain solutions (the time histories of dynamic responses) by efficiently taking into account as many high frequency wave modes as are required by using the FFT algorithm. Accordingly, the problems solved by using the spectral element model can be utilized as benchmark problems for evaluating the accuracy and performance of a newly developed solution method.
2. **Smallness of the problem size and DOFs.** This is true because one-element modeling suffices for the representation of a regular structure member of any size, without any structural and material discontinuities inside.
3. **Low computation cost.** For computing time-domain responses, this is certainly true due to the smallness of problem size as well as due to the use of the FFT algorithm. The quick computation of the time-domain responses also enables us to treat the wave propagation simulation more realistically. However, it is worth mentioning that the most significant part of time consumption in SEM is for computing eigenfrequencies by using an iteration method of root finding.
4. **Effective to deal with frequency-domain problems.** This is the situation because the SEM itself is a frequency-domain solution method. Accordingly the SEM can be successfully applied to the following situations: (i) when the characteristics of a system (e.g., constitutive equation or internal damping) are dependent on frequency (e.g., viscoelastic materials), and (ii) when the boundary conditions of a system are specified in the frequency domain (e.g., the impedance boundary conditions for the fluid-structure interaction problems).
5. **Effective to deal with the non-reflecting boundary conditions of the infinite- or semi-infinite-domain problems.** This is achieved by using the semi-infinite spectral element (often called through-off element), which can be formulated by simply removing the wave modes reflected from the boundary at the infinite from the dynamic shape functions.
6. **Locking-free method.** The SEM does not exhibit the shear locking problems often raised in the conventional FEM, because the exact wave solutions to the governing equations are used as the dynamic shape functions in the SEM.
7. **Effective to deal with digitized data.** This is true because the FFT algorithm used in SEM is an efficient tool to deal with digitized data as follows: (i) the experimental data measured digitally through the analogue-to-digital converters, and (ii) the excitation forces measured or specified in the forms of numerical values at regular intervals of time rather than in the forms of analytical functions.
8. **The system transfer functions** (the inverse of global dynamic stiffness matrix) are the by-products of the spectral element analysis. Thus, it is very straightforward to perform the inverse problems, such as the system identification (e.g., identification of system parameters, boundary conditions, structural joints, or structural damages) and the force identification.

Despite the aforementioned advantages of SEM, there are also some disadvantages as follows:

1. The exact spectral element formulation is possible for the problems where the exact wave solutions to the governing equations are available. However, unfortunately the exact wave solutions are not always available for most complex and multi-dimensional problems. Thus,

the exact spectral element models developed to-date have been mostly for one-dimensional (1-D) problems (e.g., rods and beams) or some multi-dimensional problems that can be transformed into the equivalent 1-D problems (e.g., Levy-type plates).

If the exact spectral element model is not available for a problem, the approximate spectral element modeling approach can be applied by adopting approximate dynamic shape functions obtained by using the wavenumbers computed from the approximate dispersion relation or by using another appropriate approximation method. As the assumptions possibly made for the approximate spectral element modeling will be less severe than those usually made for the conventional FEM, the approximate spectral element models may still provide very accurate solutions.

2. As the SEM is a frequency-domain analysis method based on DFT/FFT techniques, it cannot be directly applied to time-variant, non-linear systems for which the principle of superposition does not hold. Thus, for the case of non-linear systems, for instance, one may need to use an iteration method by treating non-linear terms as the pseudo-forces.
3. In the SEM, the time-domain solutions are post-processed by convolving the transfer functions with external loadings in the frequency domain via the IFFT algorithm. Thus, even though the SEM guarantees exact frequency-domain solutions, it is not true for the time-domain solutions, because errors due to aliasing or leakage are inevitable in practice. As the aliasing or leakage-induced errors depend on the FFT conditions, special care must be taken in determining the successful FFT conditions.

1.2 Historical Background

The history of spectral analysis, which is also known as the Fourier analysis or frequency-domain analysis, began with the pioneering work "*Theorie analytique de la chaleur* (*The analytical theory of heat*)" published in 1822 by Joseph Fourier. He showed how an infinite series of sine and cosine functions can be used to analyse heat conduction in solids. Owing to the distrust in the use of series then, Fourier's method did not gain acceptability during his lifetime. Since then, works by Dirichlet, Riemann and other mathematicians have resolved any doubts about the validity of the Fourier series, and spectral analysis has followed two major pathways: the continuous Fourier transform and the discrete Fourier transform (DFT).

The continuous Fourier transform has the drawback that it involves the integration of a signal (function) specified analytically over the time period, which is not always easy to carry out. This is especially true when the signal is experimentally measured in a digitized data form. Thus, the continuous Fourier transform is usually transformed in a discrete form by replacing the integration by the summation of a finite number of signal values sampled at regular intervals of time. The discrete version of the continuous Fourier transform is the DFT and it is a very effective means for transforming a periodic signal sampled in the time domain into an expression of the data in the frequency domain and vice versa.

The major difficulty in using the DFT is the requirement of extensive computation time. Although some techniques and ideas to reduce the computation time appeared early in the twentieth century, it was not until 1965 when James W. Cooley (IBM researcher) and John W. Tukey (Princeton faculty member) developed a computational algorithm that is now known as the fast Fourier transform (FFT). The FFT is simply a remarkably efficient algorithm for computing the DFT and its inversion. The FFT algorithm reduces the number of arithmetical operations for computing the DFT from the order N^2 to the order $N\log_2 N$, where N is the number of samples. This is a dramatic reduction for large values of N: the reduction ratio is over 100 : 1 when $N = 1024$ and 400 : 1 when $N = 4096$. Historically it has been recognized that Carl Friedrich Gauss developed the same FFT algorithm around 1805, but without getting any attention until Cooley and Tukey rediscovered it, independently, 160 years later. The FFT rediscovery has made spectral analysis highly efficient, with widespread applications to digital signal processing and certain areas of engineering analysis. However, the applications of DFT/FFT-based spectral analysis to the dynamic analysis of structures are fairly recent and mostly limited to 1-DOF or multi-DOFs discrete systems. Extensive reviews on the subject and related applications can be found in the books by Ginsberg [49], Humar [58] and Newland [127].

In classical DSM, most researchers have been predominantly interested in formulating exact dynamic stiffness matrices to obtain more accurate natural frequencies and modes, rather than in developing it further in the form of the SEM by combination with the DFT/FFT-based SAM. As mentioned in the previous section, the exact dynamic stiffness matrix for the classical DSM is usually formulated from exact wave solutions obtained by assuming the harmonic solutions of a single frequency to the governing differential equations.

Over the last seven decades, there have been a large number of publications related to the formulation and applications of exact dynamic stiffness matrices. The 1941 work by Kolousek [70] is probably the first to derive the dynamic stiffness matrix for the Bernoulli–Euler beam. Przemieniecki [142] introduced the formulation of the frequency-dependent mass and stiffness matrices for both bar and beam elements in his book. In contrast to the conventional finite element mass and stiffness matrices, which result in the linear eigenproblems, the exact dynamic stiffness matrices result in transcendental eigenproblems, the coefficients of which are the transcendental functions of frequency. Thus, a drawback of the DSM, is that it is not an easy task to compute all natural frequencies (without missing any one within a specified frequency range of interest) accurately by solving the transcendental eigenvalue problems. In 1971, this difficulty was resolved by Wittrick and Williams [169] by developing the well-known Wittrick–Williams algorithm for automatic calculation of undamped natural frequencies [169]. The Wittrick–Williams algorithm has certainly enhanced the applicability of the DSM.

Since then numerous exact dynamic stiffness matrices have been developed for various structures. The extensive literature up to 1976 and 1983 can be found in Ref. [4] and Ref. [115], respectively. In addition, the extensive literature up to 1993, together with elegant theories for the dynamic stiffness and substructures, can be found in a book by Leung [108]. Although not all published literature can be cited herein, exact dynamic stiffness matrices have been developed mostly for the 1-D structures including the Timoshenko beams with or without axial force [26, 30, 31, 54, 159], extension-bending-torsion coupled beams [10, 16, 18, 19, 47, 52], non-uniform beams [17, 56], curved beams and springs [61, 78, 107, 136], Rayleigh–Timoshenko beams [4, 115], composite beams [20, 44], sandwich beams [14], aircraft wing [13], the beams on elastic foundation [167], axially moving string [77], and other structures [106, 110].

Although the history of the DSM spans about seven decades, it was not until 1978 when Beskos [124] introduced the fundamental concept of the SEM for the first time. He derived an exact dynamic stiffness matrix for the beam element and employed FFT for the dynamic analysis of plane frame-works. His work was further improved and generalized in his 1982 work [148], while claiming that his 1978 work [124] was the first to employ FFT for the dynamic analysis of plane frame-works in conjunction with the DSM. It was in 1988 that Doyle at Purdue University [38] published his first work on the formulation of the spectral element for the longitudinal wave propagation in rods. He seemed to be the first to use the terminology "spectral element method" in his 1992 work [146] for the DFT/FFT-based spectral element analysis approach. Doyle and his colleagues have applied the SEM mostly to wave propagations in structures. A comprehensive list of the works by Doyle's research group and other researchers up to 1997 can be found in a book by Doyle [40]. On the other hand, the present author and his students have extensively applied the SEM to various problems in structural dynamics and the outcome of this research [33, 67, 72, 80–82, 84–101, 103, 104, 130] is the major source of the material presented in this book.

2

Spectral Analysis of Signals

In mathematics, it is well known that an arbitrary periodic signal can be decomposed into many harmonic components and, in the reverse transform, the signal can be synthesized by superposing all its harmonic components. This is the basis of spectral (Fourier) analysis. For a signal given in functional form, we may evaluate its harmonic components by analytically evaluating the continuous Fourier transform integral. However, if the signal function is so complicated, we may decide to use numerical integration instead. In this case we need to know the signal values only at a discrete set of instances. A similar situation arises in most experimental measurements, because modern data acquisition systems based on analogue-to-digital converters store digitized data sampled at discrete, uniformly spaced intervals. The discrete Fourier transform (DFT) enables us to deal with such discrete sampled data from a signal. The important advantage of DFT theory is that it allows us to use the fast Fourier transform (FFT), which is an extremely efficient algorithm for computing the DFT. To formulate spectral element models, the exact solutions to the governing equations are in general represented in the spectral forms by using the DFT. Thus this chapter provides a brief review of the theories and applications associated with DFT and FFT.

2.1 Fourier Series

If $x(t)$ is a continuous periodic function of time t, with period T, we can always represent it in the form of a Fourier series as

$$x(t) = a_0 + 2\sum_{n=1}^{\infty}(a_n \cos \omega_n t + b_n \sin \omega_n t) \tag{2.1}$$

where a_0, a_n and b_n are constant Fourier coefficients given by

$$\begin{aligned} a_0 &= \frac{1}{T}\int_0^T x(t)\mathrm{d}t \\ a_n &= \frac{1}{T}\int_0^T x(t)\cos \omega_n t \, \mathrm{d}t \\ b_n &= \frac{1}{T}\int_0^T x(t)\sin \omega_n t \, \mathrm{d}t \end{aligned} \tag{2.2}$$

Spectral Element Method in Structural Dynamics Usik Lee

In the above equations, $\omega_n = n\Delta\omega$ is the frequency of the nth harmonic and $\Delta\omega$ is the frequency spacing (fundamental frequency) defined by

$$\Delta\omega = \frac{2\pi}{T} \tag{2.3}$$

The cosine and sine functions can be decomposed by using the Euler identities as follows:

$$\begin{aligned} \cos\omega_n t &= \frac{1}{2}(e^{i\omega_n t} + e^{-i\omega_n t}) \\ \sin\omega_n t &= -\frac{i}{2}(e^{i\omega_n t} - e^{-i\omega_n t}) \end{aligned} \tag{2.4}$$

where $i = \sqrt{-1}$ is an imaginary unit. By substituting Equation 2.4 into Equation 2.1, we get an expression as follows

$$x(t) = a_0 + \sum_{n=1}^{\infty}(a_n - ib_n)e^{i\omega_n t} + \sum_{n=1}^{\infty}(a_n + ib_n)e^{-i\omega_n t} \tag{2.5}$$

Defining

$$X_n = a_n - ib_n \tag{2.6}$$

we can rewrite Equation 2.5 into a compact form

$$x(t) = \sum_{n=-\infty}^{\infty} X_n e^{i\omega_n t} \tag{2.7}$$

where X_n is the nth Fourier coefficient for a continuous periodic function $x(t)$. By substituting Equation 2.2 into Equation 2.6 and then by applying Equation 2.4, we obtain

$$X_n = \frac{1}{T}\int_0^T x(t)e^{-i\omega_n t}dt (n = 0, \pm 1, \pm 2, \ldots, \pm\infty) \tag{2.8}$$

Equations 2.7 and 2.8 constitute the Fourier transform pair for the continuous periodic function $x(t)$. Equation 2.7 is referred to as the synthesis equation and Equation 2.8 as the analysis equation for a continuous periodic function $x(t)$. In summary, the Fourier transform decomposes a periodic function into a discrete spectrum of its frequency components, and the inverse transform synthesizes a function from its spectrum of frequency components.

2.2 Discrete Fourier Transform and the FFT

2.2.1 Discrete Fourier Transform (DFT)

Although $x(t)$ is a continuous periodic function of time, with period T, it is often the case that only sampled values of the function are available. This is true, for instance, when a time history $x(t)$ is experimentally measured in the form of digitized data taken at equally spaced instances by using a modern data acquisition system. Let N be the number of samples in the

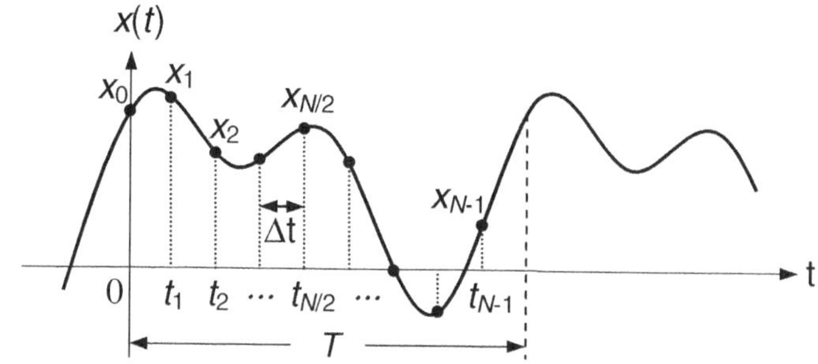

Figure 2.1 Sampling of a periodic function of period T at uniform time interval Δt

time domain and, as shown in Figure 2.1, assume that they are sampled at equally spaced time intervals given by

$$\Delta t = \frac{T}{N} \tag{2.9}$$

We express the N samples in a discrete time series $\{x_r\}$, where $x_r = x(t_r)$, $t_r = r\Delta t$ and $r = 0$, 1, 2, ..., $N-1$. By using the discrete time series $\{x_r\}$, the integral in Equation 2.8 can be approximately replaced by the summation as

$$X_n = \frac{1}{T}\sum_{r=0}^{N-1} x_r e^{-i\omega_n t_r}\Delta t \tag{2.10}$$

By using Equation 2.9, Equation. 2.10 can be rewritten as

$$X_n = \frac{1}{N}\sum_{r=0}^{N-1} x_r e^{-i\omega_n t_r} \tag{2.11}$$

Equation 2.11, the process of converting from discrete time series $\{x_r\}$ to frequency data is called the discrete Fourier transform (DFT). The frequency data X_n are called the DFT coefficients.

As the discrete time series $x(t_r)$ are real, we can readily show from Equation 2.11 that

$$\begin{aligned}
&X_0 = \text{real} && \text{(a)}\\
&X_{N/2} = X_{-N/2} = \text{real} && \text{(b)}\\
&X_{-n} = X_n^* && \text{(c)}\\
&X_{N/2+n} = X_{-N/2+n} = X_{N/2-n}^* && \text{(d)}\\
&X_{N-n} = X_n^* = X_{n-N}^* && \text{(e)}\\
&X_{N+n} = X_n && \text{(f)}
\end{aligned} \tag{2.12}$$

by using the following relationships

$$
\begin{aligned}
&e^{\pm i\omega_{p+q}t_r} = e^{\pm i\omega_p t_r} e^{\pm i\omega_q t_r} && \text{(a)}\\
&e^{\pm i\omega_{-p}t_r} = e^{\mp i\omega_p t_r} && \text{(b)}\\
&e^{\pm i\omega_{N/2}t_r} = (-1)^r && \text{(c)}\\
&e^{\pm i\omega_N t_r} = 1 && \text{(d)}
\end{aligned}
\tag{2.13}
$$

The asterisk symbol (*) is commonly used in this book to denote the complex conjugate of a complex number.

The synthesis Equation 2.7 can also be approximated by truncating the summation index n as

$$
x_r \cong \sum_{n=-N/2}^{N/2-1} X_n e^{i\omega_n t_r} \tag{2.14}
$$

We can rewrite the right-hand side of Equation 2.14 as

$$
x_r \cong \sum_{n=-N/2}^{-1} X_n e^{i\omega_n t_r} + \sum_{n=0}^{N/2-1} X_n e^{i\omega_n t_r} \tag{2.15}
$$

By using Equations 2.12 and 2.13, the first summation of Equation 2.15 can be expressed as

$$
\begin{aligned}
\sum_{n=-N/2}^{-1} X_n e^{i\omega_n t_r} &= \sum_{k=0}^{N/2-1} X_{-N/2+k} e^{i\omega_{-N/2+k} t_r}\\
&= \sum_{k=0}^{N/2-1} X_{N/2+k} e^{i\omega_{-N/2+k} t_r}\\
&= \sum_{k=0}^{N/2-1} X_{N/2+k} e^{i\omega_{N/2+k} t_r} = \sum_{n=N/2}^{N-1} X_n e^{i\omega_n t_r}
\end{aligned}
\tag{2.16}
$$

Substitution of Equation 2.16 into Equation 2.15 yields

$$
x_r \cong \sum_{n=0}^{N-1} X_n e^{i\omega_n t_r} \tag{2.17}
$$

This is the inverse discrete Fourier transform (IDFT), which is the process of converting DFT coefficients into discrete time series data.

Equations 2.11 and 2.17 constitute a DFT pair. The range of the Fourier coefficients X_n is limited to $n = 0$ to $N-1$ in order to maintain the symmetry of the DFT pair. Although the DFT pair, Equations 2.11 and 2.17, is the approximation of Equations 2.8 and 2.7, respectively, it does allow all discrete time series $\{x(t_r)\}$ to be regained exactly. In other words, the DFT processing is indeed exact. This important fact can be readily verified by showing that the discrete time series $\{x(t_r)\}$ are exactly regained by simply substituting X_n from Equation 2.11 into the right-hand side of Equation 2.17.

The $1/N$ factor before the integral in Equation 2.11 can be placed before the summation in Equation 2.17 ensuring that there is no net change in amplitude when one transforms from one domain to the other and back. Thus we can also use an alternative form of the DFT pair as

$$X_n = \sum_{r=0}^{N-1} x_r \, \mathrm{e}^{-i\omega_n t_r} \qquad (n = 0,\ 1,\ 2, \ldots,\ N-1) \tag{2.18}$$

$$x_r \cong \frac{1}{N} \sum_{n=0}^{N-1} X_n \mathrm{e}^{i\omega_n t_r} \qquad (r = 0,\ 1,\ 2, \ldots,\ N-1) \tag{2.19}$$

The DFT pair defined by Equations 2.18 and 2.19 is adopted in MATLAB®. Thus, the DFT pair defined by Equations 2.18 and 2.19 will be adopted in this book for readers who are familiar with MATLAB.

Figure 2.2a shows the DFT coefficients arrangement for an example case of $N = 2^3 = 8$. The real and imaginary parts of the DFT coefficients are arranged in the order

$$[X_{-N/2} \quad \cdots \quad X_{-1} \quad X_0 \quad X_1 \quad \cdots \quad X_{N/2-1}] \tag{2.20}$$

that is, for the example case, they are displayed in the order

$$[X_{-4} \quad X_{-3} \quad X_{-2} \quad X_{-1} \quad X_0 \quad X_1 \quad X_2 \quad X_3] \tag{2.21}$$

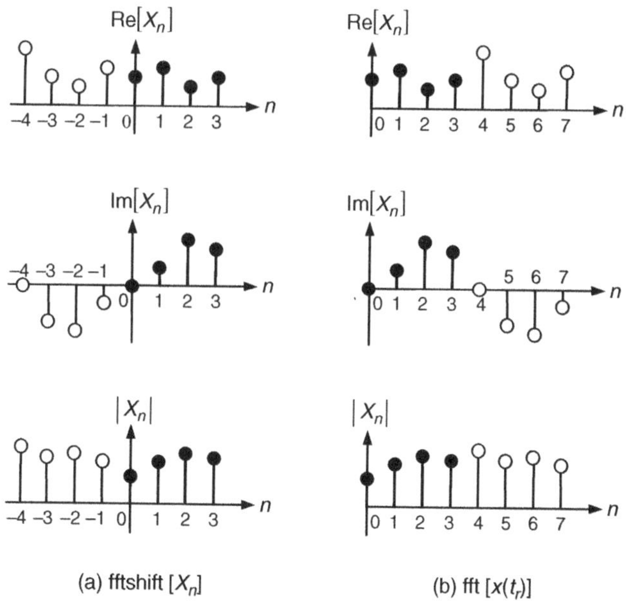

Figure 2.2 Two types of arrangement of the DFT coefficients using the MATLAB functions

However, as $X_{-N/2+k} = X_{N/2+k}$ ($k = 0, 1, \ldots, N/2-1$) is satisfied by Equation 2.12, $[X_{-N/2}, X_{-N/2+1}, \ldots, X_{-2}, X_{-1}]$ is identical with $[X_{N/2}, X_{N/2+1}, \ldots, X_{N-2}, X_{N-1}]$. Thus the arrangement of Equation 2.20 can be changed into the form

$$[X_0 \quad X_1 \quad \cdots \quad X_{N/2} \quad X_{N/2+1} \quad X_{N/2+2} \quad \cdots \quad X_{N-2} \quad X_{N-1}] \tag{2.22}$$

For the example case, the DFT coefficients are displayed in the order

$$[X_0 \quad X_1 \quad X_2 \quad X_3 \quad X_4 = X_{-4} \quad X_5 = X_{-3} \quad X_6 = X_{-2} \quad X_7 = X_{-1}] \tag{2.23}$$

which is shown in Figure 2.2b. The MATLAB fft function provides the DFT coefficients in the arrangement of Equation 2.22 and it also enables us to transform it into the arrangement of Equation 2.20. As the DFT coefficients arrangement of Equation 2.22 matches the IDFT formula of Equation 2.19, it will be adopted for use in this book.

As illustrated in Figure 2.3, owing to the properties of Equation 2.12a to Equation 2.12f, the DFT replicates periodically on the frequency axis with a period of $2\pi/\Delta t$ rad/s (or $1/\Delta t$ Hz) if the discrete time data constitutes a set of N values sampled at uniform time interval Δt (seconds). The real part is symmetrical about the middle frequency, $\omega_{N/2} = \pi/\Delta t$ rad/s (or $f_{N/2} = 1/2\Delta t$ Hz), while the imaginary part is anti-symmetric, because the DFT coefficients in the frequency range $\omega_{N/2} \le \omega \le \omega_N$ are complex conjugates of those in the frequency range of $0 \le \omega \le \omega_{N/2}$. Thus only the DFT coefficients at the frequencies below $\omega_{N/2}$ are unique, while those at the frequencies above $\omega_{N/2}$ are spurious. This means that N real discrete time data are transformed into $N/2$ complex frequency data without gaining or losing any information from a signal. Accordingly, the highest frequency we can deal with by the use of DFT is $\omega_{N/2} = \pi/\Delta t$ rad/s or $f_{N/2} = 1/2\Delta t$ Hz, which it is called the *Nyquist frequency* ω_{Nq} (rad/s) or f_{Nq} (Hz):

$$\omega_{\mathrm{Nq}} = \frac{\pi}{\Delta t}\,(\mathrm{rad/s}) \quad \text{or} \quad f_{\mathrm{Nq}} = \frac{1}{2\Delta t}\,(\mathrm{Hz}) \tag{2.24}$$

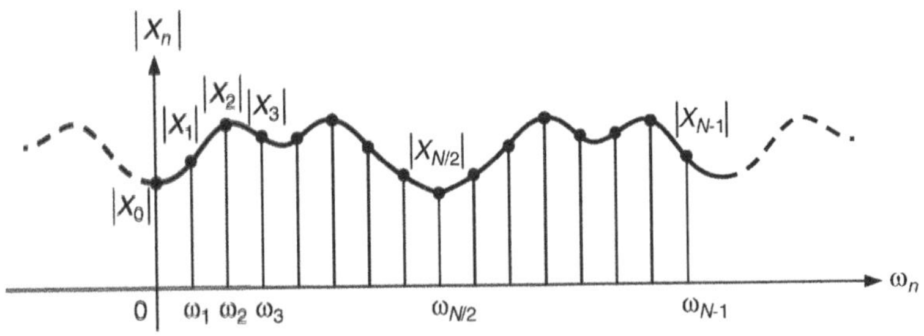

Figure 2.3 Periodicity of the DFT coefficients

2.2.2 *Fast Fourier Transform (FFT)*

The fast Fourier transform (FFT) is an ingenious highly efficient computer algorithm to compute the DFT and its inverse, IDFT, originally discovered by Gauss around 1805 and rediscovered by Cooley and Tukey at IBM in 1965. The FFT is not a different transform from the DFT, but rather just a means of computing the DFT with a considerable reduction in the computational time. The Cooley–Tukey algorithm makes the FFT extremely efficient by reducing the arithmetic operations (multiplications and additions) from something of the order of N^2 to $N/\log_2 N$. This is a dramatic reduction in computation time for large values of N. It is worth pointing out that although the FFT uses a computer, it is not a numerical method in the usual sense, because the analytical descriptions of DFT and IDFT, Equations 2.18 and 2.19, are still retained. An individual working with FFT does not necessarily need to know the details inside of the FFT algorithm, but must know how to prepare the data and how to interpret the results. More details of FFT algorithms can be found in Refs. [63, 127].

As a rule, data to be transformed by the FFT consists of N uniformly spaced discrete time series. For the maximum efficiency of computation, the optimal N is required to be an integer power of 2: that is, $N = 2^p$ where $p =$ integer. Although some FFT algorithms do not require that N be an integer power of 2, $N = 2^p$ is optimal for the maximum execution speed of the FFT algorithm. In most practical situations, the number of data sampled by using a fixed sampling time interval Δt for a fixed sampling time window T_s is unlikely to be an integer power of 2. In this case, zero padding can provide a means of making N be an integer power of 2 without losing any information, which is discussed further in the following section.

There are many computer programs available for performing the FFT. Among these, MATLAB users can use the functions listed in Table 2.1.

Table 2.1 MATLAB functions for the FFT [156]

Function	Syntax	Definition
fft	$X = \text{fft}(\boldsymbol{x})$	Returns the DFT of a vector $\boldsymbol{x}$
ifft	$\boldsymbol{x} = \text{ifft}(\boldsymbol{X})$	Returns the inverse DFT of a vector $\boldsymbol{X}$
fftshift	$Y = \text{fftshift}(\boldsymbol{X})$	Shifts zero-frequency component of DFT to the center of spectrum
ifftshift	$X = \text{ifftshift}(\boldsymbol{Y})$	Undo fftshift

2.3 Aliasing

2.3.1 Aliasing Error

According to Equation 2.12f, the DFT coefficients are symmetrically placed about the Nyquist frequency $f_{Nq} = 1/2\Delta t$ Hz when a signal is sampled at a uniform time interval Δt. In other words, the highest frequency that can be detected by DFT from the discrete time series sampled at a uniform time interval Δt is the Nyquist frequency f_{Nq}. If f_{max} is the highest frequency (called band frequency) of an actual signal, the Nyquist frequency f_{Nq} must be equal to or larger than f_{max} to accurately represent the signal as the DFT coefficients. This leads to an upper limit on the sampling time interval of

$$\Delta t \leq \frac{1}{2f_{max}} \tag{2.25}$$

This is called the Nyquist condition. As $1/f_{max}$ is the period of the highest frequency component of a signal, the Nyquist condition states that we must sample the signal more than twice within the period of the highest frequency associated with the signal. When we happen to violate the Nyquist condition (i.e., when we do not take at least two samples per period of the band frequency f_{max}) or when there are higher frequency components above the Nyquist frequency f_{Nq} in an actual signal, the higher frequency components are merged into the DFT spectrum as shown in Figure 2.4. This introduces an error into the DFT coefficients calculated for the frequencies below f_{Nq} and it is called *aliasing*.

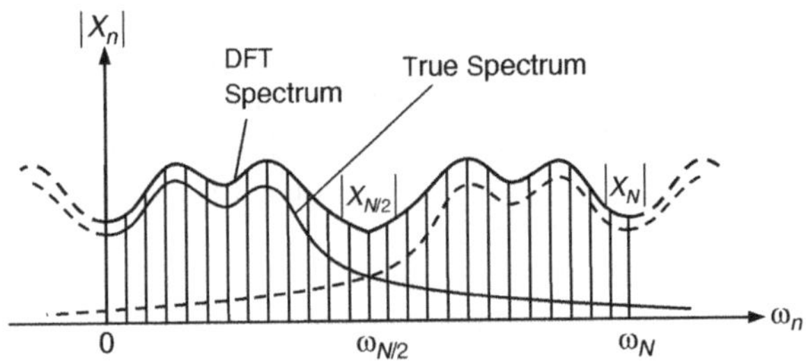

Figure 2.4 Aliasing error when the band frequency f_{max} exceeds the Nyquist frequency f_{Nq}

Example 2.1

Consider a harmonic function $x(t) = \sin 10\pi t$, which is shown in Figure 2.5a. The highest frequency of the harmonic function is $f_{max} = 5$ Hz and the required maximum sampling time interval must be smaller than $\Delta t_{req} = 0.1$ s to satisfy the Nyquist condition, Equation 2.25. Thus, when we choose the sampling time as $T = 2$ s, the required minimum number of samples must be larger than $N_{req} = T/\Delta t_{req} = 20$. We will consider the following three cases.

1. When $N = 2^3 = 8$
   Sampling time interval $\Delta t = T/N = 2/8 = 0.25\ \text{s} \geq \Delta t_{req}$
   Thus, the Nyquist frequency $f_{Nq} = 1/2\Delta t = 2\ \text{Hz} \leq f_{max}$
2. When $N = 2^4 = 16$
   Sampling time interval $\Delta t = T/N = 2/16 = 0.125\ \text{s} \geq \Delta t_{req}$
   Thus, the Nyquist frequency $f_{Nq} = 1/2\Delta t = 4\ \text{Hz} \leq f_{max}$
3. When $N = 2^5 = 32$
   Sampling time interval $\Delta t = T/N = 2/32 = 0.0625\ \text{s} \leq \Delta t_{req}$
   Thus, the Nyquist frequency $f_{Nq} = 1/2\Delta t = 8\ \text{Hz} \geq f_{max}$

The first two cases do not satisfy the Nyquist condition, while the last case satisfies it well. Hence, Figure 2.5 shows that aliasing appears for the first two cases, while it does not for the last case. In Figure 2.5, the thin lines represent the original harmonic function, while the thick lines represent the functions reconstructed from N samples by using the IDFT. The dots denote the sampled data.

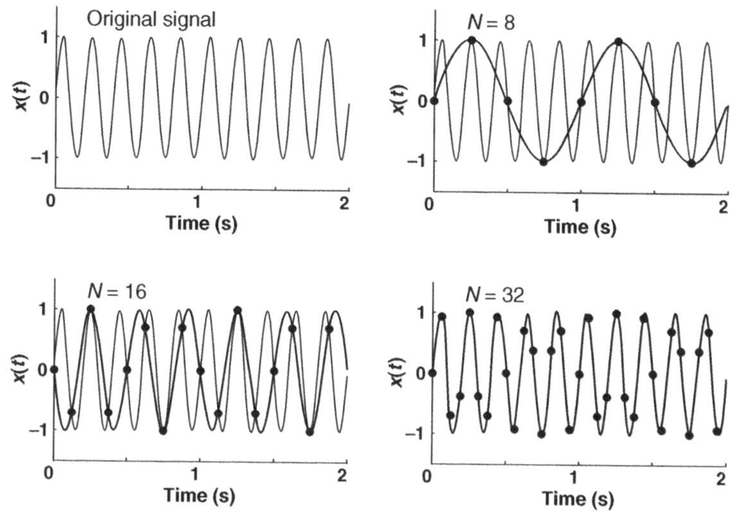

Figure 2.5 Illustration of aliasing for a harmonic function

Example 2.2

As the second example, consider a chirp signal $x(t) = \sin[2\pi(5t + 13t^2)]$ $(0 \le t \le 1)$ as shown in Figure 2.6a. The instant frequency of the chirp signal varies linearly from 5 to 31 Hz as t increase from 0 to 1 s, that is $f(t) = 5 + 26t$ Hz. Assume that the sampling time interval is given by $\Delta t = 1/36$ s. The Nyquist frequency is then given by $f_{\mathrm{Nq}} = 1/2\Delta t = 18$ Hz. Thus, the

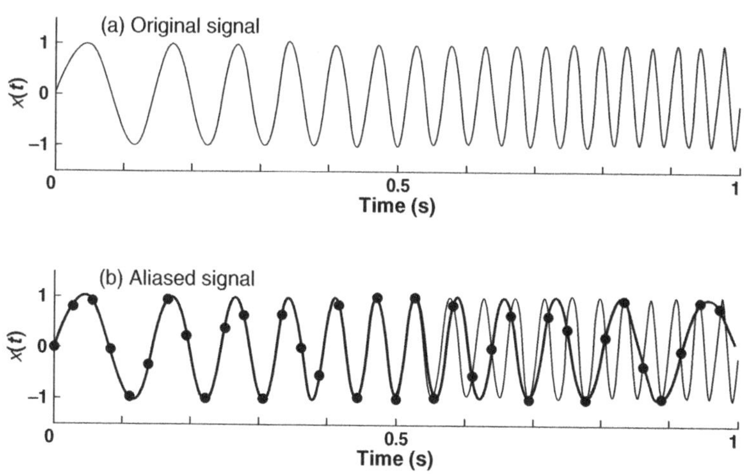

Figure 2.6 Illustration of aliasing for a chirp signal

Nyquist condition is satisfied during the time $0 \leq t \leq 1/2$ as $f_{\max} \leq f_{Nq}$ while it is not satisfied during the time $1/2 < t \leq 1$ as $f_{\max} \geq f_{Nq}$. Accordingly, aliasing is occurring as shown in Figure 2.6b, where the thin line represents the original chirp signal and the thick line represents the function reconstructed from 36 sampled data, denoted by dots.

2.3.2 *Remedy for Aliasing*

To avoid or minimize aliasing errors, we can use the following remedies.

1. **Remedy 1:** Reduce the sampling time interval Δt sufficiently so that the Nyquist condition, Equation 2.25, can be satisfied.
2. **Remedy 2:** If the first remedy does not work well, use a low-pass filter to remove significant frequency components above the Nyquist frequency before the DFT is processed.

2.4 Leakage

2.4.1 *Leakage Error*

In DFT applications, we must always deal with a restricted number of samples, such as N, in the time domain. Thus, if we set a sampling time interval Δt to a value that satisfies the Nyquist condition, the sampling time window will be fixed by $T_s = N\Delta t$. We must remember that, as the DFT theory was essentially developed for periodic functions, it can be accurately applied to the periodic time functions with period T. Thus, if the sampling time window T_s does not meet the periodicity condition of a signal under consideration, the error known as *leakage* can occur. The leakage introduces sidelobes into the frequency domain, whereas the aliasing introduces distortion of the frequency spectrum in the frequencies below the Nyquist frequency. The leakage can be observed in the following cases:

1. When the sampling time window T_s is not an integer multiple of the period T of the signal.
2. When the sampling time window T_s is not sufficiently large to see the signal decay towards zero.

Figure 2.7 shows two different sampling time windows used for a harmonic signal. When we choose a sampling time window for a signal, the DFT treats the signal as if the sampled signal repeats for ever. Thus, if we choose the sampling time window so that it exactly matches the period of the signal, the DFT will treat the original signal accurately alongwith presenting its exact DFT spectrum, as illustrated in Figure 2.7a. However, if we fail to choose the sampling time window to match with the period of the signal, the DFT misunderstands the original signal as if it is no longer a harmonic signal while presenting an inaccurate DFT spectrum, as shown in Figure 2.7b. The sidelobes appearing around the true peak spectrum represent leakage errors. The leakage errors cannot be eliminated completely because in general most signals are not periodic (or it takes an extremely long time to decay towards zero), so that there are no time windows that can cover the whole

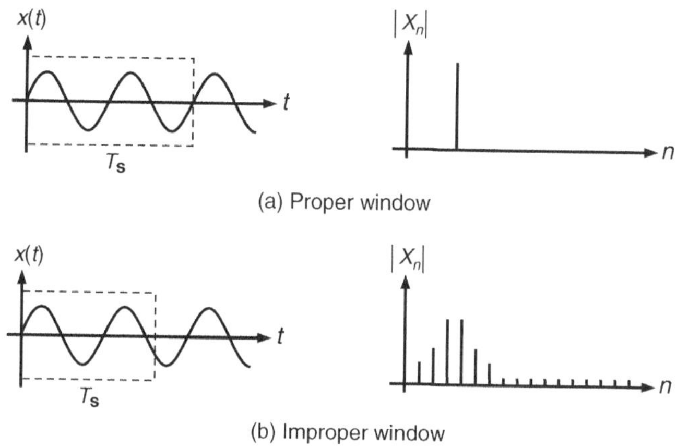

Figure 2.7 Illustration of the dependence of the leakage on the sampling time window

time signal at one time, or the maximum sampling time might be restricted by various environmental conditions.

To illustrate the leakage errors, we will consider the following two example cases.

Example 2.3

Consider a harmonic function $x(t) = \sin 6\pi t$ for which the period and maximum frequency are $T = 1/3$ s and $f_{\max} = 3$ Hz, respectively. Assume that the sampling time interval is chosen to be less than $\Delta t = 1/6$ s so that the Nyquist condition, Equation (2.25), is satisfied and the number of samples is fixed as $N = 2^4 = 16$. Now we will consider the following two cases for sampling time interval Δt.

a. When $\Delta t = 1/16$ s
   Sampling time window $T_s = N\Delta t = 16 \cdot (1/16) = 1$ s
   Thus, $T_s = 3\ T =$ integer multiple of T
b. When $\Delta t = 1/20$ s
   Sampling time window $T_s = N\Delta t = 16 \cdot (1/20) = 0.8$ s
   Thus, $T_s = 2.4\ T \neq$ integer multiple of T

When $\Delta t = 1/16$ s, the sampling time window T_s is three times the period T of signal. Thus, as shown in Figure 2.8a, the DFT provides an accurate frequency spectrum from sampled data, with a unique spectral peak at 3 Hz, and the signal reconstructed from the DFT spectrum is exactly the same as the original harmonic function $x(t)$. However, when $\Delta t = 1/20$ s, the sampling time window T_s is not the integer multiple of the period T of the signal. Thus, Figure 2.8b shows that a non-zero frequency spectrum (i.e., sidelobes) appears around the spectral peak at 3 Hz and that the signal reconstructed from the DFT spectrum is no longer the same as the original harmonic function. This occurs because the energy associated with the 3 Hz spectral peak leaks into neighboring frequencies due to the improper sampling time window.

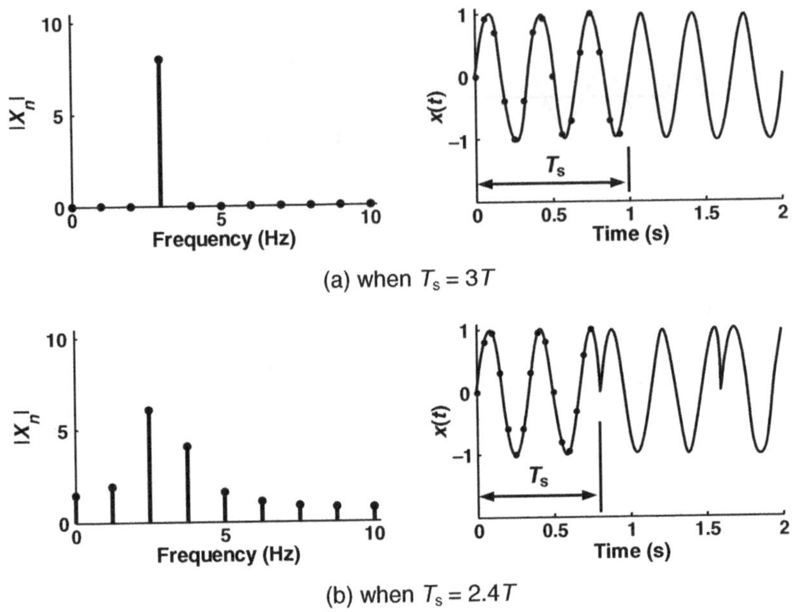

Figure 2.8 Illustrative example of leakage error

Example 2.4

Consider a damped vibration response of a one degree-of-freedom system given by $x(t) = e^{-2t} \cdot \sin 6\pi t$. In fact, the vibration response is not a periodic function. The DFT spectrum of the vibration response is shown in Figure 2.9. As the band frequency f_{max} is infinite, the sampling time interval required to satisfy the Nyquist condition, Equation (2.25), is $\Delta t = 0$ s, however this is impossible in practice. Thus, aliasing cannot be completely eliminated in this case. To investigate the leakage error as a dependence on the size of the sampling time window, we first set $\Delta t = 1/16$ s and then consider two values of N, the number of samples, as follows.

a. When $N = 2^3 = 8$
   Sampling time window $T_s = N\Delta t = 8 \cdot (1/16) = 0.5$ s
b. When $N = 2^4 = 16$
   Sampling time window $T_s = N\Delta t = 16 \cdot (1/16) = 1$ s
c. When $N = 2^5 = 32$
   Sampling time window $T_s = N\Delta t = 32 \cdot (1/16) = 2$ s

As the size of the sampling time window is increased, the DFT spectrum of the sampled signal approaches the exact continuous Fourier spectrum of the vibration response, alongwith a decrease in the leakage error.

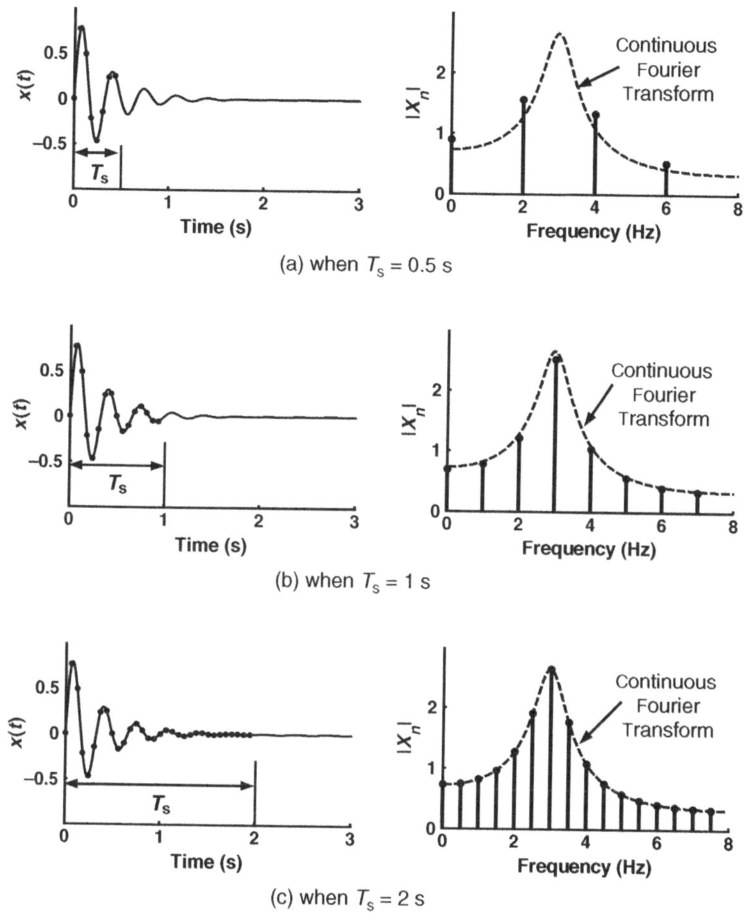

Figure 2.9 Sampling time window dependence on leakage error

2.4.2 *Artificial Damping*

When a system is lightly damped, its dynamic response takes a very long time to decay to a negligible value. If the sampling time window cannot be set up to be larger than the long decay time, leakage error will occur in the DFT processing. A possible remedy to reduce the leakage error is to use an exponential window, which will smoothly attenuate a signal at the ends of the sampled time history so that the windowed signal looks more periodic.

We will show that artificial damping assumes the role of the exponential window in order to remedy leakage. Introducing artificial damping to a system will result in a faster decay of the dynamic response, which is a good sign for the reduction of leakage error for a prefixed sampling time window. The introduction of artificial damping is only an intermediate step in the DFT processing and the final dynamic response for the original unaltered system is recovered by removing the artificial damping at the final step.

Assume that the dynamic response for a system is $x(t)$ and define the modified function by

$$\bar{x}(t) = x(t)\mathrm{e}^{-at} \tag{2.26}$$

where a is a real positive constant that assumes the role of the artificial damping. Applying the DFT to Equation 2.26, we get

$$\begin{aligned} \bar{X}_n &= \sum_{r=0}^{N-1} \bar{x}_r \mathrm{e}^{-i\omega_n t_r} = \sum_{r=0}^{N-1} [x_r \mathrm{e}^{-at_r}]\mathrm{e}^{-i\omega_n t_r} \\ &= \sum_{r=0}^{N-1} x_r \mathrm{e}^{-i(\omega_n - ia)t_r} = X_n(\omega_n - ia) \end{aligned} \tag{2.27}$$

Equation 2.27 shows that the DFT of the dynamic response $x(t)$ of a system altered by artificial damping is obtained by the frequency shifting of the DFT of the dynamic response $\bar{x}(t)$ of the original unaltered system.

For undamped or very lightly damped systems, first derive the analytical form of $X_n = X(\omega_n)$ and then compute $\bar{X}_n = X(\omega_n - ia)$ by replacing ω_n by $(\omega_n - ia)$. We then compute $\bar{x}(t)$ from $\bar{X}_n$ by using the IDFT technique. Finally, the true dynamic response is recovered by removing the artificial damping as follows:

$$x(t) = \bar{x}(t)\mathrm{e}^{at} \tag{2.28}$$

The appropriate value for the arbitrary constant a is suggested to be as follows

$$a = \frac{\ln 100}{T_s}(1/\mathrm{s}) \tag{2.29}$$

The constant a, given by Equation 2.29, ensures that the dynamic response $x(t)$ is reduced by the factor 1/100 at the end of sampling time T_s.

Example 2.5

Consider a damped one degree-of-freedom dynamic system. Assume that it is initially at rest and its motion is represented by the equation of motion as

$$\ddot{x} + \dot{x} + 36\pi^2 x = 20\delta(t)$$

where $\delta(t)$ is the delta function. Its natural frequency is 3 Hz and the dynamic response can be obtained analytically as

$$x_{\mathrm{Exact}}(t) = 1.061\mathrm{e}^{-0.5t}\sin(5.998\pi t)$$

The dynamic response will take more than 9.2 s to decay down by the factor 1/100. Figure 2.10 shows the dynamic responses obtained by using the DFT for $t \le T_s = 2$ s. The DFT spectra used for the DFT processing are given from the equation of motion by

$$X_n = \frac{20}{\Delta t}\left(\frac{1}{36\pi^2 + i\omega_n - \omega_n^2}\right) \qquad (n = 0, 1, 2, \ldots, N-1)$$

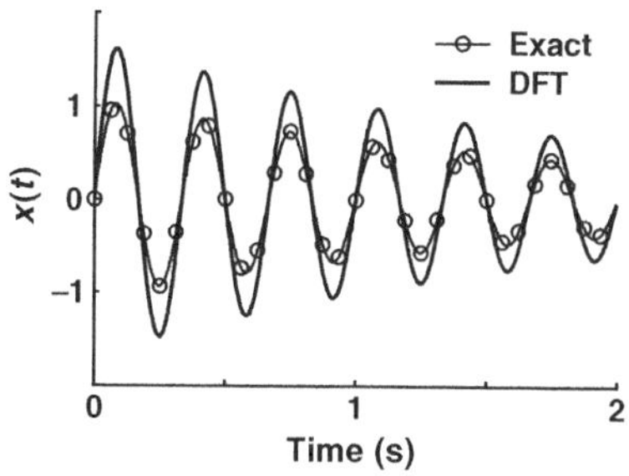

(a) Artificial damping not used

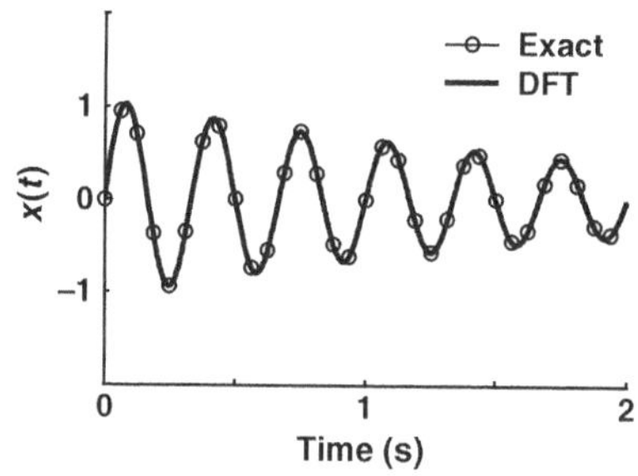

(b) Artificial damping used

Figure 2.10 Artificial damping for leakage reduction

where $N = 1024$ and $\Delta t = 1/512\,\text{s}$ are used. Figure 2.10a shows that the dynamic response obtained without using the artificial damping is not accurate enough when compared with the exact dynamic response $x_{\text{Exact}}(t)$. However, it is obvious from Figure 2.10b that the use of artificial damping improves the accuracy of the DFT application.

2.5 Picket-Fence Effect

The DFT spectrum is a discrete frequency spectrum consisting of estimates of what the spectral level is at specific discrete frequencies, which are limited to integer multiples of the fundamental frequency defined by the reciprocal of the sampling time window. Thus the DFT cannot observe the frequency spectrum as a continuous function. This means that there may be, and probably are, peaks and valleys in the true frequency spectrum of a signal that are between two of the discrete frequencies. As the observation of the frequency spectrum with the DFT is analogous to looking at a mountain range through a picket fence, the true peaks and valleys may not be accurately observed by the DFT, known as the picket-fence effect, or often, the resolution bias error.

The possible remedy for the picket-fence effect is to increase the number of DFT points by using zero padding while holding the original sampling time window fixed. The zero padding will be discussed in the next section.

2.6 Zero Padding

Zero padding is a popular technique used in FFT applications. As the name implies, it means appending a string of zeros at the end, at the beginning, or split between the beginning and end of the finite number of sampled time series data. The common purposes of this process are:

a. to extend the sampled data so that the number of samples N becomes an integer power of 2 (i.e., $N = 2^p$ where $p = \text{integer}$);
b. to improve interpolation in the transformed domain;
c. to remedy the wraparound error.

As discussed in the preceding section, making the number of samples N an integer power of 2 will maximize the computation efficiency of FFT. In the following, further discussion is given for the purposes (b) and (c).

2.6.1 *Improving Interpolation in the Transformed Domain*

The spacing of data in the frequency domain, Δf, is inversely proportional to the number of samples N in the time domain, that is, $\Delta f = 1/T = 1/N \cdot \Delta t$ where T is the period of the signal and Δt is the sampling time interval. Thus, when we increase N by using zero padding, the frequency spectrum will contain more data points in the same frequency range. In other words, zero padding will introduce more points into the same frequency range to provide interpolation between points associated with the unpadded points. Hence, by increasing the number of zeros, we can improve the frequency spectrum using FFT to get closer to those from the continuous Fourier transform, alongwith curing the picket-fence effect. The zero padding can also be applied in the frequency domain. The inverse FFT will increase the data points in the time domain to provide better interpolation between points associated with the unpadded points.

Example 2.6

Consider a signal given by $x(t) = e^{-2t} \cdot \sin 6\pi t$. Assume that the signal is sampled at the sampling time interval $\Delta t = 1/16$ s for the first one second (i.e., $T_s = 1$ s) so that the number of original samples becomes $N = 16$. Figure 2.11 shows the effect of tripling the number of samples in the time domain. The FFT for the zero-padded data (white dots) contains the same information as the unpadded data (black dots), alongwith providing interpolation with two additional data between every three points.

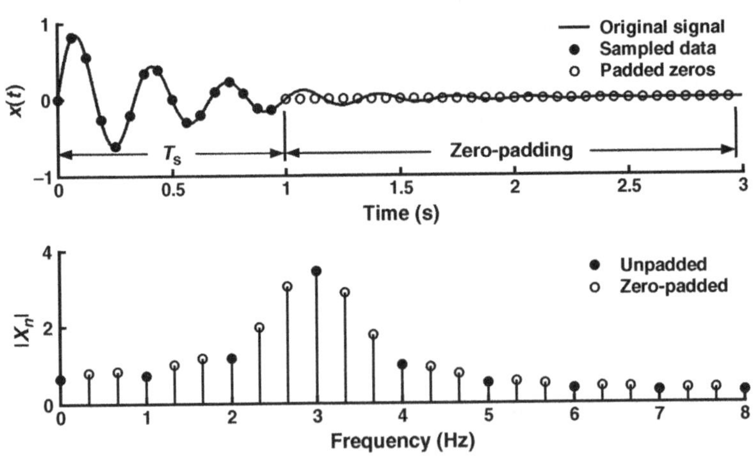

Figure 2.11 The effect of zero padding in the time domain on the frequency-domain data

Example 2.7

Consider the unpadded case of Example 2.6. The frequency spectrum for the unpadded case is zero padded to three times its original length. Figure 2.12 shows the effect of zero padding in the frequency domain. We see that the original unpadded sampled data (black dots) are interpolated by the inverse FFT of the zero-padded frequency spectrum denoted by white dots.

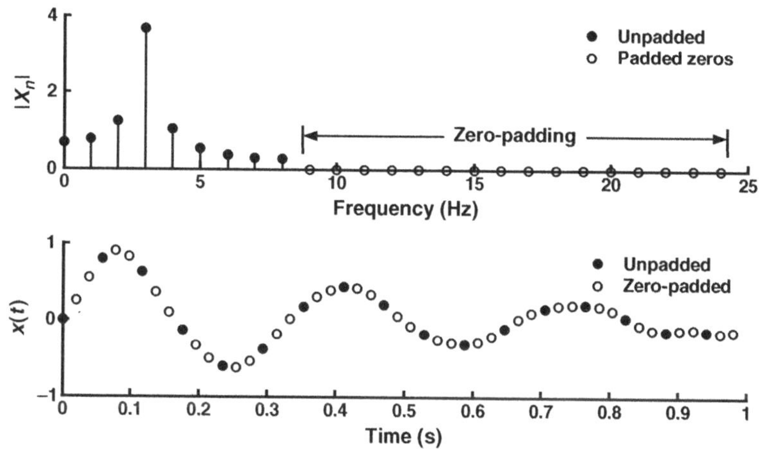

Figure 2.12 The effect of zero padding in the frequency domain on the time-domain data

2.6.2 Remedy for Wraparound Error

The dynamic response $x(t)$ of a system can be obtained from the convolution integral (or Duhamel's integral), as given by

$$x(t) = \int_0^t h(t-\tau)f(\tau)\mathrm{d}\tau \tag{2.30}$$

where $h(t)$ is the unit impulse response function and $f(t)$ is the input forcing function. They satisfy the causality conditions $h(t-\tau)=0$ if $t<\tau$ and $f(t)=0$ if $t<0$.

The DFT representation of Equation 2.30 is given by simple multiplication as

$$X_n = H_n F_n \qquad (n = 0, 1, 2, \ldots, N-1) \tag{2.31}$$

where X_n, H_n and F_n are the DFT coefficients of $x(t)$, $h(t)$ and $f(t)$, respectively. They are defined by

$$(X_n, H_n, F_n) = \sum_{r=0}^{N-1} (x_r, h_r, f_r)\mathrm{e}^{-i\omega_n t_r} \tag{2.32}$$

where $x_r = x(t_r)$, $h_r = h(t_r)$ and $f_r = f(t_r)$. From Equation 2.31, we obtain

$$x_r \cong \frac{1}{N}\sum_{n=0}^{N-1} X_n \mathrm{e}^{i\omega_n t_r} = \frac{1}{N}\sum_{n=0}^{N-1} H_n F_n \mathrm{e}^{i\omega_n t_r} \tag{2.33}$$

By applying Equation 2.32 to Equation 2.33, we are able to obtain the discrete convolution theorem as [50]

$$x_r \sim \sum_{n=0}^{r} h_n f_{r-n} + \sum_{n=r+1}^{N-1} h_n f_{r-n+N} \qquad (r = 0,\ 1,\ 2, \ldots, N-1) \tag{2.34}$$

or

$$\begin{aligned} x_0 &\sim [h_0 f_0] + [h_1 f_{N-1} + \cdots + h_{N/2} f_{N/2} + \cdots + h_{N-1} f_1] \\ x_1 &\sim [h_0 f_1 + h_1 f_0] + [h_2 f_{N-1} + \cdots + h_{N/2} f_{N/2+1} + h_{N/2+1} f_{N/2} + \cdots + h_{N-1} f_2] \\ &\vdots \\ x_{N-2} &\sim [h_0 f_{N-2} + h_1 f_{N-3} + \cdots + h_{N/2-1} f_{N/2-1} + \cdots + h_{N-2} f_0] + [h_{N-1} f_{N-1}] \\ x_{N-1} &\sim [h_0 f_{N-1} + h_1 f_{N-2} + \cdots + h_{N/2-1} f_{N/2} + h_{N/2} f_{N/2-1} + \cdots + h_{N-1} f_0] \end{aligned} \tag{2.35}$$

Equations 2.34 and 2.35 show that the DFT processing requires the values of $h(t)$ and $f(t)$ at the later times rather than the instant at which the value of $x(t)$ is to be computed, which reveals an anomaly when compared with Equation 2.30. Thus, if the DFT requires the data outside of the window $0 \le t \le T$, it will shift the data periodically into the window to cause a so called wraparound error.

The remedy for wraparound error is to use the zero-padding technique. A fail-safe process is to pad the original N sampled data with N zeros in the time domain so that the total number of samples becomes $2N$, which corresponds to an effective time window of $2T$. In this case, all terms in the second bracket, that is, the second sum in Equation 2.34 indeed vanish because they all contain the zero pads of h_r and f_r.

Example 2.8

Revisit the signal $x(t) = e^{-2t} \cdot \sin 6\pi t$ considered in Example 2.6. If we set the sampling time window as $T_s = 1$ s, the DFT will treat the signal as if it is not zero for $t < 0$, which obviously violates the causality condition. To remedy this anomaly, we pad the sampling data with zeros to extend the sampling time window to $2T_s = 2$ s. Figure 2.13 compares the signals for the unpadded and zero-padded cases. It is obvious from Figure 2.13 that the signal with zero padding has been treated by the DFT as being a signal that was quiescent before the action of input force.

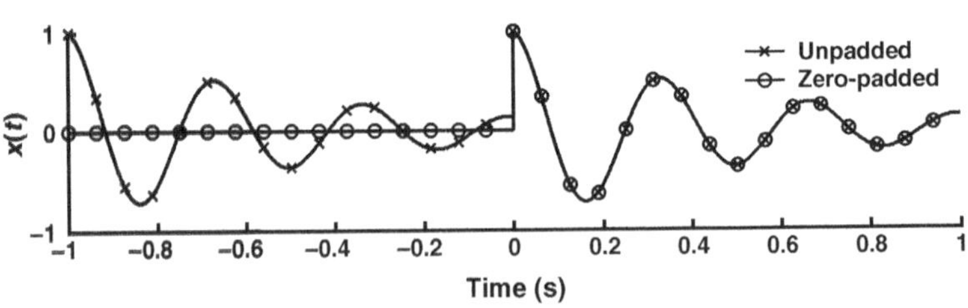

Figure 2.13 Zero padding used to satisfy the causality condition

2.7 Gibbs Phenomenon

The Gibbs phenomenon is a mathematical phenomenon that always occurs when we apply the Fourier series to a function with discontinuities. To numerically illustrate the Gibbs phenomenon, we will consider a periodic function given by

$$x(t) = \begin{cases} -1 & -1 < x < 0 \\ 0 & x = 0 \\ 1 & 0 < x < 1 \end{cases} \tag{2.36}$$

A Fourier series representation of $x(t)$ is given by

$$x(t) = \sum_{n=1}^{\infty} \frac{2(1-\cos n\pi)}{n\pi} \sin n\pi t \tag{2.37}$$

Let us approximate Equation 2.37 with a partial (finite) sum of N terms as

$$x(t)_{\text{approx}} = \sum_{n=1}^{N} \frac{2(1-\cos n\pi)}{n\pi} \sin n\pi t \tag{2.38}$$

By increasing the number of terms N, we expect that the approximate signal $x(t)_{\text{approx}}$ must converge to the original signal $x(t)$. However, this is not true. As shown in Figure 2.14, the signal always oscillates with overshoot near the discontinuity. Even at the limit of an infinite number of terms, the overshoot does not die out, but decreases the extent of the width of oscillation. The overshoot (about 10% of the height of the discontinuity) is a mathematical phenomenon, which is called the Gibbs phenomenon after the American physicist J. Willard Gibbs (1899).

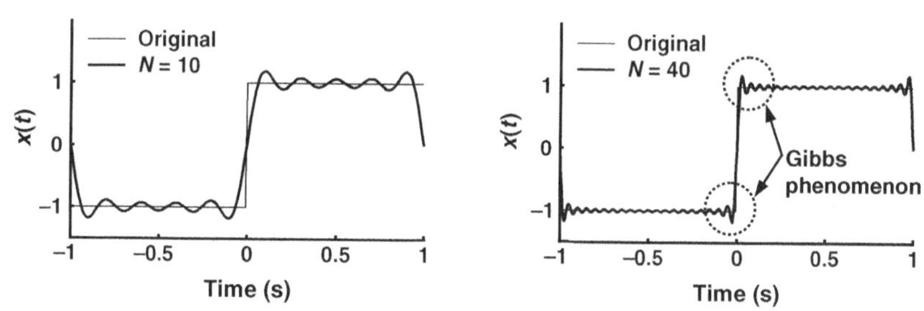

Figure 2.14 Gibbs phenomenon for various numbers of samples N

As the DFT is based on the Fourier series representation of a periodic signal, the Gibbs phenomenon may occur when the DFT is used to reconstruct a signal with discontinuity at which the time increment is much shorter than required by the Nyquist condition. It is worth noting that at a discontinuity, the DFT will provide the average of the values on either side of the discontinuity. The conventional ways to treat the Gibbs phenomenon are as follows:

1. Increase the number of samples to reduce the width of oscillation (ringing).
2. Apply a windowing function to reduce the amplitude of overshoot.

The amplitude of overshoot is proportional to the difference between the first and last sampled data (i.e., the height of the discontinuity of a periodic function). Thus it can be reduced by applying a windowing function to the signal train to taper it to zero at the end points. However, the use of windowing may distort the waveform of the original signal and introduce the spurious frequency components into the signal, that is, aliasing.

2.8 General Procedure of DFT Processing

The dynamic response $x(t)$ of a system is in general determined by its unit impulse response $h(t)$ and the input force $f(t)$, which can be represented by the convolution integral in the time domain as

$$x(t) = \int_0^t h(t-\tau)f(\tau)d\tau \tag{2.39}$$

By using the DFT theory, Equation 2.39 can be transformed into the frequency domain as

$$X_n = H_n F_n \qquad (n = 0, 1, 2, \ldots, N-1) \tag{2.40}$$

where X_n, H_n and F_n are the frequency spectrum (DFT coefficients) of $x(t)$, $h(t)$ and $f(t)$, respectively. The input force $f(t)$ can be considered as a known function because in general it is pre-specified. Once we formulate the spectral element model for a dynamic system, then we can compute the frequency spectrum H_n $(n = 0, 1, \ldots, N-1)$ by the inverse of the spectral element matrix (exact dynamic stiffness matrix). Thus we will assume that the input force $f(t)$ and the frequency spectrum $\{H_n\}$ are all known quantities. Once we compute $\{X_n\}$ from Equation 2.40 by using $\{H_n\}$ and $\{F_n\}$, we can readily compute the dynamic response $x(t)$ by simply taking the inverse FFT of $\{X_n\}$.

In the following, the general procedure of DFT processing to compute the dynamic response of a system is introduced. The general procedure of DFT processing may possibly cure the DFT-induced errors such as aliasing, leakage and wraparound error.

Step 1: Identify the times for the unit impulse response and the input force to decay down to some negligible values and they are represented by T_h and T_f, respectively. Then we determine the sampling time window T_s to satisfy

$$T_s \geq \text{maximum } (T_h \text{ and } T_f) \tag{2.41}$$

Step 2: Identify the highest frequency (band frequency) of the unit impulse response and input force as f_{max}. Then we determine the sampling time interval Δt to satisfy the Nyquist condition as

$$\Delta t \leq \frac{1}{2f_{max}} \tag{2.42}$$

Step 3: Determine the number of samples N to satisfy

$$N = 2^p \geq \frac{T_s}{\Delta t} \qquad (p = \text{integer}) \tag{2.43}$$

Step 4: Compute the artificial damping factor a from

$$a = \frac{\ln 100}{T_s} \cong \frac{4.6}{T_s}(1/s) \tag{2.44}$$

Step 5: Sample N values of input force at the sampling time intervals Δt for $0 \leq t \leq T_s$ as

$$\{f_n\} = \{f_0 \quad f_1, \cdots, f_{N-1}\} \tag{2.45}$$

Step 6: Introduce artificial damping to the sampled data of input force as

$$\{\bar{f}_n\} = \{\bar{f}_0 \quad \bar{f}_1, \cdots, \bar{f}_{N-1}\} \tag{2.46}$$

where

$$\bar{f}_n = f_n e^{-at_n} \qquad (n = 0, 1, \ldots, N-1) \tag{2.47}$$

Step 7: Double the number of samples as

$$N' = 2N \tag{2.48}$$

Step 8: Increase the number of samples of the (artificially damped) input force up to N' by padding zeros between N and $N' - 1$ as

$$\{\bar{f}'_n\} = \{\bar{f}_0 \quad \bar{f}_1 \quad \cdots \quad \bar{f}_{N-1} \quad 0_N \quad 0_{N+1} \quad \cdots \quad 0_{N'-1}\} \tag{2.49}$$

Step 9: Use the FFT algorithm to compute the DFT coefficients from the zero-padded sampled data as

$$\{\bar{F}'_n\} = \text{fft}[\{\bar{f}'_n\}] \qquad (n = 0, 1, \ldots, N' - 1) \tag{2.50}$$

Step 10: Introduce artificial damping to the frequency spectrum of the unit impulse response by the frequency shifting of the original frequency spectrum H_n $(n = 0, 1, \ldots, N' - 1)$ as

$$\bar{H}'_n = H_n(\omega_n - ia), \quad \bar{H}'_{N'-n} = \bar{H}'^{*}_n \qquad (n = 0, 1, \ldots, N'/2) \tag{2.51}$$

In SEM, the frequency spectrum $\{H_n\}$ can be obtained in analytical forms from the inverse of the global exact dynamic stiffness matrix. However, if the time-domain unit impulse responses are provided instead of the frequency spectrum, one may need to follow the steps 5–9 to compute the frequency spectrum $\{H_n\}$ from the unit impulse responses.

Step 11: Compute the (artificially damped) frequency-domain response as

$$\bar{X}'_n = \bar{H}'_n \bar{F}'_n, \quad \bar{X}'_{N-n} = \bar{X}'^{*}_n \qquad (n = 0, 1, \ldots, N'/2) \tag{2.52}$$

Step 12: Use the inverse FFT algorithm to compute the (artificially damped) time domain response as

$$\{\bar{x}'_n\} = \text{ifft}[\{\bar{X}'_n\}] \qquad (n = 0, 1, \ldots, N' - 1) \tag{2.53}$$

Step 13: Discard the values of $\bar{x}'_n$ for $n \geq N$, which are probably contaminated by the wraparound error, to obtain

$$\{\bar{x}_n\} = \{\bar{x}_0 \quad \bar{x}_1 \quad \cdots \quad \bar{x}_{N-1}\} \tag{2.54}$$

Step 14: Finally, remove the artificial damping effect from Equation 2.54 to obtain the true time domain response as

$$x_n = \bar{x}_n e^{at_n} \qquad (n = 0, 1, \ldots, N-1) \tag{2.55}$$

The steps (1), (4), (6) and (10) are to cure the leakage error, step (2) to cure the aliasing error and Gibbs phenomenon, steps (7) and (8) to cure the wraparound error and the picket-fence effect and to improve the interpolation, and step (3) is to maximize the computation efficiency of the FFT algorithm.

Example 2.9

Consider a damped 1-DOF dynamic system which is initially at rest as

$$m\ddot{x} + c\dot{x} + kx = f(t)$$

where $f(t)$ is the square impulse force given by

$$f(t) = \begin{cases} f_0 & \text{for} \quad 0 \leq t \leq t_1 \text{ s} \\ 0 & \text{for} \quad \text{elsewhere} \end{cases}$$

The exact dynamic response can be obtained in a closed form as

$$\frac{k}{f_0} x(t)_{\text{Exact}} = \left[1 - \frac{e^{-\xi\omega_N t}}{\sqrt{1-\xi^2}} \cos(\omega_d t - \phi)\right] u(t)$$
$$- \left[1 - \frac{e^{-\xi\omega_N (t-t_1)}}{\sqrt{1-\xi^2}} \cos[\omega_d (t-t_1) - \phi]\right] u(t-t_1)$$

where

$$\omega_N = \sqrt{\frac{k}{m}}, \qquad \omega_d = \omega_N \sqrt{1-\xi^2}$$

$$\xi = \frac{c}{2m\omega_N}, \qquad \phi = \tan^{-1}\left(\frac{\xi}{\sqrt{1-\xi^2}}\right)$$

and $u(t)$ is the unit step function. In the following, the exact dynamic response shown above is compared with the dynamic response obtained by using the DFT processing. For numerical computations, we used the data as follows: $m = 1$ kg, $c = 0.1$ kg/s, $k = 36\pi^2$ N/m, $f_0 = 20$ N, and $t_1 = 1$ s.

To compute the dynamic response by using the DFT processing, we conduct the following process.

Step 1: The unit impulse response of the dynamic system is given by

$$h(t) = 0.053e^{-0.05t}\sin(5.999\pi t)$$

which is found to take more than 92 s to decay down by a factor of 1/100. Thus the recommended sampling time window should be larger than the values $T_h = 92$ s and $T_f = 1$ s to make the aliasing error sufficiently small. Thus, from Equation 2.41, T_s should be larger than 92 s. However, we will set

$$T_s = 2\ \text{s}$$

to clearly investigate the effect of artificial damping on the dynamic responses obtained by using the DFT processing.

Step 2: Figure 2.15 shows the continuous Fourier spectrums for the unit impulse response $h(t)$ and the square impulse force $f(t)$. As the Fourier spectrums range from zero to infinite, we cannot help to truncate the Fourier spectrums at a specific frequency. Thus, we set

$$f_{\max} = 25\ \text{Hz}$$

Accordingly the sampling time interval is determined from Equation 2.42 as

$$\Delta t = 0.01\ \text{s} \le 1/(2f_{\max}) = 0.02\ \text{s}$$

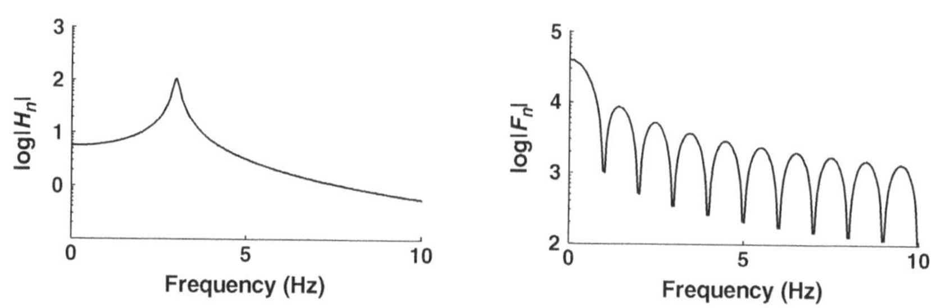

Figure 2.15 The continuous Fourier spectrums of $h(t)$ and $f(t)$

Step 3: Determine the number of samples from Equation 2.43 as

$$N = 2^8 = 256 > 2/\Delta t = 200$$

Step 4: Compute the artificial damping factor a from Equation 2.44 as

$$a = 4.6/T_s = 2.3(1/\text{s})$$

We then follow the remaining steps 5–14 to compute the dynamic response in the time domain. In Figure 2.16, the dynamic responses computed by using the DFT processing, with and without considering artificial damping (AD), are compared with the exact dynamic

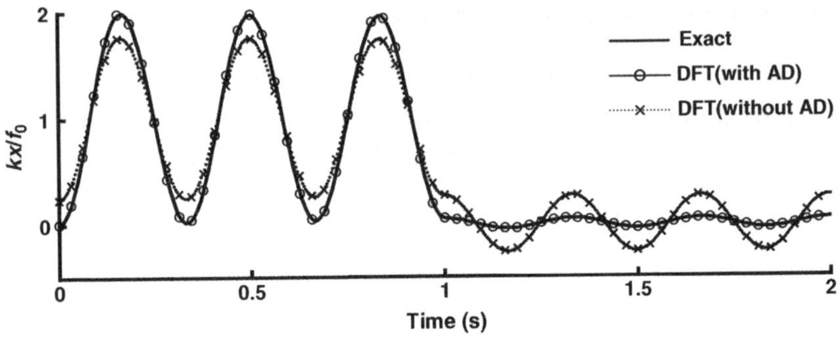

Figure 2.16 Dynamic responses obtained by the DFT processing with and without considering artificial damping (AD)

response provided in a closed form. It shows that artificial damping may enable us to obtain a satisfactory result even though the sampling time window T_s chosen for DFT processing is not sufficiently large.

2.9 DFTs of Typical Functions

2.9.1 *Product of Two Functions*

Consider a function $z(t)$ that is defined by a product of two functions $x(t)$ and $y(t)$ as

$$z(t) = x(t)y(t) \tag{2.56}$$

It is obvious from Equation 2.56 that the discrete time series $\{z_r\}$ for $z(t)$ are given by

$$z_r = x_r y_r \qquad (r = 0,\ 1,\ 2, \ldots,\ N-1) \tag{2.57}$$

The discrete time series x_r, y_r and z_r are related to the corresponding DFT spectrum X_n, Y_n and Z_n as follows:

$$\begin{aligned} x_r &= \frac{1}{N}\sum_{n=0}^{N-1} X_n e^{i\omega_n t_r} \\ y_r &= \frac{1}{N}\sum_{n=0}^{N-1} Y_n e^{i\omega_n t_r} \qquad (r = 0,\ 1, 2, \ldots,\ N-1) \\ z_r &= \frac{1}{N}\sum_{n=0}^{N-1} Z_n e^{i\omega_n t_r} \end{aligned} \tag{2.58}$$

Substituting Equation 2.58 into Equation 2.57 and using Equations 2.12 and 2.13, we can get a discrete convolution formula as

$$Z_n = \frac{1}{N}\sum_{j=0}^{N-1} X_j Y_{n-j} \qquad (n = 0,\ 1,\ 2, \ldots,\ N-1) \tag{2.59a}$$

or

$$Z_n = \frac{1}{N}\left(\sum_{j=0}^{n} X_j Y_{n-j} + \sum_{j=n+1}^{N-1} X_j Y_{N+n-j}\right) \qquad (n = 0,\ 1,\ 2, \ldots,\ N-1) \tag{2.59b}$$

or

$$Z_n = \frac{1}{N}\left(\sum_{j=0}^{n} X_j Y_{n-j} + \sum_{j=n+1}^{N/2} X_j Y^*_{j-n} + \sum_{j=1}^{N/2-1} X^*_j Y_{n+j}\right) \qquad (n = 0,\ 1,\ 2, \ldots, N/2)$$

$$Z_{N-n} = Z^*_n \tag{2.59c}$$

where the symbol (*) denotes the complex conjugate. Thus, the DFT coefficients for a function $z(t)$ can be readily computed from the DFT coefficients for $x(t)$ and $y(t)$ by using the discrete convolution formula of Equation 2.59.

Similarly to the time-domain discrete convolution theorem represented by Equation 2.34, the DFT coefficients computed by Equation 2.59 can be contaminated by the wraparound error in the frequency domain (i.e., aliasing). This product aliasing error might be cured by using the zero padding in the frequency domain.

Example 2.10

Consider a function $z(t) = x(t)\ y(t)$ that is determined by two functions given by

$$x(t) = \sin 2\pi t + 0.5 \sin 6\pi t + \cos 12\pi t + 2 \cos 14\pi t$$
$$y(t) = 2 \cos 4\pi t + \cos 8\pi t + 0.5 \sin 10\pi t + 1.5 \sin 16\pi t$$

Figure 2.17 shows that the DFT coefficients obtained by using the formula in Equation 2.59 are identical with those obtained by applying the FFT algorithm directly to $z(t)$.

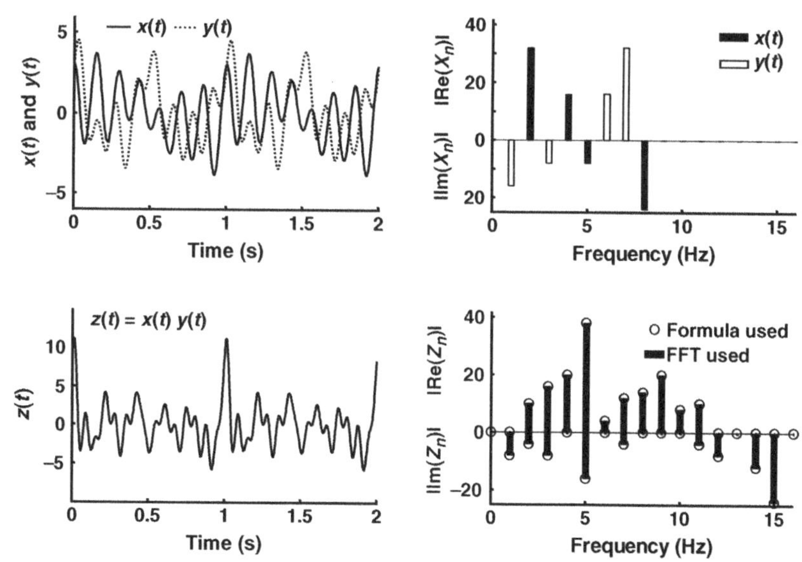

Figure 2.17 DFT coefficients computed by using the discrete convolution formula and the FFT

2.9.2 Derivative of a Function

Consider a Fourier series for a function $x(t)$ as

$$x(t) = \sum_{n=-\infty}^{\infty} X_n e^{i\omega_n t} \tag{2.60}$$

where X_n is the nth Fourier coefficient. The Fourier series for the derivative of $x(t)$ and its inverse Fourier transform are given by Equation 2.60 as

$$\dot{x}(t) = \sum_{n=-\infty}^{\infty} i\omega_n X_n e^{i\omega_n t} \equiv \sum_{n=-\infty}^{\infty} \dot{X}_n e^{i\omega_n t} \tag{2.61}$$

$$\dot{X}_n = \frac{1}{T}\int_0^T \dot{x}(t)e^{-i\omega_n t}dt \qquad (n = 0, \pm 1, \pm 2, \ldots, \pm\infty) \tag{2.62}$$

where $\dot{X}_n = i\omega_n X_n$ represent the Fourier coefficients of $\dot{x}(t)$. Following the DFT theory introduced in Section 2.2.1, we can write Equations 2.61 and 2.62 into the DFT pair as

$$\dot{x}_r \cong \frac{1}{N}\sum_{n=0}^{N-1} \dot{X}_n e^{i\omega_n t_r} \qquad (r = 0, 1, 2, \ldots, N-1) \tag{2.63}$$

$$\dot{X}_n = \sum_{r=0}^{N-1} \dot{x}_r e^{-i\omega_n t_r} \qquad (n = 0, 1, 2, \ldots, N-1) \tag{2.64}$$

Thus, the DFT coefficients for the derivative of a function $x(t)$ are given by

$$\begin{aligned} \dot{X}_n &= i\omega_n X_n \\ \dot{X}_{N-n} &= \dot{X}_n^* \end{aligned} \qquad (n = 0, 1, 2, \ldots, N/2) \tag{2.65}$$

Repeating the above process, the DFT coefficients for the kth derivative of a function $x(t)$ are given by

$$\begin{aligned} X_n^{(k)} &= (i\omega_n)^k X_n \\ X_{N-n}^{(k)} &= \dot{X}_n^{(k)*} \end{aligned} \qquad (n = 0, 1, 2, \ldots, N/2) \tag{2.66}$$

2.9.3 Other Typical Functions

Table 2.2 lists the DFTs for some typical functions such as the delta function (unit impulse function) $\delta(t)$, the unit step function $u(t)$, and the sine and cosine functions, which we often meet during the DFT/FFT-based spectral analysis. By using the formula given in Table 2.2, we can obtain the DFT coefficients X_n for a typical function as follows:

$$\begin{aligned} &X_n = \text{gvien by the formula of Table 2.2} \\ &X_{N-n} = X_n^* \end{aligned} \qquad (n = 0, 1, \ldots, N/2) \tag{2.67}$$

For the functions (5), (6) and (7) in Table 2.2, the unit step function $u(t)$ is introduced to emphasize that they have no values before the time $t = 0$ and the constant parameter a plays a

Table 2.2 The DFT coefficients of typical functions

Function $x(t)$		Sampled data x_r	DFT coefficients X_n
(1)	delta function $\delta(t)$	$\begin{cases} 1/\Delta t & \text{if } t_r = 0 \\ 0 & \text{otherwise} \end{cases}$	$\dfrac{1}{\Delta t}$
(2)	unit step function $u(t)$	1	$\begin{cases} N & \text{if } n = 0 \\ 0 & \text{otherwise} \end{cases}$
(3)	$\delta(t - t_{\text{shift}})$	$\begin{cases} 1/\Delta t & \text{if } t_r = t_{\text{shift}} \\ 0 & \text{otherwise} \end{cases}$	$\dfrac{1}{\Delta t} e^{-i\omega_n t_{\text{shift}}}$
(4)	$u(t) - u(t - t_{\text{shift}})$	$\begin{cases} 1 & \text{if } 0 < t_r < t_{\text{shift}} \\ 0 & \text{otherwise} \end{cases}$	$\dfrac{t_{\text{shift}}}{\Delta t} \operatorname{sinc}\left(\dfrac{\omega_n t_{\text{shift}}}{2\pi}\right) e^{-i\frac{\omega_n t_{\text{shift}}}{2}}$
(5)	$e^{-at} u(t)$	e^{-at_r}	$-\dfrac{i}{\Delta t(\omega_n - ia)}$
(6)	$e^{-at} \sin \bar{\omega} t\, u(t)$	$e^{-at_r} \sin \bar{\omega} t_r$	$\dfrac{\bar{\omega}}{\Delta t\left[\bar{\omega}^2 - (\omega_n - ia)^2\right]}$
(7)	$e^{-at} \cos \bar{\omega} t\, u(t)$	$e^{-at_r} \cos \bar{\omega} t_r$	$\dfrac{i(\omega_n - ia)}{\Delta t\left[\bar{\omega}^2 - (\omega_n - ia)^2\right]}$

Note: $\Delta t = T/N$, where T is the sampling time (or period) and N is the number of samples.

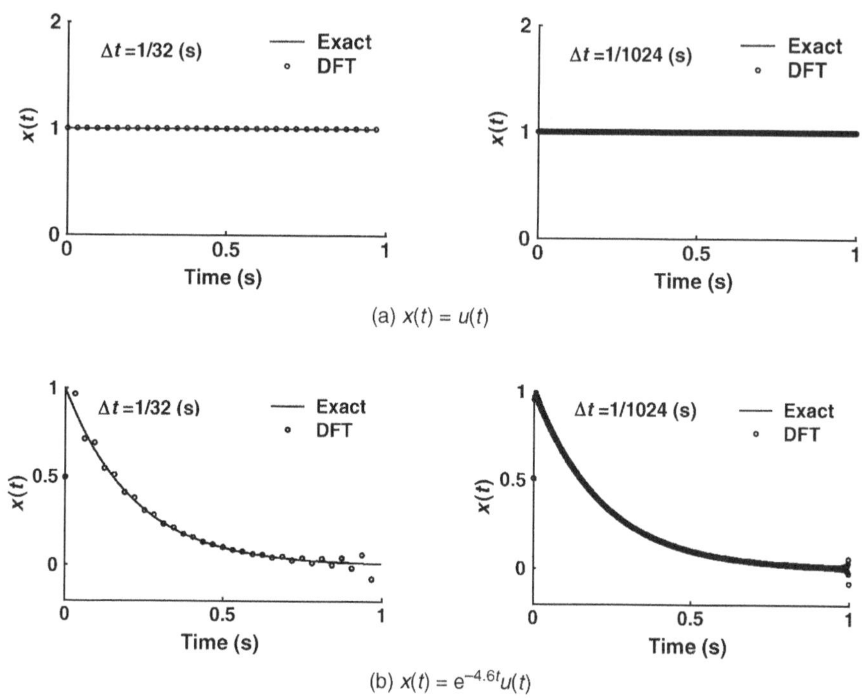

Figure 2.18 Time histories reconstructed by using the DFT formula from Table 2.2

role as the damping. Even if the constant parameter a is initially zero (i.e., no damping), it is still recommended to compute the DFT coefficients by using the formula given in the last column of the Table 2.2 by now considering the constant parameter a as the artificial damping parameter. In this instance, the original functions with $a = 0$ should be recovered by taking off the artificial damping effect through using Equation 2.28.

Figure 2.18 shows the time histories of $x(t) = e^{-4.6t}u(t)$ and $x(t) = u(t)$ by using the DFT formula (5) from Table 2.2 when the sampling time window is set as $T_s = 1$ s. Owing to the discontinuities at $t = 0$ and $t = 1$ s, we can observe the Gibbs phenomenon near the discontinuities. The Gibbs phenomenon might be alleviated by increasing the number of samples N, and by increasing the sampling time window, for instance, twice as $2T_s$ and then ignoring the reconstructed time history for $t \geq T_s$ at the end.

Part Two

Theory of Spectral Element Method

3

Methods of Spectral Element Formulation

The formulation of the spectral element model for a structure element is fairly similar to the formulation of the conventional finite element models. However, the major difference is that in general the spectral element formulation begins with transforming the governing partial differential equations of motion from the time domain into the frequency domain by using the DFT. For one-dimensional (1-D) problems, as a result of the DFT processing, the time variable disappears and the frequency becomes a parameter to transform the original time-domain partial differential equations into the frequency-domain ordinary differential equations. The frequency-domain ordinary differential equations are then solved exactly and the exact wave solutions are used to derive frequency-dependent dynamic shape functions. The exact dynamic stiffness matrix called the spectral element matrix is finally formulated by using the dynamic shape functions in an analogous way to that commonly used in conventional FEM. In this chapter, three methods of spectral element formulation are introduced: the force-displacement relation method, the variational method, and the state-vector equation method. The methods of spectral element formulation are then applied to some representative 1-D structures, such as the axial rod, Bernoulli–Euler beam, Timoshenko beam, and extended Timoshenko beam.

3.1 Force-Displacement Relation Method

In this section, the twin concepts of equilibrium and compatibility with exact solutions to the governing equations of motion, that is the direct relationship between nodal forces and nodal displacements, are used to formulate the spectral element model for a structure.

The governing equations of motion of a structure undergoing forced vibration can be symbolically represented as

$$\boldsymbol{\mathcal{L}}\,\boldsymbol{u}(x,t)+\boldsymbol{M}\ddot{\boldsymbol{u}}(x,t)=\boldsymbol{p}(x,t) \tag{3.1}$$

Spectral Element Method in Structural Dynamics Usik Lee

where $\boldsymbol{\mathcal{L}}$ is the linear differential structural operator in the time t and spatial coordinate x domains and $\boldsymbol{M}$ is the inertial operator. The dots $(\cdot)$ denote the derivatives with respective to time t. $\boldsymbol{u}(x, t)$ and $\boldsymbol{p}(x, t)$ are the displacement fields vectors and external forces vector, respectively. Assume that the external forces can be represented in the spectral form by

$$\boldsymbol{p}(x,t) = \frac{1}{N}\sum_{n=0}^{N-1} \boldsymbol{P}_n(x;\omega_n)e^{i\omega_n t} \tag{3.2}$$

where $\boldsymbol{P}_n(x;\omega)$ are the spectral components (or Fourier coefficients) of the external forces $\boldsymbol{p}(x,t)$. The solution to Equation 3.1 can be then assumed in a form similar to

$$\boldsymbol{u}(x,t) = \frac{1}{N}\sum_{n=0}^{N-1} \boldsymbol{U}_n(x;\omega_n)e^{i\omega_n t} \tag{3.3}$$

where $\boldsymbol{U}_n(x;\omega)$ are the spectral components of the displacement fields $\boldsymbol{u}(x,t)$. By substituting Equations 3.2 and 3.3 into Equation 3.1 and assuming that the spectral components $\boldsymbol{U}_n$ and $\boldsymbol{P}_n$ satisfy Equation 3.1 at each discrete frequency ω_n, we can obtain

$$\boldsymbol{L}\,\boldsymbol{U}_n(x;\omega_n) - \omega_n^2 \boldsymbol{M}\,\boldsymbol{U}_n(x;\omega_n) = \boldsymbol{P}_n(x;\omega_n) \tag{3.4}$$

where $\boldsymbol{L}$ is the linear differential structural operator in the spatial coordinate x domain.

The frequency-dependent shape functions, which are termed dynamic shape functions and used to formulate the spectral element matrix, are obtained from the homogeneous equation

$$\boldsymbol{L}\,\boldsymbol{U}_n(x) - \omega_n^2 \boldsymbol{M}\,\boldsymbol{U}_n(x) = \boldsymbol{0} \tag{3.5}$$

Equation 3.5 holds for all discrete frequencies and each discrete frequency ω_n then becomes aparameter. For brevity, the subscripts n that appeared in Equation 3.5 will be omitted in the following:

$$\boldsymbol{L}\,\boldsymbol{U}(x) - \omega^2 \boldsymbol{M}\,\boldsymbol{U}(x) = \boldsymbol{0} \tag{3.6}$$

The same form of equation as Equation 3.6 can be derived by simply assuming harmonic solutions for the homogeneous Equation 3.1, which has been popularl in the literature associated with the dynamic stiffness method.

The general solution to Equation 3.6 can be assumed in the form

$$\boldsymbol{U}(x) = \boldsymbol{c}e^{-ikx} \tag{3.7}$$

where $\boldsymbol{c}$ is a constant vector and k is the wavenumber. Substituting Equation 3.7 into Equation 3.6 yields an eigenvalue problem as

$$\boldsymbol{A}(k;\omega)\boldsymbol{c} = \boldsymbol{0} \tag{3.8}$$

The conditions for the existence of the nontrivial solution $\boldsymbol{c}$ of Equation 3.8 requires that the determinant of $\boldsymbol{A}$ $(k;\omega)$ must be zero as

$$\det \boldsymbol{A}(k;\omega) = 0 \tag{3.9}$$

which yields a characteristic equation for the wavenumber k in the form

$$k^p + \alpha_{(p-1)}(\omega)k^{(p-1)} + \alpha_{(p-2)}(\omega)k^{(p-2)} + \cdots + \alpha_1(\omega)k + \alpha_0(\omega) = 0 \tag{3.10}$$

Equation 3.10 is called a dispersion relation or spectrum relation. Assume that $k_1, k_2, \ldots, k_p$ are the distinct roots of Equation 3.10 at a discrete frequency ω. The associated eigenvector $\boldsymbol{c}_i$ can be then computed from Equation 3.8 as

$$\boldsymbol{c}_i = a_i \begin{Bmatrix} 1 \\ \boldsymbol{\beta}_i \end{Bmatrix} = a_i \boldsymbol{\phi}_i \quad (i = 1, 2, \ldots, p) \tag{3.11}$$

where the eigenvector $\boldsymbol{c}_i$ is normalized so that a component of the normalized vector $\boldsymbol{\phi}_i$ becomes unity and the other components are collected as the vector $\boldsymbol{\beta}_i$. The constants a_i will be determined to satisfy the associated boundary conditions. Once the eigensolutions k_i and $\boldsymbol{c}_i$ are obtained to satisfy the eigenvalue problem in Equation 3.8, the general solutions to Equation 3.6 can be written as

$$\boldsymbol{U}(x) = \sum_{i=1}^{p} \boldsymbol{\phi}_i e^{-ik_i(\omega)x} a_i \equiv \boldsymbol{E}(x;\omega)\boldsymbol{a} \tag{3.12}$$

where

$$\begin{aligned} \boldsymbol{E}(x;\omega) &= [\boldsymbol{\phi}_1\, \boldsymbol{\phi}_2\, \boldsymbol{\phi}_3\, \ldots\, \boldsymbol{\phi}_p]\, \boldsymbol{\Lambda}(x;\omega) \\ \boldsymbol{\Lambda}(x;\omega) &= diag\, [e^{-i\,k_i(\omega)x}] \\ \boldsymbol{a} &= \{a_1\, a_2\, a_3\, \ldots\, a_p\}^{\mathrm{T}} \end{aligned} \tag{3.13}$$

For a finite element of length L, Equation 3.12 must satisfy the geometric and natural boundary conditions at two end-nodes $x = 0$ and $x = L$: the spectral nodal displacements and slopes (simply, nodal degrees-of-freedom, DOFs) vector $\boldsymbol{d}$ and the spectral nodal forces (simply, nodal forces) vector $\boldsymbol{f}$.

The displacement and slope are generally called the primary variables and their specification on the boundaries constitutes the geometric boundary conditions. The primary variables can be related to the displacement fields in the form

$$\boldsymbol{D}(x) = \boldsymbol{L}_{\mathrm{GB}}\, \boldsymbol{U}(x) \tag{3.14}$$

where $\boldsymbol{L}_{\mathrm{GB}}$ is the linear differential operator for the geometric boundary conditions. Substituting Equation 3.12 into Equation 3.14 and then considering the nodal DOFs $\boldsymbol{d}$ specified at two end-nodes of the finite element gives

$$\boldsymbol{d} = \begin{Bmatrix} \boldsymbol{D}(0) \\ \boldsymbol{D}(L) \end{Bmatrix} = \begin{bmatrix} \boldsymbol{L}_{\mathrm{GB}}\boldsymbol{E}(x;\omega)|_{x=0} \\ \boldsymbol{L}_{\mathrm{GB}}\boldsymbol{E}(x;\omega)|_{x=L} \end{bmatrix} \boldsymbol{a} = \boldsymbol{H}(\omega)\, \boldsymbol{a} \tag{3.15}$$

By eliminating the constant vector $\boldsymbol{a}$ from Equation 3.12 by using Equation 3.15, the general solutions can be rewritten in terms of the nodal DOFs vector as

$$\boldsymbol{U}(x) = \boldsymbol{N}(x;\omega)\boldsymbol{d} \tag{3.16}$$

where $\boldsymbol{N}(x,\omega)$ is the dynamic shape function defined by

$$\boldsymbol{N}(x;\omega) = \boldsymbol{E}(x;\omega)\boldsymbol{H}^{-1}(\omega) \tag{3.17}$$

The internal forces and moment are generally called the secondary variables and their specification on the boundary constitutes the natural boundary conditions. The strength of materials provides the relationship between the secondary variables and the displacement fields in the form of

$$\boldsymbol{F}(x) = \boldsymbol{L}_{\mathrm{NB}}\boldsymbol{U}(x) \tag{3.18}$$

where $\boldsymbol{L}_{\mathrm{NB}}$ is the linear differential operator for the natural boundary conditions. Substituting Equation 3.16 into Equation 3.18 and then considering the nodal forces $\boldsymbol{f}_{\mathrm{c}}$, which are the concentrated loads specified at two end-nodes of the finite element, yields the spectral element equation

$$\boldsymbol{f}_{\mathrm{c}} \equiv \begin{Bmatrix} -\boldsymbol{F}(0) \\ +\boldsymbol{F}(L) \end{Bmatrix} = \begin{bmatrix} -\boldsymbol{L}_{\mathrm{NB}}\boldsymbol{N}(x;\omega)|_{x=0} \\ +\boldsymbol{L}_{\mathrm{NB}}\boldsymbol{N}(x;\omega)|_{x=L} \end{bmatrix} \boldsymbol{d} \equiv \boldsymbol{S}(\omega)\boldsymbol{d} \tag{3.19}$$

where

$$\boldsymbol{S}(\omega) = \begin{bmatrix} -\boldsymbol{L}_{\mathrm{NB}}\boldsymbol{N}(x;\omega)|_{x=0} \\ +\boldsymbol{L}_{\mathrm{NB}}\boldsymbol{N}(x;\omega)|_{x=L} \end{bmatrix} = \boldsymbol{G}(\omega)\,\boldsymbol{H}^{-1}(\omega) \tag{3.20}$$

with

$$\boldsymbol{G}(\omega) = \begin{bmatrix} -\boldsymbol{L}_{\mathrm{NB}}\boldsymbol{E}(x;\omega)|_{x=0} \\ +\boldsymbol{L}_{\mathrm{NB}}\boldsymbol{E}(x;\omega)|_{x=L} \end{bmatrix} \tag{3.21}$$

The matrix $\boldsymbol{S}(\omega)$ is the frequency-dependent exact dynamic stiffness matrix, which is termed the spectral element matrix in the literature. In Equation 3.19, the plus (+) and minus (−) signs reflect the difference between the sign conventions used in the theory of the strength of materials and those commonly adopted in the element methods, such as the conventional FEM and the SEM.

The recent development of symbolic computing software makes it easier to obtain the explicit analytical expression of the spectral element matrix. However, when this is formidable, computer implementation of the spectral element matrix formulation procedure can be accomplished numerically. In the following, the spectral element matrices are derived for the axial rod, the Bernoulli–Euler beam (simply the B-beam), the Timoshenko beam (simply the T-beam), and the extended Timoshenko beam (simply the ET-beam) elements as illustrative examples.

Example 3.1 Rod

The free longitudinal vibration of a uniform rod is represented by

$$EAu'' - \rho A\ddot{u} = 0 \tag{3.22}$$

where $u(x,t)$ is the longitudinal displacement, E is the Young's modulus, A is the cross-sectional area, and ρ is the mass density. The prime ($\prime$) denotes the derivatives with respective to the spatial coordinate x. The internal axial force is given by

$$N_t(x,t) = EAu'(x,t) \tag{3.23}$$

where the subscript t is used to denote the quantity in the time domain.

The solution to Equation 3.22 is assumed in the spectral form as

$$u(x,t) = \frac{1}{N}\sum_{n=0}^{N-1} U_n(x;\omega_n)\, \mathrm{e}^{i\omega_n t} \tag{3.24}$$

Substituting Equation 3.24 into Equation 3.22 yields an eigenvalue problem for a specific discrete frequency (for example, $\omega = \omega_n$) as

$$EAU'' + \omega^2 \rho A U = 0 \tag{3.25}$$

The general solution to Equation 3.25 is assumed to be

$$U(x) = a\mathrm{e}^{-ik(\omega)x} \tag{3.26}$$

By substituting Equation 3.26 into Equation 3.25, a dispersion relation can be obtained as

$$k^2 - k_{\mathrm{L}}^2 = 0 \tag{3.27}$$

where k_{L} is the wavenumber for the pure longitudinal wavemode and it is defined by

$$k_{\mathrm{L}} = \omega\sqrt{\frac{\rho A}{EA}} \tag{3.28}$$

Equation 3.27 gives two real roots as

$$k_1 = -k_2 = k_{\mathrm{L}} \tag{3.29}$$

For a finite rod element of length L, the general solution to Equation 3.25 can be then obtained in the form

$$U(x) = a_1\mathrm{e}^{-ik_{\mathrm{L}}x} + a_2\mathrm{e}^{+ik_{\mathrm{L}}x} = \boldsymbol{e}(x;\omega)\boldsymbol{a} \tag{3.30}$$

where

$$\begin{aligned} \boldsymbol{e}(x;\omega) &= [\mathrm{e}^{-ik_{\mathrm{L}}x} \quad \mathrm{e}^{+ik_{\mathrm{L}}x}] \\ \boldsymbol{a} &= \{a_1 a_2\}^{\mathrm{T}} \end{aligned} \tag{3.31}$$

The spectral nodal displacements of the finite rod element (see Figure 3.1) can be related to the displacement field as

$$\boldsymbol{d} = \begin{Bmatrix} U_1 \\ U_2 \end{Bmatrix} = \begin{Bmatrix} U(0) \\ U(L) \end{Bmatrix} \tag{3.32}$$

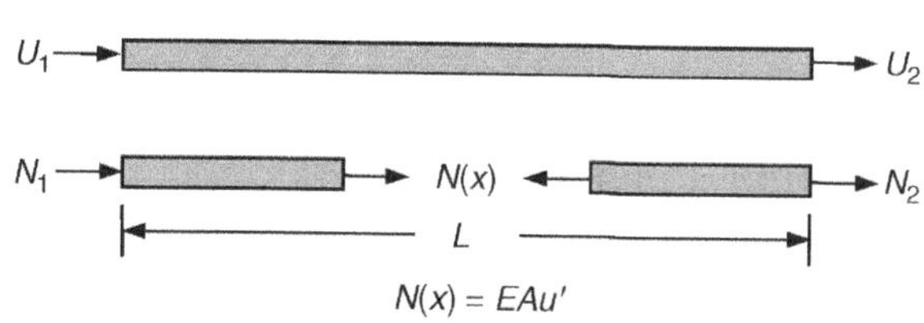

Figure 3.1 Sign convention for the rod element

Substituting Equation 3.30 into the right-hand side of Equation 3.32 gives

$$\boldsymbol{d} = \begin{bmatrix} \boldsymbol{e}(0;\omega) \\ \boldsymbol{e}(L;\omega) \end{bmatrix} \boldsymbol{a} = \boldsymbol{H}_R(\omega)\boldsymbol{a} \tag{3.33}$$

where

$$\boldsymbol{H}_R(\omega) = \begin{bmatrix} 1 & 1 \\ \mathrm{e}^{-ik_{\mathrm{L}}L} & \mathrm{e}^{+ik_{\mathrm{L}}L} \end{bmatrix} \tag{3.34}$$

By eliminating the constant vector $\boldsymbol{a}$ from Equation 3.30 by using Equation 3.33, we can represent the displacement field in the finite rod element in terms of the nodal DOFs as

$$U(x) = \boldsymbol{N}_R(x;\omega)\boldsymbol{d} \tag{3.35}$$

where

$$\begin{aligned} \boldsymbol{N}_R(x;\omega) &= \boldsymbol{e}(x;\omega)\boldsymbol{H}_R^{-1}(\omega) = [N_{R1} \quad N_{R2}] \\ N_{R1}(x;\omega) &= \csc(k_{\mathrm{L}}L)\sin[k_{\mathrm{L}}(L-x)] \\ N_{R2}(x;\omega) &= \csc(k_{\mathrm{L}}L)\sin(k_{\mathrm{L}}x) \end{aligned} \tag{3.36}$$

From Equation 3.23, the spectral components of the axial force are related to $U(x)$ by

$$N(x) = EAU'(x) \tag{3.37}$$

The spectral nodal axial forces defined for the finite rod element (see Figure 3.1) can be related to the forces defined by the strength of materials as

$$\boldsymbol{f}_{\mathrm{c}}(\omega) = \begin{Bmatrix} N_1 \\ N_2 \end{Bmatrix} = \begin{Bmatrix} -N(0) \\ +N(L) \end{Bmatrix} \tag{3.38}$$

Substituting Equations 3.35 and 3.37 into the right-hand side of Equation 3.38 gives

$$\boldsymbol{S}_R(\omega)\boldsymbol{d} = \boldsymbol{f}_{\mathrm{c}}(\omega) \tag{3.39}$$

where $\boldsymbol{S}_R(\omega)$ is the spectral element matrix for the finite rod element and it is given by

$$\boldsymbol{S}_R(\omega) = \frac{EA}{L}\begin{bmatrix} s_{R11} & s_{R12} \\ s_{R12} & s_{R22} \end{bmatrix} = \boldsymbol{S}_R(\omega)^{\mathrm{T}} \tag{3.40}$$

where

$$\begin{aligned} s_{R11} = s_{R22} &= (k_{\mathrm{L}}L)\cot(k_{\mathrm{L}}L) \\ s_{R12} &= -(k_{\mathrm{L}}L)\csc(k_{\mathrm{L}}L) \end{aligned} \tag{3.41}$$

Example 3.2 Bernoulli–Euler Beam

The free bending (flexural) vibration of a Bernoulli–Euler beam (simply, B-beam) is represented by

$$EIw'''' + \rho A\ddot{w} = 0 \tag{3.42}$$

where $w(x,t)$ is the transverse displacement, E is the Young's modulus, A is the cross-sectional area, I is the area moment of inertia about the neutral axis, and ρ is the mass density. The internal transverse shear force and bending moment are given by

$$M_t(x,t) = EIw''(x,t), \quad Q_t(x,t) = -EIw'''(x,t) \tag{3.43}$$

Assume the solution to Equation 3.42 in the spectral form to be

$$w(x,t) = \frac{1}{N}\sum_{n=0}^{N-1} W_n(x;\omega_n)\,e^{i\omega_n t} \tag{3.44}$$

Substituting Equation 3.44 into Equation 3.42 gives an eigenvalue problem for a specific discrete frequency (such as $\omega = \omega_n$) as

$$EIW'''' - \omega^2 \rho A W = 0 \tag{3.45}$$

Assume the general solution to Equation 3.45 as

$$W(x) = ae^{-ik(\omega)x} \tag{3.46}$$

Substituting Equation 3.46 into Equation 3.45 yields a dispersion relation

$$k^4 - k_F^4 = 0 \tag{3.47}$$

where k_F is the wavenumber for the pure bending (flexural) wavemode defined by

$$k_F = \sqrt{\omega}\left(\frac{\rho A}{EI}\right)^{1/4} \tag{3.48}$$

Equation 3.47 gives two pure real roots and two pure imaginary roots as

$$k_1 = -k_2 = k_F, \quad k_3 = -k_4 = ik_F \tag{3.49}$$

For the finite B-beam element of length L, the general solution of Equation 3.45 can be then obtained in the form

$$W(x;\omega) = a_1 e^{-ik_F x} + a_2 e^{-k_F x} + a_3 e^{+ik_F x} + a_4 e^{+k_F x} = \boldsymbol{e}(x;\omega)\,\boldsymbol{a} \tag{3.50}$$

where

$$\begin{aligned} \boldsymbol{e}(x;\omega) &= [e^{-ik_F x} \quad e^{-k_F x} \quad e^{+ik_F x} \quad e^{+k_F x}] \\ \boldsymbol{a} &= \{a_1 \quad a_2 \quad a_3 \quad a_4\}^{\mathrm{T}} \end{aligned} \tag{3.51}$$

The spectral nodal displacements and slopes of the finite B-beam element (see Figure 3.2) can be related to the displacement field by

$$\boldsymbol{d} = \begin{Bmatrix} W_1 \\ \Theta_1 \\ W_2 \\ \Theta_2 \end{Bmatrix} = \begin{Bmatrix} W(0) \\ W'(0) \\ W(L) \\ W'(L) \end{Bmatrix} \tag{3.52}$$

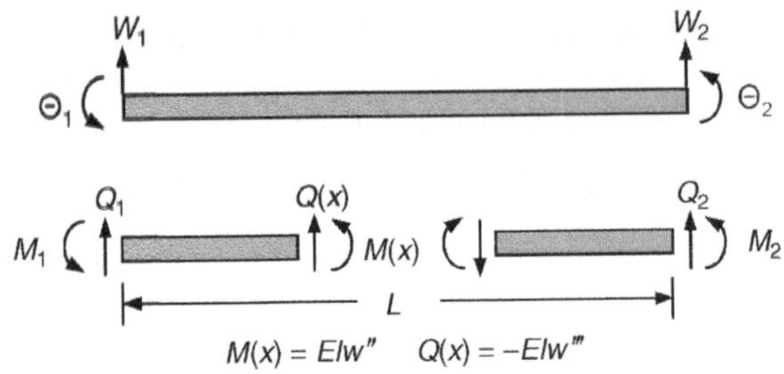

Figure 3.2 Sign convention for the Bernoulli–Euler beam element

Substituting Equation 3.50 into the right-hand side of Equation 3.52 gives

$$\boldsymbol{d} = \begin{bmatrix} e(0;\omega) \\ e'(0;\omega) \\ e(L;\omega) \\ e'(L;\omega) \end{bmatrix} \boldsymbol{a} = \boldsymbol{H}_B(\omega)\boldsymbol{a} \tag{3.53}$$

where

$$\boldsymbol{H}_B(\omega) = \begin{bmatrix} 1 & 1 & 1 & 1 \\ -ik_F & -k_F & ik_F & k_F \\ e^{-ik_FL} & e^{-k_FL} & e^{+ik_FL} & e^{+k_FL} \\ -ik_Fe^{-ik_FL} & -k_Fe^{-k_FL} & ik_Fe^{+ik_FL} & k_Fe^{+k_FL} \end{bmatrix} \tag{3.54}$$

The displacement field within the finite B-beam element can be represented in terms of the nodal DOFs vector $\boldsymbol{d}$ by eliminating the constant vector $\boldsymbol{a}$ from Equation 3.50 by using Equation 3.53, thus

$$W(x) = \boldsymbol{N}_B(x;\omega)\boldsymbol{d} \tag{3.55}$$

where

$$\begin{aligned}
\boldsymbol{N}_B(x;\omega) &= \boldsymbol{e}(x;\omega)\boldsymbol{H}_B^{-1}(\omega) = [N_{B1}\ N_{B2}\ N_{B3}\ N_{B4}] \\
N_{B1}(x) &= \eta^{-1}k_F[\cos\bar{x} - \cos(\bar{L}-\bar{x})\cosh\bar{L} - \cos\bar{L}\cosh(\bar{L}-\bar{x}) \\
&\quad + \cosh\bar{x} + \sin(\bar{L}-\bar{x})\sinh\bar{L} - \sin\bar{L}\sinh(\bar{L}-\bar{x})] \\
N_{B2}(x) &= \eta^{-1}[-\cosh(\bar{L}-\bar{x})\sin\bar{L} + \cosh\bar{L}\sin(\bar{L}-\bar{x}) + \sin\bar{x} \\
&\quad - \cos(\bar{L}-\bar{x})\sinh\bar{L} + \cos\bar{L}\sinh(\bar{L}-\bar{x}) + \sinh\bar{x}] \\
N_{B3}(x) &= \eta^{-1}k_F[\cos(\bar{L}-\bar{x}) - \cos\bar{x}\cosh\bar{L} - \cos\bar{L}\cosh\bar{x} \\
&\quad + \cosh(\bar{L}-\bar{x}) + \sin\bar{x}\sinh\bar{L} - \sin\bar{L}\sinh\bar{x}] \\
N_{B4}(x) &= -\eta^{-1}[-\cosh\bar{x}\sin\bar{L} + \cosh\bar{L}\sin\bar{x} + \sin(\bar{L}-\bar{x}) \\
&\quad - \cos\bar{x}\sinh\bar{L} + \cos\bar{L}\sinh\bar{x} + \sinh(\bar{L}-\bar{x})] \\
\eta &= 2k_F(1-\cos\bar{L}\cosh\bar{L}) \\
\bar{x} &= k_Fx, \bar{L} = k_FL
\end{aligned} \tag{3.56}$$

From Equation 3.43, the spectral components of the bending moment and transverse shear force can be related to $W(x)$ as

$$Q(x) = -EIW'''(x), \quad M(x) = EIW''(x) \tag{3.57}$$

The spectral nodal transverse shear forces and bending moments defined for the finite B-beam element (see Figure 3.2) can be related to the corresponding forces and moments defined by the strength of materials by

$$f_c = \begin{Bmatrix} Q_1 \\ M_1 \\ Q_2 \\ M_2 \end{Bmatrix} = \begin{Bmatrix} -Q(0) \\ -M(0) \\ +Q(L) \\ +M(L) \end{Bmatrix} \tag{3.58}$$

Substituting Equation 3.55 into Equation 3.57 and the results into the right-hand side of Equation 3.58 gives

$$\boldsymbol{S}_B(\omega)\boldsymbol{d} = \boldsymbol{f}_c(\omega) \tag{3.59}$$

where $\boldsymbol{S}_B(\omega)$ is the spectral element matrix for the B-beam element given by

$$\boldsymbol{S}_B(\omega) = \frac{EI}{L^3}\begin{bmatrix} s_{B11} & s_{B12} & s_{B13} & s_{B14} \\ s_{B12} & s_{B22} & s_{B23} & s_{B24} \\ s_{B13} & s_{B23} & s_{B33} & s_{B34} \\ s_{B14} & s_{B24} & s_{B34} & s_{B44} \end{bmatrix} = \boldsymbol{S}_B(\omega)^{\mathrm{T}} \tag{3.60}$$

where

$$\begin{aligned}
s_{B11} &= s_{B33} = \Delta_B \bar{L}^3(\cos\bar{L}\sinh\bar{L} + \sin\bar{L}\cosh\bar{L}) \\
s_{B22} &= s_{B44} = \Delta_B \bar{L}^3 k_F^{-2}(-\cos\bar{L}\sinh\bar{L} + \sin\bar{L}\cosh\bar{L}) \\
s_{B12} &= -s_{B34} = \Delta_B \bar{L}^3 k_F^{-1}\sin\bar{L}\sinh\bar{L} \\
s_{B13} &= -\Delta_B \bar{L}^3(\sin\bar{L} + \sinh\bar{L}) \\
s_{B14} &= -s_{B23} = \Delta_B \bar{L}^3 k_F^{-1}(-\cos\bar{L} + \cosh\bar{L}) \\
s_{B24} &= \Delta_B \bar{L}^3 k_F^{-2}(-\sin\bar{L} + \sinh\bar{L}) \\
\Delta_B &= \frac{1}{1 - \cos\bar{L}\cosh\bar{L}} \\
\bar{L} &= k_F L
\end{aligned} \tag{3.61}$$

Example 3.3 Timoshenko Beam

The free vibration of a uniform Timoshenko beam (simply, T-beam) is represented by

$$\begin{aligned}
&\kappa GA(w'' - \theta') - \rho A\ddot{w} = 0 \\
&EI\theta'' + \kappa GA(w' - \theta) - \rho I\ddot{\theta} = 0
\end{aligned} \tag{3.62}$$

where $w(x,t)$ and $\theta(x,t)$ are the transverse displacement and the slope due to bending, respectively. E is the Young's modulus, G is the shear modulus, κ is the shear correction factor, which depends on the shape of the cross-section, A is the cross-sectional area, and I is the area moment of inertia about the neutral axis. The internal bending moment and transverse shear force are given by

$$\begin{aligned} Q_t(x,t) &= \kappa GA\left[w'(x,t) - \theta(x,t)\right] \\ M_t(x,t) &= EI\theta'(x,t) \end{aligned} \tag{3.63}$$

Assume the solution of Equation 3.62 in the spectral form is

$$w(x,t) = \frac{1}{N}\sum_{n=0}^{N-1} W_n(x;\omega_n)e^{i\omega_n t}, \quad \theta(x,t) = \frac{1}{N}\sum_{n=0}^{N-1} \Theta_n(x;\omega_n)\, e^{i\omega_n t} \tag{3.64}$$

Substituting Equation 3.64 into Equation 3.62 yields an eigenvalue problem as

$$\begin{aligned} &\kappa GA(W'' - \Theta') + \rho A\omega^2 W = 0 \\ &EI\Theta'' + \kappa GA(W' - \Theta) + \rho I\omega^2\Theta = 0 \end{aligned} \tag{3.65}$$

Assume the general solutions to Equation 3.65 are

$$W(x) = ae^{-ik(\omega)x}, \quad \Theta(x) = \beta ae^{-ik(\omega)x} \tag{3.66}$$

Substitute Equation 3.66 into Equation 3.65 to obtain an eigenvalue problem as

$$\begin{bmatrix} \kappa GAk^2 - \rho A\omega^2 & -ik\kappa GA \\ ik\kappa GA & EIk^2 + \kappa GA - \rho I\omega^2 \end{bmatrix} \begin{Bmatrix} 1 \\ \beta \end{Bmatrix} = \begin{Bmatrix} 0 \\ 0 \end{Bmatrix} \tag{3.67}$$

Equation 3.67 gives a dispersion relation as

$$k^4 - \eta k_F^4 k^2 - k_F^4(1 - \eta_1 k_G^4) = 0 \tag{3.68}$$

where

$$k_F = \sqrt{\omega}\left(\frac{\rho A}{EI}\right)^{1/4}, \quad k_G = \sqrt{\omega}\left(\frac{\rho A}{\kappa GA}\right)^{1/4} \tag{3.69}$$

and

$$\eta = \eta_1 + \eta_2, \quad \eta_1 = \frac{\rho I}{\rho A}, \quad \eta_2 = \frac{EI}{\kappa GA} \tag{3.70}$$

Solving Equation 3.68 gives four roots as

$$\begin{aligned} k_1 &= -k_2 = \frac{1}{\sqrt{2}}k_F\sqrt{\eta k_F^2 + \sqrt{\eta^2 k_F^4 + 4(1 - \eta_1 k_G^4)}} \equiv k_t \\ k_3 &= -k_4 = \frac{1}{\sqrt{2}}k_F\sqrt{\eta k_F^2 - \sqrt{\eta^2 k_F^4 + 4(1 - \eta_1 k_G^4)}} \equiv k_e \end{aligned} \tag{3.71}$$

From the first line of Equation 3.67, we can obtain the wavemode ratio as

$$\beta_p(\omega) = \frac{1}{ik_p}(k_p^2 - k_G^4) = -ir_p(\omega) \quad (p = 1,\, 2,\, 3,\, 4) \tag{3.72}$$

where

$$r_p(\omega) = \frac{1}{k_p}(k_p^2 - k_G^4) \tag{3.73}$$

Note that when $\eta = 0$ and $k_G = 0$, Equation 3.68 is reduced to the disperison equation for the B-beam (see Equation 3.47) and the wavenumbers of Equation 3.71 become identical with those for the B-beam (see Equation 3.49).

By using the four wavenumbers given by Equation 3.71, the general solutions of Equation 3.65 can be written as

$$\begin{aligned} W(x) &= a_1 e^{-ik_t x} + a_2 e^{+ik_t x} + a_3 e^{-ik_e x} + a_4 e^{+ik_e x} = \boldsymbol{e}_w(x;\omega)\boldsymbol{a} \\ \Theta(x) &= \beta_1 a_1 e^{-ik_t x} + \beta_2 a_2 e^{+ik_t x} + \beta_3 a_3 e^{-ik_e x} + \beta_4 a_4 e^{+ik_e x} = \boldsymbol{e}_\theta(x;\omega)\boldsymbol{a} \end{aligned} \tag{3.74}$$

where

$$\boldsymbol{a} = \{a_1 \quad a_2 \quad a_3 \quad a_4\}^{\mathrm{T}} \tag{3.75}$$

and

$$\begin{aligned} \boldsymbol{e}_w(x;\omega) &= [e^{-ik_t x} \quad e^{+ik_t x} \quad e^{-ik_e x} \quad e^{+ik_e x}] \\ \boldsymbol{e}_\theta(x;\omega) &= \boldsymbol{e}_w(x;\omega)\boldsymbol{B}(\omega) \\ \boldsymbol{B}(\omega) &= diag[\beta_p(\omega)] \end{aligned} \tag{3.76}$$

The spectral nodal displacements and slopes of the finite T-beam element of length L (see Figure 3.3) can be related to the displacement fields by

$$\boldsymbol{d} = \begin{Bmatrix} W_1 \\ \Theta_1 \\ W_2 \\ \Theta_2 \end{Bmatrix} = \begin{Bmatrix} W(0) \\ \Theta(0) \\ W(L) \\ \Theta(L) \end{Bmatrix} \tag{3.77}$$

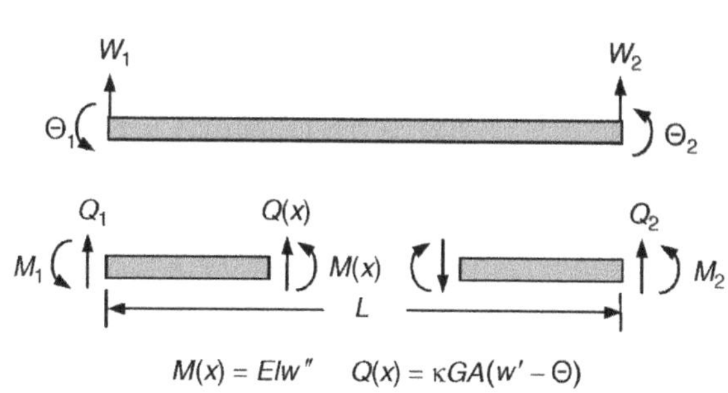

Figure 3.3 Sign convention for the Timoshenko-beam element

Substituting Equation 3.74 into the right-hand side of Equation 3.77 gives

$$d = \begin{bmatrix} e_w(0;\omega) \\ e_\theta(0;\omega) \\ e_w(L;\omega) \\ e_\theta(L;\omega) \end{bmatrix} a = H_T(\omega) a \tag{3.78}$$

where

$$H_T(\omega) = \begin{bmatrix} 1 & 1 & 1 & 1 \\ -ir_t & ir_t & -ir_e & ir_e \\ \mathrm{e}_t & \mathrm{e}_t^{-1} & \mathrm{e}_e & \mathrm{e}_e^{-1} \\ -ir_t\mathrm{e}_t & ir_t\mathrm{e}_t^{-1} & -ir_e\mathrm{e}_e & ir_e\mathrm{e}_e^{-1} \end{bmatrix} \tag{3.79}$$

with the use of following definitions:

$$\begin{aligned} \mathrm{e}_t &= \mathrm{e}^{-ik_tL}, \quad \mathrm{e}_e = \mathrm{e}^{-ik_eL} \\ r_t &= \frac{1}{k_t}(k_t^2 - k_G^4), \quad r_e = \frac{1}{k_e}(k_e^2 - k_G^4) \end{aligned} \tag{3.80}$$

By using Equation 3.78, the constant vector a can be eliminated from Equation 3.74 to express the general solutions as

$$W(x) = N_w(x;\omega)d, \quad \Theta(x) = N_\theta(x;\omega)d \tag{3.81}$$

where

$$\begin{aligned} N_w(x;\omega) &= e_w(x;\omega)H_T^{-1}(\omega) \\ N_\theta(x;\omega) &= e_\theta(x;\omega)H_T^{-1}(\omega) = e_w(x;\omega)B(\omega)H_T^{-1}(\omega) \end{aligned} \tag{3.82}$$

From Equation 3.63, the spectral components of the bending moment and transverse shear force are related to $W(x)$ and $\Theta(x)$ by

$$Q = \kappa GA(W' - \Theta), \quad M = EI\Theta' \tag{3.83}$$

The spectral nodal transverse shear forces and bending moments defined for the finite T-beam element (see Figure 3.3) can be related to the corresponding forces and moments defined by the strength of materials as

$$f_c(\omega) = \begin{Bmatrix} Q_1 \\ M_1 \\ Q_2 \\ M_2 \end{Bmatrix} = \begin{Bmatrix} -Q(0) \\ -M(0) \\ +Q(L) \\ +M(L) \end{Bmatrix} \tag{3.84}$$

Applying Equation 3.81 to Equation 3.83 and the results to the right-hand side of Equation 3.84 gives

$$S_T(\omega)d = f_c(\omega) \tag{3.85}$$

where $S_T(\omega)$ is the spectral element matrix given by

$$S_T(\omega) = EI\begin{bmatrix} s_{T11} & s_{T12} & s_{T13} & s_{T14} \\ s_{T12} & s_{T22} & s_{T23} & s_{T24} \\ s_{T13} & s_{T23} & s_{T33} & s_{T34} \\ s_{T14} & s_{T24} & s_{34} & s_{T44} \end{bmatrix} = S_T(\omega)^{\mathrm{T}} \tag{3.86}$$

where

$$\begin{aligned}
s_{T11} &= s_{T33} = i\Delta_T A[(\mathrm{e}_t^2-1)(\mathrm{e}_e^2+1)r_t - (\mathrm{e}_t^2+1)(\mathrm{e}_e^2-1)r_e](k_t r_t - k_e r_e) \\
s_{T22} &= s_{T44} = \Delta_T B[-(\mathrm{e}_t^2+1)(\mathrm{e}_e^2-1)r_t + (\mathrm{e}_t^2-1)(\mathrm{e}_e^2+1)r_e] \\
s_{T12} &= -s_{T34} = \Delta_T A\{(\mathrm{e}_t^2-1)(\mathrm{e}_e^2-1)(k_t r_e + k_e r_t) \\
&\quad -[(\mathrm{e}_t^2+1)(\mathrm{e}_e^2+1) - 4\mathrm{e}_t\mathrm{e}_e](k_t r_t + k_e r_e)\} \\
s_{T13} &= 2i\Delta_T A[(\mathrm{e}_t^2-1)\mathrm{e}_e r_t - (\mathrm{e}_e^2-1)\mathrm{e}_t r_e](-k_t r_t + k_e r_e) \\
s_{T14} &= -s_{T23} = -2\Delta_T A(\mathrm{e}_t - \mathrm{e}_e)(1-\mathrm{e}_t\mathrm{e}_e)(-k_t r_t + k_e r_e) \\
s_{T24} &= 2\Delta_T B[(\mathrm{e}_e^2-1)\mathrm{e}_t r_t - (\mathrm{e}_t^2-1)\mathrm{e}_e r_e] \\
A &= (1/8)\mathrm{e}_F(1-\eta_1 k_G^4)^{-1/2} \\
B &= (1/8)\mathrm{e}_F[\eta^2 k_F^4 + 4(1-\eta_1 k_G^4)]^{1/2} \\
\Delta_T &= \frac{8k_F^2}{i\mathrm{e}_F[2r_t r_e\{(\mathrm{e}_t^2+1)(\mathrm{e}_e^2+1) - 4\mathrm{e}_t\mathrm{e}_e\} - (r_t^2+r_e^2)(\mathrm{e}_t^2-1)(\mathrm{e}_e^2-1)]}
\end{aligned} \tag{3.87}$$

Example 3.4 Extended Timoshenko Beam

The plane motion of an extended Timoshenko beam (simply, an ET-beam) is represented by [79]

$$\begin{Bmatrix} N_t' \\ Q_t' \\ M_t' \end{Bmatrix} = \begin{bmatrix} \rho A & 0 & \rho R \\ 0 & \rho A & 0 \\ \rho R & 0 & \rho I \end{bmatrix}\begin{Bmatrix} \ddot{u} \\ \ddot{w} \\ \ddot{\theta} \end{Bmatrix} - \begin{Bmatrix} 0 \\ 0 \\ Q_t \end{Bmatrix} \tag{3.88}$$

where $N_t(x, t)$, $Q_t(x, t)$, and $M_t(x, t)$ are the axial tensile force, transverse shear force, and bending moment, respectively, and $u(x, t)$, $w(x, t)$, and $\theta(x, t)$ are the axial displacement, transverse displacement, and slope, respectively. ρA, ρR, and ρI are the effective mass per length, the first-order moment of inertia, and the second-order moment of inertia defined by

$$\rho A = \int \rho \mathrm{d}A, \quad \rho R = \int \rho z \mathrm{d}A, \quad \rho I = \int \rho z^2 \mathrm{d}A \tag{3.89}$$

where ρ is the mass density per unit volume. The force-displacement relationship for an ET-beam is given by

$$\begin{Bmatrix} N_t \\ Q_t \\ M_t \end{Bmatrix} = \begin{bmatrix} EA & C_1 & C_2 \\ C_1 & \kappa GA & C_3 \\ C_2 & C_3 & EI \end{bmatrix}\begin{Bmatrix} u' \\ w' - \theta \\ \theta' \end{Bmatrix} \tag{3.90}$$

where C_1, C_2, and C_3 are the coupling rigidities, which represent the couplings between axial, transverse shear, and bending deformations. When all the ρR, C_1, C_2, and C_3 vanish, the decoupled equations of motion for the classical Timoshenko beam and axial rod are recovered.

Represent the displacement fields in the spectral forms as

$$\begin{aligned} u(x,t) &= \frac{1}{N}\sum_{n=0}^{N-1} U_n(x;\omega_n)\mathrm{e}^{i\omega_n t} \\ w(x,t) &= \frac{1}{N}\sum_{n=0}^{N-1} W_n(x;\omega_n)\mathrm{e}^{i\omega_n t} \\ \theta(x,t) &= \frac{1}{N}\sum_{n=0}^{N-1} \Theta_n(x;\omega_n)\mathrm{e}^{i\omega_n t} \end{aligned} \tag{3.91}$$

Similarly, we represent the axial tensile force N_t, the transverse shear force Q_t, and the bending moment M_t in the spectral forms as

$$\begin{aligned} N_t(x,t) &= \frac{1}{N}\sum_{n=0}^{N-1} N_n(x;\omega_n)\mathrm{e}^{i\omega_n t} \\ Q_t(x,t) &= \frac{1}{N}\sum_{n=0}^{N-1} Q_n(x;\omega_n)\mathrm{e}^{i\omega_n t} \\ M_t(x,t) &= \frac{1}{N}\sum_{n=0}^{N-1} M_n(x;\omega_n)\mathrm{e}^{i\omega_n t} \end{aligned} \tag{3.92}$$

Substitution of Equations 3.91 and 3.92 into Equation 3.88 and 3.90 gives

$$\begin{Bmatrix} N' \\ Q' \\ M' \end{Bmatrix} = -\omega^2 \begin{bmatrix} \rho A & 0 & \rho R \\ 0 & \rho A & 0 \\ \rho R & 0 & \rho I \end{bmatrix} \begin{Bmatrix} U \\ W \\ \Theta \end{Bmatrix} - \begin{Bmatrix} 0 \\ 0 \\ Q \end{Bmatrix} \tag{3.93}$$

and

$$\begin{Bmatrix} N \\ Q \\ M \end{Bmatrix} = \begin{bmatrix} EA & C_1 & C_2 \\ C_1 & \kappa GA & C_3 \\ C_2 & C_3 & EI \end{bmatrix} \begin{Bmatrix} U' \\ W' - \Theta \\ \Theta' \end{Bmatrix} \tag{3.94}$$

By substituting Equation 3.94 into Equation 3.93, we can derive the governing equations in the frequency domain as

$$\begin{aligned} &\begin{bmatrix} EA & C_1 & C_2 \\ C_1 & \kappa GA & C_3 \\ C_2 & C_3 & EI \end{bmatrix} \begin{Bmatrix} U'' \\ W'' \\ \Theta'' \end{Bmatrix} + \begin{bmatrix} 0 & 0 & -C_1 \\ 0 & 0 & -\kappa GA \\ C_1 & \kappa GA & 0 \end{bmatrix} \begin{Bmatrix} U' \\ W' \\ \Theta' \end{Bmatrix} \\ &- \begin{bmatrix} 0 & 0 & 0 \\ 0 & 0 & 0 \\ 0 & 0 & \kappa GA \end{bmatrix} \begin{Bmatrix} U \\ W \\ \Theta \end{Bmatrix} + \omega^2 \begin{bmatrix} \rho A & 0 & \rho R \\ 0 & \rho A & 0 \\ \rho R & 0 & \rho I \end{bmatrix} \begin{Bmatrix} U \\ W \\ \Theta \end{Bmatrix} = \begin{Bmatrix} 0 \\ 0 \\ 0 \end{Bmatrix} \end{aligned} \tag{3.95}$$

Assume the general solutions of Equation 3.95 to be

$$U(x) = \alpha a e^{-ikx}, \quad W(x) = a e^{-ikx}, \quad \Theta(x) = \beta a e^{-ikx} \tag{3.96}$$

where k is the wavenumber. Substitution of Equation 3.96 into Equation 3.95 gives an eigenvalue problem as

$$\boldsymbol{X}(k)\begin{Bmatrix} \alpha \\ 1 \\ \beta \end{Bmatrix} = \begin{Bmatrix} 0 \\ 0 \\ 0 \end{Bmatrix} \tag{3.97}$$

where

$$\boldsymbol{X}(k) = \begin{bmatrix} -k^2EA + \omega^2\rho A & -k^2C_1 & ikC_1 - k^2C_2 + \omega^2\rho R \\ -k^2C_1 & -k^2\kappa GA + \omega^2\rho A & -k^2C_3 + ik\kappa GA \\ -ikC_1 - k^2C_2 + \omega^2\rho R & -k^2C_3 - ik\kappa GA & -k^2EI + \omega^2\rho I - \kappa GA \end{bmatrix} \tag{3.98}$$

From Equation 3.97, we get a dispersion relation

$$c_1k^6 + c_2k^4 + c_3k^2 + c_4 = 0 \tag{3.99}$$

where

$$\begin{aligned} c_1 &= (C_2^2 - EA \cdot EI)\kappa GA + EIC_1^2 - 2C_1C_2C_3 + C_3^2EA \\ c_2 &= \omega^2\rho A\,[EI(EA + \kappa GA) - C_2^2 - C_3^2 + 2(C_1C_3 - \kappa GA \cdot C_2)(\rho R/\rho A) \\ &\quad + (EA \cdot \kappa GA - C_1^2)\,(\rho I/\rho A)] \\ c_3 &= \omega^4\rho A^2\,[EI + (EA + \kappa GA)(\rho I/\rho A) - 2C_2(\rho R/\rho A) - \kappa GA(\rho R/\rho A)^2] \\ &\quad + \omega^2\rho A\,(C_1^2 - EA \cdot \kappa GA) \\ c_4 &= \omega^6\rho A^3[(\rho I/\rho A) - (\rho R/\rho A)^2] - \omega^4\kappa GA \cdot \rho A^2 \end{aligned} \tag{3.100}$$

We can solve Equation 3.99 for six wavenumbers

$$\begin{aligned} k_1 &= -k_2 = \sqrt{y_1 - (1/3)(c_2/c_1)} \\ k_3 &= -k_4 = \sqrt{y_2 - (1/3)(c_2/c_1)} \\ k_5 &= -k_6 = \sqrt{y_3 - (1/3)(c_2/c_1)} \end{aligned} \tag{3.101}$$

where

$$\begin{aligned} p &= (1/3)c_3c_1^{-1} - (1/9)c_2^2c_1^{-2} \\ q &= (2/27)c_2^3c_1^{-3} - (1/3)c_2c_3c_1^{-2} + c_4c_1^{-1} \\ y_1 &= \sqrt[3]{(1/2)(-q + \sqrt{q^2 + 4p^3})} + \sqrt[3]{(1/2)(-q - \sqrt{q^2 + 4p^3})} \\ y_2 &= -(-1)^{1/3}y_1 \\ y_3 &= (-1)^{2/3}y_1 \end{aligned} \tag{3.102}$$

From Equation 3.97, α_i and β_i for each wavenumber k_i ($i = 1, 2, \ldots, 6$) are determined to be

$$\begin{aligned}\alpha_i &= \chi_i^{-1}[-\rho A\rho R\omega^4 + \{\rho R + k_i(-iC_1\rho A + C_2\rho Ak_i + \kappa GA\rho Rk_i)\}\omega^2 \\ &\quad + k_i\{(C_1C_3 - C_2\kappa GA)k_i^3 + (C_3 - C_2)k_i + iC_1 - i\kappa GA\}] \\ \beta_i &= \chi_i^{-1}[-C_1^2k_i^4 - C_1k_i^2 + (EAk_i^2 - \rho A\omega^2)(\kappa GAk_i^2 - \rho A\omega^2 + 1)]\end{aligned} \tag{3.103}$$

where

$$\begin{aligned}\chi_i &= (-ik_i\kappa GA\rho A + k_i^2C_3\rho A - k_i^2C_1\rho R)\omega^2 + iEA\kappa GAk_i^3 - iC_1^2k_i^3 \\ &\quad + C_1C_2k_i^4 - C_3EAk_i^4\end{aligned} \tag{3.104}$$

By using the six wavenumbers computed from Equation 3.101, we can write the general solutions as

$$\begin{aligned}U(x) &= \sum_{i=1}^{6}\alpha_i a_i e^{-ik_ix} = \boldsymbol{e}_u(x;\omega)\,\boldsymbol{a} \\ W(x) &= \sum_{i=1}^{6} a_i e^{-ik_ix} = \boldsymbol{e}_w(x;\omega)\,\boldsymbol{a} \\ \Theta(x) &= \sum_{i=1}^{6}\beta_i a_i e^{-ik_ix} = \boldsymbol{e}_\theta(x;\omega)\,\boldsymbol{a}\end{aligned} \tag{3.105}$$

where

$$\boldsymbol{a} = \{a_1 \quad a_2 \quad a_3 \quad a_4 \quad a_5 \quad a_6\}^{\mathrm{T}} \tag{3.106}$$

and

$$\begin{aligned}\boldsymbol{e}_w(x;\omega) &= [e^{-ik_1x} \quad e^{-ik_2x} \quad e^{-ik_3x} \quad e^{-ik_4x} \quad e^{-ik_5x} \quad e^{-ik_6x}] \\ \boldsymbol{e}_u(x;\omega) &= \boldsymbol{e}_w(x;\omega)\boldsymbol{A}(\omega) \\ \boldsymbol{e}_\theta(x;\omega) &= \boldsymbol{e}_w(x;\omega)\boldsymbol{B}(\omega)\end{aligned} \tag{3.107}$$

with

$$\begin{aligned}\boldsymbol{A}(\omega) &= diag[\alpha_j(\omega)], \quad \alpha_j(\omega) = \alpha[k_j(\omega)] \\ \boldsymbol{B}(\omega) &= diag[\beta_j(\omega)], \quad \beta_j(\omega) = \beta[k_j(\omega)]\end{aligned} \qquad (j = 1, 2, \ldots, 6) \tag{3.108}$$

The spectral components of the nodal DOFs, nodal forces and nodal moments specified in Figure 3.4 are defined by

$$\begin{aligned}\boldsymbol{d} &= \{U_1 \quad W_1 \quad \Theta_1 \quad U_2 \quad W_2 \quad \Theta_2\}^{\mathrm{T}} \\ &= \{U(0) \quad W(0) \quad \Theta(0) \quad U(L) \quad W(L) \quad \Theta(L)\}^{\mathrm{T}}\end{aligned} \tag{3.109}$$

$$\begin{aligned}\boldsymbol{f}_c &= \{N_1 \quad Q_1 \quad M_1 \quad N_2 \quad Q_2 \quad M_2\}^{\mathrm{T}} \\ &= \{-N(0) \quad -Q(0) \quad -M(0) \quad N(L) \quad Q(L) \quad M(L)\}^{\mathrm{T}}\end{aligned} \tag{3.110}$$

Substitution of Equation 3.105 into Equation 3.109 yields the relationship to be

$$\boldsymbol{d} = \boldsymbol{H}_{ET}(\omega)\boldsymbol{a} \tag{3.111}$$

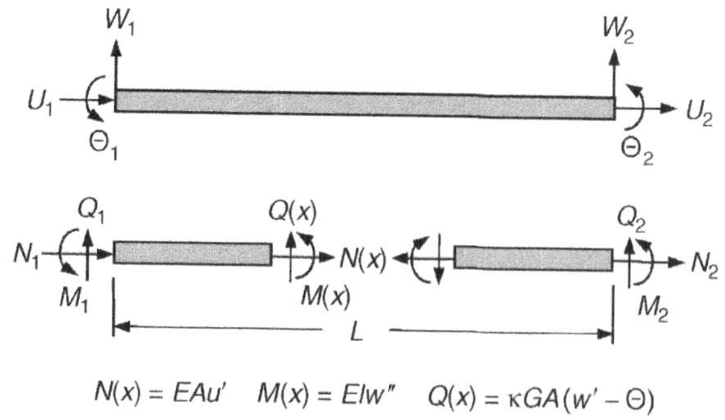

Figure 3.4 Sign convention for the extended Timoshenko-beam element

where

$$\boldsymbol{H}_{ET}(\omega) = \begin{bmatrix} \alpha_1 & \alpha_2 & \alpha_3 & \alpha_4 & \alpha_5 & \alpha_6 \\ 1 & 1 & 1 & 1 & 1 & 1 \\ \beta_1 & \beta_2 & \beta_3 & \beta_4 & \beta_5 & \beta_6 \\ \alpha_1 e_1 & \alpha_2 e_2 & \alpha_3 e_3 & \alpha_4 e_4 & \alpha_5 e_5 & \alpha_6 e_6 \\ e_1 & e_2 & e_3 & e_4 & e_5 & e_6 \\ \beta_1 e_1 & \beta_2 e_2 & \beta_3 e_3 & \beta_4 e_4 & \beta_5 e_5 & \beta_6 e_6 \end{bmatrix} \tag{3.112}$$

with the use of the definition

$$e_i = \mathrm{e}^{-ik_iL} \quad (i = 1, 2, \ldots, 6) \tag{3.113}$$

By using Equation 3.111, the constant vector $\boldsymbol{a}$ can be eliminated from Equation 3.105 to obtain

$$\begin{aligned} U(x) &= \boldsymbol{N}_u(x;\omega)\,\boldsymbol{d} \\ W(x) &= \boldsymbol{N}_w(x;\omega)\,\boldsymbol{d} \\ \Theta(x) &= \boldsymbol{N}_\theta(x;\omega)\,\boldsymbol{d} \end{aligned} \tag{3.114}$$

where $\boldsymbol{N}_u$, $\boldsymbol{N}_w$, and $\boldsymbol{N}_\theta$ are the dynamic shape functions defined by

$$\begin{aligned} \boldsymbol{N}_u(x;\omega) &= \boldsymbol{e}_u(x;\omega)\boldsymbol{H}_{ET}^{-1}(\omega) = \boldsymbol{e}_w(x;\omega)\,\boldsymbol{A}(\omega)\boldsymbol{H}_{ET}^{-1}(\omega) \\ \boldsymbol{N}_w(x;\omega) &= \boldsymbol{e}_w(x;\omega)\boldsymbol{H}_{ET}^{-1}(\omega) \\ \boldsymbol{N}_\theta(x;\omega) &= \boldsymbol{e}_\theta(x;\omega)\boldsymbol{H}_{ET}^{-1}(\omega) = \boldsymbol{e}_w(x;\omega)\boldsymbol{B}(\omega)\boldsymbol{H}_{ET}^{-1}(\omega) \end{aligned} \tag{3.115}$$

Substitution of Equation 3.114 into Equation 3.94 and the results into Equation 3.110 gives the relationship between the nodal forces and moments vector $\boldsymbol{f}_c$ and the nodal DOFs vector $\boldsymbol{d}$ as

$$\boldsymbol{S}_{ET}(\omega)\boldsymbol{d} = \boldsymbol{f}_c(\omega) \tag{3.116}$$

where the spectral element matrix $\boldsymbol{S}_{ET}(\omega)$ is given by

$$\boldsymbol{S}_{ET}(\omega) = \begin{bmatrix} \boldsymbol{C} & \boldsymbol{0} \\ \boldsymbol{0} & \boldsymbol{C} \end{bmatrix} \boldsymbol{R}(\omega)\,\boldsymbol{H}_{ET}^{-1}(\omega) \tag{3.117}$$

where

$$\boldsymbol{C} = \begin{bmatrix} EA & C_1 & C_2 \\ C_1 & \kappa GA & C_3 \\ C_2 & C_3 & EI \end{bmatrix} \tag{3.118}$$

$$\boldsymbol{R}(\omega) = \begin{bmatrix} ik_1\alpha_1 & ik_2\alpha_2 & ik_3\alpha_3 & ik_4\alpha_4 & ik_5\alpha_5 & ik_6\alpha_6 \\ \gamma_1 & \gamma_2 & \gamma_3 & \gamma_4 & \gamma_5 & \gamma_6 \\ ik_1\beta_1 & ik_2\beta_2 & ik_3\beta_3 & ik_4\beta_4 & ik_5\beta_5 & ik_6\beta_6 \\ -ik_1\alpha_1\varepsilon_1 & -ik_2\alpha_2\varepsilon_2 & -ik_3\alpha_3\varepsilon_3 & -ik_4\alpha_4\varepsilon_4 & -ik_5\alpha_5\varepsilon_5 & -ik_6\alpha_6\varepsilon_6 \\ -\gamma_1\varepsilon_1 & -\gamma_2\varepsilon_2 & -\gamma_3\varepsilon_3 & -\gamma_4\varepsilon_4 & -\gamma_5\varepsilon_5 & -\gamma_6\varepsilon_6 \\ -ik_1\beta_1\varepsilon_1 & -ik_2\beta_2\varepsilon_2 & -ik_3\beta_3\varepsilon_3 & -ik_4\beta_4\varepsilon_4 & -ik_5\beta_5\varepsilon_5 & -ik_6\beta_6\varepsilon_6 \end{bmatrix}$$

where

$$\gamma_i = ik_i + \beta_i \quad (i = 1, 2, \ldots, 6) \tag{3.119}$$

3.2 Variational Method

In this section, the frequency-domain variational method is introduced to formulate the spectral element matrix and the equivalent nodal forces for distributed dynamic forces. By use of the DFT, time no longer appears explicitly in the governing equations and in the associated boundary conditions. Thus, the original dynamic problem can be considered as a pseudo-static problem with the frequency as a parameter. In the frequency domain, the variational method can be applied in a similar manner as is applied in the conventional time-domain finite element formulation. As the variational method can be readily extended to multi-dimensional problems, the discussion in this section will be confined to the 1-D problems.

The governing equations of motion in the frequency domain are given by Equation 3.4, that is

$$\boldsymbol{L}\,\boldsymbol{U}(x) - \omega^2 \boldsymbol{M}\,\boldsymbol{U}(x) = \boldsymbol{P}(x) \tag{3.120}$$

The relationship between the displacement fields $\boldsymbol{U}$ and the primary variables $\boldsymbol{D}$ (i.e., displacements and slopes) and the relationship between the displacement fields $\boldsymbol{U}$ and the secondary variables $\boldsymbol{F}$ (i.e., internal forces and moments) are given by Equations 3.14 and 3.18, that is

$$\begin{aligned} \boldsymbol{D}(x) &= \boldsymbol{L}_{\mathrm{GB}}\,\boldsymbol{U}(x) \\ \boldsymbol{F}(x) &= \boldsymbol{L}_{\mathrm{NB}}\,\boldsymbol{U}(x) \end{aligned} \tag{3.121}$$

where $\boldsymbol{L}_{\mathrm{GB}}$ and $\boldsymbol{L}_{\mathrm{NB}}$ are the linear differential operators for the geometric and natural boundary conditions, respectively.

The weak form of the governing equation can be constructed from the weighted-integral statement given by

$$\int_0^L \delta \boldsymbol{U}^{\mathrm{T}} \{\boldsymbol{L}\boldsymbol{U}(x) - \omega^2 \boldsymbol{M}\,\boldsymbol{U}(x) - \boldsymbol{P}(x)\} \mathrm{d}x = 0 \tag{3.122}$$

where $\delta\boldsymbol{U}$ are the weight functions, which are the virtual changes of $\boldsymbol{U}$. Integrating the first term of Equation 3.122 by parts to distribute differentiation evenly between dependent variables and weighted functions and then applying Equation 3.121, we can obtain

$$\int_0^L \delta \boldsymbol{U}^{\mathrm{T}} \boldsymbol{L}\, \boldsymbol{U}(x)\mathrm{d}x = \int_0^L \boldsymbol{A}(\delta \boldsymbol{U}, \boldsymbol{U})\mathrm{d}x - (\boldsymbol{L}_{\mathrm{GB}}\delta \boldsymbol{U})^{\mathrm{T}}(\boldsymbol{L}_{\mathrm{NB}}\boldsymbol{U})\Big|_{x=0}^{x=L} \tag{3.123}$$

or

$$\int_0^L \delta \boldsymbol{U}^{\mathrm{T}} \boldsymbol{L} \boldsymbol{U}(x)\mathrm{d}x = \int_0^L \boldsymbol{A}(\delta \boldsymbol{U}, \boldsymbol{U})\mathrm{d}x - \delta \boldsymbol{D}^{\mathrm{T}} \boldsymbol{F}\Big|_{x=0}^{x=L} \tag{3.124}$$

Substituting Equation 3.124 into Equation 3.122 gives

$$\int_0^L \boldsymbol{A}(\delta \boldsymbol{U}, \boldsymbol{U})\mathrm{d}x - \omega^2 \int_0^L \delta \boldsymbol{U}^{\mathrm{T}} \boldsymbol{M} \boldsymbol{U}\,\mathrm{d}x - \int_0^L \delta \boldsymbol{U}^{\mathrm{T}} \boldsymbol{P}(x)\mathrm{d}x - \delta \boldsymbol{D}^{\mathrm{T}} \boldsymbol{F}\Big|_{x=0}^{x=L} = 0 \tag{3.125}$$

For a conservative system, the function $\boldsymbol{A}(\delta \boldsymbol{U}, \boldsymbol{U})$ will be symmetric in $\delta \boldsymbol{U}$ and $\boldsymbol{U}$, that is $\boldsymbol{A}(\delta \boldsymbol{U}, \boldsymbol{U}) = \boldsymbol{A}(\boldsymbol{U}, \delta \boldsymbol{U})$.

The general solutions to the homogeneous governing equations are given by Equation 3.16, that is

$$\boldsymbol{U}(x) = \boldsymbol{N}(x; \omega)\, \boldsymbol{d} \tag{3.126}$$

where $\boldsymbol{d}$ is the nodal DOFs vector defined by Equation 3.15, that is

$$\boldsymbol{d} = \begin{Bmatrix} \boldsymbol{D}(0) \\ \boldsymbol{D}(L) \end{Bmatrix} \tag{3.127}$$

Substituting Equation 3.126 into Equation 3.125 and using Equation 3.127 gives

$$\delta \boldsymbol{d}^{\mathrm{T}}\left(\left[\int_0^L \boldsymbol{A}(\boldsymbol{N}, \boldsymbol{N})\mathrm{d}x - \omega^2 \int_0^L \boldsymbol{N}^{T} \boldsymbol{M} \boldsymbol{N}\,\mathrm{d}x\right]\boldsymbol{d} - \int_0^L \boldsymbol{N}^{\mathrm{T}} \boldsymbol{P}(x)\mathrm{d}x - \boldsymbol{f}_{\mathrm{c}}\right) = 0 \tag{3.128}$$

where $\boldsymbol{f}_{\mathrm{c}}$ is the nodal forces defined by

$$\boldsymbol{f}_{\mathrm{c}}(\omega) \equiv \begin{Bmatrix} -\boldsymbol{F}(0) \\ +\boldsymbol{F}(L) \end{Bmatrix} \tag{3.129}$$

As $\delta \boldsymbol{d}$ is arbitrary by definition, Equation 3.128 yields the spectral element equation to be

$$\boldsymbol{S}(\omega)\boldsymbol{d} = \boldsymbol{f}_{\mathrm{c}}(\omega) + \boldsymbol{f}_{\mathrm{d}}(\omega) \tag{3.130}$$

where

$$\begin{aligned} \boldsymbol{S}(\omega) &= \int_0^L \boldsymbol{A}(\boldsymbol{N}, \boldsymbol{N})\mathrm{d}x - \omega^2 \int_0^L \boldsymbol{N}^{\mathrm{T}} \boldsymbol{M} \boldsymbol{N}\,\mathrm{d}x \\ \boldsymbol{f}_{\mathrm{d}}(\omega) &= \int_0^L \boldsymbol{N}^{\mathrm{T}} \boldsymbol{P}(x)\mathrm{d}x \end{aligned} \tag{3.131}$$

The matrix $\boldsymbol{S}(\omega)$ is the frequency-dependent spectral element matrix, $\boldsymbol{f}_{\mathrm{c}}$ is the nodal forces vector associated with the concentrated dynamic forces applied at the nodes and $\boldsymbol{f}_{\mathrm{d}}$ is the equivalent nodal forces vector associated with the distributed dynamic forces.

The advantages of the variational method over the other approaches introduced in this chapter are: (i) the boundary conditions can be consistently applied during the weak form formulation as shown in Equation 3.125, (ii) the distributed dynamic forces can be automatically transformed into the equivalent nodal forces and moments as shown by the second line of

Equation 3.131, (iii) the symmetry or asymmetry of the spectral element matrix can be readily inferred from $\boldsymbol{A}(\boldsymbol{N}, \boldsymbol{N})$ prior to the computation of the spectral element matrix, and (iv) as the spectral element formulation from the weak form statement in the frequency domain is similar to that for the conventional finite element formulation in the time domain, we can readily transform the spectral element formulation into the finite element formulation by simply replacing the frequency-dependent dynamic shape functions used for SEM with the simple polynomials commonly used for FEM and vice versa. Owing to the apparent advantages of the variational method, one may prefer to use the variational method to formulate spectral element models.

In the following the spectral element matrices are derived for the axial rod, Bernoulli–Euler beam, Timoshenko beam and extended Timoshenko beam elements as illustrative examples.

Example 3.5 Rod

The forced longitudinal vibration of a uniform rod is represented by

$$EAu'' - \rho A\ddot{u} + p(x, t) = 0 \qquad (3.132)$$

where $p(x,t)$ is the distributed dynamic force. The internal axial force is defined by

$$N_t(x, t) = EAu'(x, t) \qquad (3.133)$$

Following the procedure given in Equations 3.2–3.4, we obtain an ordinary differential equation in the frequency domain as

$$EAU'' + \omega^2 \rho AU + P(x) = 0 \qquad (3.134)$$

with the relevant force-displacement relationship being

$$N = EAU' \qquad (3.135)$$

The weak form of Equation 3.134 can be derived from the weighted-integral statement given by

$$\int_0^L (EAU'' + \omega^2 \rho AU + P)\delta U \, dx \qquad (3.136)$$

where $P(x)$ represents the spectral components of $p(x,t)$ and δU is the arbitrary variation in U. Integrating the first term in Equation 3.136 by parts and applying Equation 3.135 gives

$$\int_0^L EAU''\delta U \, dx = -\int_0^L EAU'\delta U' \, dx - N\delta U|_0 + N\delta U|_L \qquad (3.137)$$

Equation 3.137 can be rewritten in terms of nodal forces and displacements, as defined in Figure 3.1

$$\int_0^L EAU''\delta U dx = -\int_0^L EAU'\delta U' dx + \delta \boldsymbol{d}^{\mathrm{T}} \boldsymbol{f}_{\mathrm{c}} \qquad (3.138)$$

where

$$\boldsymbol{d} = \{U_1 \quad U_2\}^{\mathrm{T}}, \quad \boldsymbol{f}_{\mathrm{c}} = \{N_1 \quad N_2\}^{\mathrm{T}} \qquad (3.139)$$

Substituting Equation 3.138 into Equation 3.136 gives the weak form of Equation 3.134 as

$$\int_0^L EAU'\,\delta U' \mathrm{d}x - \int_0^L \omega^2 \rho A U\,\delta U \mathrm{d}x - \int_0^L P\delta U \mathrm{d}x - \delta \boldsymbol{d}^{\mathrm{T}} \boldsymbol{f}_{\mathrm{c}} = 0 \tag{3.140}$$

Substitute Equation 3.35 into Equation 3.140 to obtain

$$\delta \boldsymbol{d}^{\mathrm{T}} \left[\left(\int_0^L EA \boldsymbol{N}_R'^{\mathrm{T}} \boldsymbol{N}_R' \mathrm{d}x - \omega^2 \int_0^L \rho A \boldsymbol{N}_R^{\mathrm{T}} \boldsymbol{N}_R \mathrm{d}x \right) \boldsymbol{d} - \int_0^L P \boldsymbol{N}_R^{\mathrm{T}} \mathrm{d}x - \boldsymbol{f}_{\mathrm{c}} \right] = 0 \tag{3.141}$$

As $\delta \boldsymbol{d}$ is arbitrary, Equation 3.141 yields the spectral element equation to be

$$\boldsymbol{S}_R(\omega) \boldsymbol{d} = \boldsymbol{f}_{\mathrm{c}}(\omega) + \boldsymbol{f}_{\mathrm{d}}(\omega) \tag{3.142}$$

where

$$\boldsymbol{S}_R(\omega) = \int_0^L EA \boldsymbol{N}_R'^{\mathrm{T}} \boldsymbol{N}_R' \mathrm{d}x - \omega^2 \int_0^L \rho A \boldsymbol{N}_R^{\mathrm{T}} \boldsymbol{N}_R \mathrm{d}x \tag{3.143}$$

and

$$\boldsymbol{f}_{\mathrm{d}}(\omega) = \int_0^L P(x;\omega) \boldsymbol{N}_R^{\mathrm{T}}(x;\omega) \mathrm{d}x \tag{3.144}$$

The spectral element matrix given by Equation 3.143 can be shown to be identical with Equation 3.40. In Equation 3.142, $\boldsymbol{f}_{\mathrm{c}}$ and $\boldsymbol{f}_{\mathrm{d}}$ are the spectral nodal forces associated with the concentrated dynamic forces applied at the nodes and the distributed dynamic force $p(x,t)$, respectively.

Example 3.6 Bernoulli–Euler Beam

The forced bending vibration of a B-beam is represented by

$$EIw'''' + \rho A \ddot{w} = f(x,t) \tag{3.145}$$

where $f(x,t)$ is the distributed dynamic force. The internal moment and transverse shear force are given by

$$M_t(x,t) = EIw''(x,t), \quad Q_t(x,t) = -EIw'''(x,t) \tag{3.146}$$

Following the DFT processing given by Equations 3.2–3.4, Equation 3.145 and Equation 3.146 can be transformed into the frequency domain as

$$EIW'''' - \rho A \omega^2 W - F(x) = 0 \tag{3.147}$$

and

$$M = EIW'', \quad Q = -EIW''' \tag{3.148}$$

where $F(x)$ represents the spectral components of $f(x,t)$.

The weak form of Equation 3.147 can be derived from the weighted-integral statement as

$$\int_0^L (EIW'''' - \omega^2 \rho A W - F)\delta W \, dx \tag{3.149}$$

where δW is the arbitrary variation in W. Integrating the first term of Equation 3.149 by parts and applying Equation 3.148 gives

$$\int_0^L EIW'''' \, \delta W dx = \int_0^L EIW'' \delta W'' dx - Q\delta W\Big|_{x=0}^{x=L} - M\delta W'\Big|_{x=0}^{x=L} \tag{3.150}$$

Considering the sign convention for the nodal DOFs, forces and moments defined by Equations 3.52 and 3.58, Equation 3.150 can be written as

$$\int_0^L EIW'''' \delta W dx = \int_0^L EIW'' \delta W'' dx - \delta \, \boldsymbol{d}^{\mathrm{T}} \boldsymbol{f}_{\mathrm{c}} \tag{3.151}$$

where

$$\begin{aligned} \boldsymbol{d} &= \{W_1 \quad \Theta_1 \quad W_2 \quad \Theta_2\}^{\mathrm{T}} \\ \boldsymbol{f}_{\mathrm{c}} &= \{Q_1 \quad M_1 \quad Q_2 \quad M_2\}^{\mathrm{T}} \end{aligned} \tag{3.152}$$

Substitution of Equation 3.151 into Equation 3.149 gives the weak form of Equation 3.147 as

$$\int_0^L EIW'' \delta W'' dx - \omega^2 \int_0^L \rho A W \delta W dx - \int_0^L F \delta W dx - \delta \boldsymbol{d}^{\mathrm{T}} \boldsymbol{f}_{\mathrm{c}} = 0 \tag{3.153}$$

Finally, substituting Equation 3.55 into Equation 3.153 gives

$$\delta \boldsymbol{d}^{\mathrm{T}} \left[\left(\left(\int_0^L EI \boldsymbol{N}_B''^{\mathrm{T}} \boldsymbol{N}_B'' dx - \omega^2 \int_0^L \rho A \boldsymbol{N}_B^{\mathrm{T}} \boldsymbol{N}_B dx \right) \boldsymbol{d} - (\boldsymbol{f}_{\mathrm{c}} + \boldsymbol{f}_{\mathrm{d}}) \right) \right] = 0 \tag{3.154}$$

where

$$\boldsymbol{f}_{\mathrm{d}}(\omega) = \int_0^L F(x;\omega) \boldsymbol{N}_B^{\mathrm{T}}(x;\omega) dx \tag{3.155}$$

Since $\delta \boldsymbol{d}$ is arbitrary by definition, the spectral element equation can be derived from Equation 3.154 as

$$\boldsymbol{S}_B(\omega) \boldsymbol{d} = \boldsymbol{f}_{\mathrm{c}}(\omega) + \boldsymbol{f}_{\mathrm{d}}(\omega) \tag{3.156}$$

where

$$\boldsymbol{S}_B(\omega) = \int_0^L [EI \boldsymbol{N}_B''^{\mathrm{T}} \boldsymbol{N}_B'' - \omega^2 \rho A \, \boldsymbol{N}_B^{\mathrm{T}} \boldsymbol{N}_B] dx \tag{3.157}$$

The matrix $\boldsymbol{S}_B(\omega)$ is the spectral element matrix for the B-beam element. We can readily prove that $\boldsymbol{S}_B(\omega)$ computed from Equation 3.157 is identical with Equation 3.60. The nodal forces vector $\boldsymbol{f}_{\mathrm{c}}$ is associated with the concentrated dynamic forces applied at the nodes and $\boldsymbol{f}_{\mathrm{d}}$ with the distributed dynamic force $f(x,t)$.

Example 3.7 Timoshenko Beam

The forced vibration of a uniform T-beam is represented by

$$\begin{aligned} \kappa GA(w'' - \theta') - \rho A\ddot{w} &= -f(x,t) \\ EI\theta'' + \kappa GA(w' - \theta) - \rho I\ddot{\theta} &= -\tau(x,t) \end{aligned} \tag{3.158}$$

where $f(x, t)$ and $\tau(x, t)$ are the distributed dynamic forces applied on the T-beam. The force-displacement relationships are given by

$$\begin{aligned} Q_t(x,t) &= \kappa GA[w'(x,t) - \theta(x,t)] \\ M_t(x,t) &= EI\theta'(x,t) \end{aligned} \tag{3.159}$$

By using the DFT processing of Equations 3.2–3.4, Equations 3.158 and 3.159 can be transformed into the frequency-domain expressions as

$$\begin{aligned} &\kappa GA(W'' - \Theta') + \rho A\omega^2 W + F(x) = 0 \\ &EI\Theta'' + \kappa GA(W' - \Theta) + \rho I\omega^2\Theta + T(x) = 0 \end{aligned} \tag{3.160}$$

and

$$Q = \kappa GA(W' - \Theta), \quad M = EI\Theta' \tag{3.161}$$

where $F(x)$ and $T(x)$ are the spectral components of $f(x, t)$ and $\tau(x, t)$, respectively.

The weak form of Equation 3.160 can be obtained from the weighted-integral statement given by

$$\begin{aligned} &\int_0^L \{[\kappa GA(W'' - \Theta') + \omega^2\rho AW + F(x)]\delta W \\ &\quad + [EI\Theta'' + \kappa GA(W' - \Theta) + \omega^2\rho I\Theta + T(x)]\delta\Theta\}\mathrm{d}x = 0 \end{aligned} \tag{3.162}$$

where δW and $\delta\Theta$ are the arbitrary variations in W and Θ, respectively. By using the integral by parts, we can show that

$$\begin{aligned} \int_0^L \kappa GA(W'' - \Theta')\delta W\mathrm{d}x &= -\int_0^L \kappa GA(W' - \Theta)\delta W'\mathrm{d}x + Q\delta W\big|_0^L \\ \int_0^L EI\Theta''\delta\Theta\mathrm{d}x &= -\int_0^L EI\Theta'\delta\Theta'\mathrm{d}x + M\delta\Theta\big|_0^L \end{aligned} \tag{3.163}$$

where Equation 3.161 is used. By applying Equation 3.163 to Equation 3.162, we obtain the weak form as

$$\begin{aligned} &\int_0^L [\kappa GA(W'\delta W' - \Theta\delta W' - W'\delta\Theta + \Theta\delta\Theta) + EI\Theta'\delta\Theta']\mathrm{d}x + \\ &\quad - \omega^2\int_0^L (\rho AW\delta W + \rho I\Theta\,\delta\Theta)\mathrm{d}x - \int_0^L (F\delta W + T\delta\Theta)\mathrm{d}x - \boldsymbol{\delta d}^{\mathrm{T}}\boldsymbol{f}_{\mathrm{c}} = 0 \end{aligned} \tag{3.164}$$

where $\boldsymbol{d}$ and $\boldsymbol{f}_c$ are the spectral nodal DOFs vector and the nodal forces and moments vector defined by Equations 3.77 and 3.84, respectively, and they are

$$\begin{aligned} \boldsymbol{d} &= \{W_1 \quad \Theta_1 \quad W_2 \quad \Theta_2\}^{\mathrm{T}} \\ \boldsymbol{f}_c &= \{Q_1 \quad M_1 \quad Q_2 \quad M_2\}^{\mathrm{T}} \end{aligned} \tag{3.165}$$

Substituting Equation 3.81 into Equation 3.164 gives

$$\begin{aligned} \delta \boldsymbol{d}^{\mathrm{T}} \Big\{ \int_0^L [\kappa GA(\boldsymbol{N}_w'^{\mathrm{T}}\boldsymbol{N}_w' - \boldsymbol{N}_w'^{\mathrm{T}}\boldsymbol{N}_\theta - \boldsymbol{N}_\theta^{\mathrm{T}}\boldsymbol{N}_w' + \boldsymbol{N}_\theta^{\mathrm{T}}\boldsymbol{N}_\theta) + EI(\boldsymbol{N}_\theta'^{\mathrm{T}}\boldsymbol{N}_\theta') \\ - \omega^2(\rho A \boldsymbol{N}_w^{\mathrm{T}}\boldsymbol{N}_w + \rho I \boldsymbol{N}_\theta^{\mathrm{T}}\boldsymbol{N}_\theta)]\mathrm{d}x\, \boldsymbol{d} - (\boldsymbol{f}_c + \boldsymbol{f}_d) \Big\} = 0 \end{aligned} \tag{3.166}$$

where

$$\boldsymbol{f}_d(\omega) = \int_0^L [F(x;\omega)\boldsymbol{N}_w^{\mathrm{T}}(x;\omega) + T(x;\omega)\boldsymbol{N}_\theta^{\mathrm{T}}(x;\omega)]\mathrm{d}x \tag{3.167}$$

As $\delta \boldsymbol{d}$ is arbitrary, we can obtain the spectral element equation from Equation 3.166 as

$$\boldsymbol{S}_T(\omega)\boldsymbol{d} = \boldsymbol{f}_c(\omega) + \boldsymbol{f}_d(\omega) \tag{3.168}$$

where

$$\begin{aligned} \boldsymbol{S}_T(\omega) = \int_0^L [\kappa GA(\boldsymbol{N}_w'^{\mathrm{T}}\boldsymbol{N}_w' - \boldsymbol{N}_w'^{\mathrm{T}}\boldsymbol{N}_\theta - \boldsymbol{N}_\theta^{\mathrm{T}}\boldsymbol{N}_w' + \boldsymbol{N}_\theta^{\mathrm{T}}\boldsymbol{N}_\theta) + EI(\boldsymbol{N}_\theta'^{\mathrm{T}}\boldsymbol{N}_\theta') \\ - \omega^2(\rho A \boldsymbol{N}_w^{\mathrm{T}}\boldsymbol{N}_w + \rho I \boldsymbol{N}_\theta^{\mathrm{T}}\boldsymbol{N}_\theta)]\mathrm{d}x \end{aligned} \tag{3.169}$$

By substituting Equation 3.82 into Equation 3.169, we get the simplified expression

$$\boldsymbol{S}_T(\omega) = \boldsymbol{H}_T^{-\mathrm{T}}(\omega)\boldsymbol{D}_T(\omega)\boldsymbol{H}_T^{-1}(\omega) \tag{3.170}$$

where $\boldsymbol{H}_T(\omega)$ is the matrix defined by Equation 3.79 and

$$\begin{aligned} \boldsymbol{D}_T(\omega) = \int_0^L [\kappa GA(\boldsymbol{e}_w'^{\mathrm{T}}\boldsymbol{e}_w' - \boldsymbol{e}_w'^{\mathrm{T}}\boldsymbol{e}_w\boldsymbol{B} - \boldsymbol{B}\boldsymbol{e}_w^{\mathrm{T}}\boldsymbol{e}_w' + \boldsymbol{B}\boldsymbol{e}_w^{\mathrm{T}}\boldsymbol{e}_w\boldsymbol{B}) \\ + EI(\boldsymbol{B}\boldsymbol{e}_w'^{\mathrm{T}}\boldsymbol{e}_w'\boldsymbol{B}) - \omega^2\{\rho A(\boldsymbol{e}_w^{\mathrm{T}}\boldsymbol{e}_w) + \rho I(\boldsymbol{B}\boldsymbol{e}_w^{\mathrm{T}}\boldsymbol{e}_w\boldsymbol{B})\}]\mathrm{d}x \end{aligned} \tag{3.171}$$

From the definition of $\boldsymbol{e}_w$ given by Equation 3.76, we can show that

$$\boldsymbol{e}_w'(x;\omega) = -i\boldsymbol{e}_w(x;\omega)\boldsymbol{K}(\omega) \tag{3.172}$$

where

$$\boldsymbol{K}(\omega) = diag[k_p(\omega)] \quad (p = 1, 2, 3, 4) \tag{3.173}$$

By using Equations 3.76 and 3.172, Equation 3.171 can be rewritten as

$$\begin{aligned} \boldsymbol{D}_T(\omega) = -\kappa GA[\boldsymbol{K}\boldsymbol{E}_T\boldsymbol{K} - i(\boldsymbol{K}\boldsymbol{E}_T\boldsymbol{B} + \boldsymbol{B}\boldsymbol{E}_T\boldsymbol{K}) - \boldsymbol{B}\boldsymbol{E}_T\boldsymbol{B}] - EI(\boldsymbol{B}\boldsymbol{K}\boldsymbol{E}_T\boldsymbol{K}\boldsymbol{B}) \\ - \omega^2[\rho A \boldsymbol{E}_T + \rho I(\boldsymbol{B}\boldsymbol{E}_T\boldsymbol{B})] = [D_{Tij}(\omega)] \end{aligned} \tag{3.174}$$

where

$$\boldsymbol{E}_T(\omega) = \int_0^L \boldsymbol{e}_w^{\mathrm{T}}(x;\omega)\boldsymbol{e}_w(x;\omega)\mathrm{d}x = [E_{Tij}(\omega)] \tag{3.175}$$

The components of $\boldsymbol{D}_T = \boldsymbol{D}_T^{\mathrm{T}}$ and $\boldsymbol{E}_T = \boldsymbol{E}_T^{\mathrm{T}}$ are given by

$$
\begin{aligned}
D_{T11} &= \frac{i\mathrm{e}_t^2(1-\bar{\mathrm{e}}_t^2)}{2k_t}[EIk_t^2\beta_t^2 + \kappa GA(k_t-\beta_t)^2 - (\rho A - \rho I\beta_t^2)\omega^2] \\
D_{T12} &= D_{T34} = L[EIk_tk_e\beta_t\beta_e + \kappa GA(k_t-\beta_t)(k_e-\beta_e) - (\rho A - \rho I\beta_t\beta_e)\omega^2] \\
D_{T13} &= \frac{i\mathrm{e}_t\mathrm{e}_e(1-\bar{\mathrm{e}}_t\bar{\mathrm{e}}_e)}{k_t+k_e}[EIk_tk_e\beta_t\beta_e + \kappa GA(k_t-\beta_t)(k_e-\beta_e) - (\rho A - \rho I\beta_t\beta_e)\omega^2] \\
D_{T14} &= \frac{i\mathrm{e}_t\bar{\mathrm{e}}_e(1-\bar{\mathrm{e}}_t\mathrm{e}_e)}{k_t+k_e}[EIk_tk_e\beta_t\beta_e - \kappa GA(k_t-\beta_t)(k_e-\beta_e) - (\rho A + \rho I\beta_t\beta_e)\omega^2] \\
D_{T22} &= \frac{i(1-\bar{\mathrm{e}}_t^2)}{2k_t}[EIk_e^2\beta_e^2 + \kappa GA(k_e-\beta_e)^2 - (\rho A - \rho I\beta_e^2)\omega^2] \\
D_{T23} &= = \frac{i(1-\bar{e}_te_e)}{k_t-k_e}[EIk_tk_e\beta_t\beta_e - \kappa GA(k_t-\beta_t)(k_e-\beta_e) - (\rho A + \rho I\beta_t\beta_e)\omega^2] \\
D_{T24} &= \frac{i(1-\bar{\mathrm{e}}_t\bar{\mathrm{e}}_e)}{k_t+k_e}[EIk_tk_e\beta_t\beta_e - \kappa GA(k_t-\beta_t)(k_e-\beta_e) - (\rho A + \rho I\beta_t\beta_e)\omega^2] \\
D_{T33} &= \frac{i\mathrm{e}_e^2(1-\bar{\mathrm{e}}_e^2)}{2k_e}[EIk_t^2\beta_t^2 + \kappa GA(k_t-\beta_t)^2 - (\rho A - \rho I\beta_t^2)\omega^2] \\
D_{T44} &= \frac{i(1-\bar{\mathrm{e}}_e^2)}{2k_e}[EIk_e^2\beta_e^2 + \kappa GA(k_e-\beta_e)^2 - (\rho A - \rho I\beta_e^2)\omega^2]
\end{aligned}
\tag{3.176}
$$

and

$$
E_{Tij} = \begin{cases} \dfrac{i}{k_i+k_j}(\mathrm{e}^{-i(k_i+k_j)L} - 1) & (k_i+k_j \neq 0) \\ L & (k_i+k_j = 0) \end{cases} \quad (i,j = 1,\ 2,\ 3,\ 4) \tag{3.177}
$$

We can numerically prove that the spectral element matrix given by Equation 3.170 is identical with that given by Equation 3.86.

Finally, by substituting Equation 3.82 into Equation 3.167, we can obtain a new expression as

$$
f_{\mathrm{d}}(\omega) = \boldsymbol{H}_T^{-\mathrm{T}}(\omega)\int_0^L [F(x;\omega)\boldsymbol{I} + T(x;\omega)\boldsymbol{B}(\omega)]\boldsymbol{e}_w^{\mathrm{T}}(x;\omega)\mathrm{d}x \tag{3.178}
$$

Example 3.8 Extended Timoshenko Beam

The forced vibration of an ET-beam is represented by

$$
\begin{Bmatrix} N_t' \\ Q_t' \\ M_t' \end{Bmatrix} = \begin{bmatrix} \rho A & 0 & \rho R \\ 0 & \rho A & 0 \\ \rho R & 0 & \rho I \end{bmatrix} \begin{Bmatrix} \ddot{u} \\ \ddot{w} \\ \ddot{\theta} \end{Bmatrix} - \begin{Bmatrix} 0 \\ 0 \\ Q_t \end{Bmatrix} - \begin{Bmatrix} p(x,t) \\ f(x,t) \\ \tau(x,t) \end{Bmatrix} \tag{3.179}
$$

where $p(x, t)$, $f(x, t)$, and $\tau(x, t)$ are the distributed dynamic forces. The force-displacement relation for the ET-beam is given by

$$\begin{Bmatrix} N_t \\ Q_t \\ M_t \end{Bmatrix} = \begin{bmatrix} EA & C_1 & C_2 \\ C_1 & \kappa GA & C_3 \\ C_2 & C_3 & EI \end{bmatrix} \begin{Bmatrix} u' \\ w' - \theta \\ \theta' \end{Bmatrix} \tag{3.180}$$

By using the DFT processing given by Equations 3.2–3.4, the time-domain governing equations can be transformed into the frequency-domain expressions as

$$\begin{aligned} &EAU'' + C_1(W'' - \Theta') + C_2\Theta'' + \omega^2\rho AU + \omega^2\rho R\Theta + P = 0 \\ &\kappa GA(W'' - \Theta') + C_1U'' + C_3\Theta'' + \omega^2\rho AW + F = 0 \\ &EI\Theta' + \kappa GA(W' - \Theta) + C_1U'' + C_2U'' + C_3(W'' - \Theta') \\ &\qquad + C_3\Theta' + \omega^2\rho RU + \omega^2\rho I\Theta + T = 0 \end{aligned} \tag{3.181}$$

with the frequency-domain force-displacement relationship

$$\begin{aligned} N &= EAU' + C_1(W' - \Theta) + C_2\Theta' \\ Q &= \kappa GA(W' - \Theta) + C_1U' + C_3\Theta' \\ M &= EI\Theta' + C_2U' + C_3(W' - \Theta) \end{aligned} \tag{3.182}$$

The weak form of Equation 3.181 can be derived from the weighted-integral statement as

$$\begin{aligned} \int_0^L &\{[EAU'' + C_1(W'' - \Theta') + C_2\Theta'' + \omega^2\rho AU + \omega^2\rho R\Theta + P]\delta U \\ &[\kappa GA(W'' - \Theta') + C_1U'' + C_3\Theta'' + \omega^2\rho AW + F]\,\delta W \\ &[EI\Theta'' + \kappa GA(W' - \Theta) + C_1U' + C_2U'' + C_3(W'' - \Theta') + C_3\Theta' \\ &+ \omega^2\rho RU + \omega^2\rho I\Theta + T]\delta\Theta\}\,\mathrm{d}x = 0 \end{aligned} \tag{3.183}$$

Integrating by parts and applying Equation 3.182 yields the weak form as

$$\begin{aligned} \int_0^L &[EA(U'\delta U') + \kappa GA(W'\delta W' + \Theta\delta\Theta - \Theta\delta W' - W'\delta\Theta) \\ &+ EI(\Theta'\delta\Theta') + C_1(W'\delta U' + U'\delta W' - \Theta\delta U' - U'\delta\Theta) \\ &+ C_2(U'\delta\Theta' + \Theta'\delta U') + C_3(\Theta'\delta W' + W'\delta\Theta' - \Theta\delta\Theta' - \Theta'\delta\Theta) \\ &- \omega^2\{\rho A(U\delta U + W\delta W) + \rho I(\Theta\delta\Theta) + \rho R(\Theta\delta U + U\delta\Theta)\} \\ &- P\delta U - F\delta W - T\delta\Theta]\,\mathrm{d}x - N\delta U|_0^L - Q\delta W|_0^L - M\delta\Theta|_0^L = 0 \end{aligned} \tag{3.184}$$

Substituting Equation 3.114 into Equation 3.184 gives

$$\begin{aligned} \delta \boldsymbol{d}^{\mathrm{T}} \Bigg\{ \int_0^L &[EA(\boldsymbol{N}_u'^{\mathrm{T}}\boldsymbol{N}_u') + \kappa GA(\boldsymbol{N}_w'^{\mathrm{T}}\boldsymbol{N}_w' + \boldsymbol{N}_\theta^{\mathrm{T}}\boldsymbol{N}_\theta - \boldsymbol{N}_w'^{\mathrm{T}}\boldsymbol{N}_\theta - \boldsymbol{N}_\theta^{\mathrm{T}}\boldsymbol{N}_w') \\ &+ EI(\boldsymbol{N}_\theta'^{\mathrm{T}}\boldsymbol{N}_\theta') + C_1(\boldsymbol{N}_u'^{\mathrm{T}}\boldsymbol{N}_w' + \boldsymbol{N}_w'^{\mathrm{T}}\boldsymbol{N}_u' - \boldsymbol{N}_u'^{\mathrm{T}}\boldsymbol{N}_\theta - \boldsymbol{N}_\theta^{\mathrm{T}}\boldsymbol{N}_u') \\ &+ C_2(\boldsymbol{N}_\theta'^{\mathrm{T}}\boldsymbol{N}_u' + \boldsymbol{N}_u'^{\mathrm{T}}\boldsymbol{N}_\theta') + C_3(\boldsymbol{N}_w'^{\mathrm{T}}\boldsymbol{N}_\theta' + \boldsymbol{N}_\theta'^{\mathrm{T}}\boldsymbol{N}_w' - \boldsymbol{N}_\theta'^{\mathrm{T}}\boldsymbol{N}_\theta - \boldsymbol{N}_\theta^{\mathrm{T}}\boldsymbol{N}_\theta') \\ &- \omega^2\{\rho A(\boldsymbol{N}_u^{\mathrm{T}}\boldsymbol{N}_u + \boldsymbol{N}_w^{\mathrm{T}}\boldsymbol{N}_w) + \rho I(\boldsymbol{N}_\theta^{\mathrm{T}}\boldsymbol{N}_\theta) + \rho R(\boldsymbol{N}_u^{\mathrm{T}}\boldsymbol{N}_\theta + \boldsymbol{N}_\theta^{\mathrm{T}}\boldsymbol{N}_u)\}]\mathrm{d}x\,\boldsymbol{d} \\ &- (\boldsymbol{f}_{\mathrm{c}} + \boldsymbol{f}_{\mathrm{d}}) \Bigg\} = 0 \end{aligned} \tag{3.185}$$

where $\boldsymbol{d}$ is the spectral nodal DOFs vector defined by Equation 3.109 and $\boldsymbol{f}_c$ is the nodal forces and moments vector defined by Equation 3.110: they are

$$\begin{aligned}\boldsymbol{d} &= \{U_1 \quad W_1 \quad \Theta_1 \quad U_2 \quad W_2 \quad \Theta_2\}^{\mathrm{T}} \\ &= \{U(0) \quad W(0) \quad \Theta(0) \quad U(L) \quad W(L) \quad \Theta(L)\}^{\mathrm{T}}\end{aligned} \tag{3.186}$$

$$\begin{aligned}\boldsymbol{f}_c &= \{N_1 \quad Q_1 \quad M_1 \quad N_2 \quad Q_2 \quad M_2\}^{\mathrm{T}} \\ &= \{-N(0) \quad -Q(0) \quad -M(0) \quad N(L) \quad Q(L) \quad M(L)\}^{\mathrm{T}}\end{aligned} \tag{3.187}$$

The vector $\boldsymbol{f}_d$ in Equation 3.185 denotes the spectral nodal forces and moments associated with the distributed dynamic forces $p(x, t)$, $f(x, t)$, and $\tau(x, t)$ as follows:

$$\boldsymbol{f}_d(\omega) = \int_0^L \{P(x;\omega)\boldsymbol{N}_u^{\mathrm{T}}(x;\omega) + F(x;\omega)\boldsymbol{N}_w^{\mathrm{T}}(x;\omega) + T(x;\omega)\boldsymbol{N}_\theta^{\mathrm{T}}(x;\omega)\}\mathrm{d}x \tag{3.188}$$

Since $\delta\boldsymbol{d}$ is arbitrary by definition, we can obtain the spectral element equation from Equation 3.185 as

$$\boldsymbol{S}_{\mathrm{ET}}(\omega)\boldsymbol{d} = \boldsymbol{f}_c(\omega) + \boldsymbol{f}_d(\omega) \tag{3.189}$$

where

$$\begin{aligned}\boldsymbol{S}_{\mathrm{ET}}(\omega) = \int_0^L & [EA(\boldsymbol{N}_u'^{\mathrm{T}}\boldsymbol{N}_u') + \kappa GA(\boldsymbol{N}_w'^{\mathrm{T}}\boldsymbol{N}_w' + \boldsymbol{N}_\theta^{\mathrm{T}}\boldsymbol{N}_\theta - \boldsymbol{N}_w'^{\mathrm{T}}\boldsymbol{N}_\theta - \boldsymbol{N}_\theta^{\mathrm{T}}\boldsymbol{N}_w') \\ & + EI(\boldsymbol{N}_\theta'^{\mathrm{T}}\boldsymbol{N}_\theta') + C_1(\boldsymbol{N}_u'^{\mathrm{T}}\boldsymbol{N}_w' + \boldsymbol{N}_w'^{\mathrm{T}}\boldsymbol{N}_u' - \boldsymbol{N}_u'^{\mathrm{T}}\boldsymbol{N}_\theta - \boldsymbol{N}_\theta^{\mathrm{T}}\boldsymbol{N}_u') \\ & + C_2(\boldsymbol{N}_\theta'^{\mathrm{T}}\boldsymbol{N}_u' + \boldsymbol{N}_u'^{\mathrm{T}}\boldsymbol{N}_\theta') + C_3(\boldsymbol{N}_w'^{\mathrm{T}}\boldsymbol{N}_\theta' + \boldsymbol{N}_\theta'^{\mathrm{T}}\boldsymbol{N}_w' - \boldsymbol{N}_\theta'^{\mathrm{T}}\boldsymbol{N}_\theta - \boldsymbol{N}_\theta^{\mathrm{T}}\boldsymbol{N}_\theta') \\ & - \omega^2\{\rho A(\boldsymbol{N}_u^{\mathrm{T}}\boldsymbol{N}_u + \boldsymbol{N}_w^{\mathrm{T}}\boldsymbol{N}_w) + \rho I(\boldsymbol{N}_\theta^{\mathrm{T}}\boldsymbol{N}_\theta) + \rho R(\boldsymbol{N}_u^{\mathrm{T}}\boldsymbol{N}_\theta + \boldsymbol{N}_\theta^{\mathrm{T}}\boldsymbol{N}_u)\}]\mathrm{d}x\end{aligned} \tag{3.190}$$

Substitution of Equation 3.115 into Equation 3.190 gives the simplified expression

$$\boldsymbol{S}_{\mathrm{ET}}(\omega) = \boldsymbol{H}_{\mathrm{ET}}^{-\mathrm{T}}(\omega)\boldsymbol{D}_{\mathrm{ET}}(\omega)\boldsymbol{H}_{\mathrm{ET}}^{-1}(\omega) \tag{3.191}$$

where

$$\begin{aligned}\boldsymbol{D}_{\mathrm{ET}}(\omega) = \int_0^L & [EA(\boldsymbol{A}\boldsymbol{e}_w'^{\mathrm{T}}\boldsymbol{e}_w'\boldsymbol{A}) + \kappa GA(\boldsymbol{e}_w'^{\mathrm{T}}\boldsymbol{e}_w' + \boldsymbol{B}\boldsymbol{e}_w^{\mathrm{T}}\boldsymbol{e}_w\boldsymbol{B} - \boldsymbol{e}_w'^{\mathrm{T}}\boldsymbol{e}_w\boldsymbol{B} - \boldsymbol{B}\boldsymbol{e}_w^{\mathrm{T}}\boldsymbol{e}_w') \\ & + EI(\boldsymbol{B}\boldsymbol{e}_w'^{\mathrm{T}}\boldsymbol{e}_w'\boldsymbol{B}) + C_1(\boldsymbol{A}\boldsymbol{e}_w'^{\mathrm{T}}\boldsymbol{e}_w' + \boldsymbol{e}_w'^{\mathrm{T}}\boldsymbol{e}_w'\boldsymbol{A} - \boldsymbol{A}\boldsymbol{e}_w'^{\mathrm{T}}\boldsymbol{e}_w\boldsymbol{B} - \boldsymbol{B}\boldsymbol{e}_w^{\mathrm{T}}\boldsymbol{e}_w'\boldsymbol{A}) \\ & + C_2(\boldsymbol{B}\boldsymbol{e}_w'^{\mathrm{T}}\boldsymbol{e}_w'\boldsymbol{A} + \boldsymbol{A}\boldsymbol{e}_w'^{\mathrm{T}}\boldsymbol{e}_w'\boldsymbol{B}) \\ & + C_3(\boldsymbol{e}_w'^{\mathrm{T}}\boldsymbol{e}_w'\boldsymbol{B} + \boldsymbol{B}\boldsymbol{e}_w'^{\mathrm{T}}\boldsymbol{e}_w' - \boldsymbol{B}\boldsymbol{e}_w'^{\mathrm{T}}\boldsymbol{e}_w\boldsymbol{B} - \boldsymbol{B}\boldsymbol{e}_w^{\mathrm{T}}\boldsymbol{e}_w'\boldsymbol{B}) \\ & - \omega^2\{\rho A(\boldsymbol{A}\boldsymbol{e}_w^{\mathrm{T}}\boldsymbol{e}_w\boldsymbol{A} + \boldsymbol{e}_w^{\mathrm{T}}\boldsymbol{e}_w) + \rho I(\boldsymbol{B}\boldsymbol{e}_w^{\mathrm{T}}\boldsymbol{e}_w\boldsymbol{B}) \\ & + \rho R(\boldsymbol{A}\boldsymbol{e}_w^{\mathrm{T}}\boldsymbol{e}_w\boldsymbol{B} + \boldsymbol{B}\boldsymbol{e}_w^{\mathrm{T}}\boldsymbol{e}_w\boldsymbol{A})\}]\mathrm{d}x\end{aligned} \tag{3.192}$$

From the definition of $\boldsymbol{e}_w(x;\omega)$ given by Equation 3.107, we obtain

$$\boldsymbol{e}_w'(x;\omega) = -i\boldsymbol{e}_w(x;\omega)\boldsymbol{K}(\omega) \tag{3.193}$$

where

$$\boldsymbol{K}(\omega) = diag[k_p(\omega)] \quad (p = 1, 2, \ldots, 6) \tag{3.194}$$

By using Equations 3.107 and 3.193, Equation 3.192 can be further simplified as

$$\begin{aligned}\boldsymbol{D}_{\mathrm{ET}}(\omega) = & -EA(\boldsymbol{AKE}_{\mathrm{ET}}\boldsymbol{KA}) - \kappa GA(\boldsymbol{KE}_{\mathrm{ET}}\boldsymbol{K} - \boldsymbol{BE}_{\mathrm{ET}}\boldsymbol{B} - i\boldsymbol{KE}_{\mathrm{ET}}\boldsymbol{B} - i\boldsymbol{BE}_{\mathrm{ET}}\boldsymbol{K})\\ & - EI(\boldsymbol{BKE}_{\mathrm{ET}}\boldsymbol{KB}) - C_1(\boldsymbol{AKE}_{\mathrm{ET}}\boldsymbol{K} + \boldsymbol{KE}_{\mathrm{ET}}\boldsymbol{KA} - i\boldsymbol{AKE}_{\mathrm{ET}}\boldsymbol{B} - i\boldsymbol{BE}_{\mathrm{ET}}\boldsymbol{KA})\\ & - C_2(\boldsymbol{BKE}_{\mathrm{ET}}\boldsymbol{KA} + \boldsymbol{AKE}_{\mathrm{ET}}\boldsymbol{KB})\\ & - C_3(\boldsymbol{KE}_{\mathrm{ET}}\boldsymbol{KB} + \boldsymbol{BKE}_{\mathrm{ET}}\boldsymbol{K} - i\boldsymbol{BKE}_{\mathrm{ET}}\boldsymbol{B} - i\boldsymbol{BE}_{\mathrm{ET}}\boldsymbol{KB})\\ & - \omega^2[\rho A(\boldsymbol{AE}_{\mathrm{ET}}\boldsymbol{A} + \boldsymbol{E}_{\mathrm{ET}}) + \rho I(\boldsymbol{BE}_{\mathrm{ET}}\boldsymbol{B}) + \rho R(\boldsymbol{AE}_{\mathrm{ET}}\boldsymbol{B} + \boldsymbol{BE}_{\mathrm{ET}}\boldsymbol{A})]\end{aligned} \tag{3.195}$$

where

$$\boldsymbol{E}_{\mathrm{ET}}(\omega) = \int_0^L \boldsymbol{e}_w^{\mathrm{T}}(x;\omega)\boldsymbol{e}_w(x;\omega)\mathrm{d}x = [E_{\mathrm{ET}ij}] \quad (i,j = 1, 2, \ldots, 6) \tag{3.196}$$

The components of the matrix $\boldsymbol{E}_{\mathrm{ET}} = [E_{\mathrm{ET}ij}]$ are given by

$$E_{\mathrm{ET}ij} = \begin{cases} \dfrac{i}{k_i + k_j}(\mathrm{e}^{-i(k_i+k_j)L} - 1) & (k_i + k_j \neq 0)\\ L & (k_i + k_j = 0)\end{cases} \quad (i,j = 1, 2, \ldots, 6) \tag{3.197}$$

We can prove numerically that the spectral element matrix given by Equation 3.191 is identical with that given by Equation 3.117.

Finally, by substituting Equation 3.115 into Equation 3.188, we can obtain a new expression

$$\boldsymbol{f}_{\mathrm{d}}(\omega) = \boldsymbol{H}_T^{-\mathrm{T}}(\omega)\int_0^L [P(x;\omega)\boldsymbol{A}(\omega) + F(x;\omega)\boldsymbol{I} + T(x;\omega)\boldsymbol{B}(\omega)]\boldsymbol{e}_w^{\mathrm{T}}(x;\omega)\mathrm{d}x \tag{3.198}$$

3.3 State-Vector Equation Method

In the force-displacement relationship method and the variational method introduced in the preceding sections, the spectral element matrices are formulated by using the exact dynamic shape functions derived from the exact wave solutions to the frequency-domain governing equations. Thus, these two methods can only be used when the exact wave solutions to the frequency-domain governing equations are available in the closed forms. In this section, an alternative approach to spectral element formulation, in which the exact wave solutions are not required, is introduced. In the alternative approach, the spectral element matrix is derived from the transfer matrix formulated directly from the frequency-domain state-vector equation of motion.

The dynamics of 1-D structures in the frequency domain can be represented by a set of ordinary differential equations (ODEs) as shown in Equation 3.6. By using the force-displacement relationship of Equation 3.18, it may be possible to transform Equation 3.6

into a system of first-order ODEs (frequency-domain state-vector equations) as

$$\frac{\mathrm{d}\boldsymbol{y}}{\mathrm{d}x} = \boldsymbol{A}(\omega)\boldsymbol{y} \quad \text{or} \quad \boldsymbol{y}' = \boldsymbol{A}(\omega)\boldsymbol{y} \quad (0 \leq x \leq L) \tag{3.199}$$

where $\boldsymbol{A}(\omega)$ is the frequency-dependent system matrix and $\boldsymbol{y}(x)$ is the (cross-sectional) state-vector. The choice of the variables in $\boldsymbol{y}(x)$ is not unique: they may represent spatial derivatives of displacement fields and internal forces as in

$$\boldsymbol{y}(x) = \begin{Bmatrix} \boldsymbol{D}(x) \\ \boldsymbol{F}(x) \end{Bmatrix} \tag{3.200}$$

where $\boldsymbol{D}$ is the displacement fields vector and $\boldsymbol{F}$ is the internal forces vector. The general solution to Equation 3.199 is given by

$$\boldsymbol{y}(x) = \mathrm{e}^{\boldsymbol{A}x}\boldsymbol{y}(0) \tag{3.201}$$

The transfer matrix is the matrix that relates the quantities in the input side ($x = 0$) to those in the output side ($x = L$). Thus, the transfer matrix for the 1-D structural member of length L can be derived from Equation 3.201 as

$$\boldsymbol{y}(L) = \mathrm{e}^{\boldsymbol{A}L}\boldsymbol{y}(0) \equiv \boldsymbol{T}\boldsymbol{y}(0) \tag{3.202}$$

where $\boldsymbol{T} = \mathrm{e}^{\boldsymbol{A}L}$ is the transfer matrix, which can be represented in a partitioned form as

$$\boldsymbol{T} = \mathrm{e}^{\boldsymbol{A}L} = \begin{bmatrix} \boldsymbol{T}_{DD} & \boldsymbol{T}_{DF} \\ \boldsymbol{T}_{FD} & \boldsymbol{T}_{FF} \end{bmatrix} \tag{3.203}$$

where $\boldsymbol{T}_{DD}$, $\boldsymbol{T}_{DF}$, $\boldsymbol{T}_{FD}$, and $\boldsymbol{T}_{FF}$ are the submatrices of $\boldsymbol{T}$. To obtain an analytical, closed form of the partitioned transfer matrix from $\mathrm{e}^{\boldsymbol{A}L}$, one may use the symbolic computation algorithm in conjunction with the Cayley–Hamiton theorem [74].

Applying Equations 3.200 and 3.203 to Equation 3.202 gives

$$\begin{Bmatrix} \boldsymbol{D}_2 \\ \boldsymbol{F}_2 \end{Bmatrix} = \begin{bmatrix} \boldsymbol{T}_{DD} & \boldsymbol{T}_{DF} \\ \boldsymbol{T}_{FD} & \boldsymbol{T}_{FF} \end{bmatrix} \begin{Bmatrix} \boldsymbol{D}_1 \\ \boldsymbol{F}_1 \end{Bmatrix} \tag{3.204}$$

where

$$\begin{Bmatrix} \boldsymbol{D}_2 \\ \boldsymbol{F}_2 \end{Bmatrix} = \begin{Bmatrix} \boldsymbol{D}(L) \\ \boldsymbol{F}(L) \end{Bmatrix}, \begin{Bmatrix} \boldsymbol{D}_1 \\ \boldsymbol{F}_1 \end{Bmatrix} = \begin{Bmatrix} \boldsymbol{D}(0) \\ \boldsymbol{F}(0) \end{Bmatrix} \tag{3.205}$$

Comparing the sign convention defined for the spectral element with that defined for the transfer matrix of Equation 3.204 (see Figure 3.5), we obtain the relations

$$\boldsymbol{f} = \begin{Bmatrix} \boldsymbol{f}_1 \\ \boldsymbol{f}_2 \end{Bmatrix} = \begin{Bmatrix} -\boldsymbol{F}_1 \\ +\boldsymbol{F}_2 \end{Bmatrix}, \quad \boldsymbol{d} = \begin{Bmatrix} \boldsymbol{d}_1 \\ \boldsymbol{d}_2 \end{Bmatrix} = \begin{Bmatrix} +\boldsymbol{D}_1 \\ +\boldsymbol{D}_2 \end{Bmatrix} \tag{3.206}$$

By using Equation 3.206, Equation 3.204 can be written as

$$\begin{Bmatrix} +\boldsymbol{d}_2 \\ +\boldsymbol{f}_2 \end{Bmatrix} = \begin{bmatrix} \boldsymbol{T}_{DD} & \boldsymbol{T}_{DF} \\ \boldsymbol{T}_{FD} & \boldsymbol{T}_{FF} \end{bmatrix} \begin{Bmatrix} +\boldsymbol{d}_1 \\ -\boldsymbol{f}_1 \end{Bmatrix} \tag{3.207}$$

Finally, Equation 3.207 can be written in the force-displacement relation form as

$$\boldsymbol{S}(\omega)\boldsymbol{d} = \boldsymbol{f} \tag{3.208}$$

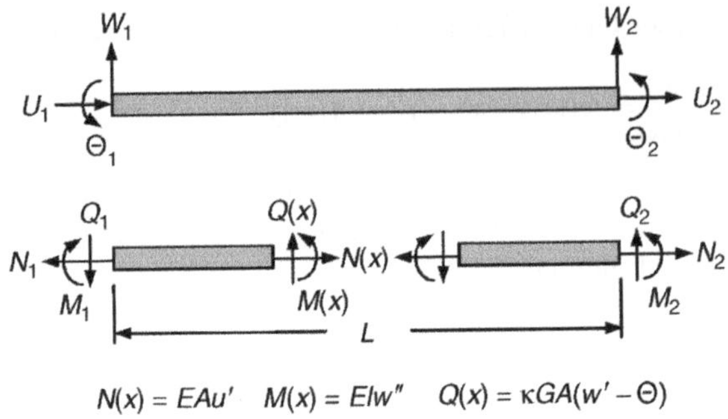

Figure 3.5 Sign convention defined for the transfer matrix

where $S(\omega)$ is the spectral element matrix given by

$$S(\omega) = \begin{bmatrix} T_{DF}^{-1} T_{DD} & -T_{DF}^{-1} \\ T_{FD} - T_{FF} T_{DF}^{-1} T_{DD} & T_{FF} T_{DF}^{-1} \end{bmatrix} \tag{3.209}$$

Rememebr that so far we have not used exact wave solutions or exact dynamic shape functions to formulate the spectral element matrix. However, for complicated higher order structure members, it will not always be easy to express the exponential matrix e^{AL} into the exact closed form of the partitioned transfer matrix T, as shown by Equation 3.203. In such a case, one may use a proper numerical method to compute T from e^{AL} at each discrete frequency.

In the following, the rod, the Bernoulli–Euler beam, the Timoshenko beam, and the extended Timoshenko beam will be considered as illustrative examples.

Example 3.9 Rod

Revisit the uniform rod considered in Example 3.1. The governing equation (*GE*) and the force-displacement relationship (*FD*) are given as

$$\begin{aligned} &(GE) \quad EAU'' + \omega^2 \rho A U = 0 \\ &(FD) \quad N = EAU' \end{aligned} \tag{3.210}$$

Equation 3.210 can be transformed into two first-order ODEs as

$$\begin{aligned} N' &= EAU'' = -\omega^2 \rho A U \\ U' &= N/EA \end{aligned} \tag{3.211}$$

The state-vector equation can be obtained from Equation 3.211 as follows:

$$y_R' = A_R(\omega) y_R \quad (0 \le x \le L) \tag{3.212}$$

where

$$y_R = \begin{Bmatrix} D \\ F \end{Bmatrix} = \begin{Bmatrix} U \\ N \end{Bmatrix} \tag{3.213}$$

and

$$A_R(\omega) = \begin{bmatrix} 0 & 1/EA \\ -EAk_L^2 & 0 \end{bmatrix} \tag{3.214}$$

where $k_L = \omega(\rho A/EA)^{1/2}$ is the wavenumber defined by Equation 3.28.

By using A_R of Equation 3.214, the transfer matrix T_R can be derived in a partitioned matrix form as

$$\begin{aligned} T_R \equiv e^{A_R L} &= \begin{bmatrix} T_{DD} & T_{DF} \\ T_{FD} & T_{FF} \end{bmatrix} \\ &= \begin{bmatrix} \cos(k_L L) & (EAk_L)^{-1}\sin(k_L L) \\ -EAk_L \sin(k_L L) & \cos(k_L L) \end{bmatrix} \end{aligned} \tag{3.215}$$

Applying Equation 3.215 to Equation 3.209 gives the spectral element matrix as

$$S_R(\omega) = \frac{EA}{L}\begin{bmatrix} (k_L L)\cot(k_L L) & -(k_L L)\csc(k_L L) \\ -(k_L L)\csc(k_L L) & (k_L L)\cot(k_L L) \end{bmatrix} \tag{3.216}$$

which is identical with Equation 3.40.

Example 3.10 Bernoulli–Euler Beam

Revisit the uniform B-beam considered in Example 3.2. The governing equation and the force-displacement relationships are given by

$$\begin{aligned} &(GE) \quad EIW'''' - \omega^2 \rho A W = 0 \\ &(FD) \quad M = EIW'' \\ &\qquad\quad\; Q = -EIW''' \end{aligned} \tag{3.217}$$

Equation 3.217 can be transformed into four first-order ODEs as

$$\begin{aligned} &W' = \Theta \\ &\Theta = W'' = M/EI \\ &Q' = -EIW''' = -\omega^2 \rho A W = -EIk_F^4 W \\ &M' = EIW''' = -Q \end{aligned} \tag{3.218}$$

Equation 3.218 provides the state-vector equation as

$$y_B' = A_B(\omega) y_B \quad (0 \le x \le L) \tag{3.219}$$

where

$$y_B = \begin{Bmatrix} D \\ F \end{Bmatrix} = \begin{Bmatrix} \begin{Bmatrix} W \\ \Theta \end{Bmatrix} \\ \begin{Bmatrix} Q \\ M \end{Bmatrix} \end{Bmatrix} \tag{3.220}$$

and

$$A_B(\omega) = \begin{bmatrix} 0 & 1 & 0 & 0 \\ 0 & 0 & 0 & 1/EI \\ -EIk_F^4 & 0 & 0 & 0 \\ 0 & 0 & -1 & 0 \end{bmatrix} \tag{3.221}$$

where $k_F = \omega(\rho A/EI)^{1/2}$ is the wavenumber defined by Equation 3.48.

By using A_B of Equation 3.221, the transfer matrix T_B can be obtained as

$$T_B \equiv e^{A_B L} = \begin{bmatrix} T_{DD} & T_{DF} \\ T_{FD} & T_{FF} \end{bmatrix}$$

$$= \frac{1}{2}\begin{bmatrix} c_1 & \dfrac{s_1}{k_F} & \dfrac{s_2}{EIk_F^3} & -\dfrac{c_2}{EIk_F^2} \\ -k_F s_2 & c_1 & \dfrac{c_2}{EIk_F^2} & \dfrac{s_1}{EIk_F} \\ -EIk_F^3 s_1 & EIk_F^2 c_2 & c_1 & k_F s_2 \\ -EIk_F^2 c_2 & -EIk_F s_2 & -\dfrac{s_1}{k_F} & c_1 \end{bmatrix} \tag{3.222}$$

where

$$\begin{aligned} c_1 &= \cos(k_F L) + \cosh(k_F L) \\ c_2 &= \cos(k_F L) - \cosh(k_F L) \\ s_1 &= \sin(k_F L) + \sinh(k_F L) \\ s_2 &= \sin(k_F L) - \sinh(k_F L) \end{aligned} \tag{3.223}$$

By substituting the submatrices of T_B into Equation 3.209, we obtain the spectral element matrix $S_B(\omega)$, which is identical with Equation 3.60.

Example 3.11 Timoshenko Beam

Revisit the uniform T-beam considered in Example 3.3. In the frequency domain, the governing equations and force-displacement relations for the T-beam are given by

$$\begin{aligned} (GE) \quad & \kappa GA(W'' - \Theta') + \rho A\omega^2 W = 0 \\ & EI\Theta'' + \kappa GA(W' - \Theta) + \rho I\omega^2\Theta = 0 \\ (FD) \quad & Q = \kappa GA(W' - \Theta) \\ & M = EI\Theta' \end{aligned} \tag{3.224}$$

Equation 3.224 can be transformed into four first-order ODEs as follows:

$$\begin{aligned} W' &= \Theta + Q/\kappa GA \\ \Theta' &= M/EI \\ Q' &= \kappa GA(W'' - \Theta') = -\rho A\omega^2 W = -EIk_F^4 W \\ M' &= EI\Theta'' = -\kappa GA(W' - \Theta) - \rho I\omega^2\Theta = -Q - EI\eta_1 k_F^4\Theta \end{aligned} \tag{3.225}$$

Thus, the above equations can be written into the form of a state-vector equation as

$$y_T' = A_T(\omega)y_T \quad (0 \le x \le L) \tag{3.226}$$

where

$$y_T = \begin{Bmatrix} \boldsymbol{D} \\ \boldsymbol{F} \end{Bmatrix} = \begin{Bmatrix} \begin{Bmatrix} W \\ \Theta \end{Bmatrix} \\ \begin{Bmatrix} Q \\ M \end{Bmatrix} \end{Bmatrix} \tag{3.227}$$

and

$$\boldsymbol{A}_T(\omega) = \begin{bmatrix} 0 & 1 & 1/\kappa GA & 0 \\ 0 & 0 & 0 & 1/EI \\ -EIk_F^4 & 0 & 0 & 0 \\ 0 & -EI\eta_1 k_F^4 & -1 & 0 \end{bmatrix} \tag{3.228}$$

where k_F and η_1 are defined by the first part of Equation 3.69 and the middle part of Equation 3.70, respectively.

By using the matrix $\boldsymbol{A}_T$ of Equation 3.228, the transfer matrix $\boldsymbol{T}_T$ can be derived as

$$\boldsymbol{T}_T \equiv e^{A_T L} = \begin{bmatrix} T_{11} & T_{12} & T_{13} & T_{14} \\ T_{21} & T_{22} & -T_{14} & T_{24} \\ T_{31} & T_{32} & T_{11} & -T_{21} \\ -T_{32} & T_{42} & -T_{12} & T_{22} \end{bmatrix} = \begin{bmatrix} \boldsymbol{T}_{DD} & \boldsymbol{T}_{DF} \\ \boldsymbol{T}_{FD} & \boldsymbol{T}_{FF} \end{bmatrix} \tag{3.229}$$

where

$$\begin{aligned}
T_{11} &= \frac{(k_G^4 - k_e^2)\cosh(k_t L) - (k_G^4 - k_t^2)\cosh(k_e L)}{k_t^2 - k_e^2} \\
T_{12} &= \frac{k_e(\eta k_F^4 - k_e^2)\sinh(k_t L) - k_t(\eta k_F^4 - k_t^2)\sinh(k_e L)}{k_t k_e (k_t^2 - k_e^2)} \\
T_{13} &= \frac{k_e[1 + \eta_2(\eta_2 k_F^4 - k_e^2)]\sinh(k_t L) - k_t[1 + \eta_2(\eta_2 k_F^4 - k_t^2)]\sinh(k_e L)}{EIk_t k_e (k_t^2 - k_e^2)} \\
T_{14} &= -\frac{\cosh(k_t L) - \cosh(k_e L)}{EI(k_t^2 - k_e^2)} \\
T_{21} &= -\frac{k_F^4[k_e \sinh(k_t L) - k_t \sinh(k_e L)]}{k_t k_e (k_t^2 - k_e^2)} \\
T_{22} &= \frac{(\eta_1 k_F^4 - k_e^2)\cosh(k_t L) - (\eta_1 k_F^4 - k_t^2)\cosh(k_e L)}{k_t^2 - k_e^2} \\
T_{24} &= \frac{k_e(\eta_1 k_F^4 - k_e^2)\sinh(k_t L) - k_t(\eta_1 k_F^4 - k_t^2)\sinh(k_e L)}{EIk_t k_e (k_t^2 - k_e^2)} \\
T_{31} &= -\frac{EIk_F^4[k_e(k_G^4 - k_e^2)\sinh(k_t L) - k_t(k_G^4 - k_t^2)\sinh(k_e L)]}{k_t k_e (k_t^2 - k_e^2)} \\
T_{32} &= \frac{EIk_F^4[\cosh(k_t L) - \cosh(k_e L)]}{k_t^2 - k_e^2} \\
T_{42} &= -\frac{EIk_F^4\{k_e[1 + \eta_1(\eta_1 k_F^4 - k_e^2)]\sinh(k_t L) - k_t[1 + \eta_1(\eta_1 k_F^4 - k_t^2)]\sinh(k_e L)\}}{k_t k_e (k_t^2 - k_e^2)}
\end{aligned} \tag{3.230}$$

In Equation 3.230, k_G is defined by the second part of Equation 3.69. By substituting the submatrices of $\boldsymbol{T}_T$ into Equation 3.209, we can obtain the spectral element matrix $\boldsymbol{S}_T(\omega)$, which is identical with Equation 3.86.

Example 3.12 Extended Timoshenko Beam

The governing equation and force-displacement relation for an ET-beam are given in the frequency domain by Equations 3.93 and 3.94, respectively. They are

$$\begin{Bmatrix} N' \\ Q' \\ M' \end{Bmatrix} = -\omega^2 \begin{bmatrix} \rho A & 0 & \rho R \\ 0 & \rho A & 0 \\ \rho R & 0 & \rho I \end{bmatrix} \begin{Bmatrix} U \\ W \\ \Theta \end{Bmatrix} - \begin{Bmatrix} 0 \\ 0 \\ Q \end{Bmatrix} \tag{3.231}$$

$$\begin{Bmatrix} N \\ Q \\ M \end{Bmatrix} = \begin{bmatrix} EA & C_1 & C_2 \\ C_1 & \kappa GA & C_3 \\ C_2 & C_3 & EI \end{bmatrix} \begin{Bmatrix} U' \\ W' - \Theta \\ \Theta' \end{Bmatrix} \tag{3.232}$$

Equations 3.231 and 3.232 can be rewritten as

$$\begin{Bmatrix} U' \\ W' \\ \Theta' \end{Bmatrix} = \begin{bmatrix} 0 & 0 & 0 \\ 0 & 0 & 1 \\ 0 & 0 & 0 \end{bmatrix} \begin{Bmatrix} U \\ W \\ \Theta \end{Bmatrix} + \begin{bmatrix} EA & C_1 & C_2 \\ C_1 & \kappa GA & C_3 \\ C_2 & C_3 & EI \end{bmatrix}^{-1} \begin{Bmatrix} N \\ Q \\ M \end{Bmatrix} \tag{3.233}$$

$$\begin{Bmatrix} N' \\ Q' \\ M' \end{Bmatrix} = \begin{bmatrix} -\omega^2 \rho A & 0 & -\omega^2 \rho R \\ 0 & -\omega^2 \rho A & 0 \\ -\omega^2 \rho R & 0 & -\omega^2 \rho I \end{bmatrix} \begin{Bmatrix} U \\ W \\ \Theta \end{Bmatrix} + \begin{bmatrix} 0 & 0 & 0 \\ 0 & 0 & 0 \\ 0 & -1 & 0 \end{bmatrix} \begin{Bmatrix} N \\ Q \\ M \end{Bmatrix} \tag{3.234}$$

A state-vector equation can be obtained from Equations 3.233 and 3.234 as

$$\boldsymbol{y}'_{ET} = \boldsymbol{A}_{ET}(\omega)\boldsymbol{y}_{ET} \tag{3.235}$$

where

$$\boldsymbol{y}_{ET} = \{U \quad W \quad \Theta \quad N \quad Q \quad M\}^{\mathrm{T}} \tag{3.236}$$

and

$$\boldsymbol{A}_{ET}(\omega) = \begin{bmatrix} 0 & 0 & 0 & & & \\ 0 & 0 & 1 & & \begin{bmatrix} EA & C_1 & C_2 \\ C_1 & \kappa GA & C_3 \\ C_2 & C_3 & EI \end{bmatrix}^{-1} & \\ 0 & 0 & 0 & & & \\ -\rho A\omega^2 & 0 & -\rho R\omega^2 & 0 & 0 & 0 \\ 0 & -\rho A\omega^2 & 0 & 0 & 0 & 0 \\ -\rho R\omega^2 & 0 & -\rho I\omega^2 & 0 & -1 & 0 \end{bmatrix} \tag{3.237}$$

The remaining step is to compute the spectral element matrix from the transfer matrix given by $\boldsymbol{T}_{\mathrm{ET}} = \mathrm{e}^{\boldsymbol{A}_{\mathrm{ET}}L}$, by following the same procedure as used in the previous examples, which is left to the readers.

3.4 Reduction from the Finite Models

There are two types of mathematical models for dynamic continuum systems: the finite models and the infinite models (or continuum models). The conventional finite element model is one of the finite models. The finite models deal with a finite number of DOFs, for example N, and they can be derived by using well-developed techniques such as the FEM and the lumped mass method. The spectral element model is one of the infinite models. Although the infinite models are formulated in terms of a finite number of DOFs, in fact they deal with an infinite number of DOFs, because some or all of the structure members between the nodal points associated with the DOFs are represented by the continuum models in which the continuous distribution of mass and stiffness of the real structure are exactly considered.

The conventional finite element model can be formulated in the standard form as

$$\boldsymbol{M}\ddot{\boldsymbol{x}}_t + \boldsymbol{C}\dot{\boldsymbol{x}}_t + \boldsymbol{K}\boldsymbol{x}_t = \boldsymbol{p}_t(t) \tag{3.238}$$

The time-domain representation of Equation 3.238 can be transformed into the frequency domain as

$$\boldsymbol{D}(\omega)\boldsymbol{x}(\omega) = \boldsymbol{p}(\omega) \tag{3.239}$$

where

$$\boldsymbol{D}(\omega) = \boldsymbol{K} + i\omega\boldsymbol{C} - \omega^2\boldsymbol{M} \tag{3.240}$$

The matrix $\boldsymbol{D}$ is the dynamic stiffness matrix for the finite model, and the matrices $\boldsymbol{M}, \boldsymbol{K}$ and $\boldsymbol{C}$ are the constant finite element mass, stiffness and damping matrices, respectively. Defining $\lambda = i\omega$, we can transform Equation 3.239 into the linearization form

$$\boldsymbol{D}_T(\omega)\boldsymbol{X}(\omega) = \boldsymbol{P}(\omega) \tag{3.241}$$

where

$$\boldsymbol{X}(\omega) = \begin{Bmatrix} \lambda\boldsymbol{x}(\omega) \\ \boldsymbol{x}(\omega) \end{Bmatrix}, \quad \boldsymbol{P}(\omega) = \begin{Bmatrix} \boldsymbol{p}(\omega) \\ \boldsymbol{0} \end{Bmatrix} \tag{3.242}$$

and

$$\boldsymbol{D}_T(\omega) = \begin{bmatrix} \boldsymbol{C} & \boldsymbol{K} \\ \boldsymbol{K} & \boldsymbol{0} \end{bmatrix} + \lambda \begin{bmatrix} \boldsymbol{M} & \boldsymbol{0} \\ \boldsymbol{0} & -\boldsymbol{K} \end{bmatrix} \tag{3.243}$$

As the dynamic stiffness matrix $\boldsymbol{D}_T(\omega)$ is a linear function of λ, we can use numerous eigensolvers available for the linear eigenvalue problems to obtain the eigensolutions of Equation 3.241. As discussed in Chapter , we can also use Sturm's theorem to accelerate the computation and to avoid the possibility of missing any frequencies of interest during the solution procedure.

The infinite model can be conceptually achieved by increasing N (the number of DOFs used for a finite model) to infinite: that is, $N \to \infty$. The infinite value will be denoted herein by N_∞ for convenience. We select a small fixed number (for example, n) of DOFs from $\boldsymbol{x}$ for use in the infinite model and represent them by $\boldsymbol{d}$. The remaining $(N - n)$ DOFs are represented by $\boldsymbol{y}$. Equation 3.239 can be then written in the partitioned form as

$$\begin{bmatrix} \boldsymbol{D}_{yy} & \boldsymbol{D}_{yd} \\ \boldsymbol{D}_{dy} & \boldsymbol{D}_{dd} \end{bmatrix} \begin{Bmatrix} \boldsymbol{y} \\ \boldsymbol{d} \end{Bmatrix} = \begin{Bmatrix} \boldsymbol{0} \\ \boldsymbol{f} \end{Bmatrix} \tag{3.244}$$

One can consider $\boldsymbol{y}$ as the DOFs associated with the internal nodes in a structure, while $\boldsymbol{d}$ as those associated with the boundary nodes of all structure members that are connected to form a complete structure. In Equation 3.244, the nodal forces associated with $\boldsymbol{y}$ are set to zero, without losing the generality of the successive discussions. Equation 3.244 can be readily condensed as

$$\boldsymbol{D}_c(\omega)\boldsymbol{d} = \boldsymbol{f}(\omega) \tag{3.245}$$

where

$$\boldsymbol{D}_c(\omega) = \boldsymbol{D}_{aa}(\omega) - \boldsymbol{D}_{dy}(\omega)\boldsymbol{D}_{yy}^{-1}(\omega)\boldsymbol{D}_{yd}(\omega) \tag{3.246}$$

The condensed matrix $\boldsymbol{D}_c(\omega)$ has the dimension of n-by-n, which is much smaller than the original matrix $\boldsymbol{D}(\omega)$. However, it is no longer a linear function of frequency ω because all of the elements of $\boldsymbol{D}_c(\omega)$ are now the quotients of two polynomials in ω. To achieve the infinite model, we increase N to infinite, N_∞. Then the dimension of $\boldsymbol{y}$ becomes infinite as $(N_\infty - n) = N_\infty$, while the dimension of $\boldsymbol{d}$ is kept constant n. In the limit of the expanding process $N \to \infty$, each element of $\boldsymbol{D}_c(\omega)$ will become infinite degrees of the polynomials, which can eventually be expressed in terms of exponential, trigonometric or other transcendental functions of ω. Thus, in the limit, we can write Equation 3.245 as

$$\boldsymbol{S}(\omega)\boldsymbol{d} = \boldsymbol{f}(\omega) \tag{3.247}$$

where

$$\boldsymbol{S}(\omega) = \lim_{N \to \infty} \boldsymbol{D}_c(\omega) \tag{3.248}$$

The matrix $\boldsymbol{S}(\omega)$ is the transcendental dynamic stiffness matrix because it is obtained by increasing N to infinite. As $\boldsymbol{S}(\omega)$ is associated with infinite DOFs, it will represent the continuous distribution of mass in a structure exactly, alongwith using only a minimum number of DOFs $\boldsymbol{d}$. Thus $\boldsymbol{S}(\omega)$ is certainly the exact dynamic stiffness matrix, termed the spectral element matrix in the SEM. Accordingly, the infinite model represented by $\boldsymbol{S}(\omega)$ is just the spectral element model. Equation 3.248 can be considered as a proof of the exactness of spectral element models.

As Equation 3.247 will provide transcendental eigenvalue problems, unfortunately the numerous eigensolvers available for the linear eigenvalue problems will no longer be useful. Thus, we must look for very reliable numerical techniques to deal with such transcendental eigenvalue problems, which will be discussed further in the next chapter.

4

Spectral Element Analysis Method

The general procedure required in spectral element analysis is very similar to that commonly used for conventional finite element analysis and it consists of the following three major steps:

- **Step 1:** formulation of spectral element equation;
- **Step 2:** assembly of spectral elements and the imposition of boundary conditions;
- **Step 3:** solution of the assembled global system equation and the post-processing of the analysis results.

Figure 4.1 shows the general procedure for spectral element analysis, which usually starts from the time-domain partial differential equations of motion for a dynamic system. The details of the spectral element formulation, as the first step in spectral element analysis, have been discussed in the previous chapter. In this chapter, each step of the spectral element analysis is briefly described, followed by the introduction of the spectral analysis methods that are useful for the computation of the dynamic responses of the discrete and continuous systems subjected to arbitrary non-null initial conditions [33, 84, 101, 117].

4.1 Formulation of Spectral Element Equation

To formulate the spectral element model for a structure subjected to excitation forces, we normally follow a procedure similar to that used in conventional finite element method (FEM). Firstly, we divide the structure into a number of finite elements and then define the nodal degrees-of-freedom (DOFs) and forces at each node of the finite elements. We then derive the frequency-dependent dynamic shape functions by solving the governing differential equations, which are transformed into the frequency domain by using the discrete Fourier transform (DFT) theory. The exact dynamic shape functions are then used to formulate the spectral element equation for a finite element of the structure, which is represented by the exact dynamic stiffness matrix. In the spectral element method (SEM), the spectrally formulated finite element model is called the spectral element model and, accordingly, the exact dynamic stiffness matrix is called the spectral element matrix. As the exactlyformulated dynamic stiffness matrix is used

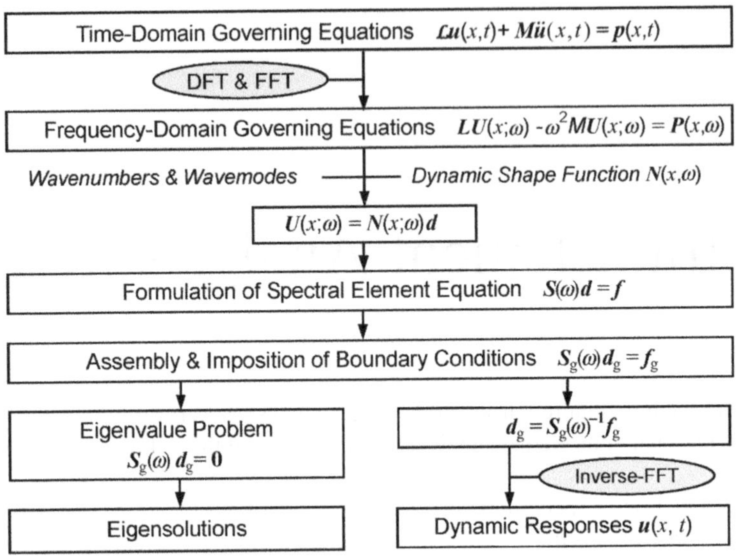

Figure 4.1 General procedure of the spectral element analysis

in the SEM, we can obtain exact solutions for a structure member without the need to divide the structure member into many finite elements. In other words, in contrast to the conventional FEM, one-element modeling is possible for a whole structure member, regardless of its length, provided the structure member does not contain any geometric or material discontinuities inside. Owing to the minimum use of finite elements, the total number of DOFs will be significantly reduced. In due course this will drastically lower the computation costs, alongwith reducing the computation errors and the discretization errors of FEM as the by-products.

To formulate the spectral element model for a given problem, one can use one of the spectral element formulation methods introduced in the previous chapter. The spectral element equation can be formulated in the standard form as

$$S(\omega)d = f \tag{4.1}$$

with

$$f = f_c + f_d \tag{4.2}$$

where $S(\omega)$ is the spectral element matrix and f is the total spectral nodal forces vector, which consists of the nodal forces vector f_c due to the concentrated (point) forces acting on the nodes and the nodal forces vector f_d due to the distributed forces.

To derive the exact dynamic shape functions required to formulate the spectral element matrix $S(\omega)$, the wavenumbers and wavemodes must be computed accurately in advance, as discussed in Section 3.1. However, it is often inevitable that a proper numerical method must be used to compute them when a dynamic system has the characteristic equation (or dispersion relation) of degree five or higher. Thus, some numerical methods that are useful for the computation of wavenumbers and wavemodes are discussed in the following sections.

4.1.1 Computation of Wavenumbers and Wavemodes

As discussed in Section 3.1, during the spectral element formulation, we will encounter an eigenvalue problem in the form of

$$\boldsymbol{A}(k)\boldsymbol{c} = \boldsymbol{0} \tag{4.3}$$

where $\boldsymbol{A}$ is the n-by-n wave matrix that is the function of wavenumber k determined by material properties and frequency ω, and $\boldsymbol{c}$ is the wave amplitude vector that is termed the wavemode. The wavenumbers (eigenvalues) are the latent roots that satisfy

$$\det \boldsymbol{A}(k) = 0 \tag{4.4}$$

or

$$P(k) = k^p + \alpha_{(p-1)}(\omega)k^{(p-1)} + \alpha_{(p-2)}(\omega)k^{(p-2)} + \cdots + \alpha_1(\omega)k + \alpha_0(\omega) = 0 \tag{4.5}$$

which is called the dispersion relation. Through the theory of linear algebra, if a wavenumber k is a latent root that satisfies the dispersion relation, there will be at least one non-trivial solution $\boldsymbol{c}$, which is the wavemode corresponding to the wavenumber. Once we compute all of the required wavenumbers and wavemodes from Equation 4.3, we use them to compute the general solutions as described by Equations 3.11–3.13. Thus, the computation of wavenumbers and wavemodes is certainly at the core of the spectral element formulation. Accordingly, the accuracy and efficiency of the SEM will in part depend on the accuracy and efficiency of the computation method adopted to compute the required wavenumbers and wavemodes. The Abel–Ruffini theorem states that there is no general formula for the latent roots of a polynomial of degree five or greater in terms of its coefficients. Thus, from a practical point of view, the approximate latent roots that are accurate enough for the desired precision will be as useful as the exact ones.

Some numerical methods that are useful for the computation of approximate latent roots (wavenumbers) of Equation 4.4 or Equation 4.5 are introduced in the following sections.

4.1.1.1 Determinant Search Method

In this method, Equation 4.4 is solved directly for k as a transcendental equation without being expanded into the polynomial form of Equation 4.5. In the first step, we commonly use a simple iteration method, such as the bisection method, to search for the intervals where the latent roots may exist. The next step is to further narrow the intervals with one or more roots to isolate each root by repeating the iteration procedure numerous times. Once the intervals containing the latent roots have been narrowed sufficiently, we can apply other elaborate iteration methods, such as the secant method and the Netwon–Raphson method, to calculate the latent roots to a desired accuracy. In MATLAB®, the function "fzero" is the subroutine that computes the roots of a function of one variable by using a combination of bisection, secant, and inverse quadratic interpolation methods.

4.1.1.2 Polynomial Root-Finding Method

In this method, Equation 4.4 is first expanded into the polynomial form of Equation 4.5 and then solved for k by using a proper iteration method. We can use the iteration methods similar to those used for the determinant search method. This polynomial root-finding method is fairly straightforward for very small eigenvalue problems because the techniques of expanding det $\mathbf{A}(k)$ into a polynomial form of Equation 4.5 and then extracting the latent roots are very well known. However, this method may not be recommended for the problems having more than four unknowns because expanding det $\mathbf{A}(k)$ into a polynomial form will no longer be efficient.

4.1.1.3 Companion Matrix Method

From the pth degree polynomial $P(k)$ represented by Equation 4.5, we can set up a companion matrix as follows [74]:

$$\mathbf{C}_a = \begin{bmatrix} 0 & 1 & 0 & \cdots & 0 \\ 0 & 0 & 1 & \cdots & 0 \\ \vdots & \vdots & \vdots & \ddots & \vdots \\ 0 & 0 & 0 & \vdots & 1 \\ -\alpha_0 & -\alpha_1 & -\alpha_2 & \cdots & -\alpha_{p-1} \end{bmatrix} \tag{4.6}$$

As det $(k\mathbf{I} - \mathbf{C}_a) = P(k)$ is satisfied, the latent roots of the polynomial $P(k)$ coincide with the eigenvalues of the companion matrix $\mathbf{C}_a$. Thus, we can compute the required eigenvalues by applying a standard eigenvalue subroutine to the companion matrix $\mathbf{C}_a$. For instance, the MATLAB function "roots" is the subroutine that provides the roots of a polynomial based on the companion matrix method.

4.1.1.4 Polynomial Eigenvalue Problem Method

Equation 4.3 can be expanded into the standard polynomial engenvalue problem (PEP) as

$$\mathbf{A}(k;\omega)\,\mathbf{c} = (\mathbf{A}_0 + k\mathbf{A}_1 + k^2\mathbf{A}_2 + \cdots + k^p\mathbf{A}_p)\mathbf{c} = 0 \tag{4.7}$$

where p is the order of the PEP and the n-by-n matrices $\mathbf{A}_i$ $(i = 1, 2, \ldots, p)$ determined by material properties and frequency ω. The PEP can be written in a linearized form as

$$\mathbf{C}_A\mathbf{x} = k\mathbf{J}\mathbf{x} \tag{4.8}$$

where

$$\mathbf{x} = \{\mathbf{c}^{\mathrm{T}} \quad k\mathbf{c}^{\mathrm{T}} \quad k^2\mathbf{c}^{\mathrm{T}} \quad \cdots \quad k^{p-1}\mathbf{c}^{\mathrm{T}}\}^{\mathrm{T}} \tag{4.9}$$

and

$$\mathbf{C}_A = \begin{bmatrix} 0 & \mathbf{I} & 0 & \cdots & 0 \\ 0 & 0 & \mathbf{I} & \cdots & 0 \\ \vdots & \vdots & \vdots & \ddots & \vdots \\ 0 & 0 & 0 & \vdots & \mathbf{I} \\ -\mathbf{A}_0 & -\mathbf{A}_1 & -\mathbf{A}_2 & \cdots & -\mathbf{A}_{p-1} \end{bmatrix}, \quad \mathbf{J} = \begin{bmatrix} \mathbf{I} & & & \\ & \mathbf{I} & & \\ & & \ddots & \\ & & & \mathbf{I} \\ & & & & \mathbf{A}_p \end{bmatrix} \tag{4.10}$$

We can show that $\mathbf{D} = \mathbf{J}^{-1}\mathbf{C}_A$ is the block companion matrix. As det $\mathbf{A}$ = det $(k\mathbf{I} - \mathbf{D}) \times$ det$\mathbf{A}_p$ with det $\mathbf{A}_p \neq 0$, the latent roots of $\mathbf{A}(k)$ coincide with the eigenvalues of $\mathbf{D}$. Thus, to compute the required eigenvalues (wavenumbers), we can apply a standard eigenvalue algorithm (for instance, the QR method) to the generalized eigenvalue problem given by Equation 4.8. The MATLAB function "polyeig", which is based on the QR method, solves the PEP of degree p to provide eigenvalues and eigenvectors all together. A disadvantage of the PEP method is that we have to work with larger matrices of the order $n \times p$, which have $n \times p$ eigenpairs. This implies that we have to check which eigenpairs satisfy the original PEP.

After the wavenumbers (eignevalues) are computed by using a proper numerical method, the corresponding wavemodes (eigenvectors) can be directly evaluated from Equation 4.3. The singular value decomposition (SVD) method is very efficient for the computation of the wavemodes. The SVD of wave matrix $\boldsymbol{A}$ has the form

$$\boldsymbol{A} = \boldsymbol{U}\Sigma\boldsymbol{V}^H \tag{4.11}$$

where the superscript H denotes the conjugate transpose. $\boldsymbol{U}$ and $\boldsymbol{V}$ are unitary matrices satisfying $\boldsymbol{U}^H\boldsymbol{U} = \boldsymbol{I}$ and $\boldsymbol{V}^H\boldsymbol{V} = \boldsymbol{I}$, respectively, and Σ is the diagonal matrix that has non-zero singular values only in the (i, i) position $(1 \le i \le r \le n)$. From Equation 4.3 it is obvious that the computation of the wavemode $\boldsymbol{c}$ is equivalent to the computation of the null-space of $\boldsymbol{A}$. As the columns of $\boldsymbol{V}$ form the basis for the null-space of $\boldsymbol{A}$, we can compute the required wavemodes from Equation 4.11. For a specific wavenumber, the wave matrix $\boldsymbol{A}$ will become singular with one or more zero singular values. The columns of $\boldsymbol{V}$ corresponding to the zero singular values are the basis for the null-space of $\boldsymbol{A}$. The MATLAB function "svd", which uses LAPACK subroutines, computes the SVD of a matrix.

4.1.2 Computation of Spectral Nodal Forces

As shown in Equation 3.130, the spectral nodal forces can be generated by two types of forces: the point (concentrated) forces and the distributed forces. The spectral nodal forces $\boldsymbol{f}_c$ due to the point forces can be readily obtained by transforming the point forces into the frequency domain by using the fast Fourier transform (FFT) algorithm. On the other hand, the spectral nodal forces $\boldsymbol{f}_d$ due to the distributed forces $\boldsymbol{p}(x, t)$ can be obtained from the virtual work δW formulated in the frequency domain. To formulate δW, we first transform $\boldsymbol{p}(x, t)$ into the frequency-domain expression $\boldsymbol{P}(x; \omega)$ by using the DFT theory. The frequency-domain virtual work is then given by

$$\delta W(\omega) = \int_0^L \delta \boldsymbol{U}^{\mathrm{T}}(x;\omega)\boldsymbol{P}(x;\omega)\mathrm{d}x \tag{4.12}$$

where $\delta\boldsymbol{U}(x; \omega)$ are the virtual displacements. As shown in Section 3.2, the displacement fields $\boldsymbol{U}(x; \omega)$ can be obtained from the frequency-domain governing equation in the form of

$$\boldsymbol{U}(x;\omega) = \boldsymbol{N}(x;\omega)\boldsymbol{d} \tag{4.13}$$

where $\boldsymbol{N}(x, \omega)$ is the dynamic shape function matrix and $\boldsymbol{d}$ is the spectral nodal DOFs vector. Substituting Equation 4.13 into Equation 4.12 gives

$$\delta W = \delta\boldsymbol{d}^{\mathrm{T}} \int_0^L \boldsymbol{N}^{\mathrm{T}}(x;\omega)\boldsymbol{P}(x;\omega)\mathrm{d}x \equiv \delta\boldsymbol{d}^{\mathrm{T}}\boldsymbol{f}_{\mathrm{d}} \tag{4.14}$$

where

$$f_d = \int_0^L N^T(x;\omega)P(x;\omega)\,dx \tag{4.15}$$

The spectral nodal forces $\boldsymbol{f}_d$ of Equation 4.15 are the equivalent frequency-domain representation of the time-domain distributed forces $\boldsymbol{p}(x, t)$. As $\boldsymbol{f}_d$ is formulated by using the same dynamic shape functions, it is natural that Equation 4.15 is identical with the second line of Equation 3.131 derived by using the variational method.

If any discontinuities exist in the distributed force $\boldsymbol{p}(x, t)$ acting on a structure member, one may need to discritize the structure member into more finite elements to improve the solution accuracy: see Example 5.1 as an illustrative example.

We can also use an alternative approximate method that is based on the superposition principle. In this method, we first represent the distributed force as a series of equivalent point forces. We then compute the vibration responses generated by each of the equivalent point forces. Lastly, by using the superposition principle, we compute the total vibration response by summing all vibration responses due to the series of equivalent point forces. The two-element method introduced in the next chapter can be used to compute the vibration response due to each equivalent point force.

4.2 Assembly and the Imposition of Boundary Conditions

The assembled global system equation for the spectral element model of a structure can be obtained by adopting the assembly procedure commonly used in the conventional FEM. The assembly procedure is based on the following two concepts:

a. The interelement continuity conditions among the local spectral nodal DOFs and the global spectral nodal DOFs.
b. The equilibrium conditions among the local force components and the globally specified force components.

After assembling all spectral elements, we impose associated boundary conditions on the assembled set of equations to obtain the global system equation in the final form as

$$S_g(\omega)d_g = f_g \tag{4.16}$$

where $\boldsymbol{S}_g$, $\boldsymbol{d}_g$, and $\boldsymbol{f}_g$ represent the exact global dynamic stiffness matrix, the global spectral nodal DOFs vector, and the global spectral nodal forces vector, respectively.

The conventional time-domain finite element models with viscous damping are commonly represented in the form

$$M_g\ddot{d}_{tg}(t) + C_g\dot{d}_{tg}(t) + K_g d_{tg}(t) = f_{tg}(t) \tag{4.17}$$

where $\boldsymbol{d}_{tg}(t)$ and $\boldsymbol{f}_{tg}(t)$ are the global nodal DOFs and forces in the time domain and the matrices $\boldsymbol{K}_g$, $\boldsymbol{C}_g$, and $\boldsymbol{M}_g$ are the constant global stiffness, damping, and mass matrices, respectively. By using the DFT theory, the time-domain ordinary differential Equations (4.17) can be transformed into the frequency-domain algebraic equations in the form of Equation 4.16.

The global dynamic stiffness matrix for a finite element model is then given by

$$\boldsymbol{S}_g(\omega) = \boldsymbol{K}_g + i\omega \boldsymbol{C}_g - \omega^2 \boldsymbol{M}_g \tag{4.18}$$

We can consider Equation 4.18 as an approximation of the exact global dynamic stiffness matrix for the spectral element model.

The fundamental differences of the SEM from the conventional FEM are as follows:

a. In SEM, $\boldsymbol{S}_g$, $\boldsymbol{d}_g$, and $\boldsymbol{f}_g$ must be assembled and computed repeatedly at each frequency step (i.e., at each FFT sampling frequency of equal interval) by adding a Do-Loop routine. The Do-Loop routine in SEM can be considered as a sequence of pseudo-static problems and it is much like the evaluation of the pseudo-static variables at each time step in the time-domain direct-integration solution method.
b. The SEM yields the transcendental eigenvalue problems, while the conventional FEM yields the linear eigenvalue problems. Although the solution methods for the linear eigenvalue problems are very extensive and have been well developed in the literature, they are no longer applicable to the nonlinear, transcendental eigenvalue problems. Thus we have to look for very proper solution methods for the eigenvalue problems derived by the SEM.

4.3 Eigenvalue Problem and Eigensolutions

The eigenvalue problem for a spectral element model can be reduced from Equation 4.16 to be

$$\boldsymbol{S}_g(\omega)\boldsymbol{d}_g = \boldsymbol{0} \tag{4.19}$$

The eigenfrequencies ω_i ($i = 1, 2, \ldots, \infty$) are determined by using the condition that the determinant of $\boldsymbol{S}_g(\omega)$ is zero at $\omega = \omega_i$, that is,

$$\det \boldsymbol{S}_g(\omega_i) = 0 \tag{4.20}$$

The real part of the ith eigenfrequency ω_i corresponds to the ith natural frequency of the system. Once an eigenfrequency is computed from Equation 4.20, its corresponding eigenvector can be computed from Equation 4.19 by substituting the eigenfrequency into Equation 4.19 and then by applying the singular value decomposition method introduced in Section 4.1.1. Other techniques to compute the eigenvectors can be found in Ref. [174].

As the global dynamic stiffness matrix $\boldsymbol{S}_g(\omega)$ is the transcendental function of frequency ω, Equation 4.19 is the transcendental eigenvalue problem. Thus, the numerous eigensolvers available for the linear eigenvalue problems are no longer applicable. The only way to compute the eigenfrequencies (or latent roots) of Equation 4.20 is to vary the frequency ω in small steps, calculating det $\boldsymbol{S}_g(\omega)$ at each step, and to seek all frequency values where det $\boldsymbol{S}_g(\omega)$ becomes zero. For such an eigenfrequency searching approach, we can use the determinant search method introduced in Section 4.1.1. To avoid the possibility of missing any eigenfrequencies below a chosen frequency, we can also use the Wittrick–Williams algorithm [169] during the eigenfrequency searching. Although the Wittrick–Williams algorithm was originally developed for computing the natural frequencies of undamped linear elastic systems, it is still valid as a guide for computing the natural frequencies of the damped linear elastic systems, because in general the effects of damping on the natural frequencies are very small.

The Wittrick–Williams algorithm, which is analogous to Sturm's theorem for the linear eigenvalue problems, states that the total number of natural frequencies of a structure below a chosen frequency ω^* is given by the formula

$$J(\omega^*) = J_c(\omega^*) + s\left[\boldsymbol{S}_g(\omega^*)\right] \tag{4.21}$$

where

$J(\omega^*)$ = total number of natural frequencies that are less than ω^*
$J_c(\omega^*)$ = total number of natural frequencies that are less than ω^* when all displacements $\boldsymbol{d}_g$ are constrained to be zero
$s[\boldsymbol{S}_g(\omega^*)]$ = the sign count of the matrix $\boldsymbol{S}_g(\omega^*)$, which is equal to the number of negative elements on the diagonal of the upper triangular matrix of $\boldsymbol{S}_g(\omega^*)$

The upper triangular matrix of $\boldsymbol{S}_g(\omega^*)$ can be obtained by using the simple Gaussian elimination method, without row interchanges.

As defined by Equation 4.21, $J_c(\omega^*)$ automatically accounts for the natural frequencies corresponding to the case of $\boldsymbol{d}_g = 0$ with det $\boldsymbol{S}_g(\omega) = \infty$. This exceptional case can occur when all nodal points corresponding to displacements $\boldsymbol{d}_g$ coincide with the nodes of some of the natural modes of a structure. If all displacements $\boldsymbol{d}_g$ are constrained to be zero, a complete structure will be divided into m structure members whose boundaries (both ends) are all fixed. Then, we can compute $J_c(\omega^*)$ from

$$J_c(\omega^*) = \sum_{j=1}^{m} J_{cj}(\omega^*) \tag{4.22}$$

where $J_{cj}(\omega^*)$ is the number of natural frequencies that are less than ω^* for the jth structure member with fixed boundaries. There can be other exceptional cases where the natural frequencies may occur when det $\boldsymbol{S}_g(\omega) \neq 0$ with $\boldsymbol{d}_g \neq 0$, which are unlikely to happen with most structural problems [168].

Example 4.1 Natural Frequencies of a Clamped–Clamped Bernoulli–Euler Beam

Consider a uniform Bernoulli–Euler beam with clamped–clamped boundaries. The number of natural frequencies below a chosen frequency ω^* is denoted by $J^c(\omega^*)$. Thus Equation 4.21 states that $J^c(\omega^*) = J_c(\omega^*) + s[\boldsymbol{S}_g(\omega^*)]$. For the beams with clamped–clamped boundaries, we cannot obtain $s[\boldsymbol{S}_g(\omega^*)]$ from the one-element model represented by the 4-by-4 spectral element matrix of Equation 3.60 because the 4-by-4 spectral element matrix will shrink to a zero-rank matrix when all clamped–clamped boundary conditions are imposed. To overcome such "rank-zero" problems, we represent the beam by the two-element model. In this case, assembling two spectral elements will give the 6-by-6 global dynamic stiffness matrix. If the clamped–clamped boundary conditions are imposed, the 6-by-6 global dynamic stiffness matrix will be reduced to the 2-by-2 dynamic stiffness matrix whose rank is no longer zero. Then we are able to obtain $s[\boldsymbol{S}_g(\omega^*)]$ from the condensed 2-by-2 dynamic stiffness matrix to eventually compute $J^c(\omega^*)$ from $J^c(\omega^*) = J_c(\omega^*) + s[\boldsymbol{S}_g(\omega^*)]$.

Instead of using the above approach, however, we will consider an alternative approach for the present example. To this end, we consider the same beam, but now subjected to simply supported boundary conditions. For the simply supported beam, the Wittrick–Williams algorithm states that the total number of natural frequencies below ω^* is given by

$$J^s(\omega^*) = J_c(\omega^*) + s\left[\boldsymbol{S}^s(\omega^*)\right] \tag{4.23}$$

where

$$\boldsymbol{S}^s(\omega) = \begin{bmatrix} s_{B22} & s_{B24} \\ s_{B24} & s_{B44} \end{bmatrix} \tag{4.24}$$

with

$$\begin{aligned} s_{B22} &= s_{B44} = \left(\frac{EI}{L}\right)\frac{\lambda(\sin\lambda\cosh\lambda - \cos\lambda\sinh\lambda)}{1-\cos\lambda\cosh\lambda} \\ s_{B24} &= \left(\frac{EI}{L}\right)\frac{\lambda(\sinh\lambda - \sin\lambda)}{1-\cos\lambda\cosh\lambda} \\ \lambda^2 &= \omega L^2\sqrt{\frac{\rho A}{EI}} \end{aligned} \tag{4.25}$$

The matrix $\boldsymbol{S}^s(\omega)$ is obtained by imposing the simply supported boundary conditions to the 4-by-4 spectral element matrix $\boldsymbol{S}(\omega)$ of Equation 3.60. By definition, $J_c(\omega^*)$ is the number of natural frequencies below ω^* when all boundaries are clamped. In other words, $J_c(\omega^*)$ is what we want to compute for the clamped–clamped beam: that is, $J_c(\omega^*) = J^c(\omega^*)$. Thus we can write Equation 4.23 as

$$J^c(\omega^*) = J^s(\omega^*) - s\left[\boldsymbol{S}^s(\omega^*)\right] \tag{4.26}$$

For the evaluation of $J^s(\omega^*)$, the characteristic equation for the simply supported beam is derived from the condition det $\boldsymbol{S}^s(\omega) = 0$. The characteristic equation is derived as $\sin\lambda = 0$, which is satisfied when

$$\lambda = n\pi \quad (n = 1, 2, \ldots, \infty) \tag{4.27}$$

From Equation 4.27, we can obtain

$$J^s(\omega^*) = i \equiv \text{the largest integer} \leq \frac{\lambda}{\pi} \tag{4.28}$$

Next, for the evaluation of the sign count $s[\boldsymbol{S}^s(\omega^*)]$, the upper triangular matrix of $\boldsymbol{S}^s(\omega)$ is obtained as

$$\boldsymbol{U}(\omega^*) = \begin{bmatrix} s_{B22} & s_{B24} \\ 0 & \bar{s}_{B44} \end{bmatrix} \tag{4.29}$$

where

$$\bar{s}_{B44} = \frac{s_{B22}s_{B44} - s_{B24}^2}{s_{B22}} = 2\left(\frac{EI}{L}\right)\frac{\lambda\sin\lambda\sinh\lambda}{\sin\lambda\cosh\lambda - \cos\lambda\sinh\lambda} \tag{4.30}$$

The sign count $s[\boldsymbol{S}^s(\omega^*)]$ can be obtained from

$$s[\boldsymbol{S}^s(\omega^*)] = s\{s_{B22}\} + s\{\bar{s}_{B44}\} \tag{4.31}$$

where $s\{\ \} = 1$ if the expression within the braces is negative, otherwise $s\{\ \} = 0$. Finally, substituting Equations 4.28 and 4.31 into Equation 4.26 gives

$$J^c(\omega^*) = i - [s\{s_{B22}\} + s\{\bar{s}_{B44}\}] \tag{4.32}$$

4.4 Dynamic Responses with Null Initial Conditions

In the SEM, the time-domain dynamic responses are represented in the spectral forms as

$$\boldsymbol{u}(x,t) = \sum_{n=0}^{N-1} \boldsymbol{U}_n(x;\omega_n)e^{i\omega_n t} \tag{4.33}$$

The spectral components $\boldsymbol{U}_n$ $(n = 0, 1, 2, \ldots, N-1)$ are then related to the local spectral nodal DOFs $\boldsymbol{d}_n$ as

$$\boldsymbol{U}_n(x;\omega_n) = \boldsymbol{N}_n(x)\,\boldsymbol{d}_n \quad (n = 0,\ 1,\ 2, \ldots,\ N-1) \tag{4.34}$$

Following the procedure for spectral element formulation discussed in the previous chapter, the spectral element model for a continuous dynamic system can be derived in the form of

$$\boldsymbol{S}_g(\omega_n)\boldsymbol{d}_{gn} = \boldsymbol{f}_{gn} \quad (n = 0,\ 1,\ 2, \ldots,\ N-1) \tag{4.35}$$

where $\boldsymbol{S}_g(\omega_n)$ is the global dynamic stiffness matrix, $\boldsymbol{d}_{gn}$ is the global spectral nodal DOFs vector, and $\boldsymbol{f}_{gn}$ is the global spectral nodal forces vector.

4.4.1 *Frequency-Domain and Time-Domain Responses*

The frequency-domain and time-domain dynamic responses of a system subjected to completely null initial conditions can be obtained as follows.

Equation 4.35 is solved for $\boldsymbol{d}_{gn}$ as

$$\boldsymbol{d}_{gn} = \boldsymbol{S}_g^{-1}(\omega_n)\boldsymbol{f}_{gn} \tag{4.36}$$

The local spectral nodal DOFs $\boldsymbol{d}_n$ associated with the finite element containing a point x of interest can be related to the global spectral nodal DOFs $\boldsymbol{d}_{gn}$ by

$$\boldsymbol{d}_n = \boldsymbol{L}\boldsymbol{d}_{gn} \tag{4.37}$$

where $\boldsymbol{L}$ is the locator matrix. Substituting Equation 4.36 into Equation 4.37 gives

$$\boldsymbol{d}_n = \boldsymbol{L}\boldsymbol{S}_g^{-1}(\omega_n)\boldsymbol{f}_{gn} \tag{4.38}$$

The spectral components of the dynamic responses at x are then obtained by substituting Equation 4.38 into Equation 4.34 hence

$$\boldsymbol{U}_n(x;\omega_n) = \boldsymbol{N}_n(x;\omega_n)\boldsymbol{L}\boldsymbol{S}_g^{-1}(\omega_n)\boldsymbol{f}_{gn} \quad (n = 0,\ 1,\ 2, \ldots,\ N-1) \tag{4.39}$$

4.4.1.1 Frequency-Response Function (FRF)

Equation 4.39 can be written in a simple form as

$$\boldsymbol{U}_n(x;\omega_n) = \boldsymbol{H}_n(x;\omega_n)\boldsymbol{f}_{gn} \quad (n = 0, 1, 2, \ldots, N-1) \tag{4.40}$$

where

$$\boldsymbol{H}_n(x;\omega_n) = \boldsymbol{N}_n(x;\omega_n)\,\boldsymbol{LS}_g^{-1}(\omega_n) \tag{4.41}$$

The matrix $\boldsymbol{H}_n(x, \omega_n)$ is the frequency response function matrix.

4.4.1.2 Time-Domain Response

The time-domain dynamic responses (time histories) can be computed from the spectral components $\boldsymbol{U}_n(x, \omega_n)$ $(n = 0, 1, 2, \ldots, N-1)$ by using the inverse-FFT algorithm as follows:

$$\boldsymbol{u}(x,t) \quad \Leftarrow \quad \mathrm{IFFT}\{\boldsymbol{U}_n(x;\omega_n)| \quad n = 0, 1, 2, \ldots, N-1\} \tag{4.42}$$

During the FFT processing, the aliasing and leakage errors can be cured or minimized by using the FFT techniques introduced in Chapter 2.

It is worth remembering that the dynamic responses computed from Equation 4.42 are the forced vibration responses of a continuous system subjected to completely null initial conditions.

4.4.2 Equivalence between Spectral Element Equation and Convolution Integral

To show that the dynamic responses computed from the spectral nodal DOFs given by Equation 4.36 are the forced vibration responses, which are independent of the initial conditions, we consider a uniform Bernoulli–Euler beam subjected to completely null initial conditions as an illustrative example. The equation of motion of the beam is given by

$$EIw'''' + \rho A\ddot{w} = f(x,t) \tag{4.43}$$

where $w(x, t)$ is the transverse displacement and $f(x, t)$ is the excitation force.

The forced vibration response of the beam initially at rest can be obtained by using the convolution integral (or Duhamel's integral) as follows:

$$w(x,t) = \int_0^L \int_0^t h(x;\xi,t-\tau) f(\xi,\tau)\mathrm{d}\tau\mathrm{d}\xi \quad (t \geq 0) \tag{4.44}$$

where $h(x; \xi, t)$ is the unit impulse response excited by a unit impulse applied at $x = \xi$ and it satisfies

$$EIh'''' + \rho A\ddot{h} = \delta(x-\xi)\delta(t) \tag{4.45}$$

By using the causality conditions $h(x; \xi, t-\tau) = 0$ for $t-\tau < 0$ and $f(\xi, \tau) = 0$ for $\tau < 0$, Equation 4.44 can be rewritten as

$$w(x,t) = \int_0^L \int_{-\infty}^{+\infty} h(x;\xi,t-\tau) f(\xi,\tau)\mathrm{d}\tau\,\mathrm{d}\xi \tag{4.46}$$

Equations 4.45 and 4.46 can be discrete Fourier transformed into the frequency domain as

$$EIH''''(x) - \omega^2 \rho AH(x) = \delta(x - \xi) \tag{4.47}$$

and

$$W(x;\xi) = \int_0^L H(x;\xi)F(\xi)\mathrm{d}\xi \tag{4.48}$$

where $H(x)$ and $F(x)$ are the spectral components of $h(x, t)$ and $f(x, t)$, respectively.

The weak form of Equation 4.47 is given by

$$\int_0^L (EIH''\delta H'' - \omega^2 \rho AH\delta H)\,\mathrm{d}x = \int_0^L \delta(x-\xi)\delta H\,\mathrm{d}x \tag{4.49}$$

The general solution to Equation 4.49 can be represented by using the dynamic shape function $\boldsymbol{N}_B(x;\omega)$ of Equation 3.56 as

$$H(x;\omega) = \boldsymbol{N}_B(x;\omega)\boldsymbol{d}_H \tag{4.50}$$

where $\boldsymbol{d}_H$ is the spectral nodal DOFs excited by the unit impulse. Substituting Equation 4.50 into Equation 4.49 gives

$$\boldsymbol{S}_B(\omega)\boldsymbol{d}_H = \boldsymbol{N}_B^{\mathrm{T}}(\xi) \tag{4.51}$$

where $\boldsymbol{S}_B(\omega)$ is the spectral element matrix defined by Equation 3.60.

We solve Equation 4.51 for $\boldsymbol{d}_H$ and then substitute the result into Equation 4.50 to obtain

$$H(x;\xi) = \boldsymbol{N}_B(x;\omega)\boldsymbol{S}_B(\omega)^{-1}\boldsymbol{N}_B^{\mathrm{T}}(\xi) \tag{4.52}$$

Substitution of Equation 4.52 into Equation 4.48 gives

$$W(x;\xi) = \boldsymbol{N}_B(x;\omega)\boldsymbol{S}_B(\omega)^{-1}\boldsymbol{f} \tag{4.53}$$

where

$$\boldsymbol{f} = \int_0^L F(\xi)\boldsymbol{N}_B^{\mathrm{T}}(\xi)\mathrm{d}\xi \tag{4.54}$$

Equation 4.53 represents the spectral components of the dynamic response derived from the convolution integral that is valid for the linear system initially at rest.

As shown in Equation 3.55, the spectral components of the dynamic response can be related to the spectral nodal DOFs $\boldsymbol{d}$ excited by the total excitation force $f(x, t)$ as

$$W(x;\omega) = \boldsymbol{N}_B(x;\omega)\boldsymbol{d} \tag{4.55}$$

By comparing Equation 4.53 with Equation 4.55, we obtain

$$\boldsymbol{S}_B(\omega)\boldsymbol{d} = \boldsymbol{f} \tag{4.56}$$

Equation 4.56 is identical with the spectral element equation (3.59) for the Bernoulli–Euler beam formulated in Example 3.2.

So far we have shown that the spectral element equation for a dynamic system can be reduced directly from the discrete Fourier transformed convolution integral. This proves that the spectral element equation is equivalent to the discrete Fourier transformed convolution integral. It is well known that the convolution integral is valid only for the linear dynamic systems that are subjected to completely null initial conditions. Thus we can conclude that the dynamic responses computed from the spectral nodal DOFs given by $\boldsymbol{d} = \boldsymbol{S}_B^{-1}(\omega)\boldsymbol{f}$ are the forced dynamic responses of the linear continuous system initially at rest.

The dynamic responses of the linear discrete and continuous systems subjected to arbitrary, non-null initial conditions are considered in the next section.

4.5 Dynamic Responses with Arbitrary Initial Conditions

This section considers the spectral analysis methods that are useful for the computation of the dynamic responses of the linear dynamic systems subjected to arbitrary, non-null initial conditions.

Before beginning this subject, we first consider a differential equation of motion for a linear discrete or continuous system subjected to the arbitrary non-null initial conditions as

$$x(t=0) = x_0, \qquad \dot{x}(t=0) = \dot{x}_0 \tag{4.57}$$

In mathematics, the complete solution of a differential equation can be obtained by summing a particular solution $x_f(t; f)$ and a complementary (homogeneous) solution $x_c(t; \boldsymbol{c})$ as follows:

$$x(t) = x_f(t; f) + x_c(t; \boldsymbol{c}) \tag{4.58}$$

The particular solution is determined by the excitation force f, while the complementary solution contains integral constants $\boldsymbol{c}$, to be determined by the initial conditions. Thus the complementary solution will consist of two parts: the first part solely depends on the excitation force f and the other part solely on the non-null initial conditions as follows.

$$x_c(t) = x_{c0}(t; x_0, \dot{x}_0) + x_{cf}(t; f) \tag{4.59}$$

By substituting Equation 4.59 into Equation 4.58, the complete solution can be rewritten as

$$x(t) = x_p(t) + x_h(t) \tag{4.60}$$

where

$$\begin{aligned} x_p(t) &= x_{cf}(t; f) + x_f(t; f) \\ x_h(t) &= x_{c0}(t; x_0, \dot{x}_0) \end{aligned} \tag{4.61}$$

$x_p(t)$ is the forced vibration response part determined solely by the excitation force $f(t)$, and $x_h(t)$ is the free vibration response part determined solely by the initial conditions x_0 and $\dot{x}_0$. It is obvious from Equation 4.61 that if a linear dynamic system initially at rest is subjected to the external force, it will have the forced vibration response part $x_p(t)$ only.

In the theory of dynamics, it is well known that the convolution integral (or the Duhamel's integral) is valid for computing the dynamic responses of the linear dynamic systems initially at rest. In the previous section, it has been shown that the discrete Fourier transform of the convolution integral is equivalent to the spectral element equation represented by

$$S(\omega)d = f \tag{4.62}$$

Thus, the dynamic responses computed from the spectral nodal DOFs given by $d = S(\omega)^{-1}f$ correspond to the forced vibration responses of the linear continuous system subjected to completely null initial conditions.

It is worth stating that, when a continuous system is approximately represented by the conventional finite element model, the spectral element matrix $S(\omega)$ will be replaced by the dynamic stiffness matrix given by $(K + i\omega C - \omega^2 M)$, where K, M, and C are the finite element stiffness, mass, and damping matrices, respectively. In the following two subsections, two types of dynamic systems are considered: the linear discrete system and the linear continuous system.

4.5.1 *Discrete Systems with Arbitrary Initial Conditions*

Consider a viscously damped linear m-DOFs system whose dynamics are represented by the equations of motion

$$M\ddot{u}(t) + C\dot{u}(t) + Ku(t) = f(t) \tag{4.63}$$

and the initial conditions

$$u(0) = u_0 \quad \text{and} \quad \dot{u}(0) = \dot{u}_0 \tag{4.64}$$

where $u(t)$ is the nodal DOFs and $f(t)$ is the excitation force. The matrices $M = [m_{ij}]$, $C = [c_{ij}]$, and $K = [k_{ij}]$ $(i, j = 1, 2, \ldots, m)$ are the finite element mass matrix, damping matrix, and stiffness matrix, respectively. The complete dynamic responses of the system can be obtained by summing the forced vibration response part $u_p(t)$ and the free vibration response part $u_h(t)$ as follows:

$$u(t) = u_p(t) + u_h(t) \tag{4.65}$$

4.5.1.1 Forced Vibration Response Part: $u_p(t)$

Represent the forced vibration response part $u_p(t)$ and the excitation forces $f(t)$ in the spectral forms as

$$u_p(t) = \frac{1}{N}\sum_{n=0}^{N-1} P_n e^{i\omega_n t} \quad \text{(a)}, \qquad f(t) = \frac{1}{N}\sum_{n=0}^{N-1} F_n e^{i\omega_n t} \quad \text{(b)} \tag{4.66}$$

Applying Equation 4.66 to Equation 4.63 yields

$$P_n = D(\omega_n)^{-1}F_n, \qquad P_{N-n} = P_n^* \quad (n = 1, 2, \ldots, N/2) \tag{4.67}$$

where

$$D(\omega_n) = K + i\omega_n C - \omega_n^2 M \tag{4.68}$$

The symbol (*) denotes the complex conjugate of a complex quantity. Once the spectral components $\boldsymbol{P}_n$ are computed from Equation 4.67 by using the spectral nodal forces $\boldsymbol{F}_n$, the time-domain forced vibration responses can be computed by using the inverse-FFT algorithm to be

$$\boldsymbol{u}_p(t) \quad \Leftarrow \quad \text{IFFT}\{P_n| \ n=0,\ 1,\ 2,\ldots,\ N-1\} \tag{4.69}$$

The time derivative of $\boldsymbol{u}_p(t)$ can readily be obtained from Equation 4.66(a) as

$$\dot{\boldsymbol{u}}_p(t) = \frac{1}{N}\sum_{n=0}^{N-1} \dot{\boldsymbol{P}}_n e^{i\omega_n t} \tag{4.70}$$

where

$$\dot{\boldsymbol{P}}_n = (i\omega_n)\boldsymbol{P}_n, \qquad \dot{\boldsymbol{P}}_{N-n} = \dot{\boldsymbol{P}}_n^* \quad (n=1,\ 2,\ldots,\ N/2) \tag{4.71}$$

4.5.1.2 Free Vibration Response Part: $\underline{u}_h(t)$

The free vibration response part $\boldsymbol{u}_h(t)$ must satisfy the homogeneous equations of motion

$$\boldsymbol{M}\ddot{\boldsymbol{u}}_h + \boldsymbol{C}\ddot{\boldsymbol{u}}_h + \boldsymbol{K}\boldsymbol{u}_h = \boldsymbol{0} \tag{4.72}$$

In the following, we consider two types of viscous damping: one is the non-proportional damping and the other is the proportional damping.

Non-Proportional Damping

For the case of the non-proportional damping $\boldsymbol{C}$, we assume the general solution to Equation 4.72 in the form of

$$\boldsymbol{u}_h(t) = \boldsymbol{h}e^{\lambda t} \tag{4.73a}$$

or

$$u_{hk}(t) = h_k e^{\lambda t} \quad (k=1,\ 2,\ldots,\ m) \tag{4.73b}$$

Substituting Equation 15.73 into Equation 4.72 yields an eigenvalue problem as

$$[\lambda^2\boldsymbol{M} + \lambda\boldsymbol{C} + \boldsymbol{K}]\boldsymbol{h} = \boldsymbol{0} \tag{4.74a}$$

or

$$\begin{bmatrix} (m_{11}\lambda^2 + c_{11}\lambda + k_{11}) & \cdots & (m_{1m}\lambda^2 + c_{1m}\lambda + k_{1m}) \\ \vdots & \ddots & \vdots \\ (m_{m1}\lambda^2 + c_{m1}\lambda + k_{m1}) & \cdots & (m_{mm}\lambda^2 + c_{mm}\lambda + k_{mm}) \end{bmatrix} \begin{Bmatrix} h_1 \\ \vdots \\ h_m \end{Bmatrix} = \boldsymbol{0} \tag{4.74b}$$

For the existence of the non-trivial solution $\boldsymbol{h}$, it follows that

$$\det\,[\lambda^2\boldsymbol{M} + \lambda\boldsymbol{C} + \boldsymbol{K}] = \boldsymbol{0} \tag{4.75}$$

Equation 4.75 yields a $2m$-degree algebraic equation with λ as the eigenvalue. In general the roots (eigenvalues) of the algebraic equation for the under-damped systems appear in the complex conjugate pairs, because all the coefficients of the algebraic equation are real. Thus, the eigenvalues computed from Equation 4.75 can be represented in the complex conjugate forms as

$$\lambda_j = \xi_j + i\omega_j, \qquad \lambda_j^* = \xi_j - i\omega_j \qquad (j = 1, 2, \ldots, m) \tag{4.76}$$

where ω_j is the jth natural frequency and ξ_j is the rate of exponential decay of the jth vibration mode. By substituting an eigenvalue λ_j into Equation 4.74, the corresponding eigenvector $\boldsymbol{h}_j$ can be computed. The ratio between the components h_{jk} of the jth eigenvector $\boldsymbol{h}_j$ are given by

$$\frac{h_{j1}}{C_{j1}} = \frac{h_{j2}}{C_{j2}} = \cdots = \frac{h_{jm}}{C_{jm}} = r_j \tag{4.77}$$

where r_j is an arbitrary complex number and C_{jk} is the cofactor of the jth row of the determinant in Equation 4.75 for a particular λ_j.

By using Equation 4.77, the kth component of the jth vibration mode corresponding to the jth eigenvalue λ_j is found from Equation 4.73 to be

$$u_{hk}^j = h_{jk} e^{\lambda_j t} = r_j C_{jk} e^{\lambda_j t} \quad (k = 1, 2, \ldots, m) \tag{4.78}$$

As there are $2m$ eigenvalues λ_j ($j = 1, 2, \ldots, 2m$), the kth component of $\boldsymbol{u}_h(t)$ can be compounded from Equation 4.78 to give

$$u_{hk}(t) = \sum_{j=1}^{2m} b_j u_{hk}^j = \sum_{j=1}^{2m} b_j r_j C_{jk} e^{\lambda_j t} \quad (k = 1, 2, \ldots, m) \tag{4.79}$$

where b_j are constants. As $2m$ eigenvalues are given by m pairs of complex conjugates, we can rewrite Equation 4.79 as

$$u_{hk}(t) = \sum_{j=1}^{m} a_j C_{jk} e^{\lambda_j t} + \sum_{j=1}^{m} a_j^* C_{jk}^* e^{\lambda_j^* t} \quad (k = 1, 2, \ldots, m) \tag{4.80}$$

where $a_j = b_j r_j$ or a_j^* ($j = 1, 2, \ldots, m$) are the constants to be determined by the initial conditions. The time derivative of Equation 4.80 is given by

$$\dot{u}_{hk}(t) = \sum_{j=1}^{m} a_j C_{jk} \lambda_j e^{\lambda_j t} + \sum_{j=1}^{m} a_j^* C_{jk}^* \lambda_j^* e^{\lambda_j^* t} \quad (k = 1, 2, \ldots, m) \tag{4.81}$$

The DFT pairs for $u_{hk}(t)$ and $\dot{u}_{hk}(t)$ are given by

$$\begin{aligned} u_{hk}(t) &= \frac{1}{N} \sum_{n=0}^{N-1} H_{kn} e^{i\omega_n t} \\ \dot{u}_{hk}(t) &= \frac{1}{N} \sum_{n=0}^{N-1} \bar{H}_{kn} e^{i\omega_n t} \end{aligned} \qquad (k = 1, 2, \ldots, m) \tag{4.82}$$

and

$$\begin{aligned} H_{kn} &= \sum_{r=0}^{N-1} u_{hk}(t_r)\mathrm{e}^{-i\omega_n t_r} \\ \bar{H}_{kn} &= \sum_{r=0}^{N-1} \dot{u}_{hk}(t_r)\mathrm{e}^{-i\omega_n t_r} \end{aligned} \qquad (k = 1, 2, \ldots, m) \tag{4.83}$$

Substituting Equations 4.80 and 4.81 into Equation 4.83, we can obtain the spectral components as

$$\begin{aligned} H_{kn} &= \sum_{j=1}^{m} (a_j X_{jkn} + a_j^* Y_{jkn}) \\ \bar{H}_{kn} &= \sum_{j=1}^{m} (a_j \lambda_j X_{jkn} + a_j^* \lambda_j^* Y_{jkn}) \end{aligned} \tag{4.84}$$

where

$$\begin{aligned} X_{jkn} &= C_{jk}\frac{1-\alpha_{jn}^N}{1-\alpha_{jn}}, \qquad \alpha_{jn} = \mathrm{e}^{(\lambda_j - i\omega_n)\Delta t} \\ Y_{jkn} &= C_{jk}^*\frac{1-\beta_{jn}^N}{1-\beta_{jn}}, \qquad \beta_{jn} = \mathrm{e}^{(\lambda_j^* - i\omega_n)\Delta t} \end{aligned} \tag{4.85}$$

where Δt is the sampling time interval.

Substitute Equation 4.84 into Equation 4.82 and express the results in the vector forms

$$\boldsymbol{u}_h(t) = \frac{1}{N}\sum_{n=0}^{N-1} \boldsymbol{H}_n \mathrm{e}^{i\omega_n t}, \qquad \dot{\boldsymbol{u}}_h(t) = \frac{1}{N}\sum_{n=0}^{N-1} \bar{\boldsymbol{H}}_n \mathrm{e}^{i\omega_n t} \tag{4.86}$$

where

$$\begin{aligned} \boldsymbol{u}_h(t) &= \{u_{h1}(t) \quad u_{h2}(t) \quad \cdots \quad u_{hm}(t)\}^{\mathrm{T}} \\ \dot{\boldsymbol{u}}_h(t) &= \{\dot{u}_{h1}(t) \quad \dot{u}_{h2}(t) \quad \cdots \quad \dot{u}_{hm}(t)\}^{\mathrm{T}} \end{aligned} \tag{4.87}$$

and

$$\begin{aligned} \boldsymbol{H}_n &= \{H_{1n} \quad H_{2n} \quad \cdots \quad H_{mn}\}^{\mathrm{T}} = \boldsymbol{X}_n \boldsymbol{a} + \boldsymbol{Y}_n \boldsymbol{a}^* \quad \text{(a)} \\ \bar{\boldsymbol{H}}_n &= \{\bar{H}_{1n} \quad \bar{H}_{2n} \quad \cdots \quad \bar{H}_{mn}\}^{\mathrm{T}} = \boldsymbol{X}_n \Lambda \boldsymbol{a} + \boldsymbol{Y}_n \Lambda^* \boldsymbol{a}^* \quad \text{(b)} \end{aligned} \tag{4.88}$$

with the use of definitions

$$\begin{aligned} \boldsymbol{a} &= \{a_1 \quad a_2 \quad \cdots \quad a_m\}^{\mathrm{T}} \\ \boldsymbol{a}^* &= \{a_1^* \quad a_2^* \quad \cdots \quad a_m^*\}^{\mathrm{T}} \end{aligned}$$

$$\boldsymbol{X}_n = \begin{bmatrix} X_{11n} & X_{21n} & \cdots & X_{m1n} \\ X_{12n} & X_{22n} & \cdots & X_{m2n} \\ \vdots & \vdots & \ddots & \vdots \\ X_{1mn} & X_{2mn} & \cdots & X_{mmn} \end{bmatrix} \tag{4.89}$$

$$\boldsymbol{Y}_n = \begin{bmatrix} Y_{11n} & Y_{21n} & \cdots & Y_{m1n} \\ Y_{12n} & Y_{22n} & \cdots & Y_{m2n} \\ \vdots & \vdots & \ddots & \vdots \\ Y_{1mn} & Y_{2mn} & \cdots & Y_{mmn} \end{bmatrix}$$

$$\Lambda = diag\,[\lambda_1 \quad \lambda_2 \quad \ldots \quad \lambda_m]$$

$$\Lambda^* = diag[\lambda_1^* \quad \lambda_2^* \quad \ldots \quad \lambda_m^*]$$

The complete dynamic responses in Equation 4.65 must satisfy the initial conditions of Equation 4.64. Thus we substitute Equation 4.65 into Equation 4.64 to obtain

$$\begin{aligned} \boldsymbol{u}_0 &= \boldsymbol{u}_p(0) + \boldsymbol{u}_h(0) = \boldsymbol{u}_h(0) \\ \dot{\boldsymbol{u}}_0 &= \dot{\boldsymbol{u}}_p(0) + \dot{\boldsymbol{u}}_h(0) = \dot{\boldsymbol{u}}_h(0) \end{aligned} \tag{4.90}$$

where we set $\boldsymbol{u}_p(0) = 0$ and $\dot{\boldsymbol{u}}_p(0) = \boldsymbol{0}$ because $\boldsymbol{u}_p(t)$ are the forced vibration responses of the system initially at rest. By applying Equation 4.86 to Equation 4.90, we obtain

$$\boldsymbol{u}_0 = \frac{1}{N}\sum_{n=0}^{N-1} \boldsymbol{H}_n, \qquad \dot{\boldsymbol{u}}_0 = \frac{1}{N}\sum_{n=0}^{N-1} \bar{\boldsymbol{H}}_n \tag{4.91}$$

Substitution of Equation 4.88 into Equation 4.91 gives

$$\bar{\boldsymbol{X}}\boldsymbol{a} + \bar{\boldsymbol{Y}}\boldsymbol{a}^* = \boldsymbol{d}, \qquad \tilde{\boldsymbol{X}}\boldsymbol{a} + \tilde{\boldsymbol{Y}}\boldsymbol{a}^* = \boldsymbol{v} \tag{4.92}$$

where

$$\boldsymbol{d} = N\boldsymbol{u}_0, \qquad \boldsymbol{v} = N\dot{\boldsymbol{u}}_0 \tag{4.93}$$

and

$$\begin{aligned} \bar{\boldsymbol{X}} &= \sum_{n=0}^{N-1} \boldsymbol{X}_n, \qquad \bar{\boldsymbol{Y}} = \sum_{n=0}^{N-1} \boldsymbol{Y}_n \\ \tilde{\boldsymbol{X}} &= \bar{\boldsymbol{X}}\Lambda, \qquad \tilde{\boldsymbol{Y}} = \bar{\boldsymbol{Y}}\Lambda^* \end{aligned} \tag{4.94}$$

The constant vectors $\boldsymbol{a}$ or $\boldsymbol{a}^*$ can be solved from Equation 4.92 as

$$\boldsymbol{a} = (\tilde{\boldsymbol{X}} - \tilde{\boldsymbol{Y}}\bar{\boldsymbol{Y}}^{-1}\bar{\boldsymbol{X}})^{-1}(\boldsymbol{v} - \tilde{\boldsymbol{Y}}\bar{\boldsymbol{Y}}^{-1}\boldsymbol{d}) \tag{4.95}$$

Once $\boldsymbol{a}$ is computed from Equation 4.95 by using the initial conditions, we can readily compute $\boldsymbol{H}_n$ from Equation 4.88(a). Next, by using the inverse-FFT algorithm, we can compute the free vibration response part $\boldsymbol{u}_h(t)$ from $\boldsymbol{H}_n$ as follows:

$$\boldsymbol{u}_h(t) \quad \Leftarrow \quad \text{IFFT}\{\boldsymbol{H}_n |\ n = 0, 1, 2, \ldots, N-1\} \tag{4.96}$$

Finally, as shown in Equation 4.65, the complete dynamic responses can be obtained by summing $\boldsymbol{u}_p(t)$ and $\boldsymbol{u}_h(t)$ computed from Equations 4.69 and 4.96, respectively.

Proportional Damping

For the case of the proportional damping $\boldsymbol{C}$, Equation 4.72 can be readily decoupled by using the modal analysis method. To this end, we assume the solution of Equation 4.72 in the form of

$$\boldsymbol{u}_h(t) = \Phi \boldsymbol{x}_h(t) \tag{4.97}$$

where Φ is the modal matrix that satisfies the orthogonality properties

$$\Phi^{\mathrm{T}} \boldsymbol{M} \Phi = \boldsymbol{I}, \quad \Phi^{\mathrm{T}} \boldsymbol{K} \Phi = \Omega^2 \tag{4.98}$$

In above equation, $\boldsymbol{I}$ is the identity matrix and $\boldsymbol{\Omega}^2$ is the diagonal matrix defined by

$$\boldsymbol{\Omega}^2 = diag\begin{bmatrix} \Omega_1^2 & \Omega_2^2 & \ldots & \Omega_m^2 \end{bmatrix} \tag{4.99}$$

where $\Omega_1 < \Omega_2 < \ldots < \Omega_m$ are the natural frequencies. Substituting Equation 4.97 into Equation 4.72 and applying Equation 4.98, we can obtain a set of decoupled modal equations as

$$\ddot{x}_{hk} + 2\xi_k \Omega_k \dot{x}_{hk} + \Omega_k^2 x_{hk} = 0 \quad (k = 1, 2, \ldots, m) \tag{4.100}$$

where ξ_k is the kth modal damping ratio. The solution to Equation 4.100 can be obtained as

$$x_{hk}(t) = a_k e^{\lambda_{k1} t} + a_k^* e^{\lambda_{k1}^* t} \quad (k = 1, 2, \ldots, m) \tag{4.101}$$

where λ_{k1} and $\lambda_{k2} = \lambda_{k1}^*$ are the conjugate roots of the characteristic equation

$$\lambda_k^2 + 2\xi_k \Omega_k \lambda_k + \Omega_k^2 = (\lambda_k - \lambda_{k1})(\lambda_k - \lambda_{k2}) = 0 \tag{4.102}$$

The constants a_k or a_k^* are determined by the initial conditions. The time derivative of Equation 4.101 is given by

$$\dot{x}_{hk}(t) = a_k \lambda_{k1} e^{\lambda_{k1} t} + a_k^* \lambda_{k1}^* e^{\lambda_{k1}^* t} \quad (k = 1, 2, \ldots, m) \tag{4.103}$$

The DFT pairs of $x_{hk}(t)$ and $\dot{x}_{hk}(t)$ are given by

$$\begin{aligned} x_{hk}(t) &= \frac{1}{N} \sum_{n=0}^{N-1} H_{kn} e^{i\omega_n t} \\ \dot{x}_{hk}(t) &= \frac{1}{N} \sum_{n=0}^{N-1} \bar{H}_{kn} e^{i\omega_n t} \end{aligned} \quad (k = 1, 2, \ldots, m) \tag{4.104}$$

and

$$\begin{aligned} H_{kn} &= \sum_{r=0}^{N-1} x_{hk}(t_r) e^{-i\omega_n t_r} \\ \bar{H}_{kn} &= \sum_{r=0}^{N-1} \dot{x}_{hk}(t_r) e^{-i\omega_n t_r} \end{aligned} \quad (k = 1, 2, \ldots, m) \tag{4.105}$$

By substituting Equations 4.101 and 4.103 into Equation 4.105, we can get

$$\begin{aligned} H_{kn} &= a_k X_{kn} + a_k^* Y_{kn} \\ \bar{H}_{kn} &= a_k \lambda_{k1} X_{kn} + a_k^* \lambda_{k1}^* Y_{kn} \end{aligned} \qquad (k = 1, 2, \ldots, m) \tag{4.106}$$

where

$$\begin{aligned} X_{kn} &= \frac{1-\alpha_{kn}^N}{1-\alpha_{kn}}, \quad \alpha_{kn} = \mathrm{e}^{(\lambda_{k1} - i\omega_n)\Delta t} \\ Y_{kn} &= \frac{1-\beta_{kn}^N}{1-\beta_{kn}}, \quad \beta_{kn} = \mathrm{e}^{(\lambda_{k1}^* - i\omega_n)\Delta t} \end{aligned} \tag{4.107}$$

Substitute Equation 4.106 into Equation 4.104 and then express the results in the vector forms of

$$\begin{aligned} \boldsymbol{x}_h(t) &= \frac{1}{N}\sum_{n=0}^{N-1} \boldsymbol{H}_n \mathrm{e}^{i\omega_n t} \\ \dot{\boldsymbol{x}}_h(t) &= \frac{1}{N}\sum_{n=0}^{N-1} \bar{\boldsymbol{H}}_n \mathrm{e}^{i\omega_n t} \end{aligned} \tag{4.108}$$

where

$$\begin{aligned} \boldsymbol{x}_h &= \{x_{h1} \quad x_{h2} \quad \cdots \quad x_{hm}\}^{\mathrm{T}} \\ \dot{\boldsymbol{x}}_h &= \{\dot{x}_{h1} \quad \dot{x}_{h2} \quad \cdots \quad \dot{x}_{hm}\}^{\mathrm{T}} \end{aligned} \tag{4.109}$$

and

$$\begin{aligned} \boldsymbol{H}_n &= \{H_{1n} \quad H_{2n} \quad \cdots \quad H_{mn}\}^{\mathrm{T}} = \boldsymbol{X}_n \boldsymbol{a} + \boldsymbol{Y}_n \boldsymbol{a}^* \quad \text{(a)} \\ \bar{\boldsymbol{H}}_n &= \{\bar{H}_{1n} \quad \bar{H}_{2n} \quad \cdots \quad \bar{H}_{mn}\}^{\mathrm{T}} = \boldsymbol{X}_n \boldsymbol{\Lambda} \boldsymbol{a} + \boldsymbol{Y}_n \boldsymbol{\Lambda}^* \boldsymbol{a}^* \quad \text{(b)} \end{aligned} \tag{4.110}$$

with the use of definitions

$$\begin{aligned} \boldsymbol{a} &= \{a_1 \quad a_2 \quad \ldots \quad a_m\}^{\mathrm{T}} \\ \boldsymbol{a}^* &= \{a_1^* \quad a_2^* \quad \ldots \quad a_m^*\}^{\mathrm{T}} \\ \boldsymbol{X}_n &= diag\,[X_{1n} \quad X_{2n} \quad \ldots \quad X_{mn}] \\ \boldsymbol{Y}_n &= diag\,[Y_{1n} \quad Y_{2n} \quad \ldots \quad Y_{mn}] \\ \Lambda &= diag\,[\lambda_{11} \quad \lambda_{21} \quad \ldots \quad \lambda_{m1}] \\ \Lambda^* &= diag\,[\lambda_{11}^* \quad \lambda_{21}^* \quad \ldots \quad \lambda_{m1}^*] \end{aligned} \tag{4.111}$$

By substituting Equation 4.108 into Equation 4.97 and its time derivative, we can obtain

$$\begin{aligned} \boldsymbol{u}_h(t) &= \frac{1}{N}\Phi\sum_{n=0}^{N-1}\boldsymbol{H}_n e^{i\omega_n t} \quad \text{(a)} \\ \dot{\boldsymbol{u}}_h(t) &= \frac{1}{N}\Phi\sum_{n=0}^{N-1}\bar{\boldsymbol{H}}_n e^{i\omega_n t} \quad \text{(b)} \end{aligned} \tag{4.112}$$

Applying the initial conditions of Equation 4.64 to the complete dynamic responses in Equation 4.65 and their time derivatives obtained by using Equations 4.69, 4.70 and 4.112 gives

$$\begin{aligned} \boldsymbol{u}_0 &= \boldsymbol{u}_p(0) + \boldsymbol{u}_h(0) = \boldsymbol{u}_h(0) = \frac{1}{N}\Phi\sum_{n=0}^{N-1}\boldsymbol{H}_n \\ \dot{\boldsymbol{u}}_0 &= \dot{\boldsymbol{u}}_p(0) + \dot{\boldsymbol{u}}_h(0) = \dot{\boldsymbol{u}}_h(0) = \frac{1}{N}\Phi\sum_{n=0}^{N-1}\bar{\boldsymbol{H}}_n \end{aligned} \tag{4.113}$$

Substituting Equation 4.110 into Equation 4.113 and then using Equation 4.98 gives

$$\bar{\boldsymbol{X}}\boldsymbol{a} + \bar{\boldsymbol{Y}}\boldsymbol{a}^* = \boldsymbol{d}, \qquad \tilde{\boldsymbol{X}}\boldsymbol{a} + \tilde{\boldsymbol{Y}}\boldsymbol{a}^* = \boldsymbol{v} \tag{4.114}$$

where

$$\boldsymbol{d} = N\Phi^{\mathrm{T}}\boldsymbol{M}\boldsymbol{u}_0, \qquad \boldsymbol{v} = N\Phi^{\mathrm{T}}\boldsymbol{M}\dot{\boldsymbol{u}} \tag{4.115}$$

and

$$\begin{aligned} \bar{\boldsymbol{X}} &= \sum_{n=0}^{N-1}\boldsymbol{X}_n, \quad \bar{\boldsymbol{Y}} = \sum_{n=0}^{N-1}\boldsymbol{Y}_n \\ \tilde{\boldsymbol{X}} &= \bar{\boldsymbol{X}}\Lambda, \qquad \tilde{\boldsymbol{Y}} = \bar{\boldsymbol{Y}}\Lambda^* \end{aligned} \tag{4.116}$$

Solving Equation 4.114 for $\boldsymbol{a}$ gives

$$\boldsymbol{a} = (\tilde{\boldsymbol{X}} - \tilde{\boldsymbol{Y}}\bar{\boldsymbol{Y}}^{-1}\bar{\boldsymbol{X}})^{-1}(\boldsymbol{v} - \tilde{\boldsymbol{Y}}\bar{\boldsymbol{Y}}^{-1}\boldsymbol{d}) \tag{4.117}$$

which can be further simplified by using Equation 4.116 as

$$\boldsymbol{a} = \frac{i}{2}\boldsymbol{Z}(\Lambda^*\boldsymbol{d} - \boldsymbol{v}) \tag{4.118}$$

where

$$\boldsymbol{Z} = diag\,[Z_1 \quad Z_2 \quad \ldots \quad Z_m] \tag{4.119}$$

with

$$Z_k = \frac{1}{\bar{X}_k \cdot \text{imaginary}\,(\lambda_{k1})}, \qquad \bar{X}_k = \sum_{n=0}^{N-1} X_{kn} \quad (k = 1, 2, \ldots, m) \tag{4.120}$$

Once $\boldsymbol{a}$ is computed from Equation 4.118 by using the initial conditions, we first compute the spectral components $\boldsymbol{H}_n$ from Equation 4.110(a). Based on Equation 4.112(a), we then use the inverse-FFT algorithm to compute the free vibration response part

$$\mathbf{u}_h(t) \quad \Leftarrow \quad \Phi \cdot \mathrm{IFFT}\{\mathbf{H}_n | \quad n = 0, 1, 2, \ldots, N-1\} \tag{4.121}$$

The remaining step is to compute the complete dynamic response $\mathbf{u}(t)$ by simply summing the forced vibration response part $\mathbf{u}_p(t)$ computed from Equation 4.69 and the free vibration response part $\mathbf{u}_h(t)$ computed from Equation 4.121.

Example 4.2

Consider a viscously damped linear 3-DOFs system

$$\begin{bmatrix} 1 & 0 & 0 \\ 0 & 1 & 0 \\ 0 & 0 & 1 \end{bmatrix} \begin{Bmatrix} \ddot{u}_1 \\ \ddot{u}_2 \\ \ddot{u}_3 \end{Bmatrix} + \begin{bmatrix} 4 & -2 & 0 \\ -2 & 4 & -1 \\ 0 & -1 & 4 \end{bmatrix} \begin{Bmatrix} \dot{u}_1 \\ \dot{u}_2 \\ \dot{u}_3 \end{Bmatrix}$$

$$+100 \begin{bmatrix} 3 & -2 & 0 \\ -2 & 3 & -1 \\ 0 & -1 & 3 \end{bmatrix} \begin{Bmatrix} u_1 \\ u_2 \\ u_3 \end{Bmatrix} = \begin{Bmatrix} 0 \\ 0 \\ f(t) \end{Bmatrix}$$

subjected to the initial conditions

$$\{u_1, u_2, u_3\} = \{1, 0, 0\} \quad \text{and} \quad \{\dot{u}_1, \dot{u}_2, \dot{u}_3\} = \{0, 1, 0\}$$

where $f(t) = 1 - u(t-1)$, where $u(t)$ is the unit step function.

Figure 4.2(a) compares the solution for u_3 obtained by using the present spectral analysis method (SAM) with the exact solution analytically obtained by using the modal analysis method (MAM) and with the solution obtained by using the Runge–Kutta method. To obtain the solutions within a 0.5% time-averaged error, the sampling time window $T = 4$ s and the number of samples $N = 1024$ are used for the SAM, while the time increment $\Delta t = 0.0039$ s is used for the Runge–Kutta method. The solutions obtained by using the non-proportional damping-based approach are also confirmed to be identical with those obtained by using the proportional damping based approach. It is obvious from Figure 4.2(a) that the present SAM provides very reliable solutions when compared with the solutions by both the MAM and the

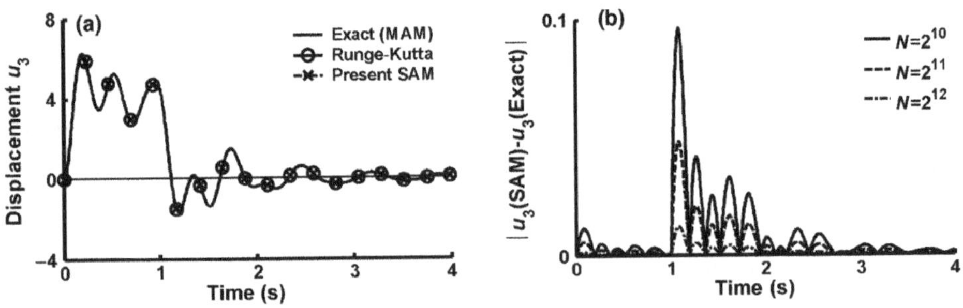

Figure 4.2 Comparison of the dynamic responses obtained by using different methods

Table 4.1 Comparison of the CPU times and time-averaged errors

Solution methods		Present SAM		Runge–Kutta method	
		$N=1024$	$N=2048$	$\Delta t=3.91$ (ms)	$\Delta t=1.95$ (ms)
CPU time (s)		0.203	0.296	0.234	0.344
Time-averaged error (%)	$u_1(t)$	0.214	0.087	0.502	0.283
	$u_2(t)$	0.232	0.120	0.265	0.236
	$u_3(t)$	0.168	0.085	0.264	0.230

Runge–Kutta method. Of course, as displayed in Figure 4.2(b), the accuracy of the SAM results can be controlled by varying the number of samples N for a chosen sampling time window $T=4$ s. Table 4.1 compares the CPU times and the time-averaged errors depending on the conditions used for the present SAM and the Runge–Kutta method.

4.5.2 Continuous Systems with Arbitrary Initial Conditions

Consider a linear continuous system whose dynamics are represented in the general form as

$$\boldsymbol{\mathcal{L}}\boldsymbol{u}(x,t)+\boldsymbol{M}\ddot{\boldsymbol{u}}(x,t)=\boldsymbol{p}(x,t) \tag{4.122}$$

where $\boldsymbol{\mathcal{L}}$ is the linear differential structural operator, $\boldsymbol{M}$ is the linear differential inertial operator, $\boldsymbol{p}(x, t)$ is the excitation force vector, and $\boldsymbol{u}(x, t)$ is the displacement vector. Assume that the linear continuous system is subjected to arbitrary, non-null initial conditions given by

$$\boldsymbol{u}(x,0)=\boldsymbol{u}_0(x), \qquad \dot{\boldsymbol{u}}(x,0)=\dot{\boldsymbol{u}}_0(x) \tag{4.123}$$

As discussed in the early part of this section, the complete dynamic response of a linear continuous system can be obtained by summing the forced vibration response part $\boldsymbol{u}_p(x, t)$ and the free vibration response part $\boldsymbol{u}_h(x, t)$ as follows:

$$\boldsymbol{u}(x,t)=\boldsymbol{u}_p(x,t;\boldsymbol{p})+\boldsymbol{u}_h(x,t;\boldsymbol{u}_0,\dot{\boldsymbol{u}}_0) \tag{4.124}$$

The forced vibration response part $\boldsymbol{u}_p(x, t)$ is determined solely by the excitation force $\boldsymbol{p}(x,t)$, while the free vibration response part $\boldsymbol{u}_h(x, t)$ is determined solely by the initial conditions $\boldsymbol{u}_0$ and $\dot{\boldsymbol{u}}_0$. Equation 4.124 implies that the linear continuous dynamic system when subjected to excitation forces, but initially at rest, will have only the forced vibration response part $\boldsymbol{u}_p(x, t)$. It is known that the convolution integral (or Duhamel's integral) provides the forced vibration responses for the linear systems initially at rest. It is also shown in Section 4.4.2 that the discrete Fourier transformed convolution integral for a linear continuous system is equivalent to the spectral element equation, $\boldsymbol{S}(\omega)\boldsymbol{d}=\boldsymbol{f}$. Thus, the dynamic responses computed from $\boldsymbol{d}=\boldsymbol{S}(\omega)^{-1}\boldsymbol{f}$ are just the forced vibration responses of the linear continuous system subjected to completely null initial conditions.

Thus, to predict the dynamic responses of a linear continuous system subjected to arbitrary non-zero initial conditions by using the SEM, we may need a strategy. The strategy introduced herein is to first transform the original governing equation, Equation 4.122, subjected to arbitrary non-null initial conditions, into the new governing equation subjected to completely null initial conditions by using the pseudo-force method, and next to apply the SEM to obtain the desired dynamic responses.

4.5.2.1 Pseudo-Force Method

Pseudo-Force for the Initial Displacement

Consider the free vibration of a continuous system, represented by the governing equation

$$\mathcal{L}\boldsymbol{u}_1(x,t)+\boldsymbol{M}\ddot{\boldsymbol{u}}_1(x,t)=\boldsymbol{0} \tag{4.125}$$

and the initial conditions

$$\boldsymbol{u}_1(x,0)=\boldsymbol{u}_0(x), \qquad \dot{\boldsymbol{u}}_1(x,0)=\boldsymbol{0} \tag{4.126}$$

Assume the solution to Equation 4.125 is

$$\boldsymbol{u}_1(x,t)=\boldsymbol{u}_0(x)+\Delta\boldsymbol{u}_1(x,t) \tag{4.127}$$

By using Equation 4.127, Equations 4.125 and 4.126 can be transformed into

$$\mathcal{L}\Delta\boldsymbol{u}_1(x,t)+\boldsymbol{M}\Delta\ddot{\boldsymbol{u}}_1(x,t)=\boldsymbol{f}_{u_0}(x,t) \tag{4.128}$$

and

$$\Delta\boldsymbol{u}_1(x,0)=\boldsymbol{0}, \qquad \Delta\dot{\boldsymbol{u}}_1(x,0)=\boldsymbol{0} \tag{4.129}$$

In Equation 4.128, $\boldsymbol{f}_{u_0}$ is the pseudo-force defined by

$$\boldsymbol{f}_{u_0}(x,t)=-\mathcal{L}\boldsymbol{u}_0(x)u(t) \tag{4.130}$$

where $u(t)$ is the unit step function.

Pseudo-Force for the Initial Velocity

Now consider a linear continuous system subjected to an impulse force

$$\mathcal{L}\boldsymbol{u}_2(x,t)+\boldsymbol{M}\ddot{\boldsymbol{u}}_2(x,t)=\boldsymbol{i}(x)\delta(t) \tag{4.131}$$

and the initial conditions

$$\boldsymbol{u}_2(x,0^-)=\boldsymbol{0}, \qquad \dot{\boldsymbol{u}}_2(x,0^-)=\boldsymbol{0} \tag{4.132}$$

where $\delta(t)$ is the Dirac delta function. The initial conditions in Equation 4.132 indicate that the linear continuous system is at rest before the application of an impulse force. The time $t=0^-$ emphasizes the instant of time just before the application of the impulse force, that is, the negative side of the time $t=0$.

Integrating both sides of Equation 4.131 from $t=0^-$ to $t=\Delta t$ yields

$$\mathcal{L}\boldsymbol{u}_{2\,\mathrm{avg}}(x)+\boldsymbol{M}[\dot{\boldsymbol{u}}_2(x,\Delta t)-\dot{\boldsymbol{u}}_2(x,0^-)]=\boldsymbol{i}(x) \tag{4.133}$$

where $\boldsymbol{u}_{2\mathrm{avg}}(x)$ is the time average of $\boldsymbol{u}_2(x,t)$ for the short time duration Δt, which will become zero as $\Delta t\to 0$. By applying the null initial conditions of Equation 4.132 to Equation 4.133 and then setting $\Delta t\to 0$, we can obtain

$$\boldsymbol{M}\dot{\boldsymbol{u}}_2(x,0)=\boldsymbol{i}(x) \tag{4.134}$$

Equation 4.134 implies that the impulse force applied at $t=0$ results in a sudden change in its velocity from zero, which is equal to $\boldsymbol{M}^{-1}\boldsymbol{i}(x)$, without resulting in an appreciable change in its displacement, because the time duration required to develop the displacement is zero from $\Delta t \rightarrow 0$. Accordingly, the original governing equation (Equation 4.131) can be expressed as

$$\boldsymbol{\mathcal{L}}\boldsymbol{u}_2(x,t)+\boldsymbol{M}\ddot{\boldsymbol{u}}_2(x,t)=\boldsymbol{0} \tag{4.135}$$

subjected to the new initial conditions

$$\boldsymbol{u}_2(x,0)=\boldsymbol{0}, \qquad \dot{\boldsymbol{u}}_2(x,0)=\boldsymbol{M}^{-1}\boldsymbol{i}(x)\equiv\dot{\boldsymbol{u}}_0(x) \tag{4.136}$$

The original set of equations 4.131 and 4.132 is equivalent to the new set of equations 4.135 and 4.136. Thus we can treat the system that has been subjected to the non-null initial velocity $\dot{\boldsymbol{u}}_0(x)$ as the system subjected to the pseudo-impulse force $\boldsymbol{f}_{\dot{u}_0}$ defined by

$$\boldsymbol{f}_{\dot{u}_0}(x,t)=\boldsymbol{M}\dot{\boldsymbol{u}}_0(x)\delta(t) \tag{4.137}$$

In other words, the set of equations 4.135 and 4.136 can be replaced with the new set of equations

$$\boldsymbol{\mathcal{L}}\boldsymbol{u}_2(x,t)+\boldsymbol{M}\ddot{\boldsymbol{u}}_2(x,t)=\boldsymbol{f}_{\dot{u}_0}(x,t) \tag{4.138}$$

and

$$\boldsymbol{u}_2(x,0)=\boldsymbol{0}, \qquad \dot{\boldsymbol{u}}_2(x,0)=\boldsymbol{0} \tag{4.139}$$

Pseudo-Force for the General Case

Finally, consider the general case when a continuous system is subjected to the excitation force $\boldsymbol{p}(x,t)$ and to the arbitrary non-null initial conditions $\boldsymbol{u}_0(x)$ and $\dot{\boldsymbol{u}}_0(x)$, as represented by the governing equation (4.122) and the initial conditions (4.123). The complete dynamic responses are assumed by

$$\boldsymbol{u}(x,t)=\boldsymbol{u}_0(x)+\Delta\boldsymbol{u}(x,t) \tag{4.140}$$

By substituting Equation 4.140 into the original governing equation (4.122) and initial conditions (4.123) and then by using the pseudo-forces developed previously, we can obtain a new set of governing equation and initial conditions for $\Delta\boldsymbol{u}(x,t)$. They are given by

$$\boldsymbol{\mathcal{L}}\Delta\boldsymbol{u}(x,t)+\boldsymbol{M}\Delta\ddot{\boldsymbol{u}}(x,t)=\boldsymbol{p}(x,t)+\boldsymbol{f}_{u_0}(x,t)+\boldsymbol{f}_{\dot{u}_0}(x,t) \tag{4.141}$$

and

$$\Delta\boldsymbol{u}(x,0)=\boldsymbol{0}, \qquad \Delta\dot{\boldsymbol{u}}(x,0)=\boldsymbol{0} \tag{4.142}$$

where $\boldsymbol{f}_{uo}$ and $\boldsymbol{f}_{\dot{u}_0}$ are the pseudo-forces defined by

$$\boldsymbol{f}_{u_0}(x,t)=-\boldsymbol{\mathcal{L}}\boldsymbol{u}_0(x)u(t) \tag{4.143}$$

$$\boldsymbol{f}_{\dot{u}_0}(x,t)=\boldsymbol{M}\dot{\boldsymbol{u}}_0(x)\delta(t) \tag{4.144}$$

As the new governing equation (4.141) is now subjected to completely null initial conditions (Equation 4.142), we can use either the convolution integral or the SEM to compute $\Delta\boldsymbol{u}(x,t)$.

Once $\Delta \boldsymbol{u}(x, t)$ is obtained from Equations 4.141 and 4.142, the complete dynamic responses can be computed from Equation 4.140 by adding $\Delta \boldsymbol{u}(x, t)$ to the initial displacements $\boldsymbol{u}_0(x)$.

4.5.2.2 Application to the Transverse Vibration of a Beam

As an illustrative example, consider a uniform Bernoulli–Euler beam subjected to an excitation force $f(x, t)$. The transverse vibration $w(x, t)$ of the beam is governed by the equation of motion

$$EIw'''' + \rho A\ddot{w} = f(x,t) \tag{4.145}$$

and the initial conditions

$$w(x,0) = w_0(x), \qquad \dot{w}(x,0) = \dot{w}_0(x) \tag{4.146}$$

where EI is the bending rigidity and ρA is the mass per unit length of the beam.

Transform of the Equation of Motion

By comparing Equations 4.145 and 4.146 with Equations 4.122 and 4.123, we find the following relationships:

$$\begin{aligned} &\boldsymbol{\mathcal{L}} \Leftrightarrow EI(\,)'''', && \boldsymbol{M} \Leftrightarrow \rho A \\ &\boldsymbol{u}(x,t) \Leftrightarrow w(x,t), && \boldsymbol{p}(x,t) \Leftrightarrow f(x,t) \\ &\boldsymbol{u}_0(x) \Leftrightarrow w_0(x), && \dot{\boldsymbol{u}}_0(x) \Leftrightarrow \dot{w}_0(x) \end{aligned} \tag{4.147}$$

Assume the complete solution of Equation 4.145 to be

$$w(x,t) = w_0(x) + \Delta w(x,t) \tag{4.148}$$

By using the relationships of Equation 4.147, the original equation of motion and initial conditions can be transformed into the new equation of motion

$$EI\Delta w''''(x,t) + \rho A\Delta\ddot{w}(x,t) = p(x,t) \tag{4.149}$$

and the new initial conditions

$$\Delta w(x,0) = \Delta\dot{w}(x,0) = 0 \tag{4.150}$$

where

$$p(x,t) = f(x,t) - EIw_0''''u(t) + \rho A\dot{w}_0\delta(t) \tag{4.151}$$

Spectral Element Model for Δw(x, t)

Represent the solution $\Delta w(x, t)$ to Equation 4.149 and the excitation force $f(x, t)$ in the spectral forms as

$$\Delta w(x,t) = \frac{1}{N}\sum_{n=0}^{N-1} \Delta W_n(x) e^{i\omega_n t} \tag{4.152}$$

$$f(x,t) = \frac{1}{N}\sum_{n=0}^{N-1} F_n(x) e^{i\omega_n t} \tag{4.153}$$

The spectral components F (omitting the subscripts n for brevity) can be derived from Equation 4.151 as

$$P(x) = F(x) - EIw_0''''(x)\delta(\omega) + \rho A\dot{w}_0(x)u(\omega) \tag{4.154}$$

where $\delta(\omega)$ is the Dirac delta function and $u(\omega)$ is the unit step function.

Substituting Equations 4.152 and 4.153 into Equation 4.149 yields

$$EI\Delta W''''(x) - \omega^2\rho A\Delta W(x) = P(x) \tag{4.155}$$

The weak form of Equation 4.155 is given by

$$\int_0^l (EI\Delta W''\delta\Delta W'' - \omega^2\rho A\Delta W\delta\Delta W)\mathrm{d}x - \int_0^l P(x)\delta\Delta W\mathrm{d}x = 0 \tag{4.156}$$

Equation 4.155 is essentially same as Equation 3.147 considered in Example 3.6. Thus we can use the same dynamic shape function $\boldsymbol{N}_B(x,\omega)$ as used in Example 3.6 to express the general solution of Equation 4.155 as

$$\Delta W(x) = \boldsymbol{N}_B(x;\omega)\Delta\boldsymbol{d} \tag{4.157}$$

where $\Delta\boldsymbol{d}$ is the spectral DOFs vector defined by

$$\Delta\boldsymbol{d} = \{\Delta W_1 \quad \Delta\Theta_1 \quad \Delta W_2 \quad \Delta\Theta_2\}^{\mathrm{T}} = \{\Delta W(0) \quad \Delta W'(0) \quad \Delta W(l) \quad \Delta W'(l)\}^{\mathrm{T}} \tag{4.158}$$

Substituting Equation 4.157 into Equation 4.156, we can derive the spectral element equation as

$$\boldsymbol{S}(\omega)\Delta\boldsymbol{d} = \boldsymbol{f} \tag{4.159}$$

where

$$\boldsymbol{S}(\omega) = \int_0^l (EI\boldsymbol{N}_B''^{\mathrm{T}}\boldsymbol{N}_B'' - \omega^2\rho A\boldsymbol{N}_B^{\mathrm{T}}\boldsymbol{N}_B)\mathrm{d}x \tag{4.160}$$

$$\boldsymbol{f} = \int_0^l P(x)\boldsymbol{N}_B^{\mathrm{T}}\mathrm{d}x \tag{4.161}$$

The spectral element matrix $\boldsymbol{S}(\omega)$ given by Equation 4.160 is identical with that given by Equation 3.156 in Example 3.6.

Solution Procedure

The spectral element equations are assembled, followed by the imposition of the boundary conditions to derive the global system dynamic equation in the form of

$$\boldsymbol{S}_{\mathrm{g}}(\omega)\Delta\boldsymbol{d}_{\mathrm{g}} = \boldsymbol{f}_{\mathrm{g}} \tag{4.162}$$

where $\boldsymbol{S}_{\mathrm{g}}(\omega)$ is the global dynamic stiffness matrix, $\Delta\boldsymbol{d}_{\mathrm{g}}$ is the global perturbed spectral nodal DOFs vector, and $\boldsymbol{f}_{\mathrm{g}}$ is the global spectral nodal forces vector. The remaining solution procedure is as follows:

a. Compute $\Delta \boldsymbol{d}_g$ from Equation 4.162.
b. Compute the spectral components $\Delta W_n(x)$ ($n = 0, 1, \ldots, N-1$) from Equation 4.157.
c. Compute the perturbed displacement $\Delta w(x, t)$ from $\Delta W_n(x)$ ($n = 0, 1, \ldots, N-1$) by using the inverse-FFT algorithm.
d. Finally compute the complete dynamic response $w(x, t)$ from Equation 4.148 by adding $\Delta w(x, t)$ to the initial displacement $w_0(x)$.

Example 4.3

Consider a uniform Bernoulli–Euler beam that is simply supported at its two ends. Assume that the beam is subjected to the initial conditions

$$w(x, 0) = w_0 \sin\frac{\pi x}{L}, \qquad \dot{w}(x, 0) = 0$$

When a point unit impulse is applied at the middle point ($x = L/2$) of the beam at time $t = t_0$, the dynamic response can be analytically obtained in a closed form by using the modal analysis method as

$$w(x, t) = w_p(x, t) + w_h(x, t)$$

where

$$w_p(x, t) = \begin{cases} 0 & (0 \le t \le t_0) \\ \displaystyle\sum_{n=1}^{\infty} \frac{1}{\omega_n} \frac{2}{\rho A L} \sin\frac{n\pi}{2} \sin\frac{n\pi x}{L} \sin\omega_n (t - t_0) & (t \ge t_0) \end{cases}$$

$$w_h(x, t) = w_0 \sin\frac{\pi x}{L} \cos\omega_1 t \quad (t \ge 0)$$

and ω_n are the natural frequencies defined by

$$\omega_n = n^2 \pi^2 \sqrt{\frac{EI}{\rho A L^4}} \quad (n = 1, 2, \ldots, \infty)$$

Figure 4.3 compares the dynamic response obtained by using the present SEM with the exact analytical solution obtained by using the modal analysis method (MAM). For a numerical comparison, $L = 0.5$ m, $\rho A = 0.5$ kg/m, $EI = 20$ N m, $t_0 = 0.003$ s and $w_0 = 5$ mm are used. For the analytical solution by MAM, 100 modes are superposed to obtain a sufficiently converged dynamic response. For the SEM solution, the sampling time window $T = 0.1024$ s and the number of samples $N = 2^{12}$ are used. Figure 4.3 shows that the dynamic response obtained by using the present SEM certainly satisfies all initial conditions and is almost identical with that obtained by using MAM.

4.6 Dynamic Responses of Nonlinear Systems

This section considers the spectral analysis method for the computation of the dynamic responses of the nonlinear systems subjected to arbitrary, non-null initial conditions.

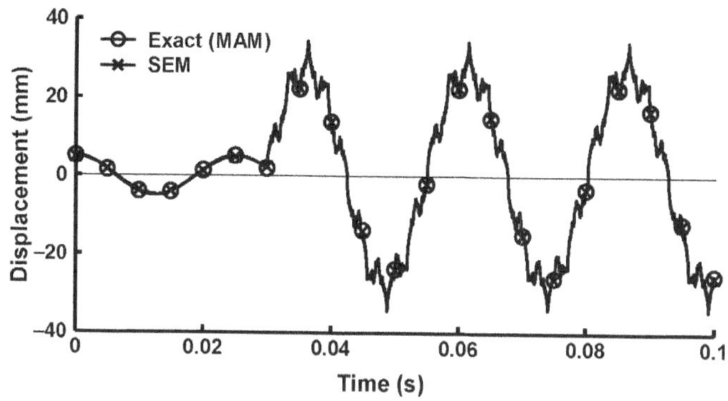

Figure 4.3 Comparison of the dynamic responses obtained by the present SEM and the modal analysis method (MAM)

4.6.1 Discrete Systems with Arbitrary Initial Conditions

Assume that a viscously damped, nonlinear m-DOFs system is represented by

$$\boldsymbol{M}\ddot{\boldsymbol{u}}(t) + \boldsymbol{C}\dot{\boldsymbol{u}}(t) + \boldsymbol{K}\boldsymbol{u}(t) + \mathcal{N}(\boldsymbol{u}(t), \dot{\boldsymbol{u}}(t)) = \boldsymbol{f}(t) \tag{4.163}$$

with the initial conditions

$$\boldsymbol{u}(0) = \boldsymbol{u}_0 \quad \text{and} \quad \dot{\boldsymbol{u}}(0) = \dot{\boldsymbol{u}}_0 \tag{4.164}$$

where $\mathcal{N}(\boldsymbol{u}, \dot{\boldsymbol{u}})$ is the nonlinear function of the unknown variable $\boldsymbol{u}(t)$. The pseudo-force concept introduced in Section 4.5.2 can be applied to the present discrete system in an analogous way. Comparing Equations 4.163 and 4.164 with Equations 4.122 and 4.123, we can find the relationships between the operators and variables used for the discrete and continuous systems as follows:

$$\begin{array}{ccc} \text{(Continuous system)} & & \text{(Discrete system)} \\ \mathcal{L} & \Leftrightarrow & \boldsymbol{C}d(\cdot)/\mathrm{d}t + \boldsymbol{K} \\ \boldsymbol{M} & \Leftrightarrow & \boldsymbol{M} \\ \boldsymbol{u}(x,t) & \Leftrightarrow & \boldsymbol{u}(t) \\ \boldsymbol{p}(x,t) & \Leftrightarrow & \boldsymbol{f}(t) \\ \boldsymbol{u}_0(x) & \Leftrightarrow & \boldsymbol{u}_0 \\ \dot{\boldsymbol{u}}_0(x) & \Leftrightarrow & \dot{\boldsymbol{u}}_0 \end{array} \tag{4.165}$$

We assume the solution of Equation 4.163 is

$$\boldsymbol{u}(t) = \boldsymbol{u}_0 + \Delta\boldsymbol{u}(t) \tag{4.166}$$

By using Equation 4.166, the nonlinear term $\mathsf{N}(\boldsymbol{u}, \dot{\boldsymbol{u}})$ can be expanded in a power series in $\Delta\boldsymbol{u}$ and $\Delta\dot{\boldsymbol{u}}$ as follows:

$$\begin{aligned} \mathcal{N}(\boldsymbol{u}(t), \dot{\boldsymbol{u}}(t)) &= \mathcal{N}(\boldsymbol{u}_0 + \Delta\boldsymbol{u}, \Delta\dot{\boldsymbol{u}}) \\ &= \boldsymbol{K}_{\mathrm{NL}}(\boldsymbol{u}_0)\Delta\boldsymbol{u}(t) + \boldsymbol{C}_{\mathrm{NL}}(\boldsymbol{u}_0)\Delta\dot{\boldsymbol{u}}(t) + \boldsymbol{f}_{\mathrm{NL}}(\Delta\boldsymbol{u}, \Delta\dot{\boldsymbol{u}}; \boldsymbol{u}_0)u(t) \end{aligned} \tag{4.167}$$

where $f_{NL}(\Delta u, \Delta \dot{u}; u_0)$ contains all constant terms and nonlinear terms in either $\Delta u(t)$ or $\Delta \dot{u}(t)$. Substituting Equation 4.166 into Equation 4.163, replacing $\mathcal{N}(u, \dot{u})$ by Equation 4.167, and then using the pseudo-force concept explained by Equations 4.143 and 4.144 in an analogous way, we can transform the original governing equation (4.163) and the initial conditions (4.164) into the perturbed governing equation

$$M\Delta\ddot{u}(t) + \bar{C}\Delta\dot{u}(t) + \bar{K}\Delta u(t) = \bar{f}(t) - f_{NL}(\Delta u, \Delta\dot{u}; u_0)u(t) \tag{4.168}$$

and the null initial conditions

$$\Delta u(0) = \mathbf{0}, \qquad \Delta\dot{u}(0) = \mathbf{0} \tag{4.169}$$

where

$$\begin{aligned} \bar{C} &= C + C_{NL}(u_0) \\ \bar{K} &= K + K_{NL}(u_0) \\ \bar{f}(t) &= f(t) - Ku_0u(t) + M\ddot{u}_0\delta(t) \end{aligned} \tag{4.170}$$

The solution $\Delta u(t)$ of the nonlinear perturbed governing equation (4.168) can be obtained by using an iteration method. Let us assume that the solution of the k-th iteration (previous iteration) is known. Then we can compute the nonlinear pseudo-force term $f_{NL}^{k}(\Delta u, \Delta\dot{u}; u_0)$ in advance so that we can obtain the new solution $\Delta u^{k+1}(t)$ from

$$M\Delta\ddot{u}^{k+1} + \bar{C}\Delta\dot{u}^{k+1} + \bar{K}\Delta u^{k+1} = \bar{f}(t) - f_{NL}^{k}(\Delta u^k, \Delta\dot{u}^k; u_0)u(t) \tag{4.171}$$

For the spectral analysis, we represent $\Delta u(t)$, $\bar{f}(t)$ and $f_{NL}(t)$ in the spectral forms as

$$\begin{aligned} \Delta u(t) &= \frac{1}{N}\sum_{n=0}^{N-1} \Delta U_n e^{i\omega_n t} \\ \bar{f}(t) &= \frac{1}{N}\sum_{n=0}^{N-1} \bar{F}_n e^{i\omega_n t} \\ f_{NL}(t) &= \frac{1}{N}\sum_{n=0}^{N-1} F_{NLn} e^{i\omega_n t} \end{aligned} \tag{4.172}$$

By using Equation 4.172, we can transform Equation 4.171 into

$$D(\omega_n)\Delta U_n^{k+1} = \bar{F}_n - F_{NLn}^{k} \tag{4.173}$$

where

$$D(\omega_n) = \bar{K} + i\omega_n\bar{C} - \omega_n^2 M \tag{4.174}$$

From Equation 4.173, we obtain

$$\Delta U_n^{k+1} = D(\omega_n)^{-1}(\bar{F}_n - F_{NLn}^{k}) \tag{4.175}$$

The new time-domain solution $\Delta \boldsymbol{u}^{k+1}$ can be computed from $\Delta \boldsymbol{U}_n^{k+1}$ ($n=0, 1, \ldots, N-1$) by using the inverse-FFT algorithm as follows:

$$\Delta \boldsymbol{u}^{k+1}(t) \quad \Leftarrow \quad \text{IFFT}\{\Delta \boldsymbol{U}_n^{k+1} |\ n=0, 1, 2, \ldots, N-1\} \tag{4.176}$$

The iteration process is repeated until the solution converges to a required accuracy.

The complete solution $\boldsymbol{u}(t)$ of the original governing equation is finally obtained by simply adding $\Delta \boldsymbol{u}(t)$ computed from Equation 4.176 to the initial displacements $\boldsymbol{u}_0$, as shown by Equation 4.166.

4.6.2 Continuous Systems with Arbitrary Initial Conditions

Consider a nonlinear continuous system represented by the governing equation

$$\mathcal{L}\boldsymbol{u}(x,t) + \boldsymbol{M}\ddot{\boldsymbol{u}}(x,t) + \mathcal{N}(\boldsymbol{u}(x,t), \dot{\boldsymbol{u}}(x,t)) = \boldsymbol{p}(x,t) \tag{4.177}$$

and the initial conditions

$$\boldsymbol{u}(x,0) = \boldsymbol{u}_0(x), \qquad \dot{\boldsymbol{u}}(x,0) = \dot{\boldsymbol{u}}_0(x) \tag{4.178}$$

where $\mathcal{L}$ and $\boldsymbol{M}$ are the linear differential structural and inertial operators, respectively, and $\mathcal{N}(\boldsymbol{u}, \dot{\boldsymbol{u}})$ is the nonlinear function of $\boldsymbol{u}(x, t)$ and $\dot{\boldsymbol{u}}(x, t)$. Following the pseudo-force approach introduced in Section 4.5.2, we assume the complete solution to be

$$\boldsymbol{u}(x,t) = \boldsymbol{u}_0(x) + \Delta \boldsymbol{u}(x,t) \tag{4.179}$$

By using Equation 4.179, the nonlinear term $\mathcal{N}(\boldsymbol{u}, \dot{\boldsymbol{u}})$ can be expanded in the form

$$\begin{aligned} \mathcal{N}(\boldsymbol{u}(x,t), \dot{\boldsymbol{u}}(x,t)) &= \mathcal{N}(\boldsymbol{u}_0 + \Delta \boldsymbol{u}, \Delta \dot{\boldsymbol{u}}) \\ &= \boldsymbol{L}_{\text{NL}} \Delta \boldsymbol{u}(x,t) + \boldsymbol{p}_{\text{NL}}(\Delta \boldsymbol{u}, \Delta \dot{\boldsymbol{u}}; \boldsymbol{u}_0) u(t) \end{aligned} \tag{4.180}$$

where $\boldsymbol{L}_{\text{NL}}$ is the linear operator and $\boldsymbol{p}_{\text{NL}}(\Delta \boldsymbol{u}, \Delta \dot{\boldsymbol{u}}; \boldsymbol{u}_0)$ contains all constant terms and nonlinear terms in either $\Delta \boldsymbol{u}(x, t)$ or $\Delta \dot{\boldsymbol{u}}(x, t)$. Substituting Equation 4.179 into Equation 4.177, replacing the nonlinear term $\mathcal{N}(\boldsymbol{u}, \dot{\boldsymbol{u}})$ by Equation 4.180, and then using the pseudo-forces given by Equations 4.143 and 4.144, we can obtain a perturbed governing equation

$$\bar{\mathcal{L}}\Delta \boldsymbol{u}(x,t) + \boldsymbol{M}\Delta \ddot{\boldsymbol{u}}(x,t) = \bar{\boldsymbol{p}}(x,t) - \boldsymbol{p}_{\text{NL}}(\Delta \boldsymbol{u}, \Delta \dot{\boldsymbol{u}}; \boldsymbol{u}_0) u(t) \tag{4.181}$$

subjected to the null initial conditions

$$\Delta \boldsymbol{u}(x,0) = \boldsymbol{0}, \qquad \Delta \dot{\boldsymbol{u}}(x,0) = \boldsymbol{0} \tag{4.182}$$

where

$$\begin{aligned} \bar{\mathcal{L}} &= \mathcal{L} + \boldsymbol{L}_{\text{NL}} \\ \bar{\boldsymbol{p}}(x,t) &= \boldsymbol{p}(x,t) - \mathcal{L}\boldsymbol{u}_0(x) u(t) + \boldsymbol{M}\dot{\boldsymbol{u}}_0(x)\delta(t) \end{aligned} \tag{4.183}$$

Similar to the nonlinear discrete systems, the iteration method can be used to solve the nonlinear perturbed governing Equation 4.181 for $\Delta \boldsymbol{u}(x, t)$. Assuming that the nonlinear pseudo-force term $\boldsymbol{p}_{\mathrm{NL}}(\Delta\boldsymbol{u}, \Delta\dot{\boldsymbol{u}}; \boldsymbol{u}_0)$ can be computed in advance by using the k-th iteration (previous iteration) solution, we can obtain the new solution $\Delta\boldsymbol{u}^{k+1}(x, t)$ from

$$\begin{aligned}&\bar{\boldsymbol{\mathcal{L}}}\Delta\boldsymbol{u}^{k+1}(x,t) + \boldsymbol{M}\Delta\ddot{\boldsymbol{u}}^{k+1}(x,t)\\&\quad = \bar{\boldsymbol{p}}(x,t) - \boldsymbol{p}_{\mathrm{NL}}^{k}(\Delta\boldsymbol{u}^{k}, \Delta\dot{\boldsymbol{u}}^{k}; \boldsymbol{u}_0)u(t)\end{aligned} \tag{4.184}$$

Represent $\Delta\boldsymbol{u}(x,t), \bar{\boldsymbol{p}}(x,t)$ and $\boldsymbol{p}_{\mathrm{NL}}(\Delta\boldsymbol{u}, \Delta\dot{\boldsymbol{u}}; \boldsymbol{u}_0)$ in the spectral forms as

$$\begin{aligned}\Delta\boldsymbol{u}(x,t) &= \frac{1}{N}\sum_{n=0}^{N-1}\Delta\boldsymbol{U}_n(x)e^{i\omega_n t}\\ \bar{\boldsymbol{p}}(x,t) &= \frac{1}{N}\sum_{n=0}^{N-1}\bar{\boldsymbol{P}}_n(x)e^{i\omega_n t}\\ \boldsymbol{p}_{\mathrm{NL}}(\Delta\boldsymbol{u}, \Delta\dot{\boldsymbol{u}}; \boldsymbol{u}_0) &= \frac{1}{N}\sum_{n=0}^{N-1}\boldsymbol{P}_{\mathrm{NL}n}(x)e^{i\omega_n t}\end{aligned} \tag{4.185}$$

Following the procedure of spectral element formulation, we transform Equation 4.184 into the frequency domain by using Equation 4.185 and then relate $\Delta\boldsymbol{U}_n(x;\omega_n)$ to the spectral nodal DOFs vector $\Delta\boldsymbol{d}_n$ by

$$\Delta\boldsymbol{U}_n(x;\omega_n) = \boldsymbol{N}(x;\omega_n)\Delta\boldsymbol{d}_n \tag{4.186}$$

where $\boldsymbol{N}(x;\ \omega_n)$ is the dynamic shape function matrix. From the weak form of the discrete Fourier transformed perturbed governing equation, we can derive the spectral element equation in the form of

$$\boldsymbol{S}(\omega_n)\Delta\boldsymbol{d}_n = \bar{\boldsymbol{F}}_n - \boldsymbol{F}_{\mathrm{NL}n} \tag{4.187}$$

where $\boldsymbol{S}(\omega_n)$ is the spectral element matrix for the perturbed governing equation and

$$\begin{aligned}\bar{\boldsymbol{F}}_n &= \int_0^L \boldsymbol{N}^{\mathrm{T}}(x;\omega_n)\bar{\boldsymbol{P}}_n(x)\mathrm{d}x\\ \boldsymbol{F}_{\mathrm{NL}n} &= \int_0^L \boldsymbol{N}^{\mathrm{T}}(x;\omega_n)\boldsymbol{P}_{NLn}(x)\mathrm{d}x\end{aligned} \tag{4.188}$$

The nonlinear equation (4.187) can be iteratively solved for $\Delta\boldsymbol{d}_n\ (n=0, 1, \ldots N-1)$ as follows:

$$\boldsymbol{S}(\omega_n)\Delta\boldsymbol{d}_n^{k+1} = \bar{\boldsymbol{F}}_n - \boldsymbol{F}_{\mathrm{NL}n}^{k} \tag{4.189}$$

From Equation 4.189, we can obtain

$$\Delta\boldsymbol{d}_n^{k+1} = \boldsymbol{S}(\omega_n)^{-1}(\bar{\boldsymbol{F}}_n - \boldsymbol{F}_{\mathrm{NL}n}^{k}) \tag{4.190}$$

By using $\Delta\boldsymbol{d}_n^{k+1}\ (n=0, 1, \ldots, N-1)$ computed from Equation 4.190, we compute the spectral components $\Delta\boldsymbol{U}_n^{k+1}(x)\ (n=0, 1, \ldots, N-1)$ from Equation 4.186. Lastly we compute the new time-domain solution $\Delta\boldsymbol{u}^{k+1}(x, t)$ from $\Delta\boldsymbol{U}_n^{k+1}(x)\ (n=0, 1, \ldots, N-1)$ by using the inverse-FFT algorithm as

$$\Delta u^{k+1}(x,t) \quad \Leftarrow \quad \mathrm{IFFT}\{\Delta U_n^{k+1}(x)|\ n=0,\ 1,\ 2, \ldots,\ N-1\} \tag{4.191}$$

The above iteration process will be repeated until the solution $\Delta \boldsymbol{u}(x, t)$ converges to the required accuracy. As explained by Equation 4.179, the complete solution $\boldsymbol{u}(x, t)$ of the original governing equation is finally obtained by adding $\Delta \boldsymbol{u}(x, t)$ computed from Equation 4.191 to the initial displacements $\boldsymbol{u}_0(x)$.

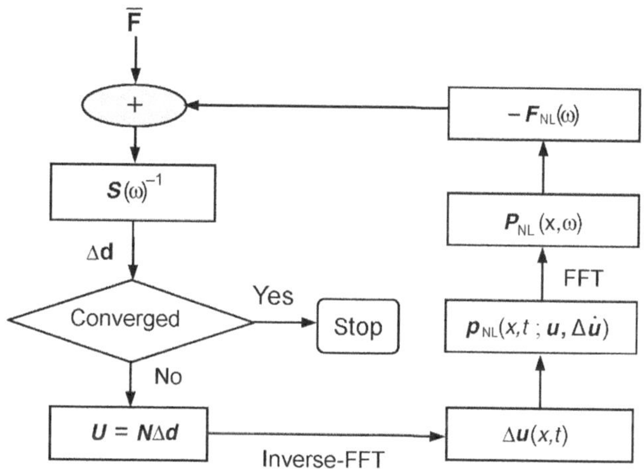

Figure 4.4 Direct iteration based on the AFT method

Figure 4.4 shows a direct iteration method where the alternating frequency-time (AFT) method [25] is used. In this approach, we evaluate the nonlinear pseudo-force term $\boldsymbol{p}_{NL}$ in the time domain by using the time history $\Delta \boldsymbol{u}(x, t)$ available at the previous iteration and then switch to the frequency domain to compute its spectral components $\boldsymbol{P}_{NL}$ by using the FFT algorithm for the use in the next iteration.

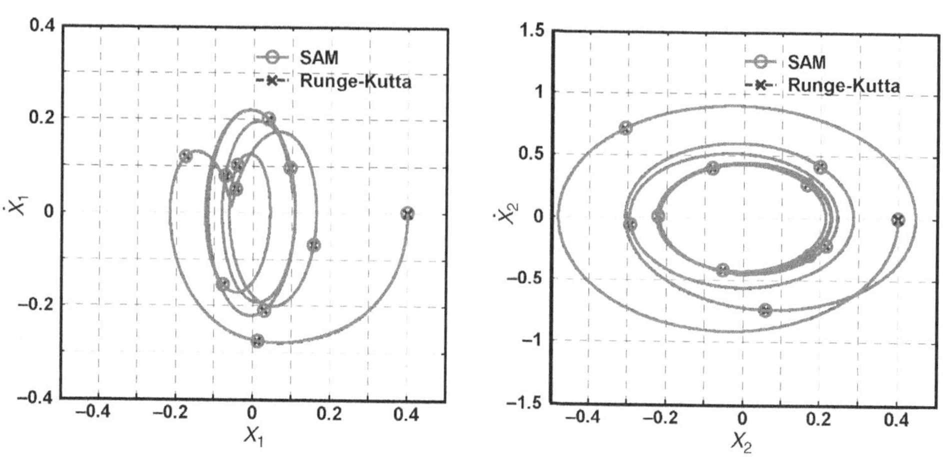

Figure 4.5 Phase plots for $x_1(t)$ and $x_2(t)$

Example 4.4

Consider a nonlinear 2-DOFs system represented by the equations of motion

$$\boldsymbol{m}\ddot{\boldsymbol{x}} + \boldsymbol{c}\dot{\boldsymbol{x}} + \boldsymbol{k}\boldsymbol{x} + \boldsymbol{k}_3\boldsymbol{x}^3 = \boldsymbol{f}(t)$$

and the initial conditions

$$\boldsymbol{x}_0 = \{0.5 \quad 0.2\}^{\mathrm{T}}, \qquad \dot{\boldsymbol{x}}_0 = \{0 \quad 0\}^{\mathrm{T}}$$

where

$$\boldsymbol{x}(t) = \begin{Bmatrix} x_1(t) \\ x_2(t) \end{Bmatrix}, \qquad \boldsymbol{f}(t) = \begin{Bmatrix} 0.5\cos 2t \\ 0 \end{Bmatrix}$$

$$\boldsymbol{m} = \begin{bmatrix} 1 & 0 \\ 0 & 1 \end{bmatrix}, \qquad \boldsymbol{c} = \begin{bmatrix} 0.4 & 0 \\ 0 & 0.2 \end{bmatrix}, \qquad \boldsymbol{k} = \begin{bmatrix} 1 & 1 \\ 1 & 4 \end{bmatrix}, \qquad \boldsymbol{k}_3 = \begin{bmatrix} 0.5 & 0 \\ 0 & 0.8 \end{bmatrix}$$

By assuming $\boldsymbol{x}(t) = \boldsymbol{x}_0 + \Delta\boldsymbol{x}(t)$, we can write the above nonlinear equations of motion in the form

$$\boldsymbol{m}\Delta\ddot{\boldsymbol{x}} + \boldsymbol{c}\Delta\dot{\boldsymbol{x}} + (\boldsymbol{k} + \boldsymbol{k}_{\mathrm{NL}})\Delta\boldsymbol{x} = \bar{\boldsymbol{f}}(t) - \boldsymbol{f}_{\mathrm{NL}}(t)$$

subjected to the null initial conditions

$$\Delta\boldsymbol{x}_0 = \{0 \quad 0\}^{\mathrm{T}}, \qquad \Delta\dot{\boldsymbol{x}}_0 = \{0 \quad 0\}^{\mathrm{T}}$$

where

$$\begin{aligned} \boldsymbol{k}_{NL} &= 3\boldsymbol{k}_3[diag(\boldsymbol{x}_0^2)] \\ \bar{\boldsymbol{f}}(t) &= \boldsymbol{f}(t) - \boldsymbol{k}\boldsymbol{x}_0 u(t) + \boldsymbol{m}\dot{\boldsymbol{x}}_0\delta(t) \\ \boldsymbol{f}_{NL} &= \boldsymbol{k}_3(\Delta\boldsymbol{x}^3 + 3[diag(\boldsymbol{x}_0)]\Delta\boldsymbol{x}^2 + \boldsymbol{x}_0^3)u(t) \end{aligned}$$

where $[diag(\boldsymbol{v})]$ denotes the diagonal matrix whose diagonal components are identical with the components of the vector $\boldsymbol{v}$. Figure 4.5 compares the phase plots for $x_1(t)$ and $x_2(t)$ computed by using the present spectral analysis method (SAM) and the Runge–Kutta (RK) method. The figure shows that the present SAM provides almost identical dynamic responses with those obtained by the Runge–Kutta method.

Part Three

Applications of Spectral Element Method

5

Dynamics of Beams and Plates

In this chapter, the spectral element method (SEM) is applied to the beams and plates subjected to distributed dynamic forces. In SEM, the distributed dynamic forces can be represented by the equivalent spectral nodal forces, which are defined by the integration of the multiplication of the discrete Fourier transformed distributed dynamic forces and the dynamic shape function matrix. As an alternative approach, the two-element method (TEM) can be applied to deal with the distributed dynamic forces acting on the beams or plates. For the TEM-based spectral element analysis, the distributed dynamic force acting on a beam is first divided into a sequence of point forces. The complete dynamic response is then obtained by superposing all dynamic responses excited by each point force. To compute the dynamic response due to a point force, the beam is represented by a two-element model so that the interconnection node becomes the point at which the point force is located. Unlike the 1-D beam structures, the plates show very complicated two-dimensional wave characteristics. Thus, developing the spectral element models for the plates with general boundary conditions is not so straightforward. Therefore, in this chapter, the spectral element modeling is introduced only for the Levy-type plates. The TEM developed for the beams is then extended to the Levy-type plates subjected to the distributed dynamic forces.

5.1 Beams

5.1.1 Spectral Element Equation

As the theory discussed in this section can be readily applied to any 1-D beam-type structure, we will consider a Bernoulli–Euler beam (B-beam) subjected to the distributed dynamic force as an illustrative example of the problem. The spectral element equation for a B-beam element is given by Equation 3.156

$$\boldsymbol{S}(\omega_n)\boldsymbol{d}_n = \boldsymbol{f}_n \qquad (n = 0,\ 1,\ 2, \ldots,\ N-1) \tag{5.1}$$

where $\boldsymbol{S}(\omega_n)$ is the spectral element matrix, $\boldsymbol{d}_n$ is the spectral nodal degrees-of-freedom (DOFs) vector, and $\boldsymbol{f}_n$ is the equivalent spectral nodal forces vector defined by

$$\boldsymbol{f}_n = \int_0^L \boldsymbol{N}^{\mathrm{T}}(x, \omega_n) F_n(x)\mathrm{d}x \tag{5.2}$$

where $F_n(x; \omega_n)$ is the spectral components of the distributed dynamic force $f(x, t)$. Once the distributed dynamic force $f(x, t)$ is specified, the corresponding equivalent spectral nodal forces f_n can readily be computed from Equation 5.2 to complete the spectral element equation.

There are two approaches to dealing with distributed dynamic forces. One approach is to represent the distributed dynamic forces by the equivalent spectral nodal forces computed from Equation 5.2. The other approach is to use the so-called two-element method that has been developed based on the linear superposition principle. In the next section, the two-element method developed for the beams is applied to the Levy-type plates.

5.1.2 Two-Element Method

As shown in Figure 5.1, a distributed dynamic force $f(x, t)$ acting on a beam can be represented by a sequence of equivalent dynamic point forces (simply, point forces). By using the linear superposition principle, the total dynamic response of the beam can be obtained by summing all dynamic responses excited by each of the point forces.

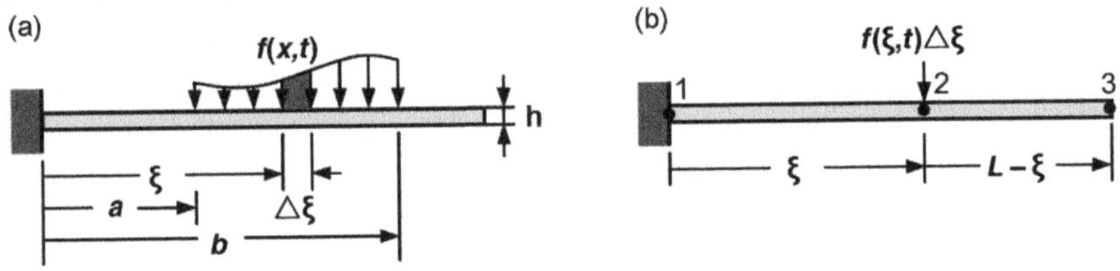

Figure 5.1 Linear superposition representation of a distributed dynamic force

Consider a beam subjected to a point force at $x = \xi$. The point force is given by

$$\Delta f(\xi, t) = f(\xi, t)\Delta\xi \tag{5.3}$$

where $\Delta\xi = (b - a)/N_d$, where N_d is the total number of increments. The dynamic response excited by $\Delta f(\xi, t)$ during the time $0 \leq \tau \leq t$ can be obtained by using the convolution integral (or Duhamel's integral) as

$$\Delta w(x, \xi, t) = \int_0^t h(x, \xi, t - \tau)\Delta f(\xi, \tau)\mathrm{d}\tau \tag{5.4}$$

where $h(x, \xi, t)$ is the unit impulse response function. Equation 5.4 can be discrete Fourier transformed as

$$\Delta W_n(x, \xi; \omega_n) = H_n(x, \xi; \omega_n)\Delta F_n(\xi; \omega_n) \qquad (n = 0, 1, 2, \ldots, N - 1) \tag{5.5}$$

where

$$\Delta F_n(\xi; \omega_n) = F_n(\xi; \omega_n)\Delta\xi \tag{5.6}$$

$H_n(x, \xi; \omega_n)$ are the spectral components of $h(x, \xi, t)$ (frequency response functions) and $F_n(\xi; \omega_n)$ are the spectral components of $f(\xi, t)$. Equation 5.5 represents the spectral components of the incremental dynamic response $\Delta w(x, \xi, t)$ excited by a point force $\Delta f(\xi, t)$ at $x = \xi$.

Thus, by using the linear superposition principle, the spectral components of the total dynamic response $w(x, t)$ excited by the whole dynamic force $f(x, t)$ distributed over the region $a \leq x \leq b$, as shown in Figure 5.1, can be obtained from

$$W_n(x;\omega_n) = \int \Delta W_n(x,\xi;\omega_n) = \int_a^b H_n(x,\xi,\omega_n)F_n(\xi;\omega_n)\mathrm{d}\xi \tag{5.7}$$

The integration of Equation 5.7 can be readily accomplished by using a proper numerical integration method. When we use the Trapezoidal algorithm, for instance, Equation 5.7 can be written in a numerical form as

$$\begin{aligned} W_n(x;\omega_n) = &\frac{1}{2}\Delta\xi[\Delta W_n(x,a;\omega_n) + 2\Delta W_n(x,a+\Delta\xi;\omega_n) \\ &+ 2\Delta W_n(x,a+2\Delta;\omega_n) + \cdots + \Delta W_n(x,b;\omega_n)] \end{aligned} \tag{5.8}$$

When we obtain spectral components $W_n(x;\omega_n)$ from either Equation 5.7 or Equation 5.8, the time-domain dynamic response $w(x, t)$ can be obtained by using the inverse-fast Fourier transform (FFT) algorithm as follows:

$$w(x,t) \Leftarrow IFFT\{W_n(x;\omega_n)|\ n = 0,\ 1,\ 2,\ldots,\ N-1\} \tag{5.9}$$

Equation 5.7 shows that we need $H_n(x, \xi; \omega_n)$ to compute spectral components $W_n(x; \omega_n)$ from either Equation 5.7 or Equation 5.8. Thus, we introduce the two-element method in the following for the computation of $H_n(x,\xi; \omega_n)$. To this end, we consider once again the same beam subjected to a point force $\Delta F_n(\xi; \omega_n)$ at $x = \xi$. As there exists a point force at $x = \xi$, we divide the beam into two segments as shown in Figure 5.1b to represent it by two spectral elements. Two spectral elements are connected at $x = \xi$ (node 2) at which the point force $\Delta F_n(\xi; \omega_n)$ is located. Assembling two spectral elements and then imposing associated boundary conditions, we can obtain a matrix equation as

$$\boldsymbol{S}_g(\xi;\omega_n)\Delta\boldsymbol{d}_{gn}(\xi) = \Delta\boldsymbol{f}_{gn}(\xi) \tag{5.10}$$

where

$$\begin{aligned} &\Delta\boldsymbol{d}_{gn} = \begin{Bmatrix} \Delta\boldsymbol{d}_{1n} \\ \Delta\boldsymbol{d}_{2n} \\ \Delta\boldsymbol{d}_{3n} \end{Bmatrix}, \qquad \Delta\boldsymbol{f}_{gn} = \begin{Bmatrix} \Delta\boldsymbol{f}_{1n} \\ \Delta\boldsymbol{f}_{2n} \\ \Delta\boldsymbol{f}_{3n} \end{Bmatrix} = \Delta F_n(\xi;\omega_n)\boldsymbol{J} \\ &\boldsymbol{J} = \{0 \quad \cdots \quad 1 \quad \cdots \quad 0\}^{\mathrm{T}} \end{aligned} \tag{5.11}$$

and where $\Delta\boldsymbol{d}_{in}$ $(i = 1, 2, 3)$ are the incremental spectral nodal DOFs at node i (remaining after imposing all boundary conditions) and $\Delta\boldsymbol{f}_{in}$ $(i = 1, 2, 3)$ are the incremental spectral nodal forces corresponding to $\Delta\boldsymbol{d}_{in}$. $\boldsymbol{J}$ is the vector for which all components are zero, except for just one component that corresponds to the nodal DOF with which the equivalent point force is associated. It is worth mentioning that $\boldsymbol{S}_g$ and $\Delta\boldsymbol{f}_{gn}$ are the functions of ξ (the location of point force) and thus $\Delta\boldsymbol{d}_{gn}$ also. From Equation 5.10, we can obtain the global incremental nodal DOFs as

$$\Delta\boldsymbol{d}_{gn}(\xi) = \boldsymbol{S}_g(\xi;\omega_n)^{-1}\Delta\boldsymbol{f}_{gn}(\xi) = \boldsymbol{S}_g(\xi;\omega_n)^{-1}\boldsymbol{J}\Delta F_n(\xi;\omega_n) \tag{5.12}$$

In the frequency domain, the dynamic response measured at x is given by Equation 3.55 as

$$W_n(x;\omega_n) = \boldsymbol{N}(x;\omega_n)\boldsymbol{d}_n \tag{5.13}$$

where $\boldsymbol{N}(x;\omega_n)$ is the dynamic shape function matrix and $\boldsymbol{d}_n$ is the local spectral nodal DOFs vector for the spectral element in which the measurement point x is located. The local spectral nodal DOFs vector $\boldsymbol{d}_n$ (and $\Delta\boldsymbol{d}_n$) can be related to the global nodal DOFs vector $\boldsymbol{d}_{gn}$ (and $\Delta\boldsymbol{d}_{gn}$) by introducing a locator matrix $\boldsymbol{L}_{x/\xi}$ as

$$\boldsymbol{d}_n = \boldsymbol{L}_{x/\xi}\boldsymbol{d}_{gn} \text{ and } \Delta\boldsymbol{d}_n = \boldsymbol{L}_{x/\xi}\Delta\boldsymbol{d}_{gn} \tag{5.14}$$

where, if $x \leq \xi$

$$\boldsymbol{d}_n = \begin{Bmatrix} \boldsymbol{d}_{1n} \\ \boldsymbol{d}_{2n} \end{Bmatrix}, \quad \boldsymbol{L}_{x/\xi} = \begin{bmatrix} \boldsymbol{I} & \boldsymbol{0} & \boldsymbol{0} \\ \boldsymbol{0} & \boldsymbol{I} & \boldsymbol{0} \end{bmatrix} \tag{5.15}$$

and, if $x \geq \xi$

$$\boldsymbol{d}_n = \begin{Bmatrix} \boldsymbol{d}_{2n} \\ \boldsymbol{d}_{3n} \end{Bmatrix}, \quad \boldsymbol{L}_{x/\xi} = \begin{bmatrix} \boldsymbol{0} & \boldsymbol{I} & \boldsymbol{0} \\ \boldsymbol{0} & \boldsymbol{0} & \boldsymbol{I} \end{bmatrix} \tag{5.16}$$

Equations 5.15 and 5.16 show that the locator matrix $\boldsymbol{L}_{x/\xi}$ depends on the location of measurement point x relative to the location of point force ξ. Substituting Equation 5.12 into the second part of Equation 5.14 gives

$$\Delta\boldsymbol{d}_n = \boldsymbol{L}_{x/\xi}\boldsymbol{S}_g(\xi;\omega_n)^{-1}\boldsymbol{J}\Delta F_n(\xi;\omega_n) \tag{5.17}$$

From Equations 5.13 and 5.17, we can obtain the incremental frequency response as

$$\Delta W_n(x;\omega_n) = \boldsymbol{N}(x;\omega_n)\Delta\boldsymbol{d}_n = \boldsymbol{N}(x;\omega_n)\boldsymbol{L}_{x/\xi}\boldsymbol{S}_g(\xi;\omega_n)^{-1}\boldsymbol{J}\Delta F_n(\xi;\omega_n) \tag{5.18}$$

Comparing Equation 5.18 with Equation 5.5, we can obtain

$$H_n(x,\xi;\omega_n) = \boldsymbol{N}(x;\omega_n)\boldsymbol{L}_{x/\xi}\boldsymbol{S}_g(\xi;\omega_n)^{-1}\boldsymbol{J} \tag{5.19}$$

Equation 5.19 provides the frequency response functions that are required to compute spectral components $W_n(x;\omega_n)$ from Equation 5.7. For Equation 5.19, the B-beam subjected to an equivalent dynamic point force is represented by two spectral elements: from here onwards, this approach is called the two-element method (TEM).

Example 5.1

As an illustrative example, consider a simply-supported B-beam subjected to a distributed dynamic force. The beam has the length $L = 2$ m, $EI = 10\,\text{kNm}^2$ and $\rho A = 10$ kg/m. The dynamic force of magnitude $F_0 = 1$ kN/m is uniformly distributed over the region $L/4 \leq x \leq 3L/4$ and it is applied from $t = 0$ to $t = t_0 = 0.1$ s. The vibration response at the middle of the beam is obtained by using four different methods:

a. TEM-based SEM: The analysis results obtained by representing the distributed dynamic force by N equivalent dynamic point forces are denoted by SEM-TEM(N).

b. One-element SEM model based on Equation 5.2: In this method, the whole beam is represented by one spectral element and the distributed dynamic force is represented by the spectral nodal forces computed from Equation 5.2. The analysis results are denoted by SEM(1)-Eq. (5.2).
c. Three-element SEM model based on Equation 5.2: In this method, the beam is represented by three spectral elements: the first element (without loading) in $0 \leq x \leq 1/4L$, the second (middle) element (with uniform loading) in $1/4L \leq x \leq 3/4L$, and the third element (without loading) in $3/4L \leq x \leq L$. The uniformly distributed dynamic force on the middle element is represented by the spectral nodal forces computed from Equation 5.2. The analysis results are denoted by SEM(3)-Eq. (5.2).
d. FEM: In the finite element method (FEM) the analysis results obtained by using N conventional finite elements are denoted by FEM(N). The finite element analysis is conducted in the frequency domain by transforming the finite element equation given in Appendix 5.A into the frequency domain as follows:

$$(\boldsymbol{K} - \omega^2 \boldsymbol{M})\boldsymbol{d} = \boldsymbol{f}$$

where $\boldsymbol{d}$ and $\boldsymbol{f}$ represent the spectral nodal DOFs and forces, respectively.
e. **Modal analysis method:** The results are denoted by MAM and they are considered as the reference (exact) solution to evaluate the accuracy of other solution methods. The dynamic response analytically obtained by MAM is given by

$$w(x,t) = \sum_{n=1}^{\infty} \phi_n(x) q_n(t)$$

where

$$\phi_n(x) = \sqrt{2} \sin \frac{n\pi x}{L}$$

$$\omega_n = \left(\frac{n\pi}{L}\right)^2 \sqrt{\frac{EI}{\rho A}}$$

$$q_n(t) = \frac{2\sqrt{2}L^4 F_0}{(n\pi)^5 EI} \sin\frac{n\pi}{2} \sin\frac{n\pi}{4} \{(1 - \cos\omega_n t) u(t) - [1 - \cos\omega_n (t - t_0)] u(t - t_0)\}$$

The function $u(t)$ is the unit step function. A sufficient number of modes are superposed to obtain fully converged, accurate dynamic responses by MAM.

Figure 5.2a compares the frequency-domain dynamic response obtained by using the TEM-based SEM with the exact reference solution obtained by MAM. Figure 5.2b shows that, as the number of equivalent point forces used to represent the distributed dynamic force is increased, the time-averaged dynamic response obtained by the TEM-based SEM indeed converges to the exact reference solution.

Figure 5.3 compares the frequency-domain dynamic responses obtained by the one- and three-element SEM models subjected to the spectral nodal forces computed from Equation 5.2 with the exact reference solution obtained by MAM. It is shown that the

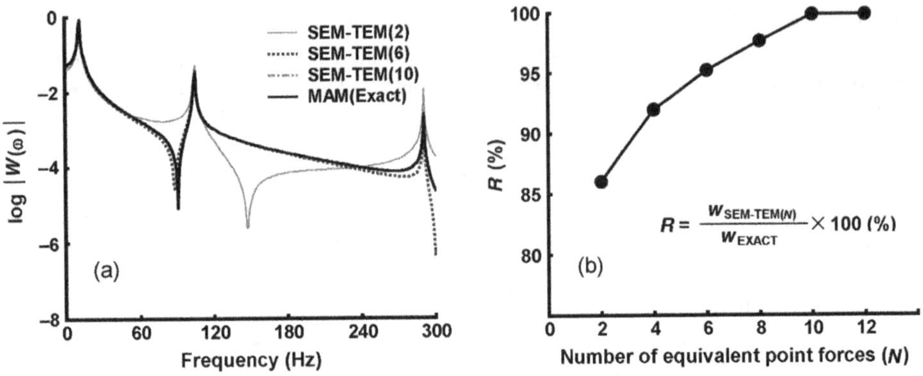

Figure 5.2 Comparison of the dynamic responses obtained by the TEM(N)-based SEM, where N is the number of equivalent point forces used in the analysis

three-element SEM model provides the better frequency-domain dynamic response when compared with the one-element SEM model. This implies that, when Equation 5.2 is used for the distributed forces acting on a structure member, we may need to mesh the structure member into more finite elements to improve the analysis results if there are discontinuities in the distributed forces.

Figure 5.4a compares the dynamic responses obtained by FEM with the exact solution obtained by MAM or three-element SEM model. The FEM results are obtained by varying the number of finite elements used for the finite element analysis. It is obvious from Figure 5.4b that the accuracy of the time-averaged dynamic response obtained by FEM improves as the number of finite elements is increased. However, the accuracy improvement of FEM seems to be restricted to a certain limit due to certain errors, such as the computational round-off errors and numerical errors, which are inevitable for most large-scale finite element models.

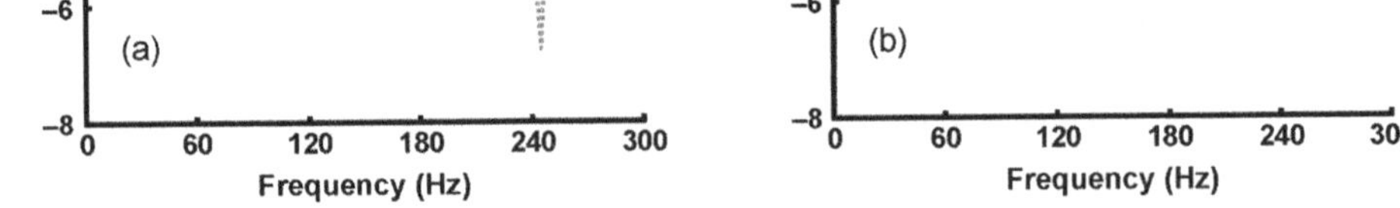

Figure 5.3 Comparison of the dynamic responses obtained by SEM(N), where N is the number of spectral elements used in the analysis

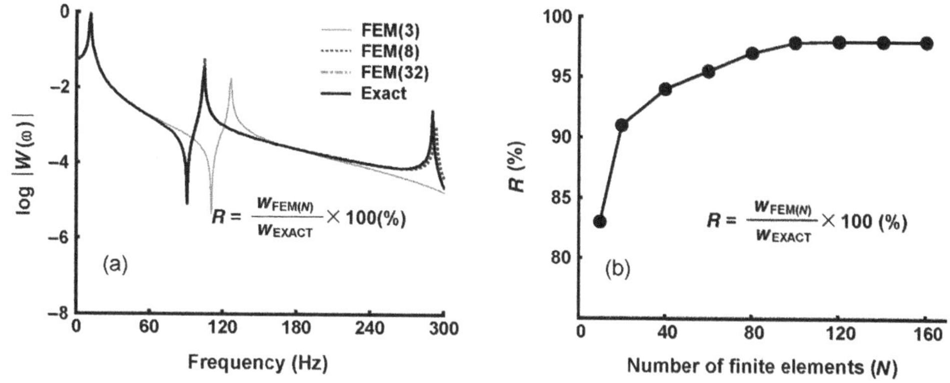

Figure 5.4 Comparison of the dynamic responses obtained by FEM(N), where N is the number of finite elements used in the analysis

5.2 Levy-Type Plates

Unlike the 1-D structures, two-dimensional (2-D) plate structures show much more complicated wave characteristics. Thus, the spectral element formulation for the plates with general boundary conditions will not be so straightforward and some approximations may be necessary. One of possible approaches is first to transform the 2-D problem (in $x-y$ domain) into an equivalent 1-D problem (in the x-domain) by determining the displacement functions of y in analytical forms so as to exactly, or approximately, satisfy the boundary conditions along two parallel sides at $y=0$ and $y=L_y$, and next to apply one of spectral element formulation methods discussed in Chapter 3 to the equivalent 1-D problem, where L_y is the dimension of the rectangular plate in the y-direction. The rectangular plates with at least two parallel sides simply supported are called Levy-type plates and it is well known in plate theory that exact solutions are available for the class of Levy-type plates. Thus the exact spectral element model will be introduced in this section for the Levy-type plates [92].

5.2.1 Equation of Motion

We will consider a thin rectangular Levy-type plate whose two opposite sides parallel to the x-axis are simply supported and the other two opposite sides parallel to the y-axis are subjected to arbitrary boundary conditions, as shown in Figure 5.5. The Levy-type plate has the dimensions

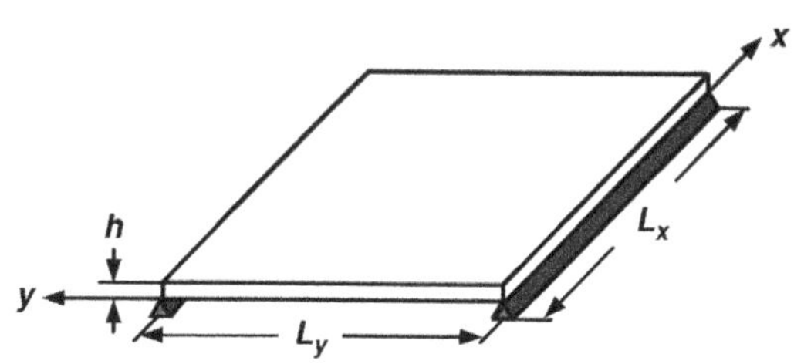

Figure 5.5 Levy-type rectangular plate

L_x and L_y in the x- and y-directions, respectively. The small transverse (out-of-plane) vibration $w(x, y, t)$ of a thin plate is represented by the classical plate equation

$$D\left(\frac{\partial^4 w}{\partial x^4}+2\frac{\partial^4 w}{\partial x^2\partial y^2}+\frac{\partial^4 w}{\partial y^4}\right)+\rho h\ddot{w}=f(x,y,t) \tag{5.20}$$

where $f(x, y, t)$ is the distributed dynamic force acting in the direction normal to the plate, h is the plate thickness, ρ is the mass density, and D is the bending rigidity of plate defined by $Eh^3/[12(1-\nu^2)]$, where E is Young's modulus and ν is the Poisson ratio. The resultant bending moments and transverse shear forces are given by

$$\begin{aligned}
M_{xx}(x,y,t) &= D\left(\frac{\partial^2 w}{\partial x^2}+\nu\frac{\partial^2 w}{\partial y^2}\right)\\
Q_{xz}(x,y,t) &= -D\left(\frac{\partial^3 w}{\partial x^3}+(2-\nu)\frac{\partial^3 w}{\partial x\partial y^2}\right)\\
M_{yy}(x,y,t) &= D\left(\frac{\partial^2 w}{\partial y^2}+\nu\frac{\partial^2 w}{\partial x^2}\right)\\
Q_{yz}(x,y,t) &= -D\left(\frac{\partial^2 w}{\partial y^3}+(2-\nu)\frac{\partial^3 w}{\partial x^2\partial y}\right)
\end{aligned} \tag{5.21}$$

The subscripts of the resultant bending moments and transverse shear forces indicate the directions of stresses that cause them. For the Levy-type plates, the simply supported boundary conditions at $y=0$ and $y=L_y$ are given by

$$\begin{aligned}
w(x,0,t)&=0, \quad M_{yy}(x,0,t)=0\\
w(x,L_y,t)&=0, \quad M_{yy}(x,L_y,t)=0
\end{aligned} \tag{5.22}$$

5.2.2 Spectral Element Modeling

Represent the dynamic response of the plate and dynamic force in the spectral forms as

$$\begin{aligned}
w(x,y,t) &= \frac{1}{N}\sum_{n=0}^{N-1} W_n(x,y;\omega_n)e^{i\omega_n t}\\
f(x,y,t) &= \frac{1}{N}\sum_{n=0}^{N-1} F_n(x,y;\omega_n)e^{i\omega_n t}
\end{aligned} \tag{5.23}$$

where $W_n(x, y)$ and $F_n(x, y)$ are the spectral components of $w(x, y, t)$ and $f(x, y, t)$, respectively, and N is the sampling number. Substitution of Equation 5.23 into Equation 5.20 gives

$$\nabla^4 W_n-\Omega_n^2 W_n=F_n(x,y;\omega_n) \tag{5.24}$$

where

$$\Omega_n^2=\omega_n\sqrt{\rho h/D} \tag{5.25}$$

Similarly, the resultant bending moments and transverse shear forces are written in the spectral forms as

$$\begin{aligned}
M_{xxn}(x,y;\omega_n) &= D\left(\frac{\partial^2 W_n}{\partial x^2}+\nu\frac{\partial^2 W_n}{\partial y^2}\right)\\
Q_{xzn}(x,y;\omega_n) &= -D\left(\frac{\partial^3 W_n}{\partial x^3}+(2-\nu)\frac{\partial^3 W_n}{\partial x\partial y^2}\right)\\
M_{yyn}(x,y;\omega_n) &= D\left(\frac{\partial^2 W_n}{\partial y^2}+\nu\frac{\partial^2 W_n}{\partial x^2}\right)\\
Q_{yzn}(x,y;\omega_n) &= -D\left(\frac{\partial^2 W_n}{\partial y^3}+(2-\nu)\frac{\partial^3 W_n}{\partial x^2\partial y}\right)
\end{aligned}\tag{5.26}$$

where M_{xxn}, Q_{xzn}, M_{yyn}, and Q_{yzn} are the spectral components of $M_{xx}(x, y, t)$, $Q_{xz}(x, y, t)$, $M_{yy}(x, y, t)$, and $Q_{yz}(x, y, t)$, respectively. The simply supported boundary conditions in Equation 5.22 at $y=0$ and $y=L_y$ are also written in the spectral forms as

$$\begin{aligned}
W_n(x,0;\omega_n)=0, \quad M_{yyn}(x,0;\omega_n)=0\\
W_n(x,L_y;\omega_n)=0, \quad M_{yyn}(x,L_y;\omega_n)=0
\end{aligned}\tag{5.27}$$

The homogeneous equation can be reduced from Equation 5.24 as

$$\nabla^4 W_n-\Omega_n^2 W_n=0 \tag{5.28}$$

Assume the general solution of Equation 5.28 in the form of

$$W_n(x,y;\omega_n)=X_n(x)Y_n(y) \tag{5.29}$$

By substituting Equation 5.29 into the boundary conditions of Equation 5.27, we can obtain the conditions for the displacement functions $Y_n(y)$ as

$$\begin{aligned}
Y_n(0)=0, \quad Y_n''(0)=0\\
Y_n(L_y)=0, \quad Y_n''(L_y)=0
\end{aligned}\tag{5.30}$$

Assume the functions $Y_n(y)$ in the form of

$$Y_n(y)=a_{n1}e^{-ik_{yn}y}+a_{n2}\,e^{-k_{yn}y}+a_{n3}e^{+ik_{yn}y}+a_{n4}\,e^{+k_{yn}y} \tag{5.31}$$

By applying the conditions (5.30) to Equation 5.31, we can obtain

$$a_{n1}=-a_{n3}, \quad a_{n2}=0, \quad a_{n4}=0 \tag{5.32}$$

and

$$k_{ynm}=\frac{m\pi}{L_y}\,(m=0,\ 1,\ 2,\ 3,\ \ldots) \tag{5.33}$$

Thus, the functions $Y_n(y)$ that satisfy the simply supported boundary conditions at $y=0$ and $y=L_y$ exactly can be obtained as

$$Y_n(y;k_{ynm})=\sum_{m=0}^{\infty}a_{nm3}(e^{ik_{ynm}y}-e^{-ik_{ynm}y})\equiv\sum_{m=0}^{\infty}b_{nm}\sin(k_{ynm}y) \tag{5.34}$$

where a_{nm3} and b_{nm} are constants. Assume that Equation 5.34 can be expressed in the spectral form as

$$Y_n(y; k_{ynm}) = \frac{1}{M}\sum_{m=0}^{M-1} Y_{nm} e^{-ik_{ynm}y} \tag{5.35}$$

where Y_{nm} are the constant spectral components of $Y_n(y)$ in the wavenumber k_y-domain and M is the sampling number.

Substituting Equation 5.35 into Equation 5.29 gives

$$W_n(x, y; \omega_n) = \frac{1}{M}\sum_{m=0}^{M-1} W_{nm}(x; k_{ynm}, \omega_n) e^{ik_{ynm}y} \tag{5.36}$$

where

$$W_{nm}(x; k_{ynm}, \omega_n) = X_n(x; k_{ynm}, \omega_n) Y_{nm} \tag{5.37}$$

$W_{nm}(x; k_{ynm}, \omega_n)$ are the spectral components of $W_n(x, y; \omega_n)$ in the wavenumber k_y-domain. Substitution of Equation 5.36 into Equation 5.28 gives

$$W_{nm}''''(x) - 2k_{ynm}^2 W_{nm}''(x) + k_{ynm}^4 - \Omega_n^2 W_{nm}(x) = 0 \tag{5.38}$$

or, by using Equation 5.37,

$$X_n''''(x) - 2k_{ynm}^2 X_n''(x) + k_{ynm}^4 - \Omega_n^2 X_n(x) = 0 \tag{5.39}$$

Assume the solution of Equation 5.38 or Equation 5.39 to be

$$W_{nm}(x) = c_{nm} e^{-ik_{xnm}x} \tag{5.40}$$

or

$$X_n(x) = \bar{c}_{nm} e^{-ik_{xnm}x} \tag{5.41}$$

where $c_{nm} = \bar{c}_{nm} Y_{nm}$. Substituting Equation 5.40 into Equation 5.38 or Equation 5.41 into Equation 5.39 yields the identical dispersion relation to be

$$k_{xnm}^4 + 2k_{ynm}^2 k_{xnm}^2 + k_{ynm}^4 - \Omega_n^2 = 0 \tag{5.42}$$

Equation 5.42 can be solved for the wavenumber k_x as

$$\begin{aligned} k_{xnm1} &= \sqrt{\Omega_n - k_{ynm}^2}, \quad k_{xnm2} = -\sqrt{\Omega_n - k_{ynm}^2} \\ k_{xnm3} &= i\sqrt{\Omega_n + k_{ynm}^2}, \quad k_{xnm4} = -i\sqrt{\Omega_n + k_{ynm}^2} \end{aligned} \tag{5.43}$$

By using four wavenumbers of Equation 5.43, the general solution to Equation 5.38 can be written as

$$W_{nm}(x; k_{ynm}, \omega_n) = c_{nm1} e^{-ik_{xnm1}x} + c_{nm2} e^{-ik_{xnm2}x} + c_{nm3} e^{-ik_{xnm3}x} + c_{nm4} e^{-ik_{xnm4}x} \tag{5.44}$$

or, simply

$$W_{nm}(x; k_{ynm}, \omega_n) = \boldsymbol{E}_{nm}(x; k_{ynm}, \omega_n) \boldsymbol{c}_{nm} \tag{5.45}$$

where

$$\boldsymbol{E}_{nm}(x; k_{ynm}, \omega_n) = [\mathrm{e}^{-ik_{xnm1}x} \quad \mathrm{e}^{-ik_{xnm2}x} \quad \mathrm{e}^{-ik_{xnm3}x} \quad \mathrm{e}^{-ik_{xnm4}x}] \tag{5.46}$$

and

$$\boldsymbol{c}_{nm} = \{c_{nm1} \quad c_{nm2} \quad c_{nm3} \quad c_{nm4}\}^{\mathrm{T}} \tag{5.47}$$

To formulate the spectral element equation for the Levy-type plate element as shown in Figure 5.5, the same procedure of spectral element formulation as used for the 1-D structures in Chapter 3 can be applied. Assume that the spectral nodal DOFs (in the frequency ω-domain) are specified on the boundary sides at $x=0$ and $x=L_x$ as follows:

$$\boldsymbol{d}_n(\omega_n) \equiv \begin{Bmatrix} W_{1n}(y;\omega_n) \\ \Theta_{1n}(y;\omega_n) \\ W_{2n}(y;\omega_n) \\ \Theta_{2n}(y;\omega_n) \end{Bmatrix} = \begin{Bmatrix} W_n(0,y;\omega_n) \\ W_n'(0,y;\omega_n) \\ W_n(L_x,y;\omega_n) \\ W_n'(L_x,y;\omega_n) \end{Bmatrix} \tag{5.48}$$

where the prime denotes the derivative with respect to x. Represent W_{1n}, Θ_{1n}, W_{2n}, and Θ_{2n} in the spectral forms in the wavenumber k_y-domain as

$$\{W_{1n}, \Theta_{1n}, W_{2n}, \Theta_{2n}\} = \frac{1}{M}\sum_{m=0}^{M-1}\{W_{1nm}, \Theta_{1nm}, W_{2nm}, \Theta_{2nm}\}\mathrm{e}^{-ik_{ynm}y} \tag{5.49}$$

By substituting Equation 5.36 and 5.49 into Equation 5.48, we can get

$$\boldsymbol{d}_{nm}(k_{ynm},\omega_n) \equiv \begin{Bmatrix} W_{1nm}(k_{ynm},\omega_n) \\ \Theta_{1nm}(k_{ynm},\omega_n) \\ W_{2nm}(k_{ynm},\omega_n) \\ \Theta_{2nm}(k_{ynm},\omega_n) \end{Bmatrix} = \begin{Bmatrix} W_{nm}(0;k_{ynm},\omega_n) \\ W_{nm}'(0;k_{ynm},\omega_n) \\ W_{nm}(L_x;k_{ynm},\omega_n) \\ W_{nm}'(L_x;k_{ynm},\omega_n) \end{Bmatrix} \tag{5.50}$$

Applying Equation 5.45 into the right-hand side of Equation 5.50 gives

$$\boldsymbol{d}_{nm}(k_{ynm},\omega_n) = \boldsymbol{\Phi}_{nm}(k_{ynm},\omega_n)\boldsymbol{c}_{nm} \tag{5.51}$$

where

$$\boldsymbol{\Phi}_{nm}(k_{ynm},\omega_n) = \begin{bmatrix} 1 & 1 & 1 & 1 \\ -ik_{xnm1} & -ik_{xnm2} & -ik_{xnm3} & -ik_{xnm4} \\ \mathrm{e}_{nm1} & \mathrm{e}_{nm2} & \mathrm{e}_{nm3} & \mathrm{e}_{nm4} \\ -ik_{xnm1}\mathrm{e}_{nm1} & -ik_{xnm2}\mathrm{e}_{nm2} & -ik_{xnm3}\mathrm{e}_{nm3} & -ik_{xnm4}\mathrm{e}_{nm4} \end{bmatrix} \tag{5.52}$$

with

$$\mathrm{e}_{nmj} = \mathrm{e}^{-ik_{xnmj}L_x} \quad (j = 1,\ 2,\ 3,\ 4) \tag{5.53}$$

By substituting Equation 5.36 into the first two equations of Equation 5.26, we obtain

$$\begin{aligned} M_{xxn}(x,y;\omega_n) &= \frac{1}{M}\sum_{m=0}^{M-1} M_{xxnm}(x;k_{ynm},\omega_n)\mathrm{e}^{-ik_{ynm}y} \\ Q_{xzn}(x,y;\omega_n) &= \frac{1}{M}\sum_{m=0}^{M-1} Q_{xznm}(x;k_{ynm},\omega_n)\mathrm{e}^{-ik_{ynm}y} \end{aligned} \tag{5.54}$$

where

$$\begin{aligned} M_{xxnm}(x; k_{ynm}, \omega_n) &= D(W''_{nm} - \nu k_{ynm}^2 W_{nm}) \\ Q_{xznm}(x; k_{ynm}, \omega_n) &= -D[W'''_{nm} - (2-\nu)k_{ynm}^2 W'_{nm}] \end{aligned} \tag{5.55}$$

Represent the spectral components (in the frequency ω-domain) of the resultant transverse shear force and bending moment specified on the side at $x = 0$ by Q_{1n} and M_{1n} and those on the side at $x = L_x$ by Q_{2n} and M_{2n}. Then, by using the sign convention for the resultant transverse shear force and bending moment, we can write

$$f_n(\omega_n) \equiv \begin{Bmatrix} Q_{1n}(y; \omega_n) \\ M_{1n}(y; \omega_n) \\ Q_{2n}(y; \omega_n) \\ M_{2n}(y; \omega_n) \end{Bmatrix} = \begin{Bmatrix} -Q_{xzn}(0, y; \omega_n) \\ +M_{xxn}(0, y; \omega_n) \\ -Q_{xzn}(L_x, y; \omega_n) \\ +M_{xxn}(L_x, y; \omega_n) \end{Bmatrix} \tag{5.56}$$

Represent Q_{1n}, M_{1n}, Q_{2n}, and M_{2n} in the spectral forms in the wavenumber k_y-domain as

$$\{Q_{1n}, M_{1n}, Q_{2n}, M_{2n}\} = \frac{1}{M} \sum_{m=0}^{M-1} \{Q_{1nm}, M_{1nm}, Q_{2nm}, M_{2nm}\} \mathrm{e}^{-ik_{ynm}y} \tag{5.57}$$

By substituting Equations 5.54 and 5.57 into Equation 5.56, we can obtain

$$f_{nm}(k_{ynm}, \omega_n) \equiv \begin{Bmatrix} Q_{1nm}(k_{ynm}, \omega_n) \\ M_{1nm}(k_{ynm}, \omega_n) \\ Q_{2nm}(k_{ynm}, \omega_n) \\ M_{2nm}(k_{ynm}, \omega_n) \end{Bmatrix} = \begin{Bmatrix} -Q_{xznm}(0; k_{ynm}, \omega_n) \\ +M_{xxnm}(0; k_{ynm}, \omega_n) \\ -Q_{xznm}(L_x; k_{ynm}, \omega_n) \\ +M_{xxnm}(L_x; k_{ynm}, \omega_n) \end{Bmatrix} \tag{5.58}$$

Substitution of Equation 5.45 into Equation 5.55 gives

$$\begin{aligned} M_{xxnm}(x; k_{ynm}, \omega_n) &= D[\boldsymbol{E}''_{nm}(x) - \nu k_{ynm}^2 \boldsymbol{E}_{nm}(x)]\boldsymbol{c}_{nm} \\ Q_{xznm}(x; k_{ynm}, \omega_n) &= -D[\boldsymbol{E}'''_{nm}(x) - (2-\nu)k_{ynm}^2 \boldsymbol{E}'_{nm}(x)]\boldsymbol{c}_{nm} \end{aligned} \tag{5.59}$$

Applying Equation 5.59 into the right-hand side of Equation 5.58 gives

$$f_{nm}(k_{ynm}, \omega_n) = D \begin{bmatrix} \boldsymbol{E}'''_{nm}(0) - (2-\nu)k_{ynm}^2 \boldsymbol{E}'_{nm}(0) \\ \boldsymbol{E}''_{nm}(0) - \nu k_{ynm}^2 \boldsymbol{E}_{nm}(0) \\ \boldsymbol{E}'''_{nm}(L_x) - (2-\nu)k_{ynm}^2 \boldsymbol{E}'_{nm}(L_x) \\ \boldsymbol{E}''_{nm}(L_x) - \nu k_{ynm}^2 \boldsymbol{E}_{nm}(L_x) \end{bmatrix} \boldsymbol{c}_{nm} \tag{5.60}$$

By using Equation 5.46, Equation 5.60 can be written as

$$f_{nm}(k_{ynm}, \omega_n) = \boldsymbol{G}_{nm}(k_{ynm}, \omega_n)\boldsymbol{c}_{nm} \tag{5.61}$$

where

$$\boldsymbol{G}_{nm} = D[g_{ij}] \qquad (i, j = 1, 2, 3, 4) \tag{5.62}$$

with

$$\begin{aligned} g_{1j} &= ik_{xnmj}[k_{xnmj}^2 + (2-\nu)k_{ynm}^2] \\ g_{2j} &= -(k_{xnmj}^2 + \nu k_{ynm}^2) \\ g_{3j} &= g_{1j}e_{nmj} \\ g_{4j} &= g_{4j}e_{nmj} \end{aligned} \qquad (j = 1, 2, 3, 4) \tag{5.63}$$

The constant vector $\boldsymbol{c}_{nm}$, which is equal to $\bar{\mathbf{c}}_{nm}Y_{nm}$, can be eliminated from Equation 5.61 by using Equation 5.51 to obtain the spectral element equation for the Levy-type thin plate element as

$$\boldsymbol{S}(k_{ynm}, \omega_n)\boldsymbol{d}_{nm} = \boldsymbol{f}_{nm} \tag{5.64}$$

where

$$\boldsymbol{S}(k_{ynm}, \omega_n) = \boldsymbol{G}_{nm}(k_{ynm}, \omega_n)\Phi_{nm}^{-1}(k_{ynm}, \omega_n) \tag{5.65}$$

The matrix $S(k_{ynm}, \omega_n)$ is the 4-by-4 spectral element matrix that is the function of frequency ω_n and wavenumber k_{ynm}. The spectral element equations for the Levy-type thin plate elements connected in the x-direction can be assembled in an analogous way to that used for the 1-D structures to form a global system equation.

5.2.3 Equivalent 1-D Structure Representation

Figure 5.6a shows a Levy-type plate subjected to a distributed dynamic force $f(x, y, t)$. The distributed dynamic force can be represented by a series of equivalent line forces that are parallel to the y-axis. If the incremental dynamic responses excited by individual equivalent line forces are available, the complete dynamic response excited by the whole distributed

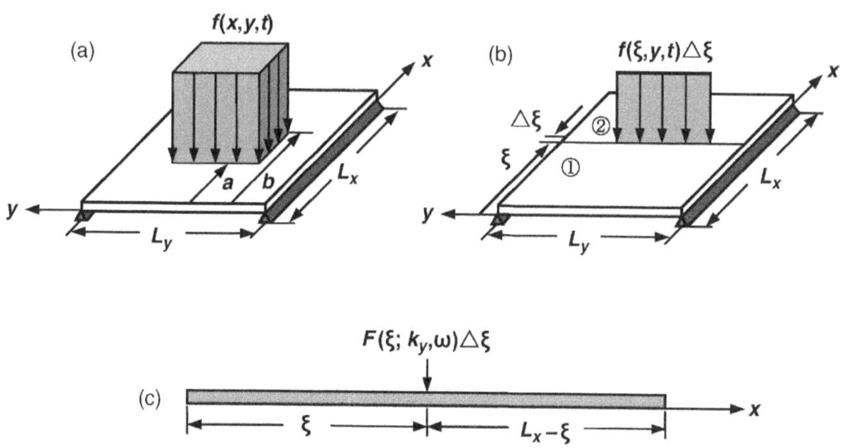

Figure 5.6 (a) A Levy-type plate subjected to a distributed dynamic force; (b) two plate-elements model with an equivalent line force, and (c) the equivalent 1-D structure representation

dynamic force can be obtained by summing all incremental dynamic responses through the use of the linear superposition principle.

For the computation of the incremental dynamic response excited by an equivalent line force, as shown in Figure 5.6b, we divide the plate into two plate elements so that the inter-connecting node line coincides with the line on which the equivalent line force is applied. The equivalent line force on $x=\xi$ has the magnitude of $f(\xi, y, t)\Delta\xi$. The equivalent line force can be discrete Fourier transformed into the frequency ω-domain and wavenumber k_y-domain to obtain

$$\begin{aligned} \Delta F_n &= F_n(\xi, y; \omega_n)\Delta\xi \Leftarrow FFT_\omega\{f(\xi, y, t)\Delta\xi\} \\ \Delta F_{nm} &= F_{nm}(\xi; k_{ynm}, \omega_n)\Delta\xi \Leftarrow FFT_{k_y}\{F_n(\xi, y; \omega_n)\Delta\xi\} \end{aligned} \tag{5.66}$$

This procedure removes the y-dependence of the equivalent line force to transform the 2-D problem into an equivalent beam-like 1-D problem, as illustrated in Figure 5.6c. We benefit from this approach in that the two-element method developed for the beams in the previous section can be equally applied to the Levy-type plate vibration.

5.2.4 Computation of Dynamic Responses

Consider a plate subjected to a typical equivalent line force. The equivalent line force $f(\xi,y,t)\Delta\xi$ is acting on the inter-connecting node line at $x=\xi$ as shown in Figure 5.6b and its spectral components ΔF_{nm} are given by Equation 5.66. Assembling two spectral plate elements and imposing the boundary conditions at $x=0$ and L_x, we can obtain the global system equation in the form

$$\mathbf{S}_g \Delta\mathbf{d}_{gnm} = \Delta\mathbf{f}_{gnm} \tag{5.67}$$

where

$$\Delta\mathbf{f}_{gnm} = \Delta F_{nm}\mathbf{J} \tag{5.68}$$

with

$$\begin{aligned} \Delta F_{nm} &= F_{nm}(\xi, k_{ynm}, \omega_n)\Delta\xi \\ \mathbf{J} &= \{0 \quad 0 \quad 1 \quad 0 \quad 0 \quad 0\}^{\mathrm{T}} \end{aligned} \tag{5.69}$$

Once ΔF_{nm} or $\Delta\mathbf{f}_{gnm}$ are computed, $\Delta\mathbf{d}_{gnm}$ can be computed from Equation 5.67 as

$$\Delta\mathbf{d}_{gnm}(\xi, k_{ynm}, \omega_n) = \mathbf{S}_g^{-1}\Delta\mathbf{f}_{gnm} = \mathbf{S}_g^{-1}\mathbf{J}\Delta F_{nm}(\xi, k_{ynm}, \omega_n) \tag{5.70}$$

The local spectral nodal DOFs vector $\mathbf{d}_{nm}$ (and $\Delta\mathbf{d}_{nm}$) can be related to the global nodal DOFs vector $\mathbf{d}_{gnm}$ (and $\Delta\mathbf{d}_{gnm}$) by introducing a locator matrix $\mathbf{L}_{x/\xi}$ as

$$\begin{aligned} \mathbf{d}_{nm}(\xi, k_{ynm}, \omega_n) &= \mathbf{L}_{x/\xi}\mathbf{d}_{gnm}(\xi, k_{ynm}, \omega_n) \\ \Delta\mathbf{d}_{nm}(\xi, k_{ynm}, \omega_n) &= \mathbf{L}_{x/\xi}\Delta\mathbf{d}_{gnm}(\xi, k_{ynm}, \omega_n) \end{aligned} \tag{5.71}$$

where, if $x \le \xi$

$$\mathbf{d}_{nm} = \begin{Bmatrix} \mathbf{d}_{1nm} \\ \mathbf{d}_{2nm} \end{Bmatrix}, \quad \mathbf{L}_{x/\xi} = \begin{bmatrix} \mathbf{I} & \mathbf{0} & \mathbf{0} \\ \mathbf{0} & \mathbf{I} & \mathbf{0} \end{bmatrix} \tag{5.72}$$

and, if $x \geq \xi$

$$\boldsymbol{d}_{nm} = \begin{Bmatrix} \boldsymbol{d}_{2nm} \\ \boldsymbol{d}_{3nm} \end{Bmatrix}, \quad \boldsymbol{L}_{x/\xi} = \begin{bmatrix} \boldsymbol{0} & \boldsymbol{I} & \boldsymbol{0} \\ \boldsymbol{0} & \boldsymbol{0} & \boldsymbol{I} \end{bmatrix} \tag{5.73}$$

The locator matrix $\boldsymbol{L}_{x/\xi}$ depends on the location of measurement point (x) relative to the location of equivalent line force (ξ). Substituting Equation 5.70 into Equation 5.71b gives

$$\Delta \boldsymbol{d}_{nm}(\xi, k_{ynm}, \omega_n) = \boldsymbol{L}_{x/\xi} \boldsymbol{S}_g^{-1} \boldsymbol{J} \Delta F_{nm}(\xi, k_{ynm}, \omega_n) \tag{5.74}$$

By using Equation 5.51, Equation 5.45 can be written as

$$W_{nm}(x, \xi; k_{ynm}, \omega_n) = \boldsymbol{N}_{nm}(x; k_{ynm}, \omega_n) \boldsymbol{d}_{nm}(\xi, k_{ynm}, \omega_n) \tag{5.75}$$

where

$$\boldsymbol{N}_{nm}(x; k_{ynm}, \omega_n) = \boldsymbol{E}_{nm}(x; k_{ynm}, \omega_n) \Phi_{nm}^{-1}(k_{ynm}, \omega_n) \tag{5.76}$$

From Equation 5.75, we can get

$$\Delta W_{nm}(x, \xi; k_{ynm}, \omega_n) = \boldsymbol{N}_{nm}(x; k_{ynm}, \omega_n) \Delta \boldsymbol{d}_{nm}(\xi, k_{ynm}, \omega_n) \tag{5.77}$$

Substituting Equation 5.74 into Equation 5.77 and using Equation 5.66, we obtain

$$\Delta W_{nm}(x, \xi; k_{ynm}, \omega_n) = H_{nm}(x; k_{ynm}, \omega_n) F_{nm}(\xi; k_{ynm}, \omega_n) \Delta \xi \tag{5.78}$$

where

$$H_{nm}(x, \xi; k_{ynm}, \omega_n) = \boldsymbol{N}_{nm}(x; k_{ynm}, \omega_n) \boldsymbol{L}_{x/\xi} \boldsymbol{S}_g^{-1} \boldsymbol{J} \tag{5.79}$$

Equation 5.78 represents the spectral components of the incremental vibration response excited by an equivalent line force acting on $x = \xi$. The spectral components of the complete vibration response excited by the whole distributed dynamic force can then be obtained by superposing all incremental responses as follows:

$$\begin{aligned} W_{nm}(x; k_{ynm}, \omega_n) &= \int \Delta W_{nm}(x, \xi; k_{ynm}, \omega_n) \\ &= \int_a^b H_{nm}(x, \xi; k_{ynm}, \omega_n) F_{nm}(\xi; k_{ynm}, \omega_n) \mathrm{d}\xi \end{aligned} \tag{5.80}$$

The integration of Equation 5.80 can be conducted by using a numerical integration method. When the trapezoidal algorithm is applied, for instance, Equation 5.80 can be written in the form of

$$\begin{aligned} W_{nm}(x; k_{ynm}, \omega_n) = \frac{1}{2} \Delta\xi \big[& \Delta W_{nm}(x, a; k_{ynm}, \omega_n) + \Delta W_{nm}(x, a + \Delta\xi; k_{ynm}, \omega_n) \\ & + 2\Delta W_{nm}(x, a + 2\Delta\xi; k_{ynm}, \omega_n) + \cdots + \Delta W_{nm}(x, b; k_{ynm}, \omega_n) \big] \end{aligned} \tag{5.81}$$

Once the spectral components W_{nm} are computed from either Equation 5.80 or Equation 5.81, the time history of the complete dynamic response can be obtained by using the inverse-FFT algorithm, first with respect to wavenumber k_{ynm}, and next with respect to frequency ω_n as follows:

$$\begin{aligned} W_n(x, y; \omega_n) &\Leftarrow IFFT_{k_y}\{W_{nm}(x; k_{ynm}, \omega_n) | m = 0, 1, 2, \ldots, M-1\} \\ w(x, y, t) &\Leftarrow IFFT_{\omega}\{W_n(x, y; \omega_n) | n = 0, 1, 2, \ldots, N-1\} \end{aligned} \tag{5.82}$$

Example 5.2

Consider a simply supported square thin plate as shown in Figure 5.7. The square plate has the side $L = 2$ m, the mass density $\rho h = 150\ \mathrm{kg/m^2}$, and the bending rigidity of plate $D = 150$ kN. To take into account the structural damping, the complex bending rigidity of plate $D^* = D(1 + i\eta)$ is used for computation, where $\eta = 0.03$ is the loss factor. Two types of dynamic forces are considered. Figure 5.7a shows the plate subjected to a point dynamic force of magnitude $F_0 = 50$ kN, which is applied at the middle of the plate. Figure 5.7b shows the plate subjected to a dynamic force uniformly distributed over a square area centered at the middle of the plate. The magnitude of the force is $F_0 = 50\ \mathrm{kN/m^2}$ and the square area has the side $l = 1$ m. Both the point and distributed dynamic forces are applied from $t = 0$ to $t_0 = 0.1$ s.

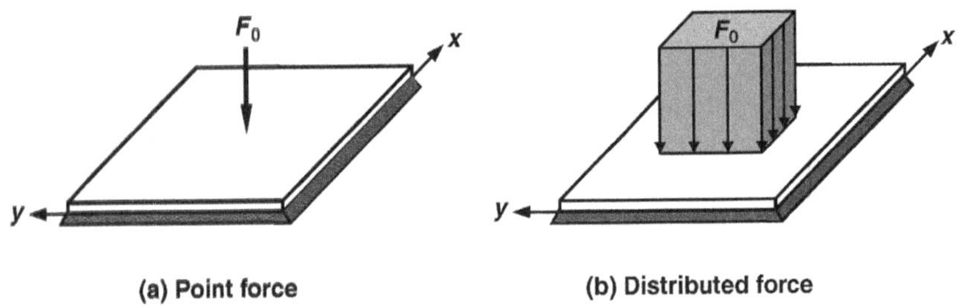

Figure 5.7 A simply supported square plate subjected to (a) a point dynamic force and (b) a partially distributed dynamic force

The dynamic responses analytically obtained by using the modal analysis method (MAM) for both the point force and the distributed dynamic force are given by

$$w(x, y, t) = \sum_{m=1}^{\infty}\sum_{n=1}^{\infty} \phi_{mn}(x, y) q_{mn}(t)$$

where

$$\phi_{mn} = \frac{2}{L\sqrt{\rho h}} \sin\frac{m\pi x}{L} \sin\frac{n\pi y}{L}, \qquad \omega_{mn} = \sqrt{\frac{D}{\rho h}}\left[\left(\frac{m\pi}{L}\right)^2 + \left(\frac{n\pi}{L}\right)^2\right]$$

$$q_{mn}(t) = \begin{cases} \dfrac{F_0}{\omega_{mn}^2} \dfrac{2}{L\sqrt{\rho h}} \sin\dfrac{m\pi}{2} \sin\dfrac{n\pi}{2} \times & \text{(point force)} \\ \quad \{(1 - \cos\omega_{mn} t) u(t) - [1 - \cos\omega_{mn}(t - t_0)] u(t - t_0)\} & \\ \dfrac{F_0}{\omega_{mn}^2} \dfrac{1}{\sqrt{\rho h}} \dfrac{8L}{mn\pi^2} \sin\dfrac{m\pi}{2} \sin\dfrac{m\pi}{4} \sin\dfrac{n\pi}{2} \sin\dfrac{n\pi}{4} \times & \text{(distributed force)} \\ \quad \{(1 - \cos\omega_{mn} t) u(t) - [1 - \cos\omega_{mn}(t - t_0)] u(t - t_0)\} & \end{cases}$$

where $u(t)$ is the unit step function. A sufficient number of modes are superposed to obtain sufficiently accurate dynamic responses by MAM. For the FEM results, the 4-node 12-DOF nonconforming rectangular plate bending element is used. The finite element stiffness and mass matrices are available in Ref. [139].

Figures 5.8 and 5.9 compare the frequency-domain dynamic responses obtained by SEM and FEM with the exact reference solution obtained by MAM. Figure 5.10a compares the time-domain undamped dynamic response with damped dynamic response obtained by the TEM-based SEM for the case of the point dynamic force, while Figure 5.10b shows the case of the distributed dynamic force. Finally, the solution accuracy and central processing unit (CPU) times are compared in Figure 5.11 for the TEM-based SEM as a dependence on the number of equivalent line forces and for the FEM as a dependence on the number of finite elements used in the analysis.

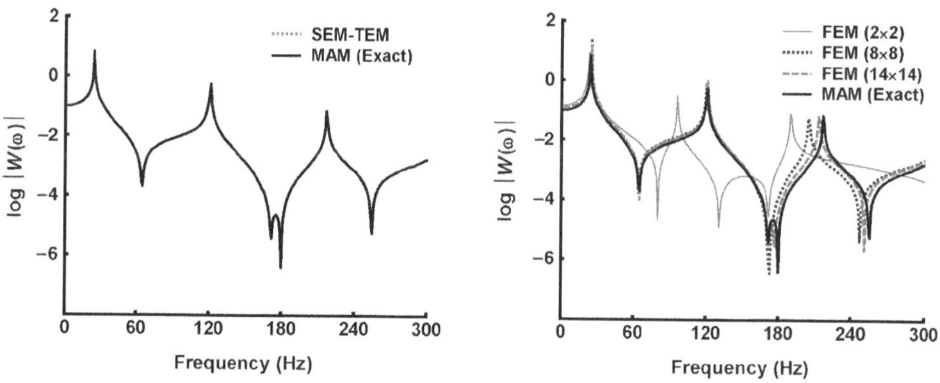

Figure 5.8 Comparison of the dynamic responses excited by a point dynamic force obtained by the TEM-based SEM and the FEM(N), where N is the number of finite elements used in the analysis

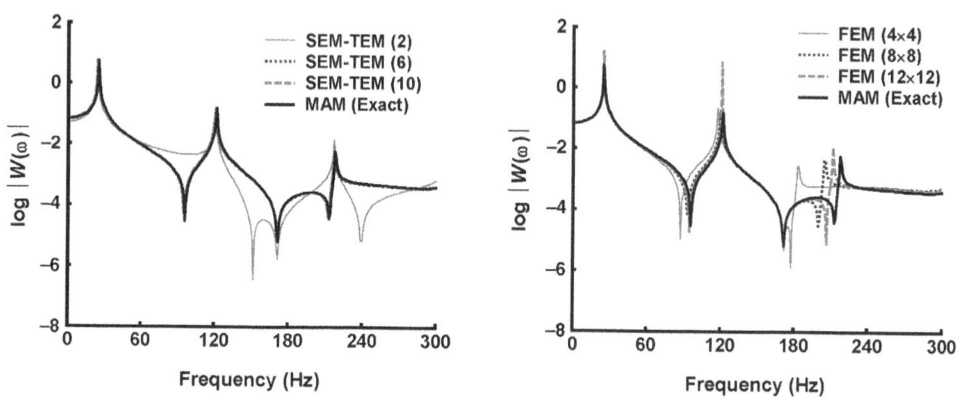

Figure 5.9 Comparison of the dynamic responses excited by a distributed dynamic force obtained by TEM(M)-based SEM and FEM(N), where M is the number of equivalent line forces and N is the number of finite elements used in the analyses

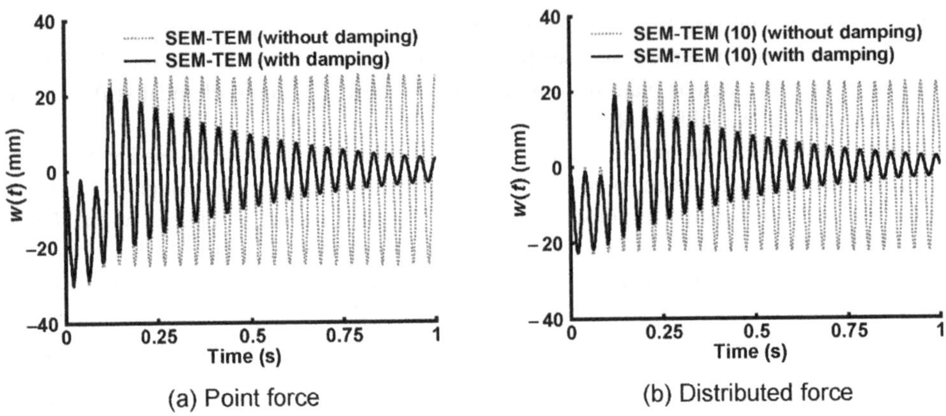

Figure 5.10 Comparison of the time-domain undamped and damped dynamic responses obtained by TEM-based SEM for the point and distributed dynamic forces

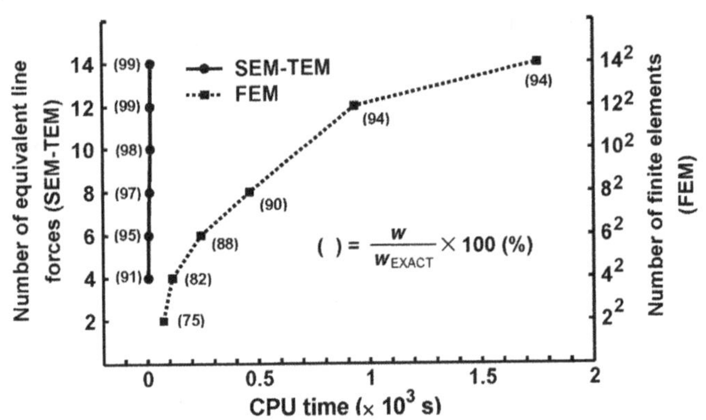

Figure 5.11 Comparison of the solution accuracy and CPU times of the TEM-based SEM and FEM

Appendix 5.A: Finite Element Model of Bernoulli–Euler Beam

The finite element equation for the Bernoulli–Euler beam model is given by

$$\boldsymbol{M}\ddot{\boldsymbol{d}} + \boldsymbol{K}\boldsymbol{d} = \boldsymbol{f}(t) \tag{5.A.1}$$

where

$$\begin{aligned} \boldsymbol{d}(t) &= \{w_1(t) \quad \theta_1(t) \quad w_2(t) \quad \theta_2(t)\}^{\mathrm{T}} \\ \boldsymbol{f}(t) &= \{Q_1(t) \quad M_1(t) \quad Q_2(t) \quad M_2(t)\}^{\mathrm{T}} \end{aligned} \tag{5.A.2}$$

The finite element mass matrix $\boldsymbol{M}$ and stiffness matrix $\boldsymbol{K}$ can be formulated by using the displacement fields

$$w(x,t) = \boldsymbol{N}(x)\boldsymbol{d}(t), \quad \theta(x,t) = w'(x,t) = \boldsymbol{N}'(x)\boldsymbol{d}(t) \tag{5.A.3}$$

where

$$N(x) = [1 - 3(x/L)^2 + 2(x/L)^3 \quad x - 2L(x/L)^2 + L(x/L)^3 \quad 3(x/L)^2 - 2(x/L)^3$$
$$- L(x/L)^2 + L(x/L)^3] \tag{5.A.4}$$

The finite element mass matrix $\boldsymbol{M}$ and stiffness matrix $\boldsymbol{K}$ are given by

$$\boldsymbol{M} = \frac{\rho AL}{420}\begin{bmatrix} 156 & 22L & 54 & -13L \\ 22L & 4L^2 & 13L & -3L^2 \\ 54 & 13L & 156 & -22L \\ -13L & -3L^2 & -22L & 4L^2 \end{bmatrix} \tag{5.A.5}$$

$$\boldsymbol{K} = \frac{EI}{L^3}\begin{bmatrix} 12 & 6L & -12 & 6L \\ 6L & 4L^2 & -6L & 2L^2 \\ -12 & -6L & 12 & -6L \\ 6L & 2L^2 & -6L & 4L^2 \end{bmatrix} \tag{5.A.6}$$

6

Flow-Induced Vibrations of Pipelines

The pipelines in chemical plants, oil pipelines, pump discharges, propellant fluid lines of liquid-filled rockets, and the human circulatory system are typical examples of the transfer of high velocity pressured fluids. The time-varying conditions imposed by pump or valve operations may change the internal fluid fields. As the internal fluid interacts with the pipe wall, the changes in the fluid fields will keep changing the dynamic behavior of a pipeline system and vice versa. Sometimes such a fluid–structure interaction phenomenon may result in serious vibrations, which can cause structural failures [8, 133]. Therefore, it is important to accurately predict the structural dynamic characteristics of a pipeline and the fluid transients inside the pipeline during the early design phase. To this end, we need to take into account more realistic fluid–structure interaction mechanism in the pipe-dynamics theory [102, 166]. In general, the structure–fluid interactions are represented by a set of coupled governing equations for the dynamics of the pipeline and for the fluid transients inside the pipeline. Thus, this chapter introduces the spectral element models [93, 95] formulated from the coupled pipe-dynamics equations of the uniform pipelines conveying internal fluids derived by using Hamilton's principle and the theory of fluid mechanics.

6.1 Theory of Pipe Dynamics

We will consider a straight uniform pipeline having a small amplitude vibration as shown in Figure 6.1. The pipeline is initially inclined at angle θ with respect to the ground. In Figure 6.1, (X, Z) represent the reference coordinates system fixed on the ground and (x, z) represent the local coordinates system fixed on the inclined pipeline. The transverse and axial displacements of the pipeline with respect to the local coordinates system (x, z) are represented by $w(x, t)$ and $u(x, t)$, respectively. Figure 6.2 shows an infinitesimal pipe–fluid element that is separated into the pipeline element and the fluid element (or control volume). In Figure 6.2b, $c(x, t)$ is the average velocity of the fluid flow and $p(x, t)$ is the average fluid pressure. The flow-induced resultant normal and tangential forces acting on the pipe wall are represented by F_N and F_T,

Spectral Element Method in Structural Dynamics Usik Lee

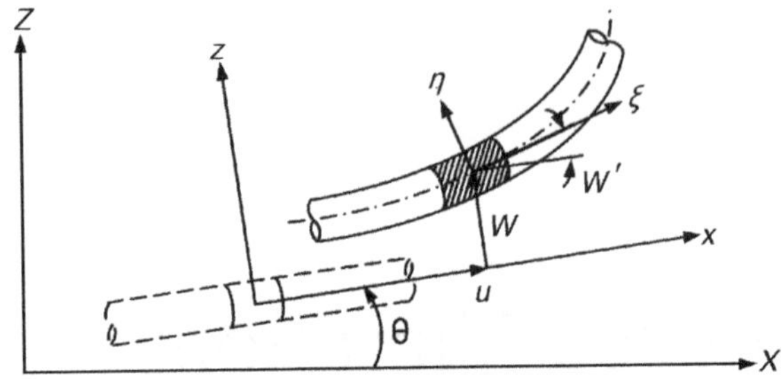

Figure 6.1 Vibration of an inclined uniform pipeline

respectively. The Hamilton principle is used to derive the equations of motion of the pipeline, while Newton's second law and the law of mass conservation are used to derive the fluid-dynamic equations for the internal fluid.

6.1.1 Equations of Motion of the Pipeline

Hamilton's principle is given by

$$\int_{t_1}^{t_2} (\delta T - \delta V + \delta W)\mathrm{d}t = 0 \tag{6.1}$$

where T and V are the kinetic and strain energies of the pipeline element given by

$$T = \frac{1}{2} m_{\mathrm{p}} \int_0^L (\dot{u}^2 + \dot{w}^2)\mathrm{d}x \tag{6.2}$$

and

$$V = \frac{1}{2}\int_0^L EA_{\mathrm{p}}(\varepsilon_{T_0} + \varepsilon_x)^2 \mathrm{d}x + \frac{1}{2}\int_0^L EI_{\mathrm{p}} w''^2 \mathrm{d}x \tag{6.3}$$

where

$$\varepsilon_{T_0} = \frac{\bar{N}}{EA_{\mathrm{p}}}, \quad \varepsilon_x = u' + \frac{1}{2}u'^2 + \frac{1}{2}w'^2 \tag{6.4}$$

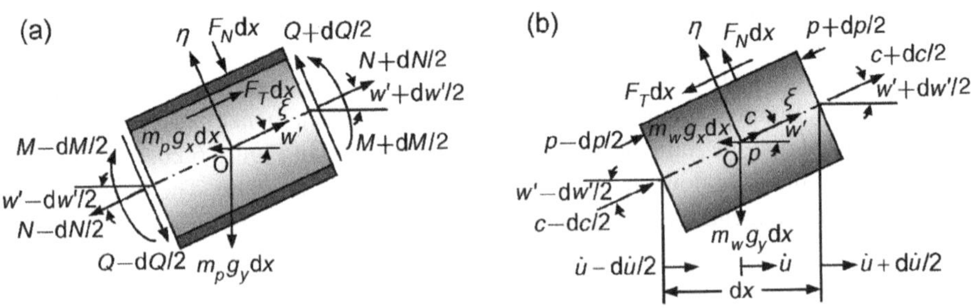

Figure 6.2 Free-body diagrams for (a) pipeline element and (b) fluid element

In the above equations, E, L, A_p, m_p, and I_p are Young's modulus, the length of pipeline element, the cross-sectional area of pipeline, the pipeline mass per unit length, and the second moment of the cross-sectional area of pipeline, respectively. The first integral of Equation 6.3 represents the stain energy due to the strains generated by a constant axial tensile force $\bar{N}$ and the longitudinal motion, and the second integral represents the strain energy due to the bending motion. By substituting Equation 6.4 into Equation 6.3, neglecting some high-order terms, and ignoring a constant term that will not eventually appear in Hamilton's principle, we obtain the strain energy as

$$V = \frac{1}{2}\int_0^L [\bar{N}(2u' + u'^2 + w'^2) + EA_p u'^2]\mathrm{d}x + \frac{1}{2}\int_0^L EI_p w''^2 \mathrm{d}x \tag{6.5}$$

The virtual work done by the flow-induced forces F_N and F_T and also by the externally applied forces and moments is given by

$$\begin{aligned}\delta W = &\int_0^L [(F_T + F_N w')\delta u + (F_T w' - F_N)\delta w]\mathrm{d}x \\ &+ M_1 \delta w'(0,t) + M_2 \delta w'(L,t) + Q_1 \delta w(0,t) + Q_2 \delta w(L,t) \\ &+ N_1 \delta u(0,t) + N_2 \delta u(L,t)\end{aligned} \tag{6.6}$$

where M_i, Q_i, and N_i $(i = 1, 2)$ are the moments, transverse shear forces, and axial forces, respectively, externally applied at the boundaries of the pipeline element, as shown in Figure 6.3.

Substituting Equation 6.2, Equations 6.5 and 6.6 into the Hamilton principle 6.1 and taking the integral by parts, we obtain

$$\begin{aligned}\int_0^t \bigg\{ &\int_0^L [(EA_{\mathrm{eff}} u'' - m_p \ddot{u} + F_N w' + F_T)\delta u - (EI_p w'''' - \bar{N} w'' + m_p \ddot{w} - F_T w' + F_N)\delta w]\mathrm{d}x \\ &- EI_p w'' \delta w' |_0^L + M_1 \delta w'(0,t) + M_2 \delta w'(L,t) \\ &+ EI_p w''' \delta w |_0^L + [Q_1 + \bar{N} w'(0,t)]\delta w(0,t) + [Q_2 - \bar{N} w'(L,t)]\delta w(L,t) \\ &- EA_{\mathrm{eff}} u' \delta u |_0^L + (N_1 + \bar{N})\delta u(0,t) + (N_2 - \bar{N})\delta u(L,t) \bigg\} \mathrm{d}t = 0\end{aligned} \tag{6.7}$$

where

$$EA_{\mathrm{eff}} = EA_p + \bar{N} \tag{6.8}$$

From Equation 6.7, we obtain the equation of motion for the axial vibration of a pipeline

$$EA_{\mathrm{eff}} u'' - m_p \ddot{u} + F_N w' + F_T = 0 \tag{6.9}$$

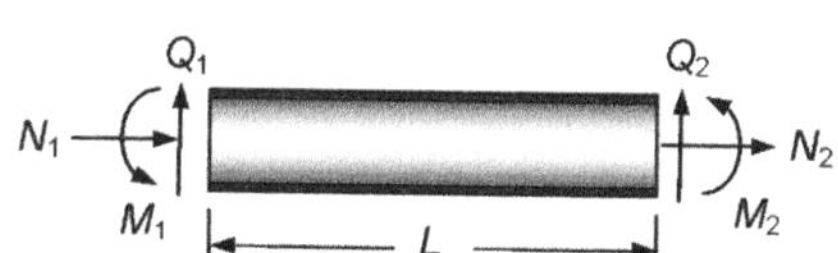

Figure 6.3 Boundary forces and moments acting on the pipeline element

and the equation of motion for the transverse vibration

$$EI_p w'''' - \bar{N}w'' + m_p \ddot{w} - F_T w' + F_N = 0 \tag{6.10}$$

with the boundary conditions

$$\begin{aligned}
M(0,t) &= -M_1(t) && \text{or} \quad w'(0,t) = \theta_1(t)\\
Q(0,t) &= -Q_1(t) - \bar{N}w'(0,t) && \text{or} \quad w(0,t) = w_1(t)\\
N(0,t) &= -N_1(t) - \bar{N} && \text{or} \quad u(0,t) = u_1(t)\\
M(L,t) &= +M_2(t) && \text{or} \quad w'(L,t) = \theta_2(t)\\
Q(L,t) &= +Q_2(t) - \bar{N}w'(L,t) && \text{or} \quad w(L,t) = w_2(t)\\
N(L,t) &= +N_2(t) - \bar{N} && \text{or} \quad u(L,t) = u_2(t)
\end{aligned} \tag{6.11}$$

For Equation 6.11, the following force-displacement relationships are applied

$$\begin{aligned}
M(x,t) &= EI_p w''\\
Q(x,t) &= -EI_p w'''\\
N(x,t) &= EA_{\text{eff}} u'
\end{aligned} \tag{6.12}$$

6.1.2 Fluid-Dynamics Equations

Now consider the fluid element shown in Figure 6.2b. Using Newton's second law, we can derive two fluid momentum equations as

$$(pA)' + F_N w' + F_T + m_w(g_X + \ddot{u} + \dot{c} + c\dot{u}' + cc') = 0 \tag{6.13}$$

$$(pAw')' + F_T w' - F_N + m_w(g_Y + \ddot{w} + 2c\dot{w}' + c^2 w'' + \dot{c}w' + cc'w') = 0 \tag{6.14}$$

where $g_X = g\sin\theta$, $g_Y = g\cos\theta$, A is the constant cross-sectional wetted area of the pipeline through which the internal fluid flows, and $m_w = \rho_w A$ where ρ_w is the density of fluid. By using the Darcy–Weisbach equation, the flow-induced tangential force F_T can be related to the velocity of the fluid flow by

$$F_T = m_w \frac{f}{2D}|c|c \equiv m_w f_D |c| c \tag{6.15}$$

where $f_D = f/(2D)$, where D is the internal diameter of the pipe and $f(\text{Re})$ is the Darcy friction factor given by the function of Reynold's number $\text{Re} = \rho_w Dc/\mu$, where μ is the viscosity of the fluid [163]. We can combine the above two fluid momentum equations into one equation by eliminating F_N, neglecting the higher order terms, and using Equation 6.15 for F_T as

$$p'A + m_w g_Y w' + m_w f_D |c| c + m_w(g_X + \ddot{u} + \dot{c} + c\dot{u}' + cc') = 0 \tag{6.16}$$

In addition to the fluid momentum in Equation 6.16, the continuity equation can be derived from the law of mass conservation, by taking into account the effects of the fluid bulk modulus of elasticity and the time rate of change of the cross-sectional area of the control volume, as

$$\dot{p}A + cp'A + a^2 m_w[c' - 2\nu(\dot{u}' + cu'')] = 0 \tag{6.17}$$

where ν is the Poisson ratio and a is the wave speed of fluid defined by

$$a = \sqrt{\frac{E_v}{\rho_w[1 + (E_v D)/(Eh)]}} \tag{6.18}$$

In above, E_v is the fluid bulk modulus of elasticity, ρ_w is the mass density of fluid, and h is the thickness of the pipeline. By neglecting convective terms, we can simplify Equation 6.17 as

$$\dot{p}A + a^2 m_w (c' - 2\nu\dot{u}') = 0 \tag{6.19}$$

6.1.3 Governing Equations for Pipe Dynamics

By using Equations 6.13 and 6.14, we can eliminate the flow-induced forces F_N and F_T from the equations of motion 6.9 and 6.10 as follows:

$$EA_{\text{eff}}u'' - m\ddot{u} - p'A - m_w(\dot{c} + c\dot{u}' + cc') - m_w g_X = 0 \tag{6.20}$$

$$\begin{aligned} EI_{\text{p}}w'''' &+ m\ddot{w} - \bar{N}w'' + p'Aw' + pAw'' \\ &+ m_w(2c\dot{w}' + c^2w'' + \dot{c}w' + cc'w') + m_w g_Y = 0 \end{aligned} \tag{6.21}$$

where

$$m = m_{\text{p}} + m_w \tag{6.22}$$

By using Equation 6.16 and ignoring the static deformation due to the pipeline weight, we can write Equation 6.20 as

$$EA_{\text{eff}}u'' - m_{\text{p}}\ddot{u} + m_w g_Y w' + m_w f_D |c| c = 0 \tag{6.23}$$

Similarly, we apply Equations 6.16–6.21. By neglecting some higher order terms and ignoring the static deformations due to the pipeline and fluid weights, we obtain

$$EI_{\text{p}}w'''' + m\ddot{w} + 2m_w c\dot{w}' - \bar{N}w'' + pAw'' + m_w c^2 w'' - m_w f_D |c| c w' = 0 \tag{6.24}$$

Thus, the governing equations of pipe dynamics consist of two equations of motion of the pipeline, Equations 6.23 and 6.24, and two fluid-dynamic equations, Equations 6.16 and 6.19.

For the linearization, we assume

$$\begin{aligned} c(x,t) &= c_0 + c_d(x,t) \quad (c_d < c_0) \\ p(x,t) &= p_0 + p_d(x,t) \quad (p_d < p_0) \end{aligned} \tag{6.25}$$

where c_0 and p_0 are constant velocity of the fluid flow and fluid pressure, respectively. By substituting Equation 6.25 into Equations 6.23 and 6.24 and neglecting nonlinear terms, we obtain the linearized equations of motion of the pipeline as

$$EI_{\text{p}}w'''' - T_0 w'' - m_w f_D c_0^2 w' + 2m_w c_0 \dot{w}' + m\ddot{w} = 0 \tag{6.26}$$

$$EA_{\text{eff}}u'' - m_{\text{p}}\ddot{u} + m_w g_Y w' + 2m_w f_D c_0 c_d + m_w f_D c_0^2 = 0 \tag{6.27}$$

where

$$T_0 = \bar{N} - p_0 A - m_w c_0^2 \tag{6.28}$$

Similarly, by substituting Equation 6.25 into Equations 6.16 and 6.19 and neglecting nonlinear terms, we obtain the linearized fluid-dynamic equations as

$$P_d' + m_w g_Y w' + m_w f_D c_0^2 + 2m_w f_D c_0 c_d + m_w(g_X + \ddot{u} + \dot{c}_d + c_0 c_d' + c_0 \dot{u}') = 0 \tag{6.29}$$

$$\dot{P}_d + a^2 m_w (c_d' - 2\nu \dot{u}') = 0 \tag{6.30}$$

where

$$P_d(x,t) = p_d(x,t)A \tag{6.31}$$

From Equation 6.30, we obtain

$$\begin{aligned} m_w c_d' &= -a^{-2}\dot{P}_d + 2m_w \nu \dot{u}' \\ m_w \dot{c}_d' &= -a^{-2}\ddot{P}_d + 2m_w \nu \ddot{u}' \\ m_w c_d'' &= -a^{-2}\dot{P}_d' + 2m_w \nu \dot{u}'' \end{aligned} \tag{6.32}$$

Applying Equation 6.32 to the derivative of Equation 6.29 with respect to x, we obtain a wave form of the equation in $P(x, t)$ as follows:

$$\begin{aligned} &P_d'' - a^{-2} c_0 \dot{P}_d' - a^{-2} \ddot{P}_d - 2a^{-2} f_D c_0 \dot{P}_d + m_w g_Y w'' \\ &+ m_w (1 + 2\nu)(\ddot{u}' + c_0 \dot{u}'') + 4m_w f_D c_0 \nu \dot{u}' = 0 \end{aligned} \tag{6.33}$$

In summary, the governing equations consist of two equations of motion of the pipeline, Equations 6.26 and 6.27, and two fluid-dynamic equations, Equations 6.33 and 6.30.

6.2 Pipelines Conveying Internal Steady Fluid

6.2.1 Governing Equations

If the velocity of the fluid flow and the fluid pressure have very small fluctuations with respect to their mean values, their effects on the motion of the pipeline may be negligible. In this situation, the perturbed values P_d and c_d are negligible and we can assume that the velocity of the fluid flow and the fluid pressure are constant. Accordingly, we can obtain the equations of motion of the pipeline conveying internal steady fluid from Equations 6.26 and 6.27 as [93]

$$EI_p w'''' - T_0 w'' - m_w f_D c_0^2 w' + 2m_w c_0 \dot{w}' + m\ddot{w} = 0 \tag{6.34}$$

$$EA_{\text{eff}} u'' - m_p \ddot{u} + m_w g_Y w' + m_w f_D c_0^2 = 0 \tag{6.35}$$

with the boundary conditions given by Equation 6.11. The force–displacement relationships are given by Equation 6.12.

6.2.2 *Spectral Element Modeling*

6.2.2.1 Governing Equations in the Frequency Domain

Represent the dynamic responses in the spectral forms as

$$w(x,t) = \frac{1}{N}\sum_{n=1}^{N} W_n(x)e^{i\omega_n t}, \quad u(x,t) = \frac{1}{N}\sum_{n=1}^{N} U_n(x)e^{i\omega_n t} \tag{6.36}$$

We also represent all boundary forces and moments in the spectral forms by

$$\begin{aligned}
\{Q(x,t), M(x,t), N(x,t)\} &= \frac{1}{N}\sum_{n=0}^{N-1}\{Q_n(x), M_n(x), N_n(x)\}e^{i\omega_n t} \\
\{Q_i(t), M_i(t), N_i(t)\} &= \frac{1}{N}\sum_{n=0}^{N-1}\{Q_{in}(t), M_{in}(t), N_{in}(t)\}e^{i\omega_n t} \\
\{w_i(t), \theta_i(t), u_i(t)\} &= \frac{1}{N}\sum_{n=0}^{N-1}\{W_{in}, \Theta_{in}, U_{in}\}e^{i\omega_n t}
\end{aligned} \tag{6.37}$$

where $i = 1, 2$. By substituting Equation 6.36 into the governing Equations 6.34 and 6.35 and applying the DFT formula of the unit step function (see Table 2.2) to the constant term in Equation 6.35, we obtain

$$EI_p W'''' - T_0 W'' - m_w f_D c_0^2 W' + 2i\omega m_w c_0 W' - \omega^2 m W = 0 \tag{6.38}$$

$$EA_{\text{eff}} U'' + \omega^2 m_p U + m_w g_Y W' + m_w f_D c_0^2 N \delta_{0n} = 0 \tag{6.39}$$

where Kronecker delta δ_{0n} has unit value when $n = 0$, otherwise it is zero: that is, a constant in the time domain has only one spectral component at zero frequency (i.e., $\omega_0 = 0$). Similarly, substitution of Equations 6.36 and 6.37 into the boundary conditions of Equation 6.11 gives

$$\begin{aligned}
Q(0) &= -Q_1 - \bar{N}W'(0) && \text{or} \quad W(0) = W_1 \\
M(0) &= -M_1 && \text{or} \quad W'(0) = \Theta_1 \\
N(0) &= -N_1 - \bar{N}N\delta_{0n} && \text{or} \quad U(0) = U_1 \\
Q(L) &= +Q_2 - \bar{N}W'(L) && \text{or} \quad W(L) = W_2 \\
M(L) &= +M_2 && \text{or} \quad W'(L) = \Theta_2 \\
N(L) &= +N_2 - \bar{N}N\delta_{0n} && \text{or} \quad U(L) = U_2
\end{aligned} \tag{6.40}$$

and substitution of Equations 6.36 and 6.37 into the force-displacement relationships 6.12 gives

$$Q(x) = -EI_p W'''(x), \quad M(x) = EI_p W''(x), \quad N(x) = EA_{\text{eff}} U'(x) \tag{6.41}$$

In the above Equations 6.38–6.41, the subscripts n are omitted for brevity.

6.2.2.2 Spectral Nodal DOFs, Forces and Moments

The spectral nodal displacements, slopes, forces, and moments are specified in Figure 6.4. The spectral nodal degrees-of-freedom (DOFs) vector is defined by

$$\begin{aligned}
\boldsymbol{d} &= \{W_1 \quad \Theta_1 \quad U_1 \quad W_2 \quad \Theta_2 \quad U_2\}^{\mathrm{T}} \\
&= \{W(0) \quad W'(0) \quad U(0) \quad W(L) \quad W'(L) \quad U(L)\}^{\mathrm{T}}
\end{aligned} \tag{6.42}$$

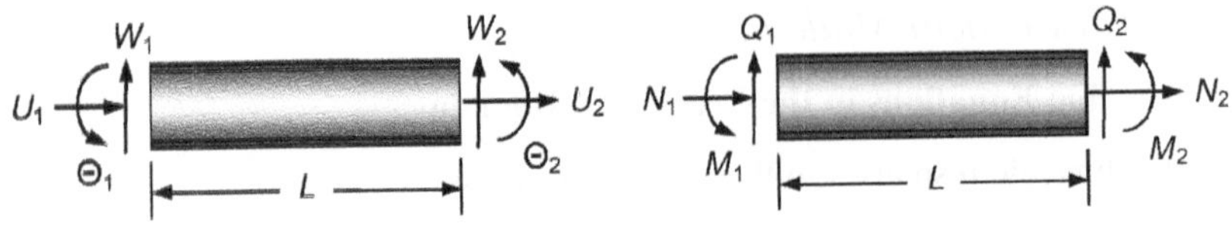

Figure 6.4 Sign convention for the pipeline element

and the spectral nodal forces and moments vector by

$$\begin{aligned} f_c &= \{Q_1 \quad M_1 \quad N_1 \quad Q_2 \quad M_2 \quad N_2\}^{\mathrm{T}} \\ &= \{-Q(0) - \bar{N}W'(0) \quad -M(0) \quad -N(0) - \bar{N}N\delta_{0n} \\ &\qquad Q(L) + \bar{N}W'(L) \quad M(L) \quad N(L) + \bar{N}N\delta_{0n}\}^{\mathrm{T}} \end{aligned} \tag{6.43}$$

6.2.2.3 Dynamic Shape Functions

To derive the frequency-dependent dynamic shape functions, consider the homogeneous equation reduced from Equations 6.38 and 6.39 as

$$EI_p W'''' - T_0 W'' - m_w f_D c_0^2 W' + 2i\omega m_w c_0 W' - \omega^2 m W = 0 \tag{6.44}$$

$$EA_{\mathrm{eff}} U'' + \omega^2 m_p U + m_w g_Y W' = 0 \tag{6.45}$$

Assume the general solutions of the homogeneous equation to be

$$W(x) = \beta a e^{-ikx}, \quad U(x) = a e^{-ikx} \tag{6.46}$$

Substituting Equation 6.46 into Equations 6.44 and 6.45, we obtain an eigenvalue problem

$$\begin{bmatrix} K_{11} & 0 \\ K_{21} & K_{22} \end{bmatrix} \begin{Bmatrix} \beta \\ 1 \end{Bmatrix} = \begin{Bmatrix} 0 \\ 0 \end{Bmatrix} \tag{6.47}$$

where

$$\begin{aligned} K_{11} &= EI_p k^4 + T_0 k^2 + i(m_w f_D c_0^2 - 2i\omega m_w c_0)k - \omega^2 m \\ K_{21} &= i m_w g_Y k \\ K_{22} &= EA_{\mathrm{eff}} k^2 - \omega^2 m_p \end{aligned} \tag{6.48}$$

Equation 6.47 provides the dispersion relation

$$EI_p k^4 + T_0 k^2 + i(m_w f_D c_0^2 - 2i\omega m_w c_0)k - \omega^2 m = 0 \tag{6.49}$$

and

$$EA_{\mathrm{eff}} k^2 - \omega^2 m_p = 0 \tag{6.50}$$

We can numerically compute four wavenumbers k_1, k_2, k_3, and k_4 from Equation 6.49 and two wavenumbers k_5 and k_6 from Equation 6.50 as

$$k_5 = -k_6 = \omega\sqrt{\frac{m_p}{EA_{\mathrm{eff}}}} \tag{6.51}$$

For each wavenumber k_j ($j = 1, 2, \ldots, 6$), we can compute the corresponding β_j from Equation 6.47 as

$$\beta_j = \begin{cases} -\dfrac{EA_{\text{eff}}k_j^2 - \omega^2 m_{\text{p}}}{ik_j m_w g_Y} & (j = 1, 2, 3, 4) \\ 0 & (j = 5, 6) \end{cases} \tag{6.52}$$

Define a diagonal matrix as

$$\boldsymbol{B} = diag[\beta_j] \quad (j = 1, 2, \ldots, 6) \tag{6.53}$$

By using six wavenumbers k_j ($j = 1, 2, \ldots, 6$) and the diagonal matrix $\boldsymbol{B}$, we can write the general solutions of Equations 6.44 and 6.45 as

$$\begin{aligned} W(x) &= \sum_{j=1}^{6} \beta_j a_j \mathrm{e}^{-ik_j x} = \boldsymbol{e}(x)\boldsymbol{Ba} \\ U(x) &= \sum_{j=1}^{6} a_j \mathrm{e}^{-ik_j x} = \boldsymbol{e}(x)\boldsymbol{a} \end{aligned} \tag{6.54}$$

where

$$\boldsymbol{e}(x;\omega) = [\mathrm{e}^{-ik_1 x} \quad \mathrm{e}^{-ik_2 x} \quad \mathrm{e}^{-ik_3 x} \quad \mathrm{e}^{-ik_4 x} \quad \mathrm{e}^{-ik_5 x} \quad \mathrm{e}^{-ik_6 x}] \tag{6.55}$$

and

$$\boldsymbol{a} = \{a_1 \quad a_2 \quad a_3 \quad a_4 \quad a_5 \quad a_6\}^{\mathrm{T}} \tag{6.56}$$

Substituting Equation 6.54 into the right-hand side of Equation 6.42, we obtain

$$\boldsymbol{d} = \boldsymbol{H}(\omega)\boldsymbol{a} \tag{6.57}$$

where

$$\boldsymbol{H}(\omega) = \begin{bmatrix} \beta_1 & \beta_2 & \beta_3 & \beta_4 & 0 & 0 \\ -ik_1\beta_1 & -ik_2\beta_2 & -ik_3\beta_3 & -ik_4\beta_4 & 0 & 0 \\ 1 & 1 & 1 & 1 & 1 & 1 \\ \beta_1 \mathrm{e}_1 & \beta_2 \mathrm{e}_2 & \beta_3 \mathrm{e}_3 & \beta_4 \mathrm{e}_4 & 0 & 0 \\ -ik_1\beta_1\mathrm{e}_1 & -ik_2\beta_2\mathrm{e}_2 & -ik_3\beta_3\mathrm{e}_3 & -ik_4\beta_4\mathrm{e}_4 & 0 & 0 \\ \mathrm{e}_1 & \mathrm{e}_2 & \mathrm{e}_3 & \mathrm{e}_4 & \mathrm{e}_5 & \mathrm{e}_6 \end{bmatrix} \tag{6.58}$$

with

$$\mathrm{e}_j = \mathrm{e}^{-ik_j L} \quad (j = 1, 2, \ldots, 6) \tag{6.59}$$

By using Equation 6.57, Equation 6.54 can be written as

$$W(x) = \boldsymbol{N}_w(x;\omega)\boldsymbol{d}, \quad U(x) = \boldsymbol{N}_u(x;\omega)\boldsymbol{d} \tag{6.60}$$

where $\boldsymbol{N}_w$ and $\boldsymbol{N}_u$ are the dynamic shape functions defined by

$$\begin{aligned} \boldsymbol{N}_w(x;\omega) &= \boldsymbol{e}(x)\boldsymbol{BH}^{-1}(\omega) \\ \boldsymbol{N}_u(x;\omega) &= \boldsymbol{e}(x)\boldsymbol{H}^{-1}(\omega) \end{aligned} \tag{6.61}$$

6.2.2.4 Weak Form of Governing Equation

The weak form of the non-homogeneous governing Equations 6.38 and 6.39 in the frequency domain can be obtained from

$$\begin{aligned}&\int_0^L [\{EI_{\mathrm{p}}W'''' - T_0W'' - m_w f_D c_0^2 W' + 2i\omega m_w c_0 W' - \omega^2 mW\}\delta W\\&- \{EA_{\mathit{eff}}U'' + \omega^2 m_{\mathrm{p}}U + m_w g_Y W' + m_w f_D c_0^2 N\delta_{0n}\}\delta U]\mathrm{d}x = 0\end{aligned} \tag{6.62}$$

By applying the integral by parts, we obtain the weak form as

$$\begin{aligned}&\int_0^L [EI_{\mathrm{p}}W''\delta W'' + T_0W'\delta W' - m_w f_D c_0^2 W'\delta W - \omega^2 mW\delta W\\&+ i\omega m_w c_0(W'\delta W - W\delta W') + EA_{\mathrm{eff}}U'\delta U' - \omega^2 m_{\mathrm{p}}U\delta U\\&- m_w g_Y W'\delta U - m_w f_D c_0^2 N\delta_{0n}\delta U + \bar{N}N\delta_{0n}\delta U'\}]\mathrm{d}x\\&+ i\omega m_w c_0 W\delta W|_0^L - (T_0 - \bar{N})W'\delta W|_0^L - (-EI_{\mathrm{p}}W''' + \bar{N}W')\delta W|_0^L\\&- EI_{\mathrm{p}}W''\delta W'|_0^L - (EA_{\mathrm{eff}}U' + \bar{N}N\delta_{0n})\delta U|_0^L = 0\end{aligned} \tag{6.63}$$

where the following relationship is used.

$$\int_0^L \bar{N}N\delta_{0n}\delta U'\mathrm{d}x - \bar{N}N\delta_{0n}\delta U|_0^L = 0 \tag{6.64}$$

By using Equation 6.41 for $Q(x)$, $M(x)$, and $N(x)$, Equation 6.63 can be written as

$$\begin{aligned}&\int_0^L [EI_{\mathrm{p}}W''\delta W'' + T_0W'\delta W' - m_w f_D c_0^2 W'\delta W - \omega^2 mW\delta W\\&+ i\omega m_w c_0(W'\delta W - W\delta W') + EA_{\mathrm{eff}}U'\delta U' - \omega^2 m_{\mathrm{p}}U\delta U\\&- m_w g_Y W'\delta U - m_w f_D c_0^2 N\delta_{0n}\delta U + \bar{N}N\delta_{0n}\delta U'\}]\mathrm{d}x\\&+ i\omega m_w c_0 W\delta W|_0^L - (T_0 - \bar{N})W'\delta W|_0^L\\&- [Q(x) + \bar{N}W']\delta W|_0^L - M(x)\delta W'|_0^L - [N(x) + \bar{N}N\delta_{0n}]\delta U|_0^L = 0\end{aligned} \tag{6.65}$$

6.2.2.5 Spectral Element Equation

Substituting Equation 6.60 into Equation 6.65, we obtain

$$\begin{aligned}&\delta \boldsymbol{d}^{\mathrm{T}}\left\{\int_0^L [EI_{\mathrm{p}}\boldsymbol{N}_w''^{\mathrm{T}}\boldsymbol{N}_w'' + T_0\boldsymbol{N}_w'^{\mathrm{T}}\boldsymbol{N}_w' - m_w f_D c_0^2 \boldsymbol{N}_w^{\mathrm{T}}\boldsymbol{N}_w' - \omega^2 m\boldsymbol{N}_w^{\mathrm{T}}\boldsymbol{N}_w\right.\\&+ i\omega m_w c_0(\boldsymbol{N}_w^{\mathrm{T}}\boldsymbol{N}_w' - \boldsymbol{N}_w'^{\mathrm{T}}\boldsymbol{N}_w) + EA_{\mathrm{eff}}\boldsymbol{N}_u'^{\mathrm{T}}\boldsymbol{N}_u' - \omega^2 m_{\mathrm{p}}\boldsymbol{N}_u^{\mathrm{T}}\boldsymbol{N}_u\\&\left.- m_w g_Y \boldsymbol{N}_u^{\mathrm{T}}\boldsymbol{N}_w']\mathrm{d}x + \left[i\omega m_w c_0 \boldsymbol{N}_w^{\mathrm{T}}\boldsymbol{N}_w - (T_0 - \bar{N})\boldsymbol{N}_w^{\mathrm{T}}\boldsymbol{N}_w'\right]\Big|_0^L\right\}\boldsymbol{d}\\&= \boldsymbol{f}_{\mathrm{c}} + \boldsymbol{f}_{\mathrm{d}}\end{aligned} \tag{6.66}$$

where

$$\boldsymbol{f}_{\mathrm{d}} = N\delta_{0n}\int_0^L (m_w f_D c_0^2 \boldsymbol{N}_u^{\mathrm{T}} - \bar{N}\boldsymbol{N}_u'^{\mathrm{T}})\mathrm{d}x \tag{6.67}$$

and $\boldsymbol{f}_c$ is defined by Equation 6.43. As $\delta\boldsymbol{d}$ is arbitrary by definition, from Equation 6.66 we obtain the spectral element equation as

$$\boldsymbol{S}(\omega)\boldsymbol{d} = \boldsymbol{f}_c + \boldsymbol{f}_d \tag{6.68}$$

where

$$\begin{aligned}\boldsymbol{S}(\omega) = \int_0^L &[EI_p\boldsymbol{N}_w''^{\mathrm{T}}\boldsymbol{N}_w'' + T_0\boldsymbol{N}_w'^{\mathrm{T}}\boldsymbol{N}_w' - m_w f_D c_0^2\boldsymbol{N}_w^{\mathrm{T}}\boldsymbol{N}_w' - \omega^2 m\boldsymbol{N}_w^{\mathrm{T}}\boldsymbol{N}_w \\ &+ i\omega m_w c_0(\boldsymbol{N}_w^{\mathrm{T}}\boldsymbol{N}_w' - \boldsymbol{N}_w'^{\mathrm{T}}\boldsymbol{N}_w) + EA_{\mathrm{eff}}\boldsymbol{N}_u'^{\mathrm{T}}\boldsymbol{N}_u' - \omega^2 m_p\boldsymbol{N}_u^{\mathrm{T}}\boldsymbol{N}_u \\ &- m_w g_Y\boldsymbol{N}_u^{\mathrm{T}}\boldsymbol{N}_w']\mathrm{d}x + [i\omega m_w c_0\boldsymbol{N}_w^{\mathrm{T}}\boldsymbol{N}_w - (T_0 - \bar{N})\boldsymbol{N}_w^{\mathrm{T}}\boldsymbol{N}_w']_0^L\end{aligned} \tag{6.69}$$

From Equation 6.61, we can get

$$\begin{aligned}\boldsymbol{N}_w'(x;\omega) &= -i\boldsymbol{e}(x)\boldsymbol{KBH}^{-1}(\omega) \\ \boldsymbol{N}_u'(x;\omega) &= -i\boldsymbol{e}(x)\boldsymbol{KH}^{-1}(\omega) \\ \boldsymbol{N}_w''(x;\omega) &= -\boldsymbol{e}(x)\boldsymbol{K}^2\boldsymbol{BH}^{-1}(\omega)\end{aligned} \tag{6.70}$$

where

$$\boldsymbol{K} = diag[k_i], \quad \boldsymbol{K}^2 = diag[k_i^2] \tag{6.71}$$

Substitution of Equations 6.61 and 6.70 into Equation 6.69 yields

$$\boldsymbol{S}(\omega) = \boldsymbol{H}^{-\mathrm{T}}(\omega)\boldsymbol{D}(\omega)\boldsymbol{H}^{-1}(\omega) \tag{6.72}$$

where

$$\begin{aligned}\boldsymbol{D}(\omega) = &EI_p\boldsymbol{BK}^2\boldsymbol{EK}^2\boldsymbol{B} - T_0\boldsymbol{BKEKB} + im_w f_D c_0^2\boldsymbol{BEKB} - \omega^2 m\boldsymbol{BEB} \\ &+ \omega m_w c_0\boldsymbol{B}(\boldsymbol{EK} - \boldsymbol{KE})\boldsymbol{B} - EA_{eff}\boldsymbol{KEK} - \omega^2 m_p\boldsymbol{E} + im_w g_Y\boldsymbol{EKB} \\ &+ i\omega m_w c_0\boldsymbol{B}\bar{\boldsymbol{E}}\boldsymbol{B} + i(T_0 - \bar{N})\boldsymbol{B}\bar{\boldsymbol{E}}\boldsymbol{KB}\end{aligned} \tag{6.73}$$

and where

$$\begin{aligned}\boldsymbol{E}(\omega) &= \int_0^L \boldsymbol{e}^{\mathrm{T}}(x;\omega)\boldsymbol{e}(x;\omega)\mathrm{d}x = [E_{rs}(\omega)] \\ \bar{\boldsymbol{E}}(\omega) &= \boldsymbol{e}^{\mathrm{T}}(x;\omega)\boldsymbol{e}(x;\omega)|_0^L = [\bar{E}_{rs}(\omega)]\end{aligned} \qquad (r, s = 1, 2, \ldots 6) \tag{6.74}$$

with

$$\begin{aligned}E_{rs} &= \begin{cases}\dfrac{i}{k_r + k_s}\bar{E}_{rs} & \text{if } k_r + k_s \neq 0 \\ L & \text{if } k_r + k_s = 0\end{cases} \\ \bar{E}_{rs} &= \mathrm{e}^{-i(k_r + k_s)L} - 1\end{aligned} \tag{6.75}$$

The spectral element equations can be assembled to form a global system equation, after imposing the relevant boundary conditions, in the form of

$$\boldsymbol{S}_g(\omega)\boldsymbol{d}_g = \boldsymbol{f}_g \tag{6.76}$$

where $\boldsymbol{S}_g(\omega)$ is the assembled global dynamic stiffness matrix, $\boldsymbol{d}_g$ is the global spectral nodal DOFs vector, and $\boldsymbol{f}_g$ is the global spectral nodal forces vector.

6.2.3 Finite Element Model

The variational approach is applied to formulate the finite element equation. The weak form of the governing Equations 6.34 and 6.35 can be constructed from the weighted-integral statement given by

$$\begin{aligned}&\int_0^L [(EI_\mathrm{p} w'''' - T_0 w'' - m_w f_D c_0^2 w' + 2m_w c_0 \dot{w}' + m\ddot{w})\delta w \\ &- (EA_\mathrm{eff} u'' - m_\mathrm{p}\ddot{u} + m_w g_Y w' + m_w f_D c_0^2)\delta u]\mathrm{d}x = 0\end{aligned} \tag{6.77}$$

Integrating Equation 6.77 by parts and then applying the force-displacement relationships 6.12 yields the weak form as

$$\begin{aligned}&\int_0^L [EI_\mathrm{p} w''\delta w'' + T_0 w'\delta w' - m_w f_D c_0^2 w'\delta w + m_w c_0(\dot{w}'\delta w - \dot{w}\delta w') \\ &+ m\ddot{w}\delta w + EA_\mathrm{eff} u'\delta u' + m_\mathrm{p}\ddot{u}\delta u - m_w g_Y w'\delta u - m_w f_D c_0^2 \delta u + \bar{N}\delta u']\mathrm{d}x \\ &+ m_w c_0 \dot{w}\delta w|_0^L - (T_0 - \bar{N})w'\delta w|_0^L \\ &- [Q(x,t) + \bar{N}w']\delta w|_0^L - M(x,t)\delta w'|_0^L - [N(x,t) + \bar{N}]\delta u|_0^L = 0\end{aligned} \tag{6.78}$$

The spectral nodal DOFs vector and the spectral forces and moments vector are defined by

$$\begin{aligned}\boldsymbol{d}(t) &= \{w_1 \quad \theta_1 \quad u_1 \quad w_2 \quad \theta_2 \quad u_2\}^\mathrm{T} \\ &= \{w(0,t)\ w'(0,t)\ u(0,t)\ w(L,t)\ w'(L,t)\ u(L,t)\}^\mathrm{T}\end{aligned} \tag{6.79}$$

$$\begin{aligned}\boldsymbol{f}_\mathrm{c}(t) &= \{Q_1 \quad M_1 \quad N_1 \quad Q_2 \quad M_2 \quad N_2\}^\mathrm{T} \\ &= \{-Q(0,t) - \bar{N}W'(0,t) \quad -M(0,t) \quad -N(0,t) - \bar{N}N\delta_{0n} \\ &\qquad Q(L,t) + \bar{N}W'(L,t) \quad M(L,t) \quad N(L,t) + \bar{N}N\delta_{0n}\}^\mathrm{T}\end{aligned} \tag{6.80}$$

The displacement fields are assumed by

$$w(x,t) = \boldsymbol{N}_w(x)\boldsymbol{d}(t), \quad u(x,t) = \boldsymbol{N}_u(x)\boldsymbol{d}(t) \tag{6.81}$$

where the shape functions are defined by

$$\begin{aligned}\boldsymbol{N}_w(x) &= [N_{w1}(x) \quad N_{w2}(x) \quad 0 \quad N_{w3}(x) \quad N_{w4}(x) \quad 0] \\ \boldsymbol{N}_u(x) &= [0 \quad 0 \quad N_{u1}(x) \quad 0 \quad 0 \quad N_{u2}(x)]\end{aligned} \tag{6.82}$$

with

$$\begin{aligned}N_{w1}(x) &= 1 - 3(x/L)^2 + 2(x/L)^3 \\ N_{w2}(x) &= L(x/L)[1 - 2(x/L) + (x/L)^2] \\ N_{w3}(x) &= (x/L)^2[3 - 2(x/L)] \\ N_{w4}(x) &= -L(x/L)^2[1 - (x/L)] \\ N_{u1}(x) &= 1 - (x/L) \\ N_{u2}(x) &= x/L\end{aligned} \tag{6.83}$$

Substituting Equation 6.81 into the weak form of Equation 6.78, we obtain the finite element equation as

$$\boldsymbol{M}\ddot{\boldsymbol{d}}(t) + \boldsymbol{C}\dot{\boldsymbol{d}}(t) + \boldsymbol{K}\boldsymbol{d}(t) = \boldsymbol{f}_\mathrm{c}(t) + \boldsymbol{f}_\mathrm{d} \tag{6.84}$$

where

$$f_{\mathrm{d}}=\int_{0}^{L}(m_{w}f_{D}c_{0}^{2}N_{u}^{\mathrm{T}}-\bar{N}N_{u}^{\prime\mathrm{T}})\mathrm{d}x \tag{6.85}$$

The finite element matrices are given in Appendix 6.A.

Example 6.1

Consider a straight uniform pipeline conveying internal steady fluid. The pipeline is simply supported at both its ends. The axial motions of the pipeline are fixed at $x=0$ and free at $x=L$. The geometric and material properties of the pipeline are: length $L=5$ m, internal diameter $D=30$ mm, thickness $h=1.5$ mm, Young's modulus $E=70$ GPa, Poisson's ratio $\nu=0.33$, mass density $\rho_{\mathrm{p}}=2500\,\mathrm{kg/m^3}$, and the axial tension $\bar{N}=0$ N. The properties of the fluid are the mass density $\rho_w=1000\,\mathrm{kg/m^3}$, the fluid bulk modulus of elasticity $E_v=2.2$ GPa, and the viscosity of fluid $\mu=1.003\times10^{-3}\,\mathrm{N\,s/m^2}$.

Table 6.1 compares the natural frequencies obtained by the spectral element method (SEM) and the finite element method (FEM). When $c_0=0$ m/s, the SEM results are identical with the exact theoretical results. As the number of finite elements used in FEM is increased, the FEM results certainly converge towards the SEM results. It is worth mentioning that the first natural frequency vanishes first when the fluid velocity is increased up to a critical value c_{D}, known as the divergence fluid velocity, at which the static instability "divergence" occurs. Figure 6.5 shows the first three eigenvalues $\lambda=i\omega$ as functions of the fluid velocity c_0. The stability of the pipeline is determined by the signs of the real and imaginary parts of the eigenvalues as follows: (a) "stable" if the real part of λ is negative; (b) "divergence" if the real part of λ is positive and the imaginary part of λ is zero; (c) "flutter" if the real part of λ is positive and the imaginary part

Table 6.1 Comparison of the natural frequencies obtained by SEM and FEM

Fluid velocity c_0 (m/s)	Method	Natural frequency (Hz)			
		$f_1^{(b)}$	$f_3^{(b)}$	$f_5^{(b)}$	$f_{14}^{(a)}$
0	Theory [24]	2.1750	19.575	54.375	264.58
	SEM(1)	2.1750	19.575	54.375	264.58
	FEM(2)	2.1752	19.468	60.352	265.66
	FEM(5)	2.1750	19.586	54.590	264.85
	FEM(10)	2.1750	19.576	54.389	264.64
15	SEM(1)	1.7680	19.299	54.119	264.58
	FEM(2)	1.7685	19.468	60.152	265.66
	FEM(5)	1.7681	19.310	54.340	264.85
	FEM(10)	1.7680	19.300	54.133	264.64
26.94 (c_{D})	SEM(1)	0.0000	18.677	53.546	264.58
	FEM(2)	0.0000	18.877	59.704	265.66
	FEM(5)	0.0000	18.690	53.783	264.85
	FEM(10)	0.0000	18.677	53.562	264.64

Note: () = total number of the equal-length finite elements used in the analysis. [a] = axial modes, [b] = bending modes.

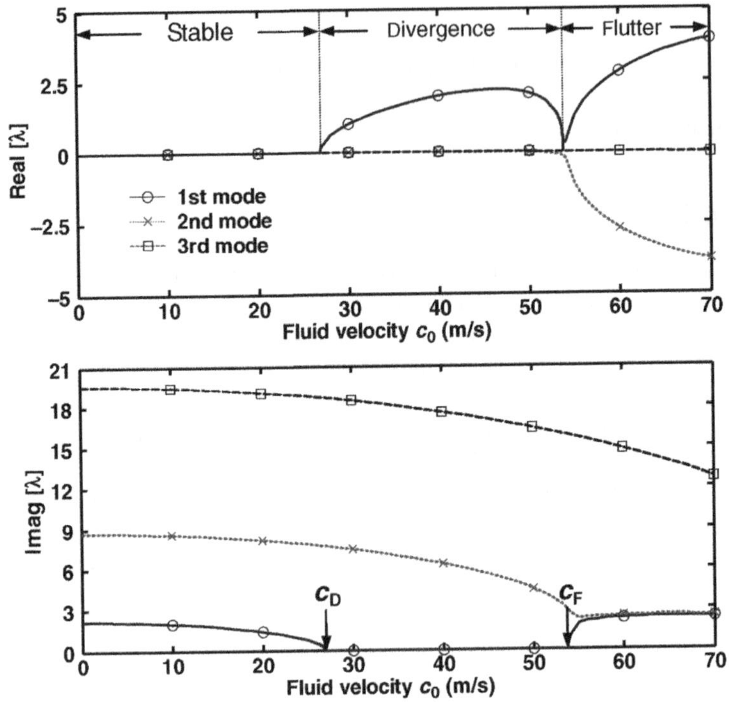

Figure 6.5 The lowest three eigenvalues λ (Hz) versus the fluid velocity c_0 (m/s)

of λ is non-zero. As shown in Figure 6.5, the pipeline is stable below a critical fluid velocity $c_D = 26.94$ m/s at which the divergence occurs. The divergence instability lasts until the fluid velocity reaches $c_F = 53.74$ m/s at which the flutter instability occurs. At a fluid velocity higher than c_F, the first and second natural frequencies are merged together resulting in the merging frequency or coupled-mode flutter.

6.3 Pipelines Conveying Internal Unsteady Fluid

6.3.1 Governing Equations

The governing equations for the transverse and axial vibrations of a pipeline conveying internal unsteady fluid are given by Equations 6.26 and 6.27, respectively, and the fluid dynamics equation for the internal fluid is given by Equation 6.33. They are written as

$$EI_p w'''' - T_0 w'' - m_w f_D c_0^2 w' + 2m_w c_0 \dot{w}' + m\ddot{w} = 0 \tag{6.86}$$

$$EA_{\text{eff}} u'' - m_p \ddot{u} + m_w g_Y w' + 2m_w f_D c_0 c_d + m_w f_D c_0^2 = 0 \tag{6.87}$$

$$\begin{aligned} &P_d'' - a^{-2} c_0 \dot{P}_d' - a^{-2} \ddot{P}_d - 2a^{-2} f_D c_0 \dot{P}_d + m_w g_Y w'' \\ &+ m_w (1 + 2\nu)(\ddot{u}' + c_0 \dot{u}'') + 4 m_w f_D c_0 \nu \dot{u}' = 0 \end{aligned} \tag{6.88}$$

The continuity Equation 6.30 is used as the complimentary equation to relate the fluid pressure to the fluid velocity. It is given by

$$\dot{P}_d + a^2 m_w (c_d' - 2\nu \dot{u}') = 0 \tag{6.89}$$

The associated boundary conditions are given by Equation 6.11, with the force-displacement relations given by Equation 6.12.

6.3.2 Spectral Element Modeling

6.3.2.1 Governing Equations in the Frequency Domain

Represent the dynamic responses of the pipeline and the pressure and velocity of the internal fluid in the spectral forms as

$$\begin{aligned} w(x,t) &= \frac{1}{N}\sum_{n=1}^{N} W_n(x)e^{i\omega_n t}, \quad u(x,t) = \frac{1}{N}\sum_{n=1}^{N} U_n(x)e^{i\omega_n t} \\ P_d(x,t) &= \frac{1}{N}\sum_{n=1}^{N} P_n(x)e^{i\omega_n t}, \quad c_d(x,t) = \frac{1}{N}\sum_{n=1}^{N} C_n(x)e^{i\omega_n t} \end{aligned} \tag{6.90}$$

By using Equation 6.90, the governing Equations 6.86–6.88 can be written in the spectral forms as

$$EI_p W'''' - T_0 W'' - m_w f_D c_0^2 W' + 2i\omega m_w c_0 W' - \omega^2 m W = 0 \tag{6.91}$$

$$EA_{\text{eff}} U'' + \omega^2 m_p U + m_w g_Y W' + 2m_w f_D c_0 C + m_w f_D c_0^2 N \delta_{0n} = 0 \tag{6.92}$$

$$\begin{aligned} &P'' - i\omega a^{-2} c_0 P' - i\omega a^{-2}(i\omega + 2f_D c_0)P + m_w g_Y W'' \\ &\quad - m_w \omega^2 U' + 2i\omega\nu m_w (i\omega + 2f_D c_0)U' + i\omega m_w (1+2\nu)c_0 U'' = 0 \end{aligned} \tag{6.93}$$

Similarly Equation 6.89 can be written in the spectral form as

$$i\omega P + a^2 m_w (C' - 2i\omega\nu U') = 0 \tag{6.94}$$

By using the same spectral representations of the forces and moments as defined by Equation 6.37, the boundary conditions 6.11 can be written in the spectral forms as

$$\begin{aligned} Q(0) &= -Q_1 - \bar{N}W'(0) \quad &\text{or}\quad W(0) &= W_1 \\ M(0) &= -M_1 \quad &\text{or}\quad W'(0) &= \Theta_1 \\ N(0) &= -N_1 - \bar{N}N\delta_{0n} \quad &\text{or}\quad U(0) &= U_1 \\ Q(L) &= +Q_2 - \bar{N}W'(L) \quad &\text{or}\quad W(L) &= W_2 \\ M(L) &= +M_2 \quad &\text{or}\quad W'(L) &= \Theta_2 \\ N(L) &= +N_2 - \bar{N}N\delta_{0n} \quad &\text{or}\quad U(L) &= U_2 \end{aligned} \tag{6.95}$$

and the force-displacement relationships 6.12 in the spectral forms as

$$Q(x) = -EI_p W'''(x), \quad M(x) = EI_p W''(x), \quad N(x) = EA_{\text{eff}} U'(x) \tag{6.96}$$

The subscripts n are omitted in the above Equations 6.91–6.96 for brevity.

6.3.2.2 Spectral Nodal DOFs, Forces and Moments

The spectral nodal DOFs are defined in Figure 6.6. Thus we introduce the spectral nodal DOFs vector as

$$\begin{aligned} \boldsymbol{d} &= \{W_1 \quad \Theta_1 \quad U_1 \quad P_1 \quad W_2 \quad \Theta_2 \quad U_2 \quad P_2\}^{\mathrm{T}} \\ &= \{W(0) \quad W'(0) \quad U(0) \quad P(0) \quad W(L) \quad W'(L) \quad U(L) \quad P(L)\}^{\mathrm{T}} \end{aligned} \tag{6.97}$$

and the spectral nodal forces, moments, and flow velocities are defined by

$$\begin{aligned} Q_1 &= -Q(0) - \bar{N}W'(0), & Q_2 &= +Q(L) + \bar{N}W'(L) \\ M_1 &= -M(0), & M_2 &= +M(L) \\ N_1 &= -N(0) - \bar{N}N\delta_{0n}, & N_2 &= +N(L) + \bar{N}N\delta_{0n} \\ C_1 &= +C(0), & C_2 &= -C(L) \end{aligned} \tag{6.98}$$

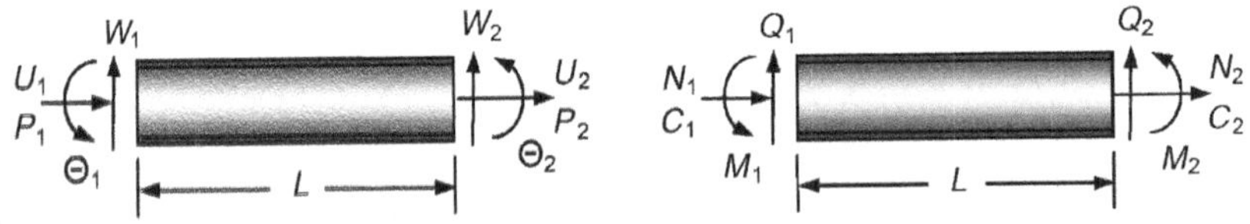

Figure 6.6 Sign convention for the pipeline element

6.3.2.3 Dynamic Shape Functions

The homogeneous governing equations reduced from Equations 6.91–6.93 are given by

$$\begin{aligned} & EI_pW'''' - T_0W'' - m_w f_D c_0^2 W' + 2i\omega m_w c_0 W' - \omega^2 mW = 0 \\ & EA_{\text{eff}}U'' + \omega^2 m_p U + m_w g_Y W' + 2m_w f_D c_0 C = 0 \\ & P'' - i\omega a^{-2} c_0 P' - i\omega a^{-2}(i\omega + 2f_D c_0)P + m_w g_Y W'' \\ & \quad - m_w\omega^2 U' + 2i\omega\nu m_w(i\omega + 2f_D c_0)U' + i\omega m_w(1+2\nu)c_0 U'' = 0 \end{aligned} \tag{6.99}$$

with

$$i\omega P + a^2 m_w(C' - 2i\omega\nu U') = 0 \tag{6.100}$$

Assume the general solutions of Equations 6.99 and 6.100 as

$$\begin{aligned} W(x) &= q\alpha e^{-ikx}, & U(x) &= \alpha e^{-ikx} \\ P(x) &= t\alpha e^{-ikx}, & C(x) &= r\alpha e^{-ikx} \end{aligned} \tag{6.101}$$

Substitution of Equation 6.101 into Equation 6.100 yields the relationship

$$r = \frac{\omega t}{a^2 m_w k} + 2i\omega\nu \tag{6.102}$$

By substituting Equation 6.101 into the governing Equation 6.99 and then applying Equation 6.102, we can obtain an eigenvalue problem as

$$\begin{bmatrix} K_{11} & 0 & 0 \\ K_{21} & K_{22} & K_{23} \\ K_{31} & K_{32} & K_{33} \end{bmatrix} \begin{Bmatrix} q \\ 1 \\ t \end{Bmatrix} = \begin{Bmatrix} 0 \\ 0 \\ 0 \end{Bmatrix} \tag{6.103}$$

where

$$\begin{aligned} K_{11} &= EI_{\mathrm{p}}k^4 + T_0k^2 + i(m_w f_D c_0^2 - 2i\omega m_w c_0)k - \omega^2 m \\ K_{21} &= im_w g_Y k \\ K_{22} &= EA_{\mathrm{eff}}k^2 - \omega^2 m_{\mathrm{p}} - 4i\omega m_w f_D c_0 \nu \\ K_{23} &= -2a^{-2} f_D c_0 \omega k^{-1} \\ K_{31} &= m_w g_Y k^2 \\ K_{32} &= i\omega m_w c_0(1+2\nu)k^2 - \omega m_w[i\omega(1+2\nu) + 4 f_D c_0 \nu]k \\ K_{33} &= k^2 + \omega a^{-2} c_0 k - \omega^2 a^{-2} + 2i\omega a^{-2} f_D c_0 \end{aligned} \tag{6.104}$$

From Equation 6.103, we can obtain two dispersion relation as

$$K_{11} = 0 \tag{6.105}$$

$$K_{22}K_{33} - K_{23}K_{32} = 0 \tag{6.106}$$

or

$$EI_{\mathrm{p}}k^4 + T_0k^2 + (2\omega c_0 + if_D c_0^2)m_w k - \omega^2 m = 0 \tag{6.107}$$

$$b_4k^4 + b_3k^3 + b_2k^2 + b_1k + b_0 = 0 \tag{6.108}$$

where

$$\begin{aligned} b_4 &= a^2 EA_{\mathrm{eff}} \\ b_3 &= \omega c_0 EA_{\mathrm{eff}} \\ b_2 &= -EA_{\mathrm{eff}}\omega^2 - \omega^2 m_p a^2 - 4i\omega m_w a^2 f_D c_0 \nu + 2i\omega EA_{\mathrm{eff}} f_D c_0 \\ b_1 &= 2i\omega^2 m_w f_D c_0^2 - \omega^3 m_{\mathrm{p}} c_0 \\ b_0 &= \omega^4 m_{\mathrm{p}} - 2i\omega^3(m_{\mathrm{p}} + m_w) f_D c_0 \end{aligned} \tag{6.109}$$

We can numerically compute four wavenumbers k_1, k_2, k_3, and k_4 from the first dispersion relation 6.107 and the other four wavenumbers k_5, k_6, k_7, and k_8 from the second dispersion relation 6.108.

By substituting each wavenumber k_j ($j = 1, 2, \ldots, 8$) into Equation 6.103 and using Equations 6.105 and 6.106, we can obtain the corresponding q_j, t_j, and r_j as follows:

$$\begin{aligned} q_j &= \begin{cases} \dfrac{K_{23}K_{32} - K_{22}K_{33}}{K_{21}K_{33} - K_{23}K_{31}} & (j = 1, 2, 3, 4) \\ 0 & (j = 5, 6, 7, 8) \end{cases} \\ t_j &= \begin{cases} -(K_{21}q_j + K_{22})K_{23}^{-1} & (j = 1, 2, 3, 4) \\ -K_{22}K_{23}^{-1} & (j = 5, 6, 7, 8) \end{cases} \\ r_j &= \frac{\omega}{a^2 m_w k} t_j + 2i\omega\nu \qquad (j = 1, 2, \cdots, 8) \end{aligned} \tag{6.110}$$

For k_j, q_j, t_j, and r_j ($j = 1, 2, \ldots, 8$), we define diagonal matrices as

$$\begin{aligned} \boldsymbol{K}(\omega) &= diag[k_j(\omega)], \quad \boldsymbol{Q}(\omega) = diag[q_j(\omega)] \\ \boldsymbol{T}(\omega) &= diag[t_j(\omega)], \quad \boldsymbol{R}(\omega) = diag[r_j(\omega)] \end{aligned} \tag{6.111}$$

By using eight wavenumbers k_j ($j = 1, 2, \ldots, 8$) and the diagonal matrices defined by Equation 6.111, we can write the general solutions in the forms of

$$\begin{aligned} W(x) &= \sum_{j=1}^{8} q_j \alpha_j \mathrm{e}^{-ik_j x} = \boldsymbol{e}(x)\boldsymbol{Q}\boldsymbol{\alpha} \\ U(x) &= \sum_{j=1}^{8} \alpha_j \mathrm{e}^{-ik_j x} = \boldsymbol{e}(x)\boldsymbol{\alpha} \\ P(x) &= \sum_{j=1}^{8} t_j \alpha_j \mathrm{e}^{-ik_j x} = \boldsymbol{e}(x)\boldsymbol{T}\boldsymbol{\alpha} \\ C(x) &= \sum_{j=1}^{8} r_j \alpha_j \mathrm{e}^{-ik_j x} = \boldsymbol{e}(x)\boldsymbol{R}\boldsymbol{\alpha} \end{aligned} \tag{6.112}$$

where

$$\boldsymbol{e}(x) = \left[\mathrm{e}^{-ik_1 x} \quad \mathrm{e}^{-ik_2 x} \quad \mathrm{e}^{-ik_3 x} \quad \mathrm{e}^{-ik_4 x} \quad \mathrm{e}^{-ik_5 x} \quad \mathrm{e}^{-ik_6 x} \quad \mathrm{e}^{-ik_7 x} \quad \mathrm{e}^{-ik_8 x}\right] \tag{6.113}$$

and

$$\boldsymbol{\alpha} = \{\alpha_1 \quad \alpha_2 \quad \alpha_3 \quad \alpha_4 \quad \alpha_5 \quad \alpha_6 \quad \alpha_7 \quad \alpha_8\}^{\mathrm{T}} \tag{6.114}$$

Substitution of Equation 6.112 into the definitions of spectral nodal DOFs 6.97 gives the relationship

$$\boldsymbol{H}(\omega)\boldsymbol{\alpha} = \boldsymbol{d} \tag{6.115}$$

where

$$\boldsymbol{H}(\omega) = \begin{bmatrix} q_1 & q_2 & q_3 & q_4 & 0 & 0 & 0 & 0 \\ -ik_1 q_1 & -ik_2 q_2 & -ik_3 q_3 & -ik_4 q_4 & 0 & 0 & 0 & 0 \\ 1 & 1 & 1 & 1 & 1 & 1 & 1 & 1 \\ t_1 & t_2 & t_3 & t_4 & t_5 & t_6 & t_7 & t_8 \\ q_1 \mathrm{e}_1 & q_2 \mathrm{e}_2 & q_3 \mathrm{e}_3 & q_4 \mathrm{e}_4 & 0 & 0 & 0 & 0 \\ -ik_1 q_1 \mathrm{e}_1 & -ik_2 q_2 \mathrm{e}_2 & -ik_3 q_3 \mathrm{e}_3 & -ik_4 q_4 \mathrm{e}_4 & 0 & 0 & 0 & 0 \\ \mathrm{e}_1 & \mathrm{e}_2 & \mathrm{e}_3 & \mathrm{e}_4 & \mathrm{e}_5 & \mathrm{e}_6 & \mathrm{e}_7 & \mathrm{e}_8 \\ t_1 \mathrm{e}_1 & t_2 \mathrm{e}_2 & t_3 \mathrm{e}_3 & t_4 \mathrm{e}_4 & t_5 \mathrm{e}_5 & t_6 \mathrm{e}_6 & t_7 \mathrm{e}_7 & t_8 \mathrm{e}_8 \end{bmatrix} \tag{6.116}$$

By using Equation 6.115, we can eliminate the constant vector $\boldsymbol{\alpha}$ from Equation 6.112 to write the general solutions as

$$\begin{aligned} W(x) &= \boldsymbol{N}_w(x;\omega)\boldsymbol{d}, \quad U(x) = \boldsymbol{N}_u(x;\omega)\boldsymbol{d} \\ P(x) &= \boldsymbol{N}_p(x;\omega)\boldsymbol{d}, \quad C(x) = \boldsymbol{N}_c(x;\omega)\boldsymbol{d} \end{aligned} \tag{6.117}$$

where $\boldsymbol{N}_w(x;\omega)$, $\boldsymbol{N}_u(x;\omega)$, $\boldsymbol{N}_p(x;\omega)$, and $\boldsymbol{N}_c(x;\omega)$ are the dynamic shape functions defined by

$$\begin{aligned} \boldsymbol{N}_w(x;\omega) &= \boldsymbol{e}(x)\boldsymbol{Q}\boldsymbol{H}^{-1}(\omega), \quad \boldsymbol{N}_u(x;\omega) = \boldsymbol{e}(x)\boldsymbol{H}^{-1}(\omega) \\ \boldsymbol{N}_p(x;\omega) &= \boldsymbol{e}(x)\boldsymbol{T}\boldsymbol{H}^{-1}(\omega), \quad \boldsymbol{N}_c(x;\omega) = \boldsymbol{e}(x)\boldsymbol{R}\boldsymbol{H}^{-1}(\omega) \end{aligned} \tag{6.118}$$

6.3.2.4 Weak Form of Governing Equations

The weak form of the governing Equations 6.91–6.93 are obtained from

$$\begin{aligned}\int_0^L &\{(EI_p W'''' - T_0 W'' - m_w f_D c_0^2 W' + 2i\omega m_w c_0 W' - \omega^2 m W)\delta W \\ &- (EA_{\text{eff}} U'' + \omega^2 m_p U + m_w g_Y W' + 2m_w f_D c_0 C + m_w f_D c_0^2 N\delta_{0n})\delta U \\ &- [P'' - i\omega a^{-2} c_0 P' - i\omega a^{-2}(i\omega + 2f_D c_0)P + m_w g_Y W'' \\ &+ 2i\omega m_w \nu(i\omega + 2f_D c_0)U' - \omega^2 m_w U' + i\omega m_w(1+2\nu)c_0 U'']\delta P\}\mathrm{d}x = 0\end{aligned} \tag{6.119}$$

Equation 6.29 can be transformed into the frequency domain by using Equation 6.90 as

$$\begin{aligned}P' = &- m_w(i\omega + 2f_D c_0)C - m_w(f_D c_0^2 + g_X)N\delta_{0n} \\ &+ i\omega a^{-2} c_0 P - m_w g_Y W' + \omega^2 m_w U - i\omega m_w c_0(1+2\nu)U'\end{aligned} \tag{6.120}$$

Then we can obtain

$$\begin{aligned}P'\delta P|_0^L = &- m_w(i\omega + 2f_D c_0)C\delta P|_0^L - m_w(f_D c_0^2 + g_X)N\delta_{0n}\delta P|_0^L \\ &+ [i\omega a^{-2} c_0 P - m_w g_Y W' + \omega^2 m_w U - i\omega m_w c_0(1+2\nu)U']\delta P|_0^L\end{aligned} \tag{6.121}$$

By integrating Equation 6.119 by parts and using the force-displacement relationships 6.96 and Equation 6.121, we obtain the weak form as

$$\begin{aligned}\int_0^L &[EI_p W''\delta W'' + T_0 W'\delta W' - m_w f_D c_0^2 W'\delta W - \omega^2 m W\delta W \\ &+ i\omega m_w c_0(W'\delta W - W\delta W') + EA_{\text{eff}} U'\delta U' - \omega^2 m_p U\delta U - m_w g_Y W'\delta U \\ &- 2m_w f_D c_0 C\delta U + P'\delta P' + (1/2)i\omega a^{-2} c_0(P'\delta P - P\delta P') \\ &+ m_w g_Y W'\delta P' + i\omega a^{-2}(i\omega + 2f_D c_0)P\delta P - \omega^2 m_w U\delta P' \\ &- 2i\omega m_w \nu(i\omega + 2f_D c_0)U'\delta P + i\omega m_w c_0(1+2\nu)U'\delta P' \\ &+ N\delta_{0n}(-m_w f_D c_0^2 \delta U + \bar{N}\delta U')]\mathrm{d}x \\ &+ i\omega m_w c_0 W\delta W|_0^L - (T_0 - \bar{N})W'\delta W|_0^L + i\omega m_w C\delta P|_0^L - (1/2)i\omega a^{-2} c_0 P\delta P|_0^L \\ &+ m_w(f_D c_0^2 + g_X)N\delta_{0n}\delta P|_0^L - [Q(x) + \bar{N}W']\delta W|_0^L - M(x)\delta W'|_0^L \\ &- [N(x) + \bar{N}N\delta_{0n}]\delta U|_0^L + 2m_w f_D c_0 C(x)\delta P|_0^L = 0\end{aligned} \tag{6.122}$$

6.3.2.5 Spectral Element Equation

Substitution of Equation 6.117 into Equation 6.122 gives

$$\begin{aligned}\delta \boldsymbol{d}^{\mathrm{T}}\Bigg\{\int_0^L &[EI_p \boldsymbol{N}_w''^{\mathrm{T}} \boldsymbol{N}_w'' + T_0 \boldsymbol{N}_w'^{\mathrm{T}} \boldsymbol{N}_w' - m_w f_D c_0^2 \boldsymbol{N}_w^{\mathrm{T}} \boldsymbol{N}_w' - \omega^2 m \boldsymbol{N}_w^{\mathrm{T}} \boldsymbol{N}_w \\ &+ i\omega m_w c_0(\boldsymbol{N}_w^{\mathrm{T}} \boldsymbol{N}_w' - \boldsymbol{N}_w'^{\mathrm{T}} \boldsymbol{N}_w) + EA_{\text{eff}} \boldsymbol{N}_u'^{\mathrm{T}} \boldsymbol{N}_u' - \omega^2 m_p \boldsymbol{N}_u^{\mathrm{T}} \boldsymbol{N}_u \\ &- m_w g_Y \boldsymbol{N}_u^{\mathrm{T}} \boldsymbol{N}_w' - 2m_w f_D c_0 \boldsymbol{N}_u^{\mathrm{T}} \boldsymbol{N}_c + \boldsymbol{N}_p'^{\mathrm{T}} \boldsymbol{N}_p' \\ &+ (1/2)i\omega a^{-2} c_0(\boldsymbol{N}_p^{\mathrm{T}} \boldsymbol{N}_p' - \boldsymbol{N}_p'^{\mathrm{T}} \boldsymbol{N}_p) + m_w g_Y \boldsymbol{N}_p'^{\mathrm{T}} \boldsymbol{N}_w' \\ &+ i\omega a^{-2}(i\omega + 2f_D c_0)\boldsymbol{N}_p^{\mathrm{T}} \boldsymbol{N}_p - \omega^2 m_w \boldsymbol{N}_p'^{\mathrm{T}} \boldsymbol{N}_u\end{aligned}$$

$$
\begin{aligned}
&- 2i\omega m_w \nu(i\omega + 2f_D c_0)\boldsymbol{N}_p^{\mathrm{T}}\boldsymbol{N}_u' + i\omega m_w c_0(1+2\nu)\boldsymbol{N}_{\mathrm{p}}'^{\mathrm{T}}\boldsymbol{N}_u']\mathrm{d}x \\
&+ [i\omega m_w c_0 \boldsymbol{N}_w^{\mathrm{T}}\boldsymbol{N}_w - (T_0 - \bar{N})\boldsymbol{N}_w^{\mathrm{T}}\boldsymbol{N}_w' + i\omega m_w \boldsymbol{N}_p^{\mathrm{T}}\boldsymbol{N}_c \\
&- (1/2)i\omega a^{-2}c_0\boldsymbol{N}_p^{\mathrm{T}}\boldsymbol{N}_p]|_0^L \Big\}\boldsymbol{d} \\
&+ \delta\boldsymbol{d}^{\mathrm{T}}\left\{ N\delta_{0n}\int_0^L (-m_w f_D c_0^2 \boldsymbol{N}_u^{\mathrm{T}} + \bar{N}\boldsymbol{N}_u'^{\mathrm{T}})\mathrm{d}x + m_w(f_D c_0^2 + g_X)N\delta_{0n}\boldsymbol{N}_p^{\mathrm{T}}|_0^L \right\} \\
&- [Q(x) + \bar{N}W']\delta W|_0^L - M(x)\delta W'|_0^L - [N(x) + \bar{N}N\delta_{0n}]\delta U|_0^L \\
&+ 2m_w f_D c_0 C(x)\delta P|_0^L = 0
\end{aligned}
\tag{6.123}
$$

Because $\delta\boldsymbol{d}$ is arbitrary, we can obtain the spectral element equation from Equation 6.123 as follows:

$$
\boldsymbol{S}(\omega)\boldsymbol{d} = \boldsymbol{f}_{\mathrm{c}} + \boldsymbol{f}_{\mathrm{d}} \tag{6.124}
$$

where

$$
\begin{aligned}
\boldsymbol{S}(\omega) = \int_0^L &[EI_{\mathrm{p}}\boldsymbol{N}_w''^{\mathrm{T}}\boldsymbol{N}_w'' + T_0\boldsymbol{N}_w'^{\mathrm{T}}\boldsymbol{N}_w' - m_w f_D c_0^2 \boldsymbol{N}_w^{\mathrm{T}}\boldsymbol{N}_w' - \omega^2 m\boldsymbol{N}_w^{\mathrm{T}}\boldsymbol{N}_w \\
&+ i\omega m_w c_0(\boldsymbol{N}_w^{\mathrm{T}}\boldsymbol{N}_w' - \boldsymbol{N}_w'^{\mathrm{T}}\boldsymbol{N}_w) + EA_{\mathrm{eff}}\boldsymbol{N}_u'^{\mathrm{T}}\boldsymbol{N}_u' - \omega^2 m_p \boldsymbol{N}_u^{\mathrm{T}}\boldsymbol{N}_u \\
&- m_w g_Y \boldsymbol{N}_u^{\mathrm{T}}\boldsymbol{N}_w' - 2m_w f_D c_0 \boldsymbol{N}_u^{\mathrm{T}}\boldsymbol{N}_c + \boldsymbol{N}_p'^{\mathrm{T}}\boldsymbol{N}_p' \\
&+ (1/2)i\omega a^{-2}c_0(\boldsymbol{N}_p^{\mathrm{T}}\boldsymbol{N}_p' - \boldsymbol{N}_p'^{\mathrm{T}}\boldsymbol{N}_p) + m_w g_Y \boldsymbol{N}_p'^{\mathrm{T}}\boldsymbol{N}_w' \\
&+ i\omega a^{-2}(i\omega + 2f_D c_0)\boldsymbol{N}_p^{\mathrm{T}}\boldsymbol{N}_p - \omega^2 m_w \boldsymbol{N}_p'^{\mathrm{T}}\boldsymbol{N}_u \\
&- 2i\omega m_w \nu(i\omega + 2f_D c_0)\boldsymbol{N}_p^{\mathrm{T}}\boldsymbol{N}_u' + i\omega m_w c_0(1+2\nu)\boldsymbol{N}_p'^{\mathrm{T}}\boldsymbol{N}_u']\mathrm{d}x \\
&+ [i\omega m_w c_0 \boldsymbol{N}_w^{\mathrm{T}}\boldsymbol{N}_w - (T_0 - \bar{N})\boldsymbol{N}_w^{\mathrm{T}}\boldsymbol{N}_w' + i\omega m_w \boldsymbol{N}_p^{\mathrm{T}}\boldsymbol{N}_c \\
&- (1/2)i\omega a^{-2}c_0\boldsymbol{N}_p^{\mathrm{T}}\boldsymbol{N}_p]|_0^L
\end{aligned}
\tag{6.125}
$$

and

$$
\begin{aligned}
\boldsymbol{f}_{\mathrm{c}} &= \{Q_1 \quad M_1 \quad T_1 \quad \mu C_1 \quad Q_2 \quad M_2 \quad T_2 \quad \mu C_2\}^{\mathrm{T}} \\
&= \{-Q(0) - \bar{N}W'(0) \quad -M(0) \quad -N(0) - \bar{N}N\delta_{0n} \quad \mu C(0) \\
&\qquad Q(L) + \bar{N}W'(L) \quad M(L) \quad N(L) + \bar{N}N\delta_{0n} \quad -\mu C(L)\}^{\mathrm{T}}
\end{aligned}
\tag{6.126}
$$

$$
\boldsymbol{f}_{\mathrm{d}} = N\delta_{0n}\int_0^L (m_w f_D c_0^2 \boldsymbol{N}_u^{\mathrm{T}} - \bar{N}\boldsymbol{N}_u'^{\mathrm{T}})\mathrm{d}x - m_w(f_D c_o^2 + g_X)N\delta_{0n}\boldsymbol{N}_p^{\mathrm{T}}|_0^L \tag{6.127}
$$

with

$$
\mu = 2m_w f_D c_0 \tag{6.128}
$$

By substituting Equation 6.118 into Equation 6.125, the spectral element matrix can be written as

$$
\boldsymbol{S}(\omega) = \boldsymbol{H}^{-\mathrm{T}}(\omega)\boldsymbol{D}(\omega)\boldsymbol{H}^{-1}(\omega) \tag{6.129}
$$

where

$$\begin{aligned}
\boldsymbol{D} = & EI_{\mathrm{p}}\boldsymbol{QK}^2\boldsymbol{EK}^2\boldsymbol{Q} - T_0\boldsymbol{QKEKQ} + im_w f_D c_0^2\boldsymbol{QEKQ} - \omega^2 m\boldsymbol{QEQ} \\
& + \omega m_w c_0(\boldsymbol{QEKQ} - \boldsymbol{QKEQ}) - EA_{\mathrm{eff}}\boldsymbol{KEK} - \omega^2 m_p\boldsymbol{E} + im_w g_Y\boldsymbol{EKQ} \\
& - 2m_w f_D c_0\boldsymbol{ER} - \boldsymbol{TKEKT} + (1/2)\omega a^{-2}c_0(\boldsymbol{TEKT} - \boldsymbol{TKET}) \\
& - m_w g_Y\boldsymbol{TKEKQ} + i\omega a^{-2}(i\omega + 2f_D c_0)\boldsymbol{TET} + i\omega^2 m_w\boldsymbol{TKE} \\
& - 2\omega m_w\nu(i\omega + 2f_D c_0)\boldsymbol{TEK} - i\omega m_w c_0(1+2\nu)\boldsymbol{TKEK} \\
& + i\omega m_w c_0\boldsymbol{Q\bar{E}Q} + i(T_0 - \bar{N})\boldsymbol{Q\bar{E}KQ} + i\omega m_w\boldsymbol{T\bar{E}R} - (1/2)i\omega a^{-2}c_0\boldsymbol{T\bar{E}T}
\end{aligned} \tag{6.130}$$

and where

$$\begin{aligned}
\boldsymbol{E}(\omega) &= \int_0^L \boldsymbol{e}^{\mathrm{T}}(x;\omega)\boldsymbol{e}(x;\omega)\mathrm{d}x = [E_{rs}(\omega)] \\
\bar{\boldsymbol{E}}(\omega) &= \boldsymbol{e}^{\mathrm{T}}(x;\omega)\boldsymbol{e}(x;\omega)\big|_0^L = [\bar{E}_{rs}(\omega)]
\end{aligned} \qquad (r, s = 1, 2, \ldots 8) \tag{6.131}$$

with

$$\begin{aligned}
E_{rs} &= \begin{cases} \dfrac{i}{k_r + k_s}\bar{E}_{rs} & \text{if } k_r + k_s \neq 0 \\ L & \text{if } k_r + k_s = 0 \end{cases} \\
\bar{E}_{rs} &= \mathrm{e}^{-i(k_r + k_s)L} - 1
\end{aligned} \tag{6.132}$$

The spectral elements represented by Equation 6.124 can be assembled to form a global system equation, after imposing relevant boundary conditions, in the form of

$$\boldsymbol{S}_{\mathrm{g}}(\omega)\boldsymbol{d}_{\mathrm{g}} = \boldsymbol{f}_{\mathrm{g}} \tag{6.133}$$

where $\boldsymbol{S}_{\mathrm{g}}(\omega)$ is the global dynamic stiffness matrix, $\boldsymbol{d}_{\mathrm{g}}$ is the global spectral nodal DOFs vector, and $\boldsymbol{f}_{\mathrm{g}}$ is the global spectral nodal forces vector.

6.3.3 Finite Element Model

The weak form of the governing Equations 86 can be constructed from the weighted-integral statement as

$$\begin{aligned}
\int_0^L & [(EI_{\mathrm{p}}w'''' - T_0 w'' - m_w f_D c_0^2 w' + 2m_w c_0\dot{w}' + m\ddot{w})\delta w \\
& - (EA_{\mathrm{eff}}u'' - m_p\ddot{u} + m_w g_Y w' + 2m_w f_D c_0 c_d + m_w f_D c_0^2)\delta u \\
& - \{P_{\mathrm{d}}'' - a^{-2}c_0\dot{P}_d' - a^{-2}\ddot{P}_d - 2a^{-2}f_D c_0\dot{P}_d + m_w g_Y w'' \\
& + m_w(1+2\nu)(\ddot{u}' + c_0\dot{u}'') + 4m_w f_D c_0\nu\dot{u}'\}\delta P_d]\mathrm{d}x = 0
\end{aligned} \tag{6.134}$$

From Equation 6.29, we obtain

$$\begin{aligned}
P_d'\delta P_d\big|_0^L = & -m_w\dot{c}_d\delta P_d\big|_0^L - 2m_w f_D c_0 c_d\delta P_d\big|_0^L - m_w f_D c_0^2\delta P_d\big|_0^L - m_w g_X\delta P_d\big|_0^L \\
& + a^{-2}c_0\dot{P}_d\delta P_d\big|_0^L - m_w g_Y w'\delta P_d\big|_0^L - m_w\ddot{u}\delta P_d\big|_0^L - m_w c_0(1+2\nu)\dot{u}'\delta P_d\big|_0^L
\end{aligned} \tag{6.135}$$

By integrating Equation 6.134 by parts and then applying Equation 6.135 and the force-displacement relationships 6.12, we can obtain the weak form as

$$\begin{aligned}
&\int_0^L [EI_p w''\delta w'' + T_0 w'\delta w' - m_w f_D c_0^2 w'\delta w + m\ddot{w}\delta w + m_w c_0 \dot{w}'\delta w \\
&- m_w c_0 \dot{w}\delta w' + EA_{\text{eff}} u'\delta u' + m_p \ddot{u}\delta u - m_w g_Y w'\delta u - 2m_w f_D c_0 c_d \delta u + P_d'\delta P_d' \\
&+ (1/2)a^{-2}c_0(\dot{P}_d'\delta P_d - \dot{P}_d\delta P_d') + m_w g_Y w'\delta P_d' + a^{-2}\ddot{P}_d\delta P_d \\
&+ 2a^{-2} f_D c_0 \dot{P}_d \delta P_d + m_w \ddot{u}\delta P_d' - 2m_w \nu \ddot{u}'\delta P_d + m_w c_0(1+2\nu)\dot{u}'\delta P_d' \\
&- 4m_w f_D c_0 \nu \dot{u}'\delta P_d - m_w f_D c_0^2 \delta u + \bar{N}\delta u']\mathrm{d}x \\
&+ m_w c_0 \dot{w}\delta w|_0^L - (T_0 - \bar{N})w'\delta w|_0^L + m_w \dot{c}_d \delta P_d|_0^L - (1/2)a^{-2}c_0 \dot{P}_d \delta P_d|_0^L \\
&+ m_w(f_D c_0^2 + g_X)\delta P_d|_0^L - [Q(x,t) + \bar{N}w']\delta w|_0^L - M(x,t)\delta w'|_0^L \\
&- [N(x,t) + \bar{N}]\delta u|_0^L + 2m_w f_D c_0 c_d(x,t)\delta P_d|_0^L = 0
\end{aligned} \tag{6.136}$$

For the three-node finite element, we define the nodal DOFs for the displacement fields and fluid pressure as follows:

$$\begin{aligned}
&w(0,t) = w_1(t), \quad w(L/2,t) = w_2(t), \quad w(L,t) = w_3(t) \\
&w'(0,t) = \theta_1(t), \quad w'(L/2,t) = \theta_2(t), \quad w'(L,t) = \theta_3(t) \\
&u(0,t) = u_1(t), \quad u(L/2,t) = u_2(t), \quad u(L,t) = u_3(t) \\
&P_d(0,t) = p_1(t), \quad P_d(L/2,t) = p_2(t), \quad P_d(L,t) = p_3(t)
\end{aligned} \tag{6.137}$$

Similarly, the nodal forces, moments, and fluid velocities are defined by

$$\begin{aligned}
&Q(0,t) = -Q_1(t) - \bar{N}w'(0,t), \quad && Q(L,t) = Q_3(t) - \bar{N}w'(L,t) \\
&M(0,t) = -M_1(t), && M(L,t) = M_3(t) \\
&N(0,t) = -N_1(t) - \bar{N}, && N(L,t) = N_3(t) - \bar{N} \\
&c_d(0,t) = C_1(t), && c_d(L,t) = -C_3(t)
\end{aligned} \tag{6.138}$$

The displacement fields and fluid fields are assumed from

$$\begin{aligned}
&w(x,t) = \mathbf{N}_w(x)\mathbf{d}(t), \quad u(x,t) = \mathbf{N}_u(x)\mathbf{d}(t) \\
&P_d(x,t) = \mathbf{N}_p(x)\mathbf{d}(t), \quad c_d(x,t) = \mathbf{N}_c(x)\mathbf{d}_c(t)
\end{aligned} \tag{6.139}$$

where

$$\begin{aligned}
&\mathbf{d}(t) = \{w_1 \;\; \theta_1 \;\; u_1 \;\; p_1 \;\; w_2 \;\; \theta_2 \;\; u_2 \;\; p_2 \;\; w_3 \;\; \theta_3 \;\; u_3 \;\; p_3\}^{\mathrm{T}} \\
&\mathbf{d}_c(t) = \{C_1(t)\; C_3(t)\}^{\mathrm{T}}
\end{aligned} \tag{6.140}$$

and

$$\begin{aligned}
&\mathbf{N}_w(x) = [N_{w1} \;\; N_{\theta 1} \;\; 0 \;\; 0 \;\; N_{w2} \;\; N_{\theta 2} \;\; 0 \;\; 0 \;\; N_{w3} \;\; N_{\theta 3} \;\; 0 \;\; 0] \\
&\mathbf{N}_u(x) = [0 \;\; 0 \;\; N_{u1} \;\; 0 \;\; 0 \;\; 0 \;\; N_{u2} \;\; 0 \;\; 0 \;\; 0 \;\; N_{u3} \;\; 0] \\
&\mathbf{N}_p(x) = [0 \;\; 0 \;\; 0 \;\; N_{p1} \;\; 0 \;\; 0 \;\; 0 \;\; N_{p2} \;\; 0 \;\; 0 \;\; 0 \;\; N_{p3}] \\
&\mathbf{N}_c(x) = [N_{c1} \;\; N_{c3}]
\end{aligned} \tag{6.141}$$

with

$$
\begin{aligned}
N_{w1}(x) &= 1-23\xi^2+66\xi^3-68\xi^4+24\xi^5\\
N_{\theta 1}(x) &= L\xi(1-6\xi+13\xi^2-12\xi^3+4\xi^4)\\
N_{w2}(x) &= 16\xi^2(1-2\xi+\xi^2)\\
N_{\theta 2}(x) &= 8L\xi^2(-1+4\xi-5\xi^2+2\xi^3)\\
N_{w3}(x) &= \xi^2(7-34\xi+52\xi^2-24\xi^3)\\
N_{\theta 3}(x) &= L\xi^2(-1+5\xi-8\xi^2+4\xi^3)\\
N_{u1}(x) &= N_{p1}(x)=N_{c1}(x)=1-3\xi+2\xi^2\\
N_{u2}(x) &= N_{p2}(x)=4\xi(1-\xi)\\
N_{u3}(x) &= N_{p3}(x)=-N_{c3}(x)=-\xi(1-2\xi)\\
\xi &= x/L
\end{aligned}
\tag{6.142}
$$

The weighted-integral statement of Equation 6.89 can be written as

$$\int_0^L [\dot{P}_d + a^2 m_w (c'_d - 2\nu \dot{u}')]\delta c_d \mathrm{d}x = 0 \tag{6.143}$$

By substituting Equation 6.139 into Equation 6.143, we obtain the relationship

$$\boldsymbol{d}_c(t) = \boldsymbol{R}\dot{\boldsymbol{d}}(t) \tag{6.144}$$

where

$$
\begin{aligned}
\boldsymbol{R} &= \left(a^2 m_w \int_0^L \mathbf{N}_c^{\mathrm{T}}\mathbf{N}'_c dx\right)^{-1}\left(2a^2 m_w \nu \int_0^L \mathbf{N}_c^{\mathrm{T}}\mathbf{N}'_u dx - \int_0^L \mathbf{N}_c^{\mathrm{T}}\mathbf{N}_p \mathrm{d}x\right)\\
&= \begin{bmatrix} 0 & 0 & 2\nu & 11\phi & 0 & 0 & -2\nu & 8\phi & 0 & 0 & 0 & \phi \\ 0 & 0 & 0 & \phi & 0 & 0 & 2\nu & 8\phi & 0 & 0 & -2\nu & 11\phi \end{bmatrix}
\end{aligned}
\tag{6.145}
$$

with the use of definition as

$$\phi = \frac{L}{40a^2 m_w} \tag{6.146}$$

By substituting Equation 6.139 into Equation 6.136 and then using Equation 6.144, we can obtain the finite element equation in the form of

$$\boldsymbol{M}\ddot{\boldsymbol{d}}(t) + \boldsymbol{C}\dot{\boldsymbol{d}}(t) + \boldsymbol{K}\boldsymbol{d}(t) = \boldsymbol{f}_{\mathrm{c}}(t) + \boldsymbol{f}_{\mathrm{d}} \tag{6.147}$$

where

$$
\begin{aligned}
\boldsymbol{f}_{\mathrm{c}}(t) = &\{Q_1 \quad M_1 \quad N_1 \quad \mu C_1 \quad 0 \quad 0 \quad 0 \quad 0 \quad Q_3 \quad M_3 \quad N_3 \quad \mu C_3\}^{\mathrm{T}}\\
= &\{-Q(0,t)-\bar{N}w'(0,t) \quad -M(0,t) \quad -N(0,t)-\bar{N} \quad \mu c_d(0,t)\\
&\; 0 \quad 0 \quad 0 \quad 0\\
&\; Q(L,t)+\bar{N}w'(L,t) \quad M(L,t) \quad N(L,t)+\bar{N} \quad \mu c_d(L,t)\}^{\mathrm{T}}
\end{aligned}
\tag{6.148}
$$

$$f_{\mathrm{d}} = \int_0^L (m_w f_D c_0^2 N_u^{\mathrm{T}} - \bar{N} N_u'^{\mathrm{T}})\mathrm{d}x - m_w (f_D c_0^2 + g x) N_p^{\mathrm{T}} \Big|_0^L \tag{6.149}$$

where μ is defined by Equation 6.128. The finite element matrices $\boldsymbol{M}$, $\boldsymbol{C}$, $\boldsymbol{K}$, and nodal force vector $\boldsymbol{f}_{\mathrm{g}}$ are given in Appendix 6.B.

Example 6.2

Revisit the same pipeline as considered in the previous Example 6.1 and assume that the unsteady fluid is flowing through the simply-supported uniform pipeline. For the spectral element analyses, we use the same geometric and material properties of the pipeline and internal fluid as used in the previous example.

Table 6.2 shows the eigenfrequencies of the first, third and fifth bending modes, the first fluid pressure mode, and the first axial mode for six different fluid velocities c_0. When $c_0 = 0$ m/s, the eigenfrequencies obtained by the present spectral element model are shown to be identical with the exact theoretical results for the bending and axial modes [24]. Similar to the case of a steady fluid, Table 6.2 shows that all eigenfrequencies tend to decrease as the fluid velocity c_0 is increased gradually and, as a result, the eigenfrequency of the first bending mode vanishes for the first time as the fluid velocity c_0 reaches a certain value, the divergence fluid velocity ($c_{\mathrm{D}} = 26.94$ m/s), where the divergence (static) instability may occur.

Table 6.2 Eigenfrequencies versus fluid velocity c_0

Fluid velocity c_0 (m/s)	$f_1^{(b)}$	$f_3^{(b)}$	$f_5^{(b)}$	$f_1^{(p)}$	$f_1^{(a)}$
0	2.1750 (2.1750)	19.575 (19.575)	54.375 (54.375)	58.114 (n/a)	264.58 (264.58)
5	2.1323	19.544	54.347	58.113 + 0.26i	264.57 + 0.45i
10	2.0007	19.453	54.261	58.112 + 0.48i	264.57 + 0.79i
15	1.7680	19.299	54.119	58.110 + 0.68i	264.57 + 1.09i
20	1.4002	19.083	53.919	58.107 + 0.88i	264.56 + 1.39i
26.94 (c_{D})	0.0000	18.677	53.546	58.102 + 1.15i	264.56 + 1.77i

Note: () = exact theory [24] ^{a} = axial modes, ^{b} = bending modes, and ^{p} = pressure modes.

Figure 6.7 shows the time histories of the axial displacement $u(x, t)$, fluid pressure $p_d(x, t)$, and the fluid velocity $c_d(x, t)$, all predicted in the middle of the pipeline when the fluid velocity at the pipeline inlet is controlled by

$$c_d(0, t) = c_0 \delta \sin \omega t \ (\mathrm{m/s})$$

where $c_0 = 15$ m/s, $\delta = 0.01$, and $\omega = 58.11\pi$ rad/s. As the inlet fluid velocity oscillates at a frequency equal to the first natural frequency of the fluid pressure, Figure 6.7 certainly shows that all the time histories of $u(x, t)$, $p_d(x, t)$ and $c_d(x, t)$ increase rapidly to demonstrate resonance phenomenon.

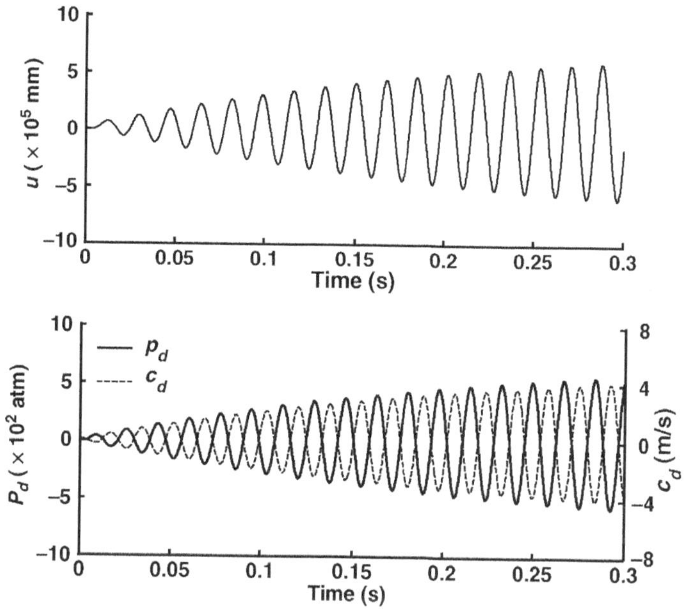

Figure 6.7 Time histories of the axial displacement, perturbed fluid pressure, and velocity at the middle of the pipeline

Appendix 6.A: Finite Element Matrices: Steady Fluid

The finite element matrices defined in Equation 6.84 are given by

$$\begin{aligned} \boldsymbol{M} &= \boldsymbol{M}_{ww} + \boldsymbol{M}_{uu} \\ \boldsymbol{C} &= \boldsymbol{G} + \boldsymbol{C}_G \\ \boldsymbol{K} &= \boldsymbol{K}_{ww1} + \boldsymbol{K}_{ww2} + \boldsymbol{K}_{ww3} + \boldsymbol{K}_{ww4} + \boldsymbol{K}_{uu} + \boldsymbol{K}_{uw} \end{aligned} \tag{6.A.1}$$

where

$$\boldsymbol{M}_{uu} = m_p \int_0^L \boldsymbol{N}_u^{\mathrm{T}} \boldsymbol{N}_u \mathrm{d}x = \frac{m_{\mathrm{p}} L}{6} \begin{bmatrix} 0 & 0 & 0 & 0 & 0 & 0 \\ 0 & 0 & 0 & 0 & 0 & 0 \\ 0 & 0 & 2 & 0 & 0 & 1 \\ 0 & 0 & 0 & 0 & 0 & 0 \\ 0 & 0 & 0 & 0 & 0 & 0 \\ 0 & 0 & 1 & 0 & 0 & 2 \end{bmatrix}$$

$$\boldsymbol{M}_{ww} = m \int_0^L \boldsymbol{N}_w^{\mathrm{T}} \boldsymbol{N}_w \mathrm{d}x = \frac{mL}{420} \begin{bmatrix} 156 & 22L & 0 & 54 & -13L & 0 \\ 22L & 4L^2 & 0 & 13L & -3L^2 & 0 \\ 0 & 0 & 0 & 0 & 0 & 0 \\ 54 & 13L & 0 & 156 & -22L & 0 \\ -13L & -3L^2 & 0 & -22L & 4L^2 & 0 \\ 0 & 0 & 0 & 0 & 0 & 0 \end{bmatrix}$$

$$\boldsymbol{G} = m_w c_0 \int_0^L (\boldsymbol{N}_w^{\mathrm{T}} \boldsymbol{N}_w' - \boldsymbol{N}_w'^{T} \boldsymbol{N}_w) \mathrm{d}x = \frac{m_w c_0}{30} \begin{bmatrix} 0 & 6L & 0 & 30 & -6L & 0 \\ -6L & 0 & 0 & 6L & -L^2 & 0 \\ 0 & 0 & 0 & 0 & 0 & 0 \\ -30 & -6L & 0 & 0 & 6L & 0 \\ 6L & L^2 & 0 & -6L & 0 & 0 \\ 0 & 0 & 0 & 0 & 0 & 0 \end{bmatrix}$$

$$\boldsymbol{C}_G = m_w c_0 \boldsymbol{N}_w^{\mathrm{T}} \boldsymbol{N}_w \big|_0^L = m_w c_0 \begin{bmatrix} -1 & 0 & 0 & 0 & 0 & 0 \\ 0 & 0 & 0 & 0 & 0 & 0 \\ 0 & 0 & 0 & 0 & 0 & 0 \\ 0 & 0 & 0 & 1 & 0 & 0 \\ 0 & 0 & 0 & 0 & 0 & 0 \\ 0 & 0 & 0 & 0 & 0 & 0 \end{bmatrix} \tag{6.A.2}$$

$$\boldsymbol{K}_{ww1} = EI_{\mathrm{p}} \int_0^L \boldsymbol{N}_w''^{\mathrm{T}} \boldsymbol{N}_w'' \mathrm{d}x = \frac{EI_{\mathrm{p}}}{L^3} \begin{bmatrix} 12 & 6L & 0 & -12 & 6L & 0 \\ 6L & 4L^2 & 0 & -6L & 2L^2 & 0 \\ 0 & 0 & 0 & 0 & 0 & 0 \\ -12 & -6L & 0 & 12 & -6L & 0 \\ 6L & 2L^2 & 0 & -6L & 4L^2 & 0 \\ 0 & 0 & 0 & 0 & 0 & 0 \end{bmatrix}$$

$$\boldsymbol{K}_{ww2} = T_0 \int_0^L \boldsymbol{N}_w'^{\mathrm{T}} \boldsymbol{N}_w' \mathrm{d}x = \frac{T_0}{30L} \begin{bmatrix} 36 & 3L & 0 & -36 & 3L & 0 \\ 3L & 4L^2 & 0 & -3L & -L^2 & 0 \\ 0 & 0 & 0 & 0 & 0 & 0 \\ -36 & -3L & 0 & 36 & -3L & 0 \\ 3L & -L^2 & 0 & -3L & 4L^2 & 0 \\ 0 & 0 & 0 & 0 & 0 & 0 \end{bmatrix}$$

$$\boldsymbol{K}_{ww3} = -m_w f_D c_0^2 \int_0^L \boldsymbol{N}_w^{\mathrm{T}} \boldsymbol{N}_w' \mathrm{d}x = -\frac{m_w f_D c_0^2}{60} \begin{bmatrix} -30 & 6L & 0 & 30 & -6L & 0 \\ -6L & 0 & 0 & 6L & -L^2 & 0 \\ 0 & 0 & 0 & 0 & 0 & 0 \\ -30 & -6L & 0 & 30 & 6L & 0 \\ 6L & L^2 & 0 & -6L & 0 & 0 \\ 0 & 0 & 0 & 0 & 0 & 0 \end{bmatrix}$$

$$\boldsymbol{K}_{ww4} = -\left(T_0 - \bar{N}\right) \bar{\boldsymbol{N}}_w^{\mathrm{T}} \bar{\boldsymbol{N}}_w' \Big|_0^L = (T_0 - \bar{N}) \begin{bmatrix} 0 & 1 & 0 & 0 & 0 & 0 \\ 0 & 0 & 0 & 0 & 0 & 0 \\ 0 & 0 & 0 & 0 & 0 & 0 \\ 0 & 0 & 0 & 0 & -1 & 0 \\ 0 & 0 & 0 & 0 & 0 & 0 \\ 0 & 0 & 0 & 0 & 0 & 0 \end{bmatrix}$$

$$K_{uu} = EA_{eff}\int_0^L N_u'^{\mathrm{T}} N_u' \mathrm{d}x = \frac{EA_{\text{eff}}}{L}\begin{bmatrix} 0 & 0 & 0 & 0 & 0 & 0 \\ 0 & 0 & 0 & 0 & 0 & 0 \\ 0 & 0 & 1 & 0 & 0 & -1 \\ 0 & 0 & 0 & 0 & 0 & 0 \\ 0 & 0 & 0 & 0 & 0 & 0 \\ 0 & 0 & -1 & 0 & 0 & 1 \end{bmatrix}$$

$$K_{uw} = -m_w g_Y \int_0^L N_u^{\mathrm{T}} N_w' \mathrm{d}x = \frac{m_w g_y}{12}\begin{bmatrix} 0 & 0 & 0 & 0 & 0 & 0 \\ 0 & 0 & 0 & 0 & 0 & 0 \\ 6 & -L & 0 & -6 & L & 0 \\ 0 & 0 & 0 & 0 & 0 & 0 \\ 0 & 0 & 0 & 0 & 0 & 0 \\ 6 & L & 0 & -6 & -L & 0 \end{bmatrix}$$

Appendix 6.B: Finite Element Matrices: Unsteady Fluid

The finite element matrices in Equation 6.147 are given by

$$\begin{aligned} M &= M_{\text{symm}} + M_{\text{asymm}} \\ C &= C_{\text{skew}} + C_{\text{symm}} + C_{\text{asymm}} \\ K &= K_{\text{symm}} + K_{\text{asymm}} \end{aligned} \tag{6.B.1}$$

where

$$M_{\text{symm}} = m\int_0^L N_w^{\mathrm{T}} N_w \mathrm{d}x + m_{\mathrm{p}}\int_0^L N_u^{\mathrm{T}} N_u \mathrm{d}x + \frac{1}{a^2}\int_0^L N_p^{\mathrm{T}} N_p \mathrm{d}x = [M_{ij}] = [M_{ji}]$$

$$\begin{aligned}
&M_{1,1} = M_{9,9} = 2092\mu_1, && M_{1,2} = -M_{9,10} = 114L\mu_1 \\
&M_{1,5} = M_{5,9} = 880\mu_1, && M_{1,6} = -M_{6,9} = -160L\mu_1 \\
&M_{1,9} = 262\mu_1, && M_{1,10} = -M_{2,9} = -29L\mu_1 \\
&M_{2,2} = M_{10,10} = 8L^2\mu_1, && M_{2,5} = -M_{5,10} = 88L\mu_1 \\
&M_{2,6} = M_{6,10} = -12L^2\mu_1, && M_{2,10} = -3L^2\mu_1 \\
&M_{3,3} = M_{11,11} = 4\mu_2, && M_{3,7} = M_{7,11} = 2\mu_2 \\
&M_{3,11} = -\mu_2, && M_{5,5} = 5632\mu_1 \\
&M_{4,4} = M_{12,12} = 4\mu_3, && M_{4,8} = M_{8,12} = 2\mu_3 \\
&M_{4,12} = -\mu_3, && M_{6,6} = 128L^2\mu_1 \\
&M_{7,7} = 16\mu_2, && M_{8,8} = 16\mu_3
\end{aligned}$$

Others = 0

$$\bar{\boldsymbol{M}}_{\text{asymm}} = m_w \int_0^L \boldsymbol{N}_p'^{\text{T}} \boldsymbol{N}_u \text{d}x - 2m_w \nu \int_0^L \boldsymbol{N}_p^{\text{T}} \boldsymbol{N}_u' \text{d}x + m_w \boldsymbol{N}_p^{\text{T}} \boldsymbol{N}_c \Big|_0^L \boldsymbol{R} = [\bar{M}_{ij}]$$

$$\begin{aligned}
&\bar{M}_{4,3} = -\bar{M}_{12,11} = -3(\mu_4 - \mu_5) - 80 m_w a^2 \nu \mu_5, && \bar{M}_{4,4} = \bar{M}_{12,12} = -11L\mu_5 \\
&\bar{M}_{4,7} = -\bar{M}_{12,7} = -4(\mu_4 + \mu_5) + 80 m_w a^2 \nu \mu_5, && \bar{M}_{4,8} = \bar{M}_{12,8} = -8L\mu_5 \\
&\bar{M}_{4,11} = -\bar{M}_{12,3} = \mu_4 + \mu_5, && \bar{M}_{4,12} = \bar{M}_{12,4} = -L\mu_5 \\
&\bar{M}_{8,3} = -\bar{M}_{8,11} = 4(\mu_4 + \mu_5), && \text{Others} = 0
\end{aligned}$$

$$\boldsymbol{C}^*_{\text{skew}} = m_w c_0 \int_0^L (\boldsymbol{N}_w^{\text{T}} \boldsymbol{N}_w' - \boldsymbol{N}_w'^{\text{T}} \boldsymbol{N}_w) \text{d}x + \frac{c_0}{2a^2} \int_0^L (\boldsymbol{N}_p^{\text{T}} \boldsymbol{N}_c' - \boldsymbol{N}_p'^{\text{T}} \boldsymbol{N}_p) \text{d}x = [C^*_{ij}] = [-C^*_{ji}]$$

$$\begin{aligned}
&C^*_{1,2} = C^*_{9,10} = 46L\chi_1, && C^*_{1,5} = C^*_{5,9} = 480\chi_1 \\
&C^*_{1,6} = -C^*_{6,9} = -128L\chi_1, && C^*_{1,9} = 150\chi_1 \\
&C^*_{1,10} = -14L\chi_1, && C^*_{2,5} = -C^*_{5,10} = 32L\chi_1 \\
&C^*_{2,6} = C^*_{6,10} = -8L^2\chi_1, && C^*_{2,9} = 14L\chi_1 \\
&C^*_{2,10} = -L^2\chi_1, && C^*_{4,8} = C^*_{8,12} = 4\chi_2 \\
&C^*_{4,12} = -\chi_2, && C^*_{5,6} = 256L\chi_1 \\
&\text{Others} = 0
\end{aligned}$$

$$\boldsymbol{C}_{\text{symm}} = m_w c_0 \boldsymbol{N}_w^{\text{T}} \boldsymbol{N}_w \Big|_0^L + \frac{2 f_D c_0}{a^2} \int_0^L \boldsymbol{N}_p^{\text{T}} \boldsymbol{N}_p \text{d}x + \frac{c_0}{2a^2} \boldsymbol{N}_p^{\text{T}} \boldsymbol{N}_p \Big|_0^L = [C_{ij}] = [C_{ji}]$$

$$\begin{aligned}
&C_{1,1} = -C_{9,9} = -m_w c_o, && C_{4,4} = 4\chi_3 + \chi_4 \\
&C_{4,8} = C_{8,12} = 2\chi_3, && C_{4,12} = -\chi_3 \\
&C_{8,8} = 16\chi_3, && C_{12,12} = 4\chi_3 - \chi_4 \\
&\text{Others} = 0
\end{aligned} \tag{6.B.2}$$

$$\boldsymbol{C}_{\text{asymm}} = m_w c_0 (1 + 2\nu) \int_0^L \boldsymbol{N}_p'^{\text{T}} \boldsymbol{N}_u' \text{d}x - 4 m_w f_D c_0 \nu \int_0^L \boldsymbol{N}_p^{\text{T}} \boldsymbol{N}_u' \text{d}x - 2 m_w f_D c_0 \int_0^L \boldsymbol{N}_u^{\text{T}} \boldsymbol{N}_c \text{d}x \boldsymbol{R} = [\bar{C}_{ij}]$$

$$\begin{aligned}
&\bar{C}_{3,3} = -\bar{C}_{7,7} = \bar{C}_{11,11} = -64L m_w f_D c_0 a^2 \nu \chi_7, && \bar{C}_{3,4} = -\bar{C}_{11,12} = -9L^2 f_D c_0 \chi_7 \\
&\bar{C}_{3,7} = \bar{C}_{11,7} = 48L m_w f_D c_0 a^2 \nu \chi_7, && \bar{C}_{3,8} = -\bar{C}_{11,8} = -8L^2 f_D c_0 \chi_7 \\
&\bar{C}_{3,11} = \bar{C}_{11,3} = 16L m_w f_D c_0 a^2 \nu \chi_7, && \bar{C}_{3,12} = -\bar{C}_{11,4} = -3L^2 f_D c_0 \chi_7 \\
&\bar{C}_{7,3} = \bar{C}_{7,11} = -32L m_w f_D c_0 a^2 \nu \chi_7, && \bar{C}_{7,4} = -\bar{C}_{7,12} = -4L^2 f_D c_0 \chi_7 \\
&\bar{C}_{4,3} = 7\chi_5 + 3\chi_6, && \bar{C}_{4,7} = \bar{C}_{8,11} = -8\chi_5 - 4\chi_6 \\
&\bar{C}_{4,11} = \chi_5 + \chi_6, && \bar{C}_{8,3} = \bar{C}_{12,7} = -8\chi_5 + 4\chi_6 \\
&\bar{C}_{8,7} = 16\chi_5, && \bar{C}_{12,3} = \chi_5 - \chi_6 \\
&\bar{C}_{12,11} = 7\chi_5 - 3\chi_6, && \text{Others} = 0
\end{aligned}$$

$$\boldsymbol{K}_{\text{symm}} = EI_{\text{p}} \int_0^L \boldsymbol{N}_w''^{\text{T}} \boldsymbol{N}_w'' \mathrm{d}x + T_0 \int_0^L \boldsymbol{N}_w'^{\text{T}} \boldsymbol{N}_w' \mathrm{d}x + EA_{eff} \int_0^L \boldsymbol{N}_u'^{\text{T}} \boldsymbol{N}_u' \mathrm{d}x + \int_0^L \boldsymbol{N}_p'^{\text{T}} \boldsymbol{N}_c' \mathrm{d}x = [K_{ij}] = [K_{ji}]$$

$K_{1,1} = K_{9,9} = 2546\kappa_1 + 1668\kappa_2$, $K_{1,2} = -K_{9,10} = 569L\kappa_1 + 39L\kappa_2$

$K_{1,5} = K_{5,9} = -1792\kappa_1 - 1536\kappa_2$, $K_{1,6} = -K_{6,9} = 960L\kappa_1 + 240L\kappa_2$

$K_{1,9} = -754\kappa_1 - 132\kappa_2$, $K_{1,10} = 121L\kappa_1 - 9L\kappa_2$

$K_{2,2} = K_{10,10} = 166L^2\kappa_1 + 28L^2\kappa_2$, $K_{2,5} = -K_{5,10} = -448L\kappa_1 - 48L\kappa_2$

$K_{2,6} = K_{6,10} = 160L^2\kappa_1 - 8L^2\kappa_2$, $K_{2,9} = -121L\kappa_1 + 9L\kappa_2$

$K_{2,10} = 19L^2\kappa_1 - 5L^2\kappa_2$, $K_{3,3} = K_{11,11} = 7\kappa_3$

$K_{3,7} = K_{7,11} = -8\kappa_3$ $K_{3,11} = \kappa_3$

$K_{4,4} = K_{12,12} = 7\kappa_4$ $K_{4,8} = K_{8,12} = -8\kappa_4$

$K_{4,12} = \kappa_4$, $K_{5,5} = 3584\kappa_1 + 3072\kappa_2$

$K_{6,6} = 640L^2\kappa_1 + 256L\kappa_2$, $K_{7,3} = 16\kappa_3$

$K_{8,8} = 16\kappa_4$, Others $= 0$

$$\boldsymbol{K}_{\text{asymm}} = -m_w f_D c_0^2 \int_0^L \boldsymbol{N}_w^{\text{T}} \boldsymbol{N}_w' \mathrm{d}x - (T_0 - \bar{N}) \boldsymbol{N}_w^{\text{T}} \boldsymbol{N}_w' \big|_0^L - m_w g_Y \int_0^L \boldsymbol{N}_u^{\text{T}} \boldsymbol{N}_w' \mathrm{d}x + m_w g_Y \int_0^L \boldsymbol{N}_p'^{\text{T}} \boldsymbol{N}_w' \mathrm{d}x = [\bar{K}_{ij}]$$

$\bar{K}_{1,1} = -\bar{K}_{9,9} = 630\kappa_5$, $\bar{K}_{1,2} = -46L\kappa_5 + (T_0 - \bar{N})$

$\bar{K}_{1,5} = \bar{K}_{5,9} = -\bar{K}_{5,1} = -\bar{K}_{9,5} = -480\kappa_5$, $\bar{K}_{1,6} = -\bar{K}_{6,1} = \bar{K}_{9,6} = -\bar{K}_{6,9} = 128L\kappa_5$

$\bar{K}_{1,9} = -\bar{K}_{9,1} = -150\kappa_5$, $\bar{K}_{1,10} = \bar{K}_{9,2} = -\bar{K}_{10,1} = -\bar{K}_{2,9} = 14L\kappa_5$

$\bar{K}_{2,1} = \bar{K}_{10,9} = 46L\kappa_5$, $\bar{K}_{2,5} = -\bar{K}_{5,10} = -\bar{K}_{5,2} = \bar{K}_{10,5} = -32L\kappa_5$

$\bar{K}_{2,6} = -\bar{K}_{6,2} = -\bar{K}_{10,6} = \bar{K}_{6,10} = 8L^2\kappa_5$, $\bar{K}_{2,10} = -\bar{K}_{10,2} = L^2\kappa_5$

$\bar{K}_{3,1} = -\bar{K}_{11,9} = 202\kappa_6$, $\bar{K}_{3,2} = \bar{K}_{11,10} = -13L\kappa_6$

$\bar{K}_{3,5} = -\bar{K}_{11,5} = -224\kappa_6$, $\bar{K}_{3,6} = \bar{K}_{11,6} = 32L\kappa_6$

$\bar{K}_{3,9} = -\bar{K}_{11,1} = 22\kappa_6$, $\bar{K}_{3,10} = \bar{K}_{11,2} = L\kappa_6$

$\bar{K}_{4,1} = \bar{K}_{12,9} = 31\kappa_7$, $\bar{K}_{4,2} = -\bar{K}_{4,10} = \bar{K}_{12,2} = -\bar{K}_{12,10} = -L\kappa_7$

$\bar{K}_{4,5} = \bar{K}_{8,1} = \bar{K}_{8,9} = \bar{K}_{12,5} = -32\kappa_7$, $\bar{K}_{4,9} = \bar{K}_{12,1} = \kappa_7$

$\bar{K}_{5,6} = -256L\kappa_5$, $\bar{K}_{6,5} = 256L\kappa_5$

$\bar{K}_{7,1} = -\bar{K}_{7,9} = 240\kappa_6$, $\bar{K}_{7,2} = \bar{K}_{7,10} = 12L\kappa_6$

$\bar{K}_{7,6} = -64L\kappa_6$, $\bar{K}_{8,2} = -\bar{K}_{8,10} = 2L\kappa_7$

$\bar{K}_{8,5} = 64\kappa_7$, $\bar{K}_{9,10} = -46L\kappa_5 - (T_0 - \bar{N})$

Others $= 0$

In above equations, $i, j = 1, 2, \ldots 8$ and the following definitions are used.

$$\begin{aligned}
&\mu_1 = \frac{mL}{13860}, \quad \mu_2 = \frac{m_p L}{30}, \quad \mu_3 = \frac{L}{30a^2} \\
&\mu_4 = \frac{m_w}{6}, \quad \mu_5 = \frac{m_w \nu}{3}, \quad \mu_5 = \frac{1}{40a^2} \\
&\chi_1 = \frac{m_w c_o}{630}, \quad \chi_2 = \frac{c_0}{6a^2}, \quad \chi_3 = \frac{f_D c_0 L}{15a^2}, \quad \chi_4 = \frac{c_0}{2a^2} \\
&\chi_5 = \frac{m_w c_0(1+2\nu)}{3L}, \quad \chi_6 = \frac{2m_w f_D c_0 \nu}{3}, \quad \chi_7 = \frac{1}{120a^2} \\
&\kappa_1 = \frac{2EI_p}{35L^3}, \quad \kappa_2 = \frac{T_0}{630L}, \quad \kappa_3 = \frac{EA_{\text{eff}}}{3L}, \quad \kappa_4 = \frac{1}{3L} \\
&\kappa_5 = \frac{m_w f_D c_0^2}{1260}, \quad \kappa_6 = \frac{m_w g_y}{420}, \quad \kappa_7 = \frac{m_w g_y}{15L}
\end{aligned} \tag{6.B.3}$$

The nodal forces vector $\boldsymbol{f}_g$ defined by Equation 6.149 is given by

$$f_g = \{0 \quad 0 \quad f_{g3} \quad f_{g4} \quad 0 \quad 0 \quad f_{g7} \quad 0 \quad 0 \quad 0 \quad f_{g11} \quad f_{g12}\}^T \tag{6.B.4}$$

where

$$\begin{aligned}
&f_{g3} = \frac{1}{6} L m_w f_D c_0^2 + \bar{N}, \quad f_{g4} = -f_{g12} = m_w(f_D c_0^2 + g_x) \\
&f_{g7} = \frac{2}{3} L m_w f_D c_0^2, \qquad f_{g11} = \frac{1}{6} L m_w f_D c_0^2 - \bar{N}
\end{aligned} \tag{6.B.5}$$

7

Dynamics of Axially Moving Structures

Axially moving structures are of technological importance and present in a wide range of engineering problems: typical examples are the thread-lines in the textile industry, chain and belt drives, high-speed paper and magnetic tapes, band saw blades, filaments, aerial cable tramways, cooling tower strips and the like. The axially moving speed of a structure can significantly affect its dynamic characteristics, giving rise to variations in natural frequencies and vibration modes. Above a certain critical moving speed, the axially moving structure may experience severe vibration or structural instability resulting in structural failure. Thus, it is important to accurately predict the dynamic characteristics of such axially moving structures to achieve successful designs. Extensive reviews on this subject can be found in Ref. [164]. If the bending stiffness of an axially moving one-dimensional structure is negligible, the structure can be represented by the string model [77]. However, if the bending stiffness is not negligible, the axially moving structure is usually represented by a beam model [99, 130, 137, 165]. For the axially moving structures with a wide width, such as the wide band saw blades and cooling tower strips, one needs to use two-dimensional plate models [67, 76, 113]. Thus, in this chapter, spectral element modeling is considered for axially moving strings, Bernoulli–Euler beams, Timoshenko beams, and thin plates.

7.1 Axially Moving String

7.1.1 Equation of Motion

Figure 7.1 shows a uniform string that travels in its axial direction at a constant moving speed c under a constant tension N_x. The string travels between two simple supports that are separated by a distance L and it has the mass density ρ and the uniform cross-sectional area A. Assume that the axially moving string takes a small transverse vibration and that its bending (flexural) rigidity is negligibly small. As the mass transports continuously across the boundaries at $x = 0$ and L, the extended form of Hamilton's principle developed by McIver [120] for the systems of changing mass is used to derive the governing equation of motion and associated boundary

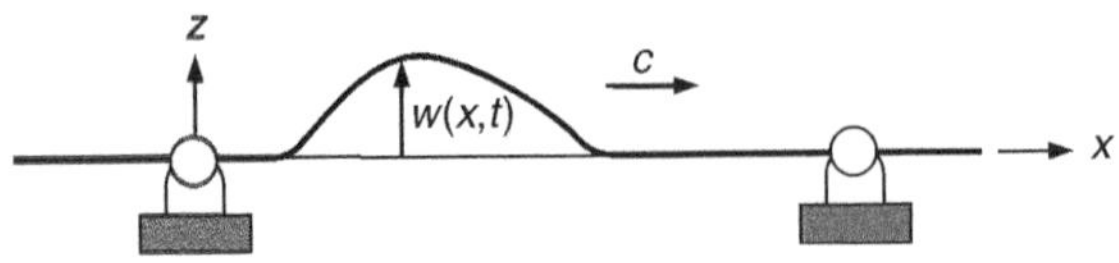

Figure 7.1 Axially moving string

conditions. The extended form of Hamilton's principle is given by

$$\int_{t_1}^{t_2} (\delta T - \delta V + \delta W_{NC} - \delta W_{MT})dt = 0 \tag{7.1}$$

where T is the kinetic energy, V is the strain energy, δW_{NC} is the virtual work done by non-conservative external forces, and δW_{MT} is the virtual momentum transport of mass across the boundaries at $x = 0$ and L.

The kinetic and potential energies are given by

$$\begin{aligned} T &= \frac{1}{2}\int_0^L \rho A[c^2 + (\dot{w} + cw')^2]dx \\ V &= \frac{1}{2}\int_0^L N_x w'^2 dx \end{aligned} \tag{7.2}$$

where $w(x, t)$ is the transverse displacement. The dot $(\cdot)$ and prime $(')$ denote the derivatives with respective to time t and spatial coordinate x, respectively. The virtual work δW_{NC} is given by

$$\delta W = \int_0^L f(x,t)\delta w dx + Q_1 \delta w(0,t) + Q_2 \delta w(L,t) \tag{7.3}$$

where $f(x, t)$ is the distributed dynamic force per length and $Q_1(t)$ and $Q_2(t)$ are the transverse forces applied at $x = 0$ and $x = L$, respectively. The virtual momentum transport of mass across the boundaries at $x = 0$ and L is given by

$$\delta W_{MT} = \int_A \rho c(\boldsymbol{v}\cdot\delta\boldsymbol{r})|_0^L dA = \rho A c(\boldsymbol{v}\cdot\delta\boldsymbol{r})|_0^L \tag{7.4}$$

where $\boldsymbol{r}$ is the position vector of a point at x, $\delta\boldsymbol{r}$ is the virtual displacement, and $\boldsymbol{v} = d\boldsymbol{r}/dt$ is the velocity field of the moving beam at time t. They are given by

$$\begin{aligned} \boldsymbol{r} &= x\boldsymbol{n}_x + w(x,t)\boldsymbol{n}_z, \qquad \delta\boldsymbol{r} = \delta w(x,t)\boldsymbol{n}_z \\ \boldsymbol{v} &= c\boldsymbol{n}_x + (\dot{w} + cw')\boldsymbol{n}_z \end{aligned} \tag{7.5}$$

where $\boldsymbol{n}_x$ and $\boldsymbol{n}_z$ are the unit direction vectors in the axial (x) and transverse (z) directions, respectively. Substitution of Equation 7.5 into Equation 7.4 gives

$$\delta W_{MT} = \rho A c(\dot{w} + cw')\delta w|_0^L \tag{7.6}$$

Substituting Equations 7.2, 7.3 and 7.6 into Equation 7.1 and applying the integral by parts gives

$$\int_{t_1}^{t_2}\left\{\int_0^L [-\rho A(c^2 w'' + 2c\dot{w}' + \ddot{w}) + N_x w'' + f(x,t)]\delta w \mathrm{d}x + Q_1 \delta w(0,t) + Q_2 \delta w(L,t) - (N_x w')\delta w\big|0L\right\}\mathrm{d}t = 0 \tag{7.7}$$

As δw is the arbitrary variation in $w(x, t)$, we can obtain the equation of motion from Equation 7.7

$$\rho A\ddot{w} + 2\rho Ac\dot{w}' + \rho Ac^2 w'' - N_x w'' = f(x,t) \tag{7.8}$$

and the boundary conditions

$$\begin{aligned} Q(0,t) &= -Q_1(t) \quad \text{or} \quad w(0,t) = w_1(t) \\ Q(L,t) &= Q_2(t) \quad \text{or} \quad w(L,t) = w_2(t) \end{aligned} \tag{7.9}$$

where $Q(x, t)$ is the transverse force defined by

$$Q(x,t) = N_x w'(x,t) \tag{7.10}$$

In Equation 7.8, the first term is related to the local acceleration, the second term to the Coriolis acceleration, and the third term to the centrifugal acceleration.

7.1.2 Spectral Element Modeling

7.1.2.1 Governing Equations in the Frequency Domain

Assume the dynamic response in the spectral form

$$w(x,t) = \frac{1}{N}\sum_{n=0}^{N-1} W_n(x)\mathrm{e}^{i\omega_n t} \tag{7.11}$$

The distributed dynamic force $f(x, t)$, the transverse force $Q(x, t)$, the boundary transverse forces $Q_1(t)$ and $Q_2(t)$, and the boundary displacements $w_1(t)$ and $w_2(t)$ are also assumed to be as follows in the spectral forms

$$\begin{aligned} \{f(x,t), Q(x,t)\} &= \frac{1}{N}\sum_{n=0}^{N-1}\{F_n(x), Q_n(x)\}\mathrm{e}^{i\omega_n t} \\ \{Q_1(t), Q_2(t)\} &= \frac{1}{N}\sum_{n=0}^{N-1}\{Q_{1n}(t), Q_{2n}(t)\}\mathrm{e}^{i\omega_n t} \\ \{w_1(t), w_2(t)\} &= \frac{1}{N}\sum_{n=0}^{N-1}\{W_{1n}, W_{2n}\}\mathrm{e}^{i\omega_n t} \end{aligned} \tag{7.12}$$

By substituting Equations 7.11 and 7.12 into Equation 7.8, we obtain the frequency-domain governing equation as

$$-\omega^2 \rho AW + 2i\omega\rho AcW' - (N_x - \rho Ac^2)W'' = F(x) \tag{7.13}$$

where the subscripts n are omitted for brevity. Similarly, by substituting Equations 7.11 and 7.12 into Equations 7.9 and 7.10, we obtain the frequency-domain boundary conditions

$$\begin{aligned} Q(0) &= -Q_1 \quad \text{or} \quad W(0) = W_1 \\ Q(L) &= +Q_2 \quad \text{or} \quad W(L) = W_2 \end{aligned} \tag{7.14}$$

and the frequency-domain force-displacement relationship

$$Q(x) = N_x W' \tag{7.15}$$

7.1.2.2 Spectral Nodal DOFs, Forces and Moments

The spectral nodal displacements, slopes, forces, and moments are indicated in Figure 7.2. The spectral nodal degrees-of-freedom (DOFs) vector is defined by

$$\boldsymbol{d} = \{W_1 \;\; W_2\}^{\mathrm{T}} = \{W(0) \;\; W(L)\}^{\mathrm{T}} \tag{7.16}$$

and the spectral nodal forces vector by

$$f_c = \{Q_1 \;\; Q_2\}^{\mathrm{T}} = \{-Q(0) \;\; +Q(L)\}^{\mathrm{T}} \tag{7.17}$$

Figure 7.2 Spectral element of an axially moving string

7.1.2.3 Dynamic Shape Functions

Consider the homogeneous equation reduced from Equation 7.13 to be

$$-\omega^2 \rho A W + 2i\omega\rho A c W' - (N_x - \rho A c^2) W'' = 0 \tag{7.18}$$

Assume the general solution of Equation 7.18 to be

$$W(x) = a e^{-ikx} \tag{7.19}$$

where k is the wavenumber. Substituting Equation 7.19 into Equation 7.18 gives a dispersion relation

$$(N_x - \rho A c^2)k^2 + 2\rho A c \omega k - \rho A \omega^2 = 0 \tag{7.20}$$

Equation 7.20 can be solved for two wavenumbers

$$\begin{aligned} k_1 &= -\frac{\rho A c - \sqrt{N_x \rho A}}{N_x - \rho A c^2}\omega \\ k_2 &= -\frac{\rho A c + \sqrt{N_x \rho A}}{N_x - \rho A c^2}\omega \end{aligned} \tag{7.21}$$

k_1 is the wavenumber for the wave traveling in the x-direction, while k_2 is in the opposite direction. By using two wavenumbers from Equation 7.21, the general solution to Equation 7.18

can be written as

$$W(x) = \sum_{r=1}^{2} a_r e^{-ik_r x} = \boldsymbol{e}(x;\omega)\boldsymbol{a} \tag{7.22}$$

where

$$\boldsymbol{e}(x;\omega) = [e^{-ik_1 x} \quad e^{-ik_2 x}] \tag{7.23}$$

and

$$\boldsymbol{a} = \{a_1 \quad a_2\}^{\mathrm{T}} \tag{7.24}$$

Substituting Equation 7.22 into the right-hand side of Equation 7.16 gives

$$\boldsymbol{d} = \boldsymbol{H}_S(\omega)\boldsymbol{a} \tag{7.25}$$

where

$$\boldsymbol{H}_S(\omega) = \begin{bmatrix} 1 & 1 \\ e^{-ik_1 L} & e^{-ik_2 L} \end{bmatrix} \tag{7.26}$$

By using Equation 7.25, the constant vector $\boldsymbol{a}$ can be eliminated from Equation 7.22 as

$$W(x) = \boldsymbol{N}(x;\omega)\boldsymbol{d} \tag{7.27}$$

where $\boldsymbol{N}(x;\omega)$ is the dynamic shape function defined by

$$\begin{aligned} \boldsymbol{N}(x;\omega) &= \boldsymbol{e}(x;\omega)\boldsymbol{H}_S^{-1}(\omega) \\ &= \frac{1}{e^{-ik_2 L} - e^{-ik_1 L}}\left[e^{-i(k_2 L + k_1 x)} - e^{-i(k_1 L + k_2 x)} \quad -e^{-ik_1 x} + e^{-ik_2 x}\right] \end{aligned} \tag{7.28}$$

7.1.2.4 Weak Form of Governing Equation

The frequency-domain weak form of Equation 7.13 can be obtained from

$$\int_0^L [-\omega^2 \rho A W + 2i\omega\rho A c W' - (N_x - \rho A c^2)W'' - F(x)]\delta W \mathrm{d}x = 0 \tag{7.29}$$

Applying integral by parts to the third term and applying Equation 7.15 gives the weak form as

$$\begin{aligned} &\int_0^L [(N_x - \rho A c^2)W'\delta W' - \omega^2 \rho A W \delta W + i\omega\rho A c(W'\delta W - W\delta W')]\mathrm{d}x \\ &\quad + (i\omega\rho A c W + \rho A c^2 W')\delta W\big|_0^L - Q(x)\delta W\big|_0^L - \textstyle\int_0^L F(x)\delta W \mathrm{d}x = 0 \end{aligned} \tag{7.30}$$

7.1.2.5 Spectral Element Equation

Substitution of Equation 7.28 into Equation 7.30 gives

$$\begin{aligned} \delta\boldsymbol{d}^{\mathrm{T}}\Bigg\{ &\int_0^L [(N_x - \rho A c^2)\boldsymbol{N}'^{\mathrm{T}}\boldsymbol{N}' - \omega^2 \rho A \boldsymbol{N}^{\mathrm{T}}\boldsymbol{N} + i\omega\rho A c(\boldsymbol{N}^{\mathrm{T}}\boldsymbol{N}' - \boldsymbol{N}'^{\mathrm{T}}\boldsymbol{N})]\mathrm{d}x \\ &+ (i\omega\rho A c \boldsymbol{N}^{\mathrm{T}}\boldsymbol{N} + \rho A c^2 \boldsymbol{N}'^{\mathrm{T}}\boldsymbol{N})\Big|_0^L \Bigg\}\boldsymbol{d} = \delta\boldsymbol{d}^{\mathrm{T}}\{\boldsymbol{f}_c + \boldsymbol{f}_d\} \end{aligned} \tag{7.31}$$

where

$$f_{\mathrm{d}}(\omega) = \int_0^L F(x)\boldsymbol{N}^{\mathrm{T}}(x;\omega)\mathrm{d}x \tag{7.32}$$

and $\boldsymbol{f}_{\mathrm{c}}$ is defined by Equation 7.17. As $\delta\boldsymbol{d}$ is arbitrary by definition, we can obtain the spectral element equation for an axially moving string from Equation 7.31 as being

$$\boldsymbol{S}_S(\omega)\boldsymbol{d} = \boldsymbol{f}(\omega) \tag{7.33}$$

where

$$\begin{aligned}\boldsymbol{S}_S(\omega) = &\int_0^L [(N_x - \rho A c^2)\boldsymbol{N}'^{\mathrm{T}}\boldsymbol{N}' - \omega^2 \rho A \boldsymbol{N}^{\mathrm{T}}\boldsymbol{N} - i\omega\rho A c(\boldsymbol{N}'^{\mathrm{T}}\boldsymbol{N} - \boldsymbol{N}^{\mathrm{T}}\boldsymbol{N}')]\mathrm{d}x \\ &+ (i\omega\rho A c\boldsymbol{N}^{\mathrm{T}}\boldsymbol{N} + \rho A c^2\boldsymbol{N}'^{\mathrm{T}}\boldsymbol{N})|_0^L\end{aligned} \tag{7.34}$$

and

$$\boldsymbol{f}(\omega) = \boldsymbol{f}_{\mathrm{c}}(\omega) + \boldsymbol{f}_{\mathrm{d}}(\omega) \tag{7.35}$$

From Equation 7.28, we can obtain

$$\boldsymbol{N}'(x;\omega) = -i\boldsymbol{e}(x;\omega)\boldsymbol{K}\boldsymbol{H}_S^{-1}(\omega) \tag{7.36}$$

where

$$\boldsymbol{K} = \begin{bmatrix} k_1 & 0 \\ 0 & k_2 \end{bmatrix} \tag{7.37}$$

By substituting Equations 7.28 and 7.36 into Equation 7.34, the spectral element matrix can be written as

$$\boldsymbol{S}_S(\omega) = \boldsymbol{H}_S^{-\mathrm{T}}(\omega)\boldsymbol{D}_S(\omega)\boldsymbol{H}_S^{-1}(\omega) \tag{7.38}$$

where

$$\begin{aligned}\boldsymbol{D}_S(\omega) = &-(N_x - \rho A c^2)\boldsymbol{K}\boldsymbol{E}_S\boldsymbol{K} - \rho A\omega^2\boldsymbol{E}_S \\ &- \omega\rho A c(\boldsymbol{K}\boldsymbol{E}_S - \boldsymbol{E}_S\boldsymbol{K}) + (i\omega\rho A c\boldsymbol{I} - i\rho A c^2\boldsymbol{K})\bar{\boldsymbol{E}}_S\end{aligned} \tag{7.39}$$

and where

$$\begin{aligned}\boldsymbol{E}_S(\omega) &= \int_0^L \boldsymbol{e}^{\mathrm{T}}(x;\omega)\boldsymbol{e}(x;\omega)\mathrm{d}x = [E_{Srs}(\omega)] \qquad (r, s = 1,\ 2) \\ \bar{\boldsymbol{E}}_S(\omega) &= \boldsymbol{e}^{\mathrm{T}}(x;\omega)\boldsymbol{e}(x;\omega)|_0^L = [\bar{E}_{Srs}(\omega)]\end{aligned} \tag{7.40}$$

with

$$\begin{aligned}E_{Srs} &= \begin{cases} \dfrac{i}{k_r + k_s}\bar{E}_{Srs} & \text{if } k_r + k_s \neq 0 \\ L & \text{if } k_r + k_s = 0 \end{cases} \\ \bar{E}_{Srs} &= \mathrm{e}^{-i(k_r + k_s)L} - 1\end{aligned} \tag{7.41}$$

Equation 7.21 implies that $k_r + k_s \neq 0$ (r, $s = 1$, 2) if $c \neq 0$. If $c = 0$, $k_r + k_s = 0$ and the spectral element matrix reduces to that of a vibrating stationary tensioned string.

By substituting Equations 7.26 and 7.39 into Equation 7.38 and then using the definitions from Equations 7.37 and 7.40, the spectral element matrix can be expressed in the form of

$$\boldsymbol{S}_S(\omega) = \begin{bmatrix} S_{S11} & S_{S12} \\ S_{S21} & S_{S22} \end{bmatrix} \tag{7.42}$$

where

$$\begin{aligned} S_{S11} &= -(N_x - \rho A c^2)\alpha_{11} - \rho A \omega^2 \beta_{11} - \omega \rho A c \varepsilon_{11} + i\omega \rho A c \lambda_{11} - i\rho A c^2 \eta_{11} \\ S_{S12} &= -(N_x - \rho A c^2)\alpha_{12} - \rho A \omega^2 \beta_{12} - \omega \rho A c \varepsilon_{12} + i\omega \rho A c \lambda_{12} - i\rho A c^2 \eta_{12} \\ S_{S21} &= -(N_x - \rho A c^2)\alpha_{21} - \rho A \omega^2 \beta_{21} - \omega \rho A c \varepsilon_{21} + i\omega \rho A c \lambda_{21} - i\rho A c^2 \eta_{21} \\ S_{S22} &= -(N_x - \rho A c^2)\alpha_{22} - \rho A \omega^2 \beta_{22} - \omega \rho A c \varepsilon_{22} + i\omega \rho A c \lambda_{22} - i\rho A c^2 \eta_{22} \end{aligned} \tag{7.43}$$

and where

$$\begin{aligned} \alpha_{11} &= i\frac{e_2^2(k_1+k_2)[k_2 e_1^2 + k_1(e_1^2-1)] - k_2[(k_1+k_2)e_1^2 + 4k_2 e_{12}(e_{12}-1)]}{2(k_1+k_2)(e_1-e_2)} \\ \alpha_{12} &= \alpha_{21} = -i\frac{k_1^2 e_2(e_1^2-1) + k_1 k_2(e_1+e_2)(1-e_{12}) + k_2^2 e_1(e_2^2-1)}{2(k_1+k_2)(e_1-e_2)^2} \\ \alpha_{22} &= i\frac{k_1^2(e_1^2-1) + k_1 k_2(e_1^2+e_2^2-4e_{12}+2) + k_2^2(e_2^2-1)}{2(k_1+k_2)(e_1-e_2)^2} \\ \beta_{11} &= i\frac{e_1^2(k_1+k_2)[k_2 e_2^2 + k_1(e_2^2-1)] - k_2[4k_1(e_{12}-1) + e_2^2(k_1+k_2)]}{2k_1 k_2(k_1+k_2)(e_1-e_2)^2} \\ \beta_{12} &= \beta_{21} = -i\frac{k_1^2 e_1(e_2^2-1) + k_1 k_2(e_1+e_2)(1-e_{12}) + k_2^2 e_2(e_1^2-1)}{2k_1 k_2(k_1+k_2)(e_1-e_2)^2} \\ \beta_{22} &= i\frac{k_1^2(e_2^2-1) + k_1 k_2(e_1^2+e_2^2-4e_{12}+2) + k_2^2(e_1^2-1)}{2k_1 k_2(k_1+k_2)(e_1-e_2)^2} \\ \varepsilon_{11} &= \varepsilon_{22} = 0 \\ \varepsilon_{21} &= -\varepsilon_{12} = i\frac{(k_1-k_2)(e_{12}-1)}{(k_1+k_2)(e_1-e_2)} \\ \lambda_{11} &= -\lambda_{22} = -1 \\ \lambda_{12} &= \lambda_{21} = 0 \\ \eta_{11} &= -\eta_{22} = \frac{(k_1+k_2)e_{12} - k_1 e_2^2 - k_2 e_1^2}{(e_1-e_2)^2} \\ \eta_{12} &= -\frac{e_{12}(k_1-k_2)}{e_1-e_2} \\ \eta_{21} &= -\frac{k_1-k_2}{e_1-e_2} \end{aligned} \tag{7.44}$$

with the use of the definitions

$$e_1 = e^{-ik_1 L}, \qquad e_2 = e^{-ik_2 L}, \qquad e_{12} = e_1 e_2 \tag{7.45}$$

The last two terms in the right-hand side of Equation 7.43 represent the effect of the mass transport across the boundaries.

Assembling the spectral elements and applying the associated boundary conditions, we can obtain a global system equation in the form of

$$\boldsymbol{S}_{Sg}(\omega)\boldsymbol{d}_g = \boldsymbol{f}_g \tag{7.46}$$

where $\boldsymbol{S}_{Sg}(\omega)$ is the global dynamic stiffness matrix, $\boldsymbol{d}_g$ is the global spectral nodal DOFs vector, and $\boldsymbol{f}_g$ is the global spectral nodal forces vector. Equation 7.30 or Equation 7.34 implies that the mass transport terms are determined only by the inlet and outlet boundary conditions. Thus, during the assembly procedure, the mass transport terms in Equation 7.43 expressed in terms of λ_{ij} and η_{ij} $(i, j = 1, 2)$ will be considered only for the spectral elements associated with the inlet and outlet boundaries. The natural frequencies ω_{NAT} can be computed from the condition where the determinant of $\boldsymbol{S}_{Sg}(\omega)$ becomes zero as follows:

$$\det \boldsymbol{S}_{Sg}(\omega_{NAT}) = 0 \tag{7.47}$$

7.1.3 Finite Element Model

The weak form of the equation of motion (7.8) in the time domain is given by

$$\begin{aligned}\int_0^L &[(N_x - \rho A c^2)w'\delta w' + \rho A\ddot{w}\delta w + \rho A c(\dot{w}'\delta w - \dot{w}\delta w')]\mathrm{d}x \\ &+ \rho A c(\dot{w} + c w')\delta w|_0^L - Q(x,t)\delta w|_0^L - \textstyle\int_0^L f(x,t)\delta w \mathrm{d}x = 0\end{aligned} \tag{7.48}$$

The sign convention defined in Figure 7.2 is valid for the nodal DOFs, forces and moments in the time domain. Thus, we define the nodal DOFs vector as

$$\boldsymbol{d}(t) = \{w_1(t) \;\; w_2(t)\}^{\mathrm{T}} = \{w(0,t) \;\; w(L,t)\}^{\mathrm{T}} \tag{7.49}$$

and the spectral nodal forces and moments vector as

$$\boldsymbol{f}_c(t) = \{Q_1(t) \;\; Q_2(t)\}^{\mathrm{T}} = \{-Q(0,t) \;\; +Q(L,t)\}^{\mathrm{T}} \tag{7.50}$$

The displacement field $w(x, t)$ within a finite element is assumed to be

$$w(x,t) = \boldsymbol{N}(x)\boldsymbol{d}(t) \tag{7.51}$$

where

$$\boldsymbol{N}(x) = [1 - x/L \quad x/L] \tag{7.52}$$

Substituting Equation 7.51 into Equation 7.48 and using Equation 7.50, we can obtain the finite element equation as

$$\boldsymbol{M}\ddot{\boldsymbol{d}}(t) + (\boldsymbol{G} + \boldsymbol{C}_{MT})\dot{\boldsymbol{d}}(t) + (\boldsymbol{K} + \boldsymbol{K}_{MT})\boldsymbol{d}(t) = \boldsymbol{f}_c(t) + \boldsymbol{f}_d(t) \tag{7.53}$$

where

$$\boldsymbol{f}_d(t) = \int_0^L f(x,t)\boldsymbol{N}^{\mathrm{T}}(x)\mathrm{d}x \tag{7.54}$$

and the finite element matrices are given in Appendix 7.A. The matrices $\boldsymbol{C}_{MT}$ and $\boldsymbol{K}_{MT}$ represent the effects of the mass transport across two end boundaries. As hinted at in Equation 7.48, the mass transport terms are determined only by the inlet and outlet boundary conditions. Thus, during the assembly procedure, the mass transport matrices $\boldsymbol{C}_{MT}$ and $\boldsymbol{K}_{MT}$ will be considered only for the finite elements associated with the inlet and outlet boundaries.

Example 7.1

Consider a uniform string that is axially moving over two fixed supports where the boundary conditions are represented by

$$w(0,t) = w(L,t) = 0$$

When $f(x,t) = 0$, the free vibration response of the axially moving string can be obtained from Equation 7.8 as

$$w(x,t) = C_1 \cos\omega(t+\alpha_1 x) + C_2 \cos\omega(t+\alpha_2 x)$$

where C_1 and C_2 are constants and

$$\alpha_1 = \frac{\rho A c + \sqrt{\rho A N_x}}{N_x - \rho A c^2}, \qquad \alpha_2 = \frac{\rho A c - \sqrt{\rho A N_x}}{N_x - \rho A c^2}$$

Applying the fixed boundary conditions gives

$$\cos\omega(t+\alpha_1 L) = \cos\omega(t+\alpha_2 L)$$

Thus we can obtain the natural frequencies as

$$f_n = \frac{\omega_n}{2\pi} = \frac{n}{(\alpha_1 - \alpha_2)L} = \frac{n(N_x - \rho A c^2)}{2L\sqrt{\rho A N_x}}$$

If $c = 0$, the natural frequencies reduce to those for the stationary tensioned string. The critical speed at which the natural frequency vanishes is equal to the wave speed given by

$$c_{\text{cr}} = \sqrt{\frac{N_x}{\rho A}}$$

Table 7.1 compares the natural frequencies for different moving speeds obtained by the present spectral element method (SEM) with those obtained by the above exact theory, and with those obtained by the finite element method (FEM) by varying the number of finite elements used in the analysis. For computation of the natural frequencies, $L = 2$ m, $\rho A = 1$ kg/m and $N_x = 10$ kN are used. Table 7.1 shows that the natural frequencies obtained by the SEM are identical with those obtained by the above exact theory. As expected, the results obtained by FEM are shown to converge towards those obtained by the exact theory or SEM as the number of finite elements used in FEM is increased. Table 7.1 also shows that the natural frequencies tend to decrease as the moving speed of the string increases.

Table 7.1 Natural frequencies of an axially moving string

Moving speed (m/s)	Method	Natural frequency (Hz)			
		f_1	f_2	f_3	f_5
0	Exact theory	25.000	50.000	75.000	125.00
	SEM (2)	25.000	50.000	75.000	125.00
	FEM(10)	25.103	50.826	77.797	137.83
	FEM(20)	25.026	50.206	75.696	128.23
	FEM(100)	25.001	50.008	75.028	125.13
20	Exact theory	24.000	48.000	72.000	120.00
	SEM (2)	24.000	48.000	72.000	120.00
	FEM(10)	24.313	49.342	75.803	135.61
	FEM(50)	24.043	48.116	72.247	120.80
	FEM(300)	24.007	48.014	72.023	120.05
50	Exact theory	18.750	37.500	56.250	93.750
	SEM (2)	18.750	37.500	56.250	93.750
	FEM(10)	19.939	41.095	64.679	123.26
	FEM(50)	18.947	37.942	57.033	95.695
	FEM(300)	18.782	37.564	56.350	93.934

Note: () = total number of the equal-length finite elements used in the analysis.

7.2 Axially Moving Bernoulli–Euler Beam

7.2.1 Equation of Motion

Consider a Bernoulli–Euler beam that travels in its axial direction at a constant moving speed c. The beam is subjected to a constant axial tensile force N_x as shown in Figure 7.3. The beam has the length L (the span between two end supports), the bending (flexural) rigidity EI, and the mass per length ρA. To take into account the continuing mass transport across the boundaries at $x = 0$ and L, the extended form of Hamilton's principle is used for the derivation of the governing equations of motion and associated boundary conditions. The extended form of Hamilton's principle is given by Equation 7.1.

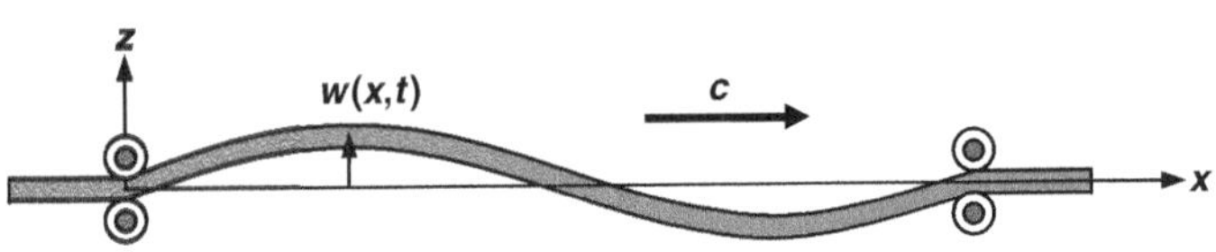

Figure 7.3 Axially moving Bernoulli–Euler beam

The kinetic and potential energies are given by

$$T = \frac{1}{2}\int_0^L \rho A[c^2 + (\dot{w} + cw')^2]\mathrm{d}x$$
$$V = \frac{1}{2}\int_0^L (EI\, w''^2 + N_x w'^2)\mathrm{d}x \tag{7.55}$$

where $w(x, t)$ is the transverse deflection. The virtual work δW_{NC} is given by

$$\delta W = \int_0^L f(x,t)\delta w \mathrm{d}x + M_1\delta\theta(0,t) + M_2\delta\theta(L,t) + Q_1\delta w(0,t) + Q_2\delta w(L,t) \tag{7.56}$$

where $f(x, t)$ is the excitation force, $\theta(x, t) = w'(x, t)$, and $M_1(t)$ and $Q_1(t)$ are the bending moment and transverse shear force applied at $x = 0$, and $M_2(t)$ and $Q_2(t)$ are at $x = L$. The virtual momentum transport of mass across the boundaries at $x = 0$ and L is given by

$$\delta W_{\mathrm{MT}} = \int_A \rho c(\boldsymbol{v}\cdot\delta\boldsymbol{r})|_0^L \mathrm{d}A \tag{7.57}$$

where $\boldsymbol{r}$ is the position vector of a point at x, $\delta\boldsymbol{r}$ is the virtual displacement, $\boldsymbol{v} = \mathrm{d}\boldsymbol{r}/\mathrm{d}t$ is the velocity field of the moving beam at time t, and A is the cross-sectional area of the beam. They are given by

$$\begin{aligned} &\boldsymbol{r} = x\boldsymbol{n}_x + w(x,t)\boldsymbol{n}_z, \qquad \delta\boldsymbol{r} = \delta w(x,t)\boldsymbol{n}_z \\ &\boldsymbol{v} = c\boldsymbol{n}_x + (\dot{w} + cw')\boldsymbol{n}_z \end{aligned} \tag{7.58}$$

where $\boldsymbol{n}_x$ and $\boldsymbol{n}_z$ are the unit direction vectors in the x- and z-directions, respectively. By substituting Equation 7.58 into Equation 7.57, we get

$$\delta W_{\mathrm{MT}} = \rho Ac(\dot{w} + cw')\delta w|_0^L \tag{7.59}$$

Substituting Equations 7.55, 7.56 and 7.59 into the extended form of Hamilton's principle, Equation 7.1, and applying the integral by parts yields

$$\begin{aligned} \int_{t_1}^{t_2} \Bigg\{ &\int_0^L [-EIw'''' - \rho A(c^2 w'' + 2c\dot{w}' + \ddot{w}) + N_x w'' + f(x,t)]\delta w \,\mathrm{d}x + EI\, w'''\delta w\Big|_0^L \\ &+ [Q_1 + N_x w'(0,t)]\delta w(0,t) + [Q_2 - N_x w'(L,t)]\delta w(L,t) - (EI\, w'')\delta w'\Big|_0^L \\ &+ M_1\delta w'(0,t) + M_2\delta w'(L,t)\Bigg\}\mathrm{d}t = 0 \end{aligned} \tag{7.60}$$

From Equation 7.60, the equation of motion can be obtained as

$$EIw'''' - (N_x - \rho Ac^2)w'' + \rho A\ddot{w} + 2\rho Ac\dot{w}' = f(x,t) \tag{7.61}$$

with the relevant boundary conditions

$$\begin{aligned} &Q(0,t) = -Q_1(t) - N_x w'(0,t) \quad &&\text{or} \quad w(0,t) = w_1(t) \\ &M(0,t) = -M_1(t) &&\text{or} \quad w'(0,t) = \theta_1(t) \\ &Q(L,t) = Q_2(t) - N_x w'(L,t) &&\text{or} \quad w(L,t) = w_2(t) \\ &M(L,t) = M_2(t) &&\text{or} \quad w'(L,t) = \theta_2(t) \end{aligned} \tag{7.62}$$

where $Q(x, t)$ and $M(x, t)$ are the resultant shear force and bending moment defined by

$$\begin{aligned} Q(x,t) &= -EIw''' \\ M(x,t) &= EIw'' \end{aligned} \tag{7.63}$$

7.2.2 Spectral Element Modeling

7.2.2.1 Governing Equations in the Frequency Domain

Assume the dynamic response in the spectral form to be

$$w(x,t) = \frac{1}{N}\sum_{n=0}^{N-1} W_n(x)e^{i\omega_n t} \tag{7.64}$$

Similarly, the external force, the internal shear force and bending moment, the boundary shear forces and bending moments, and the boundary displacements and slopes are also represented in the spectral forms as

$$\begin{aligned}
&\{f(x,t), M(x,t), Q(x,t)\} \\
&\quad = \frac{1}{N}\sum_{n=0}^{N-1}\{F_n(x), M_n(x), Q_n(x)\}e^{i\omega_n t} \\
&\{Q_1(t), M_1(t), Q_2(t), M_2(t)\} \\
&\quad = \frac{1}{N}\sum_{n=0}^{N-1}\{Q_{1n}(t), M_{1n}(t), Q_{2n}(t), M_{2n}(t)\}e^{i\omega_n t} \\
&\{w_1(t), \theta_1(t), w_2(t), \theta_2(t)\} \\
&\quad = \frac{1}{N}\sum_{n=0}^{N-1}\{W_{1n}, \Theta_{1n}, W_{2n}, \Theta_{2n}\}e^{i\omega_n t}
\end{aligned} \tag{7.65}$$

Substituting Equations 7.64 and 7.65 into Equations 7.61–7.63, we obtain the governing equation in the frequency domain, omitting the subscript n for brevity, as

$$EIW'''' - (N_x - \rho A c^2)W'' - \omega^2 \rho A W + 2i\omega\rho A c W' = F(x) \tag{7.66}$$

and the boundary conditions

$$\begin{aligned}
Q(0) &= -Q_1 - N_x W'(0) \quad &\text{or} \quad W(0) = W_1 \\
M(0) &= -M_1 \quad &\text{or} \quad W'(0) = \Theta_1 \\
Q(L) &= +Q_2 - N_x W'(L) \quad &\text{or} \quad W(L) = W_2 \\
M(L) &= +M_2 \quad &\text{or} \quad W'(L) = \Theta_2
\end{aligned} \tag{7.67}$$

and the force-displacement relationships

$$\begin{aligned}
Q(x) &= -EI\,W''' \\
M(x) &= EI\,W''
\end{aligned} \tag{7.68}$$

7.2.2.2 Spectral Nodal DOFs, Forces and Moments

The spectral nodal displacements, slopes, forces and moments are indicated in Figure 7.4. The spectral nodal DOFs vector is defined by

$$\boldsymbol{d} = \{W_1 \quad \Theta_1 \quad W_2 \quad \Theta_2\}^{\mathrm{T}} = \{W(0) \quad W'(0) \quad W(L) \quad W'(L)\}^{\mathrm{T}} \tag{7.69}$$

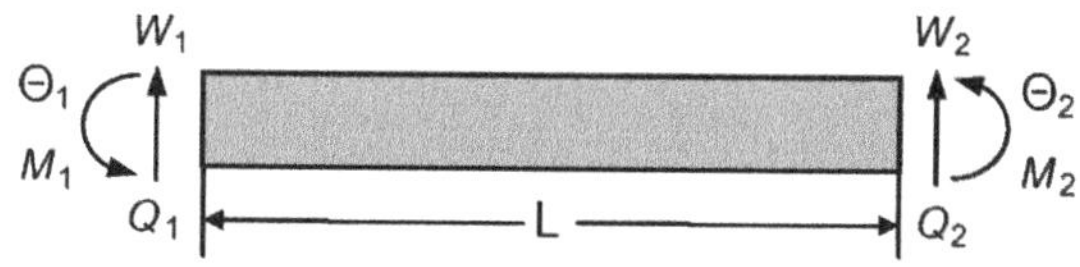

Figure 7.4 Spectral element of an axially moving Bernoulli–Euler beam

and the spectral nodal forces and moments vector by

$$\begin{aligned} f_c &= \{Q_1 \quad M_1 \quad Q_2 \quad M_2\}^T \\ &= \{-Q(0) - N_x W'(0) \quad -M(0) \quad Q(L) + N_x W'(L) \quad M(L)\}^T \end{aligned} \tag{7.70}$$

7.2.2.3 Dynamic Shape Functions

Consider the homogeneous equation reduced from Equation 7.66 to be

$$EIW'''' - (N_x - \rho A c^2)W'' - \omega^2 \rho A W + 2i\omega\rho A c W' = 0 \tag{7.71}$$

The general solution to Equation 7.71 is assumed in the form of

$$W(x) = a e^{-ikx} \tag{7.72}$$

where k is the wavenumber. Substituting Equation 7.72 into Equation 7.71 gives a dispersion relation

$$EIk^4 + (N_x - \rho A c^2)k^2 + 2\rho A c \omega k - \rho A \omega^2 = 0 \tag{7.73}$$

By using four wavenumbers k_i $(i = 1, 2, 3, 4)$ computed from Equation 7.73, we can write the general solution to Equation 7.71 as

$$W(x) = \sum_{r=1}^{4} a_r e^{-ik_r x} = \boldsymbol{e}(x;\omega)\boldsymbol{a} \tag{7.74}$$

where

$$\boldsymbol{e}(x;\omega) = [e^{-ik_1 x} \quad e^{-ik_2 x} \quad e^{-ik_3 x} \quad e^{-ik_4 x}] \tag{7.75}$$

and

$$\boldsymbol{a} = \{a_1 \quad a_2 \quad a_3 \quad a_4\}^T \tag{7.76}$$

Substituting Equation 7.74 into the right-hand side of Equation 7.69, we obtain

$$\boldsymbol{d} = \boldsymbol{H}_B(\omega)\boldsymbol{a} \tag{7.77}$$

where

$$\boldsymbol{H}_B(\omega) = \begin{bmatrix} 1 & 1 & 1 & 1 \\ -ik_1 & -ik_2 & -ik_3 & -ik_4 \\ e_1 & e_2 & e_3 & e_4 \\ -ik_1 e_1 & -ik_2 e_2 & -ik_3 e_3 & -ik_4 e_4 \end{bmatrix} \tag{7.78}$$

and where

$$e_r = e^{-ik_rL} \qquad (r = 1, 2, 3, 4) \tag{7.79}$$

The constant vector $\boldsymbol{a}$ can be eliminated from Equation 7.74, by using Equation 7.77 as

$$W(x) = \boldsymbol{N}(x;\omega)\boldsymbol{d} \tag{7.80}$$

where $\boldsymbol{N}(x;\omega)$ is the dynamic shape function defined by

$$\boldsymbol{N}(x;\omega) = \boldsymbol{e}(x;\omega)\boldsymbol{H}_B^{-1}(\omega) \tag{7.81}$$

7.2.2.4 Weak Form of Governing Equation

The weak form of the forced vibration Equation 7.66 in the frequency domain can be obtained from

$$\int_0^L [EIW'''' - (N_x - \rho Ac^2)W'' - \omega^2\rho AW + 2i\omega\rho AcW' - F(x)]\delta W \mathrm{d}x \tag{7.82}$$

Applying the integral by parts gives

$$\begin{aligned} &\int_0^L [EIW''\delta W'' + (N_x - \rho Ac^2)W'\delta W' - \omega^2\rho AW\delta W + i\omega\rho Ac(W'\delta W - W\delta W')]\mathrm{d}x \\ &\quad + (i\omega\rho AcW + \rho Ac^2W')\delta W|_0^L - M(x)\delta W'|_0^L - [Q(x) + N_xW']\delta W|_0^L \\ &\quad - \int_0^L F(x)\delta W\mathrm{d}x = 0 \end{aligned} \tag{7.83}$$

where Equation 7.68 is applied.

7.2.2.5 Spectral Element Equation

Substituting Equation 7.80 into Equation 7.83, we obtain

$$\begin{aligned} \delta\boldsymbol{d}^{\mathrm{T}}\Big\{ &\int_0^L [EI\boldsymbol{N}''^{\mathrm{T}}\boldsymbol{N}'' + (N_x - \rho Ac^2)\boldsymbol{N}'^{\mathrm{T}}\boldsymbol{N}' - \omega^2\rho A\boldsymbol{N}^{\mathrm{T}}\boldsymbol{N} + i\omega\rho Ac(\boldsymbol{N}^{\mathrm{T}}\boldsymbol{N}' - \boldsymbol{N}'^{\mathrm{T}}\boldsymbol{N})]\mathrm{d}x \\ &+ (i\omega\rho Ac\boldsymbol{N}^{\mathrm{T}}\boldsymbol{N} + \rho Ac^2\boldsymbol{N}'^{\mathrm{T}}\boldsymbol{N})\Big|_0^L\Big\}\boldsymbol{d} = \delta\boldsymbol{d}^{\mathrm{T}}\{\boldsymbol{f}_{\mathrm{c}} + \boldsymbol{f}_{\mathrm{d}}\} \end{aligned} \tag{7.84}$$

where

$$\boldsymbol{f}_{\mathrm{d}}(\omega) = \int_0^L F(x)\boldsymbol{N}^{\mathrm{T}}(x;\omega)\mathrm{d}x \tag{7.85}$$

and $\boldsymbol{f}_{\mathrm{c}}$ is defined by Equation 7.70. As $\delta\boldsymbol{d}$ is arbitrary by definition, we can obtain the spectral element equation from Equation 7.84 as

$$\boldsymbol{S}_B(\omega)\boldsymbol{d} = \boldsymbol{f}(\omega) \tag{7.86}$$

where

$$\begin{aligned} \boldsymbol{S}_B(\omega) = &\int_0^L [EI\boldsymbol{N}''^{\mathrm{T}}\boldsymbol{N}'' + (N_x - \rho Ac^2)\boldsymbol{N}'^{\mathrm{T}}\boldsymbol{N}' - \omega^2\rho A\boldsymbol{N}^{\mathrm{T}}\boldsymbol{N} \\ &- i\omega\rho Ac(\boldsymbol{N}'^{\mathrm{T}}\boldsymbol{N} - \boldsymbol{N}^{\mathrm{T}}\boldsymbol{N}')]\mathrm{d}x + (i\omega\rho Ac\boldsymbol{N}^{\mathrm{T}}\boldsymbol{N} + \rho Ac^2\boldsymbol{N}'^{T}\boldsymbol{N})|_0^L \end{aligned} \tag{7.87}$$

and

$$f(\omega) = f_c(\omega) + f_d(\omega) \tag{7.88}$$

From Equation 7.81, we get

$$N'(x;\omega) = -ie(x;\omega)KH_B^{-1}(\omega), \quad N''(x;\omega) = -e(x;\omega)K^2H_B^{-1}(\omega) \tag{7.89}$$

where

$$K = diag[k_i], \qquad K^2 = diag[k_i^2] \tag{7.90}$$

Substitution of Equations 7.81 and 7.89 into Equation 7.87 yields

$$S_B(\omega) = H_B^{-T}(\omega)D_B(\omega)H_B^{-1}(\omega) \tag{7.91}$$

where

$$\begin{aligned} D_B(\omega) &= EIK^2E_BK^2 - (N_x - \rho Ac^2)KE_BK - \rho A\omega^2 E_B \\ &\quad - \omega\rho Ac(KE_B - E_BK) + (i\omega\rho AcI - i\rho Ac^2K)\bar{E}_B \end{aligned} \tag{7.92}$$

and where

$$\begin{aligned} E_B(\omega) &= \int_0^L e^{\mathrm{T}}(x;\omega)e(x;\omega)\mathrm{d}x = [E_{Brs}(\omega)] \\ &\qquad\qquad\qquad\qquad\qquad (r, s = 1,\ 2,\ 3,\ 4) \\ \bar{E}_B(\omega) &= e^{\mathrm{T}}(x;\omega)e(x;\omega)\big|_0^L = [\bar{E}_{Brs}(\omega)] \end{aligned} \tag{7.93}$$

with

$$\begin{aligned} E_{Brs} &= \begin{cases} \dfrac{i}{k_r + k_s}\bar{E}_{Brs} & \text{if } k_r + k_s \neq 0 \\ L & \text{if } k_r + k_s = 0 \end{cases} \\ \bar{E}_{Brs} &= \mathrm{e}^{-i(k_r + k_s)L} - 1 \end{aligned} \tag{7.94}$$

The last term on the right-hand side of Equation 7.92 represents the effect of the mass transport across the boundaries at $x = 0$ and L.

Assembling the spectral elements and applying the associated boundary conditions, we can obtain a global system equation in the form as

$$S_{Bg}(\omega)d_g = f_g \tag{7.95}$$

where $S_{Bg}(\omega)$ is the global dynamic stiffness matrix, d_g is the global spectral nodal DOFs vector, and f_g is the global spectral nodal forces vector. Equation 7.83 or Equation 7.87 implies that the mass transport terms are determined only by the inlet and outlet boundary conditions. Thus, during the assembly procedure, the mass transport term expressed in terms of $\bar{E}_B$ in Equation 7.92 will be considered only for the spectral elements associated with the inlet and outlet boundaries. The natural frequencies ω_{NAT} can be computed from the condition that the determinant of $S_{Bg}(\omega)$ becomes zero as follows:

$$\det S_{Bg}(\omega_{\mathrm{NAT}}) = 0 \tag{7.96}$$

7.2.3 Finite Element Model

The weak form of the equation of motion (7.61) in the time domain is given by

$$\begin{aligned}&\int_0^L [EIw''\delta w'' + (N_x - \rho A c^2)w'\delta w' + \rho A\ddot{w}\delta w + \rho Ac(\dot{w}'\delta w - \dot{w}\delta w')]\mathrm{d}x \\ &\quad + \rho Ac(\dot{w} + cw')\delta w|_0^L - [Q(x,t) + N_x w']\delta w|_0^L - M(x,t)\delta w'|_0^L - \int_0^L f(x,t)\delta w \mathrm{d}x = 0\end{aligned} \tag{7.97}$$

The sign convention defined in Figure 7.4 is valid for the nodal DOFs, forces and moments in the time domain. Thus, we define the nodal DOFs vector as

$$\boldsymbol{d}(t) = \{w_1(t)\ \ \theta_1(t)\ \ w_2(t)\ \ \theta_2(t)\}^{\mathrm{T}} = \{w(0,t)\ \ w'(0,t)\ \ w(L,t)\ \ w'(L,t)\}^{\mathrm{T}} \tag{7.98}$$

and the spectral nodal forces and moments vector as

$$\begin{aligned}\boldsymbol{f}_{\mathrm{c}}(t) &= \{Q_1(t)\ \ M_1(t)\ \ Q_2(t)\ \ M_2(t)\}^{\mathrm{T}} \\ &= \{-Q(0,t) - N_x w'(0,t)\ \ -M(0,t)\ \ Q(L,t) + N_x w'(L,t)\ \ M(L,t)\}^{\mathrm{T}}\end{aligned} \tag{7.99}$$

The displacement field $w(x, t)$ within a finite element is assumed to be

$$w(x,t) = \boldsymbol{N}(x)\boldsymbol{d}(t) \tag{7.100}$$

where

$$\boldsymbol{N}(x) = [(1 - 3\xi^2 + 2\xi^3)\ L(\xi - 2\xi^2 + \xi^3)\ (3\xi^2 - 2\xi^3)\ L(-\xi^2 + \xi^3)] \tag{7.101}$$

with the use of the definition

$$\xi = \frac{x}{L} \tag{7.102}$$

Substituting Equation 7.100 into Equation 7.97 and using Equation 7.99, we obtain the finite element equation as

$$\boldsymbol{M}\ddot{\boldsymbol{d}}(t) + (\boldsymbol{G} + \boldsymbol{C}_{MT})\dot{\boldsymbol{d}}(t) + (\boldsymbol{K}_{EI} + \boldsymbol{K}_N + \boldsymbol{K}_{MT})\boldsymbol{d}(t) = \boldsymbol{f}_{\mathrm{d}}(t) + \boldsymbol{f}_{\mathrm{c}}(t) \tag{7.103}$$

where

$$\boldsymbol{f}_{\mathrm{d}}(t) = \int_0^L f(x,t)\boldsymbol{N}^{\mathrm{T}}(x)\mathrm{d}x \tag{7.104}$$

and the finite element matrices are given in Appendix 7.B. The matrices $\boldsymbol{C}_{MT}$ and $\boldsymbol{K}_{MT}$ represent the effects of the mass transport across two end boundaries. Equation 7.97 shows that the mass transport terms are determined only by the inlet and outlet boundary conditions. Thus, during the assembly procedure, the mass transport matrices $\boldsymbol{C}_{MT}$ and $\boldsymbol{K}_{MT}$ will be considered only for the finite elements associated with the inlet and outlet boundaries.

7.2.4 Stability Analysis

An axially moving beam may become unstable when it moves at a certain critical speed. To investigate the stability of the axially moving beam, we assume the free vibration response in the form

$$w(x,t) = W(x)\mathrm{e}^{\lambda t} \tag{7.105}$$

By substituting Equation 7.105 into the free vibration equation of Equation 7.61, we can obtain an eigenvalue problem from which the eigenvalue λ can be computed as

$$\lambda \equiv \mathrm{Re}(\lambda) + i\mathrm{Im}(\lambda) \tag{7.106}$$

where $\mathrm{Re}(\lambda)$ and $\mathrm{Im}(\lambda)$ are the real part and the imaginary part of λ, respectively. The type of instability depends on the signs of the real and imaginary parts of λ as follows:

Stable: if $\mathrm{Re}(\lambda) \leq 0$
Unstable with static instability (divergence): if $\mathrm{Re}(\lambda) > 0$ and $\mathrm{Im}(\lambda) = 0$ (7.107)
Unstable with dynamic instability (flutter): if $\mathrm{Re}(\lambda) > 0$ and $\mathrm{Im}(\lambda) \neq 0$

Thus, the eigenvalue problem to investigate the instability of an axially moving beam can be obtained from Equation 7.95, by setting $\boldsymbol{f}_g = 0$ and by replacing $i\omega$ with λ, as

$$\boldsymbol{S}_{Bg}(\lambda)\boldsymbol{d}_g = 0 \tag{7.108}$$

The types of instability at a specified moving speed c and axial tensile force N_x are then determined by investigating eigenvalues λ computed from the characteristic equation

$$\det \boldsymbol{S}_{Bg}(\lambda) = 0 \tag{7.109}$$

The wavenumbers $k_r\,(r = 1, 2, 3, 4)$ required to compute $\boldsymbol{S}_{Bg}(\lambda)$ are obtained from Equation 7.73 by replacing $i\omega$ with λ.

Hinted at in the static instability condition given by Equation 7.107, the critical speed at which the static instability (i.e., divergence speed c_D) occurs can also be obtained by considering the existence of the non-trivial equilibrium position, in other words, the static eigenvalue problem. For a beam that is axially moving over two simple supports, the static eigenvalue problem can be derived directly from Equation 7.66 by inserting $\omega = 0$ and $F = 0$ and then by assuming $W(x) = ae^{-ikx}$ as follows:

$$EIk^4 + (N_x - \rho Ac^2)k^2 = 0 \tag{7.110}$$

Equation 7.110 gives four roots as

$$k_1 = k_2 = 0, \quad k_3 = -k_4 = \bar{k} \tag{7.111}$$

where

$$\bar{k} = \sqrt{\frac{\rho Ac^2 - N_x}{EI}} \tag{7.112}$$

Thus, the non-trivial equilibrium displacement of the axially moving beam can be written as

$$w(x) = a_1 + a_2 x + a_3 e^{-i\bar{k}x} + a_4 e^{i\bar{k}x} \tag{7.113}$$

Applying the simply supported boundary conditions to Equation 7.113, we obtain

$$\bar{k}_n = \frac{n\pi}{L} \quad (n = 1, 2, 3, \ldots.) \tag{7.114}$$

Lastly, by equating Equation 7.114 to–Equation 7.112, we obtain the divergence speeds as

$$c_{Dn} = \sqrt{\frac{EI}{\rho A}\left(\bar{k}_n^2 + \frac{N_x}{EI}\right)} \tag{7.115}$$

In practice, the lowest divergence speed c_{D1} is considered as the critical speed of a moving beam. Equation 7.115 shows that, when $N=0$, the static instability occurs when a beam is axially moving at a speed that is equal to the phase speed of the bending wave, which is given by $c_p = \omega/k = k\sqrt{EI/\rho A}$.

Example 7.2

As an example problem, a uniform beam that is axially moving over two simple supports is considered. The structural and material properties of the beam are $L = 2\,\text{m}$, $EI = 10\,\text{kN m}^2$ and $\rho A = 10\,\text{kg/m}$.

Table 7.2 shows the accuracy of the present spectral element model when compared with the conventional finite element model and the effect of the moving speed c on the natural frequencies when the beam is subjected to a constant axial tensile force $N_x = 20\,\text{kN}$. Table 7.2 shows that SEM results for the stationary beam (when $c = 0\,\text{m/s}$) are identical with the exact

Table 7.2 Natural frequencies of an axially moving Bernoulli–Euler beam

		Natural frequency (Hz)			
Moving speed (m/s)	Method	f_1	f_2	f_3	f_5
0	Theory [24]	16.710	54.474	116.69	315.45
	SEM (1)	16.710	54.474	116.69	315.45
	FEM(10)	16.710	54.479	116.75	316.66
	FEM(20)	16.710	54.474	116.69	315.53
	FEM(50)	16.710	54.474	116.69	315.45
33.419 ($0.5c_{D1}$)	SEM (1)	13.913	52.217	114.70	313.69
	FEM(10)	13.913	52.224	114.77	314.98
	FEM(20)	13.913	52.217	114.71	313.77
	FEM(50)	13.913	52.217	114.70	313.69
66.838 (c_{D1})	SEM (1)	0.000	44.999	108.62	308.39
	FEM(10)	0.038	45.014	108.73	309.92
	FEM(20)	0.001	45.000	108.63	308.49
	FEM(50)	0.001	44.999	108.62	308.39

Note: () = total number of the equal-length finite elements used in the analysis.

results given by the theoretical formula [24]

$$f_n = \frac{n^2\pi}{2L^2}\sqrt{\frac{EI}{\rho A}\left(1+\frac{N_x L^2}{n^2\pi^2 EI}\right)} \quad \text{(Hz)}$$

At all moving speeds, Table 7.2 shows that FEM results converge towards the SEM results as the number of the finite elements used in FEM is increased. It is also worth mentioning that the first natural frequency indeed vanishes when the moving speed c is increased up to the lowest divergence speed $c_{D1} = 66.838$ m/s at which the static instability "divergence" occurs.

Figure 7.5 shows the lowest three eigenvalues λ defined by Equation 7.105 as the functions of the moving speed c. The static and dynamic instability zones determined by the signs of the real and imaginary parts of the eigenvalues are indicated in the figure. The divergence zone is shown to start from $c_{D1} = 66.838$ m/s, followed by the second narrow stable zone and the dynamic instability "flutter" zone from $c_{F1} = 112.372$ m/s. Figure 7.6 compares the time-domain and frequency-domain dynamic responses at flutter speed c_{F1} obtained by SEM with those obtained by FEM(n), where n is the total number of the equal-length finite elements used in the analysis. Using more finite elements certainly makes the FEM results converge towards the SEM results. The distortion of the symmetric or anti-symmetric natural mode shapes of a stationary beam due to the moving speed are displayed in Figure 7.7.

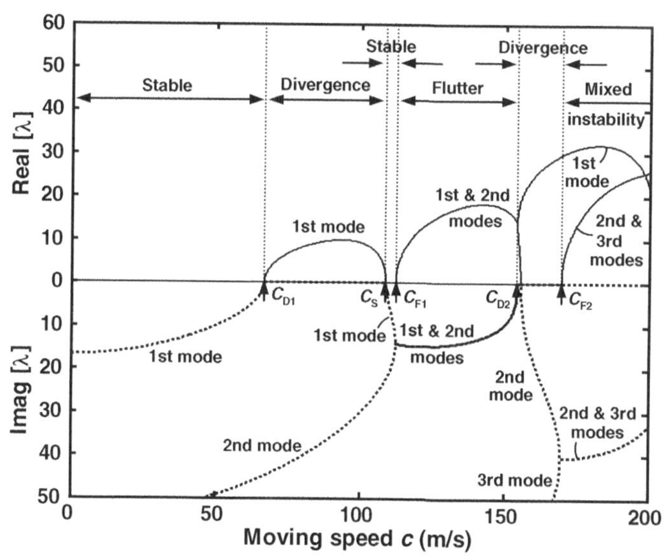

Figure 7.5 The lowest three eigenvalues λ (Hz) versus the moving speed c (m/s)

7.3 Axially Moving Timoshenko Beam

7.3.1 Equations of Motion

Consider a uniform Timoshenko beam that is traveling in its axial direction at a constant speed c, subjected to a constant axial tensile force N_x. The beam has the length L (the span between the two end supports), the bending (flexural) rigidity EI, the shear rigidity κGA, the mass per length

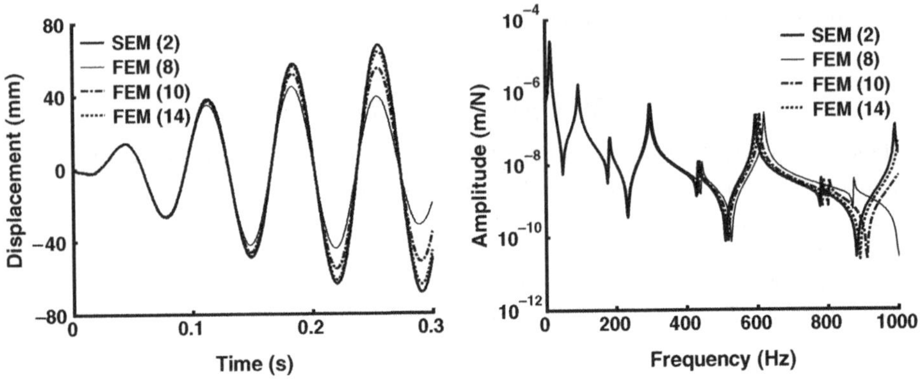

Figure 7.6 Dynamic responses at flutter speed c_{F1}

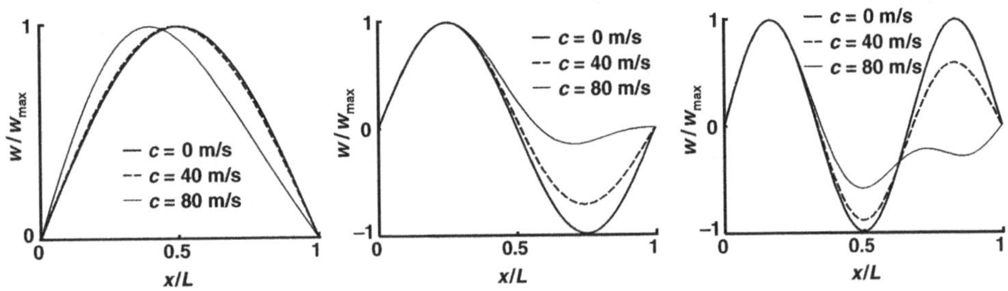

Figure 7.7 Moving speed dependence of the natural modes of an axially moving beam

ρA, and the mass moment of inertia per length ρI. Based on the Timoshenko-beam theory, the total transverse deflection and the angle of rotation due to bending are denoted by $w(x, t)$ and $\theta(x, t)$, respectively. The total kinetic and potential energies are then given by

$$T = \frac{1}{2}\int_0^L \{\rho A[c^2 + (\dot{w} + cw')^2] + \rho I(\dot{\theta} + c\theta')^2\}\mathrm{d}x$$
$$V = \frac{1}{2}\int_0^L \{EI\theta'^2 + \kappa GA(w' - \theta)^2 + N_x w'^2\}\mathrm{d}x \tag{7.116}$$

The virtual work done by the external force is given by

$$\delta W_{\mathrm{NC}} = \int_0^L f(x,t)\delta w \mathrm{d}x + M_1\delta\theta(0,t) + M_2\delta\theta(L,t) + Q_1\delta w(0,t) + Q_2\delta w(L,t) \tag{7.117}$$

where $f(x, t)$ is the excitation force, M_1 and Q_1 are the bending moment and transverse shear force applied at $x = 0$, and M_2 and Q_2 are those applied at $x = L$. Similar to the axially moving Bernoulli–Euler beam, the virtual momentum transport of mass across the boundaries at $x = 0$ and L is obtained from

$$\delta W_{MT} = \int_A \rho c(\boldsymbol{v}\cdot\delta\boldsymbol{r})|_0^L \mathrm{d}A \tag{7.118}$$

where

$$\begin{aligned} \boldsymbol{r} &= [x + z\theta(x,t)]\boldsymbol{n}_x + w(x,t)\boldsymbol{n}_z \\ \delta\boldsymbol{r} &= z\delta\theta(x,t)]\boldsymbol{n}_x + \delta w(x,t)\boldsymbol{n}_z \\ \boldsymbol{v} &= [c + z(\dot{\theta} + c\theta')]\boldsymbol{n}_x + (\dot{w} + cw')\boldsymbol{n}_z \end{aligned} \tag{7.119}$$

Substituting Equation 7.119 into Equation 7.118, we obtain

$$\delta W_{MT} = \rho Ac(\dot{w} + cw')\delta w\big|_0^L + \rho Ic(\dot{\theta} + c\theta')\delta\theta(x,t)\big|_0^L \tag{7.120}$$

Substituting Equations 7.116–7.120 into the extended form of Hamilton's principle, Equation 7.1, we obtain

$$\begin{aligned} \int_{t_1}^{t_2} \Bigg\{ \int_0^L & ([\kappa GA(w'' - \theta') - \rho A\ddot{w} - \rho Ac^2 w'' - 2\rho Ac\dot{w}' + N_x w'' + f(x,t)]\delta w \\ & + [EI\theta'' + \kappa GA(w' - \theta) - \rho I\ddot{\theta} - \rho Ic^2\theta'' - 2\rho Ic\dot{\theta}']\delta\theta)\mathrm{d}x \\ & - Q(x,t)\delta w(x,t)\Big|_0^L + [Q_1 + N_x w'(0,t)]\delta w(0,t) + [Q_2 - N_x w'(L,t)]\delta w(L,t) \\ & - M(x,t)\delta\theta(x,t)\Big|_0^L + M_1\delta\theta(0,t) + M_2\delta\theta(L,t) \Bigg\}\mathrm{d}t = 0 \end{aligned} \tag{7.121}$$

where $Q(x, t)$ and $M(x, t)$ are defined by

$$\begin{aligned} Q(x,t) &= \kappa GA(w' - \theta) \\ M(x,t) &= EI\theta' \end{aligned} \tag{7.122}$$

From Equation 7.121, the equations of motion can be derived as

$$\begin{aligned} &\kappa GA(w'' - \theta') - \rho A\ddot{w} - \rho Ac^2 w'' - 2\rho Ac\dot{w}' + N_x w'' + f(x,t) = 0 \\ &EI\theta'' + \kappa GA(w' - \theta) - \rho I\ddot{\theta} - \rho Ic^2\theta'' - 2\rho Ic\dot{\theta}' = 0 \end{aligned} \tag{7.123}$$

with the relevant boundary conditions

$$\begin{aligned} Q(0,t) &= -Q_1(t) - N_x w'(0,t) \quad &\text{or} \quad w(0,t) = w_1(t) \\ M(0,t) &= -M_1(t) &\text{or} \quad \theta(0,t) = \theta_1(t) \\ Q(L,t) &= Q_2(t) - N_x w'(L,t) &\text{or} \quad w(L,t) = w_2(t) \\ M(L,t) &= M_2(t) &\text{or} \quad \theta(L,t) = \theta_2(t) \end{aligned} \tag{7.124}$$

7.3.2 Spectral Element Modeling

7.3.2.1 Governing Equations in the Frequency Domain

Represent the dynamic responses in the spectral forms as

$$w(x,t) = \frac{1}{N}\sum_{n=1}^{N} W_n(x)\mathrm{e}^{i\omega_n t}, \qquad \theta(x,t) = \frac{1}{N}\sum_{n=1}^{N} \Theta_n(x)\mathrm{e}^{i\omega_n t} \tag{7.125}$$

Similarly, the external force, the internal shear force and bending moment, the boundary shear forces and bending moments, and the boundary displacements and slopes are also represented in the spectral forms as

$$\begin{aligned}
&\{f(x,t), M(x,t), Q(x,t)\} \\
&\quad = \frac{1}{N}\sum_{n=0}^{N-1}\{F_n(x), M_n(x), Q_n(x)\}e^{i\omega_n t} \\
&\{Q_1(t), M_1(t), Q_2(t), M_2(t)\} \\
&\quad = \frac{1}{N}\sum_{n=0}^{N-1}\{Q_{1n}, M_{1n}, Q_{2n}, M_{2n}\}e^{i\omega_n t} \\
&\{w_1(t), \theta_1(t), w_2(t), \theta_2(t)\} \\
&\quad = \frac{1}{N}\sum_{n=0}^{N-1}\{w_{1n}, \theta_{1n}, w_{2n}, \theta_{2n}\}e^{i\omega_n t}
\end{aligned} \tag{7.126}$$

Substituting Equations 7.125 and 7.126 into Equations 7.122–7.124, we obtain the governing equation in the frequency domain, omitting the subscripts n for brevity, as

$$\begin{aligned}
&(\kappa GA + N_x - \rho Ac^2)W'' - 2i\omega\rho AcW' + \omega^2\rho AW - \kappa GA\Theta' + F(x) = 0 \\
&(EI - \rho Ic^2)\Theta'' - 2i\omega\rho Ic\Theta' - (\kappa GA - \omega^2\rho I)\Theta + \kappa GAW' = 0
\end{aligned} \tag{7.127}$$

and the boundary conditions

$$\begin{aligned}
Q(0) &= -Q_1 - N_x W'(0) \quad \text{or} \quad W(0) = W_1 \\
M(0) &= -M_1 \quad \text{or} \quad W'(0) = \Theta_1 \\
Q(L) &= +Q_2 - N_x W'(L) \quad \text{or} \quad W(L) = W_2 \\
M(L) &= +M_2 \quad \text{or} \quad W'(L) = \Theta_2
\end{aligned} \tag{7.128}$$

and the force-displacement relationships

$$\begin{aligned}
Q(x) &= \kappa GA(W' - \Theta) \\
M(x) &= EI\Theta'
\end{aligned} \tag{7.129}$$

7.3.2.2 Spectral Nodal DOFs, Forces and Moments

The spectral nodal displacements, slopes, forces and moments are indicated in Figure 7.8. The spectral nodal DOFs vector is defined by

$$\mathbf{d} = \{W_1 \quad \Theta_1 \quad W_2 \quad \Theta_2\}^{\mathrm{T}} = \{W(0) \quad \Theta(0) \quad W(L) \quad \Theta(L)\}^{\mathrm{T}} \tag{7.130}$$

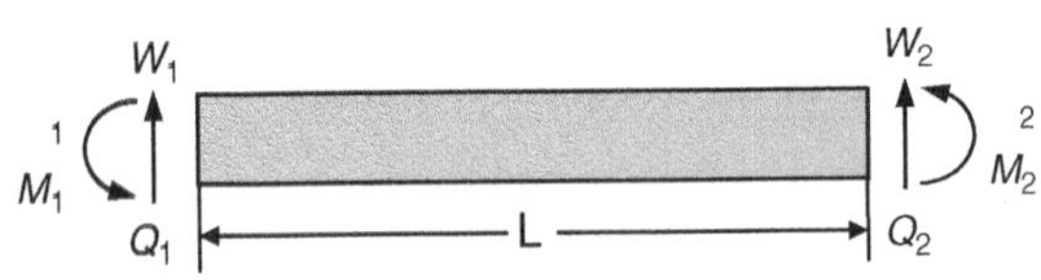

Figure 7.8 Spectral element of an axially moving Timoshenko beam

and the spectral nodal forces and moments vector by

$$\begin{aligned} f_c &= \{Q_1 \quad M_1 \quad Q_2 \quad M_2\}^{\mathrm{T}} \\ &= \{-Q(0)-N_xW'(0) \quad -M(0) \quad Q(L)+N_xW'(L) \quad M(L)\}^{\mathrm{T}} \end{aligned} \tag{7.131}$$

7.3.2.3 Dynamic Shape Functions

The homogeneous equations reduced from Equation 7.127 are given by

$$\begin{aligned} &(\kappa GA+N_x-\rho Ac^2)W''-2i\omega\rho AcW'+\omega^2\rho AW-\kappa GA\Theta'=0 \\ &(EI-\rho Ic^2)\Theta''-2i\omega\rho Ic\Theta'-(\kappa GA-\omega^2\rho I)\Theta+\kappa GAW'=0 \end{aligned} \tag{7.132}$$

Assume the general solutions to Equation 7.132 as being

$$W(x)=a\,\mathrm{e}^{-ikx}, \quad \Theta(x)=\beta a\mathrm{e}^{-ikx} \tag{7.133}$$

Substitute Equation 7.133 into Equation 7.132 to obtain an eigenvalue problem

$$\begin{bmatrix} X_{11} & X_{12} \\ X_{21} & X_{22} \end{bmatrix} \begin{Bmatrix} 1 \\ \beta \end{Bmatrix} = \begin{Bmatrix} 0 \\ 0 \end{Bmatrix} \tag{7.134}$$

where

$$\begin{aligned} X_{11} &= (\kappa GA+N_x-\rho Ac^2)k^2+2\rho Ac\omega k-\omega^2\rho A \\ X_{22} &= (EI-\rho Ic^2)k^2+2\rho Ic\omega k+(\kappa GA-\omega^2\rho I) \\ X_{12} &= -X_{21} = -ik\kappa GA \end{aligned} \tag{7.135}$$

The eigenvalue problem of Equation 7.134 provides a dispersion relation

$$k^4+c_3k^3+c_2k^2+c_1k+c_0=0 \tag{7.136}$$

where

$$\begin{aligned} \Delta &= (EI-\rho Ic^2)(\kappa GA+N_x-\rho Ac^2) \\ c_3 &= \frac{2}{\Delta}\omega\left[\rho Ic(\kappa GA+N_x-2\rho Ac^2)+EI\rho Ac\right] \\ c_2 &= \frac{1}{\Delta}\left[\kappa GA(N_x-\rho Ac^2)-\omega^2\rho I(\kappa GA+N_x-6\rho Ac^2)-\omega^2 EI\rho A\right] \\ c_1 &= \frac{2}{\Delta}\omega\rho Ac(\kappa GA-2\omega^2\rho I) \\ c_0 &= -\frac{1}{\Delta}\omega^2\rho A(\kappa GA-\omega^2\rho I) \end{aligned} \tag{7.137}$$

Four wavenumbers k_r ($r=1, 2, 3, 4$) are computed from Equation 7.136 and β is determined from Equation 7.134 for the rth wavenumber k_r as follows:

$$\beta_r = \frac{(\kappa GA+N_x-\rho Ac^2)k_r^2+2\rho Ac\omega k_r-\omega^2\rho A}{ik_r\kappa GA} \quad (r=1,\ 2,\ 3,\ 4) \tag{7.138}$$

Define a diagonal matrix

$$\boldsymbol{B} = diag[\beta_r] \quad (r = 1,\ 2,\ 3,\ 4) \tag{7.139}$$

Then the general solution of Equation 7.132 can be written as

$$\begin{aligned} W(x) &= \sum_{r=1}^{4} a_r e^{-ik_r x} = \boldsymbol{e}(x;\omega)\boldsymbol{a} \\ \Theta(x) &= \sum_{r=1}^{4} \beta_r a_r e^{-ik_r x} = \boldsymbol{e}(x;\omega)\boldsymbol{B}\boldsymbol{a} \end{aligned} \tag{7.140}$$

where

$$\boldsymbol{a} = \{a_1 \quad a_2 \quad a_3 \quad a_4\}^{\mathrm{T}} \tag{7.141}$$

and

$$\boldsymbol{e}(x;\omega) = [e^{-ik_1x} \quad e^{-ik_2x} \quad e^{-ik_3x} \quad e^{-ik_4x}] \tag{7.142}$$

Substitution of Equation 7.140 into the right-hand side of Equation 7.130 gives

$$\boldsymbol{d} = \boldsymbol{H}_T(\omega)\boldsymbol{a} \tag{7.143}$$

where

$$\boldsymbol{H}_T(\omega) = \begin{bmatrix} 1 & 1 & 1 & 1 \\ \beta_1 & \beta_2 & \beta_3 & \beta_4 \\ e_1 & e_2 & e_3 & e_4 \\ \beta_1 e_1 & \beta_2 e_2 & \beta_3 e_3 & \beta_4 e_4 \end{bmatrix} \tag{7.144}$$

with

$$e_r = e^{-ik_r L} \qquad (r = 1,\ 2,\ 3,\ 4) \tag{7.145}$$

By using Equation 7.143, the constant vector $\boldsymbol{c}$ can be eliminated from Equation 7.140 to obtain the expressions as

$$\begin{aligned} W(x) &= \boldsymbol{N}_w(x;\omega)\boldsymbol{d} \\ \Theta(x) &= \boldsymbol{N}_\theta(x;\omega)\boldsymbol{d} \end{aligned} \tag{7.146}$$

where $\boldsymbol{N}_w(x;\omega)$ and $\boldsymbol{N}_w(x;\omega)$ are the dynamic shape functions defined by

$$\begin{aligned} \boldsymbol{N}_w(x;\omega) &= \boldsymbol{e}(x;\omega)\boldsymbol{H}_T^{-1}(\omega) \\ \boldsymbol{N}_\theta(x;\omega) &= \boldsymbol{e}(x;\omega)\boldsymbol{B}\boldsymbol{H}_T^{-1}(\omega) \end{aligned} \tag{7.147}$$

7.3.2.4 Weak Form of Governing Equation

The weak form of the governing Equation 7.127 can be obtained from

$$\begin{aligned} \int_o^L &\{[(\kappa GA + N_x - \rho A c^2)W'' - 2i\omega\rho A c W' + \omega^2 \rho A W - \kappa GA\Theta' + F(x)]\delta W \\ &+ [(EI - \rho I c^2)\Theta'' - 2i\omega\rho I c\Theta' - (\kappa GA - \omega^2 \rho I)\Theta + \kappa GA W']\delta\Theta\}\mathrm{d}x = 0 \end{aligned} \tag{7.148}$$

By integrating in parts and applying Equation 7.129, we obtain the weak form as

$$\begin{aligned}
&\int_0^L [(\kappa GA + N_x - \rho Ac^2)W'\delta W' - \kappa GA(W'\delta\Theta + \Theta\delta W') + \kappa GA\Theta\delta\Theta \\
&\quad + (EI - \rho Ic^2)\Theta'\delta\Theta' + i\omega\rho Ac(W'\delta W - W\delta W') + i\omega\rho Ic(\Theta'\delta\Theta - \Theta\delta\Theta') \\
&\quad - \omega^2\rho AW\delta W - \omega^2\rho I\Theta\delta\Theta]\mathrm{d}x \\
&\quad + i\omega\rho AcW\delta W|_0^L + \rho Ac^2W'\delta W|_0^L + i\omega\rho Ic\Theta\delta\Theta|_0^L + \rho Ic^2\Theta'\delta\Theta|_0^L \\
&\quad - [Q(x) + N_xW']\delta W|_0^L - M(x)\delta\Theta|_0^L - M(x)\delta\Theta|_0^L - \int_0^L F(x)\delta W\mathrm{d}x = 0
\end{aligned} \tag{7.149}$$

7.3.2.5 Spectral Element Equation

Substituting Equation 7.146 into Equation 7.149, we obtain

$$\begin{aligned}
&\delta \boldsymbol{d}^{\mathrm{T}} \int_0^L [(\kappa GA + N_x - \rho Ac^2)\boldsymbol{N}'^{\mathrm{T}}_w\boldsymbol{N}'_w - \kappa GA(\boldsymbol{N}'^{\mathrm{T}}_w\boldsymbol{N}_\theta + \boldsymbol{N}^{\mathrm{T}}_\theta\boldsymbol{N}'_w) + \kappa GA\boldsymbol{N}^{\mathrm{T}}_\theta\boldsymbol{N}_\theta \\
&\quad + (EI - \rho Ic^2)\boldsymbol{N}'^{\mathrm{T}}_\theta\boldsymbol{N}'_\theta + i\omega\rho Ac(\boldsymbol{N}^{\mathrm{T}}_w\boldsymbol{N}'_w - \boldsymbol{N}^{\mathrm{T}}_w\boldsymbol{N}_w) + i\omega\rho Ic(\boldsymbol{N}^{\mathrm{T}}_\theta\boldsymbol{N}'_\theta - \boldsymbol{N}'^{\mathrm{T}}_\theta\boldsymbol{N}_\theta) \\
&\quad - \omega^2\rho A\boldsymbol{N}^{\mathrm{T}}_w\boldsymbol{N}_w - \omega^2\rho I\boldsymbol{N}^{\mathrm{T}}_\theta\boldsymbol{N}_\theta]\mathrm{d}x + \rho Ac^2\boldsymbol{N}^{\mathrm{T}}_w\boldsymbol{N}'_w|_0^L + i\omega\rho Ac\boldsymbol{N}^{\mathrm{T}}_w\boldsymbol{N}_w|_0^L \\
&\quad + \rho Ic^2\boldsymbol{N}^{\mathrm{T}}_\theta\boldsymbol{N}'_\theta|_0^L + i\omega\rho Ic\boldsymbol{N}^{\mathrm{T}}_\theta\boldsymbol{N}_\theta|_0^L]\boldsymbol{d} = \delta\boldsymbol{d}^{\mathrm{T}}(\boldsymbol{f}_{\mathrm{c}} + \boldsymbol{f}_{\mathrm{d}})
\end{aligned} \tag{7.150}$$

where

$$\boldsymbol{f}_{\mathrm{d}}(\omega) = \int_0^L F(x)\boldsymbol{N}^{\mathrm{T}}_w(x;\omega)\mathrm{d}x \tag{7.151}$$

and $\boldsymbol{f}_{\mathrm{c}}$ is defined by Equation 7.131. As $\delta\boldsymbol{d}$ is arbitrary, we can obtain the spectral element equation from Equation 7.150 as

$$\boldsymbol{S}_T(\omega)\boldsymbol{d} = \boldsymbol{f}(\omega) \tag{7.152}$$

where

$$\begin{aligned}
\boldsymbol{S}_T(\omega) = &\int_0^L [(\kappa GA + N_x - \rho Ac^2)\boldsymbol{N}'^{\mathrm{T}}_w\boldsymbol{N}'_w - \kappa GA(\boldsymbol{N}'^{\mathrm{T}}_w\boldsymbol{N}_\theta + \boldsymbol{N}^{\mathrm{T}}_\theta\boldsymbol{N}'_w) + \kappa GA\boldsymbol{N}^{\mathrm{T}}_\theta\boldsymbol{N}_\theta \\
&\quad + (EI - \rho Ic^2)\boldsymbol{N}'^{\mathrm{T}}_\theta\boldsymbol{N}'_\theta + i\omega\rho Ac(\boldsymbol{N}^{\mathrm{T}}_w\boldsymbol{N}'_w - \boldsymbol{N}'^{\mathrm{T}}_w\boldsymbol{N}_w) + i\omega\rho Ic(\boldsymbol{N}^{\mathrm{T}}_\theta\boldsymbol{N}'_\theta - \boldsymbol{N}'^{\mathrm{T}}_\theta\boldsymbol{N}_\theta) \\
&\quad - \omega^2\rho A\boldsymbol{N}^{\mathrm{T}}_w\boldsymbol{N}_w - \omega^2\rho I\boldsymbol{N}^{\mathrm{T}}_\theta\boldsymbol{N}_\theta]\mathrm{d}x + \rho Ac^2\boldsymbol{N}^{\mathrm{T}}_w\boldsymbol{N}'_w|_0^L + i\omega\rho Ac\boldsymbol{N}^{\mathrm{T}}_w\boldsymbol{N}_w|_0^L \\
&\quad + \rho Ic^2\boldsymbol{N}^{\mathrm{T}}_\theta\boldsymbol{N}'_\theta|_0^L + i\omega\rho Ic\boldsymbol{N}^{\mathrm{T}}_\theta\boldsymbol{N}_\theta|_0^L
\end{aligned} \tag{7.153}$$

From Equation 7.147, we obtain

$$\begin{aligned}
\boldsymbol{N}'_w(x;\omega) &= -i\boldsymbol{e}(x;\omega)\boldsymbol{K}\boldsymbol{H}_T^{-1}(\omega) \\
\boldsymbol{N}'_\theta(x;\omega) &= -i\boldsymbol{e}(x;\omega)\boldsymbol{K}\boldsymbol{B}\boldsymbol{H}_T^{-1}(\omega)
\end{aligned} \tag{7.154}$$

Substituting Equations 7.147 and 7.154 into Equation 7.153 yields

$$\boldsymbol{S}_T(\omega) = \boldsymbol{H}_T^{-\mathrm{T}}(\omega)\boldsymbol{D}_T(\omega)\boldsymbol{H}_T^{-1}(\omega) \tag{7.155}$$

where

$$
\begin{aligned}
\boldsymbol{D}_T(\omega) = & -(\kappa GA + N_x - \rho Ac^2)\boldsymbol{K E}_T\boldsymbol{K} + i\kappa GA(\boldsymbol{KE}_T\boldsymbol{B} + \boldsymbol{BE}_T\boldsymbol{K}) \\
& + \kappa GA\boldsymbol{BE}_T\boldsymbol{B} - (EI - \rho Ic^2)\boldsymbol{BKE}_T\boldsymbol{KB} + \omega\rho Ac(\boldsymbol{E}_T\boldsymbol{K} - \boldsymbol{KE}_T) \\
& + \omega\rho Ic(\boldsymbol{BE}_T\boldsymbol{KB} - \boldsymbol{BKE}_T\boldsymbol{B}) - \omega^2\rho A\boldsymbol{E}_T - \omega^2\rho I\boldsymbol{BE}_T\boldsymbol{B} \\
& - i\rho Ac^2\bar{\boldsymbol{E}}_T\boldsymbol{K} + i\omega\rho Ac\bar{\boldsymbol{E}} - i\rho Ic^2\boldsymbol{B}\bar{\boldsymbol{E}}_T\boldsymbol{KB} + i\omega\rho Ic\boldsymbol{B}\bar{\boldsymbol{E}}_T\boldsymbol{B}
\end{aligned}
\tag{7.156}
$$

and where

$$
\begin{aligned}
\boldsymbol{E}_T(\omega) &= \int_0^L \boldsymbol{e}^{\mathrm{T}}(x;\omega)\boldsymbol{e}(x;\omega)\mathrm{d}x = [E_{Trs}(\omega)] \qquad (r,s = 1,\ 2,\ 3,\ 4) \\
\bar{\boldsymbol{E}}_T(\omega) &= \boldsymbol{e}^{\mathrm{T}}(x;\omega)\boldsymbol{e}(x;\omega)|_0^L = [\bar{E}_{Trs}(\omega)]
\end{aligned}
\tag{7.157}
$$

with

$$
\begin{aligned}
E_{Trs} &= \begin{cases} \dfrac{i}{k_r + k_s}\bar{E}_{Trs} & \text{if } k_r + k_s \neq 0 \\ L & \text{if } k_r + k_s = 0 \end{cases} \\
\bar{E}_{Trs} &= \mathrm{e}^{-i(k_r + k_s)L} - 1
\end{aligned}
\tag{7.158}
$$

The last four terms in the right-hand side of Equation 7.156 represent the effects of the mass transport across the boundaries at $x=0$ and L.

Assembling the spectral elements and applying the associated boundary conditions, we can obtain a global system equation in the form as

$$
\boldsymbol{S}_{T\mathrm{g}}(\omega)\boldsymbol{d}_{\mathrm{g}} = \boldsymbol{f}_{\mathrm{g}} \tag{7.159}
$$

where $\boldsymbol{S}_{T\mathrm{g}}(\omega)$ is the global dynamic stiffness matrix, $\boldsymbol{d}_{\mathrm{g}}$ is the global spectral nodal DOFs vector, and $\boldsymbol{f}_{\mathrm{g}}$ is the global spectral nodal forces vector. It is obvious from Equation 7.149 or Equation 7.153 that the four mass transport terms are determined only by the inlet and outlet boundary conditions. Thus, during the assembly procedure, the mass transport terms expressed in terms of $\bar{\boldsymbol{E}}_T$ in Equation 7.156 will be considered only for the spectral elements associated with the inlet and outlet boundaries.

7.3.3 Finite Element Model

The finite element equation is derived from the weak form of the equations of motion (7.123), which is given by

$$
\begin{aligned}
& \int_0^L [(\kappa GA + N_x - \rho Ac^2)w'\delta w' - \kappa GA(\theta\delta w' + w'\delta\theta) + \kappa GA\theta\delta\theta \\
& + (EI - \rho Ic^2)\theta'\delta\theta' + \rho Ac(\dot{w}'\delta w - \dot{w}\delta w') + \rho Ic(\dot{\theta}'\delta\theta - \dot{\theta}\delta\theta') \\
& + \rho A\ddot{w}\delta w + \rho I\ddot{\theta}\delta\theta]\mathrm{d}x \\
& + \rho Ac\dot{w}\delta w|_0^L + \rho Ac^2 w'\delta w|_0^L + \rho Ic\dot{\theta}\delta\theta|_0^L + \rho Ic^2\theta'\delta\theta|_0^L \\
& - [Q(x) + N_x w']\delta w|_0^L - M(x)\delta\theta|_0^L - \int_0^L f(x,t)\delta w\mathrm{d}x = 0
\end{aligned}
\tag{7.160}
$$

From the sign convention defined in Figure 7.8, which is also valid in the time domain, we define the nodal DOFs vector

$$\boldsymbol{d}(t) = \{w_1(t) \quad \theta_1(t) \quad w_2(t) \quad \theta_2(t)\}^{\mathrm{T}} = \{w(0,t) \quad \theta(0,t) \quad w(L,t) \quad \theta(L,t)\}^{\mathrm{T}} \tag{7.161}$$

and the spectral nodal forces and moments vector

$$\begin{aligned} \boldsymbol{f}_{\mathrm{c}}(t) &= \{Q_1(t) \quad M_1(t) \quad Q_2(t) \quad M_2(t)\}^{\mathrm{T}} \\ &= \{-Q(0,t) - N_x w'(0,t) \quad -M(0,t) \quad Q(L,t) + N_x w'(L,t) \quad M(L,t)\}^{\mathrm{T}} \end{aligned} \tag{7.162}$$

The displacement fields $w(x, t)$ and $\theta(x, t)$ within a finite element are assumed to be

$$w(x,t) = \boldsymbol{N}_w(x)\boldsymbol{d}(t), \qquad \theta(x,t) = \boldsymbol{N}_\theta(x)\boldsymbol{d}(t) \tag{7.163}$$

where

$$\begin{aligned} \boldsymbol{N}_w(x) = [&R(1-\xi)(\xi - 2\xi^2 + R^{-1}) \qquad (1/2)RL\xi(1-\xi)(1-2\xi+R^{-1}) \\ &R\xi(-1+3\xi-2\xi^2+R^{-1}) \quad -(1/2)RL\xi(1-\xi)(-1+2\xi+R^{-1})] \\ \boldsymbol{N}_\theta(x) = [&-6RL^{-1}\xi(1-\xi) \quad -3R(1-\xi)\{\xi-(1/3)R^{-1}\} \\ &6RL^{-1}\xi(1-\xi) \quad -3R\xi\{1-\xi-(1/3)R^{-1}\}] \end{aligned} \tag{7.164}$$

with the use of definitions

$$R = \frac{L^2}{L^2 + 12r}, \quad r = \frac{EI}{\kappa GA}, \quad \xi = \frac{x}{L} \quad (0 \le \xi \le 1) \tag{7.165}$$

Substituting Equation 7.163 into Equation 7.160 and using Equation 7.162, we obtain the finite element equation as

$$\boldsymbol{M}\ddot{\boldsymbol{d}}(t) + \boldsymbol{C}\dot{\boldsymbol{d}}(t) + \boldsymbol{K}\boldsymbol{d}(t) = \boldsymbol{f}_{\mathrm{c}}(t) + \boldsymbol{f}_{\mathrm{d}}(t) \tag{7.166}$$

where

$$\boldsymbol{f}_{\mathrm{d}}(t) = \int_0^L f(x,t)\boldsymbol{N}^{\mathrm{T}}(x)\mathrm{d}x \tag{7.167}$$

and the finite element matrices are given in Appendix 7.C. The matrices $\boldsymbol{C}_{MT\rho A}$, $\boldsymbol{C}_{MT\rho I}$, $\boldsymbol{K}_{MT\rho A}$, and $\boldsymbol{K}_{MT\rho I}$ given in Appendix 7.C represent the effects of the mass transport across the boundaries at $x = 0$ and L. The four mass transport terms in Equation 7.160 are determined only by the inlet and outlet boundary conditions. Thus, during the assembly procedure, the mass transport matrices $\boldsymbol{C}_{MT\rho A}$, $\boldsymbol{C}_{MT\rho I}$, $\boldsymbol{K}_{MT\rho A}$, and $\boldsymbol{K}_{MT\rho I}$ will be considered only for the finite elements associated with the inlet and outlet boundaries.

7.3.4 Stability Analysis

As discussed in Section 7.1.4, the types of the instability for an axially moving beam can be determined by investigating the signs of the real and imaginary parts of the eigenvalues of the problem. The divergence speed at which the static instability (divergence) occurs can be obtained from the static eigenvalue problem derived from Equation 7.127 by putting $\omega = 0$ and $F = 0$ and then by assuming

$$W(x) = a\mathrm{e}^{-ikx}, \qquad \Theta(x) = \beta a\mathrm{e}^{-ikx} \tag{7.168}$$

The static eigenvalue problem is obtained as

$$\begin{bmatrix} (\kappa GA + N_x - \rho Ac^2)k^2 & -\kappa GAik \\ \kappa GAik & (EI - \rho Ic^2)k^2 + \kappa GA \end{bmatrix} \begin{Bmatrix} 1 \\ \beta \end{Bmatrix} a = \begin{Bmatrix} 0 \\ 0 \end{Bmatrix} \tag{7.169}$$

which gives

$$k_1 = 0, \quad k_2 = 0, \quad k_3 = -k_4 = \bar{k} \tag{7.170}$$

where

$$\bar{k} = \sqrt{\frac{p_2c^2 - p_3}{(1 + p_3 - p_2c^2)(p_1 - p_4c^2)}} \tag{7.171}$$

with

$$p_1 = \frac{EI}{\kappa GA}, \quad p_2 = \frac{\rho A}{\kappa GA}, \quad p_3 = \frac{N_x}{\kappa GA}, \quad p_4 = \frac{\rho I}{\kappa GA} \tag{7.172}$$

As an example case, consider a beam that is axially moving over two simple supports. The non-trivial equilibrium displacements of the beam can be written as

$$\begin{aligned} W(x) &= A_1 + A_2x + A_3e^{i\bar{k}x} + A_4e^{-i\bar{k}x} \\ \Theta(x) &= \beta_1A_1 + \beta_2A_2x + \beta_3A_3e^{i\bar{k}x} + \beta_4A_4e^{-i\bar{k}x} \end{aligned} \tag{7.173}$$

where

$$\begin{aligned} &\beta_1 = \beta_2 = 0 \\ &\beta_3 = \frac{-i\bar{k}}{1 + (p_1 - p_4c^2)\bar{k}^2}, \quad \beta_4 = \frac{i\bar{k}}{1 + (p_1 - p_4c^2)\bar{k}^2} \end{aligned} \tag{7.174}$$

Applying the simply supported boundary conditions to Equation 7.173 gives

$$\bar{k}_n = \frac{n\pi}{L} (n = 1, 2, 3, \ldots.) \tag{7.175}$$

Equating Equation 7.175 to Equation 7.171 gives the divergence speed c_{Dn} at which the divergence instability occurs in the nth mode as follows:

$$c_{Dn} = \begin{cases} \dfrac{1}{n\pi}\sqrt{\dfrac{q_1 - \sqrt{q_1^2 + q_2}}{2p_2p_4}} & \text{(for bending wave mode)} \\[2ex] \dfrac{1}{n\pi}\sqrt{\dfrac{q_1 + \sqrt{q_1^2 + q_2}}{2p_2p_4}} & \text{(for shear wave mode)} \end{cases} \tag{7.176}$$

where

$$\begin{aligned} q_1 &= n^2\pi^2(p_1p_2 + p_4 + p_3p_4) + L^2p_2 \\ q_2 &= 4n^2\pi^2p_2p_4[n^2\pi^2(p_1 + p_1p_3) + L^2p_3] \end{aligned} \tag{7.177}$$

Example 7.3

Consider a uniform Timoshenko beam that is moving axially over two simple supports as an illustrative example. The geometric and structural properties of the Timoshenko beam are given by $L = 2\,\text{m}$, $EI = 10\,\text{kN}\,\text{m}^2$, $\kappa GA = 80 \times 10^6\,\text{N}$, $\rho A = 10\,\text{kg/m}$, and $\rho I = 30 \times 10^{-6}\,\text{kg m}$.

Table 7.3 displays the natural frequencies obtained by SEM and FEM for different moving speeds and axial tensions. For the stationary Timoshenko beam (when $c = 0$ m/s), the natural frequencies can be also computed from a theoretical formula [24]

$$f_n^2\big|_{\text{Timo}} = f_n^2\big|_{\text{BE}} \frac{L^2}{2n^2\pi^2}\left(\alpha - \sqrt{\alpha^2 - \beta}\right) \quad (n = 1, 2, 3, \ldots)$$

where $f_n|_{BE}$ are the natural frequencies of a stationary Bernoulli–Euler beam

$$f_n\big|_{BE} = \frac{n^2\pi}{2L^2}\sqrt{\frac{EI}{\rho A}}$$

and

$$\alpha = \frac{\rho A}{\rho I} + \left(1 + \frac{L^2}{n^2\pi^2}\frac{\rho A}{\rho I}\right)\frac{\kappa GA}{EI}, \quad \beta = 4\frac{\rho A}{\rho I} \cdot \frac{\kappa GA}{EI}$$

Table 7.3 Natural frequencies of an axially moving Timoshenko beam

			Natural frequency (Hz)				
Moving speed (m/s)	Axial tension (kN)	Method	f_1	f_2	f_3	f_5	f_{10}
0	0	Theory [24]	11.630	46.498	104.53	289.60	1144.3
		SEM (1)	11.630	46.498	104.53	289.60	1144.3
		FEM(20)	11.630	46.498	185.65	289.71	1151.4
		FEM(100)	11.630	49.498	104.53	289.60	1144.4
0	20	SEM (1)	16.421	51.957	110.16	295.33	1150.1
		FEM(20)	16.421	51.957	110.17	295.45	1157.2
		FEM(100)	16.421	51.957	110.16	295.33	1150.2
33.092 ($0.5c_{\text{D1}}$)	20	SEM (1)	13.591	49.601	108.07	293.45	1148.3
		FEM(50)	13.774	49.818	108.30	293.68	1149.1
		FEM(300)	13.621	49.638	108.11	293.49	1148.3
66.184 (c_{D1})	20	SEM (1)	0.000	42.037	101.67	287.80	1142.9
		FEM(50)	3.386	43.035	102.60	288.70	1144.3
		FEM(300)	0.000	42.209	101.83	287.95	1143.1

Note: () = total number of the equal-length finite elements used in the analysis.

Table 7.3 shows that the present spectral element model provides very accurate natural frequencies by representing the whole beam as a single spectral element. It also shows that the first natural frequency vanishes at divergence speed $c_{\text{D1}} = 66.184$ m/s. The divergence speed c_{D1} predicted by using the Timoshenko beam theory is slightly lower than the value predicted by using the Bernoulli–Euler beam theory, and vice versa for the natural frequencies: see Table 7.2.

Figure 7.9 shows the instability boundaries as a function of the moving speed and axial tensile force. Figure 7.10 compares the dispersion curves at two different statuses: when the beam is moving axially and when the beam is not moving axially (stationary). Firstly, when the beam is stationary, k_1 is pure real at all frequency, while k_2 is pure imaginary at $0 \leq f \leq f_c$ and pure real at $f \geq f_c$, where $f_c = 2\pi\sqrt{\kappa GA/\rho I}$ is the cut-off frequency. Thus, k_1 represents the forward traveling wave at all frequencies, while k_2 represents the forward evanescent (non-propagating) wave at $f \leq f_c$ and the forward traveling wave at $f \geq f_c$. Secondly, when the beam is axially moving, k_1 keeps its value at pure real at all frequencies so that it still represents the forward traveling wave. However, k_2 changes its value from pure imaginary to pure real at $0 \leq f \leq f_{Lc}$ and from pure imaginary to complex at $f_{Lc} \leq f \leq f_c$, although it keeps its value real at $f \geq f_c$, where f_{Lc} ($< f_c$) is the lower cut-off frequency generated by the moving speed. Thus, it is interesting to investigate whether the forward evanescent wave of a stationary beam changes to the forward traveling wave at $0 \leq f \leq f_{Lc}$ when the beam starts to move axially.

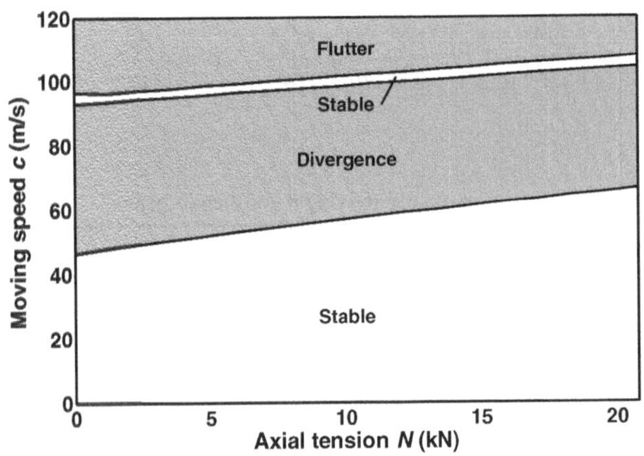

Figure 7.9 Instability boundaries of an axially moving Timoshenko beam

7.4 Axially Moving Thin Plates

7.4.1 Equation of Motion

Consider a uniform, thin plate that is moving in its length direction (x-direction) with a constant speed c, as shown in Figure 7.11. Assume that the x-direction is opposite to the gravitation g. The plate has the length L (the span between two supports), width b, and thickness h in the x-, y-, and z-directions, respectively. The material properties of the plate are the mass density ρ, Young's modulus E, and Poisson's ratio ν.

The equation of transverse vibration and the associated boundary conditions of an axially moving thin plate can be derived from the extended form of Hamilton's principle represented

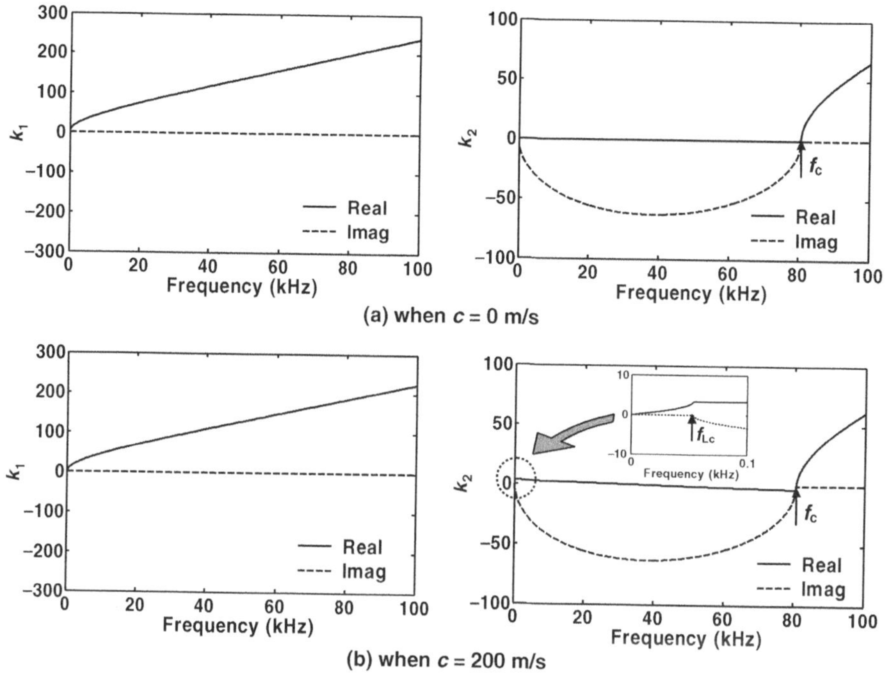

Figure 7.10 Moving speed dependence of the dispersion curves of a Timoshenko beam

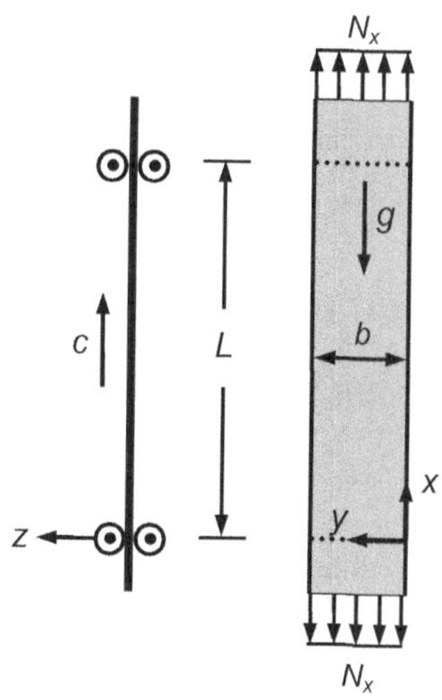

Figure 7.11 Axially moving thin plate subjected to a uniform in-plane axial tension N_x

by Equation 7.1. The kinetic energy, potential energy, and virtual work are given by [67, 76]

$$T = \frac{\rho h}{2}\int_0^b\int_0^L\left[c^2 + \left(\frac{\partial w}{\partial t} + c\frac{\partial w}{\partial x}\right)^2\right]\mathrm{d}x\mathrm{d}y$$

$$U = \frac{1}{2}\int_0^b\int_0^L\left[D\left\{\left(\frac{\partial^2 w}{\partial x^2} + \frac{\partial^2 w}{\partial y^2}\right)^2 + 2(1-\nu)\left[\left(\frac{\partial^2 w}{\partial x\partial y}\right)^2 - \frac{\partial^2 w}{\partial x^2}\frac{\partial^2 w}{\partial y^2}\right]\right\}\right.$$
$$\left. + (N_x + \rho hgx)\left(\frac{\partial w}{\partial x}\right)^2\right]\mathrm{d}x\mathrm{d}y \qquad (7.178)$$

$$\delta W = \int_0^b\int_0^L f(x,y,t)\delta w\,\mathrm{d}x\mathrm{d}y + \int_0^b V_{x1}(y,t)\delta w(0,y,t)\mathrm{d}y + \int_0^b V_{x2}(y,t)\delta w(L,y,t)\mathrm{d}y$$
$$+ \int_0^b M_{x1}(y,t)\delta\left(\frac{\partial w(0,y,t)}{\partial x}\right)\mathrm{d}y + \int_0^b M_{x2}(y,t)\delta\left(\frac{\partial w(L,y,t)}{\partial x}\right)\mathrm{d}y$$

where $w(x, y, t)$ is the transverse displacement, $f(x, y, t)$ is the external force normal to the plate surface, N_x is the constant in-plane tensile force per unit width acting in the x-direction, and $D = Eh^3/12(1-\nu^2)$ is the bending (flexural) rigidity of the plate. M_{x1} and V_{x1}, are, respectively, the bending moment and transverse shear force specified along the boundary on $x = 0$ and M_{x2} and V_{x2} on $x = L$. The virtual momentum transport of mass across the boundaries at $x = 0$ and L is given by

$$\delta W_{\mathrm{MT}} = \int_0^b \rho hc\left(\frac{\partial w}{\partial t} + c\frac{\partial w}{\partial x}\right)\delta w\Big|_0^L \mathrm{d}y \qquad (7.179)$$

Substituting Equations 7.178 and 7.179 into the extended form of Hamilton's principle, Equation 7.1, we obtain the equation of motion

$$D\nabla^4 w + \rho h\left(\frac{\partial^2 w}{\partial t^2} + 2c\frac{\partial^2 w}{\partial x\partial t} + c^2\frac{\partial^2 w}{\partial x^2}\right) - (N_x + \rho hgx)\frac{\partial^2 w}{\partial x^2} - \rho hg\frac{\partial w}{\partial x} = f(x,y,t) \qquad (7.180)$$

and the boundary conditions on $x = 0$ and $x = L$ as

$$\begin{aligned}
w(0,y,t) &= w_1(y,t) \quad \text{or} \quad V_x(0,y,t) = -V_{x1}(y,t) - N_x\frac{\partial w(0,y,t)}{\partial x}\\
\frac{\partial w(0,y,t)}{\partial x} &= \theta_1(y,t) \quad \text{or} \quad M_x(0,y,t) = +M_{x1}(y,t)\\
w(L,y,t) &= w_2(y,t) \quad \text{or} \quad V_x(L,y,t) = +V_{x2}(y,t) - N_x\frac{\partial w(L,y,t)}{\partial x}\\
\frac{\partial w(L,y,t)}{\partial x} &= \theta_2(y,t) \quad \text{or} \quad M_x(L,y,t) = -M_{x2}(y,t)
\end{aligned} \qquad (7.181)$$

and the boundary conditions on $y = 0$ and $y = b$ as

$$\begin{aligned}
V_y(x,0,t) &= 0 \quad \text{and} \quad M_y(x,0,t) = 0\\
V_y(x,b,t) &= 0 \quad \text{and} \quad M_y(x,b,t) = 0
\end{aligned} \qquad (7.182)$$

In Equation 7.180, ∇^4 is the bi-harmonic operator in rectangular coordinates. In Equations 7.181 and 7.182, V_x and V_y are the resultant Kirchhoff shear forces per unit length in the

y- and x-directions, respectively. Similarly, M_x and M_y are the resultant bending moments per unit length in the y- and x-directions, respectively. They are related to the transverse displacement $w(x, y, t)$ as follows:

$$\begin{aligned}
V_x(x,y,t) &= -D\left[\frac{\partial^3 w}{\partial x^3}+(2-\nu)\frac{\partial^3 w}{\partial x\partial y^2}\right]+\rho hgx\frac{\partial w}{\partial x}\\
V_y(x,y,t) &= -D\left[\frac{\partial^3 w}{\partial y^3}+(2-\nu)\frac{\partial^3 w}{\partial x^2 y}\right]\\
M_x(x,y,t) &= -D\left(\frac{\partial^2 w}{\partial x^2}+\nu\frac{\partial^2 w}{\partial y^2}\right)\\
M_y(x,y,t) &= -D\left(\frac{\partial^2 w}{\partial y^2}+\nu\frac{\partial^2 w}{\partial x^2}\right)
\end{aligned} \tag{7.183}$$

For the plates that are moving parallel to the ground (that is, normal to gravitation), we simply remove all gravitation-induced terms from Equations 7.180 and 7.183.

When a rectangular plate has at least two simply supported parallel edges, we call it a Levy-type plate. It is well known that closed form solutions can be obtained analytically for such Levy-type plates. The present axially moving thin plate has free boundary conditions on $y=0$ and $y=b$. Thus, provided the other two boundaries at $x=0$ and $x=L$ are not all simply supported, it is inevitable for us to adopt a proper approximate solution method such as the Rayleigh–Ritz method, finite element method (FEM), finite strip method (FSM) [32], and the Kantorovich method [64]. In the Kantorovich method, the displacements of a plate are described by the functions given as the products of assumed shape functions and of unknown functions, which reduce the partial differential equations of motion to the ordinary differential equations for unknown functions. The assumed shape functions must satisfy the boundary conditions at two parallel edges in one direction. In contrast to FSM, the unknown functions are determined to satisfy the boundary conditions at two parallel edges in the other direction by solving the reduced ordinary differential equations in an analytical way. The finite strip method is very similar in philosophy to the Kantorovich method, but it can be considered as a special form of FEM. In the FSM, the problem domain is divided into strip elements and the displacements of a strip element are described by the functions, which are given as products of trigonometric/hyperbolic series and polynomials. The series have to satisfy a priori boundary conditions at the end of the strips. Therefore, FSM will lie between the FEM and the Rayleigh–Ritz methods. In the following, we adopt the Kantorovich method to transform a two-dimensional (plate) problem into a one-dimensional-type of problem, followed by the formulation of a modal spectral element model for an axially moving thin plate.

7.4.2 Spectral Element Modeling

7.4.2.1 Governing Equation in the Frequency Domain

The vibration response of the plate is written in the spectral form as

$$w(x,y,t)=\frac{1}{N}\sum_{n=0}^{N-1}W_n(x,y)e^{i\omega_n t} \tag{7.184}$$

Similarly, we write

$$\begin{aligned}
&\{f(x,y,t), V_x(x,y,t), V_y(x,y,t), M_x(x,y,t), M_y(x,y,t)\}\\
&\quad=\frac{1}{N}\sum_{n=0}^{N-1}\{F_n(x,y), V_{xn}(x,y), V_{yn}(x,y), M_{xn}(x,y), M_{yn}(x,y)\}e^{i\omega_n t}\\
&\{V_{x1}(y,t), V_{x2}(y,t), M_{x1}(y,t), M_{x2}(y,t)\}\\
&\quad=\frac{1}{N}\sum_{n=0}^{N-1}\{V_{x1n}(y), V_{x2n}(y), M_{x1n}(y), M_{x2n}(y)\}e^{i\omega_n t}\\
&\{w_1(y,t), \theta_1(y,t), w_2(y,t), \theta_2(y,t)\}\\
&\quad=\frac{1}{N}\sum_{n=0}^{N-1}\{W_{1n}(y), \Theta_{1n}(y), W_{2n}(y), \Theta_{2n}(y)\}e^{i\omega_n t}
\end{aligned} \tag{7.185}$$

Substitution of Equations 7.184 and 7.185 into the equation of motion (7.180) gives

$$D\nabla^4 W_n - (N_x - \rho h c^2 + \rho h g x)\frac{\partial^2 W_n}{\partial x^2} + (2i\omega_n \rho h c - \rho h g)\frac{\partial W_n}{\partial x} - \rho h \omega_n^2 W_n = F_n \tag{7.186}$$

By substitution of Equations 7.184 and 7.185 into Equations 7.181 and 7.182, the boundary conditions can be also transformed into the frequency domain as

$$\begin{aligned}
W_n(0,y) &= W_{1n}(y) \quad \text{or} \quad V_{xn}(0,y) = -V_{x1n}(y) - N_x\frac{\partial W_n(0,y)}{\partial x}\\
\frac{\partial W_n(0,y)}{\partial x} &= \Theta_{1n}(y) \quad \text{or} \quad M_{xn}(0,y) = +M_{x1n}(y)\\
W_n(L,y) &= W_{2n}(y) \quad \text{or} \quad V_{xn}(L,y) = +V_{x2n}(y) - N_x\frac{\partial W_n(L,y)}{\partial x}\\
\frac{\partial W_n(L,y)}{\partial x} &= \Theta_{2n}(y) \quad \text{or} \quad M_{xn}(L,y) = -M_{x2n}(y)
\end{aligned} \tag{7.187}$$

and

$$\begin{aligned}
V_{yn}(x,0) = 0 \quad &\text{and} \quad M_{yn}(x,0) = 0\\
V_{yn}(x,b) = 0 \quad &\text{and} \quad M_{yn}(x,b) = 0
\end{aligned} \tag{7.188}$$

Similarly, the force-displacement relationships (7.183) can be transformed into the frequency domain as follows:

$$\begin{aligned}
V_{xn}(x,y) &= -D\left(\frac{\partial^3 W_n}{\partial x^3} + (2-\nu)\frac{\partial^3 W_n}{\partial x \partial y^2}\right) + \rho h g x\frac{\partial W_n}{\partial x}\\
V_{yn}(x,y) &= -D\left(\frac{\partial^3 W_n}{\partial y^3} + (2-\nu)\frac{\partial^3 W_n}{\partial x^2 \partial y}\right)\\
M_{xn}(x,y) &= -D\left(\frac{\partial^2 W_n}{\partial x^2} + \nu\frac{\partial^2 W_n}{\partial y^2}\right)\\
M_{yn}(x,y) &= -D\left(\frac{\partial^2 W_n}{\partial y^2} + \nu\frac{\partial^2 W_n}{\partial x^2}\right)
\end{aligned} \tag{7.189}$$

7.4.2.2 Dynamic Shape Functions

To formulate the spectral element, we consider the homogeneous governing equation reduced from Equation 7.186 by enforcing $F_n = 0$ as

$$D\nabla^4 W_n - (N_x - \rho hc^2 + \rho hgx)\frac{\partial^2 W_n}{\partial x^2} + (2i\omega_n \rho hc - \rho hg)\frac{\partial W_n}{\partial x} - \rho h\omega_n^2 W_n = 0 \quad (7.190)$$

By following the solution approach of the classical Kantorovich method, we assume the general solution of Equation 7.190 in the form of

$$W_n(x, y) = \sum_{p=1}^{M} X_{np}(x) Y_p(y) \quad (7.191)$$

where $Y_p(y)$ ($p = 1, 2, \ldots, M$) are the known functions chosen a priori, and $X_{np}(x)$ are the unknown functions to be determined analytically. The difference from the classical Kantorovich method is that $X_{np}(x)$ should be determined by a Do-Loop over all frequency components. We use the natural modes of the free–free Bernoulli–Euler beam (simply the "beam functions") for $Y_p(y)$. The beam functions $Y_p(y)$ satisfy the following orthogonality properties:

$$\begin{aligned}
&\int_0^b Y_p(y) Y_q(y) \mathrm{d}y = 0 \\
&\int_0^b Y_p'(y) Y_q'(y) \mathrm{d}y = 0 \\
&\int_0^b Y_p''(y) Y_q''(y) \mathrm{d}y = 0 \qquad (\text{for } p \neq q) \\
&\int_0^b Y_p''(y) Y_q(y) \mathrm{d}y = 0
\end{aligned} \quad (7.192)$$

Substituting Equation 7.191 into Equation 7.190 gives

$$\begin{aligned}
\sum_{p=1}^{M} \Bigg\{ & D\frac{d^4 X_{np}}{dx^4} Y_p + 2D\frac{d^2 X_{np}}{dx^2}\frac{d^2 Y_p}{dy^2} + DX_{nr}\frac{d^4 Y_p}{dy^4} + (\rho hc^2 - N_x - \rho hgx)\frac{d^2 X_{np}}{dx^2} Y_p \\
& + (2i\omega_n \rho hc - \rho hg)\frac{dX_{nl}}{dx} Y_p - \rho h\omega_n^2 X_{np} Y_p \Bigg\} = 0
\end{aligned} \quad (7.193)$$

Multiplying Equation 7.193 by $Y_m(y)$ and integrating from $y = 0$ to $y = b$, and then applying the orthogonal properties 7.192, we obtain

$$\begin{aligned}
& D\frac{d^4 X_{nm}}{dx^4} + 2DI_{0m}^{-1}\left(\int_0^b \frac{d^2 Y_m}{dy^2} Y_m \mathrm{d}y\right)\frac{d^2 X_{nm}}{dx^2} + DI_{0m}^{-1}\left(\int_0^b \frac{d^4 Y_m}{dy^4} Y_m \mathrm{d}y\right) X_{nm} \\
& - (N_x - \rho hc^2 + \rho hgx)\frac{d^2 X_{nm}}{dx^2} + (2i\omega_n \rho hc - \rho hg)\frac{dX_{nm}}{dx} - \rho h\omega_n^2 X_{nm} = 0
\end{aligned} \quad (7.194)$$

where

$$I_{0m} = \int_0^b Y_m^2 \mathrm{d}y \quad (7.195)$$

Applying the integral by parts to the second and third terms of Equation 7.194, we obtain

$$\begin{aligned} D\frac{d^4X_{nm}}{dx^4} &- (2DI_{0m}^{-1}I_{1m} + N_x - \rho hc^2 + \rho hgx)\frac{d^2X_{nm}}{dx^2} \\ &+ (2i\omega_n\rho hc - \rho hg)\frac{dX_{nm}}{dx} + (DI_{0m}^{-1}I_{2m} - \rho h\omega_n^2)X_{nm} \\ &+ DI_{0m}^{-1}\left(2\frac{d^2X_{nm}}{dx^2}\frac{dY_m}{dy}Y_m\bigg|_0^b + X_{nm}\frac{d^3Y_m}{dy^3}Y_m\bigg|_0^b - X_{nm}\frac{d^2Y_m}{dy^2}\frac{dY_m}{dy}\bigg|_0^b\right) = 0 \end{aligned} \tag{7.196}$$

where

$$I_{1m} = \int_0^b \left(\frac{dY_m}{dy}\right)^2 \mathrm{dy} \qquad I_{2m} = \int_0^b \left(\frac{d^2Y_m}{dy^2}\right)^2 \mathrm{dy} \tag{7.197}$$

The beam functions $Y_m(y)$ $(m = 1, 2, \ldots, M)$ used to derive Equation 7.196 do not exactly satisfy the free boundary conditions (7.188) at $y = 0$ and b. Thus we will enforce the functions $X_{nm}(x)$ to satisfy free boundary conditions at $y = 0$ and b in conjunction with $Y_m(y)$. By applying Equation 7.191 to the free boundary conditions (7.188), we can obtain

$$\begin{aligned} X_{nm}\frac{d^3Y_m}{dy^3}\bigg|_{y=0,b} &= -(2-\nu)\frac{d^2X_{nm}}{dx^2}\frac{dY_m}{dy}\bigg|_{y=0,b} \\ X_{nm}\frac{d^2Y_m}{dy^2}\bigg|_{y=0,b} &= -\nu\frac{d^2X_{nm}}{dx^2}Y_m\big|_{y=0,b} \end{aligned} \tag{7.198}$$

By applying Equation 7.198 to the last two boundary-related terms in Equation 7.196, we obtain

$$\frac{d^4X_{nm}}{dx^4} - \alpha_m\frac{d^2X_{nm}}{dx^2} + \beta_{nm}\frac{dX_{nm}}{dx} + \gamma_{nm}X_{nm} = 0 \tag{7.199}$$

where

$$\begin{aligned} \alpha_m &= \frac{N_x - \rho hc^2 + \rho hgx}{D} + \frac{2}{I_{0m}}(I_{1m} - \nu J_m) \\ \beta_{nm} &= \frac{2i\omega_n\rho hc - \rho hg}{D} \\ \gamma_{nm} &= \frac{I_{2m}}{I_{0m}} - \frac{\rho h}{D}\omega_n^2 \\ J_m &= \frac{dY_m}{dy}Y_m\bigg|_0^b \end{aligned} \tag{7.200}$$

As α_m is a function of x, we consider the following two approximation methods:

1. **Method A:** replace ρhgx by ρhgx_c.

$$\alpha_m^A = \frac{N_x - \rho hc^2 + \rho hgx_c}{D} + \frac{2}{I_{0m}}(I_{1m} - \nu J_m) \tag{7.201}$$

2. **Method B:** neglect ρhgx.

$$\alpha_m^B = \frac{N_x - \rho hc^2}{D} + \frac{2}{I_{0m}}(I_{1m} - \nu J_m) \tag{7.202}$$

where x_c is the x-wise center of the finite element.

So far, we have transformed the original two-dimensional problem represented by Equation 7.190 into a one-dimensional problem represented by Equation 7.199 by using the classical Kantorovich method. Thus, we can now readily apply the spectral element formulation technique to the newly formed one-dimensional problem of Equation 7.199. As Equation 7.199 is for a specific natural mode in the y-direction (for example, the mth natural mode of a free-free beam), the spectral element model based on Equation 7.199 is called from here onwards the "modal" spectral element model.

Assume the general solution of Equation 7.199 as

$$X_{nm}(x) = a_{nm}e^{-ik_{nm}x} \tag{7.203}$$

Substitution of Equation 7.203 into Equation 7.199 yields a dispersion relation as

$$k_{nm}^4 + \alpha_m k_{nm}^2 - i\beta_{nm}k_{nm} + \gamma_{nm} = 0 \tag{7.204}$$

where $\alpha_m = \alpha_m^A$ for Method A and $\alpha_m = \alpha_m^B$ for Method B. By using four wavenumbers k_{nmi} ($i = 1, 2, 3, 4$) computed from Equation 7.204, we can write the general solution to Equation 7.199 as

$$X_{nm}(x) = \boldsymbol{e}_{nm}(x)\boldsymbol{a}_{nm} \tag{7.205}$$

where

$$\begin{aligned} \boldsymbol{e}_{nm}(x) &= [e^{-ik_{nm1}x} \quad e^{-ik_{nm2}x} \quad e^{-ik_{nm3}x} \quad e^{-ik_{nm4}x}] \\ \boldsymbol{a}_{nm} &= \{a_{nm1} \quad a_{nm2} \quad a_{nm3} \quad a_{nm4}\}^T \end{aligned} \tag{7.206}$$

Substituting Equation 7.205 into Equation 7.191, we obtain the general solution to Equation 7.190 to be

$$W_n(x, y) = \sum_{m=1}^{M} \boldsymbol{e}_{nm}(x)Y_m(y)\boldsymbol{a}_{nm} \tag{7.207}$$

The constant vectors $\boldsymbol{a}_{nm}$ are determined to satisfy the boundary conditions at $x = 0$ and $x = L$.

Define a vector as

$$\boldsymbol{u}_n(y) = \begin{Bmatrix} W_n(0, y) \\ W'_n(0, y) \\ W_n(L, y) \\ W'_n(L, y) \end{Bmatrix} \tag{7.208}$$

where the prime (′) denotes the partial derivative with respect to x. Substitution of Equation 7.207 into Equation 7.208 gives

$$\boldsymbol{u}_n(y) = \sum_{m=1}^{M} \begin{bmatrix} e_{nm}(0) \\ e'_{nm}(0) \\ e_{nm}(L) \\ e'_{nm}(L) \end{bmatrix} Y_m(y)\boldsymbol{a}_{nm} \tag{7.209}$$

Define the generalized spectral nodal DOFs vector as

$$\boldsymbol{d}_{nm} \equiv \frac{1}{I_{0m}} \int_0^b \boldsymbol{u}_n(x) Y_m(y) \mathrm{d}y \tag{7.210}$$

By substituting Equation 7.209 into Equation 7.210 and using the properties of Equations 7.192 and 7.195, we obtain

$$\boldsymbol{d}_{nm} = \boldsymbol{H}_{nm}(\omega)\boldsymbol{a}_{nm} \tag{7.211}$$

where

$$\boldsymbol{H}_{nm}(\omega) = \begin{bmatrix} e_{nm}(0) \\ e'_{nm}(0) \\ e_{nm}(L) \\ e'_{nm}(L) \end{bmatrix} = \begin{bmatrix} 1 & 1 & 1 & 1 \\ -ik_{nm1} & -ik_{nm2} & -ik_{nm3} & -ik_{nm4} \\ \mathrm{e}_1 & \mathrm{e}_2 & \mathrm{e}_3 & \mathrm{e}_4 \\ -ik_{nm1}\mathrm{e}_1 & -ik_{nm2}\mathrm{e}_2 & -ik_{nm3}\mathrm{e}_3 & -ik_{nm4}\mathrm{e}_4 \end{bmatrix} \tag{7.212}$$

with

$$\mathrm{e}_i = \mathrm{e}^{-ik_{nmi}L} \qquad (i = 1,\ 2,\ 3,\ 4) \tag{7.213}$$

By using Equation 7.211, Equation 7.207 can be written as

$$W_n(x,y) = \sum_{m=1}^{M} \boldsymbol{N}_{nm}(x,y;\omega_n)\boldsymbol{d}_{nm} = \boldsymbol{N}_n(x,y;\omega_n)\boldsymbol{d}_n \tag{7.214}$$

where

$$\boldsymbol{d}_n = \{\boldsymbol{d}_{n1}^{\mathrm{T}} \quad \boldsymbol{d}_{n2}^{\mathrm{T}} \quad \boldsymbol{d}_{n3}^{\mathrm{T}} \ldots \boldsymbol{d}_{nM}^{\mathrm{T}}\}^{\mathrm{T}} \tag{7.215}$$

$$\boldsymbol{N}_n(x,y;\omega_n) = [\boldsymbol{N}_{n1}(x,y;\omega_n)\boldsymbol{N}_{n2}(x,y;\omega_n)\ldots\boldsymbol{N}_{nM}(x,y;\omega_n)] \tag{7.216}$$

with

$$\boldsymbol{N}_{nm}(x,y;\omega_n) = \boldsymbol{e}_{nm}(x)\boldsymbol{H}_{nm}^{-1}(\omega_n)Y_m(y) \qquad (m = 1,\ 2, \ldots,\ M) \tag{7.217}$$

$\boldsymbol{N}_n(x,y;\omega_n)$ is the 1-by-(4 × M) dynamic shape function matrix and $\boldsymbol{d}_n$ is the (4 × M)-by-1 generalized spectral nodal DOFs vector.

7.4.2.3 Weak Form of Governing Equation

The weak form of the governing Equation 7.186 in the frequency domain is obtained from

$$\int_0^b\int_0^L \Big\{D\nabla^4 W_n - (N_x - \rho hc^2 + \rho hgx)\frac{\partial^2 W_n}{\partial x^2} + (2i\omega_n\rho hc - \rho hg)\frac{\partial W_n}{\partial x} - \rho h\omega_n^2 W_n - F_n\Big\}\delta W_n \mathrm{d}x\mathrm{d}y = 0 \tag{7.218}$$

Integrating Equation 7.218 by parts and using the force-displacement relationships (7.189), we can obtain the weak form as

$$\begin{aligned}
&\int_0^b\int_0^L\left\{D\left(\frac{\partial^2 W_n}{\partial x^2}\frac{\partial^2\delta W_n}{\partial x^2}+\nu\frac{\partial^2 W_n}{\partial y^2}\frac{\partial^2\delta W_n}{\partial x^2}+\nu\frac{\partial^2 W_n}{\partial x^2}\frac{\partial^2\delta W_n}{\partial y^2}+\frac{\partial^2 W_n}{\partial y^2}\frac{\partial^2\delta W_n}{\partial y^2}\right.\right.\\
&\quad\left.+2(1-\nu)\frac{\partial^2 W_n}{\partial x\partial y}\frac{\partial^2\delta W_n}{\partial x\partial y}\right)+(N_x-\rho hc^2)\frac{\partial W_n}{\partial x}\frac{\partial\delta W_n}{\partial x}+\rho hgx\frac{\partial W_n}{\partial x}\frac{\partial\delta W_n}{\partial x}\\
&\quad\left.-\rho h\omega_n^2 W_n\delta W_n+i\omega_n\rho hc\left(\frac{\partial W_n}{\partial x}\delta W_n-W_n\frac{\partial\delta W_n}{\partial x}\right)-F_n\delta W_n\right\}\mathrm{d}x\mathrm{d}y\\
&\quad-\int_0^b\left[V_{xn}(x,y)+N_x\frac{\partial W_n}{\partial x}\right]\delta W_n\bigg|_0^L\mathrm{d}y+\int_0^b M_{xn}(x,y)\frac{\partial\delta W_n}{\partial x}\bigg|_0^L\mathrm{d}y\\
&\quad+\int_0^b\left(\rho hc^2\frac{\partial W_n}{\partial x}+i\omega_n\rho hcW_n\right)\delta W_n\bigg|_0^L\mathrm{d}y=0
\end{aligned}\tag{7.219}$$

7.4.2.4 Spectral Element Equation

By substituting Equation 7.214 into Equation 7.219 and applying the boundary conditions (7.187), we obtain

$$\begin{aligned}
&\delta\boldsymbol{d}_n^T\left[\int_0^b\int_0^L\left\{D\left[\frac{\partial^2\boldsymbol{N}_n^{\mathrm{T}}}{\partial x^2}\frac{\partial^2\boldsymbol{N}_n}{\partial x^2}+\frac{\partial^2\boldsymbol{N}_n^{\mathrm{T}}}{\partial y^2}\frac{\partial^2\boldsymbol{N}_n}{\partial y^2}+\nu\left(\frac{\partial^2\boldsymbol{N}_n^{\mathrm{T}}}{\partial x^2}\frac{\partial^2\boldsymbol{N}_n}{\partial y^2}+\frac{\partial^2\boldsymbol{N}_n^{\mathrm{T}}}{\partial y^2}\frac{\partial^2\boldsymbol{N}_n}{\partial x^2}\right)\right.\right.\right.\\
&\quad\left.+2(1-\nu)\frac{\partial^2\boldsymbol{N}_n^{\mathrm{T}}}{\partial x\partial y}\frac{\partial^2\boldsymbol{N}_n}{\partial x\partial y}\right]+(N_x-\rho hc^2)\frac{\partial\boldsymbol{N}_n^{\mathrm{T}}}{\partial x}\frac{\partial\boldsymbol{N}_n}{\partial x}+\rho hgx\frac{\partial\boldsymbol{N}_n^{\mathrm{T}}}{\partial x}\frac{\partial\boldsymbol{N}_n}{\partial x}\\
&\quad\left.-\rho h\omega_n^2\boldsymbol{N}_n^{\mathrm{T}}\boldsymbol{N}_n+i\omega_n\rho hc\left(\boldsymbol{N}_n^{\mathrm{T}}\frac{\partial\boldsymbol{N}_n}{\partial x}-\frac{\partial\boldsymbol{N}_n^{\mathrm{T}}}{\partial x}\boldsymbol{N}_n\right)\right\}\mathrm{d}x\,\mathrm{d}y\\
&\quad\left.+\int_0^b\frac{\partial\boldsymbol{N}_n^{\mathrm{T}}}{\partial x}\left(\rho hc^2\frac{\partial\boldsymbol{N}_n}{\partial x}+i\omega_n\rho hc\boldsymbol{N}_n\right)\bigg|_0^L\mathrm{d}y\right]\boldsymbol{d}_n=\delta\boldsymbol{d}_n^{\mathrm{T}}\boldsymbol{f}_n
\end{aligned}\tag{7.220}$$

where

$$\boldsymbol{f}_n(\omega_n)=\boldsymbol{f}_{\mathrm{c}n}(\omega_n)+\boldsymbol{f}_{\mathrm{d}n}(\omega_n)\tag{7.221}$$

with

$$\begin{aligned}
&\boldsymbol{f}_{\mathrm{c}n}(\omega_n)=\int_0^b\left\{\boldsymbol{N}_n^{\mathrm{T}}(0,y)V_{x1n}(y)+\frac{\partial\boldsymbol{N}_n^{\mathrm{T}}(0,y)}{\partial x}M_{x1n}(y)+\boldsymbol{N}_n^{\mathrm{T}}(L,y)V_{x2n}(y)+\frac{\partial\boldsymbol{N}_n^{\mathrm{T}}(L,y)}{\partial x}M_{x2n}(y)\right\}\mathrm{d}y\\
&\boldsymbol{f}_{\mathrm{d}n}(\omega_n)=\int_0^b\int_0^L\boldsymbol{N}_n^{\mathrm{T}}(x,y)F_n(x,y)\mathrm{d}x\mathrm{d}y
\end{aligned}\tag{7.222}$$

Since $\delta\boldsymbol{d}$ is arbitrary by definition, the spectral element equation can be derived from Equation 7.220 as

$$\boldsymbol{S}_P(\omega_n)\boldsymbol{d}_n=\boldsymbol{f}_n(\omega_n)\tag{7.223}$$

where

$$\boldsymbol{S}_P(\omega_n)=\boldsymbol{S}_I(\omega_n)+\boldsymbol{S}_{II}(\omega_n) \tag{7.224}$$

with

$$\begin{aligned}\boldsymbol{S}_I(\omega_n)=\int_0^b\int_0^L\Bigg\{&D\left[\frac{\partial^2\boldsymbol{N}_n^{\mathrm{T}}}{\partial x^2}\frac{\partial^2\boldsymbol{N}_n}{\partial x^2}+\frac{\partial^2\boldsymbol{N}_n^{\mathrm{T}}}{\partial y^2}\frac{\partial^2\boldsymbol{N}_n}{\partial y^2}+\nu\left(\frac{\partial^2\boldsymbol{N}_n^{\mathrm{T}}}{\partial x^2}\frac{\partial^2\boldsymbol{N}_n}{\partial y^2}+\frac{\partial^2\boldsymbol{N}_n^{\mathrm{T}}}{\partial y^2}\frac{\partial^2\boldsymbol{N}_n}{\partial x^2}\right)\right.\\&\left.+2(1-\nu)\frac{\partial^2\boldsymbol{N}_n^{\mathrm{T}}}{\partial x\partial y}\frac{\partial^2\boldsymbol{N}_n}{\partial x\partial y}\right]+(N_x-\rho hc^2)\frac{\partial\boldsymbol{N}_n^{\mathrm{T}}}{\partial x}\frac{\partial\boldsymbol{N}_n}{\partial x}+\rho hgx\frac{\partial\boldsymbol{N}_n^{\mathrm{T}}}{\partial x}\frac{\partial\boldsymbol{N}_n}{\partial x}\\&-\rho h\omega_n^2\boldsymbol{N}_n^{\mathrm{T}}\boldsymbol{N}_n+i\omega_n\rho hc\left(\boldsymbol{N}_n^T\frac{\partial\boldsymbol{N}_n}{\partial x}-\frac{\partial\boldsymbol{N}_n^{\mathrm{T}}}{\partial x}\boldsymbol{N}_n\right)\Bigg\}\mathrm{d}x\mathrm{d}y\end{aligned} \tag{7.225}$$

$$\begin{aligned}\boldsymbol{S}_{II}(\omega_n)=\rho hc^2&\int_0^b\left[\frac{\partial\boldsymbol{N}_n^{\mathrm{T}}(L,y)}{\partial x}\frac{\partial\boldsymbol{N}_n(L,y)}{\partial x}-\frac{\partial\boldsymbol{N}_n^{\mathrm{T}}(0,y)}{\partial x}\frac{\partial\boldsymbol{N}_n(0,y)}{\partial x}\right]\mathrm{d}y\\&+i\omega_n\rho hc\int_0^b\left[\frac{\partial\boldsymbol{N}_n^{\mathrm{T}}(L,y)}{\partial x}\boldsymbol{N}_n(L,y)-\frac{\partial\boldsymbol{N}_n^{\mathrm{T}}(0,y)}{\partial x}\boldsymbol{N}_n(0,y)\right]\mathrm{d}y\end{aligned} \tag{7.226}$$

Substitution of Equation 7.216 into Equation 7.224 will give a (4 × M)-by-(4 × M) spectral element matrix in the partitioned matrix form as

$$\boldsymbol{S}_P(\omega_n)=\begin{bmatrix}\boldsymbol{S}_{P11} & \boldsymbol{S}_{P12} & \cdots & \boldsymbol{S}_{P1j} & \cdots & \boldsymbol{S}_{P1M}\\ \boldsymbol{S}_{P21} & \boldsymbol{S}_{P22} & \cdots & \boldsymbol{S}_{P2j} & \cdots & \boldsymbol{S}_{P2M}\\ \vdots & \vdots & \ddots & \vdots & \ddots & \vdots\\ \boldsymbol{S}_{Pi1} & \boldsymbol{S}_{Pi2} & \cdots & \boldsymbol{S}_{Pij} & \cdots & \boldsymbol{S}_{PiM}\\ \vdots & \vdots & \ddots & \vdots & \ddots & \vdots\\ \boldsymbol{S}_{PM1} & \boldsymbol{S}_{PM2} & \cdots & \boldsymbol{S}_{PMj} & \cdots & \boldsymbol{S}_{PMM}\end{bmatrix} \tag{7.227}$$

In the above, $\boldsymbol{S}_{Pij}(\omega_n)$ ($i, j = 1, 2, \ldots M$) are the 4-by-4 matrices defined by

$$\boldsymbol{S}_{Pij}(\omega_n)=\boldsymbol{S}_{Iij}(\omega_n)+\boldsymbol{S}_{IIij}(\omega_n) \tag{7.228}$$

where

$$\begin{aligned}\boldsymbol{S}_{Iij}(\omega_n)=\int_0^b\int_0^L\Bigg\{&D\left[\frac{\partial^2\boldsymbol{N}_{ni}^{\mathrm{T}}}{\partial x^2}\frac{\partial^2\boldsymbol{N}_{nj}}{\partial x^2}+\frac{\partial^2\boldsymbol{N}_{ni}^{\mathrm{T}}}{\partial y^2}\frac{\partial^2\boldsymbol{N}_{nj}}{\partial y^2}\right.\\&\left.+\nu\left(\frac{\partial^2\boldsymbol{N}_{ni}^{\mathrm{T}}}{\partial x^2}\frac{\partial^2\boldsymbol{N}_{nj}}{\partial y^2}+\frac{\partial^2\boldsymbol{N}_{ni}^{\mathrm{T}}}{\partial y^2}\frac{\partial^2\boldsymbol{N}_{nj}}{\partial x^2}\right)+2(1-\nu)\frac{\partial^2\boldsymbol{N}_{ni}^{\mathrm{T}}}{\partial x\partial y}\frac{\partial^2\boldsymbol{N}_{nj}}{\partial x\partial y}\right]\\&+(N_x-\rho hc^2)\frac{\partial\boldsymbol{N}_{ni}^{\mathrm{T}}}{\partial x}\frac{\partial\boldsymbol{N}_{nj}}{\partial x}+\rho hgx\frac{\partial\boldsymbol{N}_{ni}^{\mathrm{T}}}{\partial x}\frac{\partial\boldsymbol{N}_{nj}}{\partial x}\\&-\rho h\omega_n^2\boldsymbol{N}_{ni}^{\mathrm{T}}\boldsymbol{N}_{nj}+i\omega_n\rho hc\left(\boldsymbol{N}_{ni}^{\mathrm{T}}\frac{\partial\boldsymbol{N}_{nj}}{\partial x}-\frac{\partial\boldsymbol{N}_{ni}^{\mathrm{T}}}{\partial x}\boldsymbol{N}_{nj}\right)\Bigg\}\mathrm{d}x\,\mathrm{d}y\end{aligned} \tag{7.229}$$

$$\begin{aligned} S_{IIij}(\omega_n) &= \rho h c^2 \int_0^b \left[\frac{\partial N_{ni}^{\mathrm{T}}(L,y)}{\partial x} \frac{\partial N_{nj}(L,y)}{\partial x} - \frac{\partial N_{ni}^{\mathrm{T}}(0,y)}{\partial x} \frac{\partial N_{nj}(0,y)}{\partial x} \right] \mathrm{d}y \\ &+ i\omega_n \rho h c \int_0^b \left[\frac{\partial N_{ni}^{\mathrm{T}}(L,y)}{\partial x} N_{nj}(L,y) - \frac{\partial N_{ni}^{T}(0,y)}{\partial x} N_{nj}(0,y) \right] \mathrm{d}y \end{aligned} \quad (7.230)$$

By substituting Equation 7.217 into Equations 7.229 and 7.230 and applying the orthogonality properties (7.192), we obtain the spectral element matrix in the form of

$$S_P(\omega_n) = \begin{bmatrix} S_{P11} & \mathbf{0} & \cdots & \mathbf{0} \\ \mathbf{0} & S_{P22} & \cdots & \mathbf{0} \\ \vdots & \vdots & \ddots & \vdots \\ \mathbf{0} & \mathbf{0} & \cdots & S_{PMM} \end{bmatrix} \quad (7.231)$$

where

$$S_{Pii}(\omega_n) = S_{Iii}(\omega_n) + S_{IIii}(\omega_n) \quad (i = 1, 2, 3, \ldots, M) \quad (7.232)$$

S_{Iii} and S_{IIii} are four-by-four matrices defined by

$$\begin{aligned} S_{Iii}(\omega_n) &= H_{ni}^{-\mathrm{T}} D_{Ii} H_{ni}^{-1} \\ S_{IIii}(\omega_n) &= -(\rho h c^2 A + i\omega_n \rho h c B) I_{0i} \end{aligned} \quad (7.233)$$

where

$$\begin{aligned} D_{Ii}(\omega_n) = D[&K_{ni}^2 E_{ni} K_{ni}^2 I_{0i} + E_{ni} I_{2i} - \nu(K_{ni}^2 E_{ni} + E_{ni} K_{ni}^2) I_{3i} \\ &-2(1-\nu) K_{ni} E_{ni} K_{nj} I_{1i}] - (N_x - \rho h c^2) K_{ni} E_{ni} K_{ni} I_{0i} \\ &-\rho h g K_{ni} E_{ni} K_{nj} I_{0i} - \rho h \omega_n^2 E_{ni} I_{0i} + \omega_n \rho h c (E_{ni} K_{ni} - K_{ni} E_{ni}) I_{0i} \end{aligned} \quad (7.234)$$

and

$$A = \begin{bmatrix} 0 & 0 & 0 & 0 \\ 0 & 1 & 0 & 0 \\ 0 & 0 & 0 & 0 \\ 0 & 0 & 0 & -1 \end{bmatrix}, \quad B = \begin{bmatrix} 0 & 0 & 0 & 0 \\ -1 & 0 & 0 & 0 \\ 0 & 0 & 0 & 0 \\ 0 & 0 & 1 & 0 \end{bmatrix} \quad (7.235)$$

with the use of following definitions

$$\begin{aligned} K_{ni} &= diag[k_{ni1} \;\; k_{ni2} \;\; k_{ni3} \;\; k_{ni4}] \\ K_{ni}^2 &= diag[k_{ni1}^2 \;\; k_{ni2}^2 \;\; k_{ni3}^2 \;\; k_{ni4}^2] \\ E_{ni}(\omega_n) &= \textstyle\int_0^L e_{ni}^{\mathrm{T}}(x) e_{ni}(x) \mathrm{d}x \\ \bar{E}_{ni}(\omega_n) &= e_{ni}^{\mathrm{T}}(L) e_{ni}(L) - e_{ni}^{\mathrm{T}}(0) e_{ni}(0) \end{aligned} \quad (7.236)$$

I_{0i}, I_{1i}, and I_{2i} are defined by Equation 7.195 and

$$I_{3i} = \int_0^b Y_i Y_i'' \, \mathrm{d}y \quad (7.237)$$

Similarly, by substituting Equations 7.216 and 7.217 into Equation 7.221, we obtain the generalized spectral nodal forces vector as

$$f_n(\omega_n) = \begin{Bmatrix} f_{n1} \\ f_{n2} \\ \vdots \\ f_{nM} \end{Bmatrix} = \begin{Bmatrix} f_{cn1} \\ f_{cn2} \\ \vdots \\ f_{cnM} \end{Bmatrix} + \begin{Bmatrix} f_{dn1} \\ f_{dn2} \\ \vdots \\ f_{dnM} \end{Bmatrix} \tag{7.238}$$

where

$$\begin{aligned} f_{cni}(\omega_n) = \boldsymbol{H}_{ni}^{-\mathrm{T}}\{\boldsymbol{e}_{ni}^{\mathrm{T}}(0)\int_0^b Y_m(y)V_{x1n}(y)\mathrm{d}y + \boldsymbol{e}'^{\mathrm{T}}_{ni}(0)\int_0^b Y_m(y)M_{x1n}(y)\mathrm{d}y \\ + \boldsymbol{e}_{ni}^{\mathrm{T}}(L)\int_0^b Y_m(y)V_{x2n}(y)\mathrm{d}y + \boldsymbol{e}'^{\mathrm{T}}_{ni}(L)\int_0^b Y_m(y)M_{x2n}(y)\mathrm{d}y\} \end{aligned} \tag{7.239}$$

$$f_{dni}(\omega_n) = \boldsymbol{H}_{ni}^{-\mathrm{T}}\int_0^b\int_0^L \boldsymbol{e}_{ni}^{\mathrm{T}}(x)Y_m(y)F_n(x,y)\mathrm{d}x\,\mathrm{d}y$$

Substitution of Equations 7.231 and 7.238 into Equation 7.223 yields M independent equations as follows:

$$\boldsymbol{S}_{Pii}(\omega_n)\boldsymbol{d}_{ni} = \boldsymbol{f}_{ni}(\omega_n) \quad (i = 1,\ 2,\ \ldots,\ M) \tag{7.240}$$

As Equation 7.240 represents the independent relationships between the generalized spectral nodal DOFs and the generalized spectral nodal forces for each mode (beam functions) $Y_i(y)$ $(i = 1, 2, \ldots, M)$, it is termed the "modal" spectral element equation. The modal spectral element Equation 7.240 can be assembled independently for each mode to obtain the global system equations for each mode as

$$\boldsymbol{S}_{Piig}(\omega)\boldsymbol{d}_{nig} = \boldsymbol{f}_{nig} \quad (i = 1,\ 2,\ 3,\ \ldots,\ M) \tag{7.241}$$

where $\boldsymbol{S}_{Piig}(\omega)$ is the global dynamic stiffness matrix, $\boldsymbol{d}_{nig}$ is the global spectral nodal DOFs vector, and $\boldsymbol{f}_{nig}$ is the global spectral nodal forces vector for the ith mode.

7.4.3 Finite Element Model

The weak form of the time-domain equation of motion (7.180) can be obtained from

$$\begin{aligned} \int_0^b\int_0^L \Big\{ D\nabla^4 w + \rho h\left(\frac{\partial^2 w}{\partial t^2} + 2c\frac{\partial^2 w}{\partial x \partial t} + c^2\frac{\partial^2 w}{\partial x^2}\right) \\ - (N_x + \rho hgx)\frac{\partial^2 w}{\partial x^2} - \rho hg\frac{\partial w}{\partial x} - f(x,y,t)\Big\}\delta w\mathrm{d}x\mathrm{d}y = 0 \end{aligned} \tag{7.242}$$

Integrating Equation 7.242 by parts and using the force-displacement relationships (7.183), we obtain the weak form

$$\begin{aligned}
&\int_0^b\int_0^L\left\{D\left(\frac{\partial^2 w}{\partial x^2}\frac{\partial^2\delta w}{\partial x^2}+\nu\frac{\partial^2 w}{\partial y^2}\frac{\partial^2\delta w}{\partial x^2}+\nu\frac{\partial^2 w}{\partial x^2}\frac{\partial^2\delta w}{\partial y^2}+\frac{\partial^2 w}{\partial y^2}\frac{\partial^2\delta w}{\partial y^2}\right.\right.\\
&\left.+2(1-\nu)\frac{\partial^2 w}{\partial x\partial y}\frac{\partial^2\delta w}{\partial x\partial y}\right)+(N_x-\rho hc^2)\frac{\partial w}{\partial x}\frac{\partial\delta w}{\partial x}+\rho hgx\frac{\partial w}{\partial x}\frac{\partial\delta w}{\partial x}\\
&\left.+\rho h\frac{\partial^2 w}{\partial t^2}\delta w+\rho hc\left(\frac{\partial^2 w}{\partial x\partial t}\delta w-\frac{\partial w}{\partial t}\frac{\partial\delta w}{\partial x}\right)-f(x,y,t)\delta w\right]\mathrm{d}x\,\mathrm{d}y \qquad (7.243)\\
&-\int_0^b\left[V_x(x,y,t)+N_x\frac{\partial w}{\partial x}\right]\delta w\bigg|_0^L\mathrm{d}y+\int_0^b M_x(x,y,t)\frac{\partial\delta w}{\partial x}\bigg|_0^L\mathrm{d}y\\
&+\int_0^b\left(\rho hc\frac{\partial w}{\partial t}+\rho hc^2\frac{\partial w}{\partial x}\right)\delta w\bigg|_0^L\mathrm{d}y=0
\end{aligned}$$

We will consider the four-node 12-DOF rectangular plate bending element, which is commonly referred to as the ACM element [139]. At each node, there are three DOFs: one transverse displacement DOF w, and two rotational DOFs $\theta_x=\partial w/\partial y$ and $\theta_y=-\partial w/\partial y$. The displacement field is assumed to be

$$w(x,y,t)=\boldsymbol{N}(x,y)\boldsymbol{d}(t) \qquad (7.244)$$

where

$$\begin{aligned}\boldsymbol{d}(t)=\{&w_1(t)\quad \theta_{x1}(t)\quad \theta_{y1}(t)\quad w_2(t)\quad \theta_{x2}(t)\quad \theta_{y2}(t)\quad w_3(t)\quad \theta_{x3}(t)\\ &\theta_{y3}(t)\quad w_4(t)\quad \theta_{x4}(t)\quad \theta_{y4}(t)\}^{\mathrm{T}}\end{aligned} \qquad (7.245)$$

and

$$\boldsymbol{N}(x,y)=\boldsymbol{B}(x,y)\boldsymbol{H}^{-1} \qquad (7.246)$$

with

$$\begin{aligned}
&\boldsymbol{B}(x,y)=[1\quad x\quad y\quad x^2\quad xy\quad y^2\quad x^3\quad x^2y\quad xy^2\quad y^3\quad x^3y\quad xy^3]\\
&\boldsymbol{R}(x,y)_{(3\times12)}=\begin{bmatrix}\boldsymbol{B}(x,y)\\ \partial\boldsymbol{B}(x,y)/\partial y\\ -\partial\boldsymbol{B}(x,y)/\partial x\end{bmatrix} \qquad (7.247)\\
&\boldsymbol{H}_{(12\times12)}=\begin{bmatrix}\boldsymbol{R}(0,0)\\ \boldsymbol{R}(L,0)\\ \boldsymbol{R}(L,b)\\ \boldsymbol{R}(0,b)\end{bmatrix}
\end{aligned}$$

Substitution of Equation 7.244 into Equation 7.243 gives the finite element equation in the form of

$$\boldsymbol{M}\ddot{\boldsymbol{d}}(t)+(\boldsymbol{C}_G+\boldsymbol{C}_{MT})\dot{\boldsymbol{d}}(t)+(\boldsymbol{K}_S+\boldsymbol{K}_T+\boldsymbol{K}_{\mathrm{g}}+\boldsymbol{K}_{MT})\boldsymbol{d}(t)=\boldsymbol{f}_{\mathrm{c}}(t)+\boldsymbol{f}_{\mathrm{d}}(t) \qquad (7.248)$$

where

$$
\begin{aligned}
\boldsymbol{M} &= \rho h \int_0^b \int_0^L \boldsymbol{N}^{\mathrm{T}} \boldsymbol{N} \mathrm{d}x\, \mathrm{d}y = \boldsymbol{M}^{\mathrm{T}} \\
\boldsymbol{C}_G &= \rho h c \int_0^b \int_0^L \left(\boldsymbol{N}^{\mathrm{T}} \frac{\partial \boldsymbol{N}}{\partial x} - \frac{\partial \boldsymbol{N}^{\mathrm{T}}}{\partial x} \boldsymbol{N} \right) \mathrm{d}x\, \mathrm{d}y = -\boldsymbol{C}_G^{\mathrm{T}} \\
\boldsymbol{C}_{MT} &= \rho h c \int_0^b \boldsymbol{N}^{\mathrm{T}} \boldsymbol{N} \big|_0^L \mathrm{d}y = \boldsymbol{C}_{MT}^{\mathrm{T}} \\
\boldsymbol{K}_S &= D \int_0^b \int_0^L \left[\frac{\partial^2 \boldsymbol{N}^{\mathrm{T}}}{\partial x^2} \frac{\partial^2 \boldsymbol{N}}{\partial x^2} + \nu \left(\frac{\partial^2 \boldsymbol{N}^{\mathrm{T}}}{\partial x^2} \frac{\partial^2 \boldsymbol{N}}{\partial y^2} + \frac{\partial^2 \boldsymbol{N}^{\mathrm{T}}}{\partial y^2} \frac{\partial^2 \boldsymbol{N}}{\partial x^2} \right) + \frac{\partial^2 \boldsymbol{N}^{\mathrm{T}}}{\partial y^2} \frac{\partial^2 \boldsymbol{N}}{\partial y^2} \right. \\
&\quad \left. + 2(1-\nu) \frac{\partial^2 \boldsymbol{N}^{\mathrm{T}}}{\partial x \partial y} \frac{\partial^2 \boldsymbol{N}}{\partial x \partial y} \right] \mathrm{d}x\, \mathrm{d}y = \boldsymbol{K}_S^{\mathrm{T}} \qquad (7.249) \\
\boldsymbol{K}_N &= (N_x - \rho h c^2) \int_0^b \int_0^L \frac{\partial \boldsymbol{N}^{\mathrm{T}}}{\partial x} \frac{\partial \boldsymbol{N}}{\partial x} \mathrm{d}x\, \mathrm{d}y = \boldsymbol{K}_T^{\mathrm{T}} \\
\boldsymbol{K}_g &= \rho h g \int_0^b \int_0^L x \frac{\partial \boldsymbol{N}^{\mathrm{T}}}{\partial x} \frac{\partial \boldsymbol{N}}{\partial x} \mathrm{d}x\, \mathrm{d}y = \boldsymbol{K}_g^{\mathrm{T}} \\
\boldsymbol{K}_{MT} &= \rho h c^2 \int_0^b \boldsymbol{N}^{\mathrm{T}} \frac{\partial \boldsymbol{N}}{\partial x} \bigg|_0^L \mathrm{d}y
\end{aligned}
$$

and

$$
\begin{aligned}
\boldsymbol{f}_c(t) &= \int_0^b \left\{ \boldsymbol{N}^{\mathrm{T}}(0,y) V_{x1}(y,t) + \frac{\partial \boldsymbol{N}^{\mathrm{T}}(0,y)}{\partial x} M_{x1}(y,t) + \boldsymbol{N}^{\mathrm{T}}(L,y) V_{x2}(y,t) + \frac{\partial \boldsymbol{N}^{\mathrm{T}}(L,y)}{\partial x} M_{x2}(y,t) \right\} \mathrm{d}y \\
\boldsymbol{f}_d(t) &= \int_0^b \int_0^L \boldsymbol{N}^{\mathrm{T}} f(x,y,t) \mathrm{d}x \mathrm{d}y \qquad (7.250)
\end{aligned}
$$

The finite element matrices $\boldsymbol{M}$, $\boldsymbol{C}$, and $\boldsymbol{K}$ are given in Appendix 7.D. As hinted at from Equation 7.243, the mass transport term (the last integral) is determined only by the inlet ($x = 0$) and outlet ($x = L$) boundary conditions. Thus, during the assembly procedure, the mass transport matrices $\boldsymbol{C}_{MT}$ and $\boldsymbol{K}_{MT}$ will be considered only for the finite elements associated with the inlet and outlet boundaries.

Example 7.4

Consider a uniform thin plate that is moving upward passing through two simple supports as shown in Figure 7.11. The geometric and material properties of the plate are given by $L = 15$ m, $b = 1$ m, $h = 0.003$ m, $E = 200$ GPa, $\nu = 0.333$, and $\rho = 7800\ \mathrm{kg/m^3}$.

Table 7.4 compares the natural frequencies of the stationary plate (i.e., $c = 0$ m/s) obtained by the present SEM with those obtained by FEM (m, n), where m and n denote the number of finite elements in the x- and y-directions, respectively. For SEM, the whole plate is treated as a single spectral element. It is assumed that the plate is subjected to a constant in-plane tensile force $N_x = 60$ kN/m. When the effects of the gravity (g) are not considered (or when the plate surface is parallel to the ground), the SEM results are compared with the results computed by using the theoretical formula from Ref. [24]. Table 7.4 certainly shows that the present SEM

Table 7.4 Natural frequencies (Hz) of the stationary plate subjected to $N_x = 60$ kN/m

	Without gravity ($g = 0$)		with gravity ($g = 9.8$ m/s^2)				
			FEM ($m \times n$)			SEM (1 × 1)	
Mode (p, q)	Theory [24]	SEM (1 × 1)	(15 × 1)	(60 × 4)	(120 × 8)	Method A	Method B
(1, 1)	1.688	1.688	1.458	1.711	1.712	1.710	1.713
(1, 2)	1.799	1.799	1.800	1.818	1.821	1.819	1.823
(2, 1)	3.378	3.378	2.939	3.423	3.426	3.423	3.428
(2, 2)	3.599	3.599	3.602	3.638	3.643	3.641	3.648
(3, 1)	5.071	5.072	4.456	5.139	5.143	5.141	5.145
(5, 2)	9.027	9.028	9.036	9.125	9.137	9.139	9.144

provides very accurate results when compared with the results obtained by both FEM and theoretical formula. When the effects of the gravity are considered, the "Method B" based SEM seems to slightly overestimate the natural frequencies. It also shows that the effects of the gravity tend to increase the natural frequencies. Figure 7.12 displays the natural modes of the stationary plate computed by the present SEM.

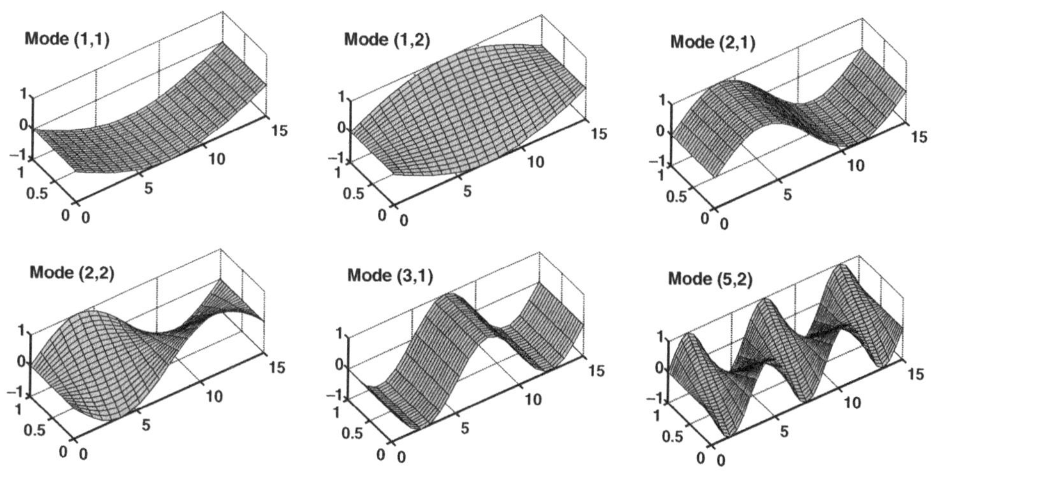

Figure 7.12 Natural modes of the stationary plate subjected to $N_x = 60$ kN/m

Table 7.5 compares the natural frequencies obtained by SEM for two different moving speeds and axial tensions, with and without considering the effects of gravity. When the effects of gravity are considered, the "Method A" based SEM is used. It is clear from Table 7.5 that the natural frequencies decrease as the moving speed increases, while they increase as the axial tension increases. Table 7.5 also shows that the natural frequencies certainly increase due to gravity.

Figure 7.13 shows the eigenvalues $\lambda = i\omega$ of the lowest four vibration modes (1, 1), (1, 2), (2, 1), and (2, 2) as functions of the moving speed c. The static and dynamic instability zones are determined by the signs of the real and imaginary parts of the eigenvalues and they are indicated in the figure. It is shown that the divergence zone starts from $c_D = 249.38$ m/s, followed by the second narrow stable zone from $c_S = 262.03$ m/s and the dynamic instability "flutter" zone from

Table 7.5 Natural frequencies (Hz) of the plate passing through two simple supports

	$c = 5$ m/s				$c = 10$ m/s			
	$N_x = 30$ kN/m		$N_x = 60$ kN/m		$N_x = 30$ kN/m		$N_x = 60$ kN/m	
Mode (p, q)	Without gravity	With gravity	Without gravity	With gravity	Without gravity	With gravity	Without gravity	With gravity
(1, 1)	1.170	1.204	1.672	1.696	1.089	1.130	1.618	1.644
(1, 2)	1.325	1.355	1.783	1.806	1.254	1.289	1.733	1.757
(2, 1)	2.344	2.413	3.345	3.394	2.181	2.240	3.238	3.296
(2, 2)	2.652	2.713	3.568	3.613	2.512	2.565	3.468	3.520
(3, 1)	3.523	3.630	5.023	5.096	3.282	3.376	4.863	4.957
(5, 2)	6.673	6.835	8.951	9.069	6.337	6.498	8.707	8.833

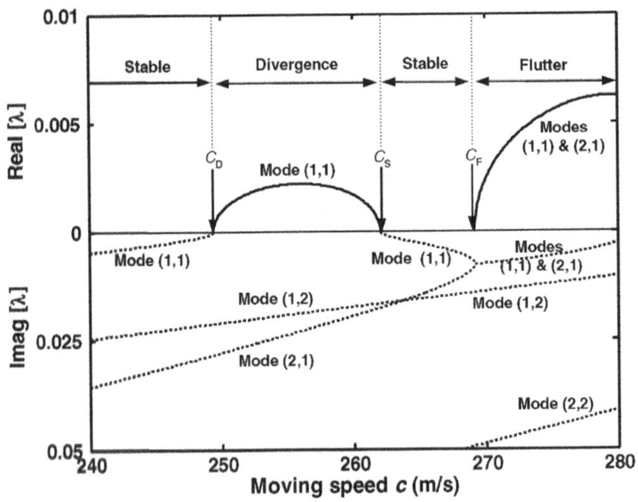

Figure 7.13 The lowest four eigenvalues λ (Hz) versus the moving speed c (m/s)

$c_F = 169.19$ m/s. The dynamic instability occurs as the coupled or mixed-mode flutter for which the natural frequencies of the mode (1, 1) and mode (2, 1) merge together near the flutter speed c_F.

Example 7.5

Consider the same uniform thin plate as considered in Example 7.4. The plate is now moving upward passing through four simple supports at an equal distance of $L/3$, where L is the distance between the bottom support and the top support as shown in Figure 7.14.

The natural frequencies obtained using the "Method A" based SEM for two different moving speeds and axial tensions are displayed in Table 7.6. The whole plate is represented by three spectral elements connected in the lengthwise direction. Table 7.6 shows that the natural frequencies decrease as the moving speed increases, while they increase as the axial tension increases. It is also shown that the effects of the gravity also increase the natural frequencies.

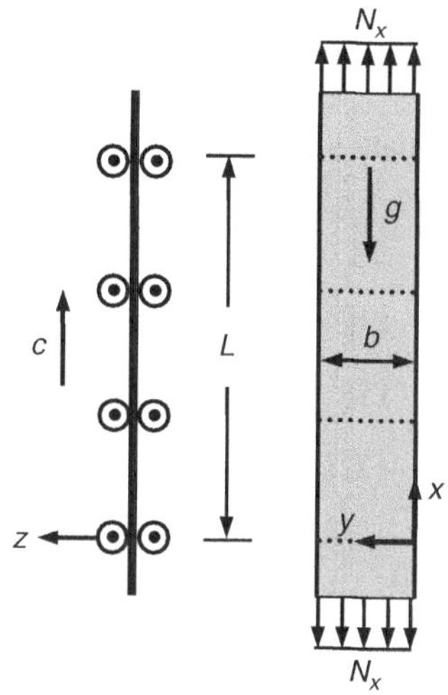

Figure 7.14 Axially moving thin plate passing through four simple supports

Table 7.6 Natural frequencies (Hz) of the plate passing through four simple supports

	$c=5$ m/s				$c=10$ m/s			
	$N_x=30$ kN/m		$N_x=60$ kN/m		$N_x=30$ kN/m		$N_x=60$ kN/m	
Mode (p, q)	Without gravity	With gravity	Without gravity	With gravity	Without gravity	With gravity	Without gravity	With gravity
(1,1)	3.520	3.555	5.021	5.045	3.211	3.259	4.835	4.863
(2,1)	3.571	3.605	5.070	5.094	3.364	3.400	4.923	4.948
(3,1)	3.668	3.703	5.166	5.191	3.459	3.495	5.018	5.042
(1,2)	3.981	4.013	5.355	5.377	3.726	3.765	5.183	5.209
(2,2)	4.032	4.062	5.404	5.426	3.848	3.879	5.265	5.288
(4,2)	8.031	8.093	10.759	10.805	7.544	7.627	10.425	10.478

Appendix 7.A: Finite Element Matrices for Axially Moving String

The finite element matrices in Equation 7.53 are given by

$$\boldsymbol{M}=\frac{\rho AL}{6}\begin{bmatrix}2 & 1\\ 1 & 2\end{bmatrix} \qquad \boldsymbol{G}=\rho Ac\begin{bmatrix}0 & 1\\ -1 & 0\end{bmatrix}$$

$$\boldsymbol{K}=\frac{N_x-\rho Ac^2}{L}\begin{bmatrix}1 & -1\\ -1 & 1\end{bmatrix}$$

$$\boldsymbol{C}_{MT}=\rho Ac\begin{bmatrix}-1 & 0\\ 0 & 1\end{bmatrix} \qquad \boldsymbol{K}_{MT}=\frac{\rho Ac^2}{L}\begin{bmatrix}1 & -1\\ -1 & 1\end{bmatrix} \tag{7.A.1}$$

Appendix 7.B: Finite Element Matrices for Axially Moving Bernoulli–Euler Beam

The finite element matrices in Equation 7.103 are given by

$$
\begin{gathered}
\boldsymbol{M} = \frac{\rho AL}{420}\begin{bmatrix} 156 & 22L & 54 & -13L \\ 22L & 4L^2 & 13L & -3L^2 \\ 54 & 13L & 156 & -22L \\ -13L & -3L^2 & -22L & 4L^2 \end{bmatrix} \\
\boldsymbol{G} = \frac{\rho Ac}{30}\begin{bmatrix} 0 & 6L & 30 & -6L \\ -6L & 0 & 6L & -L^2 \\ -30 & -6L & 0 & 6L \\ 6L & L^2 & -6L & 0 \end{bmatrix} \quad \boldsymbol{C}_{MT} = \rho Ac\begin{bmatrix} -1 & 0 & 0 & 0 \\ 0 & 0 & 0 & 0 \\ 0 & 0 & 1 & 0 \\ 0 & 0 & 0 & 0 \end{bmatrix} \\
\boldsymbol{K}_{EI} = \frac{EI}{L^3}\begin{bmatrix} 12 & 6L & -12 & 6L \\ 6L & 4L^2 & -6L & 2L^2 \\ -12 & -6L & 12 & -6L \\ 6L & 2L^2 & -6L & 4L^2 \end{bmatrix} \quad \boldsymbol{K}_{MT} = \rho Ac^2\begin{bmatrix} 0 & -1 & 0 & 0 \\ 0 & 0 & 0 & 0 \\ 0 & 0 & 0 & 1 \\ 0 & 0 & 0 & 0 \end{bmatrix} \\
\boldsymbol{K}_N = \frac{N_x - \rho Ac^2}{30L}\begin{bmatrix} 36 & 3L & -36 & 3L \\ 3L & 4L^2 & -3L & -L^2 \\ -36 & -3L & 36 & -3L \\ 3L & -L^2 & -3L & 4L^2 \end{bmatrix}
\end{gathered}
\tag{7.B.1}
$$

Appendix 7.C: Finite Element Matrices for Axially Moving Timoshenko Beam

The finite element matrices in Equation 7.166 are given by

$$
\begin{aligned}
\boldsymbol{M} &= \boldsymbol{M}_{\rho A} + \boldsymbol{M}_{\rho I} \\
\boldsymbol{C} &= \boldsymbol{G}_{\rho A} + \boldsymbol{G}_{\rho I} + \boldsymbol{C}_{MT\rho A} + \boldsymbol{C}_{\mathrm{MT}\rho I} \\
\boldsymbol{K} &= \boldsymbol{K}_1 + \boldsymbol{K}_2 + \boldsymbol{K}_3 + \boldsymbol{K}_{MT\rho A} + \boldsymbol{K}_{MT\rho I}
\end{aligned}
\tag{7.C.1}
$$

where

$$
\boldsymbol{M}_{\rho A}=\frac{\rho AL}{840}\begin{bmatrix} 4(70+7R+R^2) & L(35+7R+2R^2) & -4(-35+7R+R^2) & L(-35+7R+2R^2) \\ & L^2(7+R^2) & -L(-35+7R+2R^2) & L^2(-7+R^2) \\ & & 4(70+7R+R^2) & -L(35+7R+2R^2) \\ & \text{symmetric} & & L^2(7+R^2) \end{bmatrix}
$$

$$
\boldsymbol{M}_{\rho I}=\frac{\rho I}{30L}\begin{bmatrix} 36R^2 & 3LR(-5+6R) & -36R^2 & 3LR(-5+6R) \\ & L^2(10-15R+9R^2) & 3LR(5-6R) & L^2(5-15R+9R^2) \\ & & 36R^2 & 3LR(5-6R) \\ & \text{symmetric} & & L^2(10-15R+9R^2) \end{bmatrix}
$$

$$
\boldsymbol{G}_{\rho A}=\frac{\rho Ac}{30}\begin{bmatrix} 0 & L(5+R) & 30 & -L(5+R) \\ & 0 & L(5+R) & -L^2R \\ & & 0 & L(5+R) \\ \text{skew symmetric} & & & 0 \end{bmatrix}
$$

$$
\boldsymbol{G}_{\rho I}=\frac{\rho Ic}{L}\begin{bmatrix} 0 & 2R & 0 & -2R \\ & 0 & 2R & L(1-2R) \\ & & 0 & 2R \\ \text{skew symmetric} & & & 0 \end{bmatrix} \tag{7.C.2}
$$

$$
\boldsymbol{C}_{MT\rho A}=\rho Ac\begin{bmatrix} -1 & 0 & 0 & 0 \\ 0 & 0 & 0 & 0 \\ 0 & 0 & 1 & 0 \\ 0 & 0 & 0 & 0 \end{bmatrix} \qquad \boldsymbol{C}_{MT\rho I}=\rho Ic\begin{bmatrix} 0 & 0 & 0 & 0 \\ 0 & -1 & 0 & 0 \\ 0 & 0 & 0 & 0 \\ 0 & 0 & 0 & 1 \end{bmatrix}
$$

$$
\boldsymbol{K}_1=\frac{\kappa GA}{20L}(-1+R)^2\begin{bmatrix} 20 & 10L & -20 & 10L \\ & 5L^2 & -10L & 5L^2 \\ & & 20 & -10L \\ & \text{symmetric} & & 5L^2 \end{bmatrix}
$$

$$
\boldsymbol{K}_2=\frac{N_x-\rho Ac^2}{60L}\begin{bmatrix} 12(5+R^2) & 6LR^2 & -12(5+R^2) & 6LR^2 \\ & L^2(5+3R^2) & -6LR^2 & L^2(-5+3R^2) \\ & & 12(5+R^2) & -6LR^2 \\ & \text{symmetric} & & L^2(5+3R^2) \end{bmatrix}
$$

$$K_3 = \frac{EI - \rho I c^2}{L^3}\begin{bmatrix} 12R^2 & 6LR^2 & -12R^2 & 6LR^2 \\ & L^2(1+3R^2) & -6LR^2 & L^2(-1+3R^2) \\ & & 12R^2 & -6LR^2 \\ & \text{symmetric} & & L^2(1+3R^2) \end{bmatrix}$$

$$K_{MT\rho A} = \frac{\rho A c^2}{2L}\begin{bmatrix} 2(1-R) & -L(1+R) & -2(1-R) & L(1-R) \\ 0 & 0 & 0 & 0 \\ -2(1-R) & -L(1-R) & 2(1-R) & L(1+R) \\ 0 & 0 & 0 & 0 \end{bmatrix}$$

$$K_{MT\rho I} = \frac{\rho I c^2}{L^2}\begin{bmatrix} 0 & 0 & 0 & 0 \\ 6R & L(1+3R) & -6R & -L(1-3R) \\ 0 & 0 & 0 & 0 \\ 6R & -L(1-3R) & -6R & L(1+3R) \end{bmatrix}$$

Appendix 7.D: Finite Element Matrices for Axially Moving Plate

The finite element matrices in Equation 7.248 are given by

$$M = \frac{Lb\rho h}{25200}\begin{bmatrix} M_{11} & M_{12} & M_{13} & M_{14} \\ & M_{22} & M_{23} & M_{24} \\ & & M_{33} & M_{34} \\ & \text{symmetric} & & M_{44} \end{bmatrix} \tag{7.D.1}$$

where

$$M_{11} = \begin{bmatrix} 3454 & 461b & -461L \\ & 80b^2 & -63Lb \\ \text{symm} & & 80L^2 \end{bmatrix} \qquad M_{22} = \begin{bmatrix} 3454 & 461b & 461L \\ & 80b^2 & 63Lb \\ \text{symm} & & 80L^2 \end{bmatrix}$$

$$M_{33} = \begin{bmatrix} 3454 & -461b & 461L \\ & 80b^2 & -63Lb \\ \text{symm} & & 80L^2 \end{bmatrix} \qquad M_{44} = \begin{bmatrix} 3454 & -461b & -461L \\ & 80b^2 & 63Lb \\ \text{symm} & & 80L^2 \end{bmatrix}$$

$$M_{12} = \begin{bmatrix} 1266 & 199b & 274L \\ 199b & 40b^2 & 42Lb \\ -274L & -42Lb & -60L^2 \end{bmatrix} \qquad M_{13} = \begin{bmatrix} 394 & -166b & 116L \\ 166b & -30b^2 & 28Lb \\ -116L & 28Lb & -30L^2 \end{bmatrix}$$

$$\boldsymbol{M}_{14}=\begin{bmatrix}1226 & -274b & -199L\\ 274b & -60b^2 & -42Lb\\ -199L & 42Lb & 40L^2\end{bmatrix}\quad \boldsymbol{M}_{23}=\begin{bmatrix}1226 & -274b & 199L\\ 274b & -60b^2 & 42Lb\\ 199L & -42Lb & 40L^2\end{bmatrix}$$

$$\boldsymbol{M}_{24}=\begin{bmatrix}394 & -166b & -116L\\ 166b & -30b^2 & -28Lb\\ 116L & -28Lb & -30L^2\end{bmatrix}\quad \boldsymbol{M}_{34}=\begin{bmatrix}1226 & -199b & -274L\\ -199b & 40b^2 & 42Lb\\ 274L & -42Lb & -60L^2\end{bmatrix}$$

$$\boldsymbol{C}_G=\frac{bc\rho h}{2520}\begin{bmatrix}\boldsymbol{C}_{G11} & \boldsymbol{C}_{G12} & \boldsymbol{C}_{G13} & \boldsymbol{C}_{G14}\\ & \boldsymbol{C}_{G22} & \boldsymbol{C}_{G23} & \boldsymbol{C}_{G24}\\ & & \boldsymbol{C}_{G33} & \boldsymbol{C}_{G34}\\ \text{skew} & \text{symmetric} & & \boldsymbol{C}_{G44}\end{bmatrix} \tag{7.D.2}$$

where

$$\boldsymbol{C}_{G11}=\begin{bmatrix}0 & 0 & -175L\\ 0 & 0 & -21Lb\\ 175L & 21Lb & 0\end{bmatrix}\quad \boldsymbol{C}_{G22}=\boldsymbol{C}_{G33}=\boldsymbol{C}_{G44}=\boldsymbol{C}_{G11}$$

$$\boldsymbol{C}_{G12}=\begin{bmatrix}936 & 132b & 175L\\ 132b & 24b^2 & 21Lb\\ -175L & -21Lb & -28L^2\end{bmatrix}\quad \boldsymbol{C}_{G13}=\begin{bmatrix}324 & -78b & 77L\\ 78b & -18b^2 & 14Lb\\ -77L & 14Lb & -14L^2\end{bmatrix}$$

$$\boldsymbol{C}_{G14}=\begin{bmatrix}0 & 0 & -77L\\ 0 & 0 & -14Lb\\ 77L & -14Lb & 0\end{bmatrix}\quad \boldsymbol{C}_{G23}=\boldsymbol{C}_{G14}$$

$$\boldsymbol{C}_{G24}=\begin{bmatrix}-324 & 78b & 77L\\ -78b & 18b^2 & 14Lb\\ -77L & 14Lb & 14L^2\end{bmatrix}\quad \boldsymbol{C}_{G34}=\begin{bmatrix}-936 & 132b & 175L\\ 132b & -24b^2 & -21Lb\\ -175L & 21Lb & 28L^2\end{bmatrix}$$

$$\boldsymbol{C}_{MT}=\frac{\rho hbc}{420}\begin{bmatrix}\boldsymbol{C}_{MT11} & \mathbf{0} & \mathbf{0} & \boldsymbol{C}_{MT14}\\ & \boldsymbol{C}_{MT22} & \boldsymbol{C}_{MT23} & \mathbf{0}\\ & & \boldsymbol{C}_{MT33} & \mathbf{0}\\ & \text{symmetric} & & \boldsymbol{C}_{MT44}\end{bmatrix} \tag{7.D.3}$$

where

$$\boldsymbol{C}_{MT11}=\begin{bmatrix}-156 & -22b & 0\\ -22b & -4b^2 & 0\\ 0 & 0 & 0\end{bmatrix}\quad \boldsymbol{C}_{MT33}=\begin{bmatrix}156 & -22b & 0\\ -22b & 4b^2 & 0\\ 0 & 0 & 0\end{bmatrix}$$

$$C_{MT14} = \begin{bmatrix} -54 & 13b & 0 \\ -13b & 3b^2 & 0 \\ 0 & 0 & 0 \end{bmatrix} \quad \begin{aligned} C_{MT22} &= -C_{MT11} \\ C_{MT44} &= -C_{MT33} \\ C_{MT23} &= -C_{MT14} \end{aligned}$$

$$K_S = \frac{D}{15L^3b^3}\begin{bmatrix} K_{S11} & K_{S12} & K_{S13} & K_{S14} \\ & K_{S22} & K_{S23} & K_{S24} \\ & & K_{S33} & K_{S34} \\ & \text{symmetric} & & K_{S44} \end{bmatrix} \tag{7.D.4}$$

where

$$K_{S11} = \begin{bmatrix} 60L^4-6L^2b^2(2\nu-7)+60b^4 & 30L^4b+3L^2b^3(4\nu+1) & -30Lb^4-3L^3b^2(4\nu+1) \\ & 20L^4b^2-4L^2b^4(\nu-1) & -15L^3b^3\nu \\ \text{symmetric} & & 20L^2b^4-4L^4b^2(\nu-1) \end{bmatrix}$$

$$K_{S22} = \begin{bmatrix} 60L^4-6L^2b^2(2\nu-7)+60b^4 & 30L^2b+3L^2b^3(4\nu+1) & 30Lb^4+3L^3b^2(4\nu+1) \\ & 20L^4b^2-4L^2b^4(\nu-1) & 15L^3b^3\nu \\ \text{symmetric} & & 20L^2b^4-4L^4b^2(\nu-1) \end{bmatrix}$$

$$K_{S33} = \begin{bmatrix} 60L^4-6L^2b^2(2\nu-7)+60b^4 & -30L^4b-3L^2b^3(4\nu+1) & 30Lb^4+3L^3b^2(4\nu+1) \\ & 20L^4b^2-4L^2b^4(\nu-1) & -15L^3b^3\nu \\ \text{symmetric} & & 20L^2b^4-4L^4b^2(\nu-1) \end{bmatrix}$$

$$K_{S44} = \begin{bmatrix} 60L^4-6L^2b^2(2\nu-7)+60b^4 & -30L^4b-3L^2b^3(4\nu+1) & -30Lb^4-3L^3b^2(4\nu+1) \\ & 20L^4b^2-4L^2b^4(\nu-1) & 15L^3b^3\nu \\ \text{symmetric} & & 20L^2b^4-4L^4b^2(\nu-1) \end{bmatrix}$$

$$K_{S12} = \begin{bmatrix} 30L^4+6L^2b^2(2\nu-7)-60b^4 & 15L^4b-3L^2b^3(4\nu+1) & -30Lb^4+3L^3b^2(\nu-1) \\ & 10L^4b^2+4L^2b^4(\nu-1) & 0 \\ \text{symmetric} & & 10L^2b^4+L^4b^2(\nu-1) \end{bmatrix}$$

$$K_{S13} = \begin{bmatrix} -30L^4-6L^2b^2(2\nu-7)-30b^4 & 15L^4b+3L^2b^3(\nu-1) & -15Lb^4-3L^3b^2(\nu-1) \\ -15L^4b-3L^2b^3(\nu-1) & 5L^4b^2-4L^2b^4(\nu-1) & 0 \\ 15Lb^4+3L^3b^2(\nu-1) & 0 & 5L^2b^4-L^4b^2(\nu-1) \end{bmatrix}$$

$$\boldsymbol{K}_{S14}=\begin{bmatrix}-60L^4+6L^2b^2(2\nu-7)+30b^4 & 30L^4b-3L^2b^3(\nu-1) & -15Lb^4+3L^3b^2(4\nu+1)\\ -30L^4b+3L^2b^3(\nu-1) & 10L^4b^2+L^2b^4(\nu-1) & 0\\ -15Lb^4+3L^3b^2(4\nu+1) & 0 & 10L^2b^4+4L^4b^2(\nu-1)\end{bmatrix}$$

$$\boldsymbol{K}_{S23}=\begin{bmatrix}-60L^4+6L^2b^2(2\nu-7)+30b^4 & 30L^4b-3L^2b^3(\nu-1) & 15Lb^4-3L^3b^2(4\nu+1)\\ -30L^4b+3L^2b^3(\nu-1) & 10L^4b^2+L^2b^4(\nu-1) & 0\\ 15Lb^4-3L^3b^2(4\nu+1) & 0 & 10L^2b^4+4L^4b^2(\nu-1)\end{bmatrix}$$

$$\boldsymbol{K}_{S24}=\begin{bmatrix}-30L^4-6L^2b^2(2\nu-7)-30b^4 & 15L^4b+3L^2b^3(\nu-1) & 15Lb^4+3L^3b^2(\nu-1)\\ -15L^4b-3L^2b^3(\nu-1) & 5L^4b^2-L^2b^4(\nu-1) & 0\\ -15Lb^4-3L^3b^2(\nu-1) & 0 & 5L^2b^4-L^4b^2(\nu-1)\end{bmatrix}$$

$$\boldsymbol{K}_{S34}=\begin{bmatrix}30L^4+6L^2b^2(2\nu-7)-60b^4 & -15L^4b+3L^2b^3(4\nu+1) & 30Lb^4-3L^3b^2(\nu-1)\\ & 10L^4b^2+4L^2b^4(\nu-1) & 0\\ \text{symmetric} & & 10L^2b^4+L^4b^2(\nu-1)\end{bmatrix}$$

$$\boldsymbol{K}_N=\frac{(N_x-\rho hc^2)b}{1260L}\begin{bmatrix}\boldsymbol{K}_{N11} & \boldsymbol{K}_{N12} & \boldsymbol{K}_{N13} & \boldsymbol{K}_{N14}\\ & \boldsymbol{K}_{N22} & \boldsymbol{K}_{N23} & \boldsymbol{K}_{N24}\\ & & \boldsymbol{K}_{N33} & \boldsymbol{K}_{N34}\\ & \text{symmetric} & & \boldsymbol{K}_{N44}\end{bmatrix} \tag{7.D.5}$$

where

$$\boldsymbol{K}_{N11}=\begin{bmatrix}552 & 66b & -42L\\ & 12b^2 & 0\\ \text{symm} & & 56L^2\end{bmatrix}\qquad \boldsymbol{K}_{N22}=\begin{bmatrix}552 & 66b & 42L\\ & 12b^2 & 0\\ \text{symm} & & 56L^2\end{bmatrix}$$

$$\boldsymbol{K}_{N33}=\begin{bmatrix}552 & -66b & 42L\\ & 12b^2 & 0\\ \text{symm} & & 56L^2\end{bmatrix}\qquad \boldsymbol{K}_{N44}=\begin{bmatrix}552 & -66b & -42L\\ & 12b^2 & 0\\ \text{symm} & & 56L^2\end{bmatrix}$$

$$\boldsymbol{K}_{N12}=\begin{bmatrix}-552 & -66b & -42L\\ -66b & -12b^2 & 0\\ 42L & 0 & -14L^2\end{bmatrix}\qquad \boldsymbol{K}_{N13}=\begin{bmatrix}-204 & 39b & -21L\\ -39b & 9b^2 & 0\\ 21L & 0 & -7L^2\end{bmatrix}$$

$$\boldsymbol{K}_{N14}=\begin{bmatrix}204 & -39b & -21L\\ 39b & -9b^2 & 0\\ -21L & 0 & 28L^2\end{bmatrix}\qquad \boldsymbol{K}_{N23}=\begin{bmatrix}204 & -39b & 21L\\ 39b & -9b^2 & 0\\ 21L & 0 & 28L^2\end{bmatrix}$$

$$\boldsymbol{K}_{N24}=\begin{bmatrix}-204 & 39b & 21L\\ -39b & 9b^2 & 0\\ -21L & 0 & -7L^2\end{bmatrix}\qquad \boldsymbol{K}_{N34}=\begin{bmatrix}-552 & 66b & 42L\\ 66b & -12b^2 & 0\\ -42L & 0 & -14L^2\end{bmatrix}$$

$$\boldsymbol{K}_g=\frac{\rho hbg}{5040}\begin{bmatrix}\boldsymbol{K}_{g11} & \boldsymbol{K}_{g12} & \boldsymbol{K}_{g13} & \boldsymbol{K}_{g14}\\ & \boldsymbol{K}_{g22} & \boldsymbol{K}_{g23} & \boldsymbol{K}_{g24}\\ & & \boldsymbol{K}_{g33} & \boldsymbol{K}_{g34}\\ & \text{symmetric} & & \boldsymbol{K}_{g44}\end{bmatrix} \tag{7.D.6}$$

where

$$\boldsymbol{K}_{g11} = \begin{bmatrix} 1104 & 132b & -175L \\ & 24b^2 & -21Lb \\ \text{symm} & & 56L^2 \end{bmatrix} \qquad \boldsymbol{K}_{g22} = \begin{bmatrix} 1104 & 132b & -7L \\ & 24b^2 & -21Lb \\ \text{symm} & & 168L^2 \end{bmatrix}$$

$$\boldsymbol{K}_{g33} = \begin{bmatrix} 1104 & -132b & -7L \\ & 24b^2 & 21Lb \\ \text{symm} & & 168L^2 \end{bmatrix} \qquad \boldsymbol{K}_{g44} = \begin{bmatrix} 1104 & -132b & -175L \\ & 24b^2 & 21Lb \\ \text{symm} & & 56L^2 \end{bmatrix}$$

$$\boldsymbol{K}_{g12} = \begin{bmatrix} -1104 & -132b & 7L \\ -132b & -24b^2 & 21Lb \\ 175L & 21Lb & -28L^2 \end{bmatrix} \qquad \boldsymbol{K}_{g13} = \begin{bmatrix} -408 & 78b & -7L \\ -78b & -18b^2 & 14Lb \\ 77L & -14Lb & -14L^2 \end{bmatrix}$$

$$\boldsymbol{K}_{g14} = \begin{bmatrix} 408 & -78b & -77L \\ 78b & -18b^2 & -14Lb \\ -77L & 14Lb & 28L^2 \end{bmatrix} \qquad \boldsymbol{K}_{g23} = \begin{bmatrix} 408 & -78b & 7L \\ 78b & -18b^2 & -14Lb \\ 7L & 14Lb & 84L^2 \end{bmatrix}$$

$$\boldsymbol{K}_{g24} = \begin{bmatrix} -408 & 78b & 77L \\ -78b & 18b^2 & 14Lb \\ -7L & -14Lb & -14L^2 \end{bmatrix} \qquad \boldsymbol{K}_{g34} = \begin{bmatrix} -1104 & 132b & 175L \\ 132b & -24b^2 & -21Lb \\ 7L & -21Lb & -28L^2 \end{bmatrix}$$

$$\boldsymbol{K}_{MT} = \frac{\rho hbc^2}{420L} \begin{bmatrix} \boldsymbol{K}_{MT11} & \boldsymbol{K}_{MT12} & \boldsymbol{K}_{MT13} & \boldsymbol{K}_{MT14} \\ \boldsymbol{K}_{MT21} & \boldsymbol{K}_{MT22} & \boldsymbol{K}_{MT23} & \boldsymbol{K}_{MT24} \\ \boldsymbol{K}_{MT31} & \boldsymbol{K}_{MT32} & \boldsymbol{K}_{MT33} & \boldsymbol{K}_{MT34} \\ \boldsymbol{K}_{MT41} & \boldsymbol{K}_{MT42} & \boldsymbol{K}_{MT43} & \boldsymbol{K}_{MT44} \end{bmatrix} \tag{7.D.7}$$

where

$$\boldsymbol{K}_{MT11} = \begin{bmatrix} 9 & 22b & 147 \\ b & 4b^2 & 21b \\ 0 & 0 & 0 \end{bmatrix} \qquad \boldsymbol{K}_{MT22} = \begin{bmatrix} 9 & 22b & -147 \\ b & 4b^2 & -21b \\ 0 & 0 & 0 \end{bmatrix}$$

$$\boldsymbol{K}_{MT33} = \begin{bmatrix} 9 & -22b & -147 \\ -b & 4b^2 & 21b \\ 0 & 0 & 0 \end{bmatrix} \qquad \boldsymbol{K}_{MT44} = \begin{bmatrix} 9 & -22b & 147 \\ -b & 4b^2 & -21b \\ 0 & 0 & 0 \end{bmatrix}$$

$$\boldsymbol{K}_{MT12} = \begin{bmatrix} -9 & -22b & 0 \\ -b & -4b^2 & 0 \\ 0 & 0 & 0 \end{bmatrix} \qquad \boldsymbol{K}_{MT13} = \begin{bmatrix} 9 & 13b & 0 \\ b & 3b^2 & 0 \\ 0 & 0 & 0 \end{bmatrix}$$

$$\boldsymbol{K}_{MT14} = \begin{bmatrix} -9 & -13b & 63 \\ -b & -3b^2 & 14b \\ 0 & 0 & 0 \end{bmatrix} \qquad \boldsymbol{K}_{MT23} = \begin{bmatrix} -9 & -13b & -63 \\ -b & -3b^2 & -14b \\ 0 & 0 & 0 \end{bmatrix}$$

$$\boldsymbol{K}_{MT31}=\begin{bmatrix} 9 & -13b & 0 \\ -b & 3b^2 & 0 \\ 0 & 0 & 0 \end{bmatrix} \qquad \boldsymbol{K}_{MT32}=\begin{bmatrix} -9 & 13b & -63 \\ b & -3b^2 & 14b \\ 0 & 0 & 0 \end{bmatrix}$$

$$\boldsymbol{K}_{MT34}=\begin{bmatrix} -9 & 22b & 0 \\ b & -4b^2 & 0 \\ 0 & 0 & 0 \end{bmatrix} \qquad \boldsymbol{K}_{MT41}=\begin{bmatrix} -9 & 13b & 63 \\ b & -3b^2 & -14b \\ 0 & 0 & 0 \end{bmatrix}$$

$$\boldsymbol{K}_{MT21}=\boldsymbol{K}_{MT12} \qquad \boldsymbol{K}_{MT24}=\boldsymbol{K}_{MT13}$$

$$\boldsymbol{K}_{MT42}=\boldsymbol{K}_{MT31} \qquad \boldsymbol{K}_{MT43}=\boldsymbol{K}_{MT34}$$

8

Dynamics of Rotor Systems

Rotating machines have been used extensively in diverse engineering applications such as motors, engines, turbines, and machine tools. In general, the rotating machines consist of multiple shafts and disks (or blades) that are connected to each other to form the rotor systems supported by multiple bearings. It is very important to predict the dynamic characteristics of the rotor systems accurately early during the design phase. Thus there have been extensive studies on the modeling and analysis of such rotor systems over many decades and detailed reviews on the subject can be found in Refs. [126, 143]. This chapter introduces a spectral element model for a multi-bearing flexible rotor system with rigid disks. In the first part of the chapter, the equations of motion for the rotor systems are derived using Hamilton's principle and Lagrange's equations. The spinning circular shafts of a rotor are represented by the uniform Timoshenko shaft models and the rigid disks are considered to be wobbling about their centers. The support-bearings are regarded as the flexible bearings and they are represented by effective translational and rotational springs. The mass unbalances of disks are taken into account in the spectral element model. The finite element model is also provided in Section 8.3 for comparison with the spectral element model introduced in this chapter.

8.1 Governing Equations

A multi-bearing rotor consists of multiple flexible spinning shafts and interim disks with mass unbalances, supported by multiple supports. In this study, the spinning shafts are represented by the spinning Timoshenko shafts subjected to axial (longitudinal), torsional, and transverse vibrations and the interim disks are represented by the rigid, thin disks subjected to wobbling motions about their centers. In the following, the dynamic equations of motion for the spinning Timoshenko shafts and the wobbling rigid disks will be derived by using Hamilton's principle and Lagrange's equations, respectively.

8.1.1 Equations of Motion of the Spinning Shaft

The equations of motion and relevant boundary conditions for a uniform spinning shaft represented by the Timoshenko shaft model can be derived by using Hamilton's principle

$$\int_{t_1}^{t_2} (\delta T - \delta V + \delta W)\, \mathrm{d}t = 0 \tag{8.1}$$

where T is the kinetic energy, V is the potential energy, and δW is the virtual work perfomed by external forces and moments. We assume that the spinning Timoshenko shaft has a uniform circular cross-section and it is spinning about its central axis at a constant speed of Ω (radian/s) to take a small amplitude vibration. As shown in Figure 8.1, we define $v(x, t)$ as being the transverse displacement in the y-direction, $w(x, t)$ as the transverse displacement in the z-direction, $u(x, t)$ as the longitudinal (axial) displacement, $\theta(x, t)$ as the twist angle about the central axis x, $\phi(x, t)$ as the rotation angle about the y-axis, and $\psi(x, t)$ as the rotation angle about z-axis.

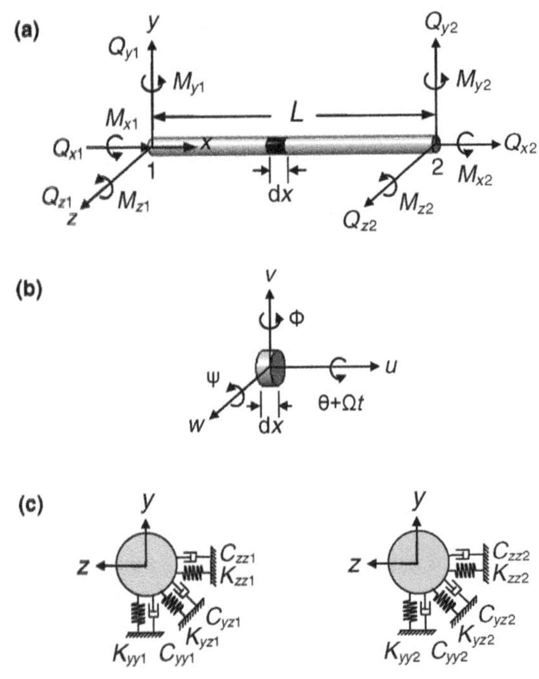

Figure 8.1 Timoshenko-beam model for a uniform spinning shaft: (a) the boundary forces and moments, (b) the six degrees of freedom for the shaft, (c) the support-bearing models

The kinetic and potential energies are then given by

$$\begin{aligned} T &= \frac{1}{2}\int_0^L \rho A(\dot{u}^2 + \dot{v}^2 + \dot{w}^2)\, \mathrm{d}x + \frac{1}{2}\int_0^L \rho I(\dot{\phi}^2 + \dot{\psi}^2)\, \mathrm{d}x \\ &+ \frac{1}{2}\int_0^L \rho J(\Omega + \dot{\theta} - \phi\dot{\psi})^2\, \mathrm{d}x \end{aligned} \tag{8.2}$$

$$V = \frac{1}{2}\int_0^L EAu'^2\,\mathrm{d}x + \frac{1}{2}\int_0^L GJ\theta'^2\,\mathrm{d}x + \frac{1}{2}\int_0^L EI(\phi'^2 + \psi'^2)\,\mathrm{d}x$$
$$+ \frac{1}{2}\int_0^L \kappa GA\,[(v' - \psi)^2 + (w' + \phi)^2]\,\mathrm{d}x + \sum_{i=1}^{2}\frac{1}{2}\{v_i w_i\}\boldsymbol{K}_{\text{support }(i)}\{v_i w_i\}^{\mathrm{T}} \tag{8.3}$$

where

$$\boldsymbol{K}_{\text{support }(i)} = \begin{bmatrix} K_{yyi} & K_{yzi} \\ K_{zyi} & K_{zzi} \end{bmatrix} \quad (i = 1,\, 2) \tag{8.4}$$

In above equations, the dot $(\cdot)$ and the prime $(')$ denote the derivatives with respect to the time t and axis x, respectively; EA is the axial rigidity, EI the bending (flexural) rigidity, GJ the torsional rigidity, κGA the transverse shear rigidity, ρA the mass per length, ρI the mass moment of inertia about the y- or z-axis, ρJ the polar mass moment of inertia, and L is the length of shaft; $\boldsymbol{K}_{\text{support}(1)}$ and $\boldsymbol{K}_{\text{support}(2)}$ represent the stiffnesses of the support-bearings at the left end "1" and at the right end "2", respectively; v_i and w_i $(i = 1, 2)$ are the transverse displacements specified at the left and right ends "1" and "2". In Equation 8.2, the first integral represents the translational kinetic energy and the remaining two integrals present the rotational kinetic energy. In Equation 8.3, the first integral represents the strain energy for axial vibration, the second integral for torsional vibration, the third integral for transverse bending vibrations, the fourth integral for transverse shear deformations, and the last term represents the strain energy stored in the bearing supports at both ends.

The virtual work δW is given by

$$\delta W = \int_0^L (p_x\delta u + p_y\delta v + p_z\delta w + \tau_x\delta\theta + \tau_y\delta\phi + \tau_z\delta\psi)\,\mathrm{d}x$$
$$+ Q_{x1}\delta u_1 + Q_{x2}\delta u_2 + Q_{y1}\delta v_1 + Q_{y2}\delta v_2 + Q_{z1}\delta w_1 + Q_{z2}\delta w_2$$
$$+ M_{x1}\delta\theta_1 + M_{x2}\delta\theta_2 + M_{y1}\delta\phi_1 + M_{y2}\delta\phi_2 + M_{z1}\delta\psi_1 + M_{z2}\delta\psi_2$$
$$- f_{\dot v_1}\delta v_1 - f_{\dot w_1}\delta w_1 - f_{\dot v_2}\delta v_2 - f_{\dot w_2}\delta w_2 \tag{8.5}$$

where Q_{xi}, Q_{yi}, Q_{zi}, M_{xi}, M_{yi}, and M_{zi} $(i = 1,\ 2)$ are the transverse shear forces, bending moments, and torques applied at two ends of the beam element, as shown in Figure 8.1, and p_x, p_y, p_z, τ_x, τ_y, and τ_z are the forces, bending moments, and torques distributed along axis x. $f_{\dot v_i}$ and $f_{\dot w_i}$ $(i = 1,\ 2)$ are the viscous damping forces due to the bearing supports and they can be computed from Rayleigh's dissipation function R as follows:

$$f_{\dot v_i} = \frac{\partial R}{\partial \dot v_i}, \qquad f_{\dot w_i} = \frac{\partial R}{\partial \dot w_i} \quad (i = 1,\ 2) \tag{8.6}$$

where

$$R = \sum_{i=1}^{2}\frac{1}{2}\{\dot v_i \dot w_i\}\boldsymbol{C}_{\text{support}(i)}\{\dot v_i \dot w_i\}^{\mathrm{T}} \tag{8.7}$$

and where

$$\boldsymbol{C}_{\text{support}(i)} = \begin{bmatrix} C_{yyi} & C_{yzi} \\ C_{zyi} & C_{zzi} \end{bmatrix} \quad (i = 1,\ 2) \tag{8.8}$$

Substituting Equations 8.2, 8.3 and 8.5 into Equation 8.1 and integrating by parts, we get the following equations of motion

$$
\begin{aligned}
&\rho A\ddot{u} - EAu'' = p_x \\
&\rho J(\ddot{\theta} - \dot{\phi}\dot{\psi} - \phi\ddot{\psi}) - GJ\theta'' = \tau_x \\
&\rho A\ddot{v} - \kappa GA(v'' - \psi') = p_y \\
&\rho I\ddot{\psi} - \rho J(\Omega + \dot{\theta} - \phi\dot{\psi})\,\dot{\phi} - \rho J(\ddot{\theta} - \dot{\phi}\dot{\psi} - \phi\ddot{\psi})\,\phi - EI\psi'' - \kappa GA(v' - \psi) = \tau_z \\
&\rho A\ddot{w} - \kappa GA(w'' + \phi') = p_z \\
&\rho I\ddot{\phi} + \rho J(\Omega + \dot{\theta} - \phi\dot{\psi})\,\dot{\psi} - EI\phi'' + \kappa GA(w' + \phi) = \tau_y
\end{aligned}
\tag{8.9}
$$

and the boundary conditions

$$
\begin{aligned}
Q_x(0,t) &= -Q_{x1} \quad \text{or} \quad u(0,t) = u_1(t) \\
Q_x(L,t) &= Q_{x2} \quad \text{or} \quad u(L,t) = u_2(t) \\
Q_y(0,t) &= K_{yy1}v_1 + (1/2)(K_{yz1} + K_{zy1})w_1 + C_{yy1}\dot{v}_1 \quad \text{or} \quad v(0,t) = v_1(t) \\
&\quad + (1/2)(C_{yz1} + C_{zy1})\dot{w}_1 - Q_{y1} \\
Q_y(L,t) &= -K_{yy2}v_2 - (1/2)(K_{yz2} + K_{zy2})w_2 - C_{yy2}\dot{v}_2 \quad \text{or} \quad v(L,t) = v_2(t) \\
&\quad - (1/2)(C_{yz2} + C_{zy2})\dot{w}_2 + Q_{y2} \\
Q_z(0,t) &= K_{zz1}w_1 + (1/2)(K_{yz1} + K_{zy1})v_1 + C_{zz1}\dot{w}_1 \quad \text{or} \quad w(0,t) = w_1(t) \\
&\quad + (1/2)(C_{yz1} + C_{zy1})\dot{v}_1 - Q_{z1} \\
Q_z(L,t) &= -K_{zz2}w_2 - (1/2)(K_{yz2} + K_{zy2})v_2 - C_{zz2}\dot{w}_2 \quad \text{or} \quad w(L,t) = w_2(t) \\
&\quad - (1/2)(C_{yz2} + C_{zy2})\dot{v}_2 + Q_{z2} \\
M_x(0,t) &= -M_{x1} \quad \text{or} \quad \theta(0,t) = \theta_1(t) \\
M_x(L,t) &= M_{x2} \quad \text{or} \quad \theta(L,t) = \theta_2(t) \\
M_y(0,t) &= -M_{y1} \quad \text{or} \quad \phi(0,t) = \phi_1(t) \\
M_y(L,t) &= M_{y2} \quad \text{or} \quad \phi(L,t) = \phi_2(t) \\
M_z(0,t) &= -M_{z1} \quad \text{or} \quad \psi(0,t) = \psi_1(t) \\
M_z(L,t) &= M_{z2} \quad \text{or} \quad \psi(L,t) = \psi_2(t)
\end{aligned}
\tag{8.10}
$$

where the force-displacement relationships are defined by

$$
\begin{aligned}
Q_x(x,t) &= EAu' \\
Q_y(x,t) &= \kappa GA(v' - \psi) \\
Q_z(x,t) &= \kappa GA(w' + \phi) \\
M_x(x,t) &= EI\theta' \\
M_y(x,t) &= EI\phi' \\
M_z(x,t) &= EI\psi'
\end{aligned}
\tag{8.11}
$$

Neglecting small coupling terms, we can further simplify Equation 8.9 as follows:

$$\rho A\ddot{u} - EAu'' = p_x \tag{8.12}$$

$$\rho J\ddot{\theta} - GJ\theta'' = \tau_x \tag{8.13}$$

$$\rho A\ddot{v} - \kappa GA(v'' - \psi') = p_y \tag{8.14}$$

$$\rho I\ddot{\psi} - \rho J\Omega\dot{\phi} - EI\psi'' - \kappa GA(v' - \psi) = \tau_z \tag{8.15}$$

$$\rho A\ddot{w} - \kappa GA(w'' + \phi') = p_z \tag{8.16}$$

$$\rho I\ddot{\phi} + \rho J\Omega\dot{\psi} - EI\phi'' + \kappa GA(w' + \phi) = \tau_y \tag{8.17}$$

The above simplified equations of motion are identical with those given in Refs. [150, 176].

8.1.2 Equations of Motion of Disks with Mass Unbalance

8.1.2.1 Thin Disk

Figure 8.2 shows a thin rigid disk with mass unbalance. The mass of the disk itself is m_D and the mass of the unbalance mass is m_e. Thus the total mass of the disk system is $M = m_D + m_e$. Assume that unbalance mass m_e is located at d_e measured from the geometric center "O" where the spinning shafts are connected. Accordingly, the mass center of the disk system is located at $d = \mu_e d_e$ measured from the geometric center "O", where $\mu_e = m_e/M$. In Figure 8.2, the disk is denoted by "D" and its mass center by "C". The global reference coordinates (x, y, z) and the moving coordinates $(1, 2, 3)$ fixed on the disk are denoted by "R" and "S", respectively. The unit direction vectors $\boldsymbol{n}_x$, $\boldsymbol{n}_y$, and $\boldsymbol{n}_z$ are defined in the x-, y-, and z-directions, while $\boldsymbol{n}_1$, $\boldsymbol{n}_2$, and $\boldsymbol{n}_3$ are in the 1-, 2-, and 3-directions, respectively. The other unit direction vectors are also defined in Figure 8.2; u, v, and w are the displacements of the geometric center "O" in the x-, y-, and z-directions, respectively, and θ, ϕ, and ψ are the rotation angles about the x-, y-, and z-axis, respectively.

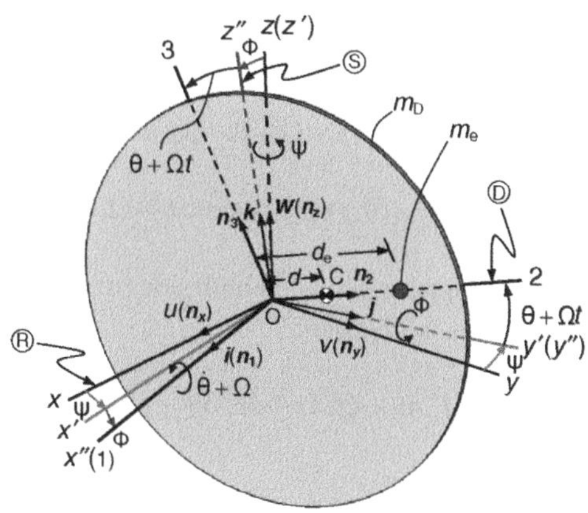

Figure 8.2 Dynamic model of a rigid disk with unbalanced mass

The position of the mass center "C" with respect to the global coordinates "R" can be then expressed by

$$\begin{aligned} {}^{R}\boldsymbol{r}^{C} &= [u + d\phi \sin(\theta + \Omega t) - d\psi \cos(\theta + \Omega t)]\boldsymbol{n}_x \\ &\quad + [v + d\cos(\theta + \Omega t)]\boldsymbol{n}_y + [w + d\sin(\theta + \Omega t)]\boldsymbol{n}_z \end{aligned} \tag{8.18}$$

From Equation 8.18, we can obtain the velocity of the mass center as

$$\begin{aligned} {}^{R}\boldsymbol{v}^{C} &= [\dot{u} + d\dot{\phi}\sin(\theta + \Omega t) - d\dot{\psi}\cos(\theta + \Omega t) \\ &\quad + d(\dot{\theta} + \Omega)\{\psi\sin(\theta + \Omega t) + \phi\cos(\theta + \Omega t)\}]\, \boldsymbol{n}_x \\ &\quad + [\dot{v} - d(\dot{\theta} + \Omega)\sin(\theta + \Omega t)]\, \boldsymbol{n}_y + [\dot{w} + d(\dot{\theta} + \Omega)\cos(\theta + \Omega t)]\, \boldsymbol{n}_z \end{aligned} \tag{8.19}$$

The angular velocity of the disk with respect to the global coordinates "R" can be obtained from Figure 8.2 as

$$\begin{aligned} {}^{R}\omega^{D} &= (\Omega + \dot{\theta})\boldsymbol{i} + \dot{\phi}\boldsymbol{j} + \dot{\psi}\, \boldsymbol{n}_z \\ &= (\Omega + \dot{\theta} - \phi\dot{\psi})\, \boldsymbol{n}_1 + [\dot{\phi}\cos(\theta + \Omega t) + \dot{\psi}\sin(\theta + \Omega t)]\, \boldsymbol{n}_2 \\ &\quad - [\dot{\phi}\sin(\theta + \Omega t) - \dot{\psi}\cos(\theta + \Omega t)]\, \boldsymbol{n}_3 \end{aligned} \tag{8.20}$$

The kinetic energy of the disk system including the unbalanced mass is given by

$$T = \frac{1}{2} M\, {}^{R}\boldsymbol{v}^{C} \cdot {}^{R}\boldsymbol{v}^{C} + \frac{1}{2}\, {}^{R}\omega^{D} \cdot \boldsymbol{I}_C^D \cdot {}^{R}\omega^{D} \tag{8.21}$$

where $\boldsymbol{I}_C^D = diag[I_{11} I_{22} I_{33}]$ is the inertia dyadic of the disk evaluated with respect to its mass center "C". Substituting Equations 8.19 and 8.20 into Equation 8.21 yields

$$\begin{aligned} T &= \frac{1}{2} M [\{\dot{u} + d\dot{\phi}\sin(\theta+\Omega t) - d\dot{\psi}\cos(\theta+\Omega t) + d(\dot{\theta}+\Omega)[\psi\sin(\theta+\Omega t) \\ &\quad + \phi\cos(\theta+\Omega t)]\}^2 + \dot{v}^2 + \dot{w}^2 + d^2(\dot{\theta}+\Omega)^2 - 2d\dot{v}(\dot{\theta}+\Omega)\sin(\theta+\Omega t) \\ &\quad + 2d\dot{w}(\dot{\theta}+\Omega)\cos(\theta+\Omega t)] + \frac{1}{2}\Big[I_{11}(\Omega+\dot{\theta}-\phi\dot{\psi})^2 + I_{22}\{\dot{\phi}\cos(\theta+\Omega t) \\ &\quad + \dot{\psi}\sin(\theta+\Omega t)\}^2 + I_{33}\{\dot{\phi}\sin(\theta+\Omega t) - \dot{\psi}\cos(\theta+\Omega t)\}^2\Big] \end{aligned} \tag{8.22}$$

Figure 8.3 shows the forces and moments acting on both sides of the disk. The virtual work δW done by these forces and moments is given by

$$\delta W = Q_x^D \delta u + M_x^D \delta\theta + Q_y^D \delta v + M_z^D \delta\psi + Q_z^D \delta w + M_y^D \delta\phi \tag{8.23}$$

where

$$Q_j^D = Q_{jl}^D + Q_{jr}^D, \qquad M_j^D = M_{jl}^D + M_{jr}^D \qquad (j = x, y, z) \tag{8.24}$$

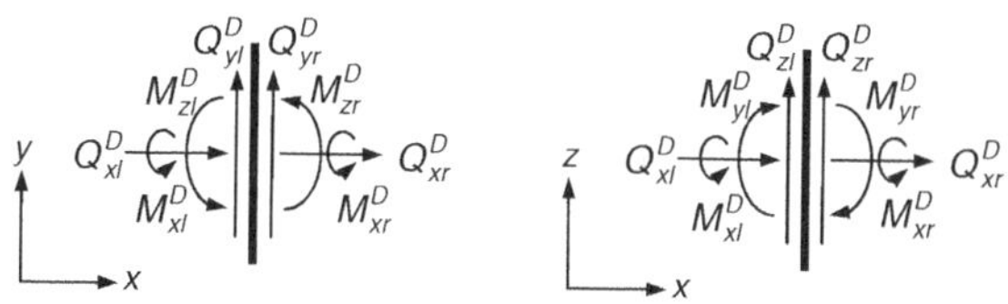

Figure 8.3 Forces and moments acting on the both sides of a rigid disk

By substituting Equations 8.22 and 8.23 into Lagrange's equations and neglecting small nonlinear terms, we can obtain the equations of motion of the disk as

$$
\begin{aligned}
&M\ddot{u} = Q_x^D \\
&J_D\ddot{\theta} = M_x^D + M_x^{m_e} \\
&M\ddot{v} = Q_y^D + f_y^{m_e} \\
&I_D\ddot{\psi} - J_D\Omega\dot{\phi} = M_z^D + M_z^{m_e} \\
&M\ddot{w} = Q_z^D + f_z^{m_e} \\
&I_D\ddot{\phi} + J_D\Omega\dot{\psi} = M_y^D + M_y^{m_e}
\end{aligned}
\tag{8.25}
$$

where $f_y^{m_e}$ and $f_z^{m_e}$ are the centrifugal unbalance forces and $M_x^{m_e}$, $M_y^{m_e}$ and $M_z^{m_e}$ are the unbalance moments, all induced by the unbalance mass. They are given by

$$
\begin{aligned}
f_y^{m_e} &= m_e d_e \Omega^2 \cos\Omega t \\
f_z^{m_e} &= m_e d_e \Omega^2 \sin\Omega t \\
M_x^{m_e} &= -m_e d_e^2 \ddot{\theta} \\
M_y^{m_e} &= -m_e d_e^2[(\ddot{\phi} - \mu_e\Omega^2\phi + 2\Omega\dot{\psi})\sin^2\Omega t - (\ddot{\psi} - \mu_e\Omega^2\psi - 2\Omega\dot{\phi})\sin\Omega t\cos\Omega t] \\
M_z^{m_e} &= -m_e d_e^2[(\ddot{\psi} - \mu_e\Omega^2\psi - 2\Omega\dot{\phi})\cos^2\Omega t - (\ddot{\phi} - \mu_e\Omega^2\phi + 2\Omega\dot{\psi})\sin\Omega t\cos\Omega t]
\end{aligned}
\tag{8.26}
$$

In most books and references associated with the theory of rotordynamics, the unbalance moments have been ignored or have not been considered. Thus, ignoring all unbalance moments, we can further simplify Equation 8.25 as follows:

$$
\begin{aligned}
&M\ddot{u} = Q_x^D \\
&J_D\ddot{\theta} = M_x^D \\
&M\ddot{v} = Q_y^D + f_y^{m_e} \\
&I_D\ddot{\psi} - J_D\Omega\dot{\phi} = M_z^D \\
&M\ddot{w} = Q_z^D + f_z^{m_e} \\
&I_D\ddot{\phi} + J_D\Omega\dot{\psi} = M_y^D
\end{aligned}
\tag{8.27}
$$

The last four equations of Equation 8.27 are identical with those derived in Ref. [42] by using the Newton's law of motion.

8.1.2.2 Thick Disk

Consider a thick rigid disk, which has the thickness $2h$ as shown in Figure 8.4. The superscripts r and l denote the translational and rotational motions at the right- and left-hand sides of the disk, respectively. The position vector ${}^R\boldsymbol{r}^C$ of the mass center "C" and the angular velocity ${}^R\omega^D$ of the thick disk are the same as in Equations 8.18 and 8.20, respectively. Thus the velocity ${}^R\boldsymbol{v}^C$ of the mass center "C" is also the same as in Equation 8.19.

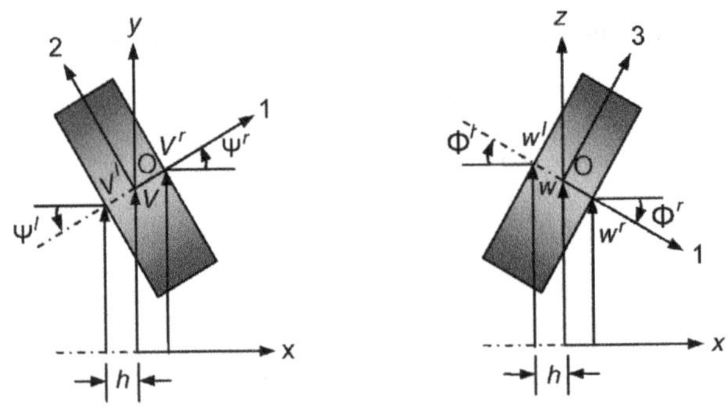

Figure 8.4 Thick rigid disk

For the rigid thick disk, the following relationships are satisfied:

$$
\begin{aligned}
u &= u^l = u^r, \qquad & v &= v^l + h\psi^l, \qquad & w &= w^l - h\phi^l \\
\theta &= \theta^l = \theta^r, \qquad & \phi &= \phi^l = \phi^r, \qquad & \psi &= \psi^l = \psi^r
\end{aligned}
\tag{8.28}
$$

and

$$
v^r = v^l + 2h\psi^l, \qquad w^r = w^l - 2h\phi^l \tag{8.29}
$$

Substituting Equation 8.28 into Equations 8.19 and 8.20 gives

$$
\begin{aligned}
{}^R\boldsymbol{v}^C &= [\{\dot{u}^l + d\dot{\phi}^l\sin(\theta^l + \Omega t) - d\dot{\psi}^l\cos(\theta^l + \Omega t) \\
&\quad + d(\dot{\theta}^l + \Omega)\{\psi^l\sin(\theta^l + \Omega t) + \phi^l\cos(\theta^l + \Omega t)\}]\,\boldsymbol{n}_x \\
&\quad + [(\dot{v}^l + h\dot{\psi}^l) - d(\dot{\theta}^l + \Omega)\sin(\theta^l + \Omega t)]\,\boldsymbol{n}_y \\
&\quad + [(\dot{w}^l - h\dot{\phi}^l) + d(\dot{\theta}^l + \Omega)\cos(\theta^l + \Omega t)]\,\boldsymbol{n}_z
\end{aligned}
\tag{8.30}
$$

and

$$
\begin{aligned}
{}^R\boldsymbol{\omega}^D &= (\Omega + \dot{\theta}^l - \phi^l\dot{\psi}^l)\,\boldsymbol{n}_1 + [\dot{\phi}^l\cos(\theta^l + \Omega t) + \dot{\psi}^l\sin(\theta^l + \Omega t)]\,\boldsymbol{n}_2 \\
&\quad - [\dot{\phi}^l\sin(\theta^l + \Omega t) - \dot{\psi}^l\cos(\theta^l + \Omega t)]\,\boldsymbol{n}_3
\end{aligned}
\tag{8.31}
$$

By using Equations 8.30 and 8.31, the kinetic energy is obtained from Equation 8.21 as

$$\begin{aligned} T = \frac{1}{2}M\Big[&\{\dot{u}^l + d\dot{\phi}^l \sin(\theta^l + \Omega t) - d\dot{\psi}^l \cos(\theta^l + \Omega t) \\ &+ d(\dot{\theta}^l + \Omega)[\psi^l \sin(\theta^l + \Omega t) + \phi^l \cos(\theta^l + \Omega t)]\}^2 \\ &+ (\dot{v}^l + h\dot{\psi}^l)^2 + (\dot{w}^l - h\dot{\phi}^l)^2 + d^2(\dot{\theta}^l + \Omega)^2 \\ &+ 2d(\dot{\theta}^l + \Omega)[(\dot{w}^l - h\dot{\phi}^l)\cos(\theta^l + \Omega t) - (\dot{v}^l + h\dot{\psi}^l)\sin(\theta^l + \Omega t)] \\ &+ \frac{1}{2}\Big[I_{11}(\Omega + \dot{\theta}^l - \phi^l\dot{\psi}^l)^2 + I_{22}\{\dot{\phi}^l \cos(\theta^l + \Omega t) + \dot{\psi}^l \sin(\theta^l + \Omega t)\}^2 \\ &+ I_{33}\{\dot{\phi}^l \sin(\theta^l + \Omega t) - \dot{\psi}^l \cos(\theta^l + \Omega t)\}^2\Big] \end{aligned} \tag{8.32}$$

The forces and moments acting on the both sides of the rigid disk are shown in Figure 8.4. The virtual work done by these forces and moments is given by

$$\begin{aligned} \delta W = \;& Q_x^D \delta u^l + Q_y^D \delta v^l + Q_z^D \delta w^l + M_x^D \delta\theta^l \\ &+ (M_y^D - 2hQ_{zr}^D)\delta\phi^l + (M_z^D + 2hQ_{yr}^D)\delta\psi^l \end{aligned} \tag{8.33}$$

where Q_j^D and M_j^D $(j = x, y, z)$ defined by Equation 8.24.

By substituting Equations 8.32 and 8.33 into the Lagrange's equations and neglecting small nonlinear terms, we obtain the equations of motion of the thick rigid disk as

$$\begin{aligned} &M\ddot{u}^l = Q_x^D \\ &J_D\ddot{\theta}^l = M_x^D + M_{xl}^{m_e} \\ &M(\ddot{v}^l + h\ddot{\psi}^l) = Q_y^D + f_y^{m_e} \\ &I_D\ddot{\psi}^l - J_D\Omega\dot{\phi}^l + Mh(\ddot{v}^l + h\ddot{\psi}^l) = M_z^D + 2hQ_{yr}^D + M_{zl}^{m_e} \\ &M(\ddot{w}^l - h\ddot{\phi}^l) = Q_z^D + f_z^{m_e} \\ &I_D\ddot{\phi}^l + J_D\Omega\dot{\psi}^l - Mh(\ddot{w}^l - h\ddot{\phi}^l) = M_y^D - 2hQ_{zr}^D + M_{yl}^{m_e} \end{aligned} \tag{8.34}$$

where

$$\begin{aligned} f_y^{m_e} &= m_e d_e \Omega^2 \cos\Omega t \\ f_z^{m_e} &= m_e d_e \Omega^2 \sin\Omega t \\ M_{xl}^{m_e} &= -m_e d_e^2 \ddot{\theta}^l \\ M_{yl}^{m_e} &= -m_e d_e h\Omega^2 \sin\Omega t - m_e d_e^2[(\ddot{\phi}^l - \mu_e\Omega^2\phi^l + 2\Omega\dot{\psi}^l)\sin^2\Omega t \\ &\qquad - (\ddot{\psi}^l - \mu_e\Omega^2\psi^l - 2\Omega\dot{\phi}^l)\sin\Omega t\cos\Omega t] \\ M_{zl}^{m_e} &= m_e d_e h\Omega^2 \cos\Omega t - m_e d_e^2[(\ddot{\psi}^l - \mu_e\Omega^2\psi^l - 2\Omega\dot{\phi}^l)\cos^2\Omega t \\ &\qquad - (\ddot{\phi}^l - \mu_e\Omega^2\phi^l + 2\Omega\dot{\psi}^l)\sin\Omega t\cos\Omega t] \end{aligned} \tag{8.35}$$

Neglecting the unbalance moments M_{mx}^l, M_{my}^l, and M_{mz}^l, the above equations (8.34) can be further simplified as

$$\begin{aligned}
&M\ddot{u}^l = Q_x^D \\
&J_D\ddot{\theta}^l = M_x^D \\
&M(\ddot{v}^l + h\ddot{\psi}^l) = Q_y^D + f_y^{m_e} \\
&I_D\ddot{\psi}^l - J_D\Omega\dot{\phi}^l + Mh(\ddot{v}^l + h\ddot{\psi}^l) = M_z^D + 2hQ_{yr}^D \\
&M(\ddot{w}^l - h\ddot{\phi}^l) = Q_z^D + f_z^{m_e} \\
&I_D\ddot{\phi}^l + J_D\Omega\dot{\psi}^l - Mh(\ddot{w}^l - h\ddot{\phi}^l) = M_y^D - 2hQ_{zr}^D
\end{aligned} \tag{8.36}$$

If we use the disk thickness $2h = 0$, the above equations (8.36) return to the equations of motion for the thin rigid disks.

8.2 Spectral Element Modeling

8.2.1 Spectral Element for the Spinning Shaft

The spectral element model for the spinning shaft is derived from the equations of motion 8.12–8.17. The equations of motion consist of two decoupled equations for longitudinal and torsional vibrations and four coupled equations for the transverse bending vibration. Thus the spectral element equations will be formulated for the longitudinal vibration, torsional vibration, and transverse bending vibration, separately, and then they will be combined to derive the spectral element equation for a spinning shaft subjected to axial, torsional, and transverse bending vibrations simultaneously.

8.2.1.1 Spectral Element for the Axial Vibration

By using the DFT theory, express the solution and external force of Equation 8.12 in the spectral forms as

$$\{u(x,t)\, p_x(x,t)\} = \frac{1}{N}\sum_{n=0}^{N-1}\{U_n(x;\omega_n)\, P_{xn}(x;\omega_n)\}e^{i\omega_n t} \tag{8.37}$$

Substitution of Equation 8.37 into Equation 8.12 yields

$$EAU'' + \omega^2\rho AU + P_x(x) = 0 \tag{8.38}$$

where the subscript n is omitted for the brevity. The homogeneous solution to Equation 8.38 can be obtained as

$$U(x) = a_{L1}e^{-ik_Lx} + a_{L2}e^{ik_Lx} = \boldsymbol{e}_L(x;\omega)\boldsymbol{a}_L \tag{8.39}$$

where $\boldsymbol{a}_L = \{a_{L1} a_{L2}\}^{\mathrm{T}}$ is the constant vector and

$$\boldsymbol{e}_L(x;\omega) = [\mathrm{e}^{-ik_L x} \mathrm{e}^{ik_L x}], \qquad k_L = \omega\sqrt{\rho A/EA} \tag{8.40}$$

The spectral nodal degrees-of-freedom (DOFs) are defined as

$$\boldsymbol{d}_L = \{U_1 \quad U_2\}^{\mathrm{T}} = \{U(0) \quad U(L)\}^{\mathrm{T}} \tag{8.41}$$

Applying Equations 8.39–8.41, we obtain

$$\boldsymbol{d}_L = \begin{bmatrix} 1 & 1 \\ \mathrm{e}^{-ik_L L} & \mathrm{e}^{ik_L L} \end{bmatrix} \begin{Bmatrix} a_1 \\ a_2 \end{Bmatrix} = \boldsymbol{H}_L(\omega)\boldsymbol{a}_L \tag{8.42}$$

By using Equation 8.42, Equation 8.39 can be rewritten as

$$U(x) = [\boldsymbol{e}_L \boldsymbol{H}_L^{-1}]\boldsymbol{d}_L = \boldsymbol{N}_L(x;\omega)\boldsymbol{d}_L \tag{8.43}$$

where $\boldsymbol{N}_L(x;\,\omega)$ is the dynamic shape function.

The weak form of Equation 8.38 can be derived from the weighted integral as

$$\int_0^L (EAU'' + \omega^2 \rho A U + P_x)\delta U \mathrm{d}x = 0 \tag{8.44}$$

where δU is the arbitrary variation in U. Applying the integral by parts to the first term of Equation 8.44, we get the weak form as

$$EA\int_0^L U'\delta U' \mathrm{d}x - \omega^2 \rho A \int_0^L U\delta U \mathrm{d}x - \int_0^L P_x \delta U \mathrm{d}x - \delta \boldsymbol{d}_L^{\mathrm{T}} \boldsymbol{f}_{\mathrm{Lc}} = 0 \tag{8.45}$$

where $\boldsymbol{f}_{\mathrm{Lc}}$ is the spectral nodal axial forces vector defined by

$$\boldsymbol{f}_{\mathrm{Lc}} = \{Q_{x1} Q_{x2}\}^{\mathrm{T}} \tag{8.46}$$

Substituting Equation 8.43 into Equation 8.45, we get

$$\delta \boldsymbol{d}_L^{\mathrm{T}}[(EA\int_0^L \boldsymbol{N'}_L^{\mathrm{T}} \boldsymbol{N'}_L \mathrm{d}x - \omega^2 \rho A \int_0^L \boldsymbol{N}_L^{\mathrm{T}} \boldsymbol{N}_L \mathrm{d}x)\boldsymbol{d}_L - \int_0^L P_x \boldsymbol{N}_L^{\mathrm{T}} \mathrm{d}x - \boldsymbol{f}_{\mathrm{Lc}}] = 0 \tag{8.47}$$

As $\delta \boldsymbol{d}_L$ is arbitrary, by definition, the spectral element equation can be obtained from Equation 8.47 as

$$\boldsymbol{S}_L(\omega)\boldsymbol{d}_L = \boldsymbol{f}_L \tag{8.48}$$

where

$$\boldsymbol{S}_L(\omega) = k_L EA \begin{bmatrix} \cot(k_L L) & -\csc(k_L L) \\ -\csc(k_L L) & \cot(k_L L) \end{bmatrix} = [s_{Lij}] \quad (i,j = 1,\, 2) \tag{8.49}$$

and

$$\boldsymbol{f}_L = \boldsymbol{f}_{\mathrm{Lc}} + \int_0^L P_x(x) \boldsymbol{N}_L^{\mathrm{T}}(x) \mathrm{d}x = \{F_{u1} F_{u2}\}^{\mathrm{T}} \tag{8.50}$$

8.2.1.2 Spectral Element for the Torsional Vibration

Represent the solution and external force of Equation 8.13 in the spectral forms as

$$\{\theta(x,t)\tau_x(x,t)\} = \frac{1}{N}\sum_{n=0}^{N-1}\{\Theta_n(x;\omega_n)T_{xn}(x;\omega_n)\}\mathrm{e}^{i\omega_n t} \tag{8.51}$$

Substituting Equation 8.51 into Equation 8.13, we get

$$GJ\Theta'' + \omega^2\rho J\Theta + T_x(x) = 0 \tag{8.52}$$

where the subscript n is again omitted for brevity. The homogeneous solution of Equation 8.52 is obtained as

$$\Theta(x) = a_{T1}\mathrm{e}^{-ik_T x} + a_{T2}\mathrm{e}^{ik_T x} = \boldsymbol{e}_T(x;\omega)\boldsymbol{a}_T \tag{8.53}$$

where $\boldsymbol{a}_T = \{a_{T1}a_{T2}\}^{\mathrm{T}}$ is the constant vector and

$$\boldsymbol{e}_T(x;\omega) = [\mathrm{e}^{-ik_T x}\mathrm{e}^{ik_T x}], \qquad k_T = \omega\sqrt{\rho J/GJ} \tag{8.54}$$

Define the spectral nodal DOFs as

$$\boldsymbol{d}_L = \{\Theta_1 \quad \Theta_2\}^{\mathrm{T}} = \{\Theta(0) \quad \Theta(L)\}^{\mathrm{T}} \tag{8.55}$$

Substitution of Equation 8.53 into Equation 8.55 gives

$$\boldsymbol{d}_T = \begin{bmatrix} 1 & 1 \\ \mathrm{e}^{-ik_T L} & \mathrm{e}^{ik_T L} \end{bmatrix} \begin{Bmatrix} a_{T1} \\ a_{T2} \end{Bmatrix} = \boldsymbol{H}_T(\omega)\boldsymbol{a}_T \tag{8.56}$$

By using Equation 8.56, Equation 8.53 can be rewritten as

$$\Theta(x) = [\boldsymbol{e}_T\boldsymbol{H}_T^{-1}]\,\boldsymbol{d}_T = \boldsymbol{N}_T(x;\omega)\,\boldsymbol{d}_T \tag{8.57}$$

where $\boldsymbol{N}_T$ is the dynamic shape function.

The weak form of Equation 8.52 can be derived from the weighted integral as

$$\int_0^L (GJ\Theta'' + \omega^2\rho J\Theta + T_x)\delta\Theta \mathrm{d}x = 0 \tag{8.58}$$

where $\delta\Theta$ is the arbitrary variation in Θ. Applying the integral by parts to the first term of Equation 8.58, we get the weak form as

$$GJ\int_0^L \Theta'\delta\Theta'\mathrm{d}x - \omega^2\rho J\int_0^L \Theta\delta\Theta\mathrm{d}x - \int_0^L T_x\delta\Theta\mathrm{d}x - \delta\boldsymbol{d}_T^{\mathrm{T}}\boldsymbol{f}_{\mathrm{Tc}} = 0 \tag{8.59}$$

where $\boldsymbol{f}_{\mathrm{Tc}}$ is the spectral nodal forces vector defined by

$$\boldsymbol{f}_{\mathrm{Tc}} = \{M_{x1} \quad M_{x2}\}^{\mathrm{T}} \tag{8.60}$$

Substituting Equation 8.57 into Equation 8.59 yields

$$\delta \boldsymbol{d}_T^{\mathrm{T}}[(GJ\int_0^L \boldsymbol{N}'^{\mathrm{T}}_T \boldsymbol{N}'_T \mathrm{d}x - \omega^2 \rho J \int_0^L \boldsymbol{N}_T^{\mathrm{T}} \boldsymbol{N}_T \mathrm{d}x)\, \boldsymbol{d}_T - \int_0^L T_x \boldsymbol{N}_T^{\mathrm{T}} \mathrm{d}x - \boldsymbol{f}_{\mathrm{Tc}}] = 0 \tag{8.61}$$

As $\delta \boldsymbol{d}_T$ is arbitrary, the spectral element equation can be derived from Equation 8.61 as

$$\boldsymbol{S}_T(\omega)\boldsymbol{d}_T = \boldsymbol{f}_T \tag{8.62}$$

where

$$S_T(\omega) = k_T GJ \begin{bmatrix} \cot(k_T L) & -\csc(k_T L) \\ -\csc(k_T L) & \cot(k_T L) \end{bmatrix} = [s_{Tij}] \qquad (i,j = 1,\, 2) \tag{8.63}$$

and

$$\boldsymbol{f}_T = \boldsymbol{f}_{\mathrm{Tc}} + \int_0^L T_x(x) \boldsymbol{N}_T^{\mathrm{T}}(x) \mathrm{d}x = \{F_{\theta 1} \quad F_{\theta 2}\}^{\mathrm{T}} \tag{8.64}$$

8.2.1.3 Spectral Element for the Transverse Bending Vibration

As previously, we represent the solutions and external forces of Equation 8.14–8.17 in the spectral forms as

$$\begin{aligned} &\{v(x,t) w(x,t) \psi(x,t) \phi(x,t)\} \\ &= \frac{1}{N}\sum_{n=0}^{N-1} \{V_n(x;\omega_n) W_n(x;\omega_n) \Psi_n(x;\omega_n) \Phi_n(x;\omega_n)\} \mathrm{e}^{i\omega_n t} \end{aligned} \tag{8.65}$$

$$\begin{aligned} &\{p_y(x,t)\; p_z(x,t)\; \tau_y(x,t)\; \tau_z(x,t)\} \\ &= \frac{1}{N}\sum_{n=0}^{N-1} \{P_{yn}(x;\omega_n)\; P_{zn}(x;\omega_n)\; T_{yn}(x;\omega_n)\; T_{zn}(x;\omega_n)\} \mathrm{e}^{i\omega_n t} \end{aligned} \tag{8.66}$$

Substituting Equations 8.65 and 8.66 into Equation 8.14–8.17, we get

$$\begin{aligned} &\kappa GA(V'' - \Psi') + \rho A \omega^2 V + P_y = 0 \\ &EI\Psi'' + i\omega\Omega\rho J\Phi + \kappa GA(V' - \Psi) + \rho I \omega^2 \Psi + T_z = 0 \\ &\kappa GA(W'' + \Phi') + \rho A \omega^2 W + P_z = 0 \\ &EI\Phi'' - i\omega\Omega\rho J\Psi - \kappa GA(W' + \Phi) + \rho I \omega^2 \Phi + T_y = 0 \end{aligned} \tag{8.67}$$

where the subscripts n are omitted for brevity. Similarly, substituting Equation 8.65 and the spectral forms of $Q_i(x, t)$ and $M_i(x, t)$ $(i = y, z)$ into the associated force-displacement relationships from Equation 8.11, we get

$$\begin{aligned} Q_y(x) &= \kappa GA(V' - \Psi), \quad M_y(x) = EI\Phi' \\ Q_z(x) &= \kappa GA(W' + \Phi), \quad M_z(x) = EI\Psi' \end{aligned} \tag{8.68}$$

Assume the homogeneous solutions of Equation 8.67 to be

$$\begin{aligned} V(x) &= ae^{-ikx}, \quad \Psi(x) = rae^{-ikx} \\ W(x) &= tae^{-ikx}, \quad \Phi(x) = t\hat{r}ae^{-ikx} \end{aligned} \tag{8.69}$$

Substitution of Equation 8.69 into the homogeneous equations reduced from Equation 8.67 yields an eigenvalue problem as

$$\begin{bmatrix} \sigma_1 & -\sigma_3 & 0 & 0 \\ \sigma_3 & \sigma_2 & 0 & -\sigma_4 \\ 0 & 0 & \sigma_1 & \sigma_3 \\ 0 & \sigma_4 & -\sigma_3 & \sigma_2 \end{bmatrix} \begin{Bmatrix} 1 \\ r \\ t \\ t\hat{r} \end{Bmatrix} = \begin{Bmatrix} 0 \\ 0 \\ 0 \\ 0 \end{Bmatrix} \tag{8.70}$$

where

$$\begin{aligned} \sigma_1 &= k^2\kappa GA - \omega^2\rho A, \quad \sigma_2 = k^2 EI + \kappa GA - \omega^2 \rho I \\ \sigma_3 &= ik\kappa GA, \qquad \sigma_4 = i\omega\Omega\rho J \end{aligned} \tag{8.71}$$

From Equation 8.70, we can get a dispersion equation as

$$\begin{aligned} & k^8 - 2(\eta_1 k_F^4 + k_G^4)k^6 + (\eta_1^2 k_F^8 + 4\eta_1 k_G^4 k_F^4 + k_G^8 - 2k_F - \eta_2^{-2}\eta_3^2\Omega^2\omega^2)k^4 \\ & +(-2\eta_1^2 k_G^4 k_F^8 - 2\eta_1 k_G^8 k_F^4 + 2\eta_1 k_F^8 + 2k_F^8 + 2k_F^4 \eta_2^{-1}\eta_3^2\Omega^2\omega^2)k^2 \\ & + \eta_1^2 k_G^8 k_F^8 - 2\eta_1 k_G^4 k_F^8 + (1 - \eta_3^2\Omega^2\omega^2)k_F^8 = 0 \end{aligned} \tag{8.72}$$

where

$$\begin{aligned} & k_F = \sqrt{\omega}\left(\frac{\rho A}{EI}\right)^{1/4}, \qquad k_G = \sqrt{\omega}\left(\frac{\rho A}{\kappa GA}\right)^{1/4} \\ & \eta_1 = \frac{\rho I}{\rho A}, \qquad \eta_2 = \frac{EI}{\kappa GA}, \qquad \eta_3 = \frac{\rho J}{\kappa GA} \end{aligned} \tag{8.73}$$

Equation 8.72 gives eight wavenumbers as

$$\begin{aligned} k_1 &= -k_2 = -\sqrt{\alpha - (1/2)\eta_2^{-1}(\sqrt{\beta} + \gamma - \eta_3\Omega\omega)} \\ k_3 &= -k_4 = -\sqrt{\alpha + (1/2)\eta_2^{-1}(\sqrt{\beta} + \gamma + \eta_3\Omega\omega)} \\ k_5 &= -k_6 = -\sqrt{\alpha - (1/2)\eta_2^{-1}(\sqrt{\beta} - \gamma + \eta_3\Omega\omega)} \\ k_7 &= -k_8 = -\sqrt{\alpha + (1/2)\eta_2^{-1}(\sqrt{\beta} - \gamma - \eta_3\Omega\omega)} \end{aligned} \tag{8.74}$$

where

$$\begin{aligned} \alpha &= (1/2)(\eta_1 + \eta_2)k_F^4 \\ \beta &= (\eta_1 - \eta_2)^2 k_G^8 + 4\eta_2 k_G^4 + \eta_3^2\Omega^2\omega^2 \\ \gamma &= 2(\eta_1 - \eta_2)k_G^4\eta_3\Omega\omega \end{aligned} \tag{8.75}$$

By substituting k_j ($i = 1, 2, \ldots, 8$) into Equation 8.70, we can obtain

$$\begin{aligned} r_j &= ik_j^{-1}(k_G^4 - k_j^2) \\ \hat{r}_j &= -r_j \qquad\qquad (j = 1, 2, \ldots, 8) \\ t_j &= -(i\eta_3\Omega\omega)^{-1} r_j^{-1}[ik_j + (\eta_2 k_j^2 - \eta_1 k_G^4) r_j] \end{aligned} \tag{8.76}$$

By using eight wavenumbers k_j ($i = 1, 2, \ldots, 8$) computed from Equation 8.74, the homogeneous solutions of Equation 8.67 can be expressed as follows:

$$\begin{aligned} V(x) &= \boldsymbol{e}_v(x)\boldsymbol{a}_B, \quad W(x) = \boldsymbol{e}_w(x)\boldsymbol{a}_B \\ \Phi(x) &= \boldsymbol{e}_\varphi(x)\boldsymbol{a}_B, \quad \Psi(x) = \boldsymbol{e}_\psi(x)\boldsymbol{a}_B \end{aligned} \tag{8.77}$$

where $\boldsymbol{a}_B = \{a_1, a_2, \ldots, a_8\}^{\mathrm{T}}$ is the constant vector and

$$\begin{aligned} \boldsymbol{e}_v(x) &= [\mathrm{e}^{-ik_1x} \quad \mathrm{e}^{ik_1x} \quad \mathrm{e}^{-ik_3x} \quad \mathrm{e}^{ik_3x} \quad \mathrm{e}^{-ik_5x} \quad \mathrm{e}^{ik_5x} \quad \mathrm{e}^{-ik_7x} \quad \mathrm{e}^{ik_7x}] \\ \boldsymbol{e}_\psi(x) &= \boldsymbol{e}_v(x)\boldsymbol{R} \\ \boldsymbol{e}_w(x) &= \boldsymbol{e}_v(x)\boldsymbol{T} \\ \boldsymbol{e}_\phi(x) &= -\boldsymbol{e}_v(x)\boldsymbol{T}\boldsymbol{R} \end{aligned} \tag{8.78}$$

where $\boldsymbol{R}$ and $\boldsymbol{T}$ are diagonal matrices defined by

$$\boldsymbol{R} = diag[r_j], \qquad \boldsymbol{T} = diag[t_j] \qquad (j = 1, 2, \ldots, 8) \tag{8.79}$$

Define the spectral nodal DOFs vector for the transverse bending vibration as

$$\begin{aligned} \boldsymbol{d}_B &= \{V_1 \quad \Psi_1 \quad W_1 \quad \Phi_1 \quad V_2 \quad \Psi_2 \quad W_2 \quad \Phi_2\}^{\mathrm{T}} \\ &\equiv \{V(0) \quad \Psi(0) \quad W(0) \quad \Phi(0) \quad V(L) \quad \Psi(L) \quad W(L) \quad \Phi(L)\}^{\mathrm{T}} \end{aligned} \tag{8.80}$$

Applying Equation 8.77 to Equation 8.80, we can obtain

$$\boldsymbol{d}_B = \boldsymbol{H}_B(\omega)\boldsymbol{a}_B \tag{8.81}$$

where

$$\boldsymbol{H}_B(\omega) = [\boldsymbol{e}_v^{\mathrm{T}}(0) \quad \boldsymbol{e}_\psi^{\mathrm{T}}(0) \quad \boldsymbol{e}_w^{\mathrm{T}}(0) \quad \boldsymbol{e}_\phi^{\mathrm{T}}(0) \quad \boldsymbol{e}_v^{\mathrm{T}}(L) \quad \boldsymbol{e}_\psi^{\mathrm{T}}(L) \quad \boldsymbol{e}_w^{\mathrm{T}}(L) \quad \boldsymbol{e}_\phi^{\mathrm{T}}(L)]^{\mathrm{T}} \tag{8.82}$$

By using Equation 8.81, Equation 8.77 can be written as

$$\begin{aligned} V(x) &= \boldsymbol{N}_v(x;\omega)\,\boldsymbol{d}_B, \quad W(x) = \boldsymbol{N}_w(x;\omega)\,\boldsymbol{d}_B \\ \Psi(x) &= \boldsymbol{N}_\psi(x;\omega)\,\boldsymbol{d}_B, \quad \Phi(x) = \boldsymbol{N}_\phi(x;\omega)\,\boldsymbol{d}_B \end{aligned} \tag{8.83}$$

where $\boldsymbol{N}_v$, $\boldsymbol{N}_\psi$, $\boldsymbol{N}_w$, and $\boldsymbol{N}_\phi$ are the dynamic shape functions defined by

$$\begin{aligned} \boldsymbol{N}_v(x;\omega) &= \boldsymbol{e}_v(x)\,\boldsymbol{H}_B^{-1}, \quad \boldsymbol{N}_w(x;\omega) = \boldsymbol{e}_w(x)\,\boldsymbol{H}_B^{-1} \\ \boldsymbol{N}_\psi(x;\omega) &= \boldsymbol{e}_\psi(x)\,\boldsymbol{H}_B^{-1}, \quad \boldsymbol{N}_\phi(x;\omega) = \boldsymbol{e}_\phi(x)\,\boldsymbol{H}_B^{-1} \end{aligned} \tag{8.84}$$

To formulate the spectral element equation, the weak form of Equation 8.67 is derived from

$$\begin{aligned}
&\int_0^L \{[\kappa GA(V'' - \Psi') + \omega^2 \rho A V + P_y]\,\delta V \\
&+ [EI\Psi'' + i\omega\Omega\rho J\Phi + \kappa GA(V' - \Psi) + \omega^2 \rho I \Psi + T_z]\delta\Psi \\
&+ [\kappa GA(W'' + \Phi') + \omega^2 \rho A W + P_z]\,\delta W \\
&+ [EI\Phi'' - i\omega\Omega\rho J\Psi - \kappa GA(W' + \Phi) + \omega^2 \rho I \Phi + T_y]\,\delta\Phi\}\,\mathrm{d}x = 0
\end{aligned} \tag{8.85}$$

On integrating by parts and then using Equation 8.68, we can obtain the weak forms as

$$\begin{aligned}
&\int_0^L EI(\Phi'\delta\Phi' + \Psi'\delta\Psi')\mathrm{d}x + \int_0^L \kappa GA(V'\delta V' + W'\delta W')\mathrm{d}x \\
&- \int_0^L \kappa GA(\Psi\delta V' + V'\delta\Psi)\mathrm{d}x + \int_0^L \kappa GA(\Phi\delta W' + W'\delta\Phi)\mathrm{d}x \\
&+ \int_0^L \kappa GA(\Phi\delta\Phi + \Psi\delta\Psi)\mathrm{d}x - \int_0^L \omega^2 \rho A(V\delta V + W\delta W)\mathrm{d}x \\
&- \int_0^L \omega^2 \rho I(\Phi\delta\Phi + \Psi\delta\Psi)\mathrm{d}x + \int_0^L i\omega\Omega\rho J(\Psi\delta\Phi - \Phi\delta\Psi)\mathrm{d}x \\
&- \int_0^L (P_y\delta V + T_z\delta\Psi + P_z\delta W + T_y\delta\Phi)\mathrm{d}x \\
&- \left(Q_y\delta V|_0^L + M_z\delta\Psi|_0^L + Q_z\delta W|_0^L + M_y\delta\Phi|_0^L\right) = 0
\end{aligned} \tag{8.86}$$

Substituting Equation 8.83 into Equation 8.86 and applying the discrete Fourier transformed boundary conditions Equation 8.10 gives

$$\begin{aligned}
\delta \boldsymbol{d}_B^{\mathrm{T}} \Bigg\{ \Bigg(&\int_0^L EI(\boldsymbol{N}'^{\mathrm{T}}_{\phi} \boldsymbol{N}'_{\phi} + \boldsymbol{N}'^{\mathrm{T}}_{\psi} \boldsymbol{N}'_{\psi})\,\mathrm{d}x + \int_0^L \kappa GA(\boldsymbol{N}'^{\mathrm{T}}_{v} \boldsymbol{N}'_{v} + \boldsymbol{N}'^{\mathrm{T}}_{w} \boldsymbol{N}'_{w})\,\mathrm{d}x \\
&- \int_0^L \kappa GA(\boldsymbol{N}'^{\mathrm{T}}_{v} \boldsymbol{N}_{\psi} + \boldsymbol{N}^{\mathrm{T}}_{\psi} \boldsymbol{N}'_{v})\,\mathrm{d}x + \int_0^L \kappa GA(\boldsymbol{N}'^{\mathrm{T}}_{w} \boldsymbol{N}_{\phi} + \boldsymbol{N}^{\mathrm{T}}_{\phi} \boldsymbol{N}'_{w})\,\mathrm{d}x \\
&+ \int_0^L \kappa GA(\boldsymbol{N}^{\mathrm{T}}_{\phi} \boldsymbol{N}_{\phi} + \boldsymbol{N}^{\mathrm{T}}_{\psi} \boldsymbol{N}_{\psi})\,\mathrm{d}x - \int_0^L \omega^2 \rho A(\boldsymbol{N}^{\mathrm{T}}_{v} \boldsymbol{N}_{v} + \boldsymbol{N}^{\mathrm{T}}_{w} \boldsymbol{N}_{w})\,\mathrm{d}x \\
&- \int_0^L \omega^2 \rho I(\boldsymbol{N}^{\mathrm{T}}_{\phi} \boldsymbol{N}_{\phi} + \boldsymbol{N}^{\mathrm{T}}_{\psi} \boldsymbol{N}_{\psi})\,\mathrm{d}x + \int_0^L i\omega\Omega\rho J(\boldsymbol{N}^{\mathrm{T}}_{\phi} \boldsymbol{N}_{\psi} - \boldsymbol{N}^{\mathrm{T}}_{\psi} \boldsymbol{N}_{\phi})\,\mathrm{d}x \\
&+ \boldsymbol{K}_{\mathrm{support}} + i\omega \boldsymbol{C}_{\mathrm{support}} \Bigg) \boldsymbol{d}_B - \int_0^L (P_y \boldsymbol{N}^{\mathrm{T}}_{v} + T_z \boldsymbol{N}^{\mathrm{T}}_{\psi} + P_z \boldsymbol{N}^{\mathrm{T}}_{w} + T_y \boldsymbol{N}^{\mathrm{T}}_{\phi})\,\mathrm{d}x - \boldsymbol{f}_{Bc} \Bigg\} = 0
\end{aligned} \tag{8.87}$$

where f_{Bc} is the spectral nodal point forces and moments vector defined by

$$f_{Bc} = \{ Q_{y1} \quad M_{z1} \quad Q_{z1} \quad M_{y1} \quad Q_{y2} \quad M_{z2} \quad Q_{z2} \quad M_{y2} \}^{T} \tag{8.88}$$

and $K_{support}$ and $C_{support}$ are the matrices defined by

$$K_{support} = \begin{bmatrix} K_1 & 0 \\ 0 & K_2 \end{bmatrix}, \qquad C_{support} = \begin{bmatrix} C_1 & 0 \\ 0 & C_2 \end{bmatrix} \tag{8.89}$$

where

$$K_i = \begin{bmatrix} K_{yyi} & 0 & \frac{1}{2}(K_{yzi}+K_{zyi}) & 0 \\ 0 & 0 & 0 & 0 \\ \frac{1}{2}(K_{yzi}+K_{zyi}) & 0 & K_{zzi} & 0 \\ 0 & 0 & 0 & 0 \end{bmatrix}$$

$$(i = 1, 2) \tag{8.90}$$

$$C_i = \begin{bmatrix} C_{yyi} & 0 & \frac{1}{2}(C_{yzi}+C_{zyi}) & 0 \\ 0 & 0 & 0 & 0 \\ \frac{1}{2}(C_{yzi}+C_{zyi}) & 0 & C_{zzi} & 0 \\ 0 & 0 & 0 & 0 \end{bmatrix}$$

As δd_B is arbitrary by definition, we can obtain the spectral element equation for the transverse bending vibration from Equation 8.87 as

$$S_B(\omega) d_B = f_B \tag{8.91}$$

where

$$\begin{aligned} S_B(\omega) = & \int_0^L EI(N_\phi'^T N_\phi' + N_\psi'^T N_\psi')\, dx + \int_0^L \kappa GA(N_v'^T N_v' + N_w'^T N_w')\, dx \\ & - \int_0^L \kappa GA(N_v'^T N_\psi + N_\psi^T N_v')\, dx + \int_0^L \kappa GA(N_w'^T N_\phi + N_\phi^T N_w')\, dx \\ & + \int_0^L \kappa GA(N_\phi^T N_\phi + N_\psi^T N_\psi)\, dx - \int_0^L \omega^2 \rho A(N_v^T N_v + N_w^T N_w)\, dx \\ & - \int_0^L \omega^2 \rho I(N_\phi^T N_\phi + N_\psi^T N_\psi)\, dx + \int_0^L i\omega\Omega\rho J(N_\phi^T N_\psi - N_\psi^T N_\phi)\, dx \\ & + K_{support} + i\omega C_{support} \end{aligned} \tag{8.92}$$

and

$$\begin{aligned} f_B &= f_{Bc} + \int_0^L [P_y(x) N_v^T(x) + T_z(x) N_\psi^T(x) + P_z(x) N_w^T(x) + T_y(x) N_\phi^T(x)]\, dx \\ &= \{F_{v1} \quad F_{\psi1} \quad F_{w1} \quad F_{v2} \quad F_{\phi1} \quad F_{\psi2} \quad F_{w2} \quad F_{\phi2}\}^T \end{aligned} \tag{8.93}$$

The last integral of Equation 8.92 forms a skew symmetric matrix, which represents the gyroscopic effect generated by the spinning of the shaft.

By substituting the dynamic shape function matrices Equation 8.84 into Equation 8.92, the spectral element matrix $S_B(\omega)$ can be written in the form of

$$S_B(\omega) = H_B^{-T} D_B H_B^{-1} + K_{\text{support}} + i\omega C_{\text{support}} \equiv \begin{bmatrix} s_{B11} & s_{B12} \\ s_{B21} & s_{B22} \end{bmatrix} \tag{8.94}$$

where

$$\begin{aligned} D_B(\omega) = & EI\int_0^L (e_\phi'^{T} e_\phi' + e_\psi'^{T} e_\psi')\,dx + \kappa GA \int_0^L (e_v'^{T} e_v' + e_w'^{T} e_w')\,dx \\ & - \kappa GA \int_0^L (e_v'^{T} e_\psi + e_\psi^{T} e_v')\,dx + \kappa GA \int_0^L (e_w'^{T} e_\phi + e_\phi^{T} e_w')\,dx \\ & + \kappa GA \int_0^L (e_\phi^{T} e_\phi + e_\psi^{T} e_\psi)\,dx - \omega^2 \rho A \int_0^L (e_v^{T} e_v + e_w^{T} e_w)\,dx \\ & - \omega^2 \rho I \int_0^L (e_\phi^{T} e_\phi + e_\psi^{T} e_\psi)\,dx + i\omega\Omega\rho J \int_0^L (e_\phi^{T} e_\psi - e_\psi^{T} e_\phi)\,dx \end{aligned} \tag{8.95}$$

and $s_{B11}, s_{B12}, s_{B21}$, and s_{B22} are 4-by-4 submatrices of $S_B(\omega)$. Substitution of Equation 8.78 into Equation 8.95 yields

$$\begin{aligned} D_B(\omega) = & -EI(RTKEKTR + RKEKR) - \kappa GA(KEK + TKEKT) \\ & + i\kappa GA(KER + REK) + i\kappa GA(TKETR + RTEKT) \\ & + \kappa GA(RTETR + RER) - \omega^2 \rho A(E + TE\,T) \\ & - \omega^2 \rho I(RTETR + RER) - i\omega\Omega\rho J(RTER - RETR) \end{aligned} \tag{8.96}$$

where

$$E(\omega) = \int_0^L e_v^{T} e_v\,dx \equiv [E_{lm}] \quad (l,\, m = 1,\, 2,\, \ldots,\, 8) \tag{8.97}$$

$$E_{lm} = \begin{cases} \dfrac{i}{k_l + k_m}\left[e^{-i(k_l+k_m)L} - 1\right] & \text{if } k_l + k_m \neq 0 \\ L & \text{if } k_l + k_m = 0 \end{cases} \tag{8.98}$$

and

$$K = diag[k_j] \quad (j = 1,\, 2,\, \ldots,\, 8) \tag{8.99}$$

Similarly, the second term in the right-hand side of Equation 8.93 can be rewritten as

$$\begin{aligned} f_{Bd} & = \int_0^L [P_y(x) N_v^{T}(x) + T_z(x) N_\psi^{T}(x) + P_z(x) N_w^{T}(x) + T_y(x) N_\phi^{T}(x)]\,dx \\ & = H_B^{-T} \textstyle\int_0^L (P_y I + T_z R^{T} + P_z R^{T} + T_y R^{T})\, e_v^{T} dx \end{aligned} \tag{8.100}$$

8.2.1.4 Spectral Element Equation for the Spinning Shaft

The spectral element equations (8.48), (8.62) and (8.91) derived for individual axial vibration, torsional vibration, and transverse bending vibration can be combined to derive the spectral element equation for the spinning shafts subjected to the longitudinal, torsional, and transverse bending vibrations simultaneously as follows:

$$\boldsymbol{S}^S(\omega)\boldsymbol{d}^S = \boldsymbol{f}^S \tag{8.101}$$

or

$$\begin{bmatrix} \boldsymbol{S}_{11}^S & \boldsymbol{S}_{12}^S \\ \boldsymbol{S}_{21}^S & \boldsymbol{S}_{22}^S \end{bmatrix} \begin{Bmatrix} \boldsymbol{d}_1^S \\ \boldsymbol{d}_2^S \end{Bmatrix} = \begin{Bmatrix} \boldsymbol{f}_1^S \\ \boldsymbol{f}_2^S \end{Bmatrix} \tag{8.102}$$

where

$$\boldsymbol{S}_{ij}^S = \begin{bmatrix} s_{Lij} & 0 & \boldsymbol{0} \\ 0 & s_{Tij} & \boldsymbol{0} \\ \boldsymbol{0} & \boldsymbol{0} & \boldsymbol{s}_{Bij} \end{bmatrix} \quad (i,j = 1,\ 2) \tag{8.103}$$

and

$$\begin{aligned} \boldsymbol{d}_j^S &= \{U_j \quad \Theta_j \quad V_j \quad \Psi_j \quad W_j \quad \Phi_j\}^{\mathrm{T}} \\ \boldsymbol{f}_j^S &= \{F_{uj} \quad F_{\theta j} \quad F_{vj} \quad F_{\psi j} \quad F_{wj} \quad F_{\phi j}\}^{\mathrm{T}} \end{aligned} \quad (j = 1,\ 2) \tag{8.104}$$

8.2.2 Spectral Element for the Disk

8.2.2.1 Thin Disk

Represent the solutions to Equation 8.27 and the associated excitation forces in the spectral forms as

$$\begin{aligned} &\{u(t) \quad \theta(t) \quad v(t) \quad \psi(t) \quad w(t) \quad \phi(t)\} \\ &= \frac{1}{N}\sum_{n=0}^{N-1}\{U_n^D \quad \Theta_n^D \quad V_n^D \quad \Psi_n^D \quad W_n^D \quad \Phi_n^D\}\,\mathrm{e}^{i\omega_n t} \\ &\{Q_x^D(t) \quad Q_y^D(t) \quad Q_z^D(t) \quad M_x^D(t) \quad M_y^D(t) \quad M_z^D(t)\} \\ &= \frac{1}{N}\sum_{n=0}^{N-1}\{Q_{xn}^D \quad Q_{yn}^D \quad Q_{zn}^D \quad M_{xn}^D \quad M_{yn}^D \quad M_{zn}^D\}\,\mathrm{e}^{i\omega_n t} \\ &\{f_y^{m_e}(t) \quad f_z^{m_e}(t)\} = \frac{1}{N}\sum_{n=0}^{N-1}\{F_{yn}^{m_e} \quad F_{zn}^{m_e}\}\mathrm{e}^{i\omega_n t} \end{aligned} \tag{8.105}$$

Substituting Equation 8.105 into Equation 8.27 yields

$$\begin{aligned} -\omega^2 M U^D &= Q_x^D \\ -\omega^2 J_D \Theta^D &= M_x^D \\ -\omega^2 M V^D &= Q_y^D + F_y^{m_e} \\ -\omega^2 I_D \Psi^D - i\omega J_D \Omega \Phi^D &= M_z^D \\ -\omega^2 M W^D &= Q_z^D + F_z^{m_e} \\ -\omega^2 I_D \Phi^D + i\omega J_D \Omega \Psi^D &= M_y^D \end{aligned} \tag{8.106}$$

where the subscripts n are omitted for brevity. From Equation 8.106, we can obtain the spectral element equation for a thin rigid disk as

$$\boldsymbol{S}^D(\omega)\boldsymbol{d}^D = \boldsymbol{f}^D \tag{8.107}$$

where

$$\boldsymbol{d}^D = \{U^D \quad \Theta^D \quad V^D \quad \Psi^D \quad W^D \quad \Phi^D\}^{\mathrm{T}}$$
$$\boldsymbol{f}^D = \{Q_x^D \quad M_x^D \quad Q_y^D + F_y^{m_e} \quad M_z^D \quad Q_z^D + F_z^{m_e} \quad M_y^D\}^{\mathrm{T}} \tag{8.108}$$

and

$$\boldsymbol{S}^D(\omega) = \begin{bmatrix} \boldsymbol{S}_{11}^D & \boldsymbol{0} \\ \boldsymbol{0} & \boldsymbol{S}_{22}^D \end{bmatrix} \tag{8.109}$$

with

$$\boldsymbol{S}_{11}^D = -\omega^2 \begin{bmatrix} M & 0 & 0 \\ 0 & J_D & 0 \\ 0 & 0 & M \end{bmatrix}$$
$$\boldsymbol{S}_{22}^D = (-1) \begin{bmatrix} \omega^2 I_D & 0 & i\omega\Omega J_D \\ 0 & \omega^2 M & 0 \\ -i\omega\Omega J_D & 0 & \omega^2 I_D \end{bmatrix} \tag{8.110}$$

8.2.2.2 Thick Disk

Represent the solutions to Equation 8.36 and the associated excitation forces in the spectral forms as

$$\{u_i(t) \quad \theta_i(t) \quad v_i(t) \quad \psi_i(t) \quad w_i(t) \quad \phi_i(t)\} = \frac{1}{N}\sum_{n=0}^{N-1} \{U_{in}^D \quad \Theta_{in}^D \quad V_{in}^D \quad \Psi_{in}^D \quad W_{in}^D \quad \Phi_{in}^D\}\, \mathrm{e}^{i\omega_n t}$$
$$\{Q_{xi}^D(t) \quad Q_{yi}^D(t) \quad Q_{zi}^D(t) \quad M_{xi}^D(t) \quad M_{yi}^D(t) \quad M_{zi}^D(t)\} = \frac{1}{N}\sum_{n=0}^{N-1} \{Q_{xin}^D \quad Q_{yin}^D \quad Q_{zin}^D \quad M_{xin}^D \quad M_{yin}^D \quad M_{zin}^D\}\, \mathrm{e}^{i\omega_n t} \tag{8.111}$$
$$\{f_y^{m_e}(t) \quad f_z^{m_e}(t)\} = \frac{1}{N}\sum_{n=0}^{N-1} \{F_{yn}^{m_e} \quad F_{zn}^{m_e}\}\, \mathrm{e}^{i\omega_n t} \tag{8.111}$$

where $i = r$ and l. Substituting Equation 8.111 into the equations of motion (8.36) and the kinematic relationships (8.28) and (8.29), we get

$$\begin{aligned}
&-\omega^2 M U_l^D = Q_x^D \\
&-\omega^2 J_D \Theta_l^D = M_x^D \\
&-\omega^2 M(V_l^D + h\Psi_l^D) = Q_y^D + F_y^{m_e} \\
&-\omega^2 I_D \Psi_l^D - i\omega J_D \Omega \Phi_l^D - \omega^2 Mh(V_l^D + h\Psi_l^D) = M_z^D + 2hQ_{yr}^D \\
&-\omega^2 M(W_l^D - h\Phi_l^D) = Q_z^D + F_z^{m_e} \\
&-\omega^2 I_D \Phi_l^D + i\omega J_D \Omega \Psi_l^D + \omega^2 Mh(W_l^D - h\Phi_l^D) = M_y^D - 2hQ_{zr}^D
\end{aligned} \tag{8.112}$$

and

$$\begin{aligned} U_r^D - U_l^D &= 0, & \Theta_r^D - \Theta_l^D &= 0 \\ V_r^D - V_l^D - 2h\Psi_l^D &= 0, & \Psi_r^D - \Psi_l^D &= 0 \\ W_r^D - W_l^D + 2h\,\Phi_l^D &= 0, & \Phi_r^D - \Phi_l^D &= 0 \end{aligned} \tag{8.113}$$

From Equations 8.112 and 8.113, we can obtain the spectral element equation for a thick rigid disk as

$$\begin{bmatrix} \mathbf{S}_{11}^D & \mathbf{0} \\ \mathbf{S}_{21}^D & \mathbf{S}_{22}^D \end{bmatrix} \begin{Bmatrix} \mathbf{d}_l^D \\ \mathbf{d}_r^D \end{Bmatrix} = \begin{Bmatrix} \mathbf{f}^D \\ \mathbf{0} \end{Bmatrix} \tag{8.114}$$

where

$$\begin{aligned} \mathbf{d}_l^D &= \{U_l^D \quad \Theta_l^D \quad V_l^D \quad \Psi_l^D \quad W_l^D \quad \Phi_l^D\}^{\mathrm{T}} \\ \mathbf{d}_r^D &= \{U_r^D \quad \Theta_r^D \quad V_r^D \quad \Psi_r^D \quad W_r^D \quad \Phi_r^D\}^{\mathrm{T}} \\ \mathbf{f}^D &= \{Q_x^D \quad M_x^D \quad Q_y^D + F_y^{m_e} \quad M_z^D + 2hQ_{yr}^D \quad Q_z^D + F_z^{m_e} \quad M_y^D - 2hQ_{zr}^D\}^{\mathrm{T}} \end{aligned} \tag{8.115}$$

and

$$\mathbf{S}_{11}^D = (-1)\begin{bmatrix} \omega^2 \mathbf{m}_1 & \mathbf{0} & \mathbf{0} \\ \mathbf{0} & \omega^2 \mathbf{m}_2 & i\omega \mathbf{g} \\ \mathbf{0} & -i\omega \mathbf{g} & \omega^2 \mathbf{m}_3 \end{bmatrix}, \quad \mathbf{S}_{21}^D = (-1)\begin{bmatrix} \mathbf{I}_2 & \mathbf{0} & \mathbf{0} \\ \mathbf{0} & \mathbf{h}_1 & \mathbf{0} \\ \mathbf{0} & \mathbf{0} & \mathbf{h}_2 \end{bmatrix}$$
$$\mathbf{S}_{22}^D = \mathbf{I}_6 \tag{8.116}$$

where $\mathbf{I}_n$ is the n-by-n identity matrix and

$$\begin{aligned} \mathbf{m}_1 &= \begin{bmatrix} M & 0 \\ 0 & J_D \end{bmatrix}, & \mathbf{m}_2 &= \begin{bmatrix} M & Mh \\ Mh & I_D + Mh^2 \end{bmatrix} \\ \mathbf{m}_3 &= \begin{bmatrix} M & -Mh \\ -Mh & I_D + Mh^2 \end{bmatrix}, & \mathbf{g} &= \begin{bmatrix} 0 & 0 \\ 0 & \Omega J_D \end{bmatrix} \\ \mathbf{h}_1 &= \begin{bmatrix} 1 & 2h \\ 0 & 1 \end{bmatrix}, & \mathbf{h}_2 &= \begin{bmatrix} 1 & -2h \\ 0 & 1 \end{bmatrix} \end{aligned} \tag{8.117}$$

8.2.3 Assembly of Spectral Elements

The rotor systems may consist of multiple spinning shafts and interim disks, supported by multiple supports. The spectral element models for such rotor systems can be derived by assembling spectral elements for each of the spinning shaft elements and disks. The assembly can be readily conducted by adopting the same assembling procedure as used in the conventional FEM. As an illustrative example, we will consider the assembly for the shaft with an interim disk as shown in Figure 8.5.

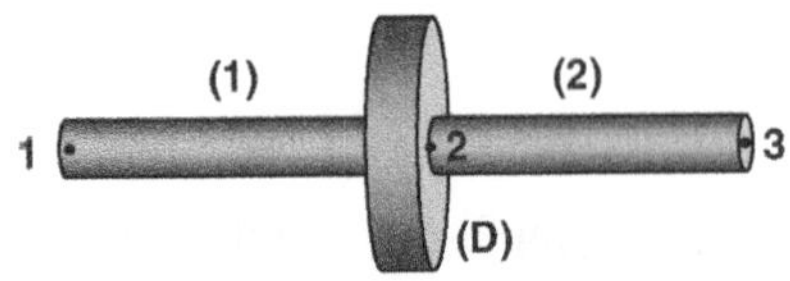

Figure 8.5 Spinning shaft with an interim rigid disk

8.2.3.1 Shaft with a Thin Disk

For a shaft with an interim thin disk, the spectral element equations for the shaft (1), the disk (D), and the shaft (2) are given by as follows:

$$\begin{bmatrix} S_{11}^1 & S_{12}^1 \\ S_{21}^1 & S_{22}^1 \end{bmatrix} \begin{Bmatrix} d_l^1 \\ d_r^1 \end{Bmatrix} = \begin{Bmatrix} f_l^1 \\ f_r^1 \end{Bmatrix} \quad \text{(shaft 1)}$$

$$S^D(\omega) d^D = f^D \quad \text{(disk D)} \tag{8.118}$$

$$\begin{bmatrix} S_{11}^2 & S_{12}^2 \\ S_{21}^2 & S_{22}^2 \end{bmatrix} \begin{Bmatrix} d_l^2 \\ d_r^2 \end{Bmatrix} = \begin{Bmatrix} f_l^2 \\ f_r^2 \end{Bmatrix} \quad \text{(shaft 2)}$$

Define the global spectral nodal DOFs as

$$\begin{aligned} d_1 &\equiv d_l^1 = \{U_1 \quad \Theta_1 \quad V_1 \quad \Psi_1 \quad W_1 \quad \Phi_1\}^{\mathrm{T}} \\ d_2 &\equiv d_r^1 = d^D = d_l^2 = \{U_2 \quad \Theta_2 \quad V_2 \quad \Psi_2 \quad W_2 \quad \Phi_2\}^{\mathrm{T}} \\ d_3 &\equiv d_r^2 = \{U_3 \quad \Theta_3 \quad V_3 \quad \Psi_3 \quad W_3 \quad \Phi_3\}^{\mathrm{T}} \end{aligned} \tag{8.119}$$

By using the above global spectral nodal DOFs, we can assemble the spectral element equations given in Equation 8.118 to obtain the global dynamic stiffness equation in the form

$$\begin{bmatrix} S_{11}^1 & S_{12}^1 & 0 \\ S_{21}^1 & S_{22}^1 + S^D + S_{11}^2 & S_{12}^2 \\ 0 & S_{21}^2 & S_{22}^2 \end{bmatrix} \begin{Bmatrix} d_1 \\ d_2 \\ d_3 \end{Bmatrix} = \begin{Bmatrix} F_1 \\ F_2 \\ F_3 \end{Bmatrix} \tag{8.120}$$

where

$$\begin{aligned} F_1 &\equiv f_l^1 = \{F_{ul}^1 \quad F_{\theta l}^1 \quad F_{vl}^1 \quad F_{\psi l}^1 \quad F_{wl}^1 \quad F_{\phi l}^1\}^{\mathrm{T}} \\ F_2 &\equiv f_r^1 + f_l^2 + f^D \\ &= \{F_{ur}^1 + F_{ul}^2 + Q_x^D \quad F_{\theta r}^1 + F_{\theta l}^2 + M_x^D \quad F_{vr}^1 + F_{vl}^2 + Q_y^D + F_y^{m_e} \\ &\qquad F_{\psi r}^1 + F_{\psi l}^2 + M_z^D \quad F_{wr}^1 + F_{wl}^2 + Q_z^D + F_z^{m_e} \quad F_{\phi r}^1 + F_{\phi l}^2 + M_y^D\}^{\mathrm{T}} \\ F_3 &\equiv f_r^2 = \{F_{ur}^2 \quad F_{\theta r}^2 \quad F_{vr}^2 \quad F_{\psi r}^2 \quad F_{wr}^2 \quad F_{\phi r}^2\}^{\mathrm{T}} \end{aligned} \tag{8.121}$$

8.2.3.2 Shaft with a Thick Disk

When the interim disk is a thick disk, the spectral element equations for the shaft (1), the disk (D), and the shaft (2) are given by as follows:

$$\begin{bmatrix} S_{11}^1 & S_{12}^1 \\ S_{21}^1 & S_{22}^1 \end{bmatrix} \begin{Bmatrix} d_l^1 \\ d_r^1 \end{Bmatrix} = \begin{Bmatrix} f_l^1 \\ f_r^1 \end{Bmatrix} \quad \text{(shaft 1)}$$
$$\begin{bmatrix} S_{11}^D & 0 \\ S_{21}^D & S_{22}^D \end{bmatrix} \begin{Bmatrix} d_l^D \\ d_r^D \end{Bmatrix} = \begin{Bmatrix} f^D \\ 0 \end{Bmatrix} \quad \text{(Disk D)} \qquad (8.122)$$
$$\begin{bmatrix} S_{11}^2 & S_{12}^2 \\ S_{21}^2 & S_{22}^2 \end{bmatrix} \begin{Bmatrix} d_l^2 \\ d_r^2 \end{Bmatrix} = \begin{Bmatrix} f_l^2 \\ f_r^2 \end{Bmatrix} \quad \text{(shaft 2)}$$

Define the global spectral nodal DOFs as

$$\begin{aligned} d_1 &\equiv d_l^1 = \{U_1 \quad \Theta_1 \quad V_1 \quad \Psi_1 \quad W_1 \quad \Phi_1\}^{\mathrm{T}} \\ d_2 &\equiv d_r^1 = d_l^D = \{U_2 \quad \Theta_2 \quad V_2 \quad \Psi_2 \quad W_2 \quad \Phi_2\}^{\mathrm{T}} \\ d_3 &\equiv d_r^D = d_l^2 = \{U_3 \quad \Theta_3 \quad V_3 \quad \Psi_3 \quad W_3 \quad \Phi_3\}^{\mathrm{T}} \\ d_4 &\equiv d_r^2 = \{U_4 \quad \Theta_4 \quad V_4 \quad \Psi_4 \quad W_4 \quad \Phi_4\}^{\mathrm{T}} \end{aligned} \qquad (8.123)$$

By using the global spectral nodal DOFs defined above, we can obtain assemble the spectral element equations given in Equation 8.122 in the form

$$\begin{bmatrix} S_{11}^1 & S_{12}^1 & 0 & 0 \\ S_{21}^1 & S_{22}^1 + S_{11}^D & 0 & 0 \\ 0 & S_{21}^D & S_{22}^2 + S_{11}^D & S_{12}^2 \\ 0 & 0 & S_{21}^2 & S_{22}^2 \end{bmatrix} \begin{Bmatrix} d_1 \\ d_2 \\ d_3 \\ d_4 \end{Bmatrix} = \begin{Bmatrix} f_l^1 \\ f_r^1 + f^D \\ f_l^2 \\ f_r^2 \end{Bmatrix} \qquad (8.124)$$

where

$$\begin{aligned} f_l^1 &= \{F_{ul}^1 \quad F_{\theta l}^1 \quad F_{vl}^1 \quad F_{\psi l}^1 \quad F_{wl}^1 \quad F_{\phi l}^1\}^{\mathrm{T}} \\ f_r^1 &= \{F_{ur}^1 \quad F_{\theta r}^1 \quad F_{vr}^1 \quad F_{\psi r}^1 \quad F_{wr}^1 \quad F_{\phi r}^1\}^{\mathrm{T}} \\ f^D &= \{Q_x^D \quad M_x^D \quad Q_x^D + F_y^{m_e} \quad M_z^D + 2hQ_{yr}^D \quad Q_z^D + F_z^{m_e} \quad M_y^D - 2hQ_{zr}^D\}^{\mathrm{T}} \\ f_l^2 &= \{F_{ul}^2 \quad F_{\theta l}^2 \quad F_{vl}^2 \quad F_{\psi l}^2 \quad F_{wl}^2 \quad F_{\phi l}^2\}^{\mathrm{T}} \\ f_r^2 &= \{F_{ur}^2 \quad F_{\theta r}^2 \quad F_{vr}^2 \quad F_{\psi r}^2 \quad F_{wr}^2 \quad F_{\phi r}^2\}^{\mathrm{T}} \end{aligned} \qquad (8.125)$$

From Equation 8.113, we obtain the relationship

$$d_r^D = T_D d_l^D \quad \text{or} \quad d_3 = T_D d_2 \qquad (8.126)$$

where

$$T_D = \begin{bmatrix} \boldsymbol{I}_2 & \boldsymbol{0} & \boldsymbol{0} \\ \boldsymbol{0} & \boldsymbol{h}_1 & \boldsymbol{0} \\ \boldsymbol{0} & \boldsymbol{0} & \boldsymbol{h}_2 \end{bmatrix} \tag{8.127}$$

where $\boldsymbol{I}_2$ is the 2-by-2 identity matrix, $\boldsymbol{h}_1$ and $\boldsymbol{h}_2$ are the 2-by-2 matrices defined by Equation 8.117. By using Equation 8.126, Equation 8.124 can be condensed into the following form

$$\begin{bmatrix} \boldsymbol{S}_{11}^1 & \boldsymbol{S}_{12}^1 & \boldsymbol{0} \\ \boldsymbol{S}_{21}^1 & \boldsymbol{S}_{22}^1 + \bar{\boldsymbol{S}}^D + \bar{\boldsymbol{S}}_{11}^2 & \bar{\boldsymbol{S}}_{12}^2 \\ \boldsymbol{0} & \bar{\boldsymbol{S}}_{21}^2 & \boldsymbol{S}_{22}^2 \end{bmatrix} \begin{Bmatrix} \boldsymbol{d}_1 \\ \boldsymbol{d}_2 \\ \boldsymbol{d}_4 \end{Bmatrix} = \begin{Bmatrix} \boldsymbol{F}_1 \\ \boldsymbol{F}_2 \\ \boldsymbol{F}_4 \end{Bmatrix} \tag{8.128}$$

where

$$\begin{aligned} &\bar{\boldsymbol{S}}_{11}^2 = \boldsymbol{T}_D^{\mathrm{T}} \boldsymbol{S}_{11}^2 \boldsymbol{T}_D, \qquad \bar{\boldsymbol{S}}_{12}^2 = \boldsymbol{T}_D^{\mathrm{T}} \boldsymbol{S}_{12}^2, \qquad \bar{\boldsymbol{S}}_{21}^2 = \boldsymbol{S}_{21}^2 \boldsymbol{T}_D \\ &\bar{\boldsymbol{S}}^D = \boldsymbol{S}_{11}^D + \boldsymbol{T}_D^{\mathrm{T}} \boldsymbol{S}_{21}^D + \boldsymbol{T}_D^{\mathrm{T}} \boldsymbol{S}_{22}^D \boldsymbol{T}_D \end{aligned} \tag{8.129}$$

and

$$\begin{aligned} &\boldsymbol{F}_1 \equiv \boldsymbol{f}_l^1 \\ &\boldsymbol{F}_2 \equiv \boldsymbol{f}_r^1 + \boldsymbol{f}^D + \boldsymbol{T}_D^T \boldsymbol{f}_l^2 = \{ F_{2U} \quad F_{2\Theta} \quad F_{2V} \quad F_{2\Psi} \quad F_{2W} \quad F_{2\Phi} \}^{\mathrm{T}} \\ &\boldsymbol{F}_4 \equiv \boldsymbol{f}_r^2 \end{aligned} \tag{8.130}$$

with

$$\begin{aligned} F_{2U} &= F_{ur}^1 + Q_x^D + F_{ul}^2 \\ F_{2\Theta} &= F_{\theta r}^1 + M_x^D + F_{\theta l}^2 \\ F_{2V} &= F_{vr}^1 + Q_y^D + F_y^{m_e} + F_{vl}^2 \\ F_{2\Psi} &= F_{\psi r}^1 + M_z^D + F_{\psi l}^2 + 2hQ_{yr}^D + 2hF_{vl}^2 \\ F_{2W} &= F_{wr}^1 + Q_z^D + F_z^{m_e} + F_{wl}^2 \\ F_{2\Phi} &= F_{\phi r}^1 + M_y^D + F_{\phi l}^2 - 2hQ_{zr}^D - 2hF_{wl}^2 \end{aligned} \tag{8.131}$$

If there are no external forces, except for the centrifugal unbalance forces, Equation 8.130 becomes

$$\begin{aligned} &\boldsymbol{F}_1 = \boldsymbol{F}_4 = 0 \\ &\boldsymbol{F}_2 = \boldsymbol{f}^D = \{0 \quad 0 \quad F_y^{m_e} \quad 0 \quad F_z^{m_e} \quad 0\}^{\mathrm{T}} \end{aligned} \tag{8.132}$$

If we set the disk thickness to $2h = 0$, Equation 8.128 for the thick rigid disk returns to Equation 8.120 for the thin rigid disk.

8.3 Finite Element Model

Similar to the spectral element modeling, the finite element modeling of a rotor system with rigid disks also consists of the finite element modeling for the axial, torsion and transverse bending vibrations of the spinning shafts and the finite element modeling for the dynamics of the disks.

8.3.1 Finite Element for the Spinning Shaft

8.3.1.1 Finite Element for the Axial Vibration

The equation of motion for the axial vibration of a shaft is given by Equation 8.12 and its weak form is given by

$$\int_0^L EAu'\delta u'\,\mathrm{d}x + \int_0^L \rho A\ddot{u}\delta u\,\mathrm{d}x = Q_{x1}\delta u_1 + Q_{x2}\delta u_2 + \int_0^L p_x\delta u\,\mathrm{d}x \tag{8.133}$$

where Q_{x1} and Q_{x2} are the nodal axial forces shown in Figure 8.1(a). The displacement field within the element is represented by

$$u(x,t) = \boldsymbol{N}_L(x)\boldsymbol{d}_L(t) \tag{8.134}$$

where

$$\boldsymbol{N}_L(x) = [1-(x/L) \quad x/L]$$

$$\boldsymbol{d}_L(t) = \{u_1(t) \quad u_2(t)\}^{\mathrm{T}} = \{u(0,t) \quad u(L,t)\}^{\mathrm{T}} \tag{8.135}$$

By substituting Equation 8.134 into Equation 8.133, we obtain

$$\boldsymbol{M}_L\ddot{\boldsymbol{d}}_L(t) + \boldsymbol{K}_L\boldsymbol{d}_L(t) = \boldsymbol{f}_L(t) \tag{8.136}$$

where

$$\begin{aligned}
\boldsymbol{M}_L &= \int_0^L \rho A\boldsymbol{N}_L^{\mathrm{T}}\boldsymbol{N}_L\mathrm{d}x = \frac{\rho AL}{6}\begin{bmatrix} 2 & 1 \\ 1 & 2 \end{bmatrix} = [m_{Lij}] \quad (i,j=1,\ 2) \\
\boldsymbol{K}_L &= \int_0^L EA\boldsymbol{N}'^{\mathrm{T}}_L\boldsymbol{N}'_L\mathrm{d}x = \frac{EA}{L}\begin{bmatrix} 1 & -1 \\ -1 & 1 \end{bmatrix} = [k_{Lij}] \quad (i,j=1,\ 2) \\
\boldsymbol{f}_L(t) &= \{Q_{x1}(t) \quad Q_{x2}(t)\}^{\mathrm{T}} + \int_0^L \boldsymbol{N}_L^T(x)p_x(x,t)\mathrm{d}x \\
&= \{f_{u1}(t) \quad f_{u2}(t)\}^{\mathrm{T}}
\end{aligned} \tag{8.137}$$

8.3.1.2 Finite Element for the Torsional Vibration

The equation of motion for the torsional vibration is given by Equation 8.13 and its weak form is given by

$$\int_0^L GJ\theta'\delta\theta'\,\mathrm{d}x + \int_0^L \rho J\ddot{\theta}\delta\theta\,\mathrm{d}x = M_{x1}\delta\theta_1 + M_{x2}\delta\theta_2 + \int_0^L \tau_x\delta\theta\,\mathrm{d}x \tag{8.138}$$

where M_{x1} and M_{x2} are the nodal torques shown in Figure 8.1(a). The displacement field $\theta(x,t)$ is represented by

$$\theta(x,t) = \boldsymbol{N}_T(x)\mathbf{d}_T(t) \tag{8.139}$$

where

$$\begin{aligned} \boldsymbol{N}_T(x) &= [1-(x/L) \quad x/L] \\ \boldsymbol{d}_T(t) &= \{\theta_1(t) \quad \theta_2(t)\}^{\mathrm{T}} = \{\theta(0,t) \quad \theta(L,t)\}^{\mathrm{T}} \end{aligned} \tag{8.140}$$

By substituting Equation 8.139 into Equation 8.138, we obtain

$$\boldsymbol{M}_T\ddot{\boldsymbol{d}}_T(t) + \boldsymbol{K}_T\boldsymbol{d}_T(t) = \boldsymbol{f}_T(t) \tag{8.141}$$

where

$$\begin{aligned} \boldsymbol{M}_T &= \int_0^L \rho J \boldsymbol{N}_T^{\mathrm{T}}\boldsymbol{N}_T \mathrm{d}x = \frac{\rho JL}{6}\begin{bmatrix} 2 & 1 \\ 1 & 2 \end{bmatrix} = [m_{Tij}] \qquad (i,j=1,\ 2) \\ \boldsymbol{K}_T &= \int_0^L GJ \boldsymbol{N}'^{\mathrm{T}}_T\boldsymbol{N}'_T \mathrm{d}x = \frac{GJ}{L}\begin{bmatrix} 1 & -1 \\ -1 & 1 \end{bmatrix} = [k_{Tij}] \qquad (i,j=1,\ 2) \\ \boldsymbol{f}_T(t) &= \{M_{x1}(t) \quad M_{x2}(t)\}^{\mathrm{T}} + \int_0^L \boldsymbol{N}_T^{\mathrm{T}}(x)\tau_x(x,t)\,\mathrm{d}x \\ &= \{f_{\theta 1}(t) \quad f_{\theta 2}(t)\}^{\mathrm{T}} \end{aligned} \tag{8.142}$$

8.3.1.3 Finite Element for the Transverse Bending Vibration

The equations of motion for the transverse bending vibration are given by Equations 8.14–8.17 and their weak form is derived as

$$\begin{aligned} &\int_0^L EI(\phi'\delta\phi' + \psi'\delta\psi')\mathrm{d}x + \int_0^L \kappa GA(v'\delta v' + w'\delta w')\,\mathrm{d}x \\ &- \int_0^L \kappa GA(\psi\delta v' + v'\delta\psi)\mathrm{d}x + \int_0^L \kappa GA(\phi\delta w' + w'\delta\phi)\,\mathrm{d}x \\ &+ \int_0^L \kappa GA(\phi\delta\phi + \psi\delta\psi)\mathrm{d}x + \int_0^L \rho A(\ddot{v}\delta v + \ddot{w}\delta w)\,\mathrm{d}x \\ &+ \int_0^L \rho I(\ddot{\phi}\delta\phi + \ddot{\psi}\delta\psi)\mathrm{d}x + \int_0^L \Omega\rho J(\dot{\psi}\delta\phi - \dot{\phi}\delta\psi)\,\mathrm{d}x \\ &- \int_0^L (p_y\delta v + \tau_z\delta\psi + p_z\delta w + \tau_y\delta\phi)\,\mathrm{d}x \\ &- (Q_{y1}\delta v_1 + M_{z1}\delta\psi_1 + Q_{z1}\delta w_1 + M_{y1}\delta\phi_1 \\ &+ Q_{y2}\delta v_2 + M_{z2}\delta\psi_2 + Q_{z2}\delta w_2 + M_{y2}\delta\phi_2) = 0 \end{aligned} \tag{8.143}$$

where Q_{yj} and M_{yj} ($j=1$, 2, 3) are the forces and moments shown in Figure 8.1(a). The displacement fields $v(x, t)$, $w(x, t)$, $\phi(x, t)$, and $\psi(x, t)$ are represented by

$$\begin{aligned} v &= \boldsymbol{N}_v(x)\boldsymbol{d}_B(t), \quad w = \boldsymbol{N}_w(x)\boldsymbol{d}_B(t) \\ \phi &= \boldsymbol{N}_\phi(x)\boldsymbol{d}_B(t), \quad \psi = \boldsymbol{N}_\psi(x)\boldsymbol{d}_B(t) \end{aligned} \tag{8.144}$$

where

$$\begin{aligned} \boldsymbol{d}_B(t) &= \{\boldsymbol{d}_{B1}(t) \quad \boldsymbol{d}_{B2}(t)\}^{\mathrm{T}} \\ \boldsymbol{d}_{Bj}(t) &= \{\, v_j(t) \quad \psi_j(t) \quad w_j(t) \quad \phi_j(t)\,\}^{\mathrm{T}} \quad (j = 1,\,2) \end{aligned} \tag{8.145}$$

and

$$\begin{aligned} \boldsymbol{N}_v(x) &= \big[\,(1-\xi)(2-\xi-\xi^2+6r)(R/4) \quad L(1-\xi^2)(1-\xi+3r)(R/8) \quad 0 \quad 0 \\ &\qquad (1+\xi)(2+\xi-\xi^2+6r)(R/4) \quad -L(1-\xi^2)(1+\xi+3r)(R/8) \quad 0 \quad 0\big] \\ \boldsymbol{N}_w(x) &= \big[0 \quad 0 \quad (1-\xi)(2-\xi-\xi^2+6r)(R/4) \quad -L(1-\xi^2)(1-\xi+3r)(R/8) \\ &\qquad 0 \quad 0 \quad (1+\xi)(2+\xi-\xi^2+6r)(R/4) \quad L(1-\xi^2)(1+\xi+3r)(R/8)\big] \\ \boldsymbol{N}_\phi(x) &= \big[0 \quad 0 \quad 3(1-\xi^2)(R/2L) \quad -(1-\xi)(1+3\xi-6r)(R/4) \\ &\qquad 0 \quad 0 \quad -3(1-\xi^2)(R/2L) \quad -(1+\xi)(1-3\xi-6r)(R/4)\,\big] \\ \boldsymbol{N}_\psi(x) &= \big[-3(1-\xi^2)(R/2L) \quad -(1-\xi)(1+3\xi-6r)(R/4) \quad 0 \quad 0 \\ &\qquad 3(1-\xi^2)(R/2L) \quad -(1+\xi)(1-3\xi-6r)(R/4) \quad 0 \quad 0\big] \end{aligned} \tag{8.146}$$

with

$$\xi = 2\left(\frac{x}{L}\right) - 1 \quad (0 \le x \le L), \qquad r = \frac{4EI}{\kappa GAL^2}, \qquad R = \frac{1}{1+3r} \tag{8.147}$$

Substitution of Equation 8.144 into Equation 8.143 gives

$$\boldsymbol{M}_B\ddot{\boldsymbol{d}}_B(t) + \boldsymbol{G}_B\dot{\boldsymbol{d}}_B(t) + \boldsymbol{K}_B\boldsymbol{d}_B(t) = \boldsymbol{f}_B(t) \tag{8.148}$$

where

$$\begin{aligned} \boldsymbol{M}_B &= [m_{Bij}] = \int_0^L \rho A(\boldsymbol{N}_v^{\mathrm{T}}\boldsymbol{N}_v + \boldsymbol{N}_w^{\mathrm{T}}\boldsymbol{N}_w)\,\mathrm{d}x + \int_0^L \rho I(\boldsymbol{N}_\phi^{\mathrm{T}}\boldsymbol{N}_\phi + \boldsymbol{N}_\psi^{\mathrm{T}}\boldsymbol{N}_\psi)\,\mathrm{d}x \\ \boldsymbol{G}_B &= [g_{Bij}] = \int_0^L \Omega\rho J(\boldsymbol{N}_\phi^{\mathrm{T}}\boldsymbol{N}_\psi - \boldsymbol{N}_\psi^{\mathrm{T}}\boldsymbol{N}_\phi)\,\mathrm{d}x \\ \boldsymbol{K}_B &= [k_{Bij}] = \int_0^L EI(\boldsymbol{N}_\phi'^{\mathrm{T}}\boldsymbol{N}_\phi' + \boldsymbol{N}_\psi'^{\mathrm{T}}\boldsymbol{N}_\psi')\,\mathrm{d}x + \int_0^L \kappa GA(\boldsymbol{N}_v'^{\mathrm{T}}\boldsymbol{N}_v' + \boldsymbol{N}_w'^{\mathrm{T}}\boldsymbol{N}_w' \\ &\qquad + \boldsymbol{N}_\phi^{\mathrm{T}}\boldsymbol{N}_\phi + \boldsymbol{N}_\psi^{\mathrm{T}}\boldsymbol{N}_\psi + \boldsymbol{N}_w'^{\mathrm{T}}\boldsymbol{N}_\phi + \boldsymbol{N}_\phi^{\mathrm{T}}\boldsymbol{N}_w' - \boldsymbol{N}_v'^{\mathrm{T}}\boldsymbol{N}_\psi - \boldsymbol{N}_\psi^{\mathrm{T}}\boldsymbol{N}_v')\,\mathrm{d}x \end{aligned} \tag{8.149}$$

and

$$\begin{aligned} \boldsymbol{f}_B(t) &= \boldsymbol{f}_{Bc}(t) + \boldsymbol{f}_{Bd}(t) \equiv \{\boldsymbol{f}_{B1}(t) \quad \boldsymbol{f}_{B2}(t)\,\}^{\mathrm{T}} \\ \boldsymbol{f}_{Bc}(t) &= \{\,Q_{y1}(t) \quad M_{z1}(t) \quad Q_{z1}(t) \quad M_{y1}(t) \quad Q_{y2}(t) \quad M_{z2}(t) \quad Q_{z2}(t) \quad M_{y2}(t)\,\}^{\mathrm{T}} \\ \boldsymbol{f}_{Bd}(t) &= \int_0^L (\boldsymbol{N}_v^{\mathrm{T}}p_y + \boldsymbol{N}_w^{\mathrm{T}}p_z + \boldsymbol{N}_\phi^{\mathrm{T}}\tau_y + \boldsymbol{N}_\psi^{\mathrm{T}}\tau_z)\,\mathrm{d}x \\ \boldsymbol{f}_{Bi}(t) &= \{f_{vi}(t) \quad f_{\psi i}(t) \quad f_{wi}(t) \quad f_{\phi i}(t)\,\}^{\mathrm{T}} \quad (i = 1,\,2) \end{aligned} \tag{8.150}$$

$\boldsymbol{M}_B$ and $\boldsymbol{K}_B$ are the 8-by-8 symmetric matrices and $\boldsymbol{G}_B$ is the 8-by-8 skew symmetric matrix. The matrices $\boldsymbol{M}_B$, $\boldsymbol{K}_B$, and $\boldsymbol{G}_B$ are given in Appendix 8.A. For the next use, the matrices $\boldsymbol{M}_B$, $\boldsymbol{K}_B$, and $\boldsymbol{G}_B$ and the vector $\boldsymbol{f}_B$ are written in the partitioned forms as

$$\boldsymbol{M}_B = \begin{bmatrix} \boldsymbol{M}_{B11} & \boldsymbol{M}_{B12} \\ \boldsymbol{M}_{B21} & \boldsymbol{M}_{B22} \end{bmatrix}, \quad \boldsymbol{G}_B = \begin{bmatrix} \boldsymbol{G}_{B11} & \boldsymbol{G}_{B12} \\ \boldsymbol{G}_{B21} & \boldsymbol{G}_{B22} \end{bmatrix}, \quad \boldsymbol{K}_B = \begin{bmatrix} \boldsymbol{K}_{B11} & \boldsymbol{K}_{B12} \\ \boldsymbol{K}_{B21} & \boldsymbol{K}_{B22} \end{bmatrix} \tag{8.151}$$

8.3.1.4 Finite Element Equation for the Spinning Shaft

The finite element for a spinning shaft can be obtained by assembling the finite element equations for the axial, torsion, and transverse bending vibrations as follows:

$$\boldsymbol{M}^S\ddot{\boldsymbol{d}}^S(t)+(\boldsymbol{G}^S+\bar{\boldsymbol{C}})\dot{\boldsymbol{d}}^S(t)+(\boldsymbol{K}^S+\bar{\boldsymbol{K}})\boldsymbol{d}^S(t)=\boldsymbol{f}^S(t) \tag{8.152}$$

where

$$\begin{aligned}\boldsymbol{d}^S(t)&=\{u_1(t)\quad\theta_1(t)\quad d_{B1}(t)\quad u_2(t)\quad\theta_2(t)\quad d_{B2}\ (t)\}^{\mathrm{T}}\\ \boldsymbol{f}^S(t)&=\{f_{u1}(t)\quad f_{\theta1}(t)\quad f_{B1}(t)\quad f_{u2}(t)\quad f_{\theta2}(t)\quad f_{B2}(t)\}^{\mathrm{T}}\end{aligned} \tag{8.153}$$

and

$$\boldsymbol{M}^S=\begin{bmatrix}\boldsymbol{M}_{11}^S & \boldsymbol{M}_{12}^S\\ \boldsymbol{M}_{21}^S & \boldsymbol{M}_{22}^S\end{bmatrix},\quad \boldsymbol{G}^S=\begin{bmatrix}\boldsymbol{G}_{11}^S & \boldsymbol{G}_{12}^S\\ \boldsymbol{G}_{21}^S & \boldsymbol{G}_{22}^S\end{bmatrix},\quad \boldsymbol{K}^S=\begin{bmatrix}\boldsymbol{K}_{11}^S & \boldsymbol{K}_{12}^S\\ \boldsymbol{K}_{21}^S & \boldsymbol{K}_{22}^S\end{bmatrix} \tag{8.154}$$

with

$$\boldsymbol{M}_{ij}^S=\begin{bmatrix}m_{Lij} & 0 & \boldsymbol{0}\\ 0 & m_{Tij} & \boldsymbol{0}\\ \boldsymbol{0} & \boldsymbol{0} & \boldsymbol{M}_{Bij}\end{bmatrix},\quad \boldsymbol{G}_{ij}^S=\begin{bmatrix}0 & 0 & \boldsymbol{0}\\ 0 & 0 & \boldsymbol{0}\\ \boldsymbol{0} & \boldsymbol{0} & \boldsymbol{G}_{Bij}\end{bmatrix},\quad \boldsymbol{K}_{ij}^S=\begin{bmatrix}k_{Lij} & 0 & \boldsymbol{0}\\ 0 & k_{Tij} & \boldsymbol{0}\\ \boldsymbol{0} & \boldsymbol{0} & \boldsymbol{K}_{Bij}\end{bmatrix} \tag{8.155}$$

In Equation 8.152, $\bar{\boldsymbol{K}}$ and $\bar{\boldsymbol{C}}$ are the stiffness and damping matrices for the bearing supports shown in Figure 8.1(c). They are given by

$$\bar{\boldsymbol{K}}=\begin{bmatrix}\boldsymbol{0} & \boldsymbol{0} & \boldsymbol{0} & \boldsymbol{0}\\ \boldsymbol{0} & \boldsymbol{K}_1 & \boldsymbol{0} & \boldsymbol{0}\\ \boldsymbol{0} & \boldsymbol{0} & \boldsymbol{0} & \boldsymbol{0}\\ \boldsymbol{0} & \boldsymbol{0} & \boldsymbol{0} & \boldsymbol{K}_2\end{bmatrix},\quad \bar{\boldsymbol{C}}=\begin{bmatrix}\boldsymbol{0} & \boldsymbol{0} & \boldsymbol{0} & \boldsymbol{0}\\ \boldsymbol{0} & \boldsymbol{C}_1 & \boldsymbol{0} & \boldsymbol{0}\\ \boldsymbol{0} & \boldsymbol{0} & \boldsymbol{0} & \boldsymbol{0}\\ \boldsymbol{0} & \boldsymbol{0} & \boldsymbol{0} & \boldsymbol{C}_2\end{bmatrix} \tag{8.156}$$

where $\boldsymbol{K}_i$ and $\boldsymbol{C}_i$ $(i=1, 2)$ are the 4-by-4 matrices defined by Equation 8.90.

8.3.2 Finite Element for the Disk

8.3.2.1 Thin Disk

The equations of motion for a thin rigid disk are given by Equation 8.27 and they can be written in the matrix equation form as

$$\boldsymbol{M}^D\ddot{\boldsymbol{d}}^D(t)+\boldsymbol{G}^D\dot{\boldsymbol{d}}^D(t)=\boldsymbol{f}^D(t) \tag{8.157}$$

where

$$\boldsymbol{M}^D=\begin{bmatrix}\boldsymbol{m}_1 & \boldsymbol{0} & \boldsymbol{0}\\ \boldsymbol{0} & \bar{\boldsymbol{m}}_2 & \boldsymbol{0}\\ \boldsymbol{0} & \boldsymbol{0} & \bar{\boldsymbol{m}}_3\end{bmatrix},\quad \boldsymbol{G}^D=\begin{bmatrix}\boldsymbol{0} & \boldsymbol{0} & \boldsymbol{0}\\ \boldsymbol{0} & \boldsymbol{0} & -\boldsymbol{g}\\ \boldsymbol{0} & \boldsymbol{g} & \boldsymbol{0}\end{bmatrix} \tag{8.158}$$

and

$$\begin{aligned} \boldsymbol{d}^D(t) &= \{u^D(t) \quad \theta^D(t) \quad v^D(t) \quad \psi^D(t) \quad w^D(t) \quad \phi^D(t)\}^{\mathrm{T}} \\ \boldsymbol{f}^D(t) &= \{Q_x^D(t) \quad M_x^D(t) \quad Q_y^D(t)+f_y^{m_e}(t) \quad M_z^D(t) \\ &\qquad Q_z^D(t)+f_z^{m_e}(t) \quad M_y^D(t)\}^{\mathrm{T}} \end{aligned} \tag{8.159}$$

The matrices $\boldsymbol{m}_1$ and $\boldsymbol{g}$ are defined by Equation 8.117 and

$$\bar{\boldsymbol{m}}_2 = \bar{\boldsymbol{m}}_3 = \begin{bmatrix} M & 0 \\ 0 & I_D \end{bmatrix} \tag{8.160}$$

8.3.2.2 Thick Disk

The equations of motion for a thick rigid disk given by Equation 8.36 and the kinematic relationships given by Equations 8.28 and 8.29 can be written in the matrix equation form as

$$\begin{bmatrix} \boldsymbol{M}^D & \boldsymbol{0} \\ \boldsymbol{0} & \boldsymbol{0} \end{bmatrix} \begin{Bmatrix} \ddot{\boldsymbol{d}}_l^D \\ \ddot{\boldsymbol{d}}_r^D \end{Bmatrix} + \begin{bmatrix} \boldsymbol{G}^D & \boldsymbol{0} \\ \boldsymbol{0} & \boldsymbol{0} \end{bmatrix} \begin{Bmatrix} \dot{\boldsymbol{d}}_l^D \\ \dot{\boldsymbol{d}}_r^D \end{Bmatrix} + \begin{bmatrix} \boldsymbol{0} & \boldsymbol{0} \\ -\boldsymbol{T}_D & \boldsymbol{I}_6 \end{bmatrix} \begin{Bmatrix} \boldsymbol{d}_l^D \\ \boldsymbol{d}_r^D \end{Bmatrix} = \begin{Bmatrix} \boldsymbol{f}^D \\ \boldsymbol{0} \end{Bmatrix} \tag{8.161}$$

where

$$\begin{aligned} \boldsymbol{d}_l^D(t) &= \{u_l^D(t) \quad \theta_l^D(t) \quad v_l^D(t) \quad \psi_l^D(t) \quad w_l^D(t) \quad \phi_l^D(t)\}^{\mathrm{T}} \\ \boldsymbol{d}_r^D(t) &= \{u_r^D(t) \quad \theta_r^D(t) \quad v_r^D(t) \quad \psi_r^D(t) \quad w_r^D(t) \quad \phi_r^D(t)\}^{\mathrm{T}} \\ \boldsymbol{f}^D(t) &= \{Q_x^D(t) \quad M_x^D(t) \quad Q_y^D(t)+f_y^{m_e}(t) \quad M_z^D(t)+2hQ_{yr}^D(t) \\ &\qquad Q_z^D(t)+f_z^{m_e}(t) \quad M_y^D(t)-2hQ_{zr}^D(t)\}^{\mathrm{T}} \end{aligned} \tag{8.162}$$

and

$$\boldsymbol{M}^D = \begin{bmatrix} \boldsymbol{m}_1 & \boldsymbol{0} & \boldsymbol{0} \\ \boldsymbol{0} & \boldsymbol{m}_2 & \boldsymbol{0} \\ \boldsymbol{0} & \boldsymbol{0} & \boldsymbol{m}_3 \end{bmatrix}, \quad \boldsymbol{G}^D = \begin{bmatrix} \boldsymbol{0} & \boldsymbol{0} & \boldsymbol{0} \\ \boldsymbol{0} & \boldsymbol{0} & -\boldsymbol{g} \\ \boldsymbol{0} & \boldsymbol{g} & \boldsymbol{0} \end{bmatrix} \tag{8.163}$$

The matrices $\boldsymbol{m}_1$, $\boldsymbol{m}_2$, $\boldsymbol{m}_3$, and $\boldsymbol{g}$ are defined by Equation 8.117 and $\boldsymbol{T}_D$ by Equation 8.127. In Equation 8.161, the following kinematic relationship obtained from Equations 8.28 and 8.29 is used

$$\boldsymbol{d}_r^D(t) = \boldsymbol{T}_D \boldsymbol{d}_l^D(t) \tag{8.164}$$

8.3.3 Assembly of Finite Elements

A thick rigid disk element "D" and its neighboring two shaft finite elements "1" and "2" can be assembled by using the same procedure as discussed in Section 8.2.3.2. The assembled finite element model can be represented by

$$\bar{\boldsymbol{M}}\ddot{\bar{\boldsymbol{d}}}(t) + \bar{\boldsymbol{G}}\dot{\bar{\boldsymbol{d}}}(t) + \bar{\boldsymbol{K}}\bar{\boldsymbol{d}}(t) = \bar{\boldsymbol{f}}(t) \tag{8.165}$$

where

$$\begin{aligned} \bar{\boldsymbol{d}}(t) &= \{\boldsymbol{d}_1(t) \quad \boldsymbol{d}_2(t) \quad \boldsymbol{d}_3(t) \quad \boldsymbol{d}_4(t)\}^{\mathrm{T}} \\ \bar{\boldsymbol{f}}(t) &= \{\boldsymbol{f}_l^1(t) \quad \boldsymbol{f}_r^1(t)+\boldsymbol{f}^D(t) \quad \boldsymbol{f}_l^2(t) \quad \boldsymbol{f}_r^2(t)\}^{\mathrm{T}} \end{aligned} \tag{8.166}$$

with

$$\begin{aligned} \boldsymbol{d}_1(t) &\equiv \boldsymbol{d}_l^1(t) = \{u_1(t) \quad \theta_1(t) \quad v_1(t) \quad \psi_1(t) \quad w_1(t) \quad \phi_1(t)\}^{\mathrm{T}} \\ \boldsymbol{d}_2(t) &\equiv \boldsymbol{d}_r^1(t) = \boldsymbol{d}_l^D(t) = \{u_2(t) \quad \theta_2(t) \quad v_2(t) \quad \psi_2(t) \quad w_2(t) \quad \phi_2(t)\}^{\mathrm{T}} \\ \boldsymbol{d}_3(t) &\equiv \boldsymbol{d}_r^D(t) = \boldsymbol{d}_l^2(t) = \{u_3(t) \quad \theta_3(t) \quad v_3(t) \quad \psi_3(t) \quad w_3(t) \quad \phi_3(t)\}^{\mathrm{T}} \\ \boldsymbol{d}_4(t) &\equiv \boldsymbol{d}_r^2(t) = \{u_4(t) \quad \theta_4(t) \quad v_4(t) \quad \psi_4(t) \quad w_4(t) \quad \phi_4(t)\}^{\mathrm{T}} \end{aligned} \tag{8.167}$$

and

$$\begin{aligned} \boldsymbol{f}_l^1(t) &= \{f_{ul}^1(t) \quad f_{\theta l}^1(t) \quad f_{vl}^1(t) \quad f_{\psi l}^1(t) \quad f_{wl}^1(t) \quad f_{\phi l}^1(t)\}^{\mathrm{T}} \\ \boldsymbol{f}_r^1(t) &= \{f_{ur}^1(t) \quad f_{\theta r}^1(t) \quad f_{vr}^1(t) \quad f_{\psi r}^1(t) \quad f_{wr}^1(t) \quad f_{\phi r}^1(t)\}^{\mathrm{T}} \\ \boldsymbol{f}^D(t) &= \{Q_x^D(t) \quad M_x^D(t) \quad Q_y^D(t) + f_y^{m_e}(t) \\ &\qquad M_z^D(t) + 2hQ_{yr}^D(t) \quad Q_z^D(t) + f_z^{m_e}(t) \quad M_y^D(t) - 2hQ_{zr}^D(t)\}^{\mathrm{T}} \\ \boldsymbol{f}_l^2(t) &= \{f_{ul}^2(t) \quad f_{\theta l}^2(t) \quad f_{vl}^2(t) \quad f_{\psi l}^2(t) \quad f_{wl}^2(t) \quad f_{\phi l}^2(t)\}^{\mathrm{T}} \\ \boldsymbol{f}_r^2(t) &= \{f_{ur}^2(t) \quad f_{\theta r}^2(t) \quad f_{vr}^2(t) \quad f_{\psi r}^2(t) \quad f_{wr}^2(t) \quad f_{\phi r}^2(t)\}^{\mathrm{T}} \end{aligned} \tag{8.168}$$

and where

$$\begin{aligned} \bar{\boldsymbol{M}} &= \begin{bmatrix} \boldsymbol{M}_{11}^1 & \boldsymbol{M}_{12}^1 & \boldsymbol{0} & \boldsymbol{0} \\ \boldsymbol{M}_{21}^1 & \boldsymbol{M}_{22}^1 + \boldsymbol{M}^D & \boldsymbol{0} & \boldsymbol{0} \\ \boldsymbol{0} & \boldsymbol{0} & \boldsymbol{M}_{11}^2 & \boldsymbol{M}_{12}^2 \\ \boldsymbol{0} & \boldsymbol{0} & \boldsymbol{M}_{21}^2 & \boldsymbol{M}_{22}^2 \end{bmatrix} \\ \bar{\boldsymbol{G}} &= \begin{bmatrix} \boldsymbol{G}_{11}^1 & \boldsymbol{G}_{12}^1 & \boldsymbol{0} & \boldsymbol{0} \\ \boldsymbol{G}_{21}^1 & \boldsymbol{G}_{22}^1 + \boldsymbol{G}^D & \boldsymbol{0} & \boldsymbol{0} \\ \boldsymbol{0} & \boldsymbol{0} & \boldsymbol{G}_{11}^2 & \boldsymbol{G}_{12}^2 \\ \boldsymbol{0} & \boldsymbol{0} & \boldsymbol{G}_{21}^2 & \boldsymbol{G}_{22}^2 \end{bmatrix} \\ \bar{\boldsymbol{K}} &= \begin{bmatrix} \boldsymbol{K}_{11}^1 & \boldsymbol{K}_{12}^1 & \boldsymbol{0} & \boldsymbol{0} \\ \boldsymbol{K}_{21}^1 & \boldsymbol{K}_{22}^1 & \boldsymbol{0} & \boldsymbol{0} \\ \boldsymbol{0} & -\boldsymbol{T}_D & \boldsymbol{I}_6 + \boldsymbol{K}_{11}^2 & \boldsymbol{K}_{12}^2 \\ \boldsymbol{0} & \boldsymbol{0} & \boldsymbol{K}_{21}^2 & \boldsymbol{K}_{22}^2 \end{bmatrix} \end{aligned} \tag{8.169}$$

By using the kinematic relationship Equation 8.164, we can obtain

$$\bar{\boldsymbol{d}}(t) = \boldsymbol{R}\boldsymbol{d}(t) \tag{8.170}$$

where

$$R = \begin{bmatrix} I_6 & 0 & 0 \\ 0 & I_6 & 0 \\ 0 & T_D & 0 \\ 0 & 0 & I_6 \end{bmatrix} \tag{8.171}$$

where I_6 is the 6-by-6 identity matrix. Equation 8.165 can be condensed by using Equation 8.170 as follows:

$$M\ddot{d}(t) + G\dot{d}(t) + Kd(t) = f(t) \tag{8.172}$$

where

$$\begin{aligned} d(t) &= \{\, d_1(t) \quad d_2(t) \quad d_4(t)\,\}^{\mathrm{T}}, \quad f(t) = R^{\mathrm{T}}\bar{f}(t) \\ M &= R^{\mathrm{T}}\bar{M}R, \quad G = R^{\mathrm{T}}\bar{G}R, \quad K = R^{\mathrm{T}}\bar{K}R \end{aligned} \tag{8.173}$$

8.4 Numerical Examples

Example 8.1 Spinning Shaft without Disks

Figure 8.6 shows a simply supported uniform Timoshenko shaft (T-shaft) without disks attached. The geometric and material properties used for numerical computation are: $L = 0.5$ m, $\rho A = 4.0$ kg/m, $\rho I = 1.5 \times 10^{-4}$ kg m, $\rho J = 3.0 \times 10^{-4}$ kg m, $EI = 4.0 \times 10^3$ N m^2, and $\kappa GA = 3.5 \times 10^7$ N. In Tables 8.1–8.3, the natural frequencies and critical speeds obtained by both SEM and FEM are compared with the results obtained by using the formulae given by

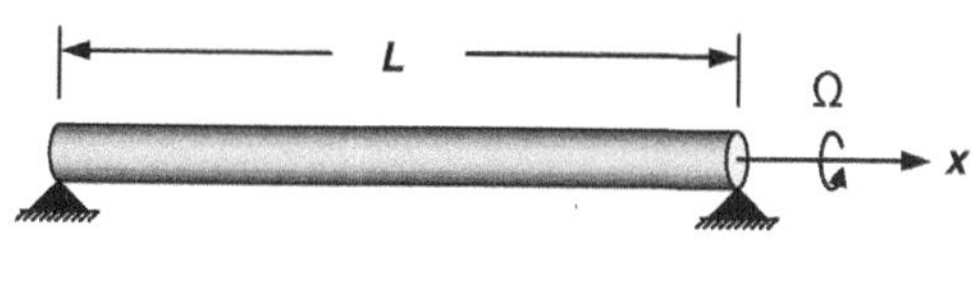

Figure 8.6 Rotating shaft without disks

Table 8.1 Natural frequencies (Hz) of the non-spinning shaft

	FEM(*N*)				
Mode	$N = 2$	$N = 10$	$N = 50$	SEM(1)	Ref. [24]
1st mode	198.96	198.10	198.10	198.10	198.10
2nd mode	879.49	785.76	785.46	785.45	785.45

Note: (*N*) = total number of finite elements used in the analysis.

Table 8.2 Natural Frequencies (Hz) of the shaft spinning at $\Omega = 3600\,\text{rpm}$

Modes		FEM(*N*) $N=2$	FEM(*N*) $N=10$	FEM(*N*) $N=50$	SEM(1)	Ref. [176]
1st mode	Backward	198.88	198.02	198.01	198.01	198.20
	Forward	199.05	198.19	198.19	198.19	
2nd mode	Backward	879.13	785.42	785.12	785.11	785.80
	Forward	879.85	786.10	785.80	785.79	

Note: (*N*) = total number of finite elements used in the analysis.

Table 8.3 Critical speeds (Hz) of the spinning shaft

Modes		FEM(*N*) $N=2$	FEM(*N*) $N=10$	FEM(*N*) $N=50$	SEM(1)	Ref. [45]
1st mode	Backward	198.96	197.82	197.81	197.81	197.81
	Forward	199.26	198.40	198.39	198.39	198.39
2nd mode	Backward	874.29	781.33	781.03	781.02	780.91
	Forward	884.78	790.26	789.96	789.95	789.99

Note: (*N*) = total number of finite elements used in the analysis.

a. Natural frequencies of the non-spinning T-shaft [24]

$$f_n = f_n|_{BE}\alpha_n\sqrt{\beta_n - \sqrt{\beta_n^2 - \eta_1^{-1}\eta_2^{-1}}} \quad \text{(Hz)}$$

b. Natural frequencies of the spinning T-shaft (at speed Ω) [176]

$$\sin\left(L\sqrt{\frac{\eta_5}{2}\left(\sigma_1 + \sqrt{\sigma_1^2 - 4\sigma_2\eta_5^{-1}}\right)}\right) = 0 \quad \text{(Hz)}$$

c. Critical speeds of the spinning T-shaft [45]

$$\Omega_n = \begin{cases} f_n|_B\sqrt{\dfrac{\alpha_n^2}{\alpha_n^2 + \eta_2 - \eta_1}} & \text{(Hz)} \quad \text{(forward whirling)} \\[2ex] f_n|_B\sqrt{\dfrac{\alpha_n^2}{\alpha_n^2 + \eta_2 + 3\eta_1}} & \text{(Hz)} \quad \text{(backward whirling)} \end{cases}$$

where η_1, η_2, and η_3 are defined by Equation 8.73 and $f_n|_B$ are the natural frequencies of the simply supported, non-spinning uniform Bernoulli–Euler beam given by

$$f_n|_B = \frac{1}{2\pi\alpha_n^2\sqrt{\eta_5}} \quad \text{(Hz)}$$

The other symbols used in above formulas are defined by

$$\eta_4 = \frac{\rho I}{\kappa GA}, \qquad \eta_5 = \frac{\rho A}{EI}, \qquad \eta_6 = \frac{\rho J}{\rho A}$$

$$\alpha_n = \frac{L}{n\pi}, \qquad \beta_n = \frac{1}{2}\left[\eta_1^{-1} + (1+\alpha_n^2\eta_1^{-1})\eta_2^{-1}\right]$$

$$\sigma_1 = (\eta_1 + \eta_2)\omega^2 - \eta_6\Omega\omega, \qquad \sigma_2 = \eta_4\omega^4 - \eta_3\Omega\omega^3 - \omega$$

Tables 8.1–8.3 show that the natural frequencies and critical speeds obtained by SEM are very close to the results obtained by using the theoretical formulae, and the FEM results certainly converge towards the SEM results as the total number of finite elements used for FEM is increased.

Example 8.2 Spinning Shaft with a Disk

Figure 8.7 shows a simply supported uniform T-shaft with a thick disk. The geometric and material properties for the T-shaft are the same as used in Example 8.1. The stiffness and damping coefficients of the bearing supports at boundaries are $K_{yy} = K_{zz} = 1.0 \times 10^9$ N/m, $K_{yz} = K_{zy} = 0$ N/m, $C_{yy} = C_{zz} = 400$ N s/m, and $C_{yz} = C_{zy} = 0$ N s/m. The disk has the thickness $2h = 20$ mm, the uniform mass $m_D = 2$ kg, the moment of inertia $I_D = 2.0 \times 10^{-3}$ kg m^2 and the polar moment of inertia $J_D = 4.0 \times 10^{-3}$ kg m^2. The unbalance mass on the disk has the mass $m_e = 0.2$ kg located at $d_e = 0.001$ m.

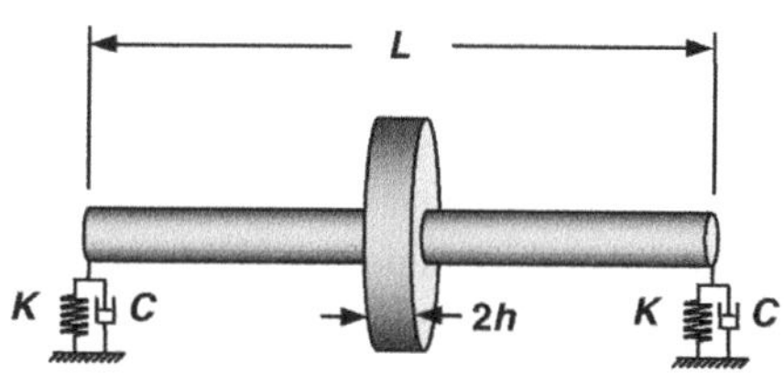

Figure 8.7 Spinning shaft with a disk

Figure 8.8 shows the spinning speed (Ω) dependence of the natural frequencies obtained by SEM. The critical speeds (Ω_c) are the spinning speeds that coincide with the natural frequencies of the backward and forward whirling modes and they are indicated in Figure 8.8. The disk can be assembled with its neighboring two-shaft elements either by ignoring disk thickness (thin disk model, Equation 8.107) or by taking into account disk thickness (thick disk model, Equation 8.114). The critical speeds obtained by using the thick disk model with varying disk thickness, but holding all inertial properties constant, are displayed in Table 8.4. If the disk thickness $2h$ is reduced from its true value of 20 mm (thick disk) to 0 mm (thin disk), the critical speeds obtained by using the thick disk model certainly converge towards the results for the thin

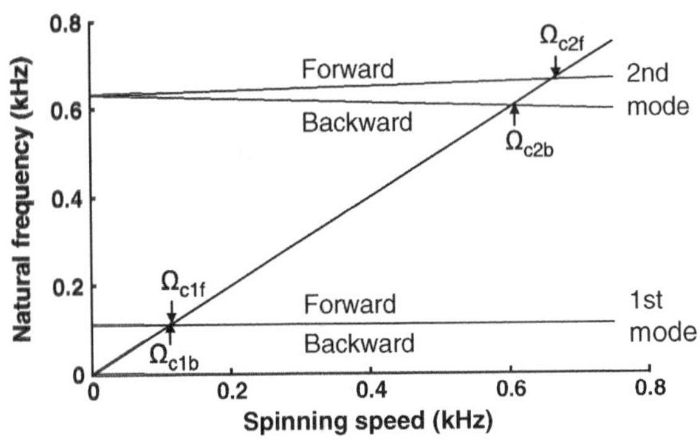

Figure 8.8 Natural frequency versus spinning speed of the shaft with a disk

Table 8.4 Disk model dependence of critical speeds (Hz)

Modes	Model 2h (mm)	Thick disk → Thin disk			
		20	10	5	0
1st mode	Backward	113.92	113.92	113.92	113.92
	Forward	114.03	114.03	114.03	114.03
2nd mode	Backward	608.41	622.67	631.32	637.09
	Forward	664.78	685.44	698.33	707.12

disk. In general, the thin disk model seems to overestimate the critical speeds when compared with the thick disk model.

Table 8.5 shows that the critical speeds obtained by FEM converge towards the SEM results as the number of finite elements used in FEM is increased. Figure 8.9 also shows that this is true for the vibration response calculated at spinning speed $\Omega = 6000$ rpm. These results prove the high accuracy of the present spectral element model.

Table 8.5 Critical speeds (Hz) of the shaft with a disk ($2h = 20$ mm)

Modes		FEM(N)			
		$N=2$	$N=10$	$N=50$	SEM(2)
1st mode	Backward	113.97	113.93	113.92	113.92
	Forward	114.08	114.04	114.03	114.03
2nd mode	Backward	633.47	608.71	608.42	608.41
	Forward	705.27	665.28	664.79	664.78

Note: (N) = total number of the finite elements used in the analysis.

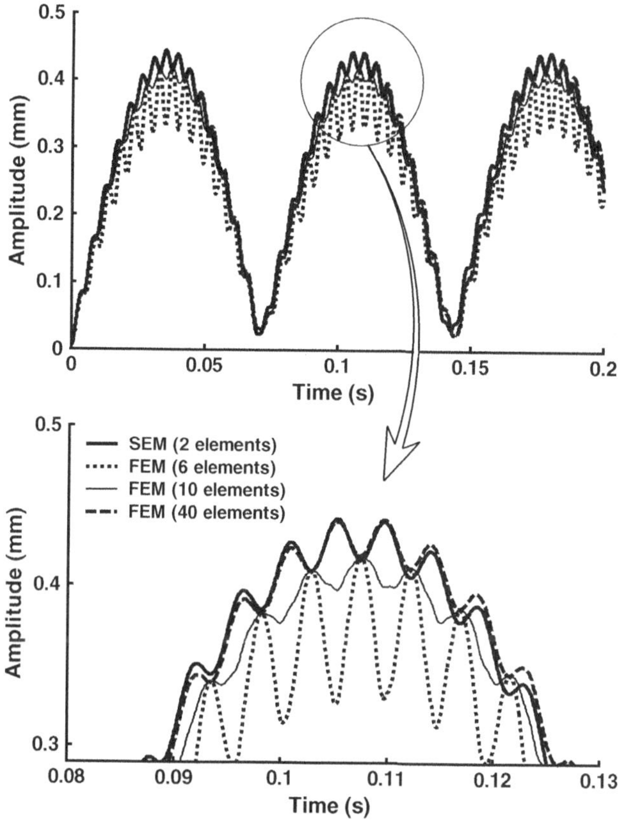

Figure 8.9 Comparison of the vibration responses obtained by SEM and FEM (at spinning speed $\Omega = 6000\,\text{rpm}$)

Appendix 8.A: Finite Element Matrices for the Transverse Bending Vibration

a. **Components of the matrix** $\boldsymbol{M}_B = \boldsymbol{M}_B^{\mathrm{T}} = [m_{Bij}]$

$$
\begin{aligned}
m_{B11} &= m_{B33} = m_{B55} = m_{B77} = 12\alpha_1(26+147r+210r^2)+36\alpha_2 \\
m_{B12} &= -m_{B34} = -m_{B56} = m_{B78} = \alpha_1 L(44+231r+315r^2)+3\alpha_2 L(1-15r) \\
m_{B15} &= m_{B37} = 36\alpha_1(3+21r+35r^2)-36\alpha_2 \\
m_{B16} &= -m_{B25} = -m_{B38} = m_{B47} = -\alpha_1 L(26+189r+315r^2)+3\alpha_2 L(1-15r) \\
m_{B22} &= m_{B44} = m_{B66} = m_{B88} = \alpha_1 L^2(8+42r+63r^2)+\alpha_2 L^2(4+15r+90r^2) \\
m_{B26} &= m_{B48} = -3\alpha_1 L^2(2+14r+21r^2)-\alpha_2 L^2(1+15r-45r^2) \\
\text{Others} &= 0 \qquad (8.\text{A}.1)
\end{aligned}
$$

where

$$\alpha_1 = \frac{R^2}{840}\rho AL, \qquad \alpha_2 = \frac{R^2}{30L}\rho I \tag{8.A.2}$$

b. **Components of the matrix $G_B = -G_B^{\mathrm{T}} = [g_{Bij}]$**

$$\begin{aligned}
g_{B13} &= -g_{B17} = g_{B35} = g_{B57} = 36\eta \\
g_{B14} &= g_{B18} = -g_{B23} = g_{B27} = g_{B36} = g_{B45} = -g_{B58} = g_{B67} = -3\eta L(1-15r) \\
g_{B24} &= g_{B68} = -\eta L^2(4+15r+90r^2) \\
g_{B28} &= -g_{B46} = \eta L^2(1+15r-45r^2) \\
&\text{Others} = 0
\end{aligned} \tag{8.A.3}$$

where

$$\eta = \frac{R^2}{30L}\rho J\Omega \tag{8.A.4}$$

c. **Components of the matrix $K_B = K_B^{\mathrm{T}} = [k_{Bij}]$**

$$\begin{aligned}
k_{B11} &= -k_{B15} = k_{B33} = -k_{B37} = k_{B55} = k_{B77} = 12\beta_1 + 540\beta_2 r^2 \\
k_{B12} &= -k_{B25} = -k_{B34} = k_{B47} = 6\beta_1 L + 270\beta_2 L r^2 \\
k_{B16} &= -k_{B38} = -k_{B56} = k_{B78} = 6\beta_1 L + 90\beta_2 L r(1+3r) \\
k_{B22} &= k_{B44} = \beta_1 L^2(4+6r+9r^2) + 135\beta_2 L^2 r^2 \\
k_{B26} &= k_{B48} = \beta_1 L^2(2-6r-9r^2) + 45\beta_2 L^2 r(1+3r) \\
k_{B66} &= k_{B88} = \beta_1 L^2(4+6r+9r^2) + \beta_2 L^2(47+210r+315r^2) \\
&\text{Others} = 0
\end{aligned} \tag{8.A.5}$$

where

$$\beta_1 = \frac{R^2}{L^3}EI, \qquad \beta_2 = \frac{R^2}{60L}\kappa GA \tag{8.A.6}$$

9

Dynamics of Multi-Layered Structures

Over the last half century, many types of multi-layered structures have been developed for various engineering applications. They include the elastic–elastic two-layer beams [53] and the passive constrained layer damping (PCLD) treated beams [9, 65, 66, 112, 122, 123, 172]. The bimetallic strip is a typical example of the elastic–elastic two-layer beam. The PCLD beam basically consists of an elastic base layer, a viscoelastic material (VEM) layer, and an elastic constraining layer and it is applied for the passive control of structural vibration and noise. For such complex multi-layer structures, more accurate analytical models and solution methods are mandatory to acquire satisfactory dynamic characteristics. In this chapter, the governing equations are derived for both an elastic–elastic two-layer beam and a PCLD beam by using Hamilton's principle. The spectral element models are then formulated by using the variational approach. The conventional finite element models are also provided to evaluate the performance of the spectral element models. The governing partial differential equations of multi-layer structures are usually fully coupled. The application of the classical modal analysis method to such coupled problems is rare in the literature. Thus, the spectral modal analysis method (SMAM) is introduced for the PCLD beam by deriving the orthogonality properties of exact damped normal modes.

9.1 Elastic–Elastic Two-Layer Beams

9.1.1 Equations of Motion

The geometry and deformations of an elastic–elastic two-layer beam of length L are shown in Figure 9.1. The following assumptions are made for the derivation of the equations of motion.

a. Geometric and material properties are uniform along the beam.
b. Perfect continuity, but no slip, at the interfaces.
c. Both layers take the same transverse displacement.

Spectral Element Method in Structural Dynamics Usik Lee

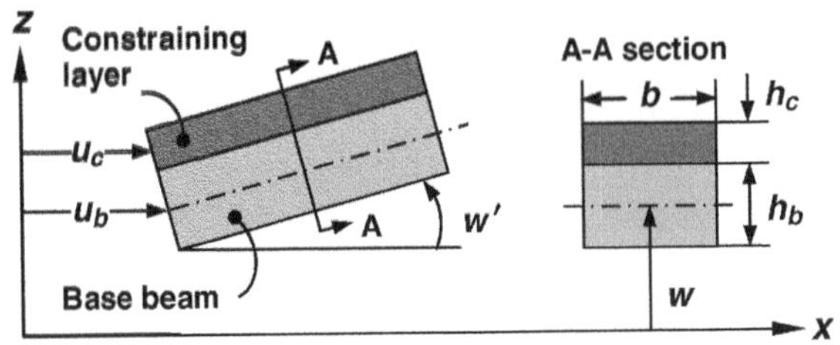

Figure 9.1 An elastic–elastic two-layer beam element

d. Small amplitude vibration.
e. Linear elasticity is valid for both layers.
f. Shear deformations and rotary inertias in both layers are negligible.

In Figure 9.1, $w(x, t)$ is the transverse displacement of the base beam, $u_b(x, t)$ is the axial displacement of the base beam, and $u_c(x, t)$ is the axial displacement of the constraining layer. h_b and h_c are the thicknesses of the base beam and constraining layer, respectively. In this chapter, the subscripts b and c are used to denote the quantities of the base beam and the constraining layer, respectively. From the perfect bonding condition, we can obtain a kinematic relationship

$$u_c = u_b - (h/2)w' \tag{9.1}$$

where

$$h = h_b + h_c \tag{9.2}$$

The prime (′) denotes the partial derivative with respect to spatial coordinate x.

The equations of motion are derived by using Hamilton's principle

$$\int_{t_1}^{t_2} (\delta T - \delta V + \delta W)\mathrm{d}t = 0 \tag{9.3}$$

where T and V are the kinetic energy and potential energy, respectively, and δW is the virtual work. The strain energy V and the kinetic energy T are given by

$$\begin{aligned} V &= \frac{1}{2}\int_0^L [E_b A_b u_b'^2 + E_b I_b w''^2 + E_c I_c w''^2 + E_c A_c u_c'^2 + F w'^2]\mathrm{d}x \\ T &= \frac{1}{2}\int_0^L [\rho_b A_b(\dot{u}_b^2 + \dot{w}^2) + \rho_c A_c(\dot{u}_c^2 + \dot{w}^2)]\mathrm{d}x \end{aligned} \tag{9.4}$$

where E, A, I and ρ (for each layer) are the Young's modulus, the cross-sectional area, the area moment of inertia about the neutral axis of each layer, and the mass density, respectively, and F is the constant axial tensile force applied to the beam. The dot (·) denotes the partial derivative with respect to time t. By eliminating u_c from Equation 9.4 by using Equation 9.1, we can write the strain and kinetic energies as

$$
\begin{aligned}
V &= \frac{1}{2}\int_0^L (EAu_b'^2 + EIw''^2 - 2\beta w'' u_b' + Fw'^2)\mathrm{d}x \\
T &= \frac{1}{2}\int_0^L (\rho A\dot{u}_b^2 + \rho A\dot{w}^2 - 2\alpha\dot{w}'\dot{u}_b + \gamma\dot{w}'^2)\mathrm{d}x
\end{aligned} \tag{9.5}
$$

where

$$
\begin{aligned}
EA &= E_bA_b + E_cA_c, \quad & EI &= E_bI_b + E_cI_c + (1/4)E_cA_ch^2 \\
\rho A &= \rho_bA_b + \rho_cA_c, \quad & \alpha &= (1/2)\rho_cA_ch \\
\beta &= (1/2)E_cA_ch, \quad & \gamma &= (1/4)\rho_cA_ch^2
\end{aligned} \tag{9.6}
$$

The virtual work δW done by the external forces, viscous damping forces, and boundary forces is given by

$$
\begin{aligned}
\delta W = &\int_0^L p(x,t)\delta w\mathrm{d}x + \int_0^L \tau(x,t)\delta u_b\mathrm{d}x - \int_0^L\int_{A_b} c_b\dot{\boldsymbol{r}}_b\cdot\delta\boldsymbol{r}_b dA_b\mathrm{d}x \\
&- \int_0^L\int_{A_c} c_c\dot{\boldsymbol{r}}_c\cdot\delta\boldsymbol{r}_c dA_c\mathrm{d}x + N_1(t)\delta u_b(0,t) + N_2(t)\delta u_b(L,t) \\
&+ M_1(t)\delta w'(0,t) + M_2(t)\delta w'(L,t) + Q_1(t)\delta w(0,t) + Q_2(t)\delta w(L,t)
\end{aligned} \tag{9.7}
$$

where

$$
\boldsymbol{r}_b = u_b\boldsymbol{n}_x + w\boldsymbol{n}_z, \quad \boldsymbol{r}_c = u_c\boldsymbol{n}_x + w\boldsymbol{n}_z \tag{9.8}
$$

and where $\boldsymbol{n}_x$ and $\boldsymbol{n}_z$ are the unit direction vectors in the x- and z-directions, respectively. N_i, Q_i, and M_i ($i=1$, 2) are the boundary forces and moments associated with u_b, w, and θ, respectively. c_b and c_c are viscous damping coefficients and $p(x, t)$ and $\tau(x, t)$ are the external forces applied along the beam. Substitute Equation 9.8 into Equation 9.7 and use Equation 9.1 to express u_c in terms of u_b. Next, apply the integral by parts to present the virtual work δW as

$$
\begin{aligned}
\delta W = &\int_0^L [p(x,t)\delta w + \tau(x,t)\delta u_b - cA\dot{u}_b\delta u_b - cA\dot{w}\delta w + c_4\dot{w}'\delta u_b \\
&+ c_4\dot{u}_b\delta w' - c_1\dot{w}'\delta w']\mathrm{d}x + N_1(t)\delta u_b(0,t) + N_2(t)\delta u_b(L,t) \\
&+ M_1(t)\delta w'(0,t) + M_2(t)\delta w'(L,t) + Q_1(t)\delta w(0,t) + Q_2(t)\delta w(L,t)
\end{aligned} \tag{9.9}
$$

where

$$
cA = c_bA_b + c_cA_c, \quad c_1 = (1/4)c_cA_ch^2, \quad c_4 = (1/2)c_cA_ch \tag{9.10}
$$

Substitution of Equations 9.5 and 9.9 into Hamilton's principle gives coupled two equations of motion

$$
\begin{aligned}
EIw'''' + \rho A\ddot{w} + cA\dot{w} &= -\alpha\ddot{u}'_b + \beta u''' + \gamma\ddot{w}'' + c_1\dot{w}'' - c_4\dot{u}'_b + Fw'' + p(x,t) \\
EAu''_b - \rho A\ddot{u}_b - cA\dot{u}_b &= -\alpha\ddot{w}' + \beta w''' - c_4\dot{w}' - \tau(x,t)
\end{aligned} \tag{9.11}
$$

and the boundary conditions

$$\begin{aligned}
Q(0,t) &= -Q_1(t) - Fw'(0,t) && \text{or} \quad w(0,t) = w_1(t) \\
M(0,t) &= -M_1(t) && \text{or} \quad w'(0,t) = \theta_1(t) \\
N(0,t) &= -N_1(t) && \text{or} \quad u_b(0,t) = u_1(t) \\
Q(L,t) &= +Q_2(t) - Fw'(L,t) && \text{or} \quad w(L,t) = w_2(t) \\
M(L,t) &= +M_2(t) && \text{or} \quad w'(L,t) = \theta_2(t) \\
N(L,t) &= +N_2(t) && \text{or} \quad u_b(L,t) = u_2(t)
\end{aligned} \tag{9.12}$$

The resultant axial force N, bending moment M, and the transverse shear force Q are related to the displacement fields as

$$\begin{aligned}
Q(x,t) &= -EI_w w''' - \alpha \ddot{u}_b + \beta u''_b + \gamma \ddot{w}' + c_1 \dot{w}' - c_4 \dot{u}_b \\
M(x,t) &= EI_w w'' - \beta u'_b \\
N(x,t) &= EAu'_b - \beta w''
\end{aligned} \tag{9.13}$$

The structural damping of the base beam layer and constraining layer can be taken into account by using the complex modulus defined by

$$E_b^* = E_b(1 + i\eta_b), \quad E_c^* = E_c(1 + i\eta_c) \tag{9.14}$$

where η_b and η_c are the loss factors of the base beam layer and the constraining layer, respectively.

9.1.2 Spectral Element Modeling

9.1.2.1 Governing Equations in the Frequency Domain

As the first step of the spectral element formulation, the governing equations are transformed into the frequency domain by using the discrete Fourier transform (DFT) theory. To this end, we represent the dynamic responses in the spectral forms as

$$\{w(x,t) \;\; u_b(x,t)\} = \frac{1}{N} \sum_{n=0}^{N-1} \{W_n(x;\omega_n) \;\; U_n(x;\omega_n)\} e^{i\omega_n t} \tag{9.15}$$

We also represent all forces and moments in the spectral forms as

$$\begin{aligned}
\{p(x,t) \;\; \tau(x,t)\} &= \frac{1}{N} \sum_{n=0}^{N-1} \{P_n(x;\omega_n) \;\; T_n(x;\omega_n)\} e^{i\omega_n t} \\
\{Q(x,t) \;\; M(x,t) \;\; N(x,t)\} &= \frac{1}{N} \sum_{n=0}^{N-1} \{Q_n(x;\omega_n) \;\; M_n(x;\omega_n) \;\; N_n(x;\omega_n)\} e^{i\omega_n t} \\
\{Q_i(x,t) \;\; M_i(x,t) \;\; N_i(x,t)\} &= \frac{1}{N} \sum_{n=0}^{N-1} \{Q_{in}(x;\omega_n) \;\; M_{in}(x;\omega_n) \;\; N_{in}(x;\omega_n)\} e^{i\omega_n t}
\end{aligned} \tag{9.16}$$

where $i = 1, 2$. By substituting Equations 9.15 and 9.16 into Equations 9.11 and 9.12, we obtain the frequency-domain expressions of the equations of motion

$$\begin{aligned} EIW'''' - \omega^2 \rho AW + i\omega cAW &= \omega^2 \alpha U' + \beta U''' - \omega^2 \gamma W'' \\ &\quad + i\omega c_1 W'' - i\omega c_4 U' + FW'' + P(x) \\ EAU'' + \omega^2 \rho AU - i\omega cAU &= \omega^2 \alpha W' + \beta W''' - i\omega c_4 W' - T(x) \end{aligned} \tag{9.17}$$

and boundary conditions

$$\begin{aligned} Q(0) &= -Q_1 - FW'(0) && \text{or} \quad W(0) = W_1 \\ M(0) &= -M_1 && \text{or} \quad W'(0) = \Theta_1 \\ N(0) &= -N_1 && \text{or} \quad U(0) = U_1 \\ Q(L) &= +Q_2 - FW'(L) && \text{or} \quad W(L) = W_2 \\ M(L) &= +M_2 && \text{or} \quad W'(L) = \Theta_2 \\ N(L) &= +N_2 && \text{or} \quad U(L) = U_2 \end{aligned} \tag{9.18}$$

Similarly, substitution of Equations 9.15 and 9.16 into the force-displacement relationships (9.13) gives

$$\begin{aligned} Q(x) &= -EIW''' + \omega^2 \alpha U + \beta U'' - \omega^2 \gamma W' + i\omega c_1 W' - i\omega c_4 U \\ M(x) &= EIW''' - \beta U' \\ N(x) &= EAU' - \beta W'' \end{aligned} \tag{9.19}$$

In above, Equations 9.17–9.19, the subscripts n are omitted for brevity.

9.1.2.2 Spectral Nodal DOFs, Forces and Moments

Figure 9.2 shows the spectral nodal degrees-of-freedom (DOFs), forces and moments. We define the spectral nodal DOFs vector as

$$\begin{aligned} \boldsymbol{d} &= \{U_1 \ W_1 \ \Theta_1 \ U_2 \ W_2 \ \Theta_2\}^{\mathrm{T}} \\ &= \{U(0) \ W(0) \ W'(0) \ U(L) \ W(L) \ W'(L)\}^{\mathrm{T}} \end{aligned} \tag{9.20}$$

and the spectral nodal forces and moments vector as

$$\begin{aligned} \boldsymbol{f}_{\mathrm{c}} &= \{N_1 \ Q_1 \ M_1 \ N_2 \ Q_2 \ M_2\}^{\mathrm{T}} \\ &= \{-N(0) \ \ -Q(0) - FW'(0) \ \ -M(0) \ \ N(L) \ \ Q(L) + FW'(L) \ \ M(L)\}^{\mathrm{T}} \end{aligned} \tag{9.21}$$

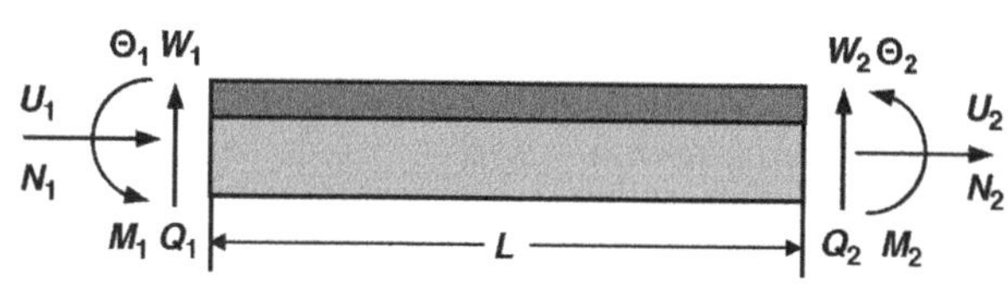

Figure 9.2 Spectral element of the elastic–elastic two-layer beam

9.1.2.3 Dynamic Shape Functions

The frequency-dependent dynamic shape functions, which are required to formulate the spectral element equation, is derived from the homogeneous governing equations reduced from Equation 9.17 by setting $P(x)=0$ and $T(x)=0$. The homogeneous governing equations are given by

$$\begin{aligned} &EIW'''' - \omega^2\rho AW + i\omega cAW \\ &\qquad = \omega^2\alpha U' + \beta U''' - \omega^2\gamma W'' + i\omega c_1 W'' - i\omega c_4 U' + FW'' \\ &EAU'' + \omega^2\rho AU - i\omega cAU = \omega^2\alpha W' + \beta W''' - i\omega c_4 W' \end{aligned} \tag{9.22}$$

The general solutions of Equation 9.22 are assumed to be

$$W(x;\omega) = Ae^{-ikx}, \quad U(x;\omega) = rAe^{-ikx} \tag{9.23}$$

Substitution of Equation 9.23 into Equation 9.22 yields an eigenvalue problem

$$\begin{bmatrix} X_{11} & X_{12} \\ X_{12} & X_{22} \end{bmatrix} \begin{Bmatrix} 1 \\ r \end{Bmatrix} = \begin{Bmatrix} 0 \\ 0 \end{Bmatrix} \tag{9.24}$$

where

$$\begin{aligned} X_{11} &= k^4 EI - \omega^2\rho A + i\omega cA - \omega^2 k^2\gamma + i\omega k^2 c_1 + k^2 F \\ X_{22} &= -k^2 EA + \omega^2\rho A - i\omega cA \\ X_{12} &= \omega k c_4 + i\omega^2 k\alpha - ik^3\beta \end{aligned} \tag{9.25}$$

The dispersion relation is obtained from Equation 9.24 as

$$a_1 k^6 + a_2 k^4 + a_3 k^2 + a_4 = 0 \tag{9.26}$$

where

$$\begin{aligned} a_1 &= \beta^2 - EAEI \\ a_2 &= \omega^2(EA\gamma + EI\rho A - 2\alpha\beta) - i\omega(EIcA + EAc_1 - 2\beta c_4) - EAF \\ a_3 &= \omega^4(\alpha^2 - \gamma\rho A) + i\omega^3(\rho A c_1 + \gamma cA - 2\alpha c_4) \\ &\quad + \omega^2(EA\rho A + cAc_1 + F\rho A - c_4^2) - i\omega cA(EA + F) \\ a_4 &= -\rho A^2\omega^4 + 2i\omega^3\rho AcA + \omega^2 cA^2 \end{aligned} \tag{9.27}$$

Six wavenumbers k_j ($j=1, 2, \ldots, 6$) can be computed from Equation 9.26. For the jth wavenumber, the corresponding r_j can be computed from Equation 9.24 as

$$r_j = \frac{k_j^2 EA - \omega^2\rho A + i\omega cA}{\omega k_j c_4 + i\omega^2 k_j\alpha - ik_j^3\beta} \quad (j = 1, 2, \ldots, 6) \tag{9.28}$$

Define a diagonal matrix as

$$\boldsymbol{R} = diag\,[r_j] \quad (j = 1, 2, \ldots, 6) \tag{9.29}$$

By using six wavenumbers k_j ($j = 1, 2, \ldots, 6$), we can write the general solutions of Equation 9.22 in the forms

$$\begin{aligned} W(x;\omega) &= \sum_{j=1}^{6} A_j \mathrm{e}^{-ik_j x} = \boldsymbol{e}(x;\omega)\boldsymbol{a} \\ U(x;\omega) &= \sum_{j=1}^{6} r_j A_j \mathrm{e}^{-ik_j x} = \boldsymbol{e}(x;\omega)\boldsymbol{R}\boldsymbol{a} \end{aligned} \tag{9.30}$$

where

$$\boldsymbol{e}(x;\omega) = [\mathrm{e}^{-ik_1 x}\ \mathrm{e}^{-ik_2 x}\ \mathrm{e}^{-ik_3 x}\ \mathrm{e}^{-ik_4 x}\ \mathrm{e}^{-ik_5 x}\ \mathrm{e}^{-ik_6 x}] \tag{9.31}$$

and

$$\boldsymbol{a} = \{A_1\ A_2\ A_3\ A_4\ A_5\ A_6\}^{\mathrm{T}} \tag{9.32}$$

Applying Equation 9.30 to the spectral nodal DOFs defined by Equation 9.20, we obtain

$$\boldsymbol{d} = \boldsymbol{H}(\omega)\boldsymbol{a} \tag{9.33}$$

where

$$\boldsymbol{H}(\omega) = \begin{bmatrix} r_1 & r_2 & r_3 & r_4 & r_5 & r_6 \\ 1 & 1 & 1 & 1 & 1 & 1 \\ -ik_1 & -ik_2 & -ik_3 & -ik_4 & -ik_5 & -ik_6 \\ r_1\mathrm{e}_1 & r_2\mathrm{e}_2 & r_3\mathrm{e}_3 & r_4\mathrm{e}_4 & r_5\mathrm{e}_5 & r_6\mathrm{e}_6 \\ e_1 & e_2 & e_3 & e_4 & e_5 & e_6 \\ -ik_1\mathrm{e}_1 & -ik_2\mathrm{e}_2 & -ik_3\mathrm{e}_3 & -ik_4\mathrm{e}_4 & -ik_5\mathrm{e}_5 & -ik_6\mathrm{e}_6 \end{bmatrix} \tag{9.34}$$

with

$$\mathrm{e}_j = \mathrm{e}^{-ik_j L} \tag{9.35}$$

The constant vector $\boldsymbol{a}$ can be eliminated from Equation 9.30 by using Equation 9.33 to write the general solutions in terms of $\boldsymbol{d}$ as follows:

$$W(x;\omega) = \boldsymbol{N}_w(x;\omega)\boldsymbol{d}, \qquad U(x;\omega) = \boldsymbol{N}_u(x;\omega)\boldsymbol{d} \tag{9.36}$$

where $\boldsymbol{N}_w$ and $\boldsymbol{N}_u$ are the dynamic shape functions defined by

$$\begin{aligned} \boldsymbol{N}_w(x;\omega) &= \boldsymbol{e}(x;\omega)\boldsymbol{H}^{-1}(\omega) \\ \boldsymbol{N}_u(x;\omega) &= \boldsymbol{e}(x;\omega)\boldsymbol{R}\boldsymbol{H}^{-1}(\omega) \end{aligned} \tag{9.37}$$

9.1.2.4 Weak Form of Governing Equations

The weak form of the governing Equation 9.17 in the frequency domain can be obtained from

$$\begin{aligned} \int_0^L &[(EIW'''' - \omega^2\rho AW + i\omega cAW - i\omega c_1 W'' + i\omega c_4 U' - \omega^2 \alpha U' \\ &- \beta U''' + \omega^2 \gamma W'' - FW'' - P)\delta W - (EAU'' + \omega^2 \rho AU \\ &- i\omega cAU - \omega^2 \alpha W' - \beta W''' + i\omega c_4 W' + T)\delta U]\mathrm{d}x = 0 \end{aligned} \tag{9.38}$$

By applying the integral by parts and using Equation 9.19 for the boundary-related terms generated by the integral by parts, we obtain the weak form as

$$\begin{aligned}
&\int_0^L [EIW''\delta W'' - \omega^2\rho AW\delta W + i\omega cAW\delta W + i\omega c_1 W'\delta W' \\
&- i\omega c_4 U\delta W' + \omega^2\alpha U\delta W' - \beta U'\delta W'' - \omega^2\gamma W'\delta W' + FW'\delta W' \\
&+ EAU'\delta U' - \omega^2\rho AU\delta U + i\omega cAU\delta U + \omega^2\alpha W'\delta U - \beta W''\delta U' \\
&- i\omega c_4 W'\delta U]\mathrm{d}x - \int_0^L [P(x)\delta W + T(x)\delta U]\mathrm{d}x \\
&- \{[Q(x) + FW']\delta W|_0^L + M(x)\delta W'|_0^L + N(x)\delta U|_0^L\} = 0
\end{aligned} \tag{9.39}$$

9.1.2.5 Spectral Element Equation

By substituting Equation 9.36 into Equation 9.39, we obtain

$$\begin{aligned}
\delta \boldsymbol{d}^{\mathrm{T}} \Bigg[&\int_0^L [EI\boldsymbol{N}''^{\mathrm{T}}_w \boldsymbol{N}''_w - \omega^2\rho A\boldsymbol{N}^{\mathrm{T}}_w\boldsymbol{N}_w + i\omega cA\boldsymbol{N}^{\mathrm{T}}_w\boldsymbol{N}_w + i\omega c_1\boldsymbol{N}'^{\mathrm{T}}_w\boldsymbol{N}'_w \\
&- i\omega c_4\boldsymbol{N}'^{\mathrm{T}}_w\boldsymbol{N}_u + \omega^2\alpha\boldsymbol{N}'^{\mathrm{T}}_w\boldsymbol{N}_u - \beta\boldsymbol{N}''^{\mathrm{T}}_w\boldsymbol{N}'_u - \omega^2\gamma\boldsymbol{N}'^{\mathrm{T}}_w\boldsymbol{N}'_w \\
&+ F\boldsymbol{N}'^{\mathrm{T}}_w\boldsymbol{N}'_w + EA\boldsymbol{N}'^{\mathrm{T}}_u\boldsymbol{N}'_u - \omega^2\rho A\boldsymbol{N}^{\mathrm{T}}_u\boldsymbol{N}_u + i\omega cA\boldsymbol{N}^{\mathrm{T}}_u\boldsymbol{N}_u \\
&+ \omega^2\alpha\boldsymbol{N}^{\mathrm{T}}_u\boldsymbol{N}'_w - \beta\boldsymbol{N}'^{\mathrm{T}}_u\boldsymbol{N}''_w - i\omega c_4\boldsymbol{N}^{\mathrm{T}}_u\boldsymbol{N}'_w]\mathrm{d}x\boldsymbol{d} - (\boldsymbol{f}_{\mathrm{c}} + \boldsymbol{f}_{\mathrm{d}})\Bigg] = 0
\end{aligned} \tag{9.40}$$

where

$$\boldsymbol{f}_{\mathrm{d}}(\omega) = \int_0^L [P(x;\omega)\boldsymbol{N}^{\mathrm{T}}_w(x;\omega) + T(x;\omega)\boldsymbol{N}^{\mathrm{T}}_u(x;\omega)]\mathrm{d}x \tag{9.41}$$

As $\delta\boldsymbol{d}$ is arbitrary by definition, we can obtain the spectral element equation from Equation 9.40 as

$$\boldsymbol{S}(\omega)\boldsymbol{d} = \boldsymbol{f}(\omega) \tag{9.42}$$

where

$$\begin{aligned}
\boldsymbol{S}(\omega) = &\int_0^L \{EA\boldsymbol{N}'^{\mathrm{T}}_u\boldsymbol{N}'_u + EI\boldsymbol{N}''^{\mathrm{T}}_w\boldsymbol{N}''_w + F\boldsymbol{N}'^{\mathrm{T}}_w\boldsymbol{N}'_w - \beta(\boldsymbol{N}''^{\mathrm{T}}_w\boldsymbol{N}'_u + \boldsymbol{N}'^{\mathrm{T}}_u\boldsymbol{N}''_w) \\
&- \omega^2[\rho A(\boldsymbol{N}^{\mathrm{T}}_w\boldsymbol{N}_w + \boldsymbol{N}^{\mathrm{T}}_u\boldsymbol{N}_u) - \alpha(\boldsymbol{N}'^{\mathrm{T}}_w\boldsymbol{N}_u + \boldsymbol{N}^{\mathrm{T}}_u\boldsymbol{N}'_w) + \gamma\boldsymbol{N}'^{\mathrm{T}}_w\boldsymbol{N}'_w] \\
&+ i\omega[cA(\boldsymbol{N}^{\mathrm{T}}_w\boldsymbol{N}_w + \boldsymbol{N}^{\mathrm{T}}_u\boldsymbol{N}_u) + c_1\boldsymbol{N}'^{\mathrm{T}}_w\boldsymbol{N}'_w - c_4(\boldsymbol{N}'^{\mathrm{T}}_w\boldsymbol{N}_u + \boldsymbol{N}^{\mathrm{T}}_u\boldsymbol{N}'_w)]\}\mathrm{d}x
\end{aligned} \tag{9.43}$$

and

$$\boldsymbol{f}(\omega) = \boldsymbol{f}_{\mathrm{c}}(\omega) + \boldsymbol{f}_{\mathrm{d}}(\omega) \tag{9.44}$$

Substitution of the dynamic shape functions (9.37) into Equation 9.43 gives

$$\boldsymbol{S}(\omega) = \boldsymbol{H}^{-\mathrm{T}}(\omega)\boldsymbol{D}(\omega)\boldsymbol{H}^{-1}(\omega) \tag{9.45}$$

where

$$\begin{aligned}
\boldsymbol{D}(\omega) = &-EA\boldsymbol{RKEKR} + EI\boldsymbol{K}^2\boldsymbol{EK}^2 - i\beta(\boldsymbol{K}^2\boldsymbol{EKR} + \boldsymbol{RKEK}^2) \\
&- \omega^2[\rho A(\boldsymbol{E} + \boldsymbol{RER}) + i\alpha(\boldsymbol{KER} + \boldsymbol{REK}) - \gamma\boldsymbol{KEK}] \\
&+ i\omega[cA(\boldsymbol{E} + \boldsymbol{RER}) - c_1\boldsymbol{KEK} + ic_4(\boldsymbol{KER} + \boldsymbol{REK})] - F\boldsymbol{RER}
\end{aligned} \tag{9.46}$$

and where

$$\boldsymbol{K} = diag[k_i] = \boldsymbol{K}^{\mathrm{T}}, \quad \boldsymbol{K}^2 = diag\,[k_i^2] = \boldsymbol{K}^{2\mathrm{T}} \quad (j = 1, 2, \ldots, 6) \tag{9.47}$$

$$\boldsymbol{E}(\omega) = \int_0^L \boldsymbol{e}^{\mathrm{T}}(x;\omega)\boldsymbol{e}(x;\omega)\mathrm{d}x = [E_{pq}(\omega)] \quad (p, q = 1, 2, \ldots, 6) \tag{9.48}$$

with

$$E_{pq} = \begin{cases} \dfrac{i}{k_p + k_q}\left[\mathrm{e}^{-i(k_p+k_q)L} - 1\right] & \text{if } k_p + k_q \neq 0 \\ L & \text{if } k_p + k_q = 0 \end{cases} \tag{9.49}$$

The spectral elements represented by Equation 9.42 can be assembled in an analogous way to that used for the conventional finite element method. After imposing the boundary conditions, the assembled global system equation can be obtained in the form

$$\boldsymbol{S}_{\mathrm{g}}(\omega)\boldsymbol{d}_{\mathrm{g}} = \boldsymbol{f}_{\mathrm{g}}(\omega) \tag{9.50}$$

where $\boldsymbol{S}_{\mathrm{g}}(\omega)$ is the global dynamic stiffness matrix, $\boldsymbol{d}_{\mathrm{g}}$ is the global spectral DOFs vector, and $\boldsymbol{f}_{\mathrm{g}}$ is the global spectral nodal forces vector.

9.1.3 Spectral Modal Analysis

For the modal analysis in the frequency domain (spectral modal analysis), the eigenfrequencies and eigenfuncions must be obtained in advance. The eigensolutions can be obtained from the eigenvalue problem given by

$$\boldsymbol{S}_{\mathrm{g}}(\omega)\boldsymbol{d}_{\mathrm{g}} = \boldsymbol{0} \tag{9.51}$$

a. **Eigenfrequencies:** (damped natural frequencies) ω_p $(p = 1, 2, \ldots, \infty)$ are obtained by numerically computing the roots of the determinant of $\boldsymbol{S}_g(\omega)$.
b. **Wavenumbers:** k_i $(i = 1, 2, \ldots, 6)$ for each eigenfrequency are computed from the dispersion relation (9.26).
c. **Eigenvectors:** (spectral nodal DOFs vector) $\boldsymbol{d}_{gp}$ $(p = 1, 2, \ldots, \infty)$ corresponding to the eigenfrequencies ω_p are obtained so as to satisfy Equation 9.51.
d. **Eigenfunctions:** (damped normal modes) W_p and U_p are finally computed from Equation 9.36 by using the six wave numbers k_i and eigenvectors $\boldsymbol{d}_{gp}$.

9.1.3.1 Orthogonality Properties of Normal Modes

For the modal analysis, the orthogonality properties of normal modes are required. Let ω_p be the pth (damped) natural frequency and W_p and U_p the corresponding (damped) normal modes of the bending and axial vibrations, respectively. The pth eigensolutions must satisfy

$$\begin{aligned} &EIW_p'''' - \omega_p^2 \rho A W_p + i\omega_p cAW_p \\ &\qquad = \omega_p^2(\alpha U_p' - \gamma W_p'') + i\omega_p(c_1 W_p'' - c_4 U_p') + \beta U_p''' + FW_p'' \\ &EAU_p'' + \omega_p^2 \rho A U_p - i\omega_p cAU_p = \omega_p^2 \alpha W'_p - i\omega_p c_4 W'_p + \beta W_p''' \end{aligned} \tag{9.52}$$

By using Equation 9.19, Equation 9.52 can be written in the simple forms of

$$\begin{aligned} \omega_p^2(\rho A W_p + \alpha U_p' - \gamma W_p'') + i\omega_p(c_1 W_p'' - c_4 U_p' - cAW_p) + FW_p'' &= M_p'' \\ \omega_p^2(\rho A U_p - \alpha W_p') + i\omega_p(c_4 W'_p - cAU_p) &= -N'_p \end{aligned} \tag{9.53}$$

We multiply the first equation from Equation 9.53 by W_q and the second equation by U_q and then integrate from $x=0$ to $x=L$, followed by integrating by parts. Finally, summing the final forms of the two equations gives

$$\begin{aligned} &\omega_p^2 \int_0^L [\rho A(W_p W_q + U_p U_q) - \alpha(U_p W_q' + U_q W_p') + \gamma W_p' W_q']\,\mathrm{d}x \\ &\quad - i\omega_p \int_0^L [cA(W_p W_q + U_p U_q) + c_1 W_p' W_q' - c_4(U_p W_q' + U_q W_p')]\,\mathrm{d}x \\ &= -N_p U_q|_0^L - (Q_p + FW'_p)W_q|_0^L - M_p W_q'|_0^L + \int_0^L N_p U_q'\,\mathrm{d}x \\ &\quad + \int_0^L M_p W_q''\,\mathrm{d}x + \int_0^L FW_p' W'_q\,\mathrm{d}x \end{aligned} \tag{9.54}$$

Similarly, for the qth eigensolutions, we obtain

$$\begin{aligned} &\omega_q^2 \int_0^L [\rho A(W_p W_q + U_p U_q) - \alpha(U_p W_q' + U_q W_p') + \gamma W_p' W_q']\mathrm{d}x \\ &\quad - i\omega_q \int_0^L [cA(W_p W_q + U_p U_q) + c_1 W_p' W'_q - c_4(U_p W_q' + U_q W_p')]\mathrm{d}x \\ &= -N_q U_p|_0^L - (Q_q + FW'_q)W_p|_0^L - M_q W_p'|_0^L + \int_0^L N_q U_p'\mathrm{d}x \\ &\quad + \int_0^L M_q W_p''\mathrm{d}x + \int_0^L FW_q' W_p'\mathrm{d}x \end{aligned} \tag{9.55}$$

Subtracting Equation 9.55 from Equation 9.54, applying homogenous boundary conditions and accomplishing a further manipulation, we obtain

$$\begin{aligned} &(\omega_p^2 - \omega_q^2) \int_0^L [\rho A(W_p W_q + U_p U_q) + \rho I \Psi_p \Psi_q - \alpha(U_p W_q' + U_q W_p') \\ &\quad + \gamma W_p' W_q']\mathrm{d}x - i(\omega_p - \omega_q) \int_0^L [cA(W_p W_q + U_p U_q) + c_1 W_p' W_q' \\ &\quad - c_4(U_p W_q' + U_q W_p')]\mathrm{d}x = 0 \end{aligned} \tag{9.56}$$

From Equation 9.56, we obtain the orthogonality properties as follows:

$$\begin{aligned} \int_0^L [\rho A(W_p W_q + U_p U_q) - \alpha(U_p W_q' + U_q W_p') + \gamma W_p' W_q']\mathrm{d}x &= m_p \delta_{pq} \\ \int_0^L [cA(W_p W_q + U_p U_q) + c_1 W_p' W_q' - c_4(U_p W_q' + U_q W_p')]\mathrm{d}x &= c_p \delta_{pq} \end{aligned} \tag{9.57}$$

where m_p and c_p are the pth modal mass and modal damping coefficient defined by

$$\begin{aligned} m_p &= \int_0^L [\rho A(U_p^2 + W_p^2) - 2\alpha U_p W_p' + \gamma {W'}_q^2]\mathrm{d}x \\ c_p &= \int_0^L [cA(U_p^2 + W_p^2) + c_1 {W'}_p^2 - 2c_4 U_p W_p']\mathrm{d}x \end{aligned} \tag{9.58}$$

By applying the above orthogonality properties to Equation 9.54, we can also obtain

$$\begin{aligned}
&-N_q U_q\big|_0^L - (Q_q + FW_q')W_q\big|_0^L - M_q\, W_q'\big|_0^L + \int_0^L N_q U_q'\, dx + \int_0^L M_q W_q''\, \mathrm{d}x \\
&\quad + \int_0^L FW_q'\, W_q' \mathrm{d}x = \omega_q^2 m_q - i\omega_q c_q \\
&-N_p U_q\big|_0^L - (Q_p + FW_p')\, W_q\big|_0^L - M_p\, W_q'\big|_0^L + \int_0^L N_p U_q' \mathrm{d}x \\
&\quad + \int_0^L M_p\, W_q'' dx + \int_0^L FW_p'\, W_q' \mathrm{d}x = 0
\end{aligned} \tag{9.59}$$

9.1.3.2 Spectral Modal Equations

Once the orthogonality properties have been derived, the forced vibration responses can be obtained by using the modal analysis method. The forced vibration responses are represented by superposing the normal modes as

$$w(x,t) = \sum_{p=1}^{\infty} W_p(x) q_p(t), \qquad u(x,t) = \sum_{p=1}^{\infty} U_p(x) q_p(t) \tag{9.60}$$

where $q_p(t)$ are the modal coordinates. The modal coordinates are written in the spectral form of

$$q_p(t) = \frac{1}{N}\sum_{n=0}^{N-1} \bar{q}_{pn} \mathrm{e}^{i\omega_n t} \tag{9.61}$$

Substitution of Equation 9.61 into Equation 9.60 gives

$$w(x,t) = \frac{1}{N}\sum_{p=1}^{\infty}\sum_{n=0}^{N-1} W_p(x) q_{pn} \mathrm{e}^{i\omega_n t}, \qquad u(x,t) = \frac{1}{N}\sum_{p=1}^{\infty}\sum_{n=0}^{N-1} U_p(x) q_{pn} \mathrm{e}^{i\omega_n t} \tag{9.62}$$

The external forces are also written in the spectral forms as

$$p(x,t) = \frac{1}{N}\sum_{n=0}^{N-1} P_n(x) \mathrm{e}^{i\omega_n t}, \qquad \tau(x,t) = \frac{1}{N}\sum_{n=0}^{N-1} T_n(x) \mathrm{e}^{i\omega_n t} \tag{9.63}$$

Substituting Equations 9.62 and 9.63 into the time-domain governing equations (9.11), we obtain

$$\begin{aligned}
&\sum_{p=1}^{\infty}(M_p'' - FW_p'')q_{pn} - \omega_n^2 \sum_{p=1}^{\infty}(\rho A W_p + \alpha U_p' - \gamma W_p'')q_{pn} \\
&\qquad + i\omega_n \sum_{p=1}^{\infty}(cAW_p - c_1 W_p'' + c_4 U_p')q_{pn} = P_n(x) \\
&\sum_{p=1}^{\infty} N_p' q_{pn} + \omega_n^2 \sum_{p=1}^{\infty}(\rho A U_p - \alpha W_p')q_{pn} \\
&\qquad - i\omega_n \sum_{p=1}^{\infty}(cAU_p - c_4 W_p')q_{pn} = -T_n(x)
\end{aligned} \tag{9.64}$$

where Equation 9.19 is used for M_p and N_p. Multiply the first equation of Equation 9.64 by W_q and the second equation by U_q and then integrate from $x = 0$ to $x = L$, followed by integrating by parts. Finally, summing the two equations gives

$$
\begin{aligned}
&\omega_n^2 \sum_{p=1}^{\infty} q_{pn} \int_0^L [\rho A(W_p W_q + U_p U_q) - \alpha U_p W_q' + \gamma W_p' W_q' - \alpha W_p' U_q] \mathrm{d}x \\
&\quad - i\omega_n \sum_{p=1}^{\infty} q_{pn} \int_0^L [cA(W_p W_q + U_p U_q) + c_1 W_p' W_q' - c_4 U_p W_q' - c_4 W_p' U_q] \mathrm{d}x \\
&= \sum_{p=1}^{\infty} q_{pn} [-(Q_p + F W_p') W_q |_0^L - M_p W_q' |_0^L - N_p U_q |_0^L \\
&\quad + \int_0^L (M_p W_q'' + F W_p' W_q' + N_p U_q') \mathrm{d}x] - \int_0^L P_n(x) W_q \mathrm{d}x - \int_0^L T_n(x) U_q \mathrm{d}x
\end{aligned}
\tag{9.65}
$$

where Equation 9.19 is used for Q_p.

Applying the orthogonality properties (9.57) through (9.59) to Equation 9.65, we obtain the modal equations in the frequency domain as

$$
(\omega_q^2 - \omega_n^2) m_q q_{qn} - i(\omega_q - \omega_n) c_q q_{qn} = f_{qn} \quad (q = 1, 2, 3, \ldots) \tag{9.66}
$$

where f_{qn} is the spectral modal force defined by

$$
f_{qn} = \int_0^L [P_n(x) W_q + T_n(x) U_q] \mathrm{d}x \quad (q = 1, 2, 3, \ldots) \tag{9.67}
$$

Equation 9.66 can be readily transformed into the time domain as

$$
m_q \ddot{q}_q(t) + c_q \dot{q}_q(t) + (\omega_q^2 m_q - i\omega_q c_q) q_q(t) = f_q(t) \tag{9.68}
$$

where f_q is the modal force given by

$$
f_q(t) = \int_0^L [p(x,t)\, W_q + \tau(x,t)\, U_q] \mathrm{d}x \tag{9.69}
$$

9.1.4 Finite Element Model

The weak form of the equations of motion (9.11) is obtained as

$$
\begin{aligned}
&\int_0^L [EI w'' \delta w'' + \rho A \ddot{w} \delta w + cA \dot{w} \delta w - \alpha \ddot{u}_b \delta w' - \beta u'_b \delta w'' + \gamma \ddot{w}' \delta w' \\
&\quad + c_1 \dot{w}' \delta w' - c_4 \dot{u}_b \delta w' + F w' \delta w' + EA u'_b \delta u'_b + \rho A \ddot{u}_b \delta u_b + cA \dot{u}_b \delta u_b \\
&\quad - \alpha \ddot{w}' \delta u_b - \beta w'' \delta u'_b - c_4 \dot{w}' \delta u'_b] \mathrm{d}x - \int_0^L [p(x,t) \delta w + \tau(x,t) \delta u_b] \mathrm{d}x \\
&\quad - \{N(x,t) \delta u_b |_0^L + M(x,t) \delta w' |_0^L + [Q(x,t) + F w'] \delta w |_0^L\} = 0
\end{aligned}
\tag{9.70}
$$

The sign conventions of the nodal DOFs, forces, and moments defined in Figure 9.2 are also valid in the time domain and we define the nodal DOFs vector as

$$
\boldsymbol{d}(t) = \{U_1(t) \;\; W_1(t) \;\; \Theta_1(t) \;\; U_2(t) \;\; W_2(t) \;\; \Theta_2(t)\}^{\mathrm{T}} \tag{9.71}
$$

and the nodal forces and moments vector by

$$\begin{aligned} f_c(t) &= \{N_1(t) \quad Q_1(t) \quad M_1(t) \quad N_2(t) \quad Q_2(t) \quad M_2(t)\}^T \\ &= \{-N(0,t) \quad -Q(0,t) - Fw'(0,t) \quad -M(0,t) \\ &\qquad N(L,t) \quad Q(L,t) + Fw'(L,t) \quad M(L,t)\}^T \end{aligned} \tag{9.72}$$

The displacement fields are assumed to be

$$u_b(x,t) = \boldsymbol{N}_u(x)\boldsymbol{d}(t), \quad w(x,t) = \boldsymbol{N}_w(x)\boldsymbol{d}(t) \tag{9.73}$$

where

$$\begin{aligned} \boldsymbol{N}_u(x) &= [N_{u1}(x) \;\; 0 \;\; 0 \;\; N_{u2}(x) \;\; 0 \;\; 0\,] \\ \boldsymbol{N}_w(x) &= [0 \;\; N_{w1}(x) \;\; N_{\theta1}(x) \;\; 0 \;\; N_{w2}(x) \;\; N_{\theta2}(x)] \end{aligned} \tag{9.74}$$

with

$$\begin{aligned} N_{u1}(x) &= 1-(x/L), & N_{u2}(x) &= x/L \\ N_{w1}(x) &= 1-3(x/L)^2+2(x/L)^3, & N_{w2}(x) &= 3(x/L)^2-2(x/L)^3 \\ N_{\theta1}(x) &= x[1-2(x/L)+(x/L)^2], & N_{\theta2}(x) &= -x[(x/L)-(x/L)^2] \end{aligned} \tag{9.75}$$

Substitution of Equation (9.73) into the weak form (9.70) gives the finite element equation as

$$\boldsymbol{M}\ddot{\boldsymbol{d}}(t) + \boldsymbol{C}\dot{\boldsymbol{d}}(t) + \boldsymbol{K}\boldsymbol{d}(t) = \boldsymbol{f}_c(t) + \boldsymbol{f}_d(t) \tag{9.76}$$

where

$$\begin{aligned} \boldsymbol{M} &= \int_0^L [\rho A(\boldsymbol{N}_w^T\boldsymbol{N}_w + \boldsymbol{N}_u^T\boldsymbol{N}_u) - \alpha(\boldsymbol{N}_w'^T\boldsymbol{N}_u + \boldsymbol{N}_u^T\boldsymbol{N}_w') + \gamma\boldsymbol{N}_w'^T\boldsymbol{N}_w']dx \\ \boldsymbol{C} &= \int_0^L [cA(\boldsymbol{N}_w^T\boldsymbol{N}_w + \boldsymbol{N}_u^T\boldsymbol{N}_u) + c_1\boldsymbol{N}_w'^T\boldsymbol{N}_w' - c_4(\boldsymbol{N}_w'^T\boldsymbol{N}_u + \boldsymbol{N}_u'^T\boldsymbol{N}_w')]dx \\ \boldsymbol{K} &= \int_0^L [EA\boldsymbol{N}_u'^T\boldsymbol{N}_u' + EI\boldsymbol{N}_w''^T\boldsymbol{N}_w'' - \beta(\boldsymbol{N}_w''^T\boldsymbol{N}'_u + \boldsymbol{N}'_u{}^T\boldsymbol{N}_w'') + F\boldsymbol{N}_w'^T\boldsymbol{N}_w']dx \end{aligned} \tag{9.77}$$

and

$$\boldsymbol{f}_d(t) = \int_0^L [p(x,t)\boldsymbol{N}_w^T(x) + \tau(x,t)\boldsymbol{N}_u^T(x)]dx \tag{9.78}$$

The finite element matrices $\boldsymbol{M}$, $\boldsymbol{C}$, and $\boldsymbol{K}$ are given in Appendix 9.A.

Example 9.1

Consider a cantilevered elastic–elastic two-layer beam as shown in Figure 9.3. The base beam is fully covered with a uniform elastic constraining layer. The geometric and material properties are given in Table 9.1. The high accuracy of the present spectral element model is shown in Table 9.2 by comparing the natural frequencies obtained by SEM with those obtained by FEM through varying the total number of finite elements used in the analysis. Figure 9.4 shows the natural frequencies as a function of the ratio of the constraining layer thickness h_c to a fixed base beam thickness $h_b = 5$ mm.

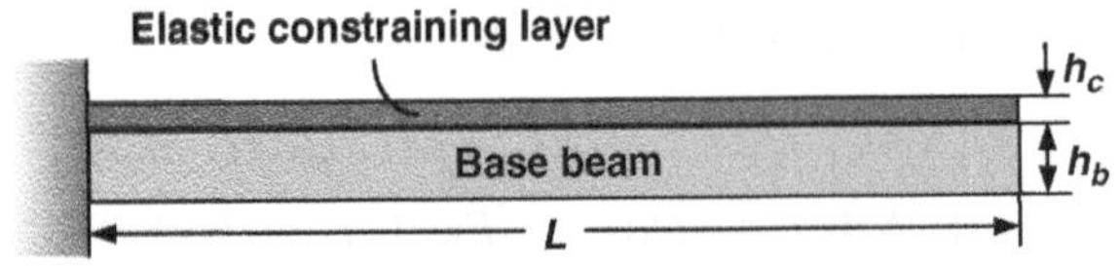

Figure 9.3 An elastic–elastic two-layer beam

Table 9.1 Geometric and material properties of the elastic–elastic two-layer beam

L	300 mm	b	20 mm	h_b	5 mm	h_c	0.75 mm
E_b	70 GPa	E_c	200 GPa	ρ_b	2700 kg/m^3	ρ_c	7850 kg/m^3

Table 9.2 Natural frequencies (Hz) of an elastic–elastic two layer beam

	SEM(*N*)	FEM(*N*)				
Mode	$N=1$	$N=5$	$N=10$	$N=20$	$N=40$	$N=100$
1st	56.5538	56.6368	56.5745	56.5590	56.5551	56.5540
2nd	354.321	358.060	355.230	354.547	354.378	354.330
3rd	991.686	1018.43	997.855	993.191	992.054	991.738
4th	1942.06	2048.77	1965.02	1947.57	1943.42	1942.28
5th	3207.71	3527.25	3270.37	3222.38	3211.32	3208.29
10th	11385.70	13097.98	12292.17	11570.39	11429.25	11392.55

Note: (*N*) = total number of the equal-length finite elements used in the analysis.

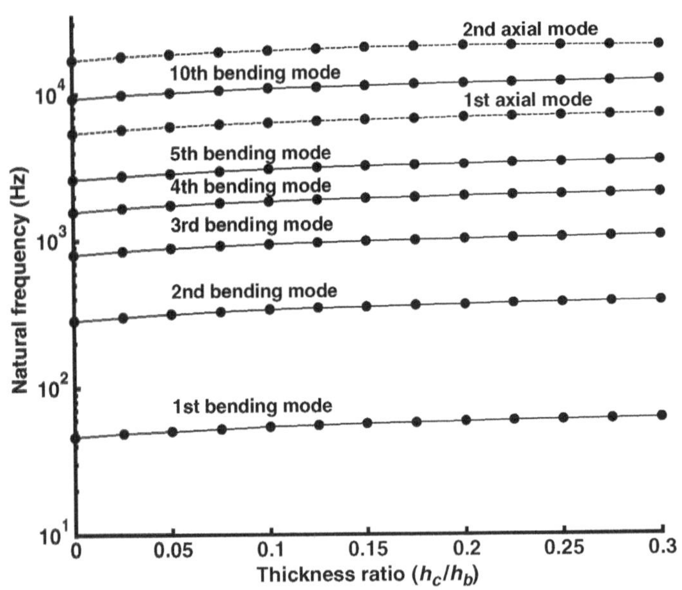

Figure 9.4 Thickness ratio dependence of the natural frequencies

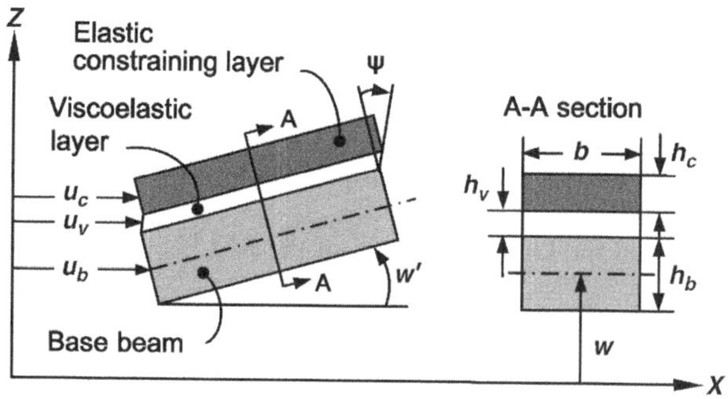

Figure 9.5 An elastic–VEM–elastic three-layer (PCLD) beam element

9.2 Elastic–Viscoelastic–elastic Three-Layer (PCLD) Beams

9.2.1 Equations of Motion

The passive constrained layered damping (PCLD) treated beams consist of three layers: an elastic base beam, a viscoelastic material (VEM) layer, and an elastic constraining layer. The geometry and deformations of a PCLD beam are shown in Figure 9.5. $w(x, t)$ is the transverse displacement of the base beam, $u_b(x, t)$ is the axial displacement of the base beam, $u_v(x, t)$ is the axial displacement of the VEM layer, and $u_c(x, t)$ is the axial displacement of the constraining layer. h_b, h_v, and h_c are the thicknesses of the base beam, VEM layer, and constraining layer, respectively, and ψ is the shear angle of the VEM layer. The subscripts b, v, and c are used to denote the quantities of the base beam, VEM layer, and constraining layer, respectively. The following assumptions are made for the derivation of the equations of motion [66].

a. Geometric and material properties are uniform along the beam.
b. Perfect continuity, but no slip, at the interfaces.
c. All layers take the same transverse displacement.
d. Small amplitude vibration.
e. Linear elasticity is valid for the base beam and constraining layer.
f. Linear viscoelasticity is valid for the VEM layer.
g. Shear deformations and rotary inertias in the base beam and constraining layer are negligible.

Based on the above assumption, we obtain the kinematics relationships from Figure 9.5 as

$$\begin{aligned} u_v &= u_b - (h_1/2)w' + (h_v/2)\psi \\ u_c &= u_b - (h_2/2)w' + h_v\psi \end{aligned} \tag{9.79}$$

where

$$h_1 = h_b + h_v, \qquad h_2 = h_b + 2h_v + h_c \tag{9.80}$$

The constitutive relationship for the one-dimensional viscoelastic materials can be represented in the following Stieltjes integral form [34]

$$\tau(x, t) = g \circ \psi = \int_{-\infty}^{t} g(t-\tau)\dot{\psi}(x, \tau)\mathrm{d}\tau \tag{9.81}$$

where τ is the shear stress and $g(t)$ is the relaxation function of VEM (the stress response to a unit-step strain input). This stress relaxation represents the energy loss from viscoelastic materials, hence damping. Equation 9.81 can be transformed in the frequency domain by using the DFT as follows:

$$DFT[g \circ \psi] = G(\omega)(i\omega)\Psi(\omega) \equiv G^*(\omega)\Psi(\omega) \tag{9.82}$$

where $G(\omega)$ and $\Psi(\omega)$ are the spectral components of $g(t)$ and $\psi(t)$, respectively, and $G^*(\omega) = i\omega G(\omega)$.

The strain energy V and the kinetic energy T of the PCLD beam are given by

$$\begin{aligned} V &= \frac{1}{2}\int_0^L [E_b A_b u_b'^2 + E_b I_b w''^2 + E_v A_v u_v'^2 + E_v I_v (w'' - \psi')^2 + A_v (g \circ \psi)\psi \\ &\quad + E_c I_c w''^2 + E_c A_c u_c'^2 + F w'^2] \mathrm{d}x \\ T &= \frac{1}{2}\int_0^L [\rho_b A_b (\dot{u}_b^2 + \dot{w}^2) + \rho_v A_v (\dot{u}_v^2 + \dot{w}^2) + \rho_v I_v (\dot{w}' - \dot{\psi})^2 + \rho_c A_c (\dot{u}_c^2 + \dot{w}^2)] \mathrm{d}x \end{aligned} \tag{9.83}$$

where E, I, A, and ρ (for each layer) are the Young's modulus, area moment of inertia about the neutral axis, cross-sectional area, and the mass density, respectively. By eliminating u_v and u_c from Equation 9.83 by using Equation 9.79, the strain and kinetic energies can be written as

$$\begin{aligned} V &= \frac{1}{2}\int_0^L [EAu_b'^2 + EI_w w''^2 + EI_\psi \psi'^2 - 2\beta w'' u_b' + 2\varepsilon_4 u'_b \psi' - 2\varepsilon_2 w'' \psi' \\ &\quad + A_v (g \circ \psi)\psi + F w'^2] \mathrm{d}x \\ T &= \frac{1}{2}\int_0^L \{\rho A \dot{u}_b^2 + \rho A \dot{w}^2 + \rho I \dot{\psi}^2 - 2\alpha \dot{w}' \dot{u}_b + 2\varepsilon_3 \dot{u}_b \dot{\psi} - 2\varepsilon_1 \dot{w}' \dot{\psi} + \gamma \dot{w}'^2\} \mathrm{d}x \end{aligned} \tag{9.84}$$

where

$$\begin{aligned} EA &= E_b A_b + E_v A_v + E_c A_c \\ EI_w &= E_b I_b + E_v I_v + E_c I_c + (1/4) E_v A_v h_1^2 + (1/4) E_c A_c h_2^2 \\ EI_\psi &= E_v I_v + (1/4) E_v A_v h_v^2 + E_c A_c h_v^2 \\ \rho A &= \rho_b A_b + \rho_v A_v + \rho_c A_c \\ \rho I &= \rho_v I_v + (1/4)\rho_v A_v h_v^2 + \rho_c A_c h_v^2 \\ \alpha &= (1/2)(\rho_v A_v h_1 + \rho_c A_c h_2) \\ \beta &= (1/2)(E_v A_v h_1 + E_c A_c h_2) \\ \gamma &= \rho_v I_v + (1/4)\rho_v A_v h_1^2 + (1/4)\rho_c A_c h_2^2 \\ \varepsilon_1 &= \rho_v I_v + (1/4)\rho_v A_v h_1 h_v + (1/2)\rho_c A_c h_2 h_v \\ \varepsilon_2 &= E_v I_v + (1/4) E_v A_v h_1 h_v + (1/2) E_c A_c h_2 h_v \\ \varepsilon_3 &= (1/2)\rho_v A_v h_v + \rho_c A_c h_v \\ \varepsilon_4 &= (1/2) E_v A_v h_v + E_c A_c h_v \end{aligned} \tag{9.85}$$

The virtual work δW done by the external forces, viscous damping forces, and boundary forces is given by

$$\begin{aligned} \delta W &= \int_0^L p(x,t)\delta w \mathrm{d}x + \int_0^L \tau(x,t)\delta u_b \mathrm{d}x + \int_0^L \gamma(x,t)\delta\psi \mathrm{d}x \\ &\quad - \int_0^L \int_{A_b} c_b \dot{\mathbf{r}}_b \cdot \delta \mathbf{r}_b \mathrm{d}A_b \mathrm{d}x - \int_0^L \int_{A_v} c_v \dot{\mathbf{r}}_v \cdot \delta \mathbf{r}_v \mathrm{d}A_v \mathrm{d}x - \int_0^L \int_{A_c} c_c \dot{\mathbf{r}}_c \cdot \delta \mathbf{r}_c \mathrm{d}A_c \mathrm{d}x \\ &\quad + N_1(t)\,\delta u_b(0,t) + N_2(t)\,\delta u_b(L,t) + M_1(t)\,\delta w'(0,t) + M_2(t)\,\delta w'(L,t) \\ &\quad + Q_1(t)\,\delta w(0,t) + Q_2(t)\,\delta w(L,t) + R_1(t)\,\delta\psi(0,t) + R_2(t)\,\delta\psi(L,t) \end{aligned} \tag{9.86}$$

with

$$\begin{aligned} \boldsymbol{r}_b &= u_b \boldsymbol{n}_x + w \boldsymbol{n}_z \\ \boldsymbol{r}_v &= [u_v - z(w' - \psi)]\boldsymbol{n}_x + w \boldsymbol{n}_z \\ \boldsymbol{r}_c &= u_c \boldsymbol{n}_x + w \boldsymbol{n}_z \end{aligned} \tag{9.87}$$

where $\boldsymbol{n}_x$ and $\boldsymbol{n}_z$ are the unit direction vectors in the x- and z-directions, respectively. In Equation 9.86, N_i, Q_i, M_i, and R_i $(i = 1, 2)$ are the boundary forces and moments associated with u_b, w, θ, and ψ, respectively. c_b, c_v, and c_c are viscous damping coefficients and $p(x, t)$, $\tau(x, t)$ and $r(x, t)$ are the external forces applied along the beam. Substitute Equation 9.87 into Equation 9.86 and then use Equation 9.79 to express u_v and u_c in terms of u_b. Next, apply the integral by parts to write Equation 9.86 in the form

$$\begin{aligned} \delta W = \int_0^L & [p(x,t)\delta w + \tau(x,t)\delta u_b + \gamma(x,t)\delta\psi \\ & - cA\dot{u}_b\delta u_b - cA\dot{w}\delta w + c_4\dot{w}'\delta u_b - c_5\dot{\psi}\delta u_b - c_5\dot{u}_b\delta\psi \\ & + c_3\dot{w}'\delta\psi - c_2\dot{\psi}\delta\psi + c_4\dot{u}_b\delta w' - c_1\dot{w}'\delta w' + c_3\dot{\psi}\delta w']\mathrm{d}x \\ & + M_1(t)\delta w'(0,t) + M_2(t)\delta w'(L,t) + Q_1(t)\delta w(0,t) \\ & + Q_2(t)\delta w(L,t) + N_1(t)\delta u_b(0,t) + N_2(t)\delta u_b(L,t) \\ & + R_1(t)\delta\psi(0,t) + R_2(t)\delta\psi(L,t) \end{aligned} \tag{9.88}$$

where

$$\begin{aligned} cA &= c_b A_b + c_v A_v + c_c A_c \\ c_1 &= c_v I_v + (1/4)(c_v A_v h_1^2 + c_c A_c h_2^2) \\ c_2 &= c_v I_v + (1/4)c_v A_v h_v^2 + c_c A_c h_v^2 \\ c_3 &= c_v I_v + (1/4)c_v A_v h_1 h_v + (1/2)c_c A_c h_2 h_v \\ c_4 &= (1/2)(c_v A_v h_1 + c_c A_c h_2) \\ c_5 &= (1/2)c_v A_v h_v + c_c A_c h_v \end{aligned} \tag{9.89}$$

Finally, by substituting Equations 9.84 and 9.88 into Hamilton's principle (9.3), we obtain a set of coupled equations of motion as follows

$$\begin{aligned} EI_w w'''' + \rho A\ddot{w} + cA\dot{w} &= -\alpha\ddot{u}_b' + \beta u''' + \gamma\ddot{w}'' \\ &\quad - \varepsilon_1\ddot{\psi}_b' + \varepsilon_2\psi''' + c_1\dot{w}'' - c_4\dot{u}_b' - c_3\dot{\psi}' + Fw'' + p(x,t) \\ EAu''_b - \rho A\ddot{u}_b - cA\dot{u}_b &= -\alpha\ddot{w}' + \beta w''' + \varepsilon_3\ddot{\psi} - \varepsilon_4\psi'' - c_4\dot{w}' + c_5\dot{\psi} - \tau(x,t) \\ EI_\psi\psi'' - \rho I\ddot{\psi} - c_2\dot{\psi} &= -\varepsilon_1\ddot{w}' + \varepsilon_2 w''' \\ &\quad + \varepsilon_3\ddot{u}_b - \varepsilon_4 u_b'' - c_3\dot{w}' + c_5\dot{u}_b + A_v(g\circ\psi) - \gamma(x,t) \end{aligned} \tag{9.90}$$

and the boundary conditions

$$\begin{aligned} Q(0,t) &= -Q_1(t) - Fw'(0,t) & \text{or} \quad w(0,t) &= w_1(t) \\ M(0,t) &= -M_1(t) & \text{or} \quad w'(0,t) &= \theta_1(t) \\ N(0,t) &= -N_1(t) & \text{or} \quad u_b(0,t) &= u_1(t) \\ R(0,t) &= -R_1(t) & \text{or} \quad \psi(0,t) &= \psi_1(t) \\ Q(L,t) &= +Q_2(t) - Fw'(L,t) & \text{or} \quad w(L,t) &= w_2(t) \\ M(L,t) &= +M_2(t) & \text{or} \quad w'(L,t) &= \theta_2(t) \\ N(L,t) &= +N_2(t) & \text{or} \quad u_b(L,t) &= u_2(t) \\ R(L,t) &= +R_2(t) & \text{or} \quad \psi(L,t) &= \psi_2(t) \end{aligned} \tag{9.91}$$

The resultant forces and moments N, M, Q, and R are related to the displacement fields as follows:

$$\begin{aligned} Q(x,t) &= -EI_w w''' - \alpha\ddot{u}_b + \beta u_b'' + \gamma\ddot{w}' - \varepsilon_1\ddot{\psi} + \varepsilon_2\psi'' + c_1\dot{w}' - c_4\dot{u}_b - c_3\dot{\psi} \\ M(x,t) &= EI_w w'' - \beta u_b' - \varepsilon_2\psi' \\ N(x,t) &= EAu'_b - \beta w'' + \varepsilon_4\psi' \\ R(x,t) &= EI_\psi\psi' - \varepsilon_2 w'' + \varepsilon_4 u_b' \end{aligned} \tag{9.92}$$

The structural damping of the base beam and elastic constraining layer can also be readily taken into account by using the complex modulus given by Equation 9.14.

If all VEM related terms (i.e., $\varepsilon_1, \varepsilon_2, \varepsilon_3, \varepsilon_4$, and the terms with the subscript v) are removed, the equations of motion (9.90) and the force-displacement relationships (9.92) return to the equations of motion (9.11) and the force-displacement relationships (9.13) for the elastic–elastic two-layer beams considered in the previous section.

9.2.2 Spectral Element Modeling

9.2.2.1 Governing Equations in the Frequency Domain

Following the same procedure of spectral element formulation as used for the two-layer beams in the previous section, we represent the dynamic responses in the spectral forms as

$$\{w(x,t)\, u_b(x,t)\psi(x,t)\} = \frac{1}{N}\sum_{n=0}^{N-1}\{\, W_n(x;\omega_n)\; U_n(x;\omega_n)\; \Psi_n(x;\omega_n)\,\}\, e^{i\omega_n t} \tag{9.93}$$

We also represent all forces and moments and the voltage in the spectral forms

$$\begin{aligned} &\{p(x,t)\;\; \tau(x,t)\;\; \gamma(x,t)\} \\ &\quad = \frac{1}{N}\sum_{n=0}^{N-1}\{P_n(x;\omega_n)T_n(x;\omega_n)\Gamma_n(x;\omega_n)\}e^{i\omega_n t} \\ &\{Q(x,t)\;\; M(x,t)\;\; N(x,t)\;\; R(x,t)\} \\ &\quad = \frac{1}{N}\sum_{n=0}^{N-1}\{Q_n(x;\omega_n)\;\; M_n(x;\omega_n)\;\; N_n(x;\omega_n)\;\; R_n(x;\omega_n)\}e^{i\omega_n t} \\ &\{Q_i(x,t)\;\; M_i(x,t)\;\; N_i(x,t)\;\; R_i(x,t)\} \\ &\quad = \frac{1}{N}\sum_{n=0}^{N-1}\{Q_{in}(x;\omega_n)\;\; M_{in}(x;\omega_n)\;\; N_{in}(x;\omega_n)\;\; R_{in}(x;\omega_n)\}e^{i\omega_n t} \end{aligned} \tag{9.94}$$

where $i, j = 1, 2$.

Substitution of Equations 9.93 and 9.94 into Equations 9.90 and 9.91 yields the discrete Fourier transformed equations of motion

$$\begin{aligned} &EI_w W'''' - \omega^2\rho AW + i\omega cAW = \omega^2\alpha U' + \beta U''' - \omega^2\gamma W'' \\ &\quad + \omega^2\varepsilon_1\Psi' + \varepsilon_2\Psi''' + i\omega c_1 W'' - i\omega c_4 U' - i\omega c_3\Psi' + FW'' + P(x) \\ &EAU'' + \omega^2\rho AU - i\omega cAU \\ &\quad = \omega^2\alpha W' + \beta W''' - \omega^2\varepsilon_3\Psi - \varepsilon_4\Psi'' - i\omega c_4 W' + i\omega c_5\Psi - T(x) \\ &EI_\psi\Psi'' + \omega^2\rho I\Psi - i\omega c_2\Psi = \omega^2\varepsilon_1 W' + \varepsilon_2 W''' - \omega^2\varepsilon_3 U - \varepsilon_4 U'' \\ &\quad - i\omega c_3 W' + i\omega c_5 U + A_v G^*\Psi - \Gamma(x) \end{aligned} \tag{9.95}$$

and the associated boundary conditions

$$\begin{aligned}
Q(0) &= -Q_1 - FW'(0) && \text{or} \quad W(0) = W_1 \\
M(0) &= -M_1 && \text{or} \quad W'(0) = \Theta_1 \\
N(0) &= -N_1 && \text{or} \quad U(0) = U_1 \\
R(0) &= -R_1 && \text{or} \quad \Psi(0) = \Psi_1 \\
Q(L) &= +Q_2 - FW'(L) && \text{or} \quad W(L) = W_2 \\
M(L) &= +M_2 && \text{or} \quad W'(L) = \Theta_2 \\
N(L) &= +N_2 && \text{or} \quad U(L) = U_2 \\
R(L) &= +R_2 && \text{or} \quad \Psi(L) = \Psi_2
\end{aligned} \tag{9.96}$$

Equation 9.82 is applied to the convolution term in Equation 9.95. Similarly, by using Equations 9.93 and 9.94, the resultant force-displacement relationships (9.92) can be discrete Fourier transformed as

$$\begin{aligned}
Q(x) &= -EI_w W''' + \omega^2 \alpha U + \beta U'' - \omega^2 \gamma W' + \omega^2 \varepsilon_1 \Psi + \varepsilon_2 \Psi'' \\
&\quad + i\omega c_1 W' - i\omega c_4 U - i\omega c_3 \Psi \\
M(x) &= EI_w W'' - \beta U' - \varepsilon_2 \Psi' \\
N(x) &= EAU' - \beta W'' + \varepsilon_4 \Psi' \\
R(x) &= EI_\psi \Psi' - \varepsilon_2 W'' + \varepsilon_4 U'
\end{aligned} \tag{9.97}$$

In Equations 9.95–9.97, the subscripts n are omitted for brevity.

9.2.2.2 Spectral Nodal DOFs, Forces, and Moments

The spectral nodal DOFs, forces, and moments are specified in Figure 9.6. The spectral nodal DOFs vector is defined by

$$\begin{aligned}
\boldsymbol{d} &= \{U_1 \ W_1 \ \Theta_1 \ \Psi_1 \ U_2 \ W_2 \ \Theta_2 \ \Psi_2\}^{\mathrm{T}} \\
&= \{U(0) \ W(0) \ W'(0) \ \Psi(0) \ U(L) \ W(L) \ W'(L) \ \Psi(L)\}^{\mathrm{T}}
\end{aligned} \tag{9.98}$$

and the spectral nodal forces and moments vector by

$$\begin{aligned}
f_c &= \{N_1 \ Q_1 \ M_1 \ R_1 \ N_2 \ Q_2 \ M_2 \ R_2\}^{\mathrm{T}} \\
&= \{-N(0) \ \ -Q(0) - FW'(0) \ \ -M(0) \ \ -R(0) \ \ N(L) \ \ Q(L) + FW'(L) \ \ M(L) \ \ R(L)\}^{\mathrm{T}}
\end{aligned} \tag{9.99}$$

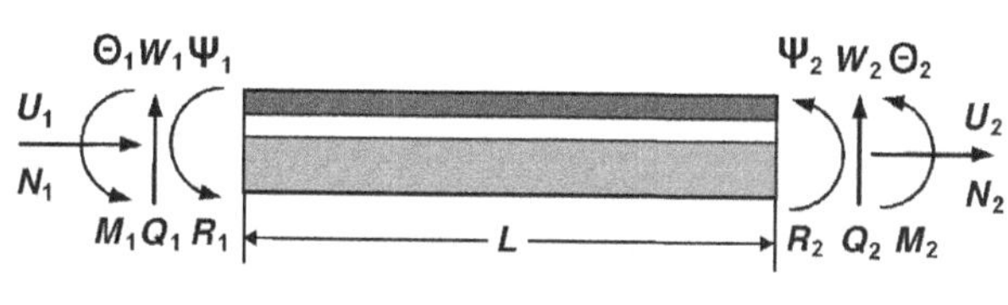

Figure 9.6 Spectral element of PCLD beam

9.2.2.3 Dynamic Shape Functions

The frequency-dependent dynamic shape functions can be derived from the homogeneous governing equations, which are reduced from Equation 9.95 as

$$\begin{aligned} & EI_w W'''' - \omega^2 \rho A W + i\omega c A W = \omega^2 \alpha U' + \beta U''' - \omega^2 \gamma W'' \\ & \qquad + \omega^2 \varepsilon_1 \Psi' + \varepsilon_2 \Psi''' + i\omega c_1 W'' - i\omega c_4 U' - i\omega c_3 \Psi' + F W'' \\ & EAU'' + \omega^2 \rho A U - i\omega c A U \\ & \qquad = \omega^2 \alpha W' + \beta W''' - \omega^2 \varepsilon_3 \Psi - \varepsilon_4 \Psi'' - i\omega c_4 W' + i\omega c_5 \Psi \\ & EI_\psi \Psi'' + \omega^2 \rho I \Psi - i\omega c_2 \Psi = \omega^2 \varepsilon_1 W' \\ & \qquad + \varepsilon_2 W''' - \omega^2 \varepsilon_3 U - \varepsilon_4 U'' - i\omega c_3 W' + i\omega c_5 U + A_v G^* \Psi \end{aligned} \tag{9.100}$$

The general solutions to Equation 9.100 are assumed to be

$$W(x;\omega) = Ae^{-ikx}, \qquad U(x;\omega) = r_1 Ae^{-ikx}, \qquad \psi(x;\omega) = r_2 Ae^{-ikx} \tag{9.101}$$

By substituting Equation 9.101 into Equation 9.100, we obtain an eigenvalue problem

$$\begin{bmatrix} X_{11} & X_{12} & X_{13} \\ X_{12} & X_{22} & X_{23} \\ X_{13} & X_{23} & X_{33} \end{bmatrix} \begin{Bmatrix} 1 \\ r_1 \\ r_2 \end{Bmatrix} = \begin{Bmatrix} 0 \\ 0 \\ 0 \end{Bmatrix} \tag{9.102}$$

where

$$\begin{aligned} & X_{11} = k^4 EI_w - \omega^2 \rho A + i\omega c A - \omega^2 k^2 \gamma + i\omega k^2 c_1 + k^2 F \\ & X_{12} = \omega k c_4 + i\omega^2 k \alpha - ik^3 \beta, \qquad X_{13} = \omega k c_3 + i\omega^2 k \varepsilon_1 - ik^3 \varepsilon_2 \\ & X_{22} = -k^2 EA + \omega^2 \rho A - i\omega c A, \qquad X_{23} = \omega^2 \varepsilon_3 - k^2 \varepsilon_4 - i\omega c_5 \\ & X_{33} = -k^2 EI_\psi + \omega^2 \rho I - i\omega c_2 - A_v G^* \end{aligned} \tag{9.103}$$

The eigenvalue problem, Equation 9.102, yields a dispersion equation as

$$a_1 k^8 + a_2 k^6 + a_3 k^4 + a_4 k^2 + a_5 = 0 \tag{9.104}$$

where

$$a_1 = EAEI_w EI_\psi - \varepsilon_4^2 EI_w + 2\beta \varepsilon_4 \varepsilon_2 - EA\varepsilon_2^2 - EI_\psi \beta^2$$

$$\begin{aligned} a_2 = {} & \omega^2 [-\rho I \beta^2 + 2\beta \varepsilon_1 \varepsilon_4 - \gamma\, \varepsilon_4^2 - 2\alpha\, EI_\psi \beta + 2\varepsilon_2 \varepsilon_3 \beta - \rho A \varepsilon_2^2 + \gamma EAEI_\psi \\ & - 2EA\varepsilon_1 \varepsilon_2 + 2\alpha \varepsilon_2 \varepsilon_4 + EI_w(\rho IEA + \rho AEI_\psi - 2\varepsilon_3 \varepsilon_4)] \\ & - i\omega\, [-c_2 \beta^2 - 2c_4 \beta EI_\psi + 2c_3 \varepsilon_4 \beta - cA\varepsilon_2^2 - \varepsilon_4^2 c_1 + c_1 EA\, EI_\psi \\ & + 2\varepsilon_2(\varepsilon_4 c_4 - c_3 EA + c_5 \beta) + EI_w(cA\, EI_\psi - 2\varepsilon_4 c_5 + EA\, c_2)] \\ & + \varepsilon_4^2 F - EAEI_w A_v G^* - EAEI_\psi F + A_v G^* \beta^2 \end{aligned}$$

$$\begin{aligned} a_3 = {} & \omega^4\, [-EA\varepsilon_1^2 - 2\rho A \varepsilon_2 \varepsilon_1 + 2\beta \varepsilon_3 \varepsilon_1 + 2\alpha \varepsilon_4 \varepsilon_1 - 2\alpha\beta\, \rho I + EA\rho I\, \gamma \\ & + EI_\psi(\gamma\, \rho A - \alpha^2) + 2\alpha \varepsilon_2 \varepsilon_3 + EI_w(\rho A \rho I - \varepsilon_3^2) - 2\gamma\, \varepsilon_3 \varepsilon_4] \\ & - i\omega^3 [-2cA\, \varepsilon_1 \varepsilon_2 + 2\varepsilon_2 \varepsilon_3 c_4 - 2\rho A \varepsilon_2 c_3 + 2\alpha \varepsilon_2 c_5 + EA\rho I c_1 - 2\varepsilon_3 \varepsilon_4 c_1 \\ & - 2c_4 \beta\, \rho I + 2\varepsilon_1 \varepsilon_4 c_4 + EI_\psi(\gamma\, cA + \rho A\, c_1 - 2\alpha c_4) - 2EA\varepsilon_1 c_3 \end{aligned}$$

$$
\begin{aligned}
&+2\beta\,\varepsilon_3 c_3 + 2\alpha\varepsilon_4 c_3 + 2\beta\varepsilon_1 c_5 - 2\gamma\varepsilon_4 c_5 - 2\alpha\beta c_2 + EA\gamma\, c_2 \\
&+ EI_w(cA\rho I - 2\varepsilon_3 c_4 + \rho A\, c_2)] \\
&+\omega^2\{\rho A\,\varepsilon_4^2 + 2F\,\varepsilon_3\varepsilon_4 - 2c_4c_3\varepsilon_4 + 2c_1c_5\varepsilon_4 + EA\,c_3^2 + 2A_vG^*\alpha\beta \\
&-\gamma EA\,A_vG^* - EA\rho I\,F - EI_\psi[-c_4^2 + \rho A\,(EA+F) + cA\,c_1] \\
&+2cA\,\varepsilon_2 c_3 - 2\varepsilon_2 c_4 c_5 - 2\beta c_3 c_5 - EA\,c_1c_2 + 2\beta\,c_5c_2 \\
&\quad + EI_w(c_5^2 - \rho A\,A_vG^* - cA\,c_2)\} \\
&+i\omega\,[-cA\,\varepsilon_4^2 - 2Fc_5\varepsilon_4 + cA\,EI_wA_vG^* + cA\,EI_\psi(EA+F) + c_1EA\,A_vG^* \\
&-2A_vG^*\beta c_4 + c_2EA\,F] + F\,EA\,A_vG^*
\end{aligned}
\tag{9.105}
$$

$$
\begin{aligned}
a_4 = &-\omega^6[\rho A\varepsilon_1^2 - 2\alpha\varepsilon_1\varepsilon_3 + \gamma\,\varepsilon_3^2 + \rho I(\alpha^2 - \gamma\,\rho A)] \\
&- i\omega^5[-c_2\alpha^2 - 2\alpha c_4\rho I + 2\alpha\varepsilon_3 c_3 - cA\varepsilon_1^2 + \gamma\,cA\,\rho I + c_1(\rho A\rho I - \varepsilon_3^2) \\
&-2\gamma\,\varepsilon_3 c_5 + 2\varepsilon_1(\varepsilon_3 c_4 - c_3\rho A + \alpha c_5) + \gamma\,c_2\rho A] \\
&+\omega^4[A_vG^*\alpha^2 - 2\alpha c_3c_5 + 2\alpha c_4c_2 + F\,\varepsilon_3^2 + \rho I\,c_4^2 + \rho A\,c_3^2 + \gamma\,c_5^2 \\
&-\gamma\,\rho A\,A_vG^* - EA\rho A\rho I - F\rho A\rho I - \rho A^2 EI_\psi - cA\rho Ic_1 + 2cA\varepsilon_1c_3 \\
&-2\varepsilon_1c_4c_5 + 2\varepsilon_3(\rho A\,\varepsilon_4 - c_4c_3 + c_1c_5) - \gamma\,cAc_2 - \rho A\,c_1c_2] \\
&+i\omega^3[-c_2c_4^2 - 2\alpha c_4A_vG^* + 2c_3c_4c_5 - cA\,c_3^2 - c_1c_5^2 + cA\gamma A_vG^* + \rho IEA\,cA \\
&+FcA\rho I + 2cA\rho AEI_\psi + \rho Ac_1A_vG^* - 2\rho A\,\varepsilon_4c_5 - 2\varepsilon_3(cA\varepsilon_4 + Fc_5) \\
&+c_2EA\rho A + F\,\rho Ac_2 + cA\,c_1c_2] \\
&+\omega^2(EI_\psi cA^2 + cAc_1A_vG^* - 2\varepsilon_4c_5cA + EAc_2cA + Fc_2cA - A_vG^*c_4^2 - Fc_5^2 \\
&+EA\rho AA_vG^* + A_vG^*F\,\rho A) - i\omega(A_vG^*\,EA\,cA + A_vG^*cA) \\
a_5 = &\;\omega^6\rho A(\varepsilon_3^2 - \rho A\rho I) + i\omega^5[-cA\varepsilon_3^2 - 2\rho Ac_5\varepsilon_3 + \rho A(2cA\,\rho I + \rho Ac_2)] \\
&+\omega^4(\rho I\,cA^2 - 2\varepsilon_3c_5cA + 2\rho Ac_2cA + A_vG^*\,\rho A^2 - \rho A\,c_5^2) \\
&-i\omega^3cA(-c_5^2 + 2A_vG^*\,\rho A + cAc_2) - \omega^2A_vG^*cA^2
\end{aligned}
$$

We can compute eight wavenumbers k_i ($i=1, 2,\ldots, 8$) from the dispersion relation, Equation 9.104. For each wavenumber k_i ($i=1, 2,\ldots, 8$), we can obtain the corresponding r_{1i} and r_{2i} from Equation 9.102 as

$$
r_{1i} = \frac{b_id_i - c_ie_i}{b_ic_i - a_ie_i}, \quad r_{2i} = \frac{a_id_i - c_i^2}{b_ic_i - a_ie_i} \quad (i=1, 2,\ldots, 8)
\tag{9.106}
$$

where

$$
\begin{aligned}
a_i &= -k_i^2EA + \omega^2\rho A - i\omega cA \\
b_i &= \omega^2\varepsilon_3 - k_i^2\varepsilon_4 - i\omega c_5 \\
c_i &= \omega k_ic_4 - i\omega^2k_i\alpha - ik_i^3\beta \\
d_i &= -k_i^4EI_w + \omega^2\rho A - i\omega cA - i\omega k_i^2c_1 + \omega^2k_i^2\gamma - k_i^2F \\
e_i &= -i\omega^2k_i\varepsilon_1 + ik_i^3\varepsilon_2 - \omega k_ic_3
\end{aligned}
\tag{9.107}
$$

Define two diagonal matrices as

$$\boldsymbol{R}_1 = diag\,[r_{1i}], \qquad \boldsymbol{R}_2 = diag\,[r_{2i}] \tag{9.108}$$

By using eight wavenumbers k_i ($i = 1, 2, \ldots, 8$) and the matrices $\boldsymbol{R}_1$ and $\boldsymbol{R}_2$, we can write the general solutions to Equation 9.100 as

$$\begin{aligned} W(x;\omega) &= \sum_{i=1}^{8} A_i \mathrm{e}^{-ik_i x} = \boldsymbol{e}(x;\omega)\boldsymbol{a} \\ U(x;\omega) &= \sum_{i=1}^{8} r_{1i} A_i \mathrm{e}^{-ik_i x} = \boldsymbol{e}(x;\omega)\boldsymbol{R}_1\boldsymbol{a} \\ \Psi(x;\omega) &= \sum_{i=1}^{8} r_{2i} A_i \mathrm{e}^{-ik_i x} = \boldsymbol{e}(x;\omega)\boldsymbol{R}_2\boldsymbol{a} \end{aligned} \tag{9.109}$$

where

$$\boldsymbol{a} = \{A_1\ A_2\ A_3\ A_4\ A_5\ A_6\ A_7\ A_8\}^{\mathrm{T}} \tag{9.110}$$

and

$$\boldsymbol{e}(x;\omega) = \left[\mathrm{e}^{-ik_1x}\ \mathrm{e}^{-ik_2x}\ \mathrm{e}^{-ik_3x}\ \mathrm{e}^{-ik_4x}\ \mathrm{e}^{-ik_5x}\ \mathrm{e}^{-ik_6x}\ \mathrm{e}^{-ik_7x}\ \mathrm{e}^{-ik_8x}\right] \tag{9.111}$$

Applying Equation 9.109 to the spectral nodal DOFs vector defined by Equation 9.98 gives

$$\boldsymbol{d} = \boldsymbol{H}(\omega)\boldsymbol{a} \tag{9.112}$$

where

$$\boldsymbol{H}(\omega) = \begin{bmatrix} r_{11} & r_{12} & r_{13} & r_{14} & r_{15} & r_{16} & r_{17} & r_{18} \\ 1 & 1 & 1 & 1 & 1 & 1 & 1 & 1 \\ -ik_1 & -ik_2 & -ik_3 & -ik_4 & -ik_5 & -ik_6 & -ik_7 & -ik_8 \\ r_{21} & r_{22} & r_{23} & r_{24} & r_{25} & r_{26} & r_{27} & r_{28} \\ r_{11}\mathrm{e}_1 & r_{12}\mathrm{e}_2 & r_{13}\mathrm{e}_3 & r_{14}\mathrm{e}_4 & r_{15}\mathrm{e}_5 & r_{16}\mathrm{e}_6 & r_{17}\mathrm{e}_7 & r_{18}\mathrm{e}_8 \\ e_1 & e_2 & e_3 & e_4 & e_5 & e_6 & e_7 & e_8 \\ -ik_1\mathrm{e}_1 & -ik_2\mathrm{e}_2 & -ik_3\mathrm{e}_3 & -ik_4\mathrm{e}_4 & -ik_5\mathrm{e}_5 & -ik_6\mathrm{e}_6 & -ik_7\mathrm{e}_7 & -ik_8\mathrm{e}_8 \\ r_{21}\mathrm{e}_1 & r_{22}\mathrm{e}_2 & r_{23}\mathrm{e}_3 & r_{24}\mathrm{e}_4 & r_{25}\mathrm{e}_5 & r_{26}\mathrm{e}_6 & r_{27}\mathrm{e}_7 & r_{28}\mathrm{e}_8 \end{bmatrix} \tag{9.113}$$

with

$$\mathrm{e}_j = \mathrm{e}^{-ik_jL} \tag{9.114}$$

The constant vector $\boldsymbol{a}$ can be eliminated from Equation 9.109 by using Equation 9.112 to write the general solutions as

$$\begin{aligned} W(x;\omega) &= \boldsymbol{N}_w(x;\omega)\boldsymbol{d} \\ U(x;\omega) &= \boldsymbol{N}_u(x;\omega)\boldsymbol{d} \\ \Psi(x;\omega) &= \boldsymbol{N}_\psi(x;\omega)\boldsymbol{d} \end{aligned} \tag{9.115}$$

where N_w, N_u, and N_ψ are the dynamic shape functions defined by

$$\begin{aligned} N_w(x;\omega) &= e(x;\omega)H^{-1}(\omega) \\ N_u(x;\omega) &= e(x;\omega)R_1H^{-1}(\omega) \\ N_\psi(x;\omega) &= e(x;\omega)R_2H^{-1}(\omega) \end{aligned} \tag{9.116}$$

9.2.2.4 Weak Form of Governing Equations

The weak form of the governing equations (9.95) can be obtained from

$$\begin{aligned} &\int_0^L [(EI_w W'''' - \omega^2\rho AW + i\omega cAW - i\omega c_1 W'' + i\omega c_4 U' - \omega^2\alpha U' \\ &- \beta U''' + \omega^2\gamma W'' - \omega^2\varepsilon_1\Psi' - \varepsilon_2\Psi''' + i\omega c_3\Psi' - FW'' - P)\delta W \\ &- (EAU'' + \omega^2\rho AU - i\omega\, cAU - \omega^2\alpha W' - \beta W''' + \omega^2\varepsilon_3\Psi + \varepsilon_4\Psi'' \\ &+ i\omega c_4 W' - i\omega c_5\Psi + T)\delta U \\ &- (EI_\psi\Psi'' + \omega^2\rho I\Psi - i\omega\xi_5\Psi - \omega^2\varepsilon_1 W' - \varepsilon_2 W''' + \omega^2\varepsilon_3 U + \varepsilon_4 U'' \\ &+ i\omega c_3 W' - i\omega c_5 U - A_v G^*\Psi + \Gamma)\delta\Psi]\mathrm{d}x = 0 \end{aligned} \tag{9.117}$$

By integrating Equation 9.117 by parts and using Equation 9.97, we can obtain the weak form as

$$\begin{aligned} &\int_0^L [EI_w W''\delta W'' - \omega^2\rho AW\delta W + i\omega cAW\delta W + i\omega c_1 W'\delta W' \\ &- i\omega c_4 U\delta W' + \omega^2\alpha U\delta W' - \beta U'\delta W'' - \omega^2\gamma W'\delta W' \\ &+ \omega^2\varepsilon_1\Psi\delta W' - \varepsilon_2\Psi'\delta W'' - i\omega c_3\Psi\delta W' + FW'\delta W' \\ &+ EAU'\delta U' - \omega^2\rho AU\delta U + i\omega cAU\delta U + \omega^2\alpha W'\delta U - \beta W''\delta U' \\ &- \omega^2\varepsilon_3\Psi\delta U + \varepsilon_4\Psi'\delta U' - i\omega c_4 W'\delta U + i\omega c_5\Psi\delta U \\ &+ EI_\psi\Psi'\delta\Psi' - \omega^2\rho I\Psi\delta\Psi + i\omega c_2\Psi\delta\Psi + \omega^2\varepsilon_1 W'\delta\Psi - \varepsilon_2 W''\delta\Psi' \\ &- \omega^2\varepsilon_3 U\delta\Psi + \varepsilon_4 U'\delta\Psi' - i\omega c_3 W'\delta\Psi + i\omega c_5 U\delta\Psi + A_v G^*\Psi\delta\Psi]\mathrm{d}x \\ &- \int_0^L [P(x)\delta W + T(x)\delta U + \Gamma(x)\delta\Psi]\mathrm{d}x \\ &- \{[Q(x) + FW']\delta W|_0^L + M(x)\delta W'|_0^L + N(x)\delta U|_0^L + R(x)\delta\Psi|_0^L\} = 0 \end{aligned} \tag{9.118}$$

9.2.2.5 Spectral Element Equation

By substituting Equation 9.115 into Equation 9.118 and using Equation 9.99, we obtain

$$\begin{aligned} \delta d^{\mathrm{T}}\Bigg[&\int_0^L [EI_w N''^{\mathrm{T}}_w N''_w - \omega^2\rho A N^{\mathrm{T}}_w N_w + i\omega cA N^{\mathrm{T}}_w N_w + i\omega c_1 N'^{\mathrm{T}}_w N'_w \\ &- i\omega c_4 N'^{\mathrm{T}}_w N_u + \omega^2\alpha N'^{\mathrm{T}}_w N_u - \beta N''^{\mathrm{T}}_w N'_u - \omega^2\gamma N'^{\mathrm{T}}_w N'_w \\ &+ \omega^2\varepsilon_1 N'^{\mathrm{T}}_w N_\psi - \varepsilon_2 N''^{\mathrm{T}}_w N'_\psi - i\omega c_3 N'^{\mathrm{T}}_w N_\psi + F N'^{\mathrm{T}}_w N'_w \\ &+ EA N'^{\mathrm{T}}_u N'_u - \omega^2\rho A N^{\mathrm{T}}_u N_u + i\omega cA N^{\mathrm{T}}_u N_u + \omega^2\alpha N^{\mathrm{T}}_u N'_w \\ &- \beta N'^{\mathrm{T}}_u N''_w - \omega^2\varepsilon_3 N^{\mathrm{T}}_u N_\psi + \varepsilon_4 N'^{\mathrm{T}}_u N'_\psi - i\omega c_4 N^{\mathrm{T}}_u N'_w \\ &+ i\omega c_5 N^{\mathrm{T}}_u N_\psi + EI_\psi N'^{\mathrm{T}}_\psi N'_\psi - \omega^2\rho I N^{\mathrm{T}}_\psi N_\psi + i\omega c_2 N^{\mathrm{T}}_\psi N_\psi \\ &+ \omega^2\varepsilon_1 N^{\mathrm{T}}_\psi N'_w - \varepsilon_2 N'^{\mathrm{T}}_\psi N''_w - \omega^2\varepsilon_3 N^{\mathrm{T}}_\psi N_u + \varepsilon_4 N'^{\mathrm{T}}_\psi N'_u \\ &- i\omega c_3 N^{\mathrm{T}}_\psi N'_w + i\omega c_5 N^{\mathrm{T}}_\psi N_u + A_v G^* N^{\mathrm{T}}_\psi N_\psi]\mathrm{d}x\, d - (f_{\mathrm{c}} + f_{\mathrm{d}})\Bigg] = 0 \end{aligned} \tag{9.119}$$

where

$$f_{\mathrm{d}}(\omega) = \int_0^L [P(x;\omega)N_w^{\mathrm{T}}(x;\omega) + T(x;\omega)N_u^{\mathrm{T}}(x;\omega) + \Gamma(x;\omega)N_\psi^{\mathrm{T}}(x;\omega)]\mathrm{d}x \tag{9.120}$$

As $\delta \boldsymbol{d}$ is arbitrary, we can obtain the spectral element equation from Equation 9.119 as

$$\boldsymbol{S}(\omega)\boldsymbol{d} = \boldsymbol{f}(\omega) \tag{9.121}$$

where

$$\begin{aligned} \boldsymbol{S}(\omega) = \int_0^L &[EA\boldsymbol{N}'^{\mathrm{T}}_u\boldsymbol{N}'_u + EI_w\boldsymbol{N}''^{\mathrm{T}}_w\boldsymbol{N}''_w + EI_\psi\boldsymbol{N}'^{\mathrm{T}}_\psi\boldsymbol{N}'_\psi + F\boldsymbol{N}'^{\mathrm{T}}_w\boldsymbol{N}'_w \\ &- \beta(\boldsymbol{N}''^{\mathrm{T}}_w\boldsymbol{N}'_u + \boldsymbol{N}'^{\mathrm{T}}_u\boldsymbol{N}''_w) - \varepsilon_2(\boldsymbol{N}''^{\mathrm{T}}_w\boldsymbol{N}'_\psi + \boldsymbol{N}'^{\mathrm{T}}_\psi\boldsymbol{N}''_w) \\ &- \omega^2\{\rho A(\boldsymbol{N}^{\mathrm{T}}_w\boldsymbol{N}_w + \boldsymbol{N}^{\mathrm{T}}_u\boldsymbol{N}_u) + \rho I\boldsymbol{N}^{\mathrm{T}}_\psi\boldsymbol{N}_\psi - \alpha(\boldsymbol{N}'^{\mathrm{T}}_w\boldsymbol{N}_u + \boldsymbol{N}^{\mathrm{T}}_u\boldsymbol{N}'_w) \\ &+ \gamma\boldsymbol{N}'^{\mathrm{T}}_w\boldsymbol{N}'_w - \varepsilon_1(\boldsymbol{N}'^{\mathrm{T}}_w\boldsymbol{N}_\psi + \boldsymbol{N}^{\mathrm{T}}_\psi\boldsymbol{N}'_w) + \varepsilon_3(\boldsymbol{N}^{\mathrm{T}}_u\boldsymbol{N}_\psi + \boldsymbol{N}^{\mathrm{T}}_\psi\boldsymbol{N}_u)\} \\ &+ \varepsilon_4(\boldsymbol{N}'^{\mathrm{T}}_u\boldsymbol{N}'_\psi + \boldsymbol{N}'^{\mathrm{T}}_\psi\boldsymbol{N}'_u) + i\omega\{cA(\boldsymbol{N}^{\mathrm{T}}_w\boldsymbol{N}_w + \boldsymbol{N}^{\mathrm{T}}_u\boldsymbol{N}_u) + c_1\boldsymbol{N}'^{\mathrm{T}}_w\boldsymbol{N}'_w \\ &+ c_2\boldsymbol{N}^{\mathrm{T}}_\psi\boldsymbol{N}_\psi - c_3(\boldsymbol{N}'^{\mathrm{T}}_w\boldsymbol{N}_\psi + \boldsymbol{N}^{\mathrm{T}}_\psi\boldsymbol{N}'_w) - c_4(\boldsymbol{N}'^{\mathrm{T}}_w\boldsymbol{N}_u + \boldsymbol{N}^{\mathrm{T}}_u\boldsymbol{N}'_w) \\ &+ c_5(\boldsymbol{N}^{\mathrm{T}}_u\boldsymbol{N}_\psi + \boldsymbol{N}^{\mathrm{T}}_\psi\boldsymbol{N}_u)\} + A_vG^*\boldsymbol{N}^{\mathrm{T}}_\psi\boldsymbol{N}_\psi]\mathrm{d}x = \boldsymbol{S}(\omega)^{\mathrm{T}} \end{aligned} \tag{9.122}$$

and

$$\boldsymbol{f}(\omega) = \boldsymbol{f}_{\mathrm{c}}(\omega) + \boldsymbol{f}_{\mathrm{d}}(\omega) \tag{9.123}$$

Substitution of the dynamic shape functions (9.116) into Equation 9.122 gives

$$\boldsymbol{S}(\omega) = \boldsymbol{H}^{-\mathrm{T}}(\omega)\boldsymbol{D}(\omega)\boldsymbol{H}^{-1}(\omega) \tag{9.124}$$

where

$$\begin{aligned} \boldsymbol{D}(\omega) = &-EA\boldsymbol{R}_1^{\mathrm{T}}\boldsymbol{KEKR}_1 + EI_w\boldsymbol{K}^2\boldsymbol{EK}^2 - EI_\psi\boldsymbol{R}_2^{\mathrm{T}}\boldsymbol{KEKR}_2 - F\boldsymbol{KEK} \\ &- i\beta(\boldsymbol{K}^2\boldsymbol{EKR}_1 + \boldsymbol{R}_1^{\mathrm{T}}\boldsymbol{KEK}^2) - i\varepsilon_2(\boldsymbol{K}^2\boldsymbol{EKR}_2 + \boldsymbol{R}_2^{\mathrm{T}}\boldsymbol{KEK}^2) \\ &- \omega^2\{\rho A(\boldsymbol{E} + \boldsymbol{R}_1^{\mathrm{T}}\boldsymbol{ER}_1) + \rho I\boldsymbol{R}_2^{\mathrm{T}}\boldsymbol{ER}_2 + i\alpha(\boldsymbol{KER}_1 + \boldsymbol{R}_1^{\mathrm{T}}\boldsymbol{EK}) \\ &- \gamma\boldsymbol{KEK} + i\varepsilon_1(\boldsymbol{KER}_2 + \boldsymbol{R}_2^{\mathrm{T}}\boldsymbol{EK}) + \varepsilon_3(\boldsymbol{R}_1^{\mathrm{T}}\boldsymbol{ER}_2 + \boldsymbol{R}_2^{\mathrm{T}}\boldsymbol{ER}_1)\} \\ &- \varepsilon_4(\boldsymbol{R}_1^{\mathrm{T}}\boldsymbol{KEKR}_2 + \boldsymbol{R}_2^{\mathrm{T}}\boldsymbol{KEKR}_1) + i\omega\{cA(\boldsymbol{E} + \boldsymbol{R}_1^{\mathrm{T}}\boldsymbol{ER}_1) - c_1\boldsymbol{KEK} \\ &+ c_2\boldsymbol{R}_2^{\mathrm{T}}\boldsymbol{ER}_2 + ic_3(\boldsymbol{KER}_2 + \boldsymbol{R}_2^{\mathrm{T}}\boldsymbol{EK}) + ic_4(\boldsymbol{KER}_1 + \boldsymbol{R}_1^{\mathrm{T}}\boldsymbol{EK}) \\ &+ c_5(\boldsymbol{R}_1^{\mathrm{T}}\boldsymbol{ER}_2 + \boldsymbol{R}_2^{\mathrm{T}}\boldsymbol{ER}_1)\} + A_vG^*\boldsymbol{R}_2^{\mathrm{T}}\boldsymbol{ER}_2 \end{aligned} \tag{9.125}$$

and where

$$\boldsymbol{K} = diag[k_i] = \boldsymbol{K}^{\mathrm{T}}, \qquad \boldsymbol{K}^2 = diag[k_i^2] = \boldsymbol{K}^{2\mathrm{T}} \tag{9.126}$$

$$E(\omega) = \int_0^L e^{\mathrm{T}}(x;\omega)e(x;\omega)\mathrm{d}x = [E_{lm}(\omega)] \quad (l,m = 1,\ 2,\ \ldots,\ 8) \tag{9.127}$$

with

$$E_{lm} = \begin{cases} \dfrac{i}{k_l + k_m}\left[\mathrm{e}^{-i(k_l+k_m)L} - 1\right] & \text{if } k_l + k_m \neq 0 \\ L & \text{if } k_l + k_m = 0 \end{cases} \tag{9.128}$$

The spectral elements represented by Equation 9.121 can be assembled, with imposing the boundary conditions, in the form

$$S_{\mathrm{g}}(\omega)d_{\mathrm{g}} = f_{\mathrm{g}}(\omega) \tag{9.129}$$

where $S_{\mathrm{g}}(\omega)$ is the global dynamic stiffness matrix, d_{g} is the global spectral DOFs vector, and f_{g} is the global spectral nodal forces vector.

9.2.3 Spectral Modal Analysis

The eigenvalue problem for the spectral PCLD element model can be reduced from Equation 9.129, by enforcing the forcing term to be zero, in the form of

$$S_{\mathrm{g}}(\omega)d_{\mathrm{g}} = 0 \tag{9.130}$$

The (damped) natural frequencies ω_p $(p = 1, 2, \ldots, \infty)$ can be computed from $\det S_{\mathrm{g}}(\omega) = 0$. Eight wavenumbers k_i $(i = 1,\ 2, \ldots,\ 8)$ can be obtained for each natural frequency from the dispersion relation Equation 9.104. The eigenvector d_{g} corresponding to each natural frequency ω_p can be then computed from Equation 9.130. By using eight wavenumbers and eigenvector d_{g} for each natural frequency, we can obtain the corresponding (damped) normal modes from Equation 9.115. It is worth remembering that the normal modes used in the present spectral modal analysis are the damped normal modes into which the effect of the VEM layer damping is smeared.

9.2.3.1 Orthogonality Properties of Normal Modes

Assume that ω_p is the pth natural frequency and W_p, U_p, and Ψ_p are the corresponding normal modes, respectively. Then, the pth eigensolutions should satisfy

$$\begin{aligned}
EI_w W_p'''' - \omega_p^2 \rho A W_p + i\omega_p c A W_p &= \omega_p^2(\alpha U'_p - \gamma W''_p + \varepsilon_1 \Psi'_p) \\
&\quad + i\omega_p(c_1 W''_p - c_4 U'_p - c_3 \Psi'_p) + \beta U'''_p + \varepsilon_2 \Psi'''_p + F W''_p \\
EA U''_p + \omega_p^2 \rho A U_p - i\omega_p c A U_p &= \omega_p^2(\alpha W'_p - \varepsilon_3 \Psi_p) \\
&\quad - i\omega_p(c_4 W'_p - c_5 \Psi_p) + \beta W''_p - \varepsilon_4 \Psi''_p \\
EI_\psi \Psi''_p + \omega_p^2 \rho I \Psi_p - i\omega_p c_2 \Psi_p &= \omega_p^2(\varepsilon_1 W'_p - \varepsilon_3 U_p) \\
&\quad - i\omega_p(c_3 W'_p - c_5 U_p) + \varepsilon_2 W'''_p - \varepsilon_4 U''_p + A_v G^* \Psi_p
\end{aligned} \tag{9.131}$$

By using Equation 9.97, Equation 9.131 can be written in the simple forms as

$$\begin{aligned}
&\omega_p^2(\rho A W_p + \alpha U_p' + \varepsilon_1 \Psi'_p - \gamma W_p'') \\
&\quad + i\omega_p(c_1 W_p'' - c_3 \Psi_p' - c_4 U_p' - cAW_p) + FW_p'' = M_p'' \\
&\omega_p^2(\rho A U_p - \alpha W_p' + \varepsilon_3 \Psi_p) + i\omega_p(c_4 W_p' - c_5 \Psi_p - cAU_p) = -N_p' \\
&\omega_p^2(\rho I \Psi_p - \varepsilon_1 W_p' + \varepsilon_3 U_p) - i\omega_p(c_2 \Psi_p - c_3 W_p' + c_5 U_p) - A_v G \Psi_p = -R_p'
\end{aligned} \tag{9.132}$$

Following the same procedure as used in Section 9.1.3, we first multiply the first equation in Equation 9.132 by W_q, the second equation by U_q, and the last equation by Ψ_q and then integrate from $x=0$ to $x=L$, followed by the application of the integral by parts. Finally, summing three equations gives

$$\begin{aligned}
&\omega_p^2 \int_0^L [\rho A(W_p W_q + U_p U_q) + \rho I \Psi_p \Psi_q - \alpha(U_p W_q' + U_q W_p') \\
&\quad - \varepsilon_1(\Psi_p W_q' + \Psi_q W_p') + \varepsilon_3(\Psi_p U_q + \Psi_q U_p) + \gamma W_p' W_q'] \mathrm{d}x \\
&\quad - i\omega_p \int_0^L [cA(W_p W_q + U_p U_q) + c_1 W_p' W_q' + c_2 \Psi_p \Psi_q - c_3(\Psi_p W_q' + \Psi_q W_p') \\
&\quad - c_4(U_p W_q' + U_q W_p') + c_5(\Psi_p U_q + \Psi_q U_p)] \mathrm{d}x \\
&= -N_p U_q|_0^L - (Q_p + FW_p')W_q|_0^L - M_p W_q'|_0^L - R_p \Psi_q|_0^L + \int_0^L N_p U_q' \mathrm{d}x \\
&\quad + \int_0^L M_p W_q'' dx + \int_0^L R_p \Psi_q' dx + \int_0^L A_v G \Psi_p \Psi_q dx + \int_0^L F W_p' W_q' \mathrm{d}x
\end{aligned} \tag{9.133}$$

Similarly, for the qth eigensolutions, we obtain

$$\begin{aligned}
&\omega_q^2 \int_0^L [\rho A(W_p W_q + U_p U_q) + \rho I \Psi_p \Psi_q - \alpha(U_p W_q' + U_q W_p') \\
&\quad - \varepsilon_1(\Psi_p W_q' + \Psi_q W_p') + \varepsilon_3(\Psi_p U_q + \Psi_q U_p) + \gamma W_p' W_q'] \mathrm{d}x \\
&\quad - i\omega_q \int_0^L [cA(W_p W_q + U_p U_q) + c_1 W_p' W_q' + c_2 \Psi_p \Psi_q \\
&\quad - c_3(\Psi_p W_q' + \Psi_q W_p') - c_4(U_p W_q' + U_q W_p') + c_5(\Psi_p U_q + \Psi_q U_p)] \mathrm{d}x \\
&= -N_q U_p|_0^L - (Q_q + FW_q')W_p|_0^L - M_q W_p'|_0^L - R_q \Psi_p|_0^L + \int_0^L N_q U_p' \mathrm{d}x \\
&\quad + \int_0^L M_q W_p'' \mathrm{d}x + \int_0^L R_q \Psi_p' \mathrm{d}x + \int_0^L A_v G \Psi_q \Psi_p \mathrm{d}x + \int_0^L F W_q' W_p' \mathrm{d}x
\end{aligned} \tag{9.134}$$

By subtracting Equation 9.134 from Equation 9.133, applying the homogenous boundary conditions and accomplishing a further manipulation, we can obtain

$$\begin{aligned}
&(\omega_p^2 - \omega_q^2) \int_0^L [\rho A(W_p W_q + U_p U_q) + \rho I \Psi_p \Psi_q - \alpha(U_p W_q' + U_q W_p') \\
&\quad - \varepsilon_1(\Psi_p W_q' + \Psi_q W_p') + \varepsilon_3(\Psi_p U_q + \Psi_q U_p) + \gamma W_p' W_q'] \mathrm{d}x \\
&\quad - i(\omega_p - \omega_q) \int_0^L [cA(W_p W_q + U_p U_q) + c_1 W_p' W_q' + c_2 \Psi_p \Psi_q - c_3(\Psi_p W_q' + \Psi_q W_p') \\
&\quad - c_4(U_p W_q' + U_q W_p') + c_5(\Psi_p U_q + \Psi_q U_p)] \mathrm{d}x = 0
\end{aligned} \tag{9.135}$$

From Equation 9.135, we can obtain the orthogonality properties as follows:

$$\begin{aligned}&\int_0^L [\rho A(W_pW_q+U_pU_q)+\rho I\Psi_p\Psi_q-\alpha(U_pW_q'+U_qW_p')\\&\quad-\varepsilon_1(\Psi_pW_q'+\Psi_qW_p')+\varepsilon_3(\Psi_pU_q+\Psi_qU_p)+\gamma W_p'W_q']\mathrm{d}x=m_p\delta_{pq}\\&\int_0^L [cA(W_pW_q+U_pU_q)+c_1W_p'W_q'+c_2\Psi_p\Psi_q-c_3(\Psi_pW_q'+\Psi_qW_p')\\&\quad-c_4(U_pW_q'+U_qW_p')+c_5(\Psi_pU_q+\Psi_qU_p)]\mathrm{d}x=c_p\delta_{pq}\end{aligned}\tag{9.136}$$

where

$$\begin{aligned}m_q&=\int_0^L[\rho A(U_q^2+W_q^2)+\rho I\Psi_q^2-2\alpha U_qW_q'-2\varepsilon_1\Psi_qW_q'+2\varepsilon_3\Psi_qU_q+\gamma W_q'^2]\mathrm{d}x\\c_q&=\int_0^L[cA(U_q^2+W_q^2)+c_1W_q'^2+c_2\Psi_q^2-2c_3\Psi_qW_q'-2c_4U_qW_q'+2c_5\Psi_qU_q]\mathrm{d}x\end{aligned}\tag{9.137}$$

The constants m_p and c_p are the pth modal mass and modal damping coefficient, respectively. By applying the above orthogonality properties to Equation 9.133, we can obtain

$$\begin{aligned}&-N_qU_q\Big|_0^L-(Q_q+FW_q')W_q\Big|_0^L-M_qW_q'\Big|_0^L-R_q\Psi_q\Big|_0^L+\int_0^L N_qU_q'\mathrm{d}x+\int_0^L M_qW_q''\mathrm{d}x\\&\quad+\int_0^L R_q\Psi_q'\mathrm{d}x+\int_0^L A_vG\Psi_q\Psi_q\mathrm{d}x+\int_0^L FW_q'W_q'\mathrm{d}x=\omega_q^2m_q-i\omega_qc_q\\&-N_pU_q\Big|_0^L-(Q_p+FW_p')W_q\Big|_0^L-M_pW_q'\Big|_0^L-R_p\Psi_q\Big|_0^L+\int_0^L N_pU_q'\mathrm{d}x+\int_0^L M_pW_q''\mathrm{d}x\\&\quad+\int_0^L R_p\Psi_q'\mathrm{d}x+\int_0^L A_vG\Psi_p\Psi_q\mathrm{d}x+\int_0^L FW_p'W_q'\mathrm{d}x=0\end{aligned}\tag{9.138}$$

9.2.3.2 Spectral Modal Equations

For the modal analysis, the forced vibration responses are represented by

$$\{w(x,t),\ u(x,t),\ \psi(x,t)\}=\sum_{p=1}^{\infty}\{W_p(x),\ U_p(x),\ \Psi_p(x)\}\ q_p(t)\tag{9.139}$$

where $q_p(t)$ are the modal coordinates. Represent the modal coordinates in the spectral form as

$$q_p(t)=\frac{1}{N}\sum_{n=0}^{N-1}\bar{q}_{pn}e^{i\omega_nt}\tag{9.140}$$

By substitution of Equation 9.140 into Equation 9.139, we obtain

$$\{w(x,t),\ u(x,t),\ \psi(x,t)\}=\frac{1}{N}\sum_{p=1}^{\infty}\sum_{n=0}^{N-1}\{W_p(x),\ U_p(x),\Psi_p(x)\}\,q_{pn}e^{i\omega_nt}\tag{9.141}$$

The external forces are also written in the spectral forms as

$$\{p(x,t),\ \tau(x,t),\ \gamma(x,t)\}=\frac{1}{N}\sum_{n=0}^{N-1}\{P_n(x),\ T_n(x),\ \Gamma_n(x)\}\,e^{i\omega_nt}\tag{9.142}$$

Substitution of Equations 9.141 and 9.142 into the equations of motion (9.90) yields three equations, for the nth spectral component, as follows:

$$\begin{aligned}
&\sum_{p=1}^{\infty}(M_p'' - FW_p'')\bar{q}_{pn} - \omega_n^2\sum_{p=1}^{\infty}(\rho A W_p + \alpha U_p' + \varepsilon_1\Psi_p' - \gamma W_p'')\bar{q}_{pn} \\
&\qquad + i\omega_n\sum_{p=1}^{\infty}(cAW_p - c_1W_p'' + c_3\Psi_p' + c_4U_p')\bar{q}_{pn} = P_n(x) \\
&\sum_{p=1}^{\infty}N_p'\bar{q}_{pn} + \omega_n^2\sum_{p=1}^{\infty}(\rho A U_p - \alpha W_p' + \varepsilon_3\Psi_p)\bar{q}_{pn} \\
&\qquad - i\omega_n\sum_{p=1}^{\infty}(cAU_p - c_4W_p' + c_5\Psi_p)\bar{q}_{pn} = -T_n(x) \\
&\sum_{p=1}^{\infty}(R_p' - A_vG\Psi_p)\bar{q}_{pn} + \omega_n^2\sum_{p=1}^{\infty}(\rho I\Psi_p - \varepsilon_1W_p' + \varepsilon_3U_p)\bar{q}_{pn} \\
&\qquad - i\omega_n\sum_{p=1}^{\infty}(c_2\Psi_p - c_3W_p' + c_5U_p)\bar{q}_{pn} = -\Gamma_n(x)
\end{aligned} \tag{9.143}$$

where Equation 9.97 is used for M_p, N_p, and R_p. Multiply the first equation of Equation 9.143 by W_q, the second equation by U_q, and the last equation by Ψ_q to integrate from $x=0$ to $x=L$. By integrating by parts and summing the final forms of three equations, we can obtain

$$\begin{aligned}
&-\omega_n^2\sum_{p=1}^{\infty}\bar{q}_{pn}\int_0^L[\rho A(W_pW_q + U_pU_q) + \rho I\Psi_p\Psi_q - \alpha(U_pW_q' + W_p'U_q) \\
&\quad + \gamma W_p'W_q' - \varepsilon_1(\Psi_pW_q' + W_p'\Psi_q) + \varepsilon_3(\Psi_pU_q + U_p\Psi_q)]\mathrm{d}x \\
&\quad + i\omega_n\sum_{p=1}^{\infty}\bar{q}_{pn}\int_0^L[cA(W_pW_q + U_pU_q) + c_1W_p'W_q' + c_2\Psi_p\Psi_q \\
&\quad - (c_3\Psi_pW_q' + W_p'\Psi_q) - c_4(U_pW_q' + W_p'U_q) + c_5(\Psi_pU_q + U_p\Psi_q)]\mathrm{d}x \\
&= \sum_{p=1}^{\infty}\bar{q}_{pn}\left[N_pU_q\Big|_0^L + (Q_p + FW_p')W_q\Big|_0^L + M_pW_q'\Big|_0^L + R_pU_q\Big|_0^L\right. \\
&\quad \left. - \int_0^L(N_pU_q' + M_pW_q'' + R_p\Psi_q' + A_vG\Psi_p\Psi_q + FW_p'W_q')\mathrm{d}x\right] \\
&\quad + \int_0^L P_n(x)W_q\mathrm{d}x + \int_0^L T_n(x)U_q\mathrm{d}x + \int_0^L \Gamma_n(x)\Psi_q\mathrm{d}x
\end{aligned} \tag{9.144}$$

where Equation 9.97 is used for Q_p.

Finally, by applying the orthogonality properties from Equations 9.136–9.138 to Equation 9.144, we can obtain the spectral modal equations as

$$(\omega_q^2 - \omega_n^2)m_q q_{qn} - i(\omega_q - \omega_n)c_q q_{qn} = f_{qn} \tag{9.145}$$

where f_{qn} is the spectral modal force defined by

$$f_{qn} = \int_0^L[P_n(x)W_q + T_n(x)U_q + \Gamma_n(x)\Psi_q]\mathrm{d}x \tag{9.146}$$

Equations 9.145 and 9.146 can be transformed into the time domain as

$$m_q\ddot{q}_q(t) + c_q\dot{q}_q(t) + (\omega_q^2 m_q - i\omega_q c_q)q_q(t) = f_q(t) \tag{9.147}$$

where f_q is the modal force given by

$$f_q(t) = \int_0^L [p(x,t)W_q + \tau(x,t)U_q + r(x,t)\Psi_q]\mathrm{d}x \tag{9.148}$$

9.2.4 Finite Element Model

The weak form of the equations of motion (9.90) is given by

$$\begin{aligned}
&\int_0^L [EI_w w''\delta w'' + \rho A\ddot{w}\delta w + cA\dot{w}\delta w - \alpha\ddot{u}_b\delta w' - \beta u'_b\delta w'' + \gamma\ddot{w}'\delta w' \\
&\quad - \varepsilon_1\ddot{\psi}\delta w' - \varepsilon_2\psi'\delta w'' + c_1\dot{w}'\delta w' - c_4\dot{u}_b\delta w' - c_3\dot{\psi}\delta w' + Fw'\delta w' \\
&\quad + EAu'_b\delta u'_b + \rho A\ddot{u}_b\delta u_b + cA\dot{u}_b\delta u_b - \alpha\ddot{w}'\delta u_b - \beta w''\delta u'_b + \varepsilon_3\ddot{\psi}\delta u_b \\
&\quad + \varepsilon_4\psi'\delta u'_b - c_4\dot{w}'\delta u'_b + c_5\dot{\psi}\delta u_b \\
&\quad + EI_\psi\psi'\delta\psi' + \rho I\ddot{\psi}\delta\psi + c_2\dot{\psi}\delta\psi - \varepsilon_1\ddot{w}'\delta\psi - \varepsilon_2 w''\delta\psi' + \varepsilon_3\ddot{u}_b\delta\psi \\
&\quad + \varepsilon_4 u'_b\delta\psi' - c_3\dot{w}'\delta\psi + c_5\dot{u}_b\delta\psi + A_v(g\circ\psi)\delta\psi]\mathrm{d}x \\
&\quad - \int_0^L [p(x,t)\delta w + \tau(x,t)\delta u_b + \gamma(x,t)\delta\psi]\mathrm{d}x \\
&\quad - \{N(x,t)\delta u_b|_0^L + M(x,t)\delta w'|_0^L + [Q(x,t) + Fw']\delta w|_0^L + R(x,t)\delta\psi|_0^L\} = 0
\end{aligned} \tag{9.149}$$

The convolution term $A_v(g\circ\psi)\delta\psi$ in Equation 9.149 is approximately represented by using the GHM model [121], which is detailed in the following.

The material behavior of a viscoelastic material can be represented by the mass–spring model as

$$m\ddot{\psi}(x,t) + (g\circ\psi) = f(x,t) \tag{9.150}$$

where the viscoelastic spring is described by a relaxation function $g(x, t)$ and the convolution $(g\circ\psi)$ is defined by Equation 9.81. Equation 9.150 can be transformed into the Laplace domain as

$$s^2 m\bar{\psi}(s) + sG(s)\bar{\psi}(s) = \bar{f}(s) \tag{9.151}$$

where s is the Laplace variable. The material modulus function $sG(s)$, which is the complex modulus in the frequency domain, can be represented by either an analytical model

$$sG(s) = G_e + \frac{G_1}{1 + c_1(if_R/f_1)^{-\beta_2} + (if_R/f_1)^{-\beta_1}} \tag{9.152}$$

or the GHM model

$$sG(s) = \kappa\left(1 + \sum_{r=1}^{R}\alpha_r\frac{s^2 + 2\hat{\xi}_r\hat{\omega}_r s}{s^2 + 2\hat{\xi}_r\hat{\omega}_r s + \hat{\omega}_r^2}\right) \tag{9.153}$$

The parameters in Equations 9.152 and 9.153 are determined so as to fit experimentally measured frequency-dependent damping data [112]. The accuracy of the GHM model may depend on R, the number of dissipation coordinates. For instance, the one-term GHM model (i.e., $R = 1$) has one dissipation coordinate as follows

$$sG(s) = \kappa\left(1 + \alpha\frac{s^2 + 2\hat{\xi}\hat{\omega}s}{s^2 + 2\hat{\xi}\hat{\omega}s + \hat{\omega}^2}\right) \tag{9.154}$$

Substituting the one-term GHM model represented by Equation 9.154 into Equation 9.151, we obtain

$$s^2 m\bar{\psi}(s) + \kappa\left(1 + \alpha\frac{s^2 + 2\hat{\xi}\hat{\omega}\, s}{s^2 + 2\hat{\xi}\hat{\omega}\, s + \hat{\omega}^2}\right)\bar{\psi}(s) = \bar{f}(s) \tag{9.155}$$

For the one-term GHM model, an auxiliary coordinate $\bar{z}$ (known as the dissipation coordinate) is introduced as

$$\bar{z}(s) = \frac{\hat{\omega}^2}{s^2 + 2\hat{\xi}\hat{\omega}\, s + \hat{\omega}^2}\bar{\psi}(s) \tag{9.156}$$

Thus, by using Equation 9.156, we can write Equation 9.155 as

$$s^2 m\bar{\psi}(s) + (1+\alpha)\kappa\bar{\psi} - \alpha\kappa\,\bar{z}(s) = \bar{f}(s) \tag{9.157}$$

Equation 9.156 can be rewritten in the form

$$s^2\bar{z}(s) + 2\hat{\xi}\hat{\omega}\, s\,\bar{z}(s) - \hat{\omega}^2\bar{\psi}(s) + \hat{\omega}^2\,\bar{z}(s) = 0 \tag{9.158}$$

Equations 9.157 and 9.158 can be inverse Laplace transformed to obtain the time-domain expressions as follows:

$$m\ddot{\psi}(x,t) + (1+\alpha)\kappa\psi(x,t) - \alpha\kappa\, z(x,t) = f(x,t) \tag{9.159}$$

$$\ddot{z}(x,t) + 2\hat{\xi}\hat{\omega}\,\dot{z}(x,t) - \hat{\omega}^2\psi(x,t) + \hat{\omega}^2 z(x,t) = 0 \tag{9.160}$$

Comparing Equation 9.159 with Equation 9.150, we obtain

$$(g\circ\psi) \cong (1+\alpha)\kappa\psi(x,t) - \alpha\kappa\, z(x,t) \tag{9.161}$$

Thus we can replace the last term in the integral of Equation 9.149 by

$$A_v(g\circ\psi)\delta\psi \cong A_v\left[(1+\alpha)\kappa\psi(x,t) - \alpha\kappa\, z(x,t)\right]\delta\psi \tag{9.162}$$

The weak form of Equation 9.160 is written as

$$\int_0^L \left[\left(\frac{A_v\alpha\kappa}{\hat{\omega}^2}\right)\ddot{z} + \left(\frac{2A_v\alpha\kappa\hat{\xi}}{\hat{\omega}}\right)\dot{z} - A_v\alpha\kappa\psi + A_v\alpha\kappa z\right]\delta z\mathrm{d}x = 0 \tag{9.163}$$

The sign convention for the nodal DOFs, forces and moments defined in Figure 9.6 are also valid in the time domain. Thus, we define the nodal DOFs vector as

$$\begin{aligned}\boldsymbol{d}(t) &= \{U_1(t)\ W_1(t)\ \Theta_1(t)\ \Psi_1(t)\ U_2(t)\ W_2(t)\ \Theta_2(t)\ \Psi_2(t)\}^{\mathrm{T}}\\ \boldsymbol{z}(t) &= \{z_1(t)\ z_2(t)\}^{\mathrm{T}}\end{aligned} \tag{9.164}$$

and the nodal forces and moments vector as

$$\begin{aligned} f_c(t) &= \{N_1(t)\ Q_1(t)\ M_1(t)\ R_1(t)\ N_2(t)\ Q_2(t)\ M_2(t)\ R_2(t)\}^T \\ &= \{-N(0,t)\ \ -Q(0,t)-Fw'(0,t)\ \ -M(0,t)\ \ -R(0,t)\} \\ &\quad\ N(L,t)\ \ \ Q(L,t)+Fw'(L,t)\ \ \ M(L,t)\ R(L,t)\}^T \end{aligned} \tag{9.165}$$

The displacement fields are assumed as

$$\begin{aligned} u_b(x,t) &= \boldsymbol{N}_u(x)\,\boldsymbol{d}(t), \quad w(x,t) = \boldsymbol{N}_w(x)\,\boldsymbol{d}(t) \\ \psi(x,t) &= \boldsymbol{N}_\psi(x)\,\boldsymbol{d}(t), \quad z(x,t) = \boldsymbol{N}_z(x)\,\boldsymbol{z}(t) \end{aligned} \tag{9.166}$$

where

$$\begin{aligned} \boldsymbol{N}_u(x) &= [N_{u1}(x)\ 0\ 0\ 0\ N_{u2}(x)\ 0\ 0\ 0] \\ \boldsymbol{N}_w(x) &= [0\ N_{w1}(x)\ N_{\theta1}(x)\ 0\ 0\ N_{w2}(x)\ N_{\theta2}(x)\ 0] \\ \boldsymbol{N}_\psi(x) &= [0\ 0\ 0\ N_{u1}(x)\ 0\ 0\ 0\ N_{u2}(x)] \\ \boldsymbol{N}_z(x) &= [N_{z1}(x)\ N_{z2}(x)] \end{aligned} \tag{9.167}$$

with

$$\begin{aligned} N_{u1}(x) &= N_{z1}(x) = 1-(x/L) \\ N_{u2}(x) &= N_{z2}(x) = x/L \\ N_{w1}(x) &= 1-3(x/L)^2+2(x/L)^3 \\ N_{w2}(x) &= 3(x/L)^2-2(x/L)^3 \\ N_{\theta1}(x) &= x[1-2(x/L)+(x/L)^2] \\ N_{\theta2}(x) &= -x[(x/L)-(x/L)^2] \end{aligned} \tag{9.168}$$

Substituting Equation 9.166 into the weak forms of Equations 9.149 and 9.163 gives the finite element equation to be

$$\begin{bmatrix} \boldsymbol{M}_d & \boldsymbol{0} \\ \boldsymbol{0} & \boldsymbol{M}_z \end{bmatrix} \begin{Bmatrix} \ddot{\boldsymbol{d}} \\ \ddot{\boldsymbol{z}} \end{Bmatrix} + \begin{bmatrix} \boldsymbol{C}_d & \boldsymbol{0} \\ \boldsymbol{0} & \boldsymbol{C}_z \end{bmatrix} \begin{Bmatrix} \dot{\boldsymbol{d}} \\ \dot{\boldsymbol{z}} \end{Bmatrix} + \begin{bmatrix} \boldsymbol{K}_d & -\boldsymbol{K}_{dz} \\ -\boldsymbol{K}_{dz}^T & \boldsymbol{K}_z \end{bmatrix} \begin{Bmatrix} \boldsymbol{d} \\ \boldsymbol{z} \end{Bmatrix} = \begin{Bmatrix} \boldsymbol{f}_c + \boldsymbol{f}_d \\ \boldsymbol{0} \end{Bmatrix} \tag{9.169}$$

where

$$\begin{aligned} \boldsymbol{M}_d &= \int_0^L [\rho A(\boldsymbol{N}_w^T\boldsymbol{N}_w + \boldsymbol{N}_u^T\boldsymbol{N}_u) + \rho I \boldsymbol{N}_\psi^T\boldsymbol{N}_\psi - \alpha(\boldsymbol{N}_w'^T\boldsymbol{N}_u + \boldsymbol{N}_u^T\boldsymbol{N}_w') \\ &\quad + \gamma\boldsymbol{N}_w'^T\boldsymbol{N}_w' - \varepsilon_1(\boldsymbol{N}_\psi^T\boldsymbol{N}_w' + \boldsymbol{N}_w'^T\boldsymbol{N}_\psi) + \varepsilon_3(\boldsymbol{N}_u^T\boldsymbol{N}_\psi + \boldsymbol{N}_\psi^T\boldsymbol{N}_u)]\mathrm{d}x \\ \boldsymbol{C}_d &= \int_0^L [cA(\boldsymbol{N}_w^T\boldsymbol{N}_w + \boldsymbol{N}_u^T\boldsymbol{N}_u) + c_1\boldsymbol{N}_w'^T\boldsymbol{N}_w' + c_2\boldsymbol{N}_\psi^T\boldsymbol{N}_\psi - c_3(\boldsymbol{N}_w'^T\boldsymbol{N}_\psi \\ &\quad + \boldsymbol{N}_\psi^T\boldsymbol{N}_w') - c_4(\boldsymbol{N}_w'^T\boldsymbol{N}_u + \boldsymbol{N}_u'^T\boldsymbol{N}_w') + c_5(\boldsymbol{N}_u^T\boldsymbol{N}_\psi + \boldsymbol{N}_\psi^T\boldsymbol{N}_u)]\mathrm{d}x \\ \boldsymbol{K}_d &= \int_0^L [EA\boldsymbol{N}_u'^T\boldsymbol{N}_u' + EI_w\boldsymbol{N}_w''^T\boldsymbol{N}_w'' + EI_\psi\boldsymbol{N}_\psi'^T\boldsymbol{N}_\psi' \\ &\quad - \beta(\boldsymbol{N}_w''^T\boldsymbol{N}_u' + \boldsymbol{N}_u'^T\boldsymbol{N}_w'') - \varepsilon_2(\boldsymbol{N}_w''^T\boldsymbol{N}_\psi' + \boldsymbol{N}_\psi'^T\boldsymbol{N}_w'') \\ &\quad + \varepsilon_4(\boldsymbol{N}_u'^T\boldsymbol{N}_\psi' + \boldsymbol{N}_\psi'^T\boldsymbol{N}_u') + F\boldsymbol{N}_w'^T\boldsymbol{N}_w' + A_v(1+\alpha)\kappa\boldsymbol{N}_\psi^T\boldsymbol{N}_\psi]\mathrm{d}x \end{aligned}$$

$$\begin{aligned}
\boldsymbol{K}_{dz} &= A_v \alpha \kappa \int_0^L \boldsymbol{N}_\psi^{\mathrm{T}} \boldsymbol{N}_z \mathrm{d}x \\
\boldsymbol{M}_z &= \frac{A_v \alpha \kappa}{\hat{\omega}^2} \int_0^L \boldsymbol{N}_z^{\mathrm{T}} \boldsymbol{N}_z \mathrm{d}x \\
\boldsymbol{C}_z &= \frac{2A_v \alpha \kappa \hat{\xi}}{\hat{\omega}} \int_0^L \boldsymbol{N}_z^{\mathrm{T}} \boldsymbol{N}_z \mathrm{d}x \\
\boldsymbol{K}_z &= A_v \alpha \kappa \int_0^L \boldsymbol{N}_z^{\mathrm{T}} \boldsymbol{N}_z \mathrm{d}x
\end{aligned} \tag{9.170}$$

and

$$\boldsymbol{f}_{\mathrm{d}}(t) = \int_0^L [p(x,t)\boldsymbol{N}_w^{\mathrm{T}}(x) + \tau(x,t)\boldsymbol{N}_u^{\mathrm{T}}(x) + \gamma(x,t)\boldsymbol{N}_\psi^{\mathrm{T}}(x)]\mathrm{d}x \tag{9.171}$$

The finite element matrices defined by Equation 9.170 are given in Appendix 9.B.

Example 9.2

Consider a cantilevered PCLD beam as shown in Figure 9.7. The base beam is fully covered with an elastic constraining layer. The geometric and material properties of the base beam and elastic constraining layer are the same as those given in Table 9.1. The VEM layer is represented by a one-term GHM model and its properties are given in Table 9.3.

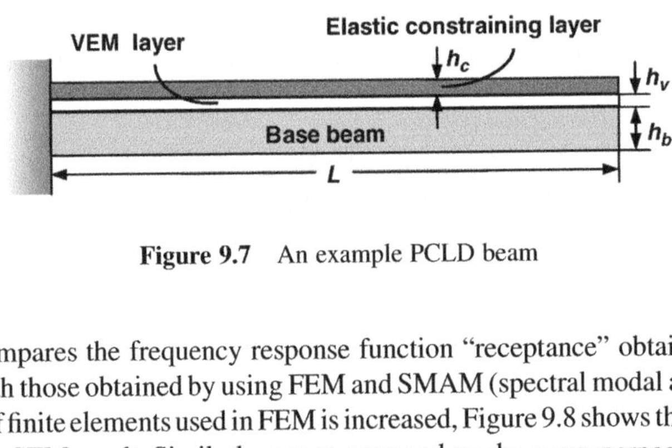

Figure 9.7 An example PCLD beam

Figure 9.8 compares the frequency response function "receptance" obtained by using the present SEM with those obtained by using FEM and SMAM (spectral modal analysis method). As the number of finite elements used in FEM is increased, Figure 9.8 shows that the FEM result gets closer to the SEM result. Similarly, as more normal modes are superposed in the spectral modal analysis, the SMAM result also converges towards the SEM result. The lowest five

Table 9.3 Geometric and material properties of the VEM layer

L	300 mm	b	20 mm	h_v	0.3 mm
E_v	0 GPa	ρ_v	1200 kg/m^3	κ	5×10^5 Pa
α	6.0	$\hat{\omega}$	1×10^5 rad/s	$\hat{\xi}$	4.0

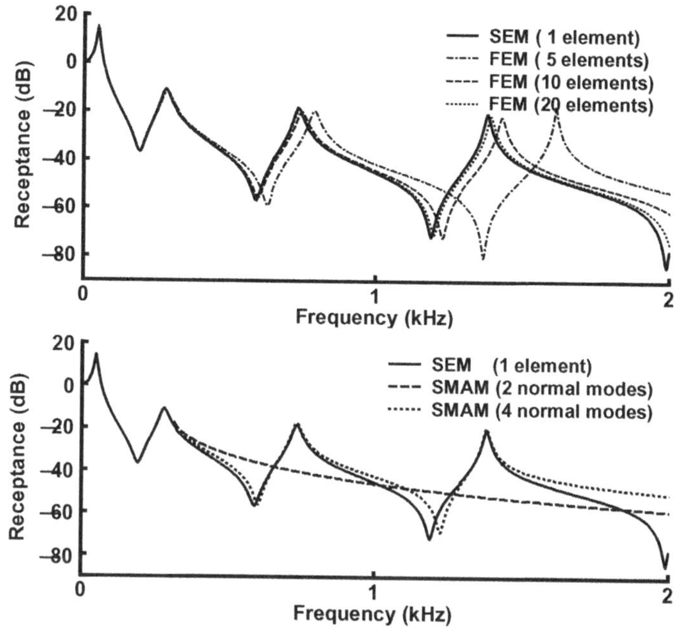

Figure 9.8 Frequency response functions obtained by SEM, FEM, and SMAM

Table 9.4 VEM-layer thickness dependence of the natural frequencies and modal damping ratios

Mode number (n)	PCLD beam $h_v = 0.3$ mm		PCLD beam $h_v = 0.03$ mm		PCLD beam $h_v = 0.003$ mm		Two-layer beam (Table 9.2)
	f_n (Hz)	ξ_n	f_n (Hz)	ξ_n	f_n (Hz)	ξ_n	f_n (Hz)
1	49.465	0.0881	55.398	0.0216	56.428	0.0028	56.554
2	280.11	0.0485	332.89	0.0249	351.53	0.0037	354.32
3	734.16	0.0164	882.14	0.0150	973.93	0.0030	991.69
4	1380.7	0.0058	1634.6	0.0045	1880.7	0.0026	1942.1
5	2237.1	0.0024	2580.2	0.0023	3055.2	0.0022	3207.7

natural frequencies and corresponding modal damping ratios are shown in Table 9.4 whilst varying the VEM layer thickness. The modal damping ratios are computed from

$$\xi_n = \frac{\omega_{nI}}{\sqrt{\omega_{n\mathrm{R}}^2 + \omega_{n\mathrm{I}}^2}}$$

where $\omega_{n\mathrm{R}}$ is the real part and $\omega_{n\mathrm{I}}$ is the imaginary part of the nth eigenfrequency ω_n. Table 9.4 shows that as the VEM layer thickness is decreased, all natural frequencies tend to increase, converging towards the natural frequencies of the elastic–elastic two-layer beam considered in Example 9.1, while the modal damping ratios decrease.

Appendix 9.A: Finite Element Matrices for the Elastic–Elastic Two-Layer Beam

The finite element matrices defined by Equation 9.76 are given by

$$\begin{aligned} \boldsymbol{M} &= \boldsymbol{M}_{\rho A} - \boldsymbol{M}_{\alpha} + \boldsymbol{M}_{\gamma} \\ \boldsymbol{C} &= \boldsymbol{C}_{cA} + \boldsymbol{C}_{c_1} - \boldsymbol{C}_{c_4} \\ \boldsymbol{K} &= \boldsymbol{K}_{EA} + \boldsymbol{K}_{EI} - \boldsymbol{K}_{\beta} + \boldsymbol{K}_{F} \end{aligned} \tag{9.A.1}$$

where

$$\begin{aligned}
\boldsymbol{M}_{\rho A} &= \rho A \int_0^L (\boldsymbol{N}_w^{\mathrm{T}} \boldsymbol{N}_w + \boldsymbol{N}_u^{\mathrm{T}} \boldsymbol{N}_u) \mathrm{d}x = \frac{\rho A L}{420} \begin{bmatrix} \boldsymbol{A}_1 & \boldsymbol{A}_2 \\ \boldsymbol{A}_2^{\mathrm{T}} & \boldsymbol{A}_3 \end{bmatrix} \\
\boldsymbol{M}_{\alpha} &= \alpha \int_0^L (\boldsymbol{N}_w'^{\mathrm{T}} \boldsymbol{N}_u + \boldsymbol{N}_u^{\mathrm{T}} \boldsymbol{N}_w') \mathrm{d}x = \frac{\alpha}{12} \begin{bmatrix} \boldsymbol{A}_4 & \boldsymbol{A}_5 \\ \boldsymbol{A}_5^{\mathrm{T}} & \boldsymbol{A}_6 \end{bmatrix} \\
\boldsymbol{M}_{\gamma} &= \gamma \int_0^L \boldsymbol{N}_w'^{\mathrm{T}} \boldsymbol{N}_w' \mathrm{d}x = \frac{\gamma}{30L} \begin{bmatrix} \boldsymbol{A}_7 & \boldsymbol{A}_8 \\ \boldsymbol{A}_8^{\mathrm{T}} & \boldsymbol{A}_9 \end{bmatrix} \\
\boldsymbol{C}_{CA} &= cA \int_0^L (\boldsymbol{N}_w^{\mathrm{T}} \boldsymbol{N}_w + \boldsymbol{N}_u^{\mathrm{T}} \boldsymbol{N}_u) \mathrm{d}x = \frac{cAL}{420} \begin{bmatrix} \boldsymbol{A}_1 & \boldsymbol{A}_2 \\ \boldsymbol{A}_2^{T} & \boldsymbol{A}_3 \end{bmatrix} \\
\boldsymbol{C}_{c_1} &= c_1 \int_0^L \boldsymbol{N}_w'^{\mathrm{T}} \boldsymbol{N}_w' \mathrm{d}x = \frac{c_1}{30L} \begin{bmatrix} \boldsymbol{A}_7 & \boldsymbol{A}_8 \\ \boldsymbol{A}_8^{\mathrm{T}} & \boldsymbol{A}_9 \end{bmatrix} \\
\boldsymbol{C}_{c_4} &= c_4 \int_0^L (\boldsymbol{N}_w'^{\mathrm{T}} \boldsymbol{N}_u + \boldsymbol{N}_u^{\mathrm{T}} \boldsymbol{N}_w') \mathrm{d}x = \frac{c_4}{12} \begin{bmatrix} \boldsymbol{A}_4 & \boldsymbol{A}_5 \\ \boldsymbol{A}_5^{\mathrm{T}} & \boldsymbol{A}_6 \end{bmatrix} \\
\boldsymbol{K}_{EA} &= EA \int_0^L \boldsymbol{N}_u'^{\mathrm{T}} \boldsymbol{N}_u' \mathrm{d}x = \frac{EA}{L} \begin{bmatrix} \boldsymbol{C}_1 & -\boldsymbol{C}_1 \\ -\boldsymbol{C}_1^{T} & \boldsymbol{C}_1 \end{bmatrix} \\
\boldsymbol{K}_{EI} &= EI \int_0^L \boldsymbol{N}_w''^{\mathrm{T}} \boldsymbol{N}_w'' \mathrm{d}x = \frac{EI}{L^3} \begin{bmatrix} \boldsymbol{A}_{10} & \boldsymbol{A}_{11} \\ \boldsymbol{A}_{11}^{\mathrm{T}} & \boldsymbol{A}_{12} \end{bmatrix} \\
\boldsymbol{K}_{\beta} &= \beta \int_0^L (\boldsymbol{N}_w''^{\mathrm{T}} \boldsymbol{N}_u' + \boldsymbol{N}_u'^{\mathrm{T}} \boldsymbol{N}_w'') \mathrm{d}x = \frac{\beta}{L} \begin{bmatrix} \boldsymbol{C}_2 & -\boldsymbol{C}_2 \\ -\boldsymbol{C}_2^{\mathrm{T}} & \boldsymbol{C}_2 \end{bmatrix} \\
\boldsymbol{K}_{F} &= F \int_0^L \boldsymbol{N}_w'^{\mathrm{T}} \boldsymbol{N}_w' \mathrm{d}x = \frac{F}{30L} \begin{bmatrix} \boldsymbol{A}_{13} & \boldsymbol{A}_{14} \\ \boldsymbol{A}_{14}^{\mathrm{T}} & \boldsymbol{A}_{15} \end{bmatrix}
\end{aligned} \tag{9.A.2}$$

with

$$
\begin{aligned}
&A_1 = \begin{bmatrix} 140 & 0 & 0 \\ 0 & 156 & 22L \\ 0 & 22L & 4L^2 \end{bmatrix} \qquad A_2 = \begin{bmatrix} 70 & 0 & 0 \\ 0 & 54 & -13L \\ 0 & 13L & -3L^2 \end{bmatrix} \\
&A_3 = \begin{bmatrix} 140 & 0 & 0 \\ 0 & 156 & -22L \\ 0 & -22L & 4L^2 \end{bmatrix} \qquad A_4 = \begin{bmatrix} 0 & -6 & L \\ -6 & 0 & 0 \\ L & 0 & 0 \end{bmatrix} \\
&A_5 = \begin{bmatrix} 0 & 6 & -L \\ -6 & 0 & 0 \\ -L & 0 & 0 \end{bmatrix} \qquad A_6 = \begin{bmatrix} 0 & 6 & L \\ 6 & 0 & 0 \\ L & 0 & 0 \end{bmatrix} \\
&A_7 = \begin{bmatrix} 0 & 0 & 0 \\ 0 & 36 & 3L \\ 0 & 3L & 4L^2 \end{bmatrix} \qquad A_8 = \begin{bmatrix} 0 & 0 & 0 \\ 0 & -36 & 3L \\ 0 & -3L & -L^2 \end{bmatrix} \\
&A_9 = \begin{bmatrix} 0 & 0 & 0 \\ 0 & 36 & -3L \\ 0 & -3L & 4L^2 \end{bmatrix} \qquad A_{10} = \begin{bmatrix} 0 & 0 & 0 \\ 0 & 12 & 6L \\ 0 & 6L & 4L^2 \end{bmatrix} \\
&A_{11} = \begin{bmatrix} 0 & 0 & 0 \\ 0 & -12 & 6L \\ 0 & -6L & 2L^2 \end{bmatrix} \qquad A_{12} = \begin{bmatrix} 0 & 0 & 0 \\ 0 & 12 & -6L \\ 0 & -6L & 4L^2 \end{bmatrix} \\
&A_{13} = \begin{bmatrix} 0 & 0 & 0 \\ 0 & 36 & 3L \\ 0 & 3L & 4L^2 \end{bmatrix} \qquad A_{14} = \begin{bmatrix} 0 & 0 & 0 \\ 0 & -36 & 3L \\ 0 & -3L & -L^2 \end{bmatrix} \\
&A_{15} = \begin{bmatrix} 0 & 0 & 0 \\ 0 & 36 & -3L \\ 0 & -3L & 4L^2 \end{bmatrix} \qquad C_1 = \begin{bmatrix} 1 & 0 & 0 \\ 0 & 0 & 0 \\ 0 & 0 & 0 \end{bmatrix} \\
&C_2 = \begin{bmatrix} 0 & 0 & 1 \\ 0 & 0 & 0 \\ 1 & 0 & 0 \end{bmatrix}
\end{aligned}
\tag{9.A.3}
$$

Appendix 9.B: Finite Element Matrices for the Elastic–VEM–Elastic Three-Layer Beam

The finite element matrices defined by Equation 9.169 are given by

$$
\begin{aligned}
&\boldsymbol{M}_d = \boldsymbol{M}_{\rho A} + \boldsymbol{M}_{\rho I} - \boldsymbol{M}_\alpha + \boldsymbol{M}_\gamma - \boldsymbol{M}_{\varepsilon_1} + \boldsymbol{M}_{\varepsilon_3} \\
&\boldsymbol{C}_d = \boldsymbol{C}_{cA} + \boldsymbol{C}_{c_1} + \boldsymbol{C}_{c_2} - \boldsymbol{C}_{c_3} - \boldsymbol{C}_{c_4} + \boldsymbol{C}_{c_5} \\
&\boldsymbol{K}_d = \boldsymbol{K}_{EA} + \boldsymbol{K}_{EI_w} + \boldsymbol{K}_{EI_\psi} - \boldsymbol{K}_\beta - \boldsymbol{K}_{\varepsilon_2} + \boldsymbol{K}_{\varepsilon_4} + \boldsymbol{K}_v + \boldsymbol{K}_F \\
&\boldsymbol{K}_{dz} = \frac{A_v \alpha \kappa L}{6} \boldsymbol{C}_7^{\mathrm{T}}, \quad \boldsymbol{M}_z = \frac{A_v \alpha \kappa L}{6\hat{\omega}^2} \boldsymbol{C}_6 \\
&\boldsymbol{C}_z = \frac{A_v \alpha \kappa \zeta L}{3\hat{\omega}} \boldsymbol{C}_6, \quad \boldsymbol{K}_z = \frac{A_v \alpha \kappa L}{6} \boldsymbol{C}_6
\end{aligned}
\tag{9.B.1}
$$

where

$$\boldsymbol{M}_{\rho A} = \rho A \int_0^L (\boldsymbol{N}_w^{\mathrm{T}} \boldsymbol{N}_w + \boldsymbol{N}_u^{\mathrm{T}} \boldsymbol{N}_u) \mathrm{d}x = \frac{\rho A L}{420} \begin{bmatrix} \bar{\boldsymbol{A}}_1 & \bar{\boldsymbol{A}}_2 \\ \bar{\boldsymbol{A}}_2^{\mathrm{T}} & \bar{\boldsymbol{A}}_3 \end{bmatrix}$$

$$\boldsymbol{M}_{\rho I} = \rho I \int_0^L \boldsymbol{N}_\psi^{\mathrm{T}} \boldsymbol{N}_\psi \mathrm{d}x = \frac{\rho I L}{6} \begin{bmatrix} 2\boldsymbol{C}_3 & \boldsymbol{C}_3 \\ \boldsymbol{C}_3^{\mathrm{T}} & 2\boldsymbol{C}_3 \end{bmatrix}$$

$$\boldsymbol{M}_{\alpha} = \alpha \int_0^L (\boldsymbol{N'}_w^{\mathrm{T}} \boldsymbol{N}_u + \boldsymbol{N}_u^{\mathrm{T}} \boldsymbol{N}'_w) \mathrm{d}x = \frac{\alpha}{12} \begin{bmatrix} \bar{\boldsymbol{A}}_4 & \bar{\boldsymbol{A}}_5 \\ \bar{\boldsymbol{A}}_5^{\mathrm{T}} & \bar{\boldsymbol{A}}_6 \end{bmatrix}$$

$$\boldsymbol{M}_{\gamma} = \gamma \int_0^L \boldsymbol{N'}_w^{\mathrm{T}} \boldsymbol{N}'_w \mathrm{d}x = \frac{\gamma}{30L} \begin{bmatrix} \bar{\boldsymbol{A}}_7 & \bar{\boldsymbol{A}}_8 \\ \bar{\boldsymbol{A}}_8^{\mathrm{T}} & \bar{\boldsymbol{A}}_9 \end{bmatrix}$$

$$\boldsymbol{M}_{\varepsilon_1} = \varepsilon_1 \int_0^L (\boldsymbol{N}_\psi^{\mathrm{T}} \boldsymbol{N}'_w + \boldsymbol{N'}_w^{\mathrm{T}} \boldsymbol{N}_\psi) \mathrm{d}x = \frac{\varepsilon_1}{12} \begin{bmatrix} \boldsymbol{A}_{16} & \boldsymbol{A}_{17} \\ \boldsymbol{A}_{17}^{\mathrm{T}} & \boldsymbol{A}_{18} \end{bmatrix}$$

$$\boldsymbol{M}_{\varepsilon_3} = \varepsilon_3 \int_0^L (\boldsymbol{N}_\psi^{\mathrm{T}} \boldsymbol{N}_u + \boldsymbol{N}_u^{\mathrm{T}} \boldsymbol{N}_\psi) \mathrm{d}x = \frac{\varepsilon_3 L}{6} \begin{bmatrix} 2\boldsymbol{C}_4 & \boldsymbol{C}_4 \\ \boldsymbol{C}_4^{\mathrm{T}} & 2\boldsymbol{C}_4 \end{bmatrix}$$

$$\boldsymbol{C}_{CA} = cA \int_0^L (\boldsymbol{N}_w^{\mathrm{T}} \boldsymbol{N}_w + \boldsymbol{N}_u^{\mathrm{T}} \boldsymbol{N}_u) \mathrm{d}x = \frac{cAL}{420} \begin{bmatrix} \bar{\boldsymbol{A}}_1 & \bar{\boldsymbol{A}}_2 \\ \bar{\boldsymbol{A}}_2^{\mathrm{T}} & \bar{\boldsymbol{A}}_3 \end{bmatrix} \tag{9.B.2}$$

$$\boldsymbol{C}_{c_1} = c_1 \int_0^L \boldsymbol{N'}_w^{\mathrm{T}} \boldsymbol{N}'_w \mathrm{d}x = \frac{c_1}{30L} \begin{bmatrix} \bar{\boldsymbol{A}}_7 & \bar{\boldsymbol{A}}_8 \\ \bar{\boldsymbol{A}}_8^{\mathrm{T}} & \bar{\boldsymbol{A}}_9 \end{bmatrix}$$

$$\boldsymbol{C}_{c_2} = c_2 \int_0^L \boldsymbol{N}_\psi{}^{\mathrm{T}} \boldsymbol{N}_\psi \mathrm{d}x = \frac{c_2 L}{6} \begin{bmatrix} 2\boldsymbol{C}_3 & \boldsymbol{C}_3 \\ \boldsymbol{C}_3^{\mathrm{T}} & 2\boldsymbol{C}_3 \end{bmatrix}$$

$$\boldsymbol{C}_{c_3} = c_3 \int_0^L (\boldsymbol{N}_\psi^{\mathrm{T}} \boldsymbol{N}'_w + \boldsymbol{N'}_w^{\mathrm{T}} \boldsymbol{N}_\psi) \mathrm{d}x = \frac{c_3}{12} \begin{bmatrix} \boldsymbol{A}_{16} & \boldsymbol{A}_{17} \\ \boldsymbol{A}_{17}^{\mathrm{T}} & \boldsymbol{A}_{18} \end{bmatrix}$$

$$\boldsymbol{C}_{c_4} = c_4 \int_0^L (\boldsymbol{N'}_w^{\mathrm{T}} \boldsymbol{N}_u + \boldsymbol{N}_u^{\mathrm{T}} \boldsymbol{N}'_w) \mathrm{d}x = \frac{c_4}{12} \begin{bmatrix} \bar{\boldsymbol{A}}_4 & \bar{\boldsymbol{A}}_5 \\ \bar{\boldsymbol{A}}_5^{\mathrm{T}} & \bar{\boldsymbol{A}}_6 \end{bmatrix}$$

$$\boldsymbol{C}_{c_5} = c_5 \int_0^L (\boldsymbol{N}_\psi^{\mathrm{T}} \boldsymbol{N}_u + \boldsymbol{N}_u^{\mathrm{T}} \boldsymbol{N}_\psi) \mathrm{d}x = \frac{c_5 L}{6} \begin{bmatrix} 2\boldsymbol{C}_4 & \boldsymbol{C}_4 \\ \boldsymbol{C}_4^{\mathrm{T}} & 2\boldsymbol{C}_4 \end{bmatrix}$$

$$\boldsymbol{K}_{EA} = EA \int_0^L \boldsymbol{N'}_u^{\mathrm{T}} \boldsymbol{N}'_u \mathrm{d}x = \frac{EA}{L} \begin{bmatrix} \bar{\boldsymbol{C}}_1 & -\bar{\boldsymbol{C}}_1 \\ -\bar{\boldsymbol{C}}_1^{\mathrm{T}} & \bar{\boldsymbol{C}}_1 \end{bmatrix}$$

$$\boldsymbol{K}_{EI_w} = EI_w \int_0^L \boldsymbol{N''}_w^{\mathrm{T}} \boldsymbol{N}''_w \mathrm{d}x = \frac{EI_w}{L^3} \begin{bmatrix} \bar{\boldsymbol{A}}_{10} & \bar{\boldsymbol{A}}_{11} \\ \bar{\boldsymbol{A}}_{11}^{\mathrm{T}} & \bar{\boldsymbol{A}}_{12} \end{bmatrix}$$

$$\boldsymbol{K}_{EI_\psi} = EI_\psi \int_0^L \boldsymbol{N}'^{\mathrm{T}}_\psi \boldsymbol{N}'_\psi \mathrm{d}x = \frac{EI_\psi}{L}\begin{bmatrix} \boldsymbol{C}_3 & -\boldsymbol{C}_3 \\ -\boldsymbol{C}_3^{\mathrm{T}} & \boldsymbol{C}_3 \end{bmatrix}$$

$$\boldsymbol{K}_\beta = \beta \int_0^L (\boldsymbol{N}''^{\mathrm{T}}_w \boldsymbol{N}'_u + \boldsymbol{N}'^{\mathrm{T}}_u \boldsymbol{N}''_w)\mathrm{d}x = \frac{\beta}{L}\begin{bmatrix} \bar{\boldsymbol{C}}_2 & -\bar{\boldsymbol{C}}_2 \\ -\bar{\boldsymbol{C}}_2^{\mathrm{T}} & \bar{\boldsymbol{C}}_2 \end{bmatrix}$$

$$\boldsymbol{K}_{\varepsilon_2} = \varepsilon_2 \int_0^L (\boldsymbol{N}'^{\mathrm{T}}_\psi \boldsymbol{N}''_w + \boldsymbol{N}''^{\mathrm{T}}_w \boldsymbol{N}'_\psi)\mathrm{d}x = \frac{\varepsilon_2}{L}\begin{bmatrix} \boldsymbol{C}_5 & -\boldsymbol{C}_5 \\ -\boldsymbol{C}_4^{\mathrm{T}} & \boldsymbol{C}_4 \end{bmatrix}$$

$$\boldsymbol{K}_v = A_v\kappa(1+\alpha) \int_0^L \boldsymbol{N}^{\mathrm{T}}_\psi \boldsymbol{N}_\psi \mathrm{d}x = \frac{A_v\kappa(1+\alpha)L}{6}\begin{bmatrix} 2\boldsymbol{C}_3 & \boldsymbol{C}_3 \\ \boldsymbol{C}_3^{\mathrm{T}} & 2\boldsymbol{C}_3 \end{bmatrix}$$

$$\boldsymbol{K}_F = F \int_0^L \boldsymbol{N}'^{\mathrm{T}}_w \boldsymbol{N}'_w \mathrm{d}x = \frac{F}{30L}\begin{bmatrix} \bar{\boldsymbol{A}}_{13} & \bar{\boldsymbol{A}}_{14} \\ \bar{\boldsymbol{A}}_{14}^{\mathrm{T}} & \bar{\boldsymbol{A}}_{15} \end{bmatrix}$$

and where

$$\bar{\boldsymbol{A}}_k = \begin{bmatrix} \boldsymbol{A}_k & \boldsymbol{0} \\ \boldsymbol{0}^{\mathrm{T}} & \boldsymbol{0} \end{bmatrix} \quad (k = 1, 2, \ldots, 15)$$

$$\bar{\boldsymbol{C}}_k = \begin{bmatrix} \boldsymbol{C}_k & \boldsymbol{0} \\ \boldsymbol{0}^{\mathrm{T}} & \boldsymbol{0} \end{bmatrix} \quad (k = 1, 2) \tag{9.B.3}$$

and

$$\boldsymbol{A}_{16} = \begin{bmatrix} 0 & 0 & 0 & 0 \\ 0 & 0 & 0 & -6 \\ 0 & 0 & 0 & L \\ 0 & -6 & L & 0 \end{bmatrix} \quad \boldsymbol{A}_{17} = \begin{bmatrix} 0 & 0 & 0 & 0 \\ 0 & 0 & 0 & -6 \\ 0 & 0 & 0 & -L \\ 0 & 6 & -L & 0 \end{bmatrix}$$

$$\boldsymbol{A}_{18} = \begin{bmatrix} 0 & 0 & 0 & 0 \\ 0 & 0 & 0 & 6 \\ 0 & 0 & 0 & L \\ 0 & 6 & L & 0 \end{bmatrix} \quad \boldsymbol{C}_3 = \begin{bmatrix} 0 & 0 & 0 & 0 \\ 0 & 0 & 0 & 0 \\ 0 & 0 & 0 & 0 \\ 0 & 0 & 0 & 1 \end{bmatrix} \tag{9.B.4}$$

$$\boldsymbol{C}_4 = \begin{bmatrix} 0 & 0 & 0 & 1 \\ 0 & 0 & 0 & 0 \\ 0 & 0 & 0 & 0 \\ 1 & 0 & 0 & 0 \end{bmatrix} \quad \boldsymbol{C}_5 = \begin{bmatrix} 0 & 0 & 0 & 0 \\ 0 & 0 & 0 & 0 \\ 0 & 0 & 0 & 1 \\ 0 & 0 & 1 & 0 \end{bmatrix}$$

$$\boldsymbol{C}_6 = \begin{bmatrix} 2 & 1 \\ 1 & 2 \end{bmatrix} \quad \boldsymbol{C}_7 = \begin{bmatrix} 0 & 0 & 0 & 2 & 0 & 0 & 0 & 1 \\ 0 & 0 & 0 & 1 & 0 & 0 & 0 & 2 \end{bmatrix}$$

The matrices A_k $(k = 1, 2, \ldots, 15)$ and C_k $(k = 1, 2)$ are given in Appendix 9.A.

10

Dynamics of Smart Structures

Piezoelectric materials have received considerable attention over recent decades due to their potential applications to the active controls of structural vibration and noise. The converse piezoelectric effects are used for actuator designs while the direct piezoelectric effect are used for sensor designs. In order to use the piezoelectric effects, the piezoelectric materials are usually bonded onto the surfaces of the structures so that the structures become the active multi-layer laminate structures. In this chapter, we consider two types of active multi-layer laminate beams: the elastic–piezoelectric two-layer beams [36, 87] and the elastic–viscoelastic–piezoelectric three-layer (active constrained layered damping (ACLD)) beams [73, 88, 105, 112, 141, 147]. By using Hamilton's principle, the coupled equations of motion are derived for both active multi-layer laminate beams. The effects of the structural damping and viscous damping of each layer are taken into account in the spectral element models formulated by using the variational method. The spectral element modeling is also introduced for an ACLD beam subjected to the feedback control. The orthogonality properties of the normal modes of a coupled structural dynamics problem are formulated to derive spectral modal equations for both active multi-layer laminate beams. The finite element models are also provided for use in the evaluation of the spectral element models.

10.1 Elastic–Piezoelectric Two-Layer Beams

10.1.1 Equations of Motion

Figure 10.1 shows an elastic–piezoelectric two-layer beam element. $w(x, t)$ is the transverse displacement of the base beam, $u_b(x, t)$ is the axial displacement of the base beam, and $u_p(x, t)$ is the axial displacement of the piezoelectric material (PZT) layer. h_b and h_p are the thicknesses of the base beam and PZT layer, respectively. In this chapter, the subscripts b and p are used to denote the quantities of the base beam and PZT layer, respectively.

For the derivation of the equations of motion, the following assumptions are made [87].

a. Geometric and material properties are uniform along the beam.
b. Perfect continuity, but no slip, at the interfaces.

Spectral Element Method in Structural Dynamics Usik Lee

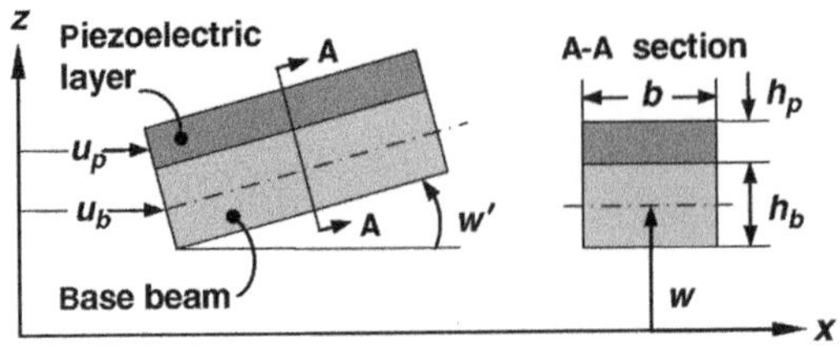

Figure 10.1 An elastic–PZT two-layer beam element

c. Both layers take the same transverse displacement.
d. Small amplitude vibration.
e. Linear elasticity and piezoelectricity are valid.
f. Shear deformations and rotary inertias in both layers are negligible.
g. Uniform voltage along the beam.

Based on the above assumptions, we can obtain a kinematic relationship as

$$u_p = u_b - (h/2)w' \tag{10.1}$$

where

$$h = h_b + h_p \tag{10.2}$$

When the PZT layer is very thin, throughout its thickness the electrical displacement D_3 can be considered to be constant. In this situation, the in-plane electrical fields will be negligible, especially when the voltage applied to the PZT layer is uniform along the axis of the beam. The constitutive relationship for a piezoelectric material under uni-axial loading is given by

$$\begin{Bmatrix} \sigma_1 \\ E_3 \end{Bmatrix} = \begin{bmatrix} C_{11}^D & -h_{31} \\ -h_{31} & \beta_{33}^S \end{bmatrix} \begin{Bmatrix} \varepsilon_1 \\ D_3 \end{Bmatrix} \tag{10.3}$$

where σ_1 and ε_1 are the mechanical stress and strain in the x-direction, respectively; D_3 is the electrical displacement (charge/area in the beam vertical direction) and E_3 is the electrical field (voltage/length along the vertical direction), all in the z-direction; C_{11}^D is the elastic stiffness; β_{33}^S is the dielectric constant; and h_{31} is the piezoelectric constant. The material constants h_{31} and β_{33}^S are related to the piezoelectric constant d_{31} through

$$h_{31}^2 \beta_{33}^{S-1} = d_{31} E_p \tag{10.4}$$

where

$$E_p = C_{11}^D - h_{31}^2 \beta_{33}^{S-1} \tag{10.5}$$

For the direct effect, applying a constant strain under the conditions of zero electric field ($E_3 = 0$) yields

$$D_3 = h_{31}^2 \beta_{33}^{S-1} \varepsilon_1 = d_{31} E_p u_p' = d_{31} E_p [u_b' - (h_2/2)w'' + h_v \psi'] \tag{10.6}$$

The equations of motion are derived by using Hamilton's principle

$$\int_{t_1}^{t_2} (\delta T - \delta V + \delta W)\mathrm{d}t = 0 \tag{10.7}$$

where T and V are the kinetic energy and potential energy, respectively, and δW is the virtual work. By using the constitutive relationship (10.3) and assuming that D_3 is constant throughout the thickness of the PZT layer, the strain energy V and the kinetic energy T are obtained as

$$V = \frac{1}{2}\int_0^L [E_bA_bu_b'^2 + E_bI_bw''^2 + C_{11}^DI_pw''^2 + C_{11}^DA_pu_p'^2 - 2A_ph_{31}D_3u_p' + A_p\beta_{33}^SD_3^2 + Fw'^2]\mathrm{d}x$$
$$T = \frac{1}{2}\int_0^L [\rho_bA_b(\dot{u}_b^2 + \dot{w}^2) + \rho_pA_p(\dot{u}_p^2 + \dot{w}^2)]\mathrm{d}x \tag{10.8}$$

where E, A, I, and ρ (for each layer) are the Young's modulus, the cross-sectional area, the area moment of inertia about the neutral axis of each layer, and the mass density, respectively, and F is the constant axial tensile force applied to the beam. By eliminating u_p and D_3 from Equation 10.8 by using Equations 10.1 and 10.6, respectively, we can write the strain and kinetic energies as

$$\begin{aligned} V &= \frac{1}{2}\int_0^L (EAu_b'^2 + EIw''^2 - 2\beta w''u_b' + Fw'^2)\mathrm{d}x \\ T &= \frac{1}{2}\int_0^L (\rho A\dot{u}_b^2 + \rho A\dot{w}^2 - 2\alpha\dot{w}'\dot{u}_b + \gamma\dot{w}'^2)\mathrm{d}x \end{aligned} \tag{10.9}$$

where

$$\begin{aligned} EA &= E_bA_b + E_pA_p, \quad & EI &= E_bI_b + C_{11}^DI_p + (1/4)E_pA_ph^2 \\ \rho A &= \rho_bA_b + \rho_pA_p, \quad & \alpha &= (1/2)\rho_pA_ph \\ \beta &= (1/2)E_pA_ph, \quad & \gamma &= (1/4)\rho_pA_ph^2 \end{aligned} \tag{10.10}$$

The virtual work δW done by the external forces, viscous damping forces, boundary forces, and applied voltage is given by

$$\begin{aligned} \delta W = & \int_0^L p(x,t)\delta w\mathrm{d}x + \int_0^L \tau(x,t)\delta u_b\mathrm{d}x - \int_0^L\int_{A_b} c_b\dot{\boldsymbol{r}}_b\cdot\delta\boldsymbol{r}_b\mathrm{d}A_b\mathrm{d}x \\ & - \int_0^L\int_{A_p} c_p\dot{\boldsymbol{r}}_p\cdot\delta\boldsymbol{r}_p\mathrm{d}A_p\mathrm{d}x + N_1(t)\delta u_b(0,t) + N_2(t)\delta u_b(L,t) \\ & + M_1(t)\delta w'(0,t) + M_2(t)\delta w'(L,t) + Q_1(t)\delta w(0,t) + Q_2(t)\delta w(L,t) \\ & + \int_0^L bV(t)\delta D_3\mathrm{d}x \end{aligned} \tag{10.11}$$

where

$$\boldsymbol{r}_b = u_b\boldsymbol{n}_x + w\boldsymbol{n}_z, \quad \boldsymbol{r}_p = u_p\boldsymbol{n}_x + w\boldsymbol{n}_z \tag{10.12}$$

where $\boldsymbol{n}_x$ and $\boldsymbol{n}_z$ are the unit direction vectors in the x- and z-directions, respectively. In Equation 10.11, $V(t)$ is the uniform voltage applied to the PZT layer and c_b and c_p are viscous damping coefficients. The definitions of the other symbols used in Equation 10.11 are the same as used in Equation 9.7. By applying Equations 10.1, 10.6 and 10.12 into Equation 10.11 and taking the integral by parts, we can write the virtual work as

$$\begin{aligned}\delta W = \int_0^L & [p(x,t)\delta w + \tau(x,t)\delta u_b - cA\dot{u}_b\delta u_b - cA\dot{w}\delta w + c_4\dot{w}'\delta u_b \\ & + c_4\dot{u}_b\delta w' - c_1\dot{w}'\delta w']dx + N_1(t)\delta u_b(0,t) + N_2(t)\delta u_b(L,t) \\ & + M_1(t)\delta w'(0,t) + M_2(t)\delta w'(L,t) + Q_1(t)\delta w(0,t) + Q_2(t)\delta w(L,t) \\ & + bd_{31}E_pV(t)\delta u_b\big|_0^L - (1/2)bhd_{31}E_pV(t)\delta w'\big|_0^L\end{aligned} \tag{10.13}$$

where

$$cA = c_bA_b + c_pA_p, \quad c_1 = (1/4)c_pA_ph^2, \quad c_4 = (1/2)c_pA_ph \tag{10.14}$$

By substituting Equations 10.8 and 10.13 into Hamilton's principle (10.7), we obtain coupled two equations of motion

$$\begin{aligned}EIw'''' + \rho A\ddot{w} + cA\dot{w} &= -\alpha\ddot{u}_b' + \beta u''' + \gamma\ddot{w}'' + c_1\dot{w}'' - c_4\dot{u}_b' + Fw'' + p(x,t) \\ EAu_b'' - \rho A\ddot{u}_b - cA\dot{u}_b &= -\alpha\ddot{w}' + \beta w''' - c_4\dot{w}' - \tau(x,t)\end{aligned} \tag{10.15}$$

and the boundary conditions

$$\begin{aligned}Q(0,t) &= -Q_1(t) - Fw'(0,t) \quad &\text{or} \quad w(0,t) &= w_1(t) \\ M(0,t) &= -M_1(t) - M^p(t) \quad &\text{or} \quad w'(0,t) &= \theta_1(t) \\ N(0,t) &= -N_1(t) + N^p(t) \quad &\text{or} \quad u_b(0,t) &= u_1(t) \\ Q(L,t) &= +Q_2(t) - Fw'(L,t) \quad &\text{or} \quad w(L,t) &= w_2(t) \\ M(L,t) &= +M_2(t) - M^p(t) \quad &\text{or} \quad w'(L,t) &= \theta_2(t) \\ N(L,t) &= +N_2(t) + N^p(t) \quad &\text{or} \quad u_b(L,t) &= u_2(t)\end{aligned} \tag{10.16}$$

where

$$M^p(t) = (1/2)bhd_{31}E_pV(t), \quad N^p(t) = bd_{31}E_pV(t) \tag{10.17}$$

where $b = b_b = b_p$ is the width of the two-layer beam. The piezoelectric force $N^p(t)$ and moment $M^p(t)$ are generated by the voltage $V(t)$ applied to the PZT layer. The resultant axial force $N(x, t)$, bending moment $M(x, t)$ and the transverse shear force $Q(x, t)$ are related to the displacement fields as follows:

$$\begin{aligned}Q(x,t) &= -EI_ww''' - \alpha\ddot{u}_b + \beta u_b'' + \gamma\ddot{w}' + c_1\dot{w}' - c_4\dot{u}_b \\ &= -M(x,t) - \alpha\ddot{u}_b + \gamma\ddot{w}' + c_1\dot{w}' - c_4\dot{u}_b \\ M(x,t) &= EI_ww'' - \beta u_b' \\ N(x,t) &= EAu_b' - \beta w''\end{aligned} \tag{10.18}$$

The structural damping of each layer can be readily taken into account by using the complex modulus as

$$E_b^* = E_b(1 + i\eta_b), \qquad E_p^* = C_{11}^{D*} - h_{31}^2\beta_{33}^{S-1}, \qquad C_{11}^{D*} = C_{11}^D(1 + i\eta_p) \tag{10.19}$$

where η_b and η_p are the loss factors of the base beam and PZT layer, respectively.

The governing Equations 10.15 for an elastic–PZT two-layer beam have exactly the same forms as the governing Equations 9.11 for an elastic–elastic two-layer beam, except for the use of the structural and damping properties (Equations 10.10 and 10.14) defined for the elastic–PZT two-layer beam. Thus, a large part of the spectral element formulation carried out in Chapter 9 for the elastic–elastic two-layer beam can be shared by the elastic–PZT two-layer beam.

10.1.2 Spectral Element Modeling

10.1.2.1 Governing Equations in the Frequency Domain

To transform the governing equations into the frequency domain, we represent all dynamic responses, forces, and moments in the spectral forms as given by Equations 9.15 and 9.16. In addition, we also represent piezoelectric forces, moments, and voltage in the spectral forms as

$$\{M^p(x,t)N^p(x,t)V(x,t)\} = \frac{1}{N}\sum_{n=0}^{N-1}\{M_n^p(x;\omega_n)N_n^p(x;\omega_n)V_n(x;\omega_n)\}e^{i\omega_n t} \tag{10.20}$$

By substituting Equations 9.15, 9.16 and 10.20 into Equations 10.15 and 10.16, we obtain the frequency-domain governing equations

$$\begin{aligned} & EIW'''' - \omega^2\rho AW + i\omega cAW \\ & \quad = \omega^2\alpha U' + \beta U''' - \omega^2\gamma W'' + i\omega c_1 W'' - i\omega c_4 U' + FW'' + P(x) \\ & EAU'' + \omega^2\rho AU - i\omega cAU = \omega^2\alpha W' + \beta W''' - i\omega c_4 W' - T(x) \end{aligned} \tag{10.21}$$

and the boundary conditions

$$\begin{aligned} Q(0) &= -Q_1 - FW'(0) \quad \text{or} \quad W(0) = W_1 \\ M(0) &= -M_1 - M^p \quad \text{or} \quad W'(0) = \Theta_1 \\ N(0) &= -N_1 + N^p \quad \text{or} \quad U(0) = U_1 \\ Q(L) &= +Q_2 - FW'(L) \quad \text{or} \quad W(L) = W_2 \\ M(L) &= +M_2 - M^p \quad \text{or} \quad W'(L) = \Theta_2 \\ N(L) &= +N_2 + N^p \quad \text{or} \quad U(L) = U_2 \end{aligned} \tag{10.22}$$

where

$$M^p = (1/2)bd_{31}E_p hV(\omega), \quad N^p = bd_{31}E_p V(\omega) \tag{10.23}$$

Similarly, we can transform the force-displacement relationships (10.18) into the frequency domain as

$$\begin{aligned} Q(x) &= -EIW''' + \omega^2\alpha U + \beta U'' - \omega^2\gamma W' + i\omega c_1 W' - i\omega c_4 U \\ M(x) &= EIW'' - \beta U' \\ N(x) &= EAU' - \beta W'' \end{aligned} \tag{10.24}$$

In the above equations, the subscripts n are omitted for brevity.

10.1.2.2 Spectral Nodal DOFs, Forces, and Moments

The spectral nodal degrees-of-freedom (DOFs) and the mechanical spectral nodal forces and moments are specified in Figure 10.2. We define the spectral nodal DOFs vector as

$$\boldsymbol{d} = \{U_1\ W_1\ \Theta_1\ U_2\ W_2\ \Theta_2\}^T = \{U(0)\ W(0)\ W'(0)\ U(L)\ W(L)\ W'(L)\}^T \tag{10.25}$$

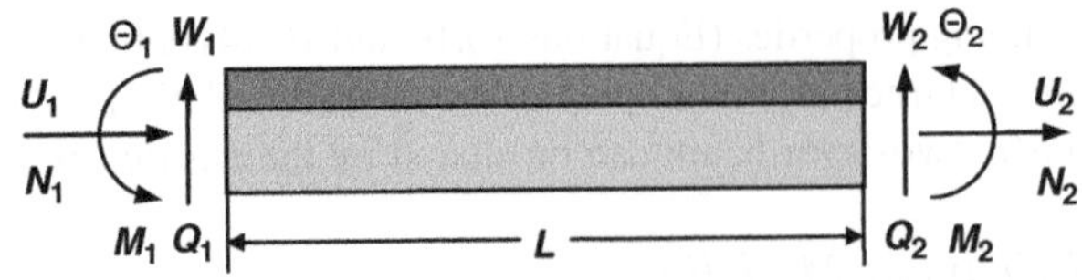

Figure 10.2 Spectral element of the elastic–PZT two-layer beam

and the mechanical spectral nodal forces and moments vector as

$$f_c^{\text{mech}} = \{N_1\ Q_1\ M_1\ N_2\ Q_2\ M_2\}^{\text{T}} \tag{10.26}$$

10.1.2.3 Dynamic Shape Functions

To formulate the spectral element equation, we consider the homogeneous governing equations reduced from Equation 10.21 to be

$$\begin{aligned} & EIW'''' - \omega^2 \rho A W + i\omega c A W \\ & \quad = \omega^2 \alpha U' + \beta U''' - \omega^2 \gamma W'' + i\omega c_1 W'' - i\omega c_4 U' + F W'' \\ & EAU'' + \omega^2 \rho A U - i\omega c A U = \omega^2 \alpha W' + \beta W''' - i\omega c_4 W' \end{aligned} \tag{10.27}$$

As Equation 10.27 is identical with Equation 9.22, except for using the structural and damping properties defined by Equations 10.10 and 10.14, we can follow the same procedure as used in Section 9.1.2.3 to derive a dispersion relation

$$a_1 k^6 + a_2 k^4 + a_3 k^2 + a_4 = 0 \tag{10.28}$$

where $a_i\ (i = 1, 2, 3, 4)$ are given by Equation 9.27 with the structural and damping properties replaced by Equations 10.10 and 10.14. By using six wavenumbers $k_i\ (i = 1, 2, \ldots, 6)$ computed from Equation 10.28, we can obtain general solutions to Equation 10.27 in the form of

$$W(x;\omega) = \boldsymbol{N}_w(x;\omega)\boldsymbol{d}, \quad U(x;\omega) = \boldsymbol{N}_u(x;\omega)\boldsymbol{d} \tag{10.29}$$

where $\boldsymbol{N}_w$ and $\boldsymbol{N}_u$ are the dynamic shape functions given by

$$\boldsymbol{N}_w(x;\omega) = \boldsymbol{e}(x;\omega)\boldsymbol{H}^{-1}(\omega), \quad \boldsymbol{N}_u(x;\omega) = \boldsymbol{e}(x;\omega)\boldsymbol{R}\boldsymbol{H}^{-1}(\omega) \tag{10.30}$$

The matrices $\boldsymbol{e}(x;\omega)$, $\boldsymbol{R}(\omega)$ and $\boldsymbol{H}(\omega)$ are defined by Equations 9.31, 9.29 and 9.34, respectively.

10.1.2.4 Weak Form of Governing Equations

As shown in Section 9.1.2.4, the weak form of the governing Equation 10.27 can be derived as

$$\begin{aligned} & \int_0^L [EIW''\delta W'' - \omega^2 \rho A W \delta W + i\omega c A W \delta W + i\omega c_1 W' \delta W' \\ & \quad - i\omega c_4 U \delta W' + \omega^2 \alpha U \delta W' - \beta U' \delta W'' - \omega^2 \gamma W' \delta W' + F W' \delta W' \\ & \quad + EAU'\delta U' - \omega^2 \rho A U \delta U + i\omega c A U \delta U + \omega^2 \alpha W' \delta U - \beta W'' \delta U' \\ & \quad - i\omega c_4 W' \delta U] \mathrm{d}x - \{[Q(x) + F W'(x)]\delta W|_0^L + M(x)\delta W'|_0^L + N(x)\delta U|_0^L\} \\ & \quad - \int_0^L [P(x)\delta W + T(x)\delta U] \mathrm{d}x = 0 \end{aligned} \tag{10.31}$$

10.1.2.5 Spectral Element Equation

As shown in Section 9.1.2.5, substitution of Equation 10.29 into Equation 10.31 gives the spectral element equation as

$$\boldsymbol{S}(\omega)\boldsymbol{d} = \boldsymbol{f}(\omega) \tag{10.32}$$

The matrix $\boldsymbol{S}(\omega)$ is the spectral element matrix defined by

$$\boldsymbol{S}(\omega) = \boldsymbol{H}^{-\mathrm{T}}(\omega)\boldsymbol{D}(\omega)\boldsymbol{H}^{-1}(\omega) \tag{10.33}$$

where

$$\begin{aligned}\boldsymbol{D}(\omega) = &- EA\boldsymbol{RKEKR} + EI\boldsymbol{K}^2\boldsymbol{EK}^2 - i\beta(\boldsymbol{K}^2\boldsymbol{EKR} + \boldsymbol{RKEK}^2)\\ &- \omega^2[\rho A(\boldsymbol{E} + \boldsymbol{RER}) + i\alpha(\boldsymbol{KER} + \boldsymbol{REK}) - \gamma\boldsymbol{KEK}]\\ &+ i\omega[cA(\boldsymbol{E} + \boldsymbol{RER}) - c_1\boldsymbol{KEK} + ic_4(\boldsymbol{KER} + \boldsymbol{REK})] - F\boldsymbol{RER}\end{aligned} \tag{10.34}$$

In Equation 10.34, the following definitions are used.

$$\boldsymbol{K} = diag\,[k_i] = \boldsymbol{K}^{\mathrm{T}}, \quad \boldsymbol{K}^2 = diag\,[k_i^2] = \boldsymbol{K}^{2\mathrm{T}} \quad (j = 1, 2, \ldots, 6) \tag{10.35}$$

$$\boldsymbol{E}(\omega) = \int_0^L \boldsymbol{e}^{\mathrm{T}}(x;\omega)\boldsymbol{e}(x;\omega)\mathrm{d}x = [E_{pq}(\omega)] \quad (p, q = 1, 2, \ldots, 6) \tag{10.36}$$

where

$$E_{pq} = \begin{cases} \dfrac{i}{k_p + k_q}\left[\mathrm{e}^{-i(k_p + k_q)L} - 1\right] & \text{if} \quad k_p + k_q \neq 0 \\ L & \text{if} \quad k_p + k_q = 0 \end{cases} \tag{10.37}$$

The vector $\boldsymbol{f}(\omega)$ is the spectral nodal forces vector defined by

$$\boldsymbol{f}(\omega) = \boldsymbol{f}_{\mathrm{c}}^{\mathrm{mech}}(\omega) + \boldsymbol{f}_{\mathrm{c}}^{\mathrm{piezo}}(\omega) + \boldsymbol{f}_{\mathrm{d}}(\omega) \tag{10.38}$$

where

$$\boldsymbol{f}_{\mathrm{c}}^{\mathrm{piezo}} = \{-N^p \quad 0 \quad M^p \quad N^p \quad 0 \quad -M^p\}^{\mathrm{T}} \tag{10.39}$$

$$\boldsymbol{f}_{\mathrm{d}}(\omega) = \int_0^L [P(x;\omega)\boldsymbol{N}_w^{\mathrm{T}}(x;\omega) + T(x;\omega)\boldsymbol{N}_u^{\mathrm{T}}(x;\omega)]\mathrm{d}x \tag{10.40}$$

In the above, $\boldsymbol{f}_{\mathrm{c}}^{\mathrm{piezo}}$ and $\boldsymbol{f}_{\mathrm{c}}^{\mathrm{mech}}$ represent the piezoelectric spectral nodal forces and moments generated by the applied voltage and the mechanical spectral nodal forces and moments vector defined by Equation 10.26, respectively.

The spectral elements represented by Equation 10.32 can be assembled to form a global system equation as

$$\boldsymbol{S}_{\mathrm{g}}(\omega)\boldsymbol{d}_{\mathrm{g}} = \boldsymbol{f}_{\mathrm{g}}(\omega) \tag{10.41}$$

where $\boldsymbol{S}_{\mathrm{g}}(\omega)$ is the global dynamic stiffness matrix, $\boldsymbol{d}_{\mathrm{g}}$ is the global spectral DOFs vector, and $\boldsymbol{f}_{\mathrm{g}}$ is the global spectral nodal forces vector.

10.1.3 Spectral Element with Active Control

Consider an elastic–PZT two-layer beam implemented with a proportional-derivative (PD) controller, as shown in Figure 10.3. Assume that the driving voltage applied to the PZT layer is determined by the vibration data measured at $x = x_0$ as

$$V(t) = -K_p w(x_0, t) - K_d \dot{w}(x_0, t) \tag{10.42}$$

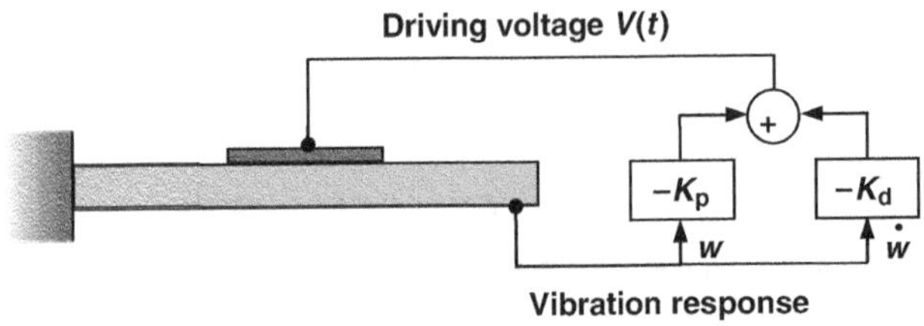

Figure 10.3 An elastic–PZT two-layer beam with a PD controller

where K_p and K_d are the proportional and derivative gains, respectively. Equation 10.42 can be written in the spectral form as

$$V(\omega) = -(K_p + i\omega K_d) W(x_0; \omega) \tag{10.43}$$

By using the first part of Equation 10.29, the measured displacement $W(x_0;\omega)$ can be related to the spectral nodal DOFs vector $\boldsymbol{d}$ as follows:

$$W(x_0; \omega) = \boldsymbol{N}_w(x_0; \omega)\boldsymbol{d} \tag{10.44}$$

The spectral nodal DOFs vector $\boldsymbol{d}$ in Equation 10.44 corresponds to the spectral element in which the sensor (i.e., x_0) is located. Substitution of Equation 10.44 into Equation 10.43 gives

$$V(\omega) = -(K_p + i\omega K_d)\boldsymbol{N}_w(x_0; \omega)\boldsymbol{d} \tag{10.45}$$

By substituting Equation 10.45 into Equation 10.23, we obtain the piezoelectric forces and moments as

$$\begin{aligned} N^p &= bd_{31}E_p V(\omega) = -\chi_N \boldsymbol{N}_w(x_0; \omega)\boldsymbol{d} \\ M^p &= (1/2)bd_{31}E_p h V(\omega) = -\chi_M \boldsymbol{N}_w(x_0; \omega)\boldsymbol{d} \end{aligned} \tag{10.46}$$

where

$$\begin{aligned} \chi_N &= bd_{31}E_p(K_p + i\omega K_d) \\ \chi_M &= (1/2)bd_{31}E_p h(K_p + i\omega K_d) \end{aligned} \tag{10.47}$$

Substitution of Equation 10.46 into Equation 10.39 gives

$$\boldsymbol{f}_c^{\text{piezo}} = -\boldsymbol{S}_{\text{piezo}}(\omega)\boldsymbol{d} \tag{10.48}$$

where

$$\boldsymbol{S}_{\text{piezo}}(\omega) = \begin{bmatrix} -\chi_N \boldsymbol{N}_w(x_0;\omega) \\ 0 \\ +\chi_M \boldsymbol{N}_w(x_0;\omega) \\ +\chi_N \boldsymbol{N}_w(x_0;\omega) \\ 0 \\ -\chi_M \boldsymbol{N}_w(x_0;\omega) \end{bmatrix} \tag{10.49}$$

By substituting Equation 10.48 into Equation 10.38 and the result into Equation 10.32, we can obtain the spectral element equation for the elastic–PZT two-layer beam element with a PD controller as follows:

$$\boldsymbol{S}(\omega)_{\text{CONTROL}}\boldsymbol{d} = \boldsymbol{f}_{\text{c}}^{\text{mech}} + \boldsymbol{f}_{\text{d}} \tag{10.50}$$

where

$$\boldsymbol{S}(\omega)_{\text{CONTROL}} = \boldsymbol{S}(\omega) + \boldsymbol{S}_{\text{piezo}}(\omega) \tag{10.51}$$

10.1.4 Spectral Modal Analysis

As the governing equations for the elastic–PZT two-layer beam have exactly the same forms as those for the elastic–elastic two-layer beam, the spectral modal analysis in Section 9.1.3 is also valid for the elastic–PZT two-layer beam.

10.1.4.1 Orthogonality Properties of Normal Modes

Let ω_p be the pth damped natural frequency and W_p and U_p the corresponding normal modes. The pth eigensolutions (ω_p, W_p, U_p) must satisfy an eigenvalue problem as

$$\begin{aligned} &EIW_p'''' - \omega_p^2\rho AW_p + i\omega_p cAW_p = \omega_p^2(\alpha U_p' - \gamma W_p'') + i\omega_p(c_1 W_p'' - c_4 U_p') + \beta U_p''' + FW_p'' \\ &EAU_p'' + \omega_p^2\rho AU_p - i\omega_p cAU_p = \omega_p^2 \alpha W_p' - i\omega_p c_4 W_p' + \beta W_p''' \end{aligned} \tag{10.52}$$

Similarly, the qth eigensolutions (ω_q, W_q, U_q) must satisfy an eigenvalue problem as

$$\begin{aligned} &EIW_q'''' - \omega_q^2\rho AW_q + i\omega_q cAW_q = \omega_q^2(\alpha U_q' - \gamma W_q'') + i\omega_q(c_1 W_q'' - c_4 U_q') + \beta U_q''' + FW_q'' \\ &EAU_q'' + \omega_q^2\rho AU_q - i\omega_q cAU_q = \omega_q^2 \alpha W_q' - i\omega_q c_4 W_q' + \beta W_q''' \end{aligned} \tag{10.53}$$

Following the procedure used in Section 9.1.3.1, we can derive the orthogonality properties of the damped normal modes from Equations 10.52 and 10.53 as follows:

$$\begin{aligned} &\int_0^L [\rho A(W_p W_q + U_p U_q) - \alpha(U_p W_q' + U_q W_p') + \gamma W_p' W_q'] \mathrm{d}x = m_p \delta_{pq} \\ &\int_0^L [cA(W_p W_q + U_p U_q) + c_1 W_p' W_q' - c_4(U_p W_q' + U_q W_p')] \mathrm{d}x = c_p \delta_{pq} \end{aligned} \tag{10.54}$$

where m_p and c_p are the modal mass and modal damping coefficients defined by

$$m_p = \int_0^L [\rho A(U_p^2 + W_p^2) - 2\alpha U_p W_p' + \gamma W_p'^2] \mathrm{d}x$$
$$c_p = \int_0^L [cA(U_p^2 + W_p^2) + c_1 W_p'^2 - 2c_4 U_p W_p'] \mathrm{d}x \tag{10.55}$$

10.1.4.2 Spectral Modal Equations

Based on the modal analysis method, the forced vibration responses are represented by superposing the normal modes as

$$w(x,t) = \sum_{p=1}^{\infty} W_p(x) q_p(t), \qquad u(x,t) = \sum_{p=1}^{\infty} U_p(x) q_p(t) \tag{10.56}$$

where $q_p(t)$ are the modal coordinates. The modal coordinates can be written in the spectral form as

$$q_p(t) = \frac{1}{N} \sum_{n=0}^{N-1} \bar{q}_{pn} \mathrm{e}^{i\omega_n t} \tag{10.57}$$

By using Equation 10.57, Equation 10.56 can be written as

$$w(x,t) = \frac{1}{N} \sum_{p=1}^{\infty} \sum_{n=0}^{N-1} W_p(x) q_{pn} \mathrm{e}^{i\omega_n t}, \qquad u(x,t) = \frac{1}{N} \sum_{p=1}^{\infty} \sum_{n=0}^{N-1} U_p(x) q_{pn} \mathrm{e}^{i\omega_n t} \tag{10.58}$$

The external forces are also written in the spectral forms as

$$p(x,t) = \frac{1}{N} \sum_{n=0}^{N-1} P_n(x) \mathrm{e}^{i\omega_n t}, \qquad \tau(x,t) = \frac{1}{N} \sum_{n=0}^{N-1} T_n(x) \mathrm{e}^{i\omega_n t} \tag{10.59}$$

By substituting Equations 10.58 and 10.59 into Equation 10.15, taking the same manipulation as carried out in Section 9.1.3.2, and applying the orthogonality properties (10.54), we can derive the modal equations in the frequency domain to be

$$(\omega_q^2 - \omega_n^2) m_q q_{qn} - i(\omega_q - \omega_n) c_q q_{qn} = f_{qn} \quad (q = 1, 2, 3, \ldots) \tag{10.60}$$

where

$$f_{qn} = \int_0^L [P_n(x) W_q + T_n(x) U_q] \mathrm{d}x \quad (q = 1, 2, 3, \ldots) \tag{10.61}$$

The spectral modal Equations 10.60 can be transformed into the time domain as

$$m_p \ddot{q}_p(t) + c_p \dot{q}_p(t) + (\omega_p^2 m_p - i\omega_p c_p) q_p(t) = f_p(t) \tag{10.62}$$

where

$$f_p(t) = \int_0^L [p(x,t) W_p + \tau(x,t) U_p] \mathrm{d}x \tag{10.63}$$

10.1.5 Finite Element Model

The finite element equation can be formulated from the weak form of the time-domain governing Equations 10.15. The weak form can be derived in the form as given by Equation 9.70. By using the same nodal DOFs vector and displacement fields as used in Section 9.1.4, we can obtain the finite element equation as

$$\boldsymbol{M}\ddot{\boldsymbol{d}}(t)+\boldsymbol{C}\dot{\boldsymbol{d}}(t)+\boldsymbol{K}\boldsymbol{d}(t)=\boldsymbol{f}_{\mathrm{c}}^{\mathrm{mech}}(t)+\boldsymbol{f}_{\mathrm{c}}^{\mathrm{piezo}}(t)+\boldsymbol{f}_{\mathrm{d}}(t) \tag{10.64}$$

where

$$\begin{aligned}\boldsymbol{M}&=\int_0^L[\rho A(\boldsymbol{N}_w^{\mathrm{T}}\boldsymbol{N}_w+\boldsymbol{N}_u^{\mathrm{T}}\boldsymbol{N}_u)-\alpha(\boldsymbol{N}_w'^{\mathrm{T}}\boldsymbol{N}_u+\boldsymbol{N}_u^{\mathrm{T}}\boldsymbol{N}_w')+\gamma\boldsymbol{N}_w'^{\mathrm{T}}\boldsymbol{N}_w']\mathrm{d}x\\ \boldsymbol{C}&=\int_0^L[cA(\boldsymbol{N}_w^{\mathrm{T}}\boldsymbol{N}_w+\boldsymbol{N}_u^{\mathrm{T}}\boldsymbol{N}_u)+c_1\boldsymbol{N}_w'^{\mathrm{T}}\boldsymbol{N}_w'-c_4(\boldsymbol{N}_w'^{\mathrm{T}}\boldsymbol{N}_u+\boldsymbol{N}_u'^{\mathrm{T}}\boldsymbol{N}_w')]\mathrm{d}x\\ \boldsymbol{K}&=\int_0^L[EA\boldsymbol{N}_u'^{\mathrm{T}}\boldsymbol{N}_u'+EI\boldsymbol{N}_w''^{\mathrm{T}}\boldsymbol{N}_w''-\beta(\boldsymbol{N}_w''^{\mathrm{T}}\boldsymbol{N}_u'+\boldsymbol{N}_u'^{\mathrm{T}}\boldsymbol{N}_w'')+F\boldsymbol{N}_w'^{\mathrm{T}}\boldsymbol{N}_w']\mathrm{d}x\end{aligned} \tag{10.65}$$

and

$$\begin{aligned}\boldsymbol{f}_c^{\mathrm{mech}}&=\{N_1(t)\quad Q_1(t)\quad M_1(t)\quad N_2(t)\quad Q_2(t)\quad M_2(t)\}^{\mathrm{T}}\\ \boldsymbol{f}_c^{\mathrm{piezo}}&=\{-N^p(t)\quad 0\quad M^p(t)\quad N^p(t)\quad 0\quad -M^p(t)\}^{\mathrm{T}}\end{aligned} \tag{10.66}$$

$$\boldsymbol{f}_{\mathrm{d}}(t)=\int_0^L[p(x,t)\boldsymbol{N}_w^{\mathrm{T}}(x)+\tau(x,t)\boldsymbol{N}_u^{\mathrm{T}}(x)]\mathrm{d}x \tag{10.67}$$

The finite element matrices $\boldsymbol{M}$, $\boldsymbol{C}$, and $\boldsymbol{K}$ are same as given in Appendix 9.A, but replacing the structural and damping properties by Equations 10.10 and 10.14 defined for the elastic–PZT two-layer beams.

Example 10.1

Consider a cantilevered elastic–PZT two-layer beam as shown in Figure 10.4. The base beam is partially covered with a uniform PZT layer. The geometric and material properties are given in Table 10.1. The high accuracy of the present spectral element model is shown in Table 10.2 by comparing the natural frequencies obtained by both SEM and FEM. Table 10.3 compares the natural frequencies and modal damping ratios of the beam with and without the PD control, which is shown in Figure 10.3. It is shown that the beam with a PD controller has larger modal damping ratios than the beam without the PD controller to reduce the vibration quicker, as displayed in Figure 10.5.

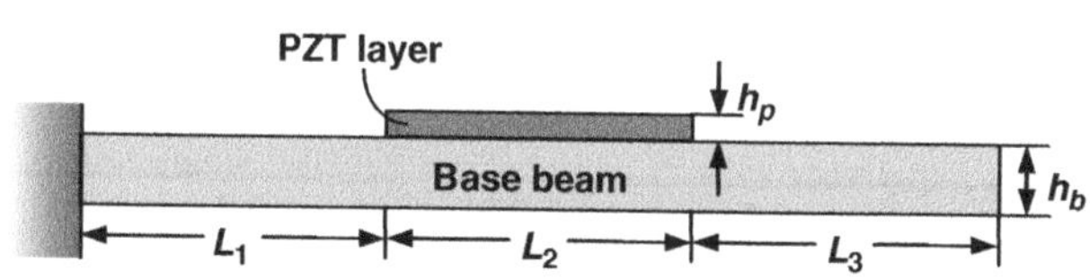

Figure 10.4 A cantilevered beam partially covered with a PZT layer

Table 10.1 Geometric and material properties of the base beam and PZT layer

L_1	100 mm	L_2	100 mm	L_3	100 mm
b	20 mm	h_b	5 mm	h_p	0.75 mm
E_b	70 GPa	E_p	65 GPa	C_{11}^D	75 GPa
d_{31}	-1.75×10^{-10} m/V	ρ_b	2700 kg/m^3	ρ_p	7500 kg/m^3

Table 10.2 Natural frequencies (Hz) of an elastic–PZT two-layer beam

	SEM(*N*)	FEM(*N*)				
Mode	$N=1$	$N=6$	$N=15$	$N=30$	$N=60$	$N=150$
1st	45.3380	45.3412	45.3385	45.3381	45.3380	45.3380
2nd	289.987	290.089	289.995	289.988	289.987	289.987
3rd	800.120	802.295	800.285	800.154	800.128	800.121
4th	1590.45	1603.74	1591.26	1590.60	1590.49	1590.46
5th	2621.51	2662.93	2623.85	2621.85	2621.58	2621.52
10th	9291.54	11438.37	9362.63	9299.39	9292.80	9291.71

Note: (*N*) = total number of the equal-length finite elements used in the analysis.

Table 10.3 Natural frequencies and modal damping ratios of the elastic–PZT two-layer beam with and without the PD controller

	Uncontrolled	Controlled			
		$K_p = 5 \times 10^7$ V, $K_d = 100$ Vs		$K_p = 5 \times 10^7$ V, $K_d = 300$ Vs	
Mode number (n)	f_n (Hz)	f_n (Hz)	ξ_n	f_n (Hz)	ξ_n
1	45.338	48.960	0.0084	49.158	0.0240
2	289.99	289.35	0.0117	289.37	0.0263
3	800.12	800.46	0.0021	800.49	0.0059
4	1590.5	1590.6	0.0012	1590.6	0.0019
5	2621.5	2618.5	0.0004	2616.9	0.0009

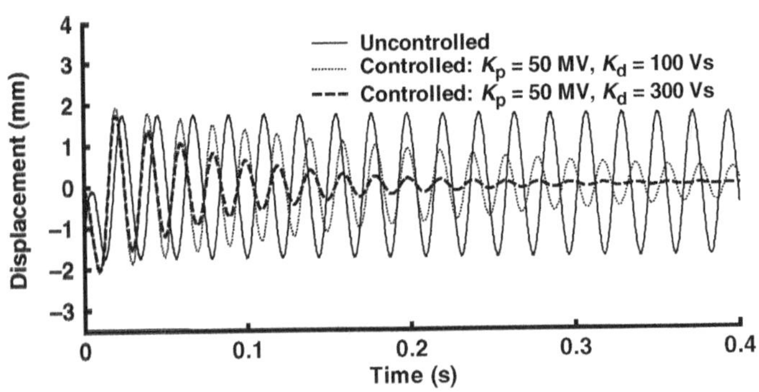

Figure 10.5 Dynamic responses of an elastic–PZT two-layer beam with a PD controller

10.2 Elastic–Viscoelastic–Piezoelectric Three-Layer (ACLD) Beams

10.2.1 Equations of Motion

The active constrained layered damping (ACLD) treated beams consist of three layers: an elastic base beam, a viscoelastic material (VEM) layer, and a piezoelectric material (PZT) layer. The geometry and deformations of an ACLD beam are shown in Figure 10.6. $u_b(x, t)$, $u_v(x, t)$, and $u_p(x, t)$ are the axial displacements of the base-beam layer, VEM layer, and PZT layer, respectively. h_b, h_v, and h_p are the thicknesses of the base beam, VEM layer, and PZT layer, respectively, and ψ is the shear angle of the VEM layer. The subscripts b, v, and p are used to denote the quantities of the base beam, VEM layer, and PZT layer, respectively. For the derivation of the equations of motion, the following assumptions are made [88].

a. Geometric and material properties are uniform along the beam.
b. Perfect continuity, but no slip, at the interfaces.
c. All layers take the same transverse displacement.
d. Small amplitude vibration.
e. Linear elasticity, viscoelasticity, and piezoelectricity are valid for the base beam, VEM layer, and PZT layer, respectively.
f. Shear deformations and rotary inertias in the base beam and PZT layer are negligible.
g. Uniform voltage along the beam.

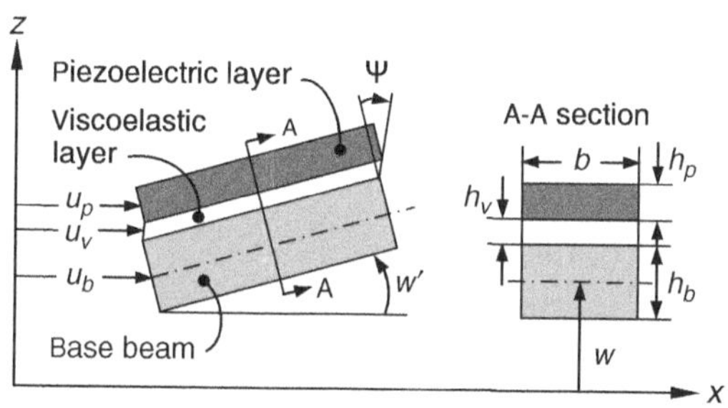

Figure 10.6 An elastic–VEM-PZT three-layer (ACLD) beam element

From the geometry of Figure 10.6, we obtain the following kinematics relationships

$$\begin{aligned} u_v &= u_b - (h_1/2)w' + (h_v/2)\psi \\ u_p &= u_b - (h_2/2)w' + h_v\psi \end{aligned} \tag{10.68}$$

where

$$h_1 = h_b + h_v, \quad h_2 = h_b + 2h_v + h_p \tag{10.69}$$

The equations of motion of the ACLD beam are derived by using Hamilton's principle

$$\int_{t_1}^{t_2} (\delta T - \delta V + \delta W)\mathrm{d}t = 0 \tag{10.70}$$

where T is the kinetic energy, V is the potential energy, and δW is the virtual work. The strain and kinetic energies are given by

$$\begin{aligned} V &= \frac{1}{2}\int_0^L [E_bA_bu_b'^2 + E_bI_bw''^2 + E_vA_vu_v'^2 + E_vI_v(w''-\psi')^2 + A_v(g\circ\psi)\psi \\ &\quad + C_{11}^DI_pw''^2 + C_{11}^DA_pu_p'^2 - 2A_ph_{31}D_3u_p' + A_p\beta_{33}^SD_3^2 + Fw'^2]\,dx \\ T &= \frac{1}{2}\int_0^L [\rho_bA_b(\dot{u}_b^2+\dot{w}^2) + \rho_vA_v(\dot{u}_v^2+\dot{w}^2) + \rho_vI_v(\dot{w}'-\dot{\psi})^2 + \rho_pA_p(\dot{u}_p^2+\dot{w}^2)]\,dx \end{aligned} \tag{10.71}$$

where the constitutive relationship in Equation 9.81 is applied for the VEM layer. By eliminating u_v and u_p from Equation 10.71 by using Equation 10.68 and also D_3 by using Equation 10.6, we can write the strain and kinetic energies as

$$\begin{aligned} V &= \frac{1}{2}\int_0^L [EAu_b'^2 + EI_ww''^2 + EI_\psi\psi'^2 - 2\beta w''u_b' + 2\varepsilon_4u_b'\psi' - 2\varepsilon_2w''\psi' \\ &\quad + A_v(g\circ\psi)\psi + Fw'^2]dx \\ T &= \frac{1}{2}\int_0^L \{\rho A\dot{u}_b^2 + \rho A\dot{w}^2 + \rho I\dot{\psi}^2 - 2\alpha\dot{w}'\dot{u}_b + 2\varepsilon_3\dot{u}_b\dot{\psi} - 2\varepsilon_1\dot{w}'\dot{\psi} + \gamma\dot{w}'^2\}dx \end{aligned} \tag{10.72}$$

where

$$\begin{aligned} EA &= E_bA_b + E_vA_v + E_pA_p \\ EI_w &= E_bI_b + E_vI_v + C_{11}^DI_p + (1/4)E_vA_vh_1^2 + (1/4)E_pA_ph_2^2 \\ EI_\psi &= E_vI_v + (1/4)E_vA_vh_v^2 + E_pA_ph_v^2 \\ \rho A &= \rho_bA_b + \rho_vA_v + \rho_pA_p \\ \rho I &= \rho_vI_v + (1/4)\rho_vA_vh_v^2 + \rho_pA_ph_v^2 \\ \alpha &= (1/2)(\rho_vA_vh_1 + \rho_pA_ph_2) \\ \beta &= (1/2)(E_vA_vh_1 + E_pA_ph_2) \\ \gamma &= \rho_vI_v + (1/4)\rho_vA_vh_1^2 + (1/4)\rho_pA_ph_2^2 \\ \varepsilon_1 &= \rho_vI_v + (1/4)\rho_vA_vh_1h_v + (1/2)\rho_pA_ph_2h_v \\ \varepsilon_2 &= E_vI_v + (1/4)E_vA_vh_1h_v + (1/2)E_pA_ph_2h_v \\ \varepsilon_3 &= (1/2)\rho_vA_vh_v + \rho_pA_ph_v \\ \varepsilon_4 &= (1/2)E_vA_vh_v + E_pA_ph_v \end{aligned} \tag{10.73}$$

The virtual work δW done by external forces, viscous damping forces, boundary forces, and the voltage applied to the PZT layer is given by

$$\begin{aligned} \delta W &= \int_0^L p(x,t)\delta w dx + \int_0^L \tau(x,t)\delta u_b dx + \int_0^L \gamma(x,t)\delta\psi dx \\ &\quad - \int_0^L\int_{A_b} c_b\dot{\mathbf{r}}_b\cdot\delta\mathbf{r}_b dA_b dx - \int_0^L\int_{A_v} c_v\dot{\mathbf{r}}_v\cdot\delta\mathbf{r}_v dA_v dx - \int_0^L\int_{A_p} c_p\dot{\mathbf{r}}_p\cdot\delta\mathbf{r}_p dA_p dx \\ &\quad + N_1(t)\delta u_b(0,t) + N_2(t)\delta u_b(L,t) + M_1(t)\delta w'(0,t) + M_2(t)\delta w'(L,t) \\ &\quad + Q_1(t)\delta w(0,t) + Q_2(t)\delta w(L,t) + R_1(t)\delta\psi(0,t) + R_2(t)\delta\psi(L,t) \\ &\quad + \int_0^L bV(t)\delta D_3 dx \end{aligned} \tag{10.74}$$

where

$$
\begin{aligned}
\boldsymbol{r}_b &= u_b \boldsymbol{n}_x + w \boldsymbol{n}_z \\
\boldsymbol{r}_v &= [u_v - z(w' - \psi)]\boldsymbol{n}_x + w \boldsymbol{n}_z \\
\boldsymbol{r}_p &= u_p \boldsymbol{n}_x + w \boldsymbol{n}_z
\end{aligned}
\tag{10.75}
$$

and where $\boldsymbol{n}_x$ and $\boldsymbol{n}_z$ are the unit direction vectors in the x- and z-directions, respectively. In Equation 10.74, $V(t)$ is the uniform voltage applied to the piezoelectric layer and c_b, c_v, and c_p are viscous damping coefficients. The definitions of the other symbols used in Equation 10.74 are the same as those used in Equation 9.86. By applying Equations 10.6, 10.68 and 10.75 into Equation 10.74 and taking the integral by parts, we can write the virtual work as

$$
\begin{aligned}
\delta W = \int_0^L & [p(x,t)\delta w + \tau(x,t)\delta u_b + \gamma(x,t)\delta\psi - cA\dot{u}_b\delta u_b - cA\dot{w}\delta w + c_4\dot{w}'\delta u_b \\
& - c_5\dot{\psi}\delta u_b - c_5\dot{u}_b\delta\psi + c_3\dot{w}'\delta\psi - c_2\dot{\psi}\delta\psi + c_4\dot{u}_b\delta w' - c_1\dot{w}'\delta w' + c_3\dot{\psi}\delta w']\,\mathrm{d}x \\
& + M_1(t)\delta w'(0,t) + M_2(t)\delta w'(L,t) + Q_1(t)\delta w(0,t) + Q_2(t)\delta w(L,t) \\
& + N_1(t)\delta u_b(0,t) + N_2(t)\delta u_b(L,t) + R_1(t)\delta\psi(0,t) + R_2(t)\delta\psi(L,t) \\
& + bd_{31}E_pV(t)\delta u_b|_0^L - (1/2)bh_2d_{31}E_pV(t)\delta w'|_0^L + (1/2)bh_vd_{31}E_pV(t)\delta\psi|_0^L
\end{aligned}
\tag{10.76}
$$

where

$$
\begin{aligned}
cA &= c_bA_b + c_vA_v + c_pA_p \\
c_1 &= c_vI_v + (1/4)(c_vA_vh_1^2 + c_pA_ph_2^2) \\
c_2 &= c_vI_v + (1/4)c_vA_vh_v^2 + c_pA_ph_v^2 \\
c_3 &= c_vI_v + (1/4)c_vA_vh_1h_v + (1/2)c_pA_ph_2h_v \\
c_4 &= (1/2)(c_vA_vh_1 + c_pA_ph_2) \\
c_5 &= (1/2)c_vA_vh_v + c_pA_ph_v
\end{aligned}
\tag{10.77}
$$

Finally, substitution of Equations 10.72 and 10.76 into Hamilton's principle (10.70) gives three coupled equations of motion

$$
\begin{aligned}
& EI_w w'''' + \rho A\ddot{w} + cA\dot{w} = -\alpha\ddot{u}_b' + \beta u''' - +\gamma\ddot{w}'' - \varepsilon_1\ddot{\psi}_b' + \varepsilon_2\psi''' + c_1\dot{w}'' - c_4\dot{u}_b' - c_3\dot{\psi}' + Fw'' + p(x,t) \\
& EAu_b'' - \rho A\ddot{u}_b - cA\dot{u}_b = -\alpha\ddot{w}' + \beta w''' + \varepsilon_3\ddot{\psi} - \varepsilon_4\psi'' - c_4\dot{w}' + c_5\dot{\psi} - \tau(x,t) \\
& EI_\psi\psi'' - \rho I\ddot{\psi} - c_2\dot{\psi} = -\varepsilon_1\ddot{w}' + \varepsilon_2 w''' + \varepsilon_3\ddot{u}_b - \varepsilon_4 u_b'' - c_3\dot{w}' + c_5\dot{u}_b + A_v(g\circ\psi) - \gamma(x,t)
\end{aligned}
\tag{10.78}
$$

and the boundary conditions

$$
\begin{aligned}
Q(0,t) &= -Q_1(t) - Fw'(0,t) \quad &\text{or}\quad & w(0,t) = w_1(t) \\
M(0,t) &= -M_1(t) - M^p(t) \quad &\text{or}\quad & w'(0,t) = \theta_1(t) \\
N(0,t) &= -N_1(t) + N^p(t) \quad &\text{or}\quad & u_b(0,t) = u_1(t) \\
R(0,t) &= -R_1(t) + R^p(t) \quad &\text{or}\quad & \psi(0,t) = \psi_1(t) \\
Q(L,t) &= +Q_2(t) - Fw'(L,t) \quad &\text{or}\quad & w(L,t) = w_2(t) \\
M(L,t) &= +M_2(t) - M^p(t) \quad &\text{or}\quad & w'(L,t) = \theta_2(t) \\
N(L,t) &= +N_2(t) + N^p(t) \quad &\text{or}\quad & u_b(L,t) = u_2(t) \\
R(L,t) &= +R_2(t) + R^p(t) \quad &\text{or}\quad & \psi(L,t) = \psi_2(t)
\end{aligned}
\tag{10.79}
$$

where

$$
\begin{aligned}
M^p(t) &= (1/2)bh_2d_{31}E_pV(t) \\
N^p(t) &= bd_{31}E_pV(t) \\
R^p(t) &= bh_vd_{31}E_pV(t)
\end{aligned}
\tag{10.80}
$$

$N^P(t)$, $M^P(t)$, and $R^P(t)$ are the piezoelectric forces and moments generated by the voltage $V(t)$ applied to the PZT layer, and $b=b_b=b_v=b_p$ is the width of the ACLD beam. The resultant force-displacement relationships are given by

$$
\begin{aligned}
Q(x,t) &= -EI_w w''' - \alpha\ddot{u}_b + \beta u_b'' + \gamma\ddot{w}' - \varepsilon_1\ddot{\psi} + \varepsilon_2\psi'' + c_1\dot{w}' - c_4\dot{u}_b - c_3\dot{\psi} \\
M(x,t) &= EI_w w'' - \beta u_b' - \varepsilon_2\psi' \\
N(x,t) &= EAu_b' - \beta w'' + \varepsilon_4\psi' \\
R(x,t) &= EI_\psi\psi' - \varepsilon_2 w'' + \varepsilon_4 u_b'
\end{aligned}
\tag{10.81}
$$

If necessary, the structural damping of the base beam and PZT layer can be taken into account by using the complex modulus given in Equation 10.19.

If all VEM related terms are removed, the equations of motion (10.78) and the force-displacement relationships (10.81) return to the equations of motion (10.15) and the force-displacement relationships (10.18) for the elastic–PZT two-layer beam, respectively. The equations of motion (10.78) are identical with the equations of motion (9.90) for the PCLD beam, except for using the structural and damping properties defined by (10.73) and (10.77) and the boundary conditions (10.79), which are different from those for the PCLD beams due to piezoelectric effects. Thus, a large part of the spectral element formulation conducted in Section 9.2 for the PCLD beams can be shared by the ACLD beams.

10.2.2 Spectral Element Modeling

10.2.2.1 Governing Equations in the Frequency Domain

As with the PCLD beams in Section 9.2.2.1, we represent all dynamic responses, forces, and moments in the spectral forms given by Equations 9.93 and 9.94. In addition, we represent piezoelectric forces, moments, and voltage in the spectral forms as

$$
\begin{aligned}
&\{M^p(x,t)\ N^p(x,t)\ R^p(x,t)\ V(t)\} \\
&= \frac{1}{N}\sum_{n=0}^{N-1}\{M_n^p(x;\omega_n)\ N_n^p(x;\omega_n)\ R_n^p(x;\omega_n)\ V_n(\omega_n)\}e^{i\omega_n t}
\end{aligned}
\tag{10.82}
$$

Substitution of Equations 9.93, 9.94 and 10.82 into Equations 10.78 and 10.79 gives the equations of motion in the frequency domain

$$
\begin{aligned}
&EI_w W'''' - \omega^2\rho A W + i\omega c A W = \omega^2\alpha U' + \beta U''' - \omega^2\gamma W'' \\
&\quad + \omega^2\varepsilon_1\Psi' + \varepsilon_2\Psi''' + i\omega c_1 W'' - i\omega c_4 U' - i\omega c_3\Psi' + FW'' + P(x) \\
&EAU'' + \omega^2\rho AU - i\omega cAU = \omega^2\alpha W' + \beta W''' \\
&\quad - \omega^2\varepsilon_3\Psi - \varepsilon_4\Psi'' - i\omega c_4 W' + i\omega c_5\Psi - T(x) \\
&EI_\psi\Psi'' + \omega^2\rho I\Psi - i\omega c_2\Psi = \omega^2\varepsilon_1 W' + \varepsilon_2 W''' - \omega^2\varepsilon_3 U \\
&\quad - \varepsilon_4 U'' - i\omega c_3 W' + i\omega c_5 U + A_v G^*\Psi - \Gamma(x)
\end{aligned}
\tag{10.83}
$$

and the associated boundary conditions

$$
\begin{aligned}
Q(0) &= -Q_1 - FW'(0) && \text{or} \quad W(0) = W_1 \\
M(0) &= -M_1 - M^p && \text{or} \quad W'(0) = \Theta_1 \\
N(0) &= -N_1 + N^p && \text{or} \quad U(0) = U_1 \\
R(0) &= -R_1 + R^p && \text{or} \quad \Psi(0) = \Psi_1 \\
Q(L) &= +Q_2 - FW'(L) && \text{or} \quad W(L) = W_2 \\
M(L) &= +M_2 - M^p && \text{or} \quad W'(L) = \Theta_2 \\
N(L) &= +N_2 + N^p && \text{or} \quad U(L) = U_2 \\
R(L) &= +R_2 + R^p && \text{or} \quad \Psi(L) = \Psi_2
\end{aligned}
\tag{10.84}
$$

where

$$
\begin{aligned}
M^p &= (1/2)bh_2 d_{31} E_p V(\omega) \\
N^p &= bd_{31} E_p V(\omega) \\
R^p &= bh_v d_{31} E_p V(\omega)
\end{aligned}
\tag{10.85}
$$

Similarly, the resultant force-displacement relationships (10.81) can be also transformed into the frequency domain as

$$
\begin{aligned}
Q(x) &= -EI_w W''' + \omega^2 \alpha U + \beta U'' - \omega^2 \gamma W' + \omega^2 \varepsilon_1 \Psi + \varepsilon_2 \Psi'' + i\omega c_1 W' - i\omega c_4 U - i\omega c_3 \Psi \\
M(x) &= EI_w W'' - \beta U' - \varepsilon_2 \Psi' \\
N(x) &= EAU' - \beta W'' + \varepsilon_4 \Psi' \\
R(x) &= EI_\psi \Psi' - \varepsilon_2 W'' + \varepsilon_4 U'
\end{aligned}
\tag{10.86}
$$

In Equations 10.83–10.86, the subscripts n are omitted for brevity.

10.2.2.2 Spectral Nodal DOFs, Forces, and Moments

By using the spectral nodal DOFs, forces, and moments specified in Figure 10.7, we define the spectral nodal DOFs vector

$$
\begin{aligned}
\boldsymbol{d} &= \{U_1\ W_1\ \Theta_1\ \Psi_1\ U_2\ W_2\ \Theta_2\ \Psi_2\}^T \\
&= \{U(0)\ W(0)\ W'(0)\ \Psi(0)\ U(L)\ W(L)\ W'(L)\ \Psi(L)\}^T
\end{aligned}
\tag{10.87}
$$

and the mechanical spectral nodal forces and moments vector

$$
f_c^{\mathrm{mech}} = \{N_1\ Q_1\ M_1\ R_1\ N_2\ Q_2\ M_2\ R_2\}^{\mathrm{T}} \tag{10.88}
$$

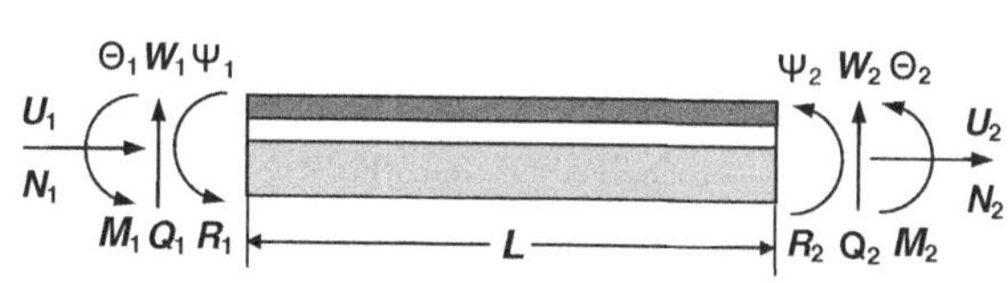

Figure 10.7 Spectral element of the ACLD beam

10.2.2.3 Dynamic Shape Functions

For the derivation of the dynamic shape function, we consider the homogeneous governing equations reduced from Equation 10.83 as

$$\begin{aligned}
&EI_w W'''' - \omega^2 \rho A W + i\omega c A W = \omega^2 \alpha U' + \beta U''' - \omega^2 \gamma W'' + \omega^2 \varepsilon_1 \Psi' + \varepsilon_2 \Psi''' \\
&\qquad + i\omega c_1 W'' - i\omega c_4 U' - i\omega c_3 \Psi' + F W'' \\
&EAU'' + \omega^2 \rho A U - i\omega c A U = \omega^2 \alpha W' + \beta W''' - \omega^2 \varepsilon_3 \Psi - \varepsilon_4 \Psi'' \\
&\quad - i\omega c_4 W' + i\omega c_5 \Psi \\
&EI_\psi \Psi'' + \omega^2 \rho I \Psi - i\omega c_2 \Psi = \omega^2 \varepsilon_1 W' + \varepsilon_2 W''' - \omega^2 \varepsilon_3 U - \varepsilon_4 U'' \\
&\quad - i\omega c_3 W' + i\omega c_5 U + A_v G^* \Psi
\end{aligned} \tag{10.89}$$

The general solutions to Equation 10.89 are assumed to be

$$W(x;\omega) = Ae^{-ikx}, \quad U(x;\omega) = r_1 Ae^{-ikx}, \quad \psi(x;\omega) = r_2 Ae^{-ikx} \tag{10.90}$$

By substituting Equation 10.90 into Equation 10.89 and following the same process as used in Section 9.2.2.3, we can derive a dispersion equation as

$$a_1 k^8 + a_2 k^6 + a_3 k^4 + a_4 k^2 + a_5 = 0 \tag{10.91}$$

where a_i ($i = 1, 2, \ldots, 5$) are given by Equation 9.105 when the structural and damping properties are replaced by the values given by Equations 10.73 and 10.77. Eight wavenumbers k_i ($i = 1, 2, \ldots, 6$) computed from Equation 10.91 can be used to obtain the general solutions to Equation 10.89 in the forms of

$$W(x;\omega) = \mathbf{N}_w(x;\omega)\mathbf{d}, \quad U(x;\omega) = \mathbf{N}_u(x;\omega)\mathbf{d}, \quad \Psi(x;\omega) = \mathbf{N}_\psi(x;\omega)\mathbf{d} \tag{10.92}$$

where $\mathbf{N}_w$, $\mathbf{N}_u$, and $\mathbf{N}_\psi$ are the dynamic shape functions defined by

$$\begin{aligned}
\mathbf{N}_w(x;\omega) &= \mathbf{e}(x;\omega)\mathbf{H}^{-1}(\omega) \\
\mathbf{N}_u(x;\omega) &= \mathbf{e}(x;\omega)\mathbf{R}_1\mathbf{H}^{-1}(\omega) \\
\mathbf{N}_\psi(x;\omega) &= \mathbf{e}(x;\omega)\mathbf{R}_2\mathbf{H}^{-1}(\omega)
\end{aligned} \tag{10.93}$$

The matrices $\mathbf{e}(x;\omega)$, $\mathbf{R}_1(\omega)$, $\mathbf{R}_2(\omega)$, and $\mathbf{H}(\omega)$ are defined in the same forms as given by Equations 9.111, 9.108 and 9.113, respectively.

10.2.2.4 Weak Form of Governing Equations

Following the procedure used in Section 9.2.2.4, we can obtain the weak form of the governing Equation 10.83 as

$$\begin{aligned}
&\int_0^L [EI_w W''\delta W'' - \omega^2 \rho A W \delta W + i\omega c A W \delta W + i\omega c_1 W' \delta W' - i\omega c_4 U \delta W' \\
&\quad + \omega^2 \alpha U \delta W' - \beta U' \delta W'' - \omega^2 \gamma W' \delta W' + \omega^2 \varepsilon_1 \Psi \delta W' - \varepsilon_2 \Psi' \delta W'' \\
&\quad - i\omega c_3 \Psi \delta W' + F W' \delta W' + EAU' \delta U' - \omega^2 \rho A U \delta U + i\omega c A U \delta U \\
&\quad + \omega^2 \alpha W' \delta U - \beta W'' \delta U' - \omega^2 \varepsilon_3 \Psi \delta U + \varepsilon_4 \Psi' \delta U' - i\omega c_4 W' \delta U \\
&\quad + i\omega c_5 \Psi \delta U + EI_\psi \Psi' \delta \Psi' - \omega^2 \rho I \Psi \delta \Psi + i\omega c_2 \Psi \delta \Psi + \omega^2 \varepsilon_1 W' \delta \Psi \\
&\quad - \varepsilon_2 W'' \delta \Psi' - \omega^2 \varepsilon_3 U \delta \Psi + \varepsilon_4 U' \delta \Psi' - i\omega c_3 W' \delta \Psi + i\omega c_5 U \delta \Psi \\
&\quad + A_v G^* \Psi \delta \Psi] dx - \int_0^L [P(x)\delta W + T(x)\delta U + \Gamma(x)\delta \Psi] dx \\
&\quad - \{[Q(x) + FW'(x)]\delta W|_0^L + M(x)\delta W'|_0^L + N(x)\delta U|_0^L + R(x)\delta \Psi|_0^L\} = 0
\end{aligned} \tag{10.94}$$

10.2.2.5 Spectral Element Equation

By substituting Equation 10.92 into Equation 10.94 and using the spectral nodal forces vector $\boldsymbol{f}_c$ defined by Equation 10.88, we obtain the spectral element equation as

$$\boldsymbol{S}(\omega)\boldsymbol{d} = \boldsymbol{f}(\omega) \tag{10.95}$$

The matrix $\boldsymbol{S}(\omega)$ is the spectral element matrix defined by

$$\boldsymbol{S}(\omega) = \boldsymbol{H}^{-\mathrm{T}}(\omega)\boldsymbol{D}(\omega)\boldsymbol{H}^{-1}(\omega) \tag{10.96}$$

where

$$\begin{aligned}\boldsymbol{D}(\omega) = & -EA\boldsymbol{R}_1^T\boldsymbol{KEKR}_1 + EI_w\boldsymbol{K}^2\boldsymbol{EK}^2 - EI_\psi\boldsymbol{R}_2^T\boldsymbol{KEKR}_2 - F\boldsymbol{KEK} \\ & - i\beta(\boldsymbol{K}^2\boldsymbol{EKR}_1 + \boldsymbol{R}_1^T\boldsymbol{KEK}^2) - i\varepsilon_2(\boldsymbol{K}^2\boldsymbol{EKR}_2 + \boldsymbol{R}_2^T\boldsymbol{KEK}^2) \\ & - \omega^2\{\rho A(\boldsymbol{E} + \boldsymbol{R}_1^T\boldsymbol{ER}_1) + \rho I\boldsymbol{R}_2^T\boldsymbol{ER}_2 + i\alpha(\boldsymbol{KER}_1 + \boldsymbol{R}_1^T\boldsymbol{EK}) \\ & - \gamma\boldsymbol{KEK} + i\varepsilon_1(\boldsymbol{KER}_2 + \boldsymbol{R}_2^T\boldsymbol{EK}) + \varepsilon_3(\boldsymbol{R}_1^T\boldsymbol{ER}_2 + \boldsymbol{R}_2^T\boldsymbol{ER}_1)\} \\ & - \varepsilon_4(\boldsymbol{R}_1^T\boldsymbol{KEKR}_2 + \boldsymbol{R}_2^T\boldsymbol{KEKR}_1) + i\omega\{cA(\boldsymbol{E} + \boldsymbol{R}_1^T\boldsymbol{ER}_1) - c_1\boldsymbol{KEK} \\ & + c_2\boldsymbol{R}_2^T\boldsymbol{ER}_2 + ic_3(\boldsymbol{KER}_2 + \boldsymbol{R}_2^T\boldsymbol{EK}) + ic_4(\boldsymbol{KER}_1 + \boldsymbol{R}_1^T\boldsymbol{EK}) \\ & + c_5(\boldsymbol{R}_1^T\boldsymbol{ER}_2 + \boldsymbol{R}_2^T\boldsymbol{ER}_1)\} + A_vG^*\boldsymbol{R}_2^T\boldsymbol{ER}_2\end{aligned} \tag{10.97}$$

In the above equation, the following definitions are used.

$$\boldsymbol{K} = diag\,[k_i] = \boldsymbol{K}^{\mathrm{T}}, \qquad \boldsymbol{K}^2 = diag\,[k_i^2] = \boldsymbol{K}^{2\mathrm{T}} \tag{10.98}$$

$$\boldsymbol{E}(\omega) = \int_0^L \boldsymbol{e}^{\mathrm{T}}(x;\omega)\boldsymbol{e}(x;\omega)\mathrm{d}x = [E_{lm}(\omega)] \quad (l,\, m = 1,\, 2, \ldots, 8) \tag{10.99}$$

where

$$E_{lm} = \begin{cases} \dfrac{i}{k_l + k_m}\left[\mathrm{e}^{-i(k_l+k_m)L} - 1\right] & \text{if} \quad k_l + k_m \neq 0 \\ L & \text{if} \quad k_l + k_m = 0 \end{cases} \tag{10.100}$$

The vector $\boldsymbol{f}(\omega)$ is the spectral nodal forces vector defined by

$$\boldsymbol{f}(\omega) = \boldsymbol{f}_{\mathrm{c}}^{\mathrm{mech}}(\omega) + \boldsymbol{f}_{\mathrm{c}}^{\mathrm{piezo}}(\omega) + \boldsymbol{f}_{\mathrm{d}}(\omega) \tag{10.101}$$

where

$$\boldsymbol{f}_{\mathrm{c}}^{\mathrm{piezo}} = \{-N^p \quad 0 \quad M^p \quad -R^p \quad N^p \quad 0 \quad -M^p \quad R^p\}^{\mathrm{T}} \tag{10.102}$$

$$\boldsymbol{f}_{\mathrm{d}}(\omega) = \int_0^L [P(x;\omega)\boldsymbol{N}_w^{\mathrm{T}}(x;\omega) + T(x;\omega)\boldsymbol{N}_u^{\mathrm{T}}(x;\omega) + \Gamma(x;\omega)\boldsymbol{N}_\psi^{\mathrm{T}}(x;\omega)]\mathrm{d}x \tag{10.103}$$

The spectral elements represented by Equation 10.95 can be assembled, alongwith imposing the boundary conditions, in the form

$$\boldsymbol{S}_{\mathrm{g}}(\omega)\boldsymbol{d}_{\mathrm{g}} = \boldsymbol{f}_{\mathrm{g}}(\omega) \tag{10.104}$$

where $\boldsymbol{S}_{\mathrm{g}}(\omega)$ is the global dynamic stiffness matrix, $\boldsymbol{d}_{\mathrm{g}}$ is the global spectral DOFs vector, and $\boldsymbol{f}_{\mathrm{g}}$ is the global spectral nodal forces vector.

10.2.3 Spectral Element with Active Control

Consider an ACLD beam element implemented with a proportional-derivative (PD) controller, as shown in Figure 10.8. Assume that the driving voltage applied to the PZT layer is determined by

$$V(t) = -K_p w - K_d \dot{w} \tag{10.105}$$

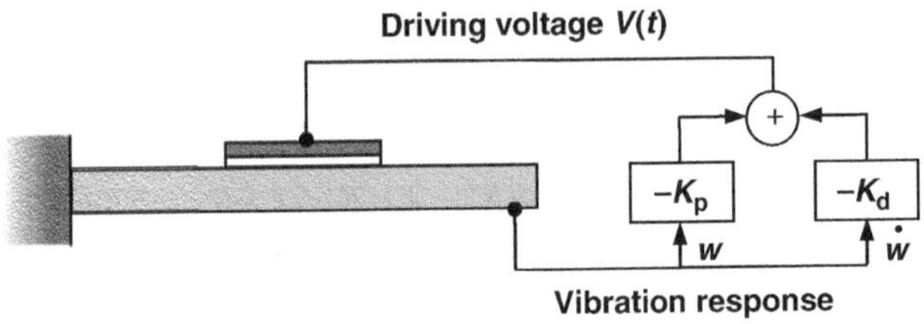

Figure 10.8 An ACLD beam with a PD controller

where K_p and K_d are the proportional and derivative gains, respectively. We can represent Equation 10.105 in the spectral form as

$$V(\omega) = -(K_p + i\omega K_d)W(x;\omega) \tag{10.106}$$

From Equation 10.92, the spectral components $W(x;\omega)$ of the displacement measured at a point $x = x_0$ within an ACLD beam element can be related to the corresponding spectral nodal DOFs vector $\boldsymbol{d}$ as follows:

$$W(x_0;\omega) = \boldsymbol{N}_w(x_0;\omega)\boldsymbol{d} \tag{10.107}$$

Substitution of Equation 10.107 into Equation 10.106 gives

$$V(\omega) = -(K_p + i\omega K_d)\boldsymbol{N}_w(x_0;\omega)\boldsymbol{d} \tag{10.108}$$

By applying Equation 10.108 to Equation 10.85, the piezoelectric forces and moments are obtained as

$$\begin{aligned} N^p &= bd_{31}E_pV(\omega) = -\chi_N\boldsymbol{N}_w(x_0;\omega)\boldsymbol{d} \\ M^p &= (1/2)bd_{31}E_phV(\omega) = -\chi_M\boldsymbol{N}_w(x_0;\omega)\boldsymbol{d} \\ R^p &= bh_vd_{31}E_pV(\omega) = -\chi_R\boldsymbol{N}_w(x_0;\omega)\boldsymbol{d} \end{aligned} \tag{10.109}$$

where

$$\begin{aligned} \chi_N &= bd_{31}E_p(K_p + i\omega K_d) \\ \chi_M &= (1/2)bd_{31}E_ph(K_p + i\omega K_d) \\ \chi_R &= bh_vd_{31}E_p(K_p + i\omega K_d) \end{aligned} \tag{10.110}$$

Now, substitute Equation 10.109 into Equation 10.102 to obtain

$$f_c^{\text{piezo}} = -\boldsymbol{S}_{\text{piezo}}(\omega)\boldsymbol{d} \tag{10.111}$$

where

$$S_{\text{piezo}}(\omega) = \begin{bmatrix} -\chi_N N_w(x_0;\omega) \\ 0 \\ +\chi_M N_w(x_0;\omega) \\ -\chi_R N_w(x_0;\omega) \\ +\chi_N N_w(x_0;\omega) \\ 0 \\ -\chi_M N_w(x_0;\omega) \\ +\chi_R N_w(x_0;\omega) \end{bmatrix} \tag{10.112}$$

Finally, by substituting Equation 10.111 into Equation 10.101 and then the result into Equation 10.95, we can obtain the spectral element equation for the ACLD beam element with a PD controller as being

$$S(\omega)_{\text{CONTROL}} d = f_c^{\text{mech}} + f_d \tag{10.113}$$

where

$$S(\omega)_{\text{CONTROL}} = S(\omega) + S_{\text{piezo}}(\omega) \tag{10.114}$$

10.2.4 Spectral Modal Analysis

The governing equations of the ACLD beam have exactly the same forms as those of the PCDL beam. Thus, we can follow the spectral modal analysis procedure introduced in Section 9.2.3.

10.2.4.1 Orthogonality Properties of Normal Modes

Let ω_p be the pth natural frequency and W_p, U_p, and Ψ_p the corresponding normal modes, respectively. Then, the pth eigensolutions must satisfy the eigenvalue problem given by

$$\begin{aligned} &EI_w W_p'''' - \omega_p^2 \rho A W_p + i\omega_p cA W_p = \omega_p^2(\alpha U_p' - \gamma W_p'' + \varepsilon_1 \Psi_p') \\ &\quad + i\omega_p(c_1 W_p'' - c_4 U_p' - c_3 \Psi_p') + \beta U_p''' + \varepsilon_2 \Psi_p''' + F W_p'' \\ &EAU_p'' + \omega_p^2 \rho A U_p - i\omega_p cA U_p = \omega_p^2(\alpha W_p' - \varepsilon_3 \Psi_p) - i\omega_p(c_4 W_p' - c_5 \Psi_p) + \beta W_p''' - \varepsilon_4 \Psi_p'' \\ &EI_\psi \Psi_p'' + \omega_p^2 \rho I \Psi_p - i\omega_p c_2 \Psi_p = \omega_p^2(\varepsilon_1 W_p' - \varepsilon_3 U_p) - i\omega_p(c_3 W_p' - c_5 U_p) + \varepsilon_2 W_p''' - \varepsilon_4 U_p'' + A_v G^* \Psi_p \end{aligned} \tag{10.115}$$

Similarly, the qth eigensolutions ω_q, W_q, U_q, and Ψ_q must satisfy the eigenvalue problem given by

$$\begin{aligned} &EI_w W_q'''' - \omega_q^2 \rho A W_q + i\omega_q cA W_q = \omega_q^2(\alpha U_q' - \gamma W_q'' + \varepsilon_1 \Psi_q') \\ &\quad + i\omega_q(c_1 W_q'' - c_4 U_q' - c_3 \Psi_q') + \beta U_q''' + \varepsilon_2 \Psi_q''' + F W_q'' \\ &EAU_q'' + \omega_q^2 \rho A U_q - i\omega_p cA U_q = \omega_q^2(\alpha W_q' - \varepsilon_3 \Psi_q) - i\omega_q(c_4 W_q' - c_5 \Psi_q) + \beta W_q''' - \varepsilon_4 \Psi_q'' \\ &EI_\psi \Psi_q'' + \omega_q^2 \rho I \Psi_q - i\omega_q c_2 \Psi_q = \omega_q^2(\varepsilon_1 W_q' - \varepsilon_3 U_q) - i\omega_q(c_3 W_q' - c_5 U_q) \\ &\quad + \varepsilon_2 W_q''' - \varepsilon_4 U_q'' + A_v G^* \Psi_q \end{aligned} \tag{10.116}$$

By following the procedure used in Section 9.2.3.1, we can derive the orthogonality properties of the normal modes from Equations 10.115 and 10.116 as

$$\begin{aligned}
&\int_0^L [\rho A(W_pW_q + U_pU_q) + \rho I\Psi_p\Psi_q - \alpha(U_pW_q' + U_qW_p') - \varepsilon_1(\Psi_pW_q' + \Psi_qW_p') \\
&\quad + \varepsilon_3(\Psi_pU_q + \Psi_qU_p) + \gamma W_p'W_q']\mathrm{d}x = m_p\delta_{pq} \\
&\int_0^L [cA(W_pW_q + U_pU_q) + c_1W_p'W_q' + c_2\Psi_p\Psi_q - c_3(\Psi_pW_q' + \Psi_qW_p') \\
&\quad - c_4(U_pW_q' + U_qW_p') + c_5(\Psi_pU_q + \Psi_qU_p)]\mathrm{d}x = c_p\delta_{pq}
\end{aligned} \tag{10.117}$$

where m_p and c_p are the modal mass and modal damping coefficients defined by

$$\begin{aligned}
m_p &= \int_0^L [\rho A(U_p^2 + W_p^2) + \rho I\Psi_p^2 - 2\alpha U_pW_p' - 2\varepsilon_1\Psi_pW_p' + 2\varepsilon_3\Psi_pU_p + \gamma W_p'^2]\mathrm{d}x \\
c_p &= \int_0^L [cA(U_p^2 + W_p^2) + c_1W_p'^2 + c_2\Psi_p^2 - 2c_3\Psi_pW_p' - 2c_4U_pW_p' + 2c_5\Psi_pU_p]\mathrm{d}x
\end{aligned} \tag{10.118}$$

10.2.4.2 Spectral Modal Equations

Based on the modal analysis method, the forced vibration responses are assumed in the forms

$$\{w(x,t), u(x,t), \psi(x,t)\} = \sum_{p=1}^{\infty}\{W_p(x), U_p(x), \Psi_p(x)\}q_p(t) \tag{10.119}$$

where $q_p(t)$ are the modal coordinates. Represent the modal coordinates in the spectral form as

$$q_p(t) = \frac{1}{N}\sum_{n=0}^{N-1}\bar{q}_{pn}\mathrm{e}^{i\omega_n t} \tag{10.120}$$

By substitution of Equation 10.120 into Equation 10.119, we obtain

$$\{w(x,t), u(x,t), \psi(x,t)\} = \frac{1}{N}\sum_{p=1}^{\infty}\sum_{n=0}^{N-1}\{W_p(x), U_p(x), \Psi_p(x)\}q_{pn}\mathrm{e}^{i\omega_n t} \tag{10.121}$$

The external forces are also written in the spectral forms as

$$\{p(x,t), \tau(x,t), \gamma(x,t)\} = \frac{1}{N}\sum_{n=0}^{N-1}\{P_n(x), T_n(x), \Gamma_n(x)\}\mathrm{e}^{i\omega_n t} \tag{10.122}$$

Substitute Equations 10.121 and 10.122 into the time-domain governing Equations 10.78 and take the same manipulation as conducted in Section 9.2.3.2. Then, we can obtain the spectral modal equations as

$$(\omega_q^2 - \omega_n^2)m_q q_{qn} - i(\omega_q - \omega_n)c_q q_{qn} = f_{qn} \tag{10.123}$$

where f_{qn} are the spectral modal forces defined by

$$f_{qn} = \int_0^L [P_n(x)W_q + T_n(x)U_q + \Gamma_n(x)\Psi_q]\mathrm{d}x \tag{10.124}$$

The spectral modal Equation 10.123 can be readily transformed into the time domain as

$$m_q\ddot{q}_q(t) + c_q\dot{q}_q(t) + (\omega_q^2 m_q - i\omega_q c_q)q_q(t) = f_q(t) \tag{10.125}$$

where

$$f_q(t) = \int_0^L [p(x,t)W_q + \tau(x,t)U_q + r(x,t)\Psi_q]\mathrm{d}x \tag{10.126}$$

10.2.5 Finite Element Model

The weak form of the equations of motion (10.78) is given by

$$\begin{aligned}
&\int_0^L [EI_w w''\delta w'' + \rho A\ddot{w}\delta w + cA\dot{w}\delta w - \alpha\ddot{u}_b\delta w' - \beta u_b'\delta w'' + \gamma\ddot{w}'\delta w' - \varepsilon_1\ddot{\psi}\delta w' \\
&\quad - \varepsilon_2\psi'\delta w'' + c_1\dot{w}'\delta w' - c_4\dot{u}_b\delta w' - c_3\dot{\psi}\delta w' + Fw'\delta w' + EAu_b'\delta u_b' + \rho A\ddot{u}_b\delta u_b \\
&\quad + cA\dot{u}_b\delta u_b - \alpha\ddot{w}'\delta u_b - \beta w''\delta u_b' + \varepsilon_3\ddot{\psi}\delta u_b + \varepsilon_4\psi'\delta u_b' - c_4\dot{w}'\delta u_b' + c_5\dot{\psi}\delta u_b \\
&\quad + EI_\psi\psi'\delta\psi' + \rho I\ddot{\psi}\delta\psi + c_2\dot{\psi}\delta\psi - \varepsilon_1\ddot{w}'\delta\psi - \varepsilon_2 w''\delta\psi' + \varepsilon_3\ddot{u}_b\delta\psi \\
&\quad + \varepsilon_4 u_b'\delta\psi' - c_3\dot{w}'\delta\psi + c_5\dot{u}_b\delta\psi + A_v(g\circ\psi)\delta\psi]\mathrm{d}x \\
&\quad - \int_0^L [p(x,t)\delta w + \tau(x,t)\delta u_b + \gamma(x,t)\delta\psi]\mathrm{d}x - \{N(x,t)\delta u_b|_0^L \\
&\quad + M(x,t)\delta w'|_0^L + [Q(x,t) + Fw'(x,t)]\delta w|_0^L + R(x,t)\delta\psi|_0^L\} = 0
\end{aligned} \tag{10.127}$$

The dynamic behavior of a viscoelastic material can be represented by using the GHM model [121]. As discussed in Section 9.2.4, if we adopt a one-term GHM model for the VEM layer, the convolution term $A_v(g^\circ\psi)\delta\psi$ in Equation 10.127 can be represented by

$$A_v(g\circ\psi)\delta\psi \cong A_v[(1+\alpha)\kappa\psi(x,t) - \alpha\kappa z(x,t)]\delta\psi \tag{10.128}$$

where $z(t)$ is the dissipation coordinate governed by

$$\ddot{z}(x,t) + 2\hat{\xi}\hat{\omega}\dot{z}(x,t) - \hat{\omega}^2\psi(x,t) + \hat{\omega}^2 z(x,t) = 0 \tag{10.129}$$

The weak form of Equation 10.129 can be written in the form of

$$\int_0^L \left[\left(\frac{A_v\alpha\kappa}{\hat{\omega}^2}\right)\ddot{z} + \left(\frac{2A_v\alpha\kappa\hat{\xi}}{\hat{\omega}}\right)\dot{z} - A_v\alpha\kappa\psi + A_v\alpha\kappa z\right]\delta z\mathrm{d}x = 0 \tag{10.130}$$

The finite element equation can be formulated from the weak forms of Equations 10.127 and 10.130. By using the same nodal DOFs vector and displacement fields as used in Section 9.2.4, we can derive the finite element equation as

$$\begin{bmatrix} \boldsymbol{M}_d & \boldsymbol{0} \\ \boldsymbol{0} & \boldsymbol{M}_z \end{bmatrix}\begin{Bmatrix} \ddot{\boldsymbol{d}} \\ \ddot{\boldsymbol{z}} \end{Bmatrix} + \begin{bmatrix} \boldsymbol{C}_d & \boldsymbol{0} \\ \boldsymbol{0} & \boldsymbol{C}_z \end{bmatrix}\begin{Bmatrix} \dot{\boldsymbol{d}} \\ \dot{\boldsymbol{z}} \end{Bmatrix} + \begin{bmatrix} \boldsymbol{K}_d & -\boldsymbol{K}_{dz} \\ -\boldsymbol{K}_{dz}^{\mathrm{T}} & \boldsymbol{K}_z \end{bmatrix}\begin{Bmatrix} \boldsymbol{d} \\ \boldsymbol{z} \end{Bmatrix} = \begin{Bmatrix} \boldsymbol{f}(t) \\ \boldsymbol{0} \end{Bmatrix} \tag{10.131}$$

where

$$
\begin{aligned}
\boldsymbol{M}_d &= \int_0^L [\rho A(\boldsymbol{N}_w^{\mathrm{T}}\boldsymbol{N}_w + \boldsymbol{N}_u^{\mathrm{T}}\boldsymbol{N}_u) + \rho I \boldsymbol{N}_\psi^{\mathrm{T}}\boldsymbol{N}_\psi - \alpha(\boldsymbol{N}'^{\mathrm{T}}_w\boldsymbol{N}_u + \boldsymbol{N}_u^{\mathrm{T}}\boldsymbol{N}'_w) \\
&\quad + \gamma \boldsymbol{N}'^{\mathrm{T}}_w\boldsymbol{N}'_w - \varepsilon_1(\boldsymbol{N}_\psi^{\mathrm{T}}\boldsymbol{N}'_w + \boldsymbol{N}'^{\mathrm{T}}_w\boldsymbol{N}_\psi) + \varepsilon_3(\boldsymbol{N}_u^{\mathrm{T}}\boldsymbol{N}_\psi + \boldsymbol{N}_\psi^{\mathrm{T}}\boldsymbol{N}_u)]\mathrm{d}x \\
\boldsymbol{M}_z &= \frac{A_\nu \alpha \kappa}{\hat{\omega}^2}\int_0^L \boldsymbol{N}_z^{\mathrm{T}}\boldsymbol{N}_z \mathrm{d}x \\
\boldsymbol{C}_d &= \int_0^L [cA(\boldsymbol{N}_w^{\mathrm{T}}\boldsymbol{N}_w + \boldsymbol{N}_u^{\mathrm{T}}\boldsymbol{N}_u) + c_1 \boldsymbol{N}'^{\mathrm{T}}_w\boldsymbol{N}'_w + c_2 \boldsymbol{N}_\psi^{\mathrm{T}}\boldsymbol{N}_\psi - c_3(\boldsymbol{N}'^{\mathrm{T}}_w\boldsymbol{N}_\psi \\
&\quad + \boldsymbol{N}_\psi^{\mathrm{T}}\boldsymbol{N}'_w) - c_4(\boldsymbol{N}'^{\mathrm{T}}_w\boldsymbol{N}_u + \boldsymbol{N}'^{\mathrm{T}}_u\boldsymbol{N}'_w) + c_5(\boldsymbol{N}_u^{\mathrm{T}}\boldsymbol{N}_\psi + \boldsymbol{N}_\psi^{\mathrm{T}}\boldsymbol{N}_u)]\mathrm{d}x \\
\boldsymbol{C}_z &= \frac{2A_\nu \alpha \kappa \hat{\xi}}{\hat{\omega}}\int_0^L \boldsymbol{N}_z^{\mathrm{T}}\boldsymbol{N}_z \mathrm{d}x \\
\boldsymbol{K}_d &= \int_0^L [EA\boldsymbol{N}'^{\mathrm{T}}_u\boldsymbol{N}'_u + EI_w \boldsymbol{N}''^{\mathrm{T}}_w\boldsymbol{N}''_w + EI_\psi \boldsymbol{N}'^{\mathrm{T}}_\psi\boldsymbol{N}'_\psi \\
&\quad - \beta(\boldsymbol{N}''^{\mathrm{T}}_w\boldsymbol{N}'_u + \boldsymbol{N}'^{\mathrm{T}}_u\boldsymbol{N}''_w) - \varepsilon_2(\boldsymbol{N}''^{\mathrm{T}}_w\boldsymbol{N}'_\psi + \boldsymbol{N}'^{\mathrm{T}}_\psi\boldsymbol{N}''_w) \\
&\quad + \varepsilon_4(\boldsymbol{N}'^{\mathrm{T}}_u\boldsymbol{N}'_\psi + \boldsymbol{N}'^{\mathrm{T}}_\psi\boldsymbol{N}'_u) + F\boldsymbol{N}'^{\mathrm{T}}_w\boldsymbol{N}'_w + A_\nu(1+\alpha)\kappa \boldsymbol{N}_\psi^{\mathrm{T}}\boldsymbol{N}_\psi]\mathrm{d}x \\
\boldsymbol{K}_{dz} &= A_\nu \alpha \kappa \int_0^L \boldsymbol{N}_\psi^{\mathrm{T}}\boldsymbol{N}_z \mathrm{d}x \\
\boldsymbol{K}_z &= A_\nu \alpha \kappa \int_0^L \boldsymbol{N}_z^{\mathrm{T}}\boldsymbol{N}_z \mathrm{d}x
\end{aligned}
\tag{10.132}
$$

and

$$
\boldsymbol{f}(t) = \boldsymbol{f}_{\mathrm{c}}^{\mathrm{mech}}(t) + \boldsymbol{f}_{\mathrm{c}}^{\mathrm{piezo}}(t) + \boldsymbol{f}_{\mathrm{d}}(t) \tag{10.133}
$$

with the use of the definitions:

$$
\begin{aligned}
\boldsymbol{f}_{\mathrm{c}}^{\mathrm{mech}}(t) &= \{N_1(t) \quad Q_1(t) \quad M_1(t) \quad R_1(t) \quad N_2(t) \quad Q_2(t) \quad M_2(t) \quad R_2(t)\}^{\mathrm{T}} \\
\boldsymbol{f}_{\mathrm{c}}^{\mathrm{piezo}}(t) &= \{-N^p(t) \quad 0 \quad M^p(t) \quad -R^p(t) \quad N^p(t) \quad 0 \quad -M^p(t) \quad R^p(t)\}^{\mathrm{T}} \\
\boldsymbol{f}_{\mathrm{d}}(t) &= \int_0^L [p(x,t)\boldsymbol{N}_w^{\mathrm{T}}(x) + \tau(x,t)\boldsymbol{N}_u^{\mathrm{T}}(x) + \gamma(x,t)\boldsymbol{N}_\psi^{\mathrm{T}}(x)]\mathrm{d}x
\end{aligned}
\tag{10.134}
$$

The finite element matrices defined by Equation 10.132 are the same as given in Appendix 9.B, except for using the structural and damping properties defined by Equations 10.73 and 10.77.

Example 10.2

Consider a cantilevered ACLD beam as shown in Figure 10.9. The middle one-third part of the base beam is covered with a uniform VEM–PZT patch. The geometric and material properties of the base beam and PZT layer are same as those given in Table 10.1. The VEM layer is represented by a one-term GHM model and its geometric and material properties are the same as given in Table 9.3, except for the length. The vibration of the ACLD beam is controlled by

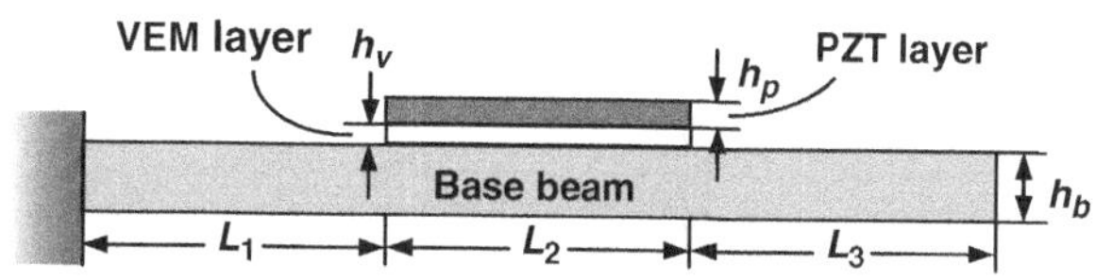

Figure 10.9 A cantilevered beam partially covered with a VEM–PZT patch

Table 10.4 Natural frequencies and modal damping ratios of an ACLD beam

	Uncontrolled		Controlled			
			$K_p = 5 \times 10^7$ V, $K_d = 100$ Vs		$K_p = 5 \times 10^7$ V, $K_d = 300$ Vs	
Mode number (n)	f_n (Hz)	ξ_n	f_n (Hz)	ξ_n	f_n (Hz)	ξ_n
1	44.098	0.0055	46.784	0.0101	46.766	0.0266
2	264.62	0.0112	263.95	0.0152	264.04	0.0286
3	760.62	0.0028	761.29	0.0028	761.36	0.0066
4	1520.2	0.0015	1520.4	0.0017	1520.3	0.0021
5	2428.5	0.0005	2429.2	0.0006	2429.2	0.0010

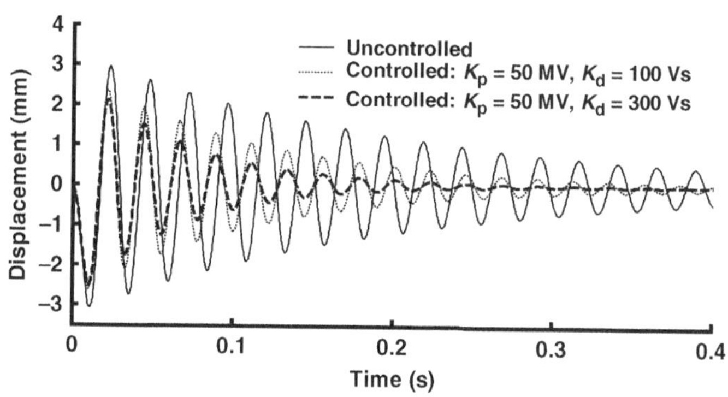

Figure 10.10 Dynamic responses of an ACLD beam with a PD controller

using a PD controller, which is shown schematically in Figure 10.8. The natural frequencies and modal damping ratios for the ACLD beam with and without the PD controller are compared in Table 10.4. It is shown that the PD controller certainly increases the modal damping ratios to suppress the vibration quicker, as shown in Figure 10.10.

11

Dynamics of Composite Laminated Structures

It is well recognized that fiber reinforced composite materials have many advantages over isotropic materials due to their high strength-to-density ratios. Thus they have been widely used in many industrial applications [12, 29, 111, 144, 161]. Typical examples of are composite aircraft wings, helicopter blades, propeller blades, turbine blades, axles of vehicles and so on. In general, the composite laminated structures are fabricated by bonding two or more laminae together. The dynamic characteristics of composite laminated structures can be enhanced by appropriately tailoring the material and structural couplings determined by the specific lay-up and fiber orientations. Of particular interest, especially for the high aspect ratio composite laminated beams, are the coupling between the extensional (axial), bending, and torsional modes of deformation. As fibrous composite materials usually have very low shear modulus, the shear deformation should also be included in the analysis and design of such composite laminated beams. Thus, this chapter covers the dynamics of the axial–bending–shear coupled composite beams and the bending–torsion–shear coupled composite beams, all based on the first-order shear deformation theory (FSDT), which is the equivalent to the Timoshenko-beam theory for composite laminated beams.

11.1 Theory of Composite Mechanics

11.1.1 Three-Dimensional Stress–Strain Relationships

For orthotropic materials, the stress–strain relationships in the coordinates aligned with its principal material coordinates (1, 2, 3) are given by

$$\begin{Bmatrix} \sigma_1 \\ \sigma_2 \\ \sigma_3 \\ \sigma_4 \\ \sigma_5 \\ \sigma_6 \end{Bmatrix} = \begin{bmatrix} C_{11} & C_{12} & C_{13} & 0 & 0 & 0 \\ C_{12} & C_{22} & C_{23} & 0 & 0 & 0 \\ C_{13} & C_{23} & C_{33} & 0 & 0 & 0 \\ 0 & 0 & 0 & C_{44} & 0 & 0 \\ 0 & 0 & 0 & 0 & C_{55} & 0 \\ 0 & 0 & 0 & 0 & 0 & C_{66} \end{bmatrix} \begin{Bmatrix} \varepsilon_1 \\ \varepsilon_2 \\ \varepsilon_3 \\ \varepsilon_4 \\ \varepsilon_5 \\ \varepsilon_6 \end{Bmatrix} \tag{11.1}$$

Spectral Element Method in Structural Dynamics Usik Lee

where $\sigma_i (i=1,2,\ldots,6)$ are the stress components, $\varepsilon_i (i=1,2,\ldots,6)$ are the strain components, C_{ij} $(i, j=1, 2, \ldots, 6)$ are the stiffness coefficients. The strain–stress relationship can be also written as

$$\begin{Bmatrix} \varepsilon_1 \\ \varepsilon_2 \\ \varepsilon_3 \\ \varepsilon_4 \\ \varepsilon_5 \\ \varepsilon_6 \end{Bmatrix} = \begin{bmatrix} S_{11} & S_{12} & S_{13} & 0 & 0 & 0 \\ S_{12} & S_{22} & S_{23} & 0 & 0 & 0 \\ S_{13} & S_{23} & S_{33} & 0 & 0 & 0 \\ 0 & 0 & 0 & S_{44} & 0 & 0 \\ 0 & 0 & 0 & 0 & S_{55} & 0 \\ 0 & 0 & 0 & 0 & 0 & S_{66} \end{bmatrix} \begin{Bmatrix} \sigma_1 \\ \sigma_2 \\ \sigma_3 \\ \sigma_4 \\ \sigma_5 \\ \sigma_6 \end{Bmatrix} \tag{11.2}$$

where C_{ij} are the compliance coefficients determined from nine independent engineering constants as follows:

$$\begin{aligned} S_{11} &= \frac{1}{E_1}, & S_{22} &= \frac{1}{E_2}, & S_{33} &= \frac{1}{E_3} \\ S_{44} &= \frac{1}{G_{23}}, & S_{55} &= \frac{1}{G_{31}}, & S_{66} &= \frac{1}{G_{12}} \\ S_{12} &= -\frac{\nu_{21}}{E_2}, & S_{13} &= -\frac{\nu_{31}}{E_3}, & S_{23} &= -\frac{\nu_{32}}{E_3} \end{aligned} \tag{11.3}$$

In the above, the subscripts are used to denote the quantities measured with respect to the principal material coordinates (1, 2, 3) and the following contracted notation for stresses and strains are used:

$$\begin{aligned} &\sigma_1 = \sigma_{11}, \quad \sigma_2 = \sigma_{22}, \quad \sigma_3 = \sigma_{33}, \quad \sigma_4 = \sigma_{23}, \quad \sigma_5 = \sigma_{31}, \quad \sigma_6 = \sigma_{12} \\ &\varepsilon_1 = \varepsilon_{11}, \quad \varepsilon_2 = \varepsilon_{22}, \quad \varepsilon_3 = \varepsilon_{33}, \quad \varepsilon_4 = 2\varepsilon_{23}, \quad \varepsilon_5 = 2\varepsilon_{31}, \quad \varepsilon_6 = 2\varepsilon_{12} \end{aligned} \tag{11.4}$$

11.1.2 Stress–Strain Relationships for an Orthotropic Lamina

For an orthotropic lamina, we assume that $\sigma_3=0$. Thus, from Equation 11.2 we can obtain

$$\varepsilon_3 = S_{13}\sigma_1 + S_{23}\sigma_2 \tag{11.5}$$

and Equation 11.2 can be reduced into the forms

$$\begin{Bmatrix} \varepsilon_1 \\ \varepsilon_2 \\ \varepsilon_6 \end{Bmatrix} = \begin{bmatrix} S_{11} & S_{12} & 0 \\ S_{12} & S_{22} & 0 \\ 0 & 0 & S_{66} \end{bmatrix} \begin{Bmatrix} \sigma_1 \\ \sigma_2 \\ \sigma_6 \end{Bmatrix} \tag{11.6}$$

and

$$\begin{Bmatrix} \varepsilon_4 \\ \varepsilon_5 \end{Bmatrix} = \begin{bmatrix} S_{44} & 0 \\ 0 & S_{55} \end{bmatrix} \begin{Bmatrix} \sigma_4 \\ \sigma_5 \end{Bmatrix} \tag{11.7}$$

The stress–strain relationships for an orthotropic lamina can be readily obtained from Equations 11.6 and 11.7 as

$$\begin{Bmatrix} \sigma_1 \\ \sigma_2 \\ \sigma_6 \end{Bmatrix} = \begin{bmatrix} Q_{11} & Q_{12} & 0 \\ Q_{12} & Q_{22} & 0 \\ 0 & 0 & Q_{66} \end{bmatrix} \begin{Bmatrix} \varepsilon_1 \\ \varepsilon_2 \\ \varepsilon_6 \end{Bmatrix} \tag{11.8}$$

and

$$\begin{Bmatrix} \sigma_4 \\ \sigma_5 \end{Bmatrix} = \begin{bmatrix} Q_{44} & 0 \\ 0 & Q_{55} \end{bmatrix} \begin{Bmatrix} \varepsilon_4 \\ \varepsilon_5 \end{Bmatrix} \tag{11.9}$$

where Q_{ij} are the reduced stiffnesses given by

$$\begin{aligned} Q_{11} &= \frac{S_{22}}{S_{11}S_{22} - S_{12}^2} = \frac{E_1}{1 - \nu_{12}\nu_{21}} \\ Q_{22} &= \frac{S_{11}}{S_{11}S_{22} - S_{12}^2} = \frac{E_2}{1 - \nu_{12}\nu_{21}} \\ Q_{12} &= -\frac{S_{12}}{S_{11}S_{22} - S_{12}^2} = \frac{\nu_{21}E_1}{1 - \nu_{12}\nu_{21}} = \frac{\nu_{12}E_2}{1 - \nu_{12}\nu_{21}} \\ Q_{44} &= \frac{1}{S_{44}} = G_{23}, \quad Q_{55} = \frac{1}{S_{55}} = G_{31}, \quad Q_{66} = \frac{1}{S_{66}} = G_{12} \end{aligned} \tag{11.10}$$

Now consider global coordinates (x, y, z), which are used in the solution to a problem and, as shown in Figure 11.1, assume that the principal material coordinates (1, 2, 3) are obtained by rotating the xy-plane counterclockwise about the z-axis by an angle ϕ. Then, we can derive the coordinate transformation rules as follows:

$$\begin{aligned} \begin{Bmatrix} \sigma_{xx} \\ \sigma_{yy} \\ \sigma_{xy} \end{Bmatrix} &= \begin{bmatrix} \cos^2\phi & \sin^2\phi & -\sin 2\phi \\ \sin^2\phi & \cos^2\phi & \sin 2\phi \\ \sin\phi\cos\phi & -\sin\phi\cos\phi & \cos^2\phi - \sin^2\phi \end{bmatrix} \begin{Bmatrix} \sigma_1 \\ \sigma_2 \\ \sigma_6 \end{Bmatrix} \\ \begin{Bmatrix} \sigma_{yz} \\ \sigma_{zx} \end{Bmatrix} &= \begin{bmatrix} \cos\phi & \sin\phi \\ -\sin\phi & \cos\phi \end{bmatrix} \begin{Bmatrix} \sigma_4 \\ \sigma_5 \end{Bmatrix} \end{aligned} \tag{11.11}$$

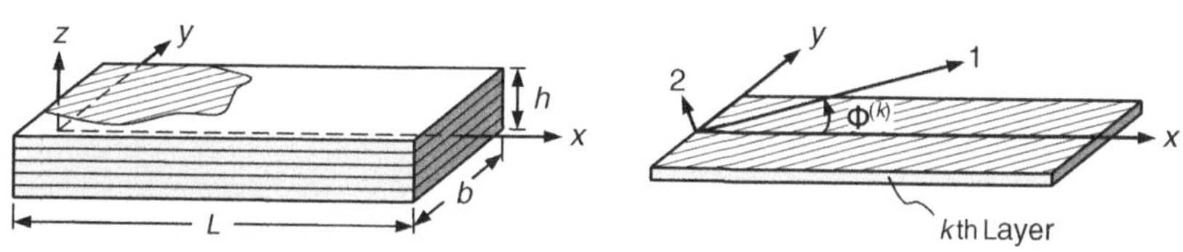

Figure 11.1 The global and principal material coordinate systems for a straight composite laminated beam

and

$$\begin{Bmatrix} \varepsilon_{xx} \\ \varepsilon_{yy} \\ \gamma_{xy} \end{Bmatrix} = \begin{bmatrix} \cos^2\phi & \sin^2\phi & -\sin\phi\cos\phi \\ \sin^2\phi & \cos^2\phi & \sin\phi\cos\phi \\ \sin 2\phi & -\sin 2\phi & \cos^2\phi - \sin^2\phi \end{bmatrix} \begin{Bmatrix} \varepsilon_1 \\ \varepsilon_2 \\ \varepsilon_6 \end{Bmatrix}$$
$$\begin{Bmatrix} \gamma_{yz} \\ \gamma_{zx} \end{Bmatrix} = \begin{bmatrix} \cos\phi & \sin\phi \\ -\sin\phi & \cos\phi \end{bmatrix} \begin{Bmatrix} \varepsilon_4 \\ \varepsilon_5 \end{Bmatrix} \tag{11.12}$$

where

$$\gamma_{xy} = 2\varepsilon_{xy}, \quad \gamma_{yz} = 2\varepsilon_{yz}, \quad \gamma_{zx} = 2\varepsilon_{zx} \tag{11.13}$$

By applying the above coordinate transformation rules to Equations 11.8 and 11.9, we can get the stress–strain relationships with respect to the global coordinates (x, y, z) as

$$\begin{Bmatrix} \sigma_{xx} \\ \sigma_{yy} \\ \sigma_{xy} \end{Bmatrix} = \begin{bmatrix} \bar{Q}_{11} & \bar{Q}_{12} & \bar{Q}_{16} \\ \bar{Q}_{12} & \bar{Q}_{22} & \bar{Q}_{26} \\ \bar{Q}_{16} & \bar{Q}_{26} & \bar{Q}_{66} \end{bmatrix} \begin{Bmatrix} \varepsilon_{xx} \\ \varepsilon_{yy} \\ \gamma_{xy} \end{Bmatrix} \tag{11.14}$$

and

$$\begin{Bmatrix} \sigma_{yz} \\ \sigma_{zx} \end{Bmatrix} = \begin{bmatrix} \bar{Q}_{44} & \bar{Q}_{45} \\ \bar{Q}_{45} & \bar{Q}_{55} \end{bmatrix} \begin{Bmatrix} \gamma_{yz} \\ \gamma_{zx} \end{Bmatrix} \tag{11.15}$$

where $\bar{Q}_{ij}$ are the transformed reduced stiffnesses defined by [144]

$$\begin{aligned}
\bar{Q}_{11} &= Q_{11}\cos^4\phi + Q_{22}\sin^4\phi + 2(Q_{12} + 2Q_{66})\sin^2\phi\cos^2\phi \\
\bar{Q}_{22} &= Q_{11}\sin^4\phi + Q_{22}\cos^4\phi + 2(Q_{12} + 2Q_{66})\sin^2\phi\cos^2\phi \\
\bar{Q}_{66} &= Q_{66}(\sin^4\phi + \cos^4\phi) + (Q_{11} + Q_{22} - 2Q_{12} - 2Q_{66})\sin^2\phi\cos^2\phi \\
\bar{Q}_{12} &= Q_{12}(\sin^4\phi + \cos^4\phi) + (Q_{11} + Q_{22} - 4Q_{66})\sin^2\phi\cos^2\phi \\
\bar{Q}_{16} &= (Q_{11} - Q_{12} - 2Q_{66})\sin\phi\cos^3\phi + (Q_{12} - Q_{22} + 2Q_{66})\sin^3\phi\cos\phi \\
\bar{Q}_{26} &= (Q_{11} - Q_{12} - 2Q_{66})\sin^3\phi\cos\phi + (Q_{12} - Q_{22} + 2Q_{66})\sin\phi\cos^3\phi \\
\bar{Q}_{44} &= Q_{44}\cos^2\phi + Q_{55}\sin^2\phi, \quad \bar{Q}_{55} = Q_{55}\cos^2\phi + Q_{44}\sin^2\phi \\
\bar{Q}_{45} &= (Q_{55} - Q_{44})\cos\phi\sin\phi
\end{aligned} \tag{11.16}$$

11.1.3 Strain–Displacement Relationships

Let $u(x, y, z, t)$, $v(x, y, z, t)$, and $w(x, y, z, t)$ represent the displacements in the x-, y-, and z-directions, respectively. The strains are given by

$$\begin{aligned}
&\varepsilon_{xx} = \frac{\partial u}{\partial x}, \quad \varepsilon_{yy} = \frac{\partial v}{\partial y}, \quad \gamma_{xy} = 2\varepsilon_{xy} = \frac{\partial v}{\partial x} + \frac{\partial u}{\partial y} \\
&\gamma_{yz} = 2\varepsilon_{yz} = \frac{\partial w}{\partial y} + \frac{\partial v}{\partial z}, \quad \gamma_{zx} = 2\varepsilon_{zx} = \frac{\partial w}{\partial x} + \frac{\partial u}{\partial z}
\end{aligned} \tag{11.17}$$

Assume that the displacements are given in the general forms as

$$\begin{aligned} u(x,y,z,t) &= u_0(x,y,t) - z\theta_x(x,y,t) \\ v(x,y,z,t) &= -z\theta_y(x,y,t) \\ w(x,y,z,t) &= w_0(x,y,t) \end{aligned} \tag{11.18}$$

where u_0 and w_0 are the displacements of a point on the reference plane $z = 0$ in the x- and z-directions, and θ_x and θ_y are the negative of the slope along the x- and y-axes, respectively. Substitution of Equation 11.18 into Equation 11.17 gives

$$\begin{aligned} &\varepsilon_{xx} = \varepsilon_{xx}^0 + z\chi_x, \qquad \varepsilon_{yy} = z\chi_y, \qquad \gamma_{xy} = \gamma_{xy}^0 + z\chi_{xy} \\ &\gamma_{yz} = \frac{\partial w_0}{\partial y} - \theta_y, \qquad \gamma_{zx} = \frac{\partial w_0}{\partial x} - \theta_x \end{aligned} \tag{11.19}$$

where ε_{xx}^0, ε_{yy}^0, and ε_{xy}^0 are the in-plane strains defined by

$$\varepsilon_{xx}^0 = \frac{\partial u_0}{\partial x}, \qquad \gamma_{xy}^0 = \frac{\partial u_0}{\partial y} \tag{11.20}$$

and χ_x, χ_y, and χ_{xy} are the curvatures defined by

$$\begin{aligned} &\chi_x = -\frac{\partial \theta_x}{\partial x}, \qquad \chi_y = -\frac{\partial \theta_y}{\partial y} \\ &\chi_{xy} = -\left(\frac{\partial \theta_y}{\partial x} + \frac{\partial \theta_x}{\partial y}\right) \end{aligned} \tag{11.21}$$

11.1.4 Resultant Forces and Moments

The resultant forces and moments can be obtained by integrating the stresses through the laminate thickness ($-h/2 \leq z \leq h/2$) as

$$\begin{Bmatrix} N_x \\ Q_x \end{Bmatrix} = \int_{-\frac{h}{2}}^{\frac{h}{2}} \begin{Bmatrix} \sigma_{xx} \\ \kappa\sigma_{zx} \end{Bmatrix} \mathrm{d}z = \sum_{k=1}^{N} \int_{z_{k-1}}^{z_k} \begin{Bmatrix} \sigma_{xx}^{(k)} \\ \kappa\sigma_{zx}^{(k)} \end{Bmatrix} \mathrm{d}z \tag{11.22}$$

$$\begin{Bmatrix} M_{xx} \\ M_{yy} \\ M_{xy} \end{Bmatrix} = \int_{-\frac{h}{2}}^{\frac{h}{2}} \begin{Bmatrix} \sigma_{xx} \\ \sigma_{yy} \\ \sigma_{xy} \end{Bmatrix} z\mathrm{d}z = \sum_{k=1}^{N} \int_{z_{k-1}}^{z_k} \begin{Bmatrix} \sigma_{xx}^{(k)} \\ \sigma_{yy}^{(k)} \\ \sigma_{xy}^{(k)} \end{Bmatrix} z\mathrm{d}z \tag{11.23}$$

where b is the width of the beam, h is the thickness of the beam, κ is the shear correction factor, and z_k and z_{k-1} are defined in Figure 11.2. The stress–strain relationships for the kth lamina are taken from Equations 11.14 and 11.15 as

$$\begin{Bmatrix} \sigma_{xx}^{(k)} \\ \sigma_{yy}^{(k)} \\ \sigma_{xy}^{(k)} \end{Bmatrix} = \begin{bmatrix} \bar{Q}_{11}^{(k)} & \bar{Q}_{12}^{(k)} & \bar{Q}_{16}^{(k)} \\ \bar{Q}_{12}^{(k)} & \bar{Q}_{22}^{(k)} & \bar{Q}_{26}^{(k)} \\ \bar{Q}_{16}^{(k)} & \bar{Q}_{26}^{(k)} & \bar{Q}_{66}^{(k)} \end{bmatrix} \begin{Bmatrix} \varepsilon_{xx}^{(k)} \\ \varepsilon_{yy}^{(k)} \\ \gamma_{xy}^{(k)} \end{Bmatrix} \tag{11.24}$$

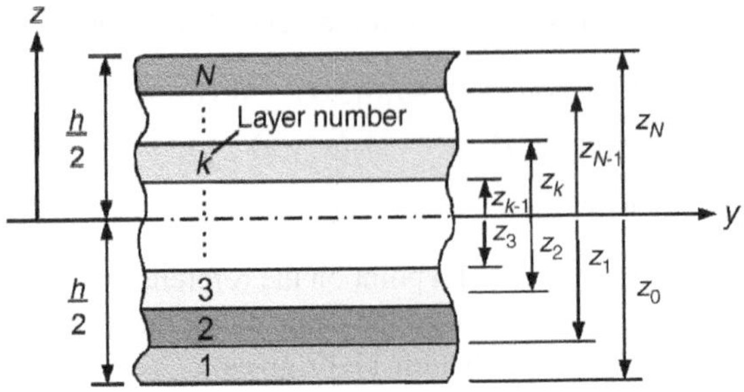

Figure 11.2 Geometry of an N-layered laminate

$$\sigma_{zx}^{(k)} = \begin{bmatrix} \bar{Q}_{45}^{(k)} & \bar{Q}_{55}^{(k)} \end{bmatrix} \begin{Bmatrix} \gamma_{yz}^{(k)} \\ \gamma_{zx}^{(k)} \end{Bmatrix} \tag{11.25}$$

By substituting Equation 11.19 into Equations 11.24 and 11.25 and the results into Equations 11.22 and 11.23, we can obtain the expressions

$$N_x = A_{11}\varepsilon_{xx}^0 + A_{16}\gamma_{xy}^0 + B_{11}\chi_x + B_{12}\chi_y + B_{16}\chi_{xy} \tag{11.26}$$

$$Q_x = \kappa A_{45}\left(\frac{\partial w_0}{\partial y} - \theta_y\right) + \kappa A_{55}\left(\frac{\partial w_0}{\partial x} - \theta_x\right) \tag{11.27}$$

$$\begin{Bmatrix} M_{xx} \\ M_{yy} \\ M_{xy} \end{Bmatrix} = \begin{bmatrix} B_{11} & B_{16} \\ B_{12} & B_{26} \\ B_{16} & B_{66} \end{bmatrix} \begin{Bmatrix} \varepsilon_{xx}^0 \\ \gamma_{xy}^0 \end{Bmatrix} + \begin{bmatrix} D_{11} & D_{12} & D_{16} \\ D_{12} & D_{22} & D_{26} \\ D_{16} & D_{26} & D_{66} \end{bmatrix} \begin{Bmatrix} \chi_x \\ \chi_y \\ \chi_{xy} \end{Bmatrix} \tag{11.28}$$

where

$$\begin{aligned} A_{ij} &= \sum_{k=1}^{N} \bar{Q}_{ij}^{(k)} (z_k - z_{k-1}) \\ B_{ij} &= \frac{1}{2}\sum_{k=1}^{N} \bar{Q}_{ij}^{(k)} (z_k^2 - z_{k-1}^2) \\ D_{ij} &= \frac{1}{3}\sum_{k=1}^{N} \bar{Q}_{ij}^{(k)} (z_k^3 - z_{k-1}^3) \end{aligned} \tag{11.29}$$

11.2 Equations of Motion for Composite Laminated Beams

The equations of motion and associated boundary conditions for composite beams are derived from Hamilton's principle based on the first-order shear deformation theory (FSDT), which is the equivalent of the Timoshenko-beam theory for composite laminated beams. The Hamilton's principle is given by

$$\int_0^t (\delta T - \delta U + \delta W)\mathrm{d}t = 0 \tag{11.30}$$

where T and U are the kinetic and strain energies, respectively, and δW is the virtual work done by external forces.

11.2.1 Axial–Bending–Shear Coupled Vibration

Consider the small amplitude axial–bending–shear coupled vibration of a straight composite laminated beam subjected to a constant axial tensile force P. The displacements of the beam are assumed as

$$\begin{aligned} w(x,y,z,t) &\cong w_0(x,t) \\ u(x,y,z,t) &\cong u(x,z,t) = u_0(x,t) - z\theta(x,t) \end{aligned} \tag{11.31}$$

Thus from Equations 11.19–11.21 we obtain

$$\begin{aligned} &\theta_x = \theta, \quad \theta_y = 0, \quad \varepsilon_{xx}^0 = u_0', \quad \gamma_{xy}^0 = 0 \\ &\chi_x = -\theta', \quad \chi_y = 0, \quad \chi_{xy} = 0 \end{aligned} \tag{11.32}$$

where the prime denotes the derivative with respective to the axial coordinate x. By using Equation 11.32, the resultant axial force, bending moment, and transverse shear force can be obtained from Equations 11.26–11.28 as follows:

$$\begin{aligned} N(x,t) &\equiv bN_x = EA_{eq}u_0' - K_{Aeq}\theta' \\ Q(x,t) &\equiv bQ_x = \kappa GA_{eq}(w_0' - \theta) \\ M(x,t) &\equiv -bM_{xx} = EI_{eq}\theta' - K_{Aeq}u_0' \end{aligned} \tag{11.33}$$

where

$$\begin{aligned} &EA_{eq} \equiv bA_{11}, \qquad EI_{eq} \equiv bD_{11} \\ &\kappa GA_{eq} \equiv b\kappa A_{55}, \quad K_{Aeq} \equiv bB_{11} \end{aligned} \tag{11.34}$$

The subscript "eq" is used to denote the apparent structural rigidities and mass inertia properties for a composite laminated beam, and the other terms are: the extensional rigidity (EA), the bending (flexural) rigidity (EI), the shear rigidity (κGA), the axial–bending material coupling rigidity (K_A), the mass per unit length (ρA), the first-order mass moment of inertia (ρR), and the second-order mass moment of inertia (ρI).

Assume that the composite beam is subjected to the external forces and moments applied at two end boundaries as shown in Figure 11.3, the distributed loads are $p(x, t)$, $f(x, t)$, and $\tau(x, t)$, and there is a constant axial tensile force P, which is assumed to act in the x-direction through the mass center of the cross-section. The strain energy U, the kinetic energy T, and the virtual

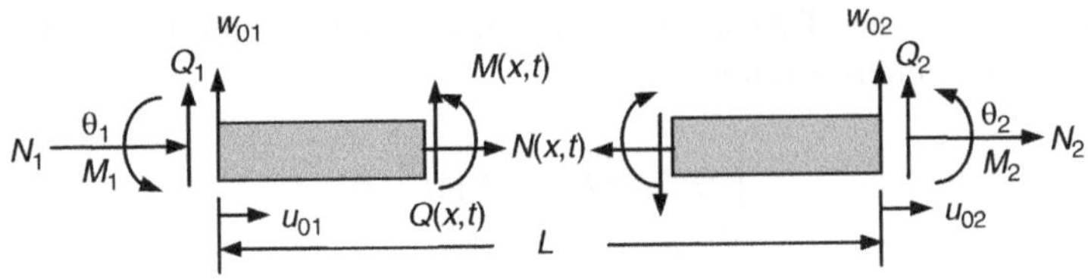

Figure 11.3 Sign convention for the composite laminated beam element

work δW done by external forces and moments are given by

$$
\begin{aligned}
U &= \frac{1}{2}\int_0^L \{Nu_0' - M\chi_x + Q(w_0' - \theta) + Pw_0'^2\}\,\mathrm{d}x \\
&= \frac{1}{2}\int_0^L \{EA_{eq}\,u_0'2 - 2K_{Aeq}u_0'\theta' + EI_{eq}\theta'^2 + \kappa GA_{eq}(w_0' - \theta)^2 + Pw_0'^2\}\,\mathrm{d}x \\
T &= \frac{1}{2}\int_0^L\int_{-\frac{h}{2}}^{\frac{h}{2}} \rho b(\dot{w}^2 + \dot{u}^2)\,\mathrm{d}z\mathrm{d}x = \frac{1}{2}\int_0^L (\rho A_{eq}\dot{w}_0^2 + \rho A_{eq}\dot{u}_0^2 - 2\rho R_{eq}\dot{u}_0\dot{\theta} + \rho I_{eq}\dot{\theta}^2)\,\mathrm{d}x \\
\delta W &= N_1\delta u_0(0) + N_2\delta u_0(L) + M_1\delta\theta(0) + M_2\delta\theta(L) + Q_1\delta w_0(0) + Q_2\delta w_0(L) \\
&\quad + \int_0^L [f_u(x,t)\delta u_0 + f_w(x,t)\delta w_0 + f_\theta(x,t)\delta\theta]\,\mathrm{d}x
\end{aligned}
\tag{11.35}
$$

where

$$
\begin{aligned}
\rho A_{eq} &\equiv b\int_{-\frac{h}{2}}^{\frac{h}{2}} \rho\mathrm{d}z = b\sum_{k=1}^{N} \rho^{(k)}(z_k - z_{k-1}) \\
\rho R_{eq} &\equiv b\int_{-\frac{h}{2}}^{\frac{h}{2}} \rho z\mathrm{d}z = \frac{b}{2}\sum_{k=1}^{N} \rho^{(k)}(z_k^2 - z_{k-1}^2) \\
\rho I_{eq} &\equiv b\int_{-\frac{h}{2}}^{\frac{h}{2}} \rho z^2\mathrm{d}z = \frac{b}{3}\sum_{k=1}^{N} \rho^{(k)}(z_k^3 - z_{k-1}^3)
\end{aligned}
\tag{11.36}
$$

where ρA_{eq} is the apparent mass of the composite laminated beam, ρR_{eq} is the apparent first-order mass moment of inertia about the y-axis and ρI_{eq} is the apparent secondorder mass moment of inertia about the y-axis, all per unit length.

By substituting Equation 11.35 into Hamilton's principle and integrating by parts, we can obtain

$$
\begin{aligned}
&\int_0^t\Big[\int_0^L \{[\kappa GA_{eq}(w_0'' - \theta') + Pw_0'' - \rho A_{eq}\ddot{w}_0 + f_w(x,t)]\delta w_0 \\
&\quad + [EI_{eq}\theta'' + \kappa GA_{eq}(w_0' - \theta) - K_{Aeq}u_0'' + \rho R_{eq}\ddot{u}_0 - \rho I_{eq}\ddot{\theta} + f_\theta(x,t)]\delta\theta \\
&\quad + [EA_{eq}u_0'' - K_{Aeq}\theta'' - \rho A_{eq}\ddot{u}_0 + \rho R_{eq}\ddot{\theta} + f_u(x,t)]\delta u_0\}\mathrm{d}x \\
&\quad - N(x,t)\delta u_0|_0^L - M(x,t)\delta\theta|_0^L - [Q(x,t) + Pw_0']\delta w_0|_0^L \\
&\quad + N_1\delta u_0(0) + N_2\delta u_0(L) + M_1\delta\theta(0) + M_2\delta\theta(L) \\
&\quad + Q_1\delta w_0(0) + Q_2\delta w_0(L)\Big]\mathrm{d}t = 0
\end{aligned}
\tag{11.37}
$$

where Equation 11.33 is used. From Equation 11.37, we get the equations of motion

$$\begin{aligned}
&\kappa GA_{eq}(w_0'' - \theta') + Pw_0'' - \rho A_{eq}\ddot{w}_0 + f_w(x,t) = 0 \\
&EI_{eq}\theta'' + \kappa GA_{eq}(w_0' - \theta) - K_{Aeq}u_0'' + \rho R_{eq}\ddot{u}_0 - \rho I_{eq}\ddot{\theta} + f_\theta(x,t) = 0 \\
&EA_{eq}u_0'' - K_{Aeq}\theta'' - \rho A_{eq}\ddot{u}_0 + \rho R_{eq}\ddot{\theta} + f_u(x,t) = 0
\end{aligned} \tag{11.38}$$

and the boundary conditions

$$\begin{aligned}
N(0,t) &= -N_1(t) && \text{or} \quad u_0(0,t) = 0 \\
N(L,t) &= +N_2(t) && \text{or} \quad u_0(L,t) = 0 \\
M(0,t) &= -M_1(t) && \text{or} \quad w_0(0,t) = 0 \\
M(L,t) &= +M_2(t) && \text{or} \quad w_0(L,t) = 0 \\
Q(0,t) &= -Q_1(t) - Pw_0'(0,t) && \text{or} \quad \theta(0,t) = 0 \\
Q(L,t) &= +Q_2(t) - Pw_0'(L,t) && \text{or} \quad \theta(L,t) = 0
\end{aligned} \tag{11.39}$$

For a laminated beam with mid-plane symmetry, K_{Aeq} and ρR_{eq} are equal to zero and the axial vibration (in-plane displacement u_0) is decoupled with the bending–shear vibration. In this case, the bending–shear vibration and the axial vibration of the composite beam can be considered separately. The spectral element models for the pure bending–shear vibration (Timoshenko-beam model) and the pure axial vibration (rod model) were introduced in Chapter 3.

11.2.2 Bending–Torsion–Shear Coupled Vibration

Now consider the small amplitude bending–torsion–shear coupled vibration of a straight composite laminated beam subjected to a constant axial tensile force P. Assume that the x-axis is chosen to coincide with the elastic axis that passes through the shear center of the beam cross-section, and the beam takes the bending deflection $h(x, t)$ along the x-axis and the twist $\alpha(x, t)$ about the x-axis. The displacements of the composite laminated beam are then assumed in the forms of

$$\begin{aligned}
w(x,y,z,t) &\cong w_0(x,y,t) \cong h(x,t) + y\alpha(x,t) \\
u(x,y,z,t) &\cong -z\theta_x(x,0,t) \equiv -z\theta(x,t) \\
v(x,y,z,t) &\cong -z\left.\frac{\partial w_0}{\partial y}\right|_{y=0} \equiv -z\alpha(x,t)
\end{aligned} \tag{11.40}$$

where $\theta(x, t)$ represents the slope along the x-axis. As the in-plane displacement u_0 is excluded from here onwards, from Equation 11.20 we get

$$\varepsilon_{xx}^0 = 0, \quad \gamma_{xy}^0 = 0 \tag{11.41}$$

Assuming the moment M_{yy} is zero and using Equation 11.41, we can eliminate the curvature χ_y from Equation 11.28 to obtain

$$\begin{Bmatrix} M_{xx} \\ M_{xy} \end{Bmatrix} = \begin{bmatrix} D_{11} - D_{12}^2 D_{22}^{-1} & D_{16} - D_{12}D_{26}D_{22}^{-1} \\ D_{16} - D_{12}D_{26}D_{22}^{-1} & D_{66} - D_{26}^2 D_{22}^{-1} \end{bmatrix} \begin{Bmatrix} \chi_x \\ \chi_{xy} \end{Bmatrix} \tag{11.42}$$

The curvatures can be obtained approximately as being

$$\chi_x = -\theta'_x = -\theta', \quad \chi_{xy} \cong -2\frac{\partial^2 w_0}{\partial x \partial y}\bigg|_{y=0} = -2\alpha' \tag{11.43}$$

The resultant bending moment (*M*) and torque (*T*) are given by [161]

$$M(x,t) \equiv -bM_{xx}, \quad T(x,t) = -2bM_{xy} \tag{11.44}$$

Thus, by applying Equations 11.42 and 11.43 into Equation 11.44, we can obtain

$$\begin{Bmatrix} M \\ T \end{Bmatrix} = \begin{bmatrix} EI_{eq} & K_{Teq} \\ K_{Teq} & GJ_{eq} \end{bmatrix} \begin{Bmatrix} \theta' \\ \alpha' \end{Bmatrix} \tag{11.45}$$

where

$$\begin{aligned} EI_{eq} &= b(D_{11} - D_{12}^2 D_{22}^{-1}) \\ GJ_{eq} &= 4b(D_{66} - D_{26}^2 D_{22}^{-1}) \\ K_{Teq} &= 2b(D_{16} - D_{12}D_{26}D_{22}^{-1}) \end{aligned} \tag{11.46}$$

and where K_{Teq} represents the bending–torsion material coupling rigidity. By defining the beam deflection along the *x*-axis by $h(x, t) = w_0(x, 0, t)$, from Equation 11.27 we can represent the transverse shear force approximately as

$$Q(x,t) \cong \kappa GA_{eq}(h' - \theta) \tag{11.47}$$

where

$$\kappa GA_{eq} = b\kappa A_{55} \tag{11.48}$$

The strain energy *U*, the kinetic energy *T*, and the virtual work δW can be obtained by

$$\begin{aligned} U &= \frac{1}{2}\int_0^L \{M\theta' + T\alpha' + Q(h' - \theta) + P(h'^2 + 2y_{CA}h'\alpha' + r_g^2\alpha'^2)\}\,dx \\ &= \frac{1}{2}\int_0^L \{EI_{eq}\theta'^2 + 2K_{T\,eq}\alpha'\theta' + GJ_{eq}\alpha'^2 + \kappa GA_{eq}(h' - \theta)^2 + P(h'^2 + 2y_{CA}h'\alpha' + r_g^2\alpha'^2)\}\,dx \\ T &= \frac{1}{2}\int_0^L \int_A \rho(\dot{w}^2 + \dot{u}^2 + \dot{v}^2)\,dA dx \\ &= \frac{1}{2}\int_0^L (\rho A_{eq}\dot{h}^2 + 2\rho A_{eq}y_{CM}\dot{h}\dot{\alpha} + \rho I_{eq}\dot{\theta}^2 + \rho J_{eq}\dot{\alpha}^2)\,dx \\ \delta W &= Q_1\delta h(0) + Q_2\delta h(L) + M_1\delta\theta(0) + M_2\delta\theta(L) + T_1\delta\alpha(0) + T_2\delta\alpha(L) \\ &\quad + \int_0^L \{f_h(x,t)\delta h + f_\theta(x,t)\delta\theta + f_\alpha(x,t)\delta\alpha - c_1\dot{h}\delta h - c_2\dot{\theta}\delta\theta - c_3\dot{\alpha}\delta\alpha\}\,dx \end{aligned} \tag{11.49}$$

where (c_1, c_2, c_3) are the viscous damping coefficients and the following definitions are used:

$$
\begin{aligned}
& A_{eq} = \int_A \mathrm{d}A, \quad J_{eq} = \int_A (y^2 + z^2)\mathrm{d}A \\
& (\rho A_{eq}, \rho I_{eq}, \rho J_{eq}) = \int_A \rho(1, z^2, y^2 + z^2)\mathrm{d}A \\
& y_{\mathrm{CA}} = \frac{1}{A}\int_A y\mathrm{d}A, \quad y_{\mathrm{CM}} = \frac{1}{\rho A_{eq}}\int_A \rho y \mathrm{d}A, \quad r_g = \sqrt{\frac{J_{eq}}{A_{eq}}}
\end{aligned}
\tag{11.50}
$$

where A_{eq} is the area of the cross-section of the composite laminate beam, J_{eq} is the polar area moment of inertia of the cross-section of the beam (about the x-axis), y_{CA} is the distance between the elastic axis (x-axis) and the center of the area of the cross-section (CA), ρA_{eq} is the apparent mass per unit length, ρI_{eq} is the mass moment of inertia per unit length about the y-axis, ρJ_{eq} is the apparent polar mass moment of inertia per unit length (about the x-axis), and y_{CM} is the distance between the elastic axis and the center of mass (CM). If the mass density is constant over the whole cross-section, one may show from Equation 11.50 that $\rho A_{eq} = \rho A$, $\rho J_{eq} = \rho I_p$, and $y_{\mathrm{CA}} = y_{\mathrm{CM}}$.

Substituting Equation 11.49 into Hamilton's principle and integrating by parts, we obtain

$$
\begin{aligned}
& \int_0^t \Big(\int_0^L [\{\kappa G A_{eq}(h'' - \theta') + P(h'' + y_{\mathrm{CA}}\alpha'') - \rho A_{eq}\ddot{h} - \rho A_{eq} y_{\mathrm{CM}}\ddot{\alpha} - c_1\dot{h} + f_h(x,t)\}\delta h \\
& \quad + \{GJ_{eq}\alpha'' + K_{Teq}\theta'' + P(y_{\mathrm{CA}}h'' + r_g^2\alpha'') - \rho J_{eq}\ddot{\alpha} - \rho A_{eq} y_{\mathrm{CM}}\ddot{h} - c_3\dot{\alpha} + f_\alpha(x,t)\}\delta\alpha \\
& \quad + \{EI_{eq}\theta'' + K_{Teq}\alpha'' + \kappa G A_{eq}(h' - \theta) - \rho I_{eq}\ddot{\theta} - c_2\dot{\theta} + f_\theta(x,t)\}\delta\theta]\mathrm{d}x \\
& \quad - \{Q(L,t) + P[h'(L) + y_{\mathrm{CA}}\alpha'(L) - Q_2]\}\delta h(L) \\
& \quad + \{Q(0,t) + P[h'(0) + y_{\mathrm{CA}}\alpha'(0)] + Q_1]\}\delta h(0) \\
& \quad - \{M(L,t) - M_2\}\delta\theta(L) + \{M(0,t) + M_1\}\delta\theta(0) \\
& \quad - \{T(L,t) + P[y_{\mathrm{CA}}h'(L) + r_g^2\alpha'(L)] - T_2\}\alpha(L) \\
& \quad + \{T(0,t) + P[y_{\mathrm{CA}}h'(0) + r_g^2\alpha'(0)] + T_1\}\delta\alpha(0)\Big)\mathrm{d}t = 0
\end{aligned}
\tag{15.51}
$$

when Equations 11.45 and 11.47 are used. From Equation 11.51, we obtain the equations of motion

$$
\begin{aligned}
& \kappa G A_{eq}(h'' - \theta') + P(h'' + y_{\mathrm{CA}}\alpha'') - \rho A_{eq}\ddot{h} - \rho A_{eq} y_{\mathrm{CM}}\ddot{\alpha} - c_1\dot{h} + f_h(x,t) = 0 \\
& GJ_{eq}\alpha'' + K_{Teq}\theta'' + P(y_{\mathrm{CA}}h'' + r_g^2\alpha'') - \rho J_{eq}\ddot{\alpha} - \rho A_{eq} y_{\mathrm{CM}}\ddot{h} - c_3\dot{\alpha} + f_\alpha(x,t) = 0 \\
& EI_{eq}\theta'' + K_{Teq}\alpha'' + \kappa G A_{eq}(h' - \theta) - \rho I_{eq}\ddot{\theta} - c_2\dot{\theta} + f_\theta(x,t) = 0
\end{aligned}
\tag{11.52}
$$

and the boundary conditions

$$
\begin{aligned}
Q(0,t) &= -Q_1(t) - P[h'(0,t) + y_{\mathrm{CA}}\alpha'(0,t)] && \text{or} \quad h(0,t) = 0 \\
Q(L,t) &= +Q_2(t) - P[h'(L,t) + y_{\mathrm{CA}}\alpha'(L,t)] && \text{or} \quad h(L,t) = 0 \\
M(0,t) &= -M_1(t) && \text{or} \quad \theta(0,t) = 0 \\
M(L,t) &= +M_2(t) && \text{or} \quad \theta(L,t) = 0 \\
T(0,t) &= -T_1(t) - P[y_{\mathrm{CA}}h'(0,t) + r_g^2\alpha'(0,t)] && \text{or} \quad \alpha(0,t) = 0 \\
T(L,t) &= +T_2(t) - P[y_{\mathrm{CA}}h'(L,t) + r_g^2\alpha'(L,t)] && \text{or} \quad \alpha(L,t) = 0
\end{aligned}
\tag{11.53}
$$

Equation 11.52 shows that the couplings between the bending, torsion, and shear displacements occurs as a result of the bending–torsion material coupling (i.e., K_{Teq}) and the non-coincidence of the shear center (x-axis) and the mass center (CM) of the beam cross-section.

11.3 Dynamics of Axial–Bending–Shear Coupled Composite Beams

11.3.1 Equations of Motion

The equations of motion and associated boundary conditions were derived in the previous section. They are given by

$$\begin{aligned} &\kappa GA_{eq}(w''-\theta')+Pw''-\rho A_{eq}\,\ddot{w}+f_w(x,t)=0\\ &EI_{eq}\theta''+\kappa GA_{eq}(w'-\theta)-K_{A\ eq}u''+\rho R_{eq}\,\ddot{u}-\rho I_{eq}\,\ddot{\theta}+f_\theta(x,t)=0\\ &EA_{eq}u''-K_{A\ eq}\theta''-\rho A_{eq}\,\ddot{u}+\rho R_{eq}\,\ddot{\theta}+f_u(x,t)=0 \end{aligned} \tag{11.54}$$

and

$$\begin{aligned} Q(0,t)&=-Q_1(t)-Pw_0'(0,t) && \text{or } \theta(0,t)=0\\ Q(L,t)&=+Q_2(t)-Pw_0'(L,t) && \text{or } \theta(L,t)=0\\ M(0,t)&=-M_1(t) && \text{or } w(0,t)=0\\ M(L,t)&=+M_2(t) && \text{or } w(L,t)=0\\ N(0,t)&=-N_1(t) && \text{or } u(0,t)=0\\ N(L,t)&=+N_2(t) && \text{or } u(L,t)=0 \end{aligned} \tag{11.55}$$

where the resultant axial tensile force N, the resultant moment M, and the resultant transverse shear force Q are defined by

$$\begin{aligned} Q(x,t)&=\kappa GA_{eq}(w'-\theta)\\ M(x,t)&=EI_{eq}\theta'-K_{A\ eq}u'\\ N(x,t)&=EA_{eq}u'-K_{A\ eq}\theta' \end{aligned} \tag{11.56}$$

In above equations, the subscripts "0" used for the in-plane displacements are omitted for brevity.

11.3.2 Spectral Element Modeling

11.3.2.1 Governing Equations in the Frequency Domain

The displacement fields are assumed in the spectral forms to be

$$\{w(x,t),\ \theta(x,t),\ u(x,t)\}=\frac{1}{N}\sum_{n=0}^{N-1}\{W_n(x;\omega_n),\ \Theta_n(x;\omega_n),\ U_n(x;\omega_n)\}\,e^{i\omega_n t} \tag{11.57}$$

Similarly, represent all forces and moments in the spectral forms as

$$\begin{aligned} \{f_w(x,t),\ f_\theta(x,t),\ f_u(x,t)\}&=\frac{1}{N}\sum_{n=0}^{N-1}\{F_{wn}(x;\omega_n),\ F_{\theta n}(x;\omega_n),\ F_{un}(x;\omega_n)\}e^{i\omega_n t}\\ \{Q(x,t),\ M(x,t),\ N(x,t)\}&=\frac{1}{N}\sum_{n=0}^{N-1}\{Q_n(x;\omega_n),\ M_n(x;\omega_n),\ N_n(x;\omega_n)\}e^{i\omega_n t}\\ \{Q_k(t),\ M_k(t),\ N_k(t)\}&=\frac{1}{N}\sum_{n=0}^{N-1}\{Q_{kn}(\omega_n),\ M_{kn}(\omega_n),\ N_{kn}(\omega_n)\}e^{i\omega_n t}\quad (k=1,\ 2) \end{aligned} \tag{11.58}$$

Substituting Equations 11.57 and 11.58 into Equations 11.54 and 11.55 gives the frequency-domain equations of motion and boundary conditions as

$$\begin{aligned}
&\kappa GA_{eq}(W'' - \Theta') + PW'' + \omega^2 \rho A_{eq} W + F_w = 0 \\
&EI_{eq}\Theta'' + \kappa GA_{eq}(W' - \Theta) - K_{Aeq}U'' - \omega^2 \rho R_{eq} U + \omega^2 \rho I_{eq}\Theta + F_\theta = 0 \\
&EA_{eq}U'' - K_{Aeq}\Theta'' + \omega^2 \rho A_{eq} U - \omega^2 \rho R_{eq}\Theta + F_u = 0
\end{aligned} \tag{11.59}$$

and

$$\begin{aligned}
Q(0) &= -Q_1 - PW'(0) && \text{or} \quad \Theta(0) = 0 \\
Q(L) &= +Q_2 - PW'(L) && \text{or} \quad \Theta(L) = 0 \\
M(0) &= -M_1 && \text{or} \quad W(0) = 0 \\
M(L) &= +M_2 && \text{or} \quad W(L) = 0 \\
N(0) &= -N_1 && \text{or} \quad U(0) = 0 \\
N(L) &= +N_2 && \text{or} \quad U(L) = 0
\end{aligned} \tag{11.60}$$

Similarly, substitution of Equations 11.57 and 11.58 into Equation 11.56 gives the frequency-domain force-displacement relationships as

$$\begin{aligned}
Q(x) &= \kappa GA_{eq}(W' - \Theta) \\
M(x) &= EI_{eq}\Theta' - K_{Aeq}U' \\
N(x) &= EA_{eq}U' - K_{Aeq}\Theta'
\end{aligned} \tag{11.61}$$

In above Equations 11.59–11.61, the subscripts n are omitted for brevity.

11.3.2.2 Spectral Nodal DOFs, Forces, and Moments

By using the spectral nodal degrees-of-freedom (DOFs), forces, and moments specified in Figure 11.4, we define the spectral nodal DOFs vector as

$$\boldsymbol{d} = \{\{U_1\, W_1\, \Theta_1\, U_2\, W_2\, \Theta_2\}^{\mathrm{T}} = \{U(0)\, W(0)\, \Theta(0)\, U(L)\, W(L)\, \Theta(L)\}^{\mathrm{T}} \tag{11.62}$$

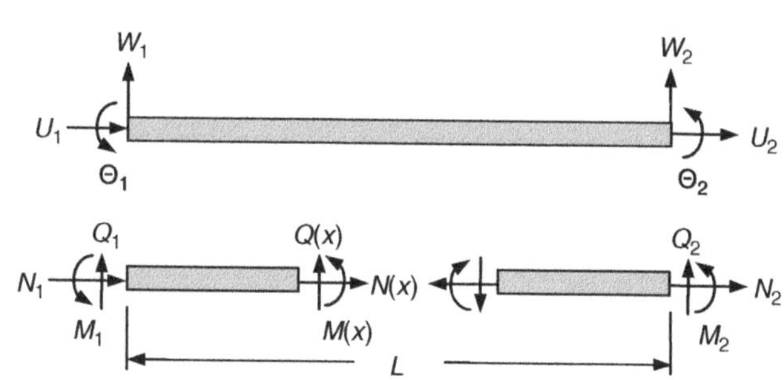

Figure 11.4 Sign convention for the axial–bending–shear coupled composite beam element

and the spectral nodal forces and moments vector as

$$f_c = \{N_1\, Q_1\, M_1\, N_2\, Q_2\, M_2\}^{\mathrm{T}} \tag{11.63}$$

11.3.2.3 Dynamic Shape Functions

The homogeneous governing equations can be reduced from Equation 11.59 to give

$$\begin{aligned}
&\kappa GA_{eq}(W''-\Theta')+PW''+\omega^2\rho A_{eq}W=0\\
&EI_{eq}\Theta''+\kappa GA_{eq}(W'-\Theta)-K_{A\,eq}U''-\omega^2\rho R_{eq}U+\omega^2\rho I_{eq}\Theta=0\\
&EA_{eq}U''-K_{A\,eq}\Theta''+\omega^2\rho A_{eq}U-\omega^2\rho R_{eq}\Theta=0
\end{aligned} \tag{11.64}$$

Assume the general solution of Equation 11.64 to be

$$W(x)=ae^{-ikx},\quad \Theta(x)=\alpha a e^{-ikx},\quad U(x)=\beta a e^{-ikx} \tag{11.65}$$

Substitution of Equation 11.65 into Equation 11.64 gives an eigenvalue problem as follows:

$$\begin{bmatrix} X_{11} & X_{12} & 0 \\ -X_{12} & X_{22} & X_{23} \\ 0 & X_{23} & X_{33} \end{bmatrix}\begin{Bmatrix} 1 \\ \alpha \\ \beta \end{Bmatrix}=\begin{Bmatrix} 0 \\ 0 \\ 0 \end{Bmatrix} \tag{11.66}$$

where

$$\begin{aligned}
X_{11} &= -(\kappa GA_{eq}+P)k^2+\omega^2\rho A_{eq}\\
X_{12} &= -i\kappa GA_{eq}k\\
X_{22} &= -EI_{eq}k^2+\omega^2\rho I_{eq}-\kappa GA_{eq}\\
X_{23} &= X_{32}=K_{A\,eq}k^2-\omega^2\rho R_{eq}\\
X_{33} &= -EA_{eq}k^2+\omega^2\rho A_{eq}
\end{aligned} \tag{11.67}$$

Equation 11.66 yields a dispersion relation as

$$c_1k_n^6+c_2k_n^4+c_3k_n^2+c_4=0 \tag{11.68}$$

where

$$\begin{aligned}
c_1 &= -(EA_{eq}EI_{eq}-K_{Aeq}^2)(\kappa GA_{eq}+P)\\
c_2 &= [-\rho A_{eq}K_{Aeq}^2+(P+\kappa GA_{eq})(EI_{eq}\rho A_{eq}-2\rho R_{eq}K_{Aeq})\\
&\quad +EA_{eq}(EI_{eq}\rho A_{eq}+\kappa GA_{eq}\rho I_{eq}+P\rho I_{eq})]\omega^2-EA_{eq}\kappa GA_{eq}P\\
c_3 &= [-EI_{eq}\rho A_{eq}^2-\kappa GA_{eq}\rho A_{eq}\rho I_{eq}+2K_{Aeq}\rho A_{eq}\rho R_{eq}+\kappa GA_{eq}\rho R_{eq}^2\\
&\quad -EA_{eq}\rho A_{eq}\rho I_{eq}+P(\rho R_{eq}^2-\rho A_{eq}\rho I_{eq})]\omega^4+P\kappa GA_{eq}\rho A_{eq}\omega^2\\
&\quad +EA_{eq}\kappa GA_{eq}\rho A_{eq}\\
c_4 &= (\rho A_{eq}^2\rho I_{eq}-\rho A_{eq}\rho R_{eq}^2)\omega^6-\kappa GA_{eq}\rho A_{eq}^2\omega^4
\end{aligned} \tag{11.69}$$

From Equation 11.68, we can obtain six wavenumbers as follows

$$
\begin{aligned}
k_{1,2} &= \pm\sqrt{B_3 + B_4 - \frac{1}{3}(c_2/c_1)} \\
k_{3,4} &= \pm\sqrt{-\frac{1}{2}(B_3 + B_4) - \frac{1}{3}(c_2/c_1) + i\tfrac{\sqrt{3}}{2}(B_3 - B_4)} \\
k_{5,6} &= \pm\sqrt{-\frac{1}{2}(B_3 + B_4) - \tfrac{1}{3}(c_2/c_1) - i\tfrac{\sqrt{3}}{2}(B_3 - B_4)}
\end{aligned}
\tag{11.70}
$$

where

$$
\begin{aligned}
B_1 &= \frac{1}{9}\left[3(c_3/c_1) - (c_2/c_1)^2\right] \\
B_2 &= \frac{1}{54}\left[9(c_2c_3/c_1^2) - 27(c_4/c_1) - 2(c_2/c_1)^3\right] \\
B_3 &= \sqrt[3]{B_2 + \sqrt{B_1^3 + B_2^2}} \\
B_4 &= \sqrt[3]{B_2 - \sqrt{B_1^3 + B_2^2}}
\end{aligned}
\tag{11.71}
$$

By substituting wavenumber k_p ($p = 1, 2, \ldots 6$) into Equation 11.66, we can obtain the corresponding α_p and β_p ($p = 1, 2, \ldots, 6$) as

$$
\begin{aligned}
\alpha_p &= i\frac{\omega^2 \rho A_{eq} - \kappa G A_{eq} k_p^2 - P k_p^2}{\kappa G A_{eq} k_p} \\
\beta_p &= \left(\frac{\omega^2 \rho R_{eq} - K_{Aeq} k_p^2}{\omega^2 \rho A_{eq} - E A_{eq} k_p^2}\right)\alpha_p
\end{aligned}
\qquad (p = 1, 2, \ldots, 6)
\tag{11.72}
$$

By using six wavenumbers given by Equation 11.70, we can write the general solution to Equation 11.64 as

$$
W(x) = \sum_{p=1}^{6} a_p e^{-ik_p x}, \quad \Theta(x) = \sum_{p=1}^{6} \alpha_p a_p e^{-ik_p x}, \quad U(x) = \sum_{i=p}^{6} \beta_p b_p e^{-ik_p x} \tag{11.73}
$$

or

$$
W(x) = \boldsymbol{e}(x;\omega)\boldsymbol{a}, \quad \Theta(x) = \boldsymbol{e}(x;\omega)\boldsymbol{A}(\omega)\boldsymbol{a}, \quad U(x) = \boldsymbol{e}(x;\omega)\boldsymbol{B}(\omega)a \tag{11.74}
$$

where

$$
\boldsymbol{e}(x;\omega) = \left[e^{-ik_1x}\ e^{-ik_2x}\ e^{-ik_3x}\ e^{-ik_4x}\ e^{-ik_5x}\ e^{-ik_6x}\right] \tag{11.75}
$$

and

$$
\begin{aligned}
&\boldsymbol{A}(\omega) = diag\,[\alpha_p(\omega)], \quad \boldsymbol{B}(\omega) = diag\,[\beta_p(\omega)] \quad (p = 1, 2, \ldots, 6) \\
&a = \{a_1\ a_2\ a_3\ a_4\ a_5\ a_6\}^{\mathrm{T}}
\end{aligned}
\tag{11.76}
$$

Applying Equation 11.74 to the definition of the spectral nodal DOFs given by Equation 11.62 gives the relationship

$$
\boldsymbol{d} = \boldsymbol{H}(\omega)\boldsymbol{a} \tag{11.77}
$$

where,

$$H(\omega) = \begin{bmatrix} \beta_1 & \beta_2 & \beta_3 & \beta_4 & \beta_5 & \beta_6 \\ 1 & 1 & 1 & 1 & 1 & 1 \\ \alpha_1 & \alpha_2 & \alpha_3 & \alpha_4 & \alpha_5 & \alpha_6 \\ \beta_1 e_1 & \beta_2 e_2 & \beta_3 e_3 & \beta_4 e_4 & \beta_5 e_5 & \beta_6 e_6 \\ e_1 & e_2 & e_3 & e_4 & e_5 & e_6 \\ \alpha_1 e_1 & \alpha_2 e_2 & \alpha_3 e_3 & \alpha_4 e_4 & \alpha_5 e_5 & \alpha_6 e_6 \end{bmatrix} \tag{11.78}$$

with

$$e_p = e^{-ik_pL} \quad (p = 1, 2, \ldots, 6) \tag{11.79}$$

By using Equation 11.77, the constant vector $\boldsymbol{a}$ can be removed from Equation 11.74 to write the general solutions in the form

$$W(x) = \boldsymbol{N}_w(x;\omega)\boldsymbol{d}, \quad \Theta(x) = \boldsymbol{N}_\theta(x;\omega)\boldsymbol{d}, \quad U(x) = \boldsymbol{N}_u(x;\omega)\boldsymbol{d} \tag{11.80}$$

where $\boldsymbol{N}_w$, $\boldsymbol{N}_\theta$, and $\boldsymbol{N}_u$ are the dynamic shape functions defined by

$$\begin{aligned} \boldsymbol{N}_w(x;\omega_n) &= \boldsymbol{e}(x;\omega_n)\boldsymbol{H}^{-1}(\omega) \\ \boldsymbol{N}_\theta(x;\omega_n) &= \boldsymbol{e}(x;\omega_n)\boldsymbol{A}(\omega)\boldsymbol{H}^{-1}(\omega) \\ \boldsymbol{N}_u(x;\omega_n) &= \boldsymbol{e}(x;\omega_n)\boldsymbol{B}(\omega)\boldsymbol{H}^{-1}(\omega) \end{aligned} \tag{11.81}$$

11.3.2.4 Weak Form of Governing Equations

The weak form of the frequency-domain governing Equations 11.59 can be obtained from

$$\begin{aligned} \int_0^L \{ & [\kappa GA_{eq}(W'' - \Theta') + PW'' + \omega^2 \rho A_{eq} W + F_w(x)]\delta W \\ & + [EI_{eq}\Theta'' + \kappa GA_{eq}(W' - \Theta) - K_{Aeq}U'' - \omega^2 \rho R_{eq} U + \omega^2 \rho I_{eq}\Theta + F_\theta(x)]\delta\Theta \\ & + [EA_{eq}U'' - K_{Aeq}\Theta'' + \omega^2 \rho A_{eq} U - \omega^2 \rho R_{eq}\Theta + F_u(x)]\delta U\} dx = 0 \end{aligned} \tag{11.82}$$

By applying the integral by parts and using Equation 11.61, we can obtain the weak form as

$$\begin{aligned} \int_0^L \{ & -\kappa GA_{eq}W'\delta W' + \kappa GA_{eq}(\Theta\delta W' + W'\delta\Theta) - \kappa GA_{eq}\Theta\delta\Theta \\ & - EI_{eq}\Theta'\delta\Theta' - EA_{eq}U'\delta U' + K_{Aeq}(U'\delta\Theta' + \Theta'\delta U') - PW'\delta W' \\ & + \omega^2 \rho A_{eq} U\delta U + \omega^2 \rho A_{eq} W\delta W + \omega^2 \rho I_{eq}\Theta\delta\Theta - \omega^2 \rho R_{eq}(U\delta\Theta + \Theta\delta U) \\ & + F_u\delta U + F_w\delta W + F_\theta\delta\Theta\} dx \\ & + N(x)\delta U\big|_0^L + \left[Q(x) + PW'\right]\delta W\big|_0^L + M(x)\delta\Theta\big|_0^L = 0 \end{aligned} \tag{11.83}$$

or, by applying the boundary conditions (11.60), Equations 11.62 and 11.63,

$$\begin{aligned}&\int_0^L \{\kappa GA_{eq}W'\delta W' - \kappa GA_{eq}(\Theta\delta W' + W'\delta\Theta) + \kappa GA_{eq}\Theta\delta\Theta \\ &\quad + EI_{eq}\Theta'\delta\Theta' + EA_{eq}U'\delta U' - K_{A\,eq}(U'\delta\Theta' + \Theta'\delta U') + PW'\delta W' \\ &\quad -\omega^2\rho A_{eq}U\delta U - \omega^2\rho A_{eq}W\delta W - \omega^2\rho I_{eq}\Theta\delta\Theta + \omega^2\rho R_{eq}(U\delta\Theta + \Theta\delta U)\}\mathrm{d}x \\ &= \delta\boldsymbol{d}^{\mathrm{T}}\boldsymbol{f}_{\mathrm{c}} + \int_0^L (F_u\delta U + F_w\delta W + F_\theta\delta\Theta)\mathrm{d}x\end{aligned} \tag{11.84}$$

11.3.2.5 Spectral Element Equation

Substituting Equation 11.80 into Equation 11.84, we obtain

$$\begin{aligned}&\delta\boldsymbol{d}^{\mathrm{T}}\int_0^L \{\kappa GA_{eq}\boldsymbol{N}_w'^{\mathrm{T}}\boldsymbol{N}_w' - \kappa GA_{eq}(\boldsymbol{N}_w'^{\mathrm{T}}\boldsymbol{N}_\theta + \boldsymbol{N}_\theta^{\mathrm{T}}\boldsymbol{N}_w') + \kappa GA_{eq}\boldsymbol{N}_\theta^{\mathrm{T}}\boldsymbol{N}_\theta \\ &\quad + EI_{eq}\boldsymbol{N}_\theta'^{\mathrm{T}}\boldsymbol{N}_\theta' + EA_{eq}\boldsymbol{N}_u'^{\mathrm{T}}\boldsymbol{N}_u' - K_{A\,eq}(\boldsymbol{N}_\theta'^{\mathrm{T}}\boldsymbol{N}_u' + \boldsymbol{N}_u'^{\mathrm{T}}\boldsymbol{N}_\theta') \\ &\quad + P\boldsymbol{N}_w'^{\mathrm{T}}\boldsymbol{N}_w' - \omega^2\rho A_{eq}(\boldsymbol{N}_u^{\mathrm{T}}\boldsymbol{N}_u + \boldsymbol{N}_w^{\mathrm{T}}\boldsymbol{N}_w) - \omega^2\rho I_{eq}\boldsymbol{N}_\theta^{\mathrm{T}}\boldsymbol{N}_\theta \\ &\quad + \omega^2\rho R_{eq}(\boldsymbol{N}_\theta^{\mathrm{T}}\boldsymbol{N}_u + \boldsymbol{N}_u^{\mathrm{T}}\boldsymbol{N}_\theta)]\}\mathrm{d}x\boldsymbol{d} = \delta\boldsymbol{d}^{\mathrm{T}}(\boldsymbol{f}_{\mathrm{c}} + \boldsymbol{f}_{\mathrm{d}})\end{aligned} \tag{11.85}$$

where

$$\boldsymbol{f}_{\mathrm{d}} = \int_0^L (F_u\boldsymbol{N}_u^{\mathrm{T}} + F_w\boldsymbol{N}_w^{\mathrm{T}} + F_\theta\boldsymbol{N}_\theta^{\mathrm{T}})\mathrm{d}x \tag{11.86}$$

As $\delta\boldsymbol{d}$ is arbitrary by definition, we can obtain the spectral element equation from Equation 11.85 as

$$\boldsymbol{S}(\omega)\boldsymbol{d} = \boldsymbol{f}_{\mathrm{c}} + \boldsymbol{f}_{\mathrm{d}} \tag{11.87}$$

where

$$\begin{aligned}\boldsymbol{S}(\omega) = &\int_0^L [\kappa GA_{eq}\boldsymbol{N}_w'^{\mathrm{T}}\boldsymbol{N}_w' - \kappa GA_{eq}(\boldsymbol{N}_w'^{\mathrm{T}}\boldsymbol{N}_\theta + \boldsymbol{N}_\theta^{\mathrm{T}}\boldsymbol{N}_w') + \kappa GA_{eq}\boldsymbol{N}_\theta^{\mathrm{T}}\boldsymbol{N}_\theta \\ &+ EI_{eq}\boldsymbol{N}_\theta'^{\mathrm{T}}\boldsymbol{N}_\theta' + EA_{eq}\boldsymbol{N}_u'^{\mathrm{T}}\boldsymbol{N}_u' - K_{Aeq}(\boldsymbol{N}_\theta'^{\mathrm{T}}\boldsymbol{N}_u' + \boldsymbol{N}_u'^{\mathrm{T}}\boldsymbol{N}_\theta') \\ &+ P\boldsymbol{N}_w'^{\mathrm{T}}\boldsymbol{N}_w' - \omega^2\rho A_{eq}(\boldsymbol{N}_u^{\mathrm{T}}\boldsymbol{N}_u + \boldsymbol{N}_w^{\mathrm{T}}\boldsymbol{N}_w) - \omega^2\rho I_{eq}\boldsymbol{N}_\theta^{\mathrm{T}}\boldsymbol{N}_\theta \\ &+ \omega^2\rho R_{eq}(\boldsymbol{N}_\theta^{\mathrm{T}}\boldsymbol{N}_u + \boldsymbol{N}_u^{\mathrm{T}}\boldsymbol{N}_\theta)]\mathrm{d}x\end{aligned} \tag{11.88}$$

By substituting Equation 11.81 into Equation 11.88, the spectral element matrix can be expressed in the form

$$\boldsymbol{S}(\omega) = \boldsymbol{H}^{-\mathrm{T}}(\omega)\boldsymbol{D}(\omega)\boldsymbol{H}^{-1}(\omega) \tag{11.89}$$

where

$$\begin{aligned}D(\omega) = \int_0^L [&\kappa GA_{eq}e'^{\mathrm{T}}e' - \kappa GA_{eq}(e'^{\mathrm{T}}eA + A^{\mathrm{T}}e^{\mathrm{T}}e') + \kappa GA_{eq}A^{\mathrm{T}}e^{\mathrm{T}}eA \\ &+ EI_{eq}A^{\mathrm{T}}e'^{\mathrm{T}}e'A + EA_{eq}B^{\mathrm{T}}e'^{\mathrm{T}}e'B - K_{A\,eq}(A^{\mathrm{T}}e'^{\mathrm{T}}e'B + B^{\mathrm{T}}e'^{\mathrm{T}}e'A) \\ &+ Pe'^{\mathrm{T}}e' - \omega^2\rho A_{eq}(B^{\mathrm{T}}e^{\mathrm{T}}eB + e^{\mathrm{T}}e) - \omega^2\rho I_{eq}A^{\mathrm{T}}e^{\mathrm{T}}eA \\ &+ \omega^2\rho R_{eq}(A^{\mathrm{T}}e^{\mathrm{T}}eB + B^{\mathrm{T}}e^{\mathrm{T}}eA)]\mathrm{d}x\end{aligned} \tag{11.90}$$

From Equation 11.75, we can obtain the derivative of $e(x;\omega)$ with respect to x as

$$e'(x;\omega) = -ie(x;\omega)K(\omega) \tag{11.91}$$

where

$$K = diag\,[k_p] \quad (p = 1,\ 2,\ \ldots,\ 6) \tag{11.92}$$

Substitution of Equations 11.75 and 11.91 into Equation 11.90 gives

$$\begin{aligned}D(\omega) = {}& \kappa GA_{eq}[-KEK + i(KEA + A^{\mathrm{T}}EK) + A^{\mathrm{T}}EA] - EI_{eq}A^{\mathrm{T}}KEKA \\ &- EA_{eq}B^{\mathrm{T}}KEKB + K_{A\,eq}(A^{\mathrm{T}}KEKB + B^{\mathrm{T}}KEKA) - PKEK \\ &- \omega^2[\rho A_{eq}(B^{\mathrm{T}}EB + E) + \rho I_{eq}A^{\mathrm{T}}EA - \rho R_{eq}(A^{\mathrm{T}}EB + B^{\mathrm{T}}EA)]\end{aligned} \tag{11.93}$$

where

$$E(\omega) = \int_0^L e^{\mathrm{T}}(x;\omega)e(x;\omega)\mathrm{d}x = [E_{rs}(\omega)] \quad (r,\ s = 1,\ 2,\ \ldots,\ 6) \tag{11.94}$$

with

$$E_{rs} = \begin{cases} \dfrac{i}{k_r + k_s}[\mathrm{e}^{-i(k_r+k_s)L} - 1] & \text{if} \quad k_r + k_s \neq 0 \\ L & \text{if} \quad k_r + k_s = 0 \end{cases} \tag{11.95}$$

11.3.3 Finite Element Model

The weak form of the equations of motion in the time domain can be obtained from Equation 11.54 as

$$\begin{aligned}&\int_0^L [\rho A_{eq}(\ddot{w}\delta w + \ddot{u}\delta u) - \rho R_{eq}(\ddot{u}\delta\theta + \ddot{\theta}\delta u) + \rho I_{eq}\ddot{\theta}\delta\theta \\ &\quad + \kappa GA_{eq}(w'\delta w' + \theta\delta\theta - w'\delta\theta - \theta\delta w') - K_{Aeq}(u'\delta\theta' + \theta'\delta u') \\ &\quad + EI_{eq}\theta'\delta\theta' + Pw'\delta w' + EA_{eq}u'\delta u']\mathrm{d}x \\ &= N(x,t)\delta u|_0^L + [Q(x,t) + Pw']\delta w|_0^L + M(x,t)\delta\theta|_0^L \\ &\quad + \int_0^L (f_u\delta u + f_w\delta w + f_\theta\delta\theta)\mathrm{d}x\end{aligned} \tag{11.96}$$

The nodal DOFs, forces, and moments in the time domain are specified in Figure 11.4 and we define the nodal DOFs vector by

$$d(t) = \{\, u_1(t) \quad w_1(t) \quad \theta_1(t) \quad u_2(t) \quad w_2(t) \quad \theta_2(t) \,\}^{\mathrm{T}} \tag{11.97}$$

and the nodal forces and moments vector by

$$f_c(t) = \{ N_1(t) \quad Q_1(t) \quad M_1(t) \quad N_2(t) \quad Q_2(t) \quad M_2(t) \}^{\mathrm{T}} \tag{11.98}$$

By applying the boundary conditions (11.55) to the right-hand side of Equation 11.96 and then by using definitions (11.97) and (11.98), Equation 11.96 can be written in the form

$$\begin{aligned}
&\int_0^L [\rho A_{eq}(\ddot{u}\delta u + \ddot{w}\delta w) + \rho I_{eq}\ddot{\theta}\delta\theta - \rho R_{eq}(\ddot{u}\delta\theta + \ddot{\theta}\delta u) \\
&\quad + \kappa G A_{eq} w'\delta w' - \kappa G A_{eq}(w'\delta\theta + \theta\delta w') + \kappa G A_{eq}\theta\delta\theta \\
&\quad + EI_{eq}\theta'\delta\theta' + EA_{eq}u'\delta u' - K_{Aeq}(u'\delta\theta' + \theta'\delta u') + Pw'\delta w']\mathrm{d}x \\
&= \delta \boldsymbol{d}^{\mathrm{T}} \boldsymbol{f}_c(t) + \int_0^L (f_u\delta u + f_w\delta w + f_\theta\delta\theta)\mathrm{d}x
\end{aligned} \tag{11.99}$$

The displacement fields within the composite beam element are assumed as being

$$u(x,t) = \boldsymbol{N}_u(x)\,\boldsymbol{d}(t), \quad w(x,t) = \boldsymbol{N}_w(x)\,\boldsymbol{d}(t), \quad \theta(x,t) = \boldsymbol{N}_\theta(x)\,\boldsymbol{d}(t) \tag{11.100}$$

where

$$\begin{aligned}
\boldsymbol{N}_u(x) &= [N_1(x)\;0\;0\;N_2(x)\;0\;0] \\
\boldsymbol{N}_w(x) &= [0\;N_3(x)\;N_4(x)\;0\;N_5(x)\;N_6(x)] \\
\boldsymbol{N}_\theta(x) &= [0\;N_7(x)\;N_8(x)\;0\;N_9(x)\;N_{10}(x)]
\end{aligned} \tag{11.101}$$

with

$$\begin{aligned}
N_1(x) &= 1-\xi, \quad N_2(x) = \xi \\
N_3(x) &= R(1-\xi)(1+\xi-2\xi^2+\delta) \\
N_4(x) &= (1/2)RL\xi(1-\xi)(2-2\xi+\delta) \\
N_5(x) &= R\xi(3\xi-2\xi^2+\delta), \qquad N_6(x) = -(1/2)RL\xi(1-\xi)(2\xi+\delta) \\
N_7(x) &= -6(R/L)\xi(1-\xi), \qquad N_8(x) = R(1-\xi)(1-3\xi+\delta) \\
N_9(x) &= 6(R/L)\xi(1-\xi), \qquad N_{10}(x) = R\xi(-2+3\xi+\delta)
\end{aligned} \tag{11.102}$$

where

$$\xi = \frac{x}{L}, \qquad R = \frac{1}{1+\delta}, \qquad \delta = \frac{12r}{L^2}, \qquad r = \frac{EI}{\kappa GA} \tag{11.103}$$

Substitution of Equation 11.100 into Equation 11.99 gives the finite element equation in the form of

$$\boldsymbol{M}\ddot{\boldsymbol{d}}(t) + \boldsymbol{K}\boldsymbol{d}(t) = \boldsymbol{f}_c(t) + \boldsymbol{f}_d(t) \tag{11.104}$$

where

$$\begin{aligned}
\boldsymbol{M} &= \int_0^L [\rho A_{eq}(\boldsymbol{N}_u^{\mathrm{T}}\boldsymbol{N}_u + \boldsymbol{N}_w^{\mathrm{T}}\boldsymbol{N}_w) + \rho I_{eq}\boldsymbol{N}_\theta^{\mathrm{T}}\boldsymbol{N}_\theta - \rho R_{eq}(\boldsymbol{N}_\theta^{\mathrm{T}}\boldsymbol{N}_u + \boldsymbol{N}_u^{\mathrm{T}}\boldsymbol{N}_\theta)]\mathrm{d}x \\
\boldsymbol{K} &= \int_0^L [\kappa G A_{eq}\boldsymbol{N}_w'^{\mathrm{T}}\boldsymbol{N}_w' - \kappa G A_{eq}(\boldsymbol{N}_w'^{\mathrm{T}}\boldsymbol{N}_\theta + \boldsymbol{N}_\theta^{\mathrm{T}}\boldsymbol{N}_w') + \kappa G A_{eq}\boldsymbol{N}_\theta^{\mathrm{T}}\boldsymbol{N}_\theta \\
&\quad + EI_{eq}\boldsymbol{N}_\theta'^{\mathrm{T}}\boldsymbol{N}_\theta' + EA_{eq}\boldsymbol{N}_u'^{\mathrm{T}}\boldsymbol{N}_u' - K_{Aeq}(\boldsymbol{N}_\theta'^{\mathrm{T}}\boldsymbol{N}_u' + \boldsymbol{N}_u'^{\mathrm{T}}\boldsymbol{N}_\theta')_u + P\boldsymbol{N}_w'^{\mathrm{T}}\boldsymbol{N}_w']\mathrm{d}x \\
\boldsymbol{f}_d(t) &= \int_0^L [f_u(x,t)\boldsymbol{N}_u^{\mathrm{T}}(x) + f_w(x,t)\boldsymbol{N}_w^{\mathrm{T}}(x) + f_\theta(x,t)\boldsymbol{N}_\theta^{\mathrm{T}}(x)]\mathrm{d}x
\end{aligned} \tag{11.105}$$

The finite element mass matrix $\boldsymbol{M}$ and stiffness matrix $\boldsymbol{K}$ are given in Appendix 11.A.

Example 11.1

As an illustrative example, we will consider a $[+\theta/-\theta/-\theta/+\theta]$ angle-ply symmetric composite beam. Each ply is made up of AS/3501-6 graphite–epoxy material and its material properties are given in Ref. [29] as being

$$\begin{aligned} &E_1 = 144.84\,\text{GPa} && E_2 = 9.65\,\text{GPa} \\ &G_{12} = G_{31} = 4.14\,\text{GPa}, && G_{23} = 3.45\,\text{GPa} \\ &\nu_{12} = 0.3 && \nu_{21} = 0.2 \\ &\rho = 1389.79\,\text{kg/m}^3 && \end{aligned}$$

The angle-ply symmetric composite beam has the length $L = 3.75\,\text{m}$ and thickness $h = 0.25\,\text{m}$. As the composite beam is symmetric with respect to its mid-plane, the bending–shear vibration is not coupled with the axial vibration. Thus, the spectral Timoshenko-beam element introduced in Chapter 3 is used to compute the natural frequencies of the composite beam. Table 11.1 compares the fundamental natural frequencies obtained by the present spectral element method (SEM) with those obtained by the conventional finite element method (FEM) and with exact analytical results cited in Ref. [29] for different four boundary conditions: simply–simply supported (SS), clamped–clamped (CC), clamped-free (CF), and clamped–simply supported (CS). Table 11.1 certainly shows that the fundamental natural frequencies obtained by the spectral element model agree totally with those given in Ref. [29].

Table 11.1 Natural frequencies (kHz) of a $[+\theta°/-\theta°/-\theta°/+\theta°]$ cross-ply composite beam

	$\theta = 30°$				$\theta = 45°$			
		FEM(N)				FEM(N)		
Boundary condition	SEM(2)	$N = 5$	$N = 50$	Ref. [29]	SEM(2)	$N = 5$	$N = 50$	Ref. [29]
SS	0.598	0.599	0.598	0.598	0.437	0.438	0.437	0.437
CC	1.165	1.172	1.165	1.165	0.904	0.908	0.905	0.904
CF	0.218	0.218	0.218	0.218	0.158	0.158	0.158	0.158
CS	0.869	0.872	0.869	0.869	0.655	0.656	0.655	0.655

Note: N = total number of the equal-length finite elements used in the analysis.

Example 11.2

As the second illustrative example, we consider a [0°/90°/0°/90°] cross-ply antisymmetric composite beam. Each ply is made up of the same graphite–epoxy material as considered in Example 11.1. The cross-ply antisymmetric composite beam also has the length $L = 3.75\,\text{m}$ and thickness $h = 0.25\,\text{m}$. The natural frequencies are computed by using the spectral element model given by Equation 11.87 and they are compared with the results obtained by FEM in Table 11.2. The FEM results are shown to converge towards the SEM results as the number of total finite elements used in the finite element analysis is increased.

Table 11.2 Natural frequencies (kHz) of a [0°/90°/0°/90°] cross-ply composite beam

	First mode				Second mode			
		FEM(N)				FEM(N)		
Boundary condition	SEM(2)	$N=5$	$N=20$	$N=100$	SEM(2)	$N=5$	$N=20$	$N=100$
SS	0.528	0.530	0.528	0.528	1.911	1.956	1.914	1.911
CC	1.055	1.070	1.056	1.055	2.510	2.617	2.517	2.510
CF	0.192	0.193	0.192	0.192	1.089	1.101	1.090	1.089
CS	0.777	0.784	0.778	0.777	2.221	2.294	2.225	2.221

Note: N = total number of the equal-length finite elements used in the analysis.

11.4 Dynamics of Bending–Torsion–Shear Coupled Composite Beams

11.4.1 Equations of Motion

The equations of motion and the associated boundary conditions are derived in the previous section. They are given by

$$\begin{aligned}
&\kappa GA_{eq}(h''-\theta')+P(h''+y_{CA}\alpha'')-\rho A_{eq}\ddot{h}-\rho A_{eq}y_{CM}\ddot{\alpha}-c_1\dot{h}+f_h(x,t)=0\\
&EI_{eq}\theta''+K_{Teq}\alpha''+\kappa GA_{eq}(h'-\theta)-\rho I_{eq}\ddot{\theta}-c_2\dot{\theta}+f_\theta(x,t)=0\\
&GJ_{eq}\alpha''+K_{Teq}\theta''+P(y_{CA}h''+r_g^2\alpha'')-\rho J_{eq}\ddot{\alpha}-\rho A_{eq}y_{CM}\ddot{h}-c_3\dot{\alpha}+f_\alpha(x,t)=0
\end{aligned} \tag{11.106}$$

and

$$\begin{aligned}
Q(0,t)&=-Q_1(t)-P[h'(0,t)+y_{CA}\alpha'(0,t)] &&\text{or}\quad h(0,t)=0\\
Q(L,t)&=+Q_2(t)-P[h'(L,t)+y_{CA}\alpha'(L,t)] &&\text{or}\quad h(L,t)=0\\
M(0,t)&=-M_1(t) &&\text{or}\quad \theta(0,t)=0\\
M(L,t)&=+M_2(t) &&\text{or}\quad \theta(L,t)=0\\
T(0,t)&=-T_1(t)-P[y_{CA}h'(0,t)+r_g^2\alpha'(0,t)] &&\text{or}\quad \alpha(0,t)=0\\
T(L,t)&=+T_2(t)-P[y_{CA}h'(L,t)+r_g^2\alpha'(L,t)] &&\text{or}\quad \alpha(L,t)=0
\end{aligned} \tag{11.107}$$

where the resultant moment M, the resultant torque T, and the resultant transverse shear force Q are defined by

$$\begin{aligned}
Q(x,t)&=\kappa GA_{eq}(h'-\theta)\\
M(x,t)&=EI_{eq}\theta'+K_{Teq}\alpha'\\
T(x,t)&=GJ_{eq}\alpha'+K_{Teq}\theta'
\end{aligned} \tag{11.108}$$

11.4.2 Spectral Element Modeling

11.4.2.1 Governing Equations in the Frequency Domain

Represent the displacement fields in the spectral forms as

$$\{h(x,t),\ \theta(x,t),\ \alpha(x,t)\}=\sum_{n=0}^{N-1}\{H_n(x;\omega_n),\ \Theta_n(x;\omega_n),\ A_n(x;\omega_n)\}e^{i\omega_n t} \tag{11.109}$$

Similarly we represent all forces and moments in the spectral forms as

$$\begin{aligned}
&\{f_h(x,t),\ f_\theta(x,t),\ f_\alpha(x,t)\} \\
&\qquad = \sum_{n=0}^{N-1}\{F_{hn}(x;\omega_n),\ F_{\theta n}(x;\omega_n),\ F_{\alpha n}(x;\omega_n)\}e^{i\omega_n t} \\
&\{Q(x,t),\ M(x,t),\ T(x,t)\} \\
&\qquad = \sum_{n=0}^{N-1}\{Q_n(x;\omega_n),\ M_n(x;\omega_n),\ T_n(x;\omega_n)\}\,e^{i\omega_n t} \\
&\{Q_k(t),\ M_k(t),\ T_k(t)\} \\
&\qquad = \sum_{n=0}^{N-1}\{Q_{kn}(\omega_n),\ M_{kn}(\omega_n),\ T_{kn}(\omega_n)\}e^{i\omega_n t} \quad (k=1,\ 2)
\end{aligned} \tag{11.110}$$

By substituting Equations 11.109 and 11.110 into Equation 11.106, we obtain the frequency-domain governing equations as

$$\begin{aligned}
&\kappa GA_{eq}(H''-\Theta')+P(H''+y_{CA}A'')+\omega^2\rho A_{eq}H+\omega^2\rho A_{eq}y_{CM}A-i\omega c_1H+F_h=0 \\
&EI_{eq}\Theta''+K_{Teq}A''+\kappa GA_{eq}(H'-\Theta)+\omega^2\rho I_{eq}\Theta-i\omega c_2\Theta+F_\theta=0 \\
&GJ_{eq}A''+K_{Teq}\Theta''+P(y_{CA}H''+r_g^2A'')+\omega^2\rho J_{eq}A+\omega^2\rho A_{eq}y_{CM}H-i\omega c_3A+F_\alpha=0
\end{aligned} \tag{11.111}$$

Similarly, we can obtain the frequency-domain boundary conditions as

$$\begin{aligned}
Q(0) &= -Q_1-P[H'(0)+y_{CA}A'(0)] &&\text{or}\quad H(0)=0 \\
Q(L) &= +Q_2-P[H'(L)+y_{CA}A'(L)] &&\text{or}\quad H(L)=0 \\
M(0) &= -M_1 &&\text{or}\quad \Theta(0)=0 \\
M(L) &= +M_2 &&\text{or}\quad \Theta(L)=0 \\
T(0) &= T_1-P[y_{CA}H'(0)+r_g^2A'(0)] &&\text{or}\quad A(0)=0 \\
T(L) &= +T_2-P[y_{CA}H'(L)+r_g^2A'(L)] &&\text{or}\quad A(L)=0
\end{aligned} \tag{11.112}$$

and the frequency-domain force-displacement relationships as

$$\begin{aligned}
Q(x) &= \kappa GA_{eq}(H'-\Theta) \\
M(x) &= EI_{eq}\Theta'+K_{Teq}A' \\
T(x) &= GJ_{eq}A'+K_{Teq}\Theta'
\end{aligned} \tag{11.113}$$

Notice that, in the above equations (Equations 11.111–11.113), the subscripts n are omitted for brevity.

11.4.2.2 Spectral Nodal DOFs, Forces, and Moments

The spectral nodal DOFs, forces, and moments for the bending–torsion–shear coupled composite beam element are specified in Figure 11.5. We define the spectral nodal DOFs vector as

$$\begin{aligned} \mathbf{d} &= \{H_1 \quad \Theta_1 \quad A_1 \quad H_2 \quad \Theta_2 \quad A_2\}^{\mathrm{T}} \\ &= \{H(0) \quad \Theta(0) \quad A(0) \quad H(L) \quad \Theta(L) \quad A(L)\}^{\mathrm{T}} \end{aligned} \tag{11.114}$$

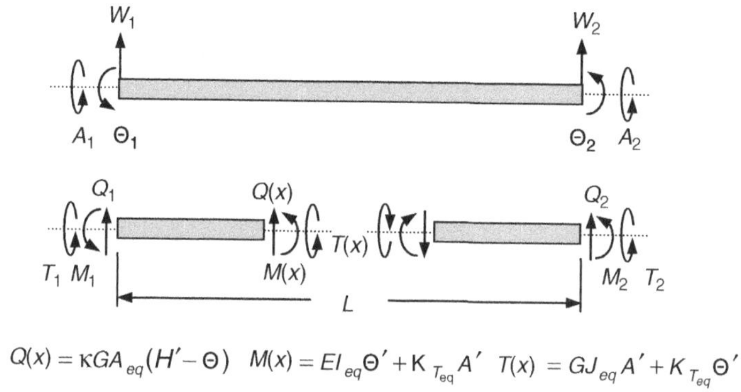

Figure 11.5 Sign convention for the bending–torsion–shear coupled composite beam element

and the spectral nodal forces and moments vector as

$$\mathbf{f}_{\mathrm{c}} = \{Q_1 \quad M_1 \quad T_1 \quad Q_2 \quad M_2 \quad T_2\}^{\mathrm{T}} \tag{11.115}$$

11.4.2.3 Dynamic Shape Functions

To formulate the dynamic shape functions, we will consider the homogeneous governing equations reduced from Equation 11.111 to be

$$\begin{aligned} &\kappa GA_{eq}(H'' - \Theta') + P(H'' + y_{\mathrm{CA}}A'') + \omega^2 \rho A_{eq} H + \omega^2 \rho A_{eq} y_{\mathrm{CM}} A - i\omega c_1 H = 0 \\ &EI_{eq}\Theta'' + K_{Teq}A'' + \kappa GA_{eq}(H' - \Theta) + \omega^2 \rho I_{eq}\Theta - i\omega c_2 \Theta = 0 \\ &GJ_{eq}A'' + K_{Teq}\Theta'' + P(y_{\mathrm{CA}}H'' + r_g^2 A'') + \omega^2 \rho J_{eq} A + \omega^2 \rho A_{eq} y_{\mathrm{CM}} H - i\omega c_3 A = 0 \end{aligned} \tag{11.116}$$

Assume the general solution to Equation 11.116 is

$$H(x) = a\mathrm{e}^{-ikx}, \qquad \Theta(x) = \alpha a\mathrm{e}^{-ikx}, \qquad A(x) = \beta a\mathrm{e}^{-ikx} \tag{11.117}$$

where k is the wave number. Substitute Equation 11.117 into Equation 11.116 to obtain an eigenvalue problem in the form of

$$\begin{bmatrix} T_{11} & T_{12} & T_{13} \\ T_{21} & T_{22} & T_{23} \\ T_{31} & T_{32} & T_{33} \end{bmatrix} \begin{Bmatrix} 1 \\ \alpha \\ \beta \end{Bmatrix} = \begin{Bmatrix} 0 \\ 0 \\ 0 \end{Bmatrix} \tag{11.118}$$

where

$$\begin{aligned}
T_{11} &= -(\kappa GA_{eq}+P)k^2+\omega^2\rho A_{eq}-i\omega c_1 \\
T_{12} &= -T_{21}=i\kappa GA_{eq}k \\
T_{13} &= T_{31}=-Py_{\mathrm{CA}}k^2+\omega^2\rho A_{eq}y_{\mathrm{CM}} \\
T_{22} &= -EI_{eq}k^2+i\kappa GA_{eq}k+\omega^2\rho I_{eq}-i\omega c_2 \\
T_{23} &= T_{32}=-K_{Teq}k^2 \\
T_{33} &= -(GJ_{eq}+\mathrm{Pr}_g^2)k^2+\omega^2\rho J_{eq}-i\omega c_3
\end{aligned} \tag{11.119}$$

Equation 11.118 yields a dispersion relation as follows:

$$b_1k_n^6+b_2k_n^5+b_3k_n^4+b_4k_n^3+b_5k_n^2+b_6k_n+b_7=0 \tag{11.120}$$

where

$$\begin{aligned}
b_1 &= EI_{eq}P(Py_{\mathrm{CA}}-Pr_g^2-GJ_{eq})+\kappa GA(K_{Teq}^2-EI_{eq}GJ_{eq}-EI_{eq}Pr_g^2)+K_{Teq}^2P \\
b_2 &= i\kappa GA[GJ_{eq}\kappa GA+Pr_g^2(\kappa GA+P)+P(GJ_{eq}-Py_{\mathrm{CA}}^2)] \\
b_3 &= \kappa GA^2(GJ_{eq}+Pr_g^2)+i\{c_1(K_{Teq}^2+EI_{eq}GJ_{eq}-EI_{eq}Pr_g^2) \\
&\quad -c_2[P(Pr_g^2+Py_{\mathrm{CA}}^2-GJ_{eq})P+\kappa GA(GJ_{eq}-Pr_g^2)]-c_3EI_{eq}(P-\kappa GA)\}\omega \\
&\quad +\{EI_{eq}(GJ_{eq}\rho A_{eq}+\rho A_{eq}Pr_g^2-Py_{\mathrm{CM}}y_{\mathrm{CA}}\rho A_{eq}+\rho J_{eq}P+\rho J_{eq}\kappa GA) \\
&\quad +\rho I_{eq}P(Pr_g^2-Py_{\mathrm{CA}}^2+\kappa GAr_g^2)+GJ_{eq}\rho I_{eq}(P+\kappa GA)-K_{Teq}^2\rho A_{eq}\}\omega^2 \\
b_4 &= -\{c_1\kappa GA(GJ_{eq}+Pr_g^2)+c_3\kappa GA(\kappa GA+P)\}\omega \\
&\quad +i\kappa GA\{\rho A_{eq}(2Py_{\mathrm{CA}}y_{\mathrm{CM}}-GJ_{eq}-Pr_g^2)-\rho J_{eq}(\kappa GA+P)\}\omega^2 \\
b_5 &= ic_3\kappa GA^2\omega \\
&\quad +\{c_1c_3EI_{eq}+c_1c_2GJ_{eq}+c_2c_3P+c_1c_2Pr_g^2+\kappa GA(c_2c_3-\kappa GA\rho J_{eq})\}\omega^2 \\
&\quad +i\{c_1(GJ_{eq}\rho I_{eq}+\rho I_{eq}Pr_g^2+EI_{eq}\rho J_{eq})+c_2[\rho A_{eq}(GJ_{eq}+Pr_g^2-2y_{\mathrm{CA}}y_{\mathrm{CM}}P) \\
&\quad +\rho J_{eq}(P+\kappa GA)]+c_3[EI_{eq}\rho A_{eq}+\rho I_{eq}(\kappa GA+P)]\}\omega^3 \\
&\quad +\{EI_{eq}\rho A_{eq}(y_{\mathrm{CM}}^2-\rho I_{eq})-\rho A_{eq}\rho I_{eq}(GJ_{eq}-Pr_g^2+2y_{\mathrm{CA}}y_{\mathrm{CM}}P]) \\
&\quad -\rho I_{eq}\rho J_{eq}(P+\kappa GA)\}\omega^4 \\
b_6 &= -ic_1c_3\kappa GA\omega^2+\kappa GA(c_1\rho J_{eq}+c_3\rho A_{eq})\omega^3+i\kappa GA\rho A_{eq}(\rho J_{eq}-iy_{\mathrm{CM}}^2\rho A_{eq})\omega^4 \\
b_7 &= ic_1c_2c_3\omega^3-(c_2c_3\rho A_{eq}-c_1c_3\rho I_{eq}-c_1c_2\rho J_{eq})\omega^4+i\{\rho A_{eq}(c_2y_{\mathrm{CM}}^2\rho A_{eq}-c_3\rho I_{eq} \\
&\quad -c_2\rho J_{eq})-c_1\rho I_{eq}\rho J_{eq}\}\omega^5+\rho A_{eq}\rho I_{eq}(\rho J_{eq}-y_{\mathrm{CM}}^2\rho A_{eq})\omega^6
\end{aligned} \tag{11.121}$$

From Equation 11.120, we can numerically compute six wavenumbers k_p ($p=1, 2, \ldots, 6$). The coefficients α_p and β_p for each wavenumbers k_p ($p=1, 2, \ldots, 6$) can be obtained from

Equation 11.118 as

$$\begin{aligned} \alpha_p &= \frac{1}{\mu_{1p}}(K_{Teq}k_p^2\beta_p + i\kappa GA_{eq}k_p) \\ \beta_p &= \frac{\mu_{2p}}{\mu_{3p}} \end{aligned} \qquad (p = 1,\ 2,\ \ldots\ 6) \tag{11.122}$$

where

$$\begin{aligned} \mu_{1p} &= -EI_{eq}k_p^2 + i\kappa GA_{eq}k_p + \omega^2\rho I_{eq} - i\omega c_2 \\ \mu_{2p} &= (\kappa GA_{eq} + P)K_{Teq}k_p^3 + i\kappa GA_{eq}Py_{CA}k_p^2 - K_{T_{eq}}(\omega^2\rho A_{eq} - i\omega c_1)k_p \\ &\quad - i\kappa GA_{eq}\omega^2\rho A_{eq}y_{CM} \\ \mu_{3p} &= -Py_{CA}K_{Teq}k_p^3 - i\kappa GA_{eq}(GJ_{eq} + Pr_g^2)k_p^2 + \omega^2\rho A_{eq}y_{CM}K_{Teq}k_p \\ &\quad + \kappa GA_{eq}\omega(i\omega\rho J_{eq} + c_3) \end{aligned} \tag{11.123}$$

The general solutions of Equation 11.116 can be then be written in the forms

$$H(x) = \sum_{p=1}^{6} a_p e^{-ik_px}, \quad \Theta(x) = \sum_{p=1}^{6} a_p\alpha_p e^{-ik_px}, \quad A(x) = \sum_{p=1}^{6} a_p\beta_p e^{-ik_px} \tag{11.124}$$

or

$$H(x) = \boldsymbol{e}(x;\omega)\boldsymbol{a}, \quad \Theta(x) = \boldsymbol{e}(x;\omega)\boldsymbol{Aa}, \quad A(x) = \boldsymbol{e}(x;\omega)\boldsymbol{Ba} \tag{11.125}$$

where

$$\boldsymbol{a} = \{a_1 \quad a_2 \quad a_3 \quad a_4 \quad a_5 \quad a_6\}^{\mathrm{T}} \tag{11.126}$$

and

$$\boldsymbol{e}(x;\omega) = [e^{-ik_1x} \quad e^{-ik_2x} \quad e^{-ik_3x} \quad e^{-ik_4x} \quad e^{-ik_5x} \quad e^{-ik_6x}] \tag{11.127}$$

$$\begin{aligned} \boldsymbol{A}(\omega) &= diag[\alpha_p(\omega)] \\ \boldsymbol{B}(\omega) &= diag[\beta_p(\omega)] \end{aligned} \qquad (p = 1,\ 2,\ \ldots,\ 6) \tag{11.128}$$

By applying Equation 11.125 to the definition of the spectral nodal DOFs (11.114), we can obtain the relationship

$$\boldsymbol{d} = \boldsymbol{G}(\omega)\boldsymbol{a} \tag{11.129}$$

where

$$\boldsymbol{G}(\omega) = \begin{bmatrix} 1 & 1 & 1 & 1 & 1 & 1 \\ \alpha_1 & \alpha_2 & \alpha_3 & \alpha_4 & \alpha_5 & \alpha_6 \\ \beta_1 & \beta_2 & \beta_3 & \beta_4 & \beta_5 & \beta_6 \\ e_1 & e_2 & e_3 & e_4 & e_5 & e_6 \\ \alpha_1 e_1 & \alpha_2 e_2 & \alpha_3 e_3 & \alpha_4 e_4 & \alpha_5 e_5 & \alpha_6 e_6 \\ \beta_1 e_1 & \beta_2 e_2 & \beta_3 e_3 & \beta_4 e_4 & \beta_5 e_5 & \beta_6 e_6 \end{bmatrix} \tag{11.130}$$

with

$$e_p = e^{-ik_pL} \qquad (p = 1, 2, \ldots, 6) \tag{11.131}$$

By using Equation 11.129, the constant vector $\boldsymbol{a}$ can be eliminated from Equation 11.125 to obtain the expressions

$$H(x) = \boldsymbol{N}_h(x;\omega)\boldsymbol{d}, \quad \Theta(x) = \boldsymbol{N}_\theta(x;\omega)\boldsymbol{d}, \quad A(x) = \boldsymbol{N}_\alpha(x;\omega)\boldsymbol{d} \tag{11.132}$$

where $N_h(x;\,\omega_n)$, $N_\theta(x;\,\omega_n)$, $N_\alpha(x;\,\omega_n)$ are the dynamic shape functions defined by

$$\begin{aligned}\boldsymbol{N}_h(x;\omega) &= \boldsymbol{e}(x;\omega)\boldsymbol{G}^{-1}(\omega)\\ \boldsymbol{N}_\theta(x;\omega) &= \boldsymbol{e}(x;\omega)\boldsymbol{A}\boldsymbol{G}^{-1}(\omega)\\ \boldsymbol{N}_\alpha(x;\omega) &= \boldsymbol{e}(x;\omega)\boldsymbol{B}\boldsymbol{G}^{-1}(\omega_n)\end{aligned} \tag{11.133}$$

11.4.2.4 Weak Form of Governing Equations

To formulate the spectral element equation in the next subsection, we first derive the weak form of Equation 11.111 from

$$\begin{aligned}&\int_0^L \{[\kappa GA_{eq}(H'' - \Theta') + P(H'' + y_{CA}A'') + \omega^2\rho A_{eq}H + \omega^2\rho A_{eq}y_{CM}A\\ &- i\omega c_1 H + F_h]\delta H + [EI_{eq}\Theta'' + K_{Teq}A'' + \kappa GA_{eq}(H' - \Theta') + \omega^2\rho I_{eq}\Theta\\ &- i\omega c_2\Theta + F_\theta]\delta\Theta + [GJ_{eq}A'' + K_{Teq}\Theta'' + P(y_{CA}H'' + r_g^2A'')\\ &+ \omega^2\rho J_{eq}A + \omega^2\rho A_{eq}y_{CM}H - i\omega c_3 A + F_\alpha)]\delta A\}\mathrm{d}x = 0\end{aligned} \tag{11.134}$$

Applying the integral by parts to some terms in Equation 11.134 and then using the force-displacement relationships (11.113), we can obtain the weak form as

$$\begin{aligned}&\int_0^L [\kappa GA_{eq}H'\delta H' + \kappa GA_{eq}\Theta\delta\Theta - \kappa GA_{eq}(\Theta\delta H' + H'\delta\Theta) + EI_{eq}\Theta'\delta\Theta'\\ &+ (GJ_{eq} + \mathrm{P}r_g^2)A'\delta A' + K_{Teq}(A'\delta\Theta' + \Theta'\delta A') + Py_{CA}(A'\delta H' + H'\delta A')\\ &+ PH'\delta H' + (i\omega c_1 - \omega^2\rho A_{eq})H\delta H + (i\omega c_2 - \omega^2\rho I_{eq})\Theta\delta\Theta\\ &+ (i\omega c_3 - \omega^2\rho J_{eq})A\delta A - \omega^2\rho A_{eq}y_{CM}(A\delta H + H\delta A)]\mathrm{d}x\\ &= M(x)\delta\Theta\Big|_0^L + [Q(x) + P(H' + y_{CA}A')]\delta H\Big|_0^L + [T(x) + P(y_{CA}H' + r_g^2A')]\delta A\Big|_0^L\\ &+ \int_0^L [F_h(x)\delta H + F_\theta(x)\delta\Theta + F_\alpha(x)\delta A]\mathrm{d}x\end{aligned} \tag{11.135}$$

or, by applying the boundary conditions (11.112) and then using the definitions of Equations 11.114 and 11.115,

$$\begin{aligned}&\int_0^L [\kappa GA_{eq}H'\delta H' + \kappa GA_{eq}\Theta\delta\Theta - \kappa GA_{eq}(\Theta\delta H' + H'\delta\Theta) + EI_{eq}\Theta'\delta\Theta'\\ &+ (GJ_{eq} + \mathrm{P}r_g^2)A'\delta A' + K_{Teq}(A'\delta\Theta' + \Theta'\delta A') + Py_{CA}(A'\delta H' + H'\delta A')\\ &+ PH'\delta H' + (i\omega c_1 - \omega^2\rho A_{eq})H\delta H + (i\omega c_2 - \omega^2\rho I_{eq})\Theta\delta\Theta\\ &+ (i\omega c_3 - \omega^2\rho J_{eq})A\delta A - \omega^2\rho A_{eq}y_{CM}(A\delta H + H\delta A)]\mathrm{d}x\\ &= \delta\boldsymbol{d}^{\mathrm{T}}\boldsymbol{f}_c + \int_0^L \left[F_h(x)\delta H + F_\theta(x)\delta\Theta + F_\alpha(x)\delta A\right]\mathrm{d}x\end{aligned} \tag{11.136}$$

11.4.2.5 Spectral Element Equation

Substitution of Equation 11.132 into Equation 11.136 gives

$$\begin{aligned}
&\delta \boldsymbol{d}^{\mathrm{T}}\left[\int_0^L \{\kappa GA_{eq}\boldsymbol{N}_h'^{\mathrm{T}}\boldsymbol{N}_h' + \kappa GA_{eq}\boldsymbol{N}_\theta^{\mathrm{T}}\boldsymbol{N}_\theta - \kappa GA_{eq}(\boldsymbol{N}_h'^{\mathrm{T}}\boldsymbol{N}_\theta + \boldsymbol{N}_\theta^{\mathrm{T}}\boldsymbol{N}_h')\right.\\
&\quad + EI_{eq}\boldsymbol{N}_\theta'^{\mathrm{T}}\boldsymbol{N}_\theta' + (GJ_{eq} + Pr_g^2)\boldsymbol{N}_\alpha'^{\mathrm{T}}\boldsymbol{N}_\alpha' + K_{Teq}(\boldsymbol{N}_\theta'^{\mathrm{T}}\boldsymbol{N}_\alpha' + \boldsymbol{N}_\alpha'^{\mathrm{T}}\boldsymbol{N}_\theta')\\
&\quad + Py_{\mathrm{CA}}(\boldsymbol{N}_h'^{\mathrm{T}}\boldsymbol{N}_\alpha' + \boldsymbol{N}_\alpha'^{\mathrm{T}}\boldsymbol{N}_h') + \boldsymbol{P}\boldsymbol{N}_h'^{\mathrm{T}}\boldsymbol{N}_h' + (i\omega_n c_1 - \omega_n^2\rho A_{eq})\boldsymbol{N}_h^{\mathrm{T}}\boldsymbol{N}_h\\
&\quad + (i\omega_n c_2 - \omega_n^2\rho I_{eq})\boldsymbol{N}_\theta^{\mathrm{T}}\boldsymbol{N}_\theta + (i\omega_n c_3 - \omega_n^2\rho J_{eq})\boldsymbol{N}_\alpha^{\mathrm{T}}\boldsymbol{N}_\alpha\\
&\quad - \omega_n^2\rho A_{eq}y_{\mathrm{CM}}(\boldsymbol{N}_h^{\mathrm{T}}\boldsymbol{N}_\alpha + \boldsymbol{N}_\alpha^{\mathrm{T}}\boldsymbol{N}_h)\}\mathrm{d}x\boldsymbol{d}\\
&= \delta \boldsymbol{d}^{\mathrm{T}}\left[\boldsymbol{f}_{\mathrm{c}} + \int_0^L (F_h(x)\boldsymbol{N}_h^{\mathrm{T}} + F_\theta(x)\boldsymbol{N}_\theta^{\mathrm{T}} + F_\alpha(x)\boldsymbol{N}_\alpha^{\mathrm{T}})\mathrm{d}x\right]
\end{aligned} \tag{11.137}$$

$\delta \boldsymbol{d}$ is arbitrary by definition. Thus, we can obtain the spectral element equation from Equation 11.137 as being

$$\boldsymbol{S}(\omega)\boldsymbol{d} = \boldsymbol{f}_{\mathrm{c}} + \boldsymbol{f}_{\mathrm{d}} \tag{11.138}$$

where

$$\begin{aligned}
\boldsymbol{S}(\omega) = &\int_0^L \{\kappa GA_{eq}\boldsymbol{N}_h'^{\mathrm{T}}\boldsymbol{N}_h' + \kappa GA_{eq}\boldsymbol{N}_\theta^{\mathrm{T}}\boldsymbol{N}_\theta - \kappa GA_{eq}(\boldsymbol{N}_h'^{\mathrm{T}}\boldsymbol{N}_\theta + \boldsymbol{N}_\theta^{\mathrm{T}}\boldsymbol{N}_h')\\
&+ EI_{eq}\boldsymbol{N}_\theta'^{\mathrm{T}}\boldsymbol{N}_\theta' + (GJ_{eq} + Pr_g^2)\boldsymbol{N}_\alpha'^{\mathrm{T}}\boldsymbol{N}_\alpha' + K_{Teq}(\boldsymbol{N}_\theta'^{\mathrm{T}}\boldsymbol{N}_\alpha' + \boldsymbol{N}_\alpha'^{\mathrm{T}}\boldsymbol{N}_\theta')\\
&+ Py_{\mathrm{CA}}(\boldsymbol{N}_h'^{\mathrm{T}}\boldsymbol{N}_\alpha' + \boldsymbol{N}_\alpha'^{\mathrm{T}}\boldsymbol{N}_h') + \boldsymbol{P}\boldsymbol{N}_h'^{\mathrm{T}}\boldsymbol{N}_h' + (i\omega_n c_1 - \omega_n^2\rho A_{eq})\boldsymbol{N}_h^{\mathrm{T}}\boldsymbol{N}_h\\
&+ (i\omega_n c_2 - \omega_n^2\rho I_{eq})\boldsymbol{N}_\theta^{\mathrm{T}}\boldsymbol{N}_\theta + (i\omega_n c_3 - \omega_n^2\rho J_{eq})\boldsymbol{N}_\alpha^{\mathrm{T}}\boldsymbol{N}_\alpha\\
&- \omega_n^2\rho A_{eq}y_{\mathrm{CM}}(\boldsymbol{N}_h^{\mathrm{T}}\boldsymbol{N}_\alpha + \boldsymbol{N}_\alpha^{\mathrm{T}}\boldsymbol{N}_h)\}\mathrm{d}x
\end{aligned} \tag{11.139}$$

and

$$\boldsymbol{f}_{\mathrm{d}} = \int_0^L [F_h(x)\boldsymbol{N}_h^{\mathrm{T}} + F_\theta(x)\boldsymbol{N}_\theta^{\mathrm{T}} + F_\alpha(x)\boldsymbol{N}_\alpha^{\mathrm{T}}]\mathrm{d}x \tag{11.140}$$

By substituting Equation 11.133 into Equation 11.139, the spectral element matrix $\boldsymbol{S}(\omega)$ can be expressed in the form

$$\boldsymbol{S}(\omega) = \boldsymbol{G}^{-\mathrm{T}}(\omega)\boldsymbol{D}(\omega)\boldsymbol{G}^{-1}(\omega) \tag{11.141}$$

where

$$\begin{aligned}
\boldsymbol{D}(\omega) = &\int_0^L [\kappa GA_{eq}\boldsymbol{e}'^{\mathrm{T}}\boldsymbol{e}' + \kappa GA_{eq}\boldsymbol{A}\boldsymbol{e}^{\mathrm{T}}\boldsymbol{e}\boldsymbol{A} - \kappa GA_{eq}(\boldsymbol{e}'^{\mathrm{T}}\boldsymbol{e}\boldsymbol{A} + \boldsymbol{A}\boldsymbol{e}^{\mathrm{T}}\boldsymbol{e}')\\
&+ EI_{eq}\boldsymbol{A}\boldsymbol{e}'^{\mathrm{T}}\boldsymbol{e}'\boldsymbol{A} + (GJ_{eq} + Pr_g^2)\boldsymbol{B}\boldsymbol{e}'^{\mathrm{T}}\boldsymbol{e}'\boldsymbol{B}\\
&+ K_{Teq}(\boldsymbol{A}\boldsymbol{e}'^{\mathrm{T}}\boldsymbol{e}'\boldsymbol{B} + \boldsymbol{B}\boldsymbol{e}'^{\mathrm{T}}\boldsymbol{e}'\boldsymbol{A}) + Py_{\mathrm{CA}}(\boldsymbol{e}'^{\mathrm{T}}\boldsymbol{e}'\boldsymbol{B} + \boldsymbol{B}\boldsymbol{e}'^{\mathrm{T}}\boldsymbol{e}') + P\boldsymbol{e}'^{\mathrm{T}}\boldsymbol{e}'\\
&+ (i\omega_n c_1 - \omega_n^2\rho A_{eq})\boldsymbol{e}^{\mathrm{T}}\boldsymbol{e} + (i\omega_n c_2 - \omega_n^2\rho I_{eq})\boldsymbol{A}\boldsymbol{e}^{\mathrm{T}}\boldsymbol{e}\boldsymbol{A}\\
&+ (i\omega_n c_3 - \omega_n^2\rho J_{eq})\boldsymbol{B}\boldsymbol{e}^{\mathrm{T}}\boldsymbol{e}\boldsymbol{B} - \omega_n^2\rho A_{eq}y_{\mathrm{CM}}(\boldsymbol{e}^{\mathrm{T}}\boldsymbol{e}\boldsymbol{B} + \boldsymbol{B}\boldsymbol{e}^{\mathrm{T}}\boldsymbol{e})]\mathrm{d}x
\end{aligned} \tag{11.142}$$

The derivative of $\boldsymbol{e}(x;\omega)$ with respect to x can be obtained from Equation 11.127 as

$$\boldsymbol{e}'(x;\omega) = -i\boldsymbol{e}(x;\omega)\boldsymbol{K}(\omega) \tag{11.143}$$

where

$$\boldsymbol{K} = diag[k_p] \qquad (p = 1, 2, \ldots, 6) \tag{11.144}$$

Substitution of Equations 11.127 and 11.143 into Equation 11.142 gives

$$\begin{aligned}
\boldsymbol{D}(\omega) = {} & \kappa G A_{eq}[-\boldsymbol{KEK}+\boldsymbol{AEA}+i(\boldsymbol{KEA}+\boldsymbol{AEK})]-EI_{eq}\boldsymbol{AKEKA} \\
& -(GJ_{eq}+Pr_g^2)\boldsymbol{BKEKB}-K_{Teq}(\boldsymbol{AKEKB}+\boldsymbol{BKEKA}) \\
& -Py_{\mathrm{CA}}(\boldsymbol{KEKB}+\boldsymbol{BKEK})-P\boldsymbol{KEK}+(i\omega_n c_1-\omega_n^2\rho A_{eq})\boldsymbol{E} \\
& +(i\omega_n c_2-\omega_n^2\rho I_{eq})\boldsymbol{AEA}+(i\omega_n c_3-\omega_n^2\rho J_{eq})\boldsymbol{BEB} \\
& -\omega_n^2\rho A_{eq}y_{\mathrm{CM}}(\boldsymbol{EB}+\boldsymbol{BE})
\end{aligned} \tag{11.145}$$

where the same definition for $\boldsymbol{E}(\omega)$ as that given by Equation 11.94 is used.

11.4.3 Finite Element Model

We can obtain the weak form of the equations of motion (11.106) in the form of

$$\begin{aligned}
& \int_0^L [\kappa GA_{eq}h'\delta h'+\kappa GA_{eq}\theta\delta\theta-\kappa GA_{eq}(\theta\delta h'+h'\delta\theta)+EI_{eq}\theta'\delta\theta' \\
& \quad +(GJ_{eq}+\mathrm{P}r_g^2)\alpha'\delta\alpha'+K_{Teq}(\alpha'\delta\theta'+\theta'\delta\alpha')+Py_{\mathrm{CA}}(\alpha'\delta h'+h'\delta\alpha') \\
& \quad +Ph'\delta h'+(c_1\dot{h}+\rho A_{eq}\ddot{h})\delta h+(c_2\dot{\theta}+\rho I_{eq}\ddot{\theta})\delta\theta+(c_3\dot{\alpha}+\rho J_{eq}\ddot{\alpha})\delta\alpha \\
& \quad +zd\rho A_{eq}y_{\mathrm{CM}}\,\ddot{\alpha}\delta h+\rho A_{eq}y_{\mathrm{CM}}\,\ddot{h}\delta\alpha]\mathrm{d}x \\
& = [Q(x,t)+P(h'+y_{\mathrm{CA}}\alpha')]\delta h\Big|_0^L+M_t\delta\theta\Big|_0^L+[T(x,t)+P(y_{\mathrm{CA}}h'+r_g^2\alpha')]\delta\alpha\Big|_0^L \\
& \quad +\int_0^L f_h\delta h\mathrm{d}x+f_\theta\delta\theta\mathrm{d}x+\int_0^L f_\alpha\delta\alpha\mathrm{d}x
\end{aligned} \tag{11.146}$$

We use the same convention for the nodal DOFs, forces, and moments in the time domain as specified in Figure 11.5 and we define the nodal DOFs vector by

$$\boldsymbol{d}(t)=\{h_1(t)\quad \theta_1(t)\quad \alpha_1(t)\quad h_2(t)\quad \theta_2(t)\quad \alpha_2(t)\}^{\mathrm{T}} \tag{11.147}$$

and the nodal forces and moments vector by

$$\boldsymbol{f}_c(t)=\{Q_1(t)\quad M_1(t)\quad T_1(t)\quad Q_2(t)\quad M_2(t)\quad T_2(t)\}^{\mathrm{T}} \tag{11.148}$$

We apply the boundary conditions (11.107) to the right-hand side of Equation 11.146 and then use the definitions (11.147) and (11.148) to write Equation 11.146 as

$$\begin{aligned}
& \int_0^L [\kappa GA_{eq}h'\delta h'+\kappa GA_{eq}\theta\delta\theta-\kappa GA_{eq}(\theta\delta h'+h'\delta\theta)+EI_{eq}\theta'\delta\theta' \\
& \quad +(GJ_{eq}+\mathrm{P}r_g^2)\alpha'\delta\alpha'+K_{Teq}(\alpha'\delta\theta'+\theta'\delta\alpha')+Py_{\mathrm{CA}}(\alpha'\delta h'+h'\delta\alpha') \\
& \quad +Ph'\delta h'+(c_1\dot{h}+\rho A_{eq}\ddot{h})\delta h+(c_2\dot{\theta}+\rho I_{eq}\ddot{\theta})\delta\theta+(c_3\dot{\alpha}+\rho J_{eq}\ddot{\alpha})\delta\alpha \\
& \quad +\rho A_{eq}y_{\mathrm{CM}}\,\ddot{\alpha}\delta h+\rho A_{eq}y_{CM}\,\ddot{h}\delta\alpha]\mathrm{d}x \\
& = \delta\boldsymbol{d}^{\mathrm{T}}\boldsymbol{f}_c(t)+\int_0^L f_h\delta h\mathrm{d}x+f_\theta\delta\theta\mathrm{d}x+\int_0^L f_\alpha\delta\alpha\mathrm{d}x
\end{aligned} \tag{11.149}$$

Assume the displacement fields within the composite beam element to be

$$h(x,t) = \boldsymbol{N}_h(x)\boldsymbol{d}(t), \quad \theta(x,t) = \boldsymbol{N}_\theta(x)\boldsymbol{d}(t), \quad \alpha(x,t) = \boldsymbol{N}_\alpha(x)\boldsymbol{d}(t) \tag{11.150}$$

where

$$\begin{aligned} \boldsymbol{N}_h(x) &= [N_1(x) \quad N_2(x) \quad 0 \quad N_3(x) \quad N_4(x) \quad 0] \\ \boldsymbol{N}_\theta(x) &= [N_5(x) \quad N_6(x) \quad 0 \quad N_7(x) \quad N_8(x) \quad 0] \\ \boldsymbol{N}_\alpha(x) &= [0 \quad 0 \quad N_9(x) \quad 0 \quad 0 \quad N_{10}(x)] \end{aligned} \tag{11.151}$$

with

$$\begin{aligned} &N_1(x) = R(1-\xi)(1+\xi-2\xi^2+\delta) \\ &N_2(x) = (1/2)RL\xi(1-\xi)(2-2\xi+\delta) \\ &N_3(x) = R\xi(3\xi-2\xi^2+\delta), \qquad N_4(x) = -(1/2)RL\xi(1-\xi)(2\xi+\delta) \\ &N_5(x) = -6(R/L)\xi(1-\xi), \qquad N_6(x) = R(1-\xi)(1-3\xi+\delta) \\ &N_7(x) = 6(R/L)\xi(1-\xi), \qquad N_8(x) = R\xi(-2+3\xi+\delta) \\ &N_9(x) = 1-\xi, \qquad N_{10}(x) = \xi \end{aligned} \tag{11.152}$$

where

$$\xi = \frac{x}{L}, \qquad R = \frac{1}{1+\delta}, \qquad \delta = \frac{12r}{L^2}, \qquad r = \frac{EI}{\kappa GA} \tag{11.153}$$

By substituting Equation 11.150 into Equation 11.149, we can obtain the finite element equation in the form of

$$\boldsymbol{M}\ddot{\boldsymbol{d}}(t) + \boldsymbol{C}\dot{\boldsymbol{d}}(t) + \boldsymbol{K}\boldsymbol{d}(t) = \boldsymbol{f}_\mathrm{c}(t) + \boldsymbol{f}_\mathrm{d}(t) \tag{11.154}$$

where

$$\begin{aligned} \boldsymbol{M} &= \rho A_{eq}\int_0^L \boldsymbol{N}_h^\mathrm{T}\boldsymbol{N}_h \mathrm{d}x + \rho I_{eq}\int_0^L \boldsymbol{N}_\theta^\mathrm{T}\boldsymbol{N}_\theta \mathrm{d}x + \rho J_{eq}\int_0^L \boldsymbol{N}_\alpha^\mathrm{T}\boldsymbol{N}_\alpha \mathrm{d}x \\ &\quad + \rho A_{eq}y_\mathrm{CM}\int_0^L (\boldsymbol{N}_h^\mathrm{T}\boldsymbol{N}_\alpha + \boldsymbol{N}_\alpha^\mathrm{T}\boldsymbol{N}_h)\mathrm{d}x \\ \boldsymbol{C} &= c_1\int_0^L \boldsymbol{N}_h^\mathrm{T}\boldsymbol{N}_h \mathrm{d}x + c_2\int_0^L \boldsymbol{N}_\theta^\mathrm{T}\boldsymbol{N}_\theta \mathrm{d}x + c_3\int_0^L \boldsymbol{N}_\alpha^\mathrm{T}\boldsymbol{N}_\alpha \mathrm{d}x \\ \boldsymbol{K} &= \kappa GA_{eq}\int_0^L \boldsymbol{N}_h'^\mathrm{T}\boldsymbol{N}_h' \mathrm{d}x + \kappa GA_{eq}\int_0^L \boldsymbol{N}_\theta^\mathrm{T}\boldsymbol{N}_\theta \mathrm{d}x - \kappa GA_{eq}\int_0^L (\boldsymbol{N}_h'^\mathrm{T}\boldsymbol{N}_\theta + \boldsymbol{N}_\theta^\mathrm{T}\boldsymbol{N}_h')\mathrm{d}x \\ &\quad + EI_{eq}\int_0^L \boldsymbol{N}_\theta'^\mathrm{T}\boldsymbol{N}_\theta' \mathrm{d}x + (GJ_{eq} + Pr_g^2)\int_0^L \boldsymbol{N}_\alpha'^\mathrm{T}\boldsymbol{N}_\alpha' \mathrm{d}x + K_{Teq}\int_0^L (\boldsymbol{N}_\theta'^\mathrm{T}\boldsymbol{N}_\alpha' + \boldsymbol{N}_\alpha'^\mathrm{T}\boldsymbol{N}_\theta')\mathrm{d}x \\ &\quad + Py_\mathrm{CA}\int_0^L (\boldsymbol{N}_h'^\mathrm{T}\boldsymbol{N}_\alpha' + \boldsymbol{N}_\alpha'^\mathrm{T}\boldsymbol{N}_h')\mathrm{d}x + P\int_0^L \boldsymbol{N}_h'^\mathrm{T}\boldsymbol{N}_h' \mathrm{d}x \end{aligned} \tag{11.155}$$

and

$$f_d(t) = \int_0^L [f_h(x,t)N_h^T(x) + f_\theta(x,t)N_\theta^T(x) + f_\alpha(x,t)N_\alpha^T(x)]dx \tag{11.156}$$

The finite element mass matrix $\boldsymbol{M}$, stiffness matrix $\boldsymbol{K}$, and damping matrix $\boldsymbol{C}$ are given in Appendix 11.B.

Example 11.3

Consider a cantilever glass–epoxy composite beam from Refs. [12, 111] as an illustrative example. The beam is of rectangular cross-section and is made up of unidirectional plies with fiber angles in each ply set to a single angle $+15°$. Thus the composite beam can be considered as a single thick ply. The beam has the length $L = 0.1905$ m and its structural and inertia properties are taken from Refs. [12, 111] and they are given by

$$\begin{aligned} EI_{eq} &= 0.2865\ \mathrm{N\,m^2} & \kappa GA_{eq} &= 6343.3\ \mathrm{N} \\ GJ_{eq} &= 0.1891\ \mathrm{N\,m^2} & K_{Teq} &= 0.1143\ \mathrm{N\,m^2} \\ \rho A_{eq} &= 0.0544\ \mathrm{kg/m} & \rho I_{eq} &= 4.584\times 10^{-8}\ \mathrm{kg\,m} \\ \rho J_{eq} &= 0.777\times 10^{-6}\ \mathrm{kg\,m} \end{aligned}$$

with $y_{CA} = y_{CM} = 0$ m and the damping coefficients set to $c_1 = c_2 = c_3 = 0$.

The first five natural frequencies of the composite beam subjected to an axial tensile force ($P = 7.5$ N) are computed by using the present spectral element model and the finite element model, and in Table 11.3 the results are compared with the results from Refs. [12, 111]. The one-element model is used for the SEM results, while the number of finite elements used for FEM is varied in order to investigate the solution convergence to the SEM results. Similarly, Table 11.4 is for the same composite beam subjected to a compressive axial force ($P = -7.5$ N). The present SEM results are shown to be almost equal to the results obtained by an analytical method in Ref. [111] and the dynamic stiffness method in Ref. [12]. The FEM results are shown to converge towards the SEM results as the number of finite elements used in the FEM is increased. As expected, Tables 11.3 and 11.4 also show that the natural frequencies of the composite beam subjected to a compressive axial force are lower than those of the composite beam subjected to a tensile axial force.

Table 11.3 Natural frequencies of the cantilever composite beam subjected to an axial tensile force

		FEM(N)						
Mode	SEM (1)	$N=2$	$N=10$	$N=30$	$N=50$	$N=200$	Ref. [111]	Ref. [12]
1	37.10	37.19	37.10	37.10	37.10	37.10	37.10	37.10
2	197.7	205.6	198.2	197.7	197.7	197.7	197.7	197.7
3	525.7	590.3	530.0	526.2	525.7	525.7	525.7	525.6
4	648.5	654.6	649.1	648.6	648.5	648.5	648.5	648.6
5	992.9	1259	1011	994.9	993.6	992.9	992.9	—

Note: N = total number of the equal-length finite elements used in the analysis.

Table 11.4 Natural frequencies of the cantilever composite beam subjected to an axial compressive force

Mode	SEM(1)	FEM(N)					Ref. [111]	Ref. [12]
		$N=2$	$N=10$	$N=50$	$N=150$	$N=200$		
1	21.99	22.07	21.99	21.99	21.99	21.99	21.98	21.99
2	181.5	189.4	182.0	181.5	181.5	181.5	181.5	181.5
3	511.8	576.9	516.2	512.0	511.8	511.8	511.8	511.9
4	648.0	653.9	648.6	648.1	648.1	648.0	648.0	648.0
5	979.5	1247	997.9	980.2	979.6	979.5	979.5	—

Note: N = total number of the equal-length finite elements used in the analysis.

Appendix 11.A: Finite Element Matrices for Axial–Bending–Shear Coupled Composite Beams

The finite element mass and stiffness matrices defined in Equation 11.104 are given by

$$\begin{aligned} \boldsymbol{M} &= \boldsymbol{M}_{\rho A} + \boldsymbol{M}_{\rho I} - \boldsymbol{M}_{\rho R} \\ \boldsymbol{K} &= \boldsymbol{K}_{\kappa GA} + \boldsymbol{K}_{EI} + \boldsymbol{K}_{EA} - \boldsymbol{K}_{K}A_{eq} + \boldsymbol{K}_{P} \end{aligned} \tag{11.A.1}$$

where

$$\boldsymbol{M}_{\rho A} = \frac{\rho A_{eq} L}{840} \begin{bmatrix} 280 & 0 & 0 & 140 & 0 & 0 \\ & 4(70+7R+R^2) & L(35+7R+2R^2) & 0 & 4(35-7R-R^2) & -L(35-7R-2R^2) \\ & & L^2(7+R^2) & 0 & L(35-7R-2R^2) & -L^2(7-R^2) \\ & & & 280 & 0 & 0 \\ & \text{Symmetric} & & & 4(70+7R+R^2) & -L(35+7R+2R^2) \\ & & & & & L^2(7+R^2) \end{bmatrix}$$

$$
\boldsymbol{M}_{\rho I} = \frac{\rho I_{eq}}{30L}\begin{bmatrix} 0 & 0 & 0 & 0 & 0 & 0 \\ & 36R^2 & -3LR(5-6R) & 0 & -36R^2 & -3LR(5-6R) \\ & & L^2(10-15R+9R^2) & 0 & 3LR(5-6R) & L^2(5-15R+9R^2) \\ & & & 0 & 0 & 0 \\ & \text{Symmetric} & & & 36R^2 & 3LR(5-6R) \\ & & & & & L^2(10-15R+9R^2) \end{bmatrix}
$$

$$
\boldsymbol{M}_3 = \frac{\rho R_{eq}}{12}\begin{bmatrix} 0 & -6R & L(4-3R) & 0 & 6R & L(2-3R) \\ & 0 & 0 & -6R & 0 & 0 \\ & & 0 & L(2-3R) & 0 & 0 \\ & & & 0 & 6R & L(4-3R) \\ & \text{Symmetric} & & & 0 & 0 \\ & & & & & 0 \end{bmatrix} \tag{11.A.2}
$$

and

$$
\boldsymbol{K}_{\kappa GA} = \frac{\kappa GA_{eq}}{4L}(I-R)^2\begin{bmatrix} 0 & 0 & 0 & 0 & 0 & 0 \\ & 4 & 2L & 0 & -4 & 2L \\ & & L^2 & 0 & -2L & L^2 \\ & & & 0 & 0 & 0 \\ & \text{Symmetric} & & & 4 & -2L \\ & & & & & L^2 \end{bmatrix}
$$

$$
\boldsymbol{K}_{EI} = \frac{EI_{eq}}{L^3}\begin{bmatrix} 0 & 0 & 0 & 0 & 0 & 0 \\ & 12R^2 & 6LR^2 & 0 & -12R^2 & 6LR^2 \\ & & L^2(1+3R^2) & 0 & -6LR^2 & -L^2(1-3R^2) \\ & & & 0 & 0 & 0 \\ & \text{Symmetric} & & & 12R^2 & -6LR^2 \\ & & & & & L^2(1+3R^2) \end{bmatrix}
$$

$$
\boldsymbol{K}_{EA} = \frac{EA_{eq}}{L}\begin{bmatrix} 1 & 0 & 0 & -1 & 0 & 0 \\ & 0 & 0 & 0 & 0 & 0 \\ & & 0 & 0 & 0 & 0 \\ & & & 1 & 0 & 0 \\ & \text{Symmetric} & & & 0 & 0 \\ & & & & & 0 \end{bmatrix} \tag{11.A.3}
$$

$$
\boldsymbol{K}_{K_A} = \frac{K_{Aeq}}{L}\begin{bmatrix} 0 & 0 & 1 & 0 & 0 & -1 \\ & 0 & 0 & 0 & 0 & 0 \\ & & 0 & -1 & 0 & 0 \\ & & & 0 & 0 & 1 \\ & \text{Symmetric} & & & 0 & 0 \\ & & & & & 0 \end{bmatrix}
$$

$$
\boldsymbol{K}_P = \frac{P}{60L}\begin{bmatrix} 0 & 0 & 0 & 0 & 0 & 0 \\ & 12(5+R^2) & 6LR^2 & 0 & -12(5+R^2) & 6LR^2 \\ & & L^2(5+3R^2) & 0 & -6LR^2 & -L^2(5-3R^2) \\ & & & 0 & 0 & 0 \\ & \text{Symmetric} & & & 12(5+R^2) & -6LR^2 \\ & & & & & L^2(5+3R^2) \end{bmatrix}
$$

Appendix 11.B: Finite Element Matrices for Bending–Torsion–Shear Coupled Composite Beams

The finite element matrices defined in Equation 11.154 are given by

$$
\begin{aligned}
\boldsymbol{M} &= \boldsymbol{M}_{\rho A} + \boldsymbol{M}_{\rho I} + \boldsymbol{M}_{\rho J} + \boldsymbol{M}_{\rho A y_{\mathrm{CM}}} \\
\boldsymbol{C} &= \boldsymbol{C}_{c_1} + \boldsymbol{C}_{c_2} + \boldsymbol{C}_{c_3} \\
\boldsymbol{K} &= \boldsymbol{K}_{\kappa GA} + \boldsymbol{K}_{EI} + \boldsymbol{K}_{GJ+Pr_g^2} + \boldsymbol{K}_{K_T} + \boldsymbol{K}_{P y_{\mathrm{CA}}} + \boldsymbol{K}_P
\end{aligned} \tag{11.B.1}
$$

where

$$
\boldsymbol{M}_{\rho A} = \frac{\rho A_{eq} L}{840}\begin{bmatrix} 4(70+7R+R^2) & L(35+7R+2R^2) & 0 & 4(35-7R-R^2) & -L(35-7R-2R^2) & 0 \\ & L^2(7+R^2) & 0 & L(35-7R-2R^2) & -L^2(7-R^2) & 0 \\ & & 0 & 0 & 0 & 0 \\ & & & 4(70+7R+R^2) & -L(35+7R+2R^2) & 0 \\ & \text{Symmetric} & & & L^2(7+R^2) & 0 \\ & & & & & 0 \end{bmatrix}
$$

$$
\boldsymbol{M}_{\rho I} = \frac{\rho I_{eq}}{30L}\begin{bmatrix} 36R^2 & -3L(5R-6R^2) & 0 & -36R^2 & -3L(5R-6R^2) & 0 \\ & L^2(10-15R+9R^2) & 0 & 3L(5R-6R^2) & L^2(5-15R+9R^2) & 0 \\ & & 0 & 0 & 0 & 0 \\ & & & 36R^2 & 3L(5R-6R^2) & 0 \\ & \text{Symmetric} & & & L^2(10-15R+9R^2) & 0 \\ & & & & & 0 \end{bmatrix}
$$

$$\boldsymbol{M}_{\rho J} = \frac{\rho J_{eq} L}{6} \begin{bmatrix} 0 & 0 & 0 & 0 & 0 & 0 \\ & 0 & 0 & 0 & 0 & 0 \\ & & 2 & 0 & 0 & 1 \\ & & & 0 & 0 & 0 \\ & \text{Symmetric} & & & 0 & 0 \\ & & & & & 2 \end{bmatrix} \tag{11.B.2}$$

$$\boldsymbol{M}_{\rho A y_{\text{CM}}} = \frac{\rho A_{eq} y_{\text{CM}} L}{120} \begin{bmatrix} 0 & 0 & 2(20+R) & 0 & 0 & 2(10-R) \\ & 0 & L(5+R) & 0 & 0 & L(5-R) \\ & & 0 & 2(10-R) & -L(5-R) & 0 \\ & & & 0 & 0 & 2(20+R) \\ & & \text{Symmetric} & & 0 & -L(5+R) \\ & & & & & 0 \end{bmatrix}$$

and

$$\boldsymbol{C}_{c_1} = \frac{c_1 L}{840} \begin{bmatrix} 4(70+7R+R^2) & L(35+7R+2R^2) & 0 & 4(35-7R-R^2) & -L(35-7R-2R^2) & 0 \\ & L^2(7+R^2) & 0 & L(35-7R-2R^2) & -L^2(7-R^2) & 0 \\ & & 0 & 0 & 0 & 0 \\ & & & 4(70+7R+R^2) & -L(35+7R+2R^2) & 0 \\ & \text{Symmetric} & & & L^2(7+R^2) & 0 \\ & & & & & 0 \end{bmatrix}$$

$$\boldsymbol{C}_{c_2} = \frac{c_2}{30L} \begin{bmatrix} 36R^2 & -3L(5R-6R^2) & 0 & -36R^2 & -3L(5R-6R^2) & 0 \\ & L^2(10-15R+9R^2) & 0 & 3L(5R-6R^2) & L^2(5-15R+9R^2) & 0 \\ & & 0 & 0 & 0 & 0 \\ & & & 36R^2 & 3L(5R-6R^2) & 0 \\ & \text{Symmetric} & & & L^2(10-15R+9R^2) & 0 \\ & & & & & 0 \end{bmatrix}$$

$$\boldsymbol{C}_{c_3} = \frac{c_3 L}{6} \begin{bmatrix} 0 & 0 & 0 & 0 & 0 & 0 \\ & 0 & 0 & 0 & 0 & 0 \\ & & 2 & 0 & 0 & 1 \\ & & & 0 & 0 & 0 \\ & \text{Symmetric} & & & 0 & 0 \\ & & & & & 2 \end{bmatrix} \tag{11.B.3}$$

and

$$\boldsymbol{K}_{\kappa GA} = \frac{\kappa GA_{eq}}{4L}(1-R)^2 \begin{bmatrix} 4 & 2L & 0 & -4 & 2L & 0 \\ & L^2 & 0 & -2L & L^2 & 0 \\ & & 0 & 0 & 0 & 0 \\ & & & 4 & -2L & 0 \\ & \text{Symmetric} & & & L^2 & 0 \\ & & & & & 0 \end{bmatrix}$$

$$\boldsymbol{K}_{EI} = \frac{EI_{eq}}{L^3} \begin{bmatrix} 12R^2 & 6LR^2 & 0 & -12R^2 & 6LR^2 & 0 \\ & L^2(1+3R^2) & 0 & -6LR^2 & -L^2(1-3R^2) & 0 \\ & & 0 & 0 & 0 & 0 \\ & & & 12R^2 & -6LR^2 & 0 \\ & \text{Symmetric} & & & L^2(1+3R^2) & 0 \\ & & & & & 0 \end{bmatrix}$$

$$\boldsymbol{K}_{GJ+Pr_g^2} = \frac{GJ_{eq}+Pr_g^2}{L} \begin{bmatrix} 0 & 0 & 0 & 0 & 0 & 0 \\ & 0 & 0 & 0 & 0 & 0 \\ & & 1 & 0 & 0 & -1 \\ & & & 0 & 0 & 0 \\ & \text{Symmetric} & & & 0 & 0 \\ & & & & & 1 \end{bmatrix}$$

$$\boldsymbol{K}_{K_T} = \frac{K_{Teq}}{L} \begin{bmatrix} 0 & 0 & 0 & 0 & 0 & 0 \\ & 0 & 1 & 0 & 0 & -1 \\ & & 0 & 0 & -1 & 0 \\ & & & 0 & 0 & 0 \\ & \text{Symmetric} & & & 0 & 1 \\ & & & & & 0 \end{bmatrix} \tag{11.B.4}$$

$$\boldsymbol{K}_{Py_{\text{CA}}} = \frac{Py_{\text{CA}}}{L} \begin{bmatrix} 0 & 0 & 1 & 0 & 0 & -1 \\ & 0 & 0 & 0 & 0 & 0 \\ & & 0 & -1 & 0 & 0 \\ & & & 0 & 0 & 1 \\ & \text{Symmetric} & & & 0 & 0 \\ & & & & & 0 \end{bmatrix}$$

$$\boldsymbol{K}_{P} = \frac{P}{60L} \begin{bmatrix} 12(5+R^2) & 6LR^2 & 0 & -12(5+R^2) & 6LR^2 & 0 \\ & L^2(5+3R^2) & 0 & -6LR^2 & -L^2(5-3R^2) & 0 \\ & & 0 & 0 & 0 & 0 \\ & & & 12(5+R^2) & -6LR^2 & 0 \\ & \text{Symmetric} & & & L^2(5+3R^2) & 0 \\ & & & & & 0 \end{bmatrix}$$

12

Dynamics of Periodic Lattice Structures

Lattice structures are usually composed of many identically constructed cell units connected end-to-end to form spatially periodic arrays, ranging in size from nanometers to many kilometers. A carbon nanotube is an example of a periodic nanostructure, while the space station is an example of a large periodic space structure. In general, most of the periodic lattice structures are very flexible, containing a huge number of elements and nodes. Thus, a significant computation cost is inevitable in order to obtain sufficiently reliable dynamic characteristics, through the use of matrix methodologies such as the finite element method (FEM). In this chapter, two alternative solution methods are introduced: the continuum modeling method and the spectral transfer matrix method. In the continuum modeling method, a periodic lattice structure is transformed into an equivalent continuum model by determining the effective continuum structural properties in an appropriate way, which enables us to apply existing solution methods directly to the equivalent continuum model. The spectral transfer matrix method is a spectral element method (SEM) based transfer matrix method useful for beam-like one-dimensional (1-D) lattice structures. In this method, we first derive transfer matrices for each of the lattice cells in a lattice structure by using SEM and then use the classical transfer matrix method to conduct dynamic analysis for the whole lattice structure.

12.1 Continuum Modeling Method

In the continuum modeling approach, a periodic lattice structure is transformed into an equivalent continuum model by determining the effective structural properties of the equivalent continuum model from the appropriate relationshipships between the geometric and material properties of a periodic lattice structure and its equivalent continuum model. Because there can be numerous concepts and approaches to determining equivalent structural properties in terms of the geometric and material properties of the original periodic lattice structure, the continuum modeling approach will not be unique in nature. In the literature, the equivalence of a periodic lattice structure and its equivalent continuum model has been established with the

Spectral Element Method in Structural Dynamics Usik Lee

requirement that both have the same kinetic and strain energies [79, 129], constitutive relationships [125], static deformation characteristics [152], or dispersion curves [1] when they take identical global dynamic behaviors. In this section, we will consider a continuum modeling method that was developed based on the transfer matrix equivalence [80]. Figure 12.1 shows some typical examples of beam-like (1-D) periodic lattice structures and their representative lattice cells. As the continuum modeling method introduced in this chapter can be readily extended to the 2-D and 3-D periodic lattice structures, we will consider a periodic plane truss, as shown in Figure 12.1(b), as an illustrative example.

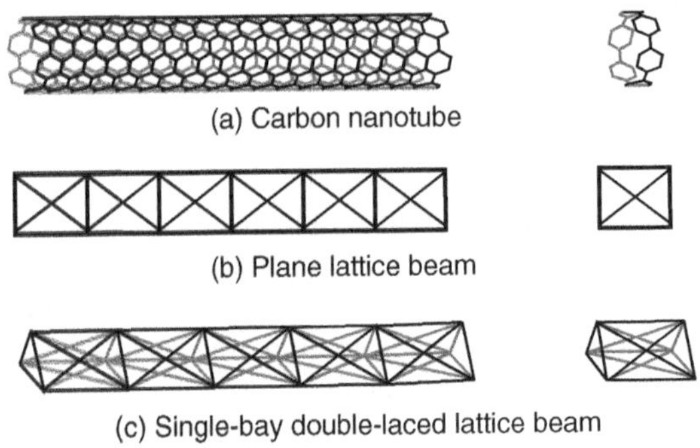

Figure 12.1 Examples of beam-like periodic lattice structures

12.1.1 Transfer Matrix for the Representative Lattice Cell (RLC)

12.1.1.1 RLC to Continuum Beam Element

Figure 12.2(a) shows a representative lattice cell (RLC) isolated from the example periodic plane truss. The RLC consists of a total of six structure members connecting four corner nodes and it has length L and height H. We will consider a continuum modeling method by which the RLC can be transformed into an equivalent continuum beam element as shown in Figure 12.2(b). The global shear deformation will be significant for discrete structures such as

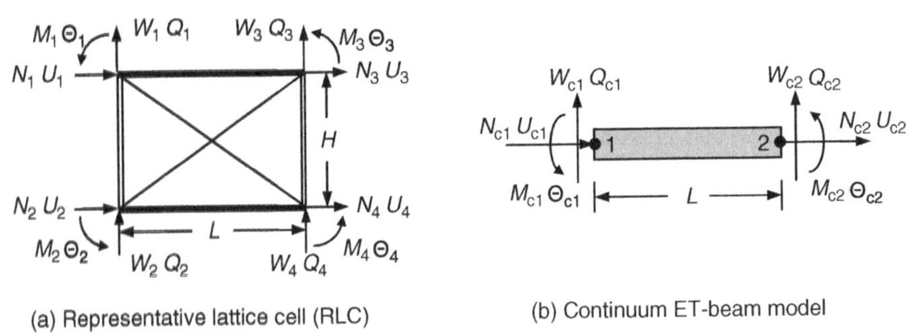

Figure 12.2 Continuum modeling of an RLC isolated from a plane truss

the periodic lattice structures, and the couplings between the axial, shear, and bending deformations may occur if a periodic lattice structure is not perfectly axisymmetric about its central axis. Thus, to take account of the global shear deformation and the couplings between the axial, shear, and bending deformations, the extended Timoshenko beam (ET-beam) model will be adopted as the equivalent continuum beam element. The ET-beam model is a generalization of the classical Timoshenko-beam model to include the couplings between the axial, shear, and bending deformations, which will be detailed in the next section.

12.1.1.2 Spectral Beam Element Representation of the Structure Members in RLC

The structure members in RLC can be represented by the rod models (axial bar models) or beam models, depending on the type of joints at each corner of the RLC. Assuming that the structural members in the RLC shown in Figure 12.2(a) are all clamped at the joints, we will represent them by the beam models, which take the longitudinal (axial) and bending vibrations simultaneously. The spectral element model for the axial–bending beam model can be obtained by combining the spectral rod element given by Equation 3.39 and the spectral beam element given by Equation 3.59 in the form

$$\bar{s}(\omega)\bar{\boldsymbol{d}} = \bar{\boldsymbol{f}} \tag{12.1}$$

where

$$\begin{aligned} \bar{\boldsymbol{d}} &= \{\bar{U}_1 \quad \bar{W}_1 \quad \bar{\Theta}_1 \quad \bar{U}_2 \quad \bar{W}_2 \quad \bar{\Theta}_2\}^{\mathrm{T}} \\ \bar{\boldsymbol{f}} &= \{\bar{N}_1 \quad \bar{Q}_1 \quad \bar{M}_1 \quad \bar{N}_2 \quad \bar{Q}_2 \quad \bar{M}_2\}^{\mathrm{T}} \end{aligned} \tag{12.2}$$

and

$$\bar{s}(\omega) = \begin{bmatrix} s_{11}^R & 0 & 0 & s_{12}^R & 0 & 0 \\ 0 & s_{11}^B & s_{12}^B & 0 & s_{13}^B & s_{14}^B \\ 0 & s_{21}^B & s_{22}^B & 0 & s_{23}^B & s_{24}^B \\ s_{21}^R & 0 & 0 & s_{22}^R & 0 & 0 \\ 0 & s_{31}^B & s_{32}^B & 0 & s_{33}^B & s_{34}^B \\ 0 & s_{41}^B & s_{42}^B & 0 & s_{43}^B & s_{44}^B \end{bmatrix} \tag{12.3}$$

where the superscripts R and B denote the components of the spectral element matrices for the rod element and Bernoulli–Euler beam element, respectively, the over-bars denote the quantities measured with respect to the local (or element) coordinates, $\bar{\boldsymbol{d}}$ is the spectral nodal degrees-of-freedom (DOFs) vector, and $\bar{\boldsymbol{f}}$ is the spectral nodal forces and moments vector: $\bar{U}_i$, $\bar{W}_i$, and $\bar{\Theta}_i$ ($i = 1, 2, 3, 4$) are the spectral components of the nodal axial displacements, transverse displacements, and rotations (slopes), respectively, and $\bar{N}_i$, $\bar{Q}_i$, and $\bar{M}_i$ ($i = 1, 2, 3, 4$) are the spectral components of the nodal resultant axial forces, transverse shear forces, and bending moments, respectively.

The spectral nodal DOFs $\bar{\boldsymbol{d}}$ in the local coordinates can be related to the spectral nodal DOFs $\boldsymbol{d}$ in the global (or reference) coordinates as

$$\bar{\boldsymbol{d}} = \boldsymbol{T}_{l/g}\boldsymbol{d} \tag{12.4}$$

where $T_{l/g}$ is the "local-to-global" coordinate transformation matrix given by

$$T_{1/g} = \begin{bmatrix} \cos\phi & \sin\phi & 0 & 0 & 0 & 0 \\ -\sin\phi & \cos\phi & 0 & 0 & 0 & 0 \\ 0 & 0 & 1 & 0 & 0 & 0 \\ 0 & 0 & 0 & \cos\phi & \sin\phi & 0 \\ 0 & 0 & 0 & -\sin\phi & \cos\phi & 0 \\ 0 & 0 & 0 & 0 & 0 & 1 \end{bmatrix} \tag{12.5}$$

where ϕ is the rotation angle of the local coordinates measured counterclockwise from the global coordinates. The coordinate transformation matrix $T_{l/g}$ is an orthogonal matrix satisfying $T_{1/g}^{-1} = T_{1/g}^{T}$. Similarly, the spectral nodal forces and moments $\bar{f}$ in the local coordinates can be related to the spectral nodal forces and moments f in the global coordinates by using the same transformation matrix T as

$$\bar{f} = T_{1/g} f \tag{12.6}$$

By substituting Equations 12.4 and 12.6 into Equation 12.1, we can obtain the spectral element equation in the global coordinates as follows:

$$s(\omega) d = f \tag{12.7}$$

where

$$s(\omega) = T_{1/g}^{T} \bar{s}(\omega) T_{1/g} \tag{12.8}$$

and

$$\begin{aligned} d &= \{U_1 \quad W_1 \quad \Theta_1 \quad U_2 \quad W_2 \quad \Theta_2\}^{T} \\ f &= \{N_1 \quad Q_1 \quad M_1 \quad N_2 \quad Q_2 \quad M_2\}^{T} \end{aligned} \tag{12.9}$$

It is worth mentioning that, if a structure member is pin connected at its joints, the structure member may be represented by the rod element. In this instance, the rotational DOFs and the associated bending moments will all be removed from Equation 12.7.

12.1.1.3 Spectral Element Modeling of the RLC in Global Nodal DOFs

The example RLC consists of six structure members: two parallel members, two vertical members, and two diagonal members. Each structure member can be modeled by using one spectral beam element represented by Equation 12.7. The spectral nodal DOFs, forces, and moments in the global coordinates are specified at each node of the RLC in Figure 12.2(a). By assembling six spectral beam elements for six structure members, we can readily obtain the global system matrix equation in the form as

$$S_{g}(\omega) d_{g} = f_{g} \tag{12.10}$$

where $S(\omega)$ is the assembled global dynamic stiffness matrix, d_g is the global spectral nodal DOFs vector, and f_g is the global spectral nodal forces and moments vector. By rearranging the order of the nodal DOFs and forces, Equation 12.10 can be written in a partitioned form as

$$\begin{bmatrix} S_{gLL}(\omega) & S_{gLR}(\omega) \\ S_{gRL}(\omega) & S_{gRR}(\omega) \end{bmatrix} \begin{Bmatrix} d_{gL} \\ d_{gR} \end{Bmatrix} = \begin{Bmatrix} f_{gL} \\ f_{gR} \end{Bmatrix} \tag{12.11}$$

where

$$\begin{aligned} \boldsymbol{d}_{\mathrm{gL}} &= \{U_1 \quad W_1 \quad \Theta_1 \quad U_2 \quad W_2 \quad \Theta_2\}^{\mathrm{T}} \\ \boldsymbol{d}_{\mathrm{gR}} &= \{U_3 \quad W_3 \quad \Theta_3 \quad U_4 \quad W_4 \quad \Theta_4\}^{\mathrm{T}} \\ \boldsymbol{f}_{\mathrm{gL}} &= \{N_1 \quad Q_1 \quad M_1 \quad N_2 \quad Q_2 \quad M_2\}^{\mathrm{T}} \\ \boldsymbol{f}_{\mathrm{gR}} &= \{N_3 \quad Q_3 \quad M_3 \quad N_4 \quad Q_4 \quad M_4\}^{\mathrm{T}} \end{aligned} \tag{12.12}$$

The subscripts L and R denote the spectral values at the nodes on the left-section and the right-section of the RLC, respectively.

12.1.1.4 Spectral Element Modeling of the RLC in Continuum Nodal DOFs

The RLC will be transformed in an equivalent continuum ET-beam element. We assume that the centerline of the continuum ET-beam element is aligned with the centerline of the RLC. The spectral nodal DOFs for the continuum ET-beam element are defined in Figure 12.2(b) and are denoted by

$$\boldsymbol{d}_{\mathrm{cL}} = \{U_{\mathrm{c1}} \quad W_{\mathrm{c1}} \quad \Theta_{\mathrm{c1}}\}^{\mathrm{T}} \qquad \boldsymbol{d}_{\mathrm{cR}} = \{U_{\mathrm{c2}} \quad W_{\mathrm{c2}} \quad \Theta_{\mathrm{c2}}\}^{\mathrm{T}} \tag{12.13}$$

where $\boldsymbol{d}_{\mathrm{cL}}$ and $\boldsymbol{d}_{\mathrm{cR}}$ are the continuum nodal DOFs specified at the left and right nodes of the continuum ET-beam element, respectively. Similarly, the spectral nodal forces and moments are defined by

$$\boldsymbol{f}_{\mathrm{cL}} = \{N_{\mathrm{c1}} \quad Q_{\mathrm{c1}} \quad M_{\mathrm{c1}}\}^{\mathrm{T}} \qquad \boldsymbol{f}_{\mathrm{cR}} = \{N_{\mathrm{c2}} \quad Q_{\mathrm{c2}} \quad M_{\mathrm{c2}}\}^{\mathrm{T}} \tag{12.14}$$

where $\boldsymbol{f}_{\mathrm{cL}}$ and $\boldsymbol{f}_{\mathrm{cR}}$ are the continuum nodal forces and moments specified at the right and left nodes of the continuum ET-beam element, respectively.

The spectral nodal DOFs vector $\boldsymbol{d}_{\mathrm{gL}}$ can be related to the continuum nodal DOFs $\boldsymbol{d}_{\mathrm{cL}}$ as

$$\boldsymbol{d}_{\mathrm{gL}} = \boldsymbol{T}_{\mathrm{g/c}} \boldsymbol{d}_{\mathrm{cL}} \tag{12.15}$$

where

$$\boldsymbol{T}_{\mathrm{g/c}} = \begin{bmatrix} 1 & 0 & -0.5H \\ 0 & 1 & 0 \\ 0 & 0 & 1 \\ 1 & 0 & +0.5H \\ 0 & 1 & 0 \\ 0 & 0 & 1 \end{bmatrix} \tag{12.16}$$

Similarly, $\boldsymbol{d}_{\mathrm{gR}}$ can be related to the continuum nodal DOFs $\boldsymbol{d}_{\mathrm{cR}}$ as

$$\boldsymbol{d}_{\mathrm{gR}} = \boldsymbol{T}_{\mathrm{g/c}} \boldsymbol{d}_{\mathrm{cR}} \tag{12.17}$$

From Equations 12.15 and 12.17, we obtain

$$\begin{Bmatrix} \boldsymbol{d}_{\mathrm{gL}} \\ \boldsymbol{d}_{\mathrm{gR}} \end{Bmatrix} = \begin{bmatrix} \boldsymbol{T}_{\mathrm{g/c}} & 0 \\ 0 & \boldsymbol{T}_{\mathrm{g/c}} \end{bmatrix} \begin{Bmatrix} \boldsymbol{d}_{\mathrm{cL}} \\ \boldsymbol{d}_{\mathrm{cR}} \end{Bmatrix} \tag{12.18}$$

The spectral nodal forces and moments of the continuum ET-beam element can be related to the spectral nodal forces and moments of the RLC as

$$\mathbf{f}_{\text{cL}} = \mathbf{T}^{\text{T}}_{\text{g/c}} \mathbf{f}_{\text{gL}} \tag{12.19}$$

and

$$\mathbf{f}_{\text{cR}} = \mathbf{T}^{\text{T}}_{\text{g/c}} \mathbf{f}_{\text{gR}} \tag{12.20}$$

From Equations 12.19 and 12.20, we obtain

$$\begin{Bmatrix} \mathbf{f}_{\text{cL}} \\ \mathbf{f}_{\text{cR}} \end{Bmatrix} = \begin{bmatrix} \mathbf{T}^{\text{T}}_{\text{g/c}} & \mathbf{0} \\ \mathbf{0} & \mathbf{T}^{\text{T}}_{\text{g/c}} \end{bmatrix} \begin{Bmatrix} \mathbf{f}_{\text{gL}} \\ \mathbf{f}_{\text{gR}} \end{Bmatrix} \tag{12.21}$$

By applying Equations 12.18 and 12.21 to Equation 12.11, we can obtain

$$\begin{bmatrix} \mathbf{S}_{\text{cLL}}(\omega) & \mathbf{S}_{\text{cLR}}(\omega) \\ \mathbf{S}_{\text{cRL}}(\omega) & \mathbf{S}_{\text{cRR}}(\omega) \end{bmatrix} \begin{Bmatrix} \mathbf{d}_{\text{cL}} \\ \mathbf{d}_{\text{cR}} \end{Bmatrix} = \begin{Bmatrix} \mathbf{f}_{\text{cL}} \\ \mathbf{f}_{\text{cR}} \end{Bmatrix} \tag{12.22}$$

where

$$\begin{bmatrix} \mathbf{S}_{\text{cLL}}(\omega) & \mathbf{S}_{\text{cLR}}(\omega) \\ \mathbf{S}_{\text{cRL}}(\omega) & \mathbf{S}_{\text{cRR}}(\omega) \end{bmatrix} = \begin{bmatrix} \mathbf{T}^{\text{T}}_{\text{g/c}} & \mathbf{0} \\ \mathbf{0} & \mathbf{T}^{\text{T}}_{\text{g/c}} \end{bmatrix} \begin{bmatrix} \mathbf{S}_{\text{gLL}}(\omega) & \mathbf{S}_{\text{gLR}}(\omega) \\ \mathbf{S}_{\text{gRL}}(\omega) & \mathbf{S}_{\text{gRR}}(\omega) \end{bmatrix} \begin{bmatrix} \mathbf{T}_{\text{g/c}} & \mathbf{0} \\ \mathbf{0} & \mathbf{T}_{\text{g/c}} \end{bmatrix} \tag{12.23}$$

12.1.1.5 Transfer Matrix for the RLC in Continuum Nodal DOFs

The sign convention defined for the transfer matrix is the same as that used in the strength of materials and it is shown in Figure 12.3. Comparing the sign convention used for the spectral ET-beam element with that for the transfer matrix, we find that only $\mathbf{f}_{\text{cL}}$ is negative to each other. Thus, by changing the sign of $\mathbf{f}_{\text{cL}}$ only, Equation 12.22 can be written in terms of the continuum nodal DOFs and forces defined for the transfer matrix as

$$\begin{bmatrix} \mathbf{S}_{\text{cLL}}(\omega) & \mathbf{S}_{\text{cLR}}(\omega) \\ \mathbf{S}_{\text{cRL}}(\omega) & \mathbf{S}_{\text{cRR}}(\omega) \end{bmatrix} \begin{Bmatrix} \mathbf{d}_{\text{cL}} \\ \mathbf{d}_{\text{cR}} \end{Bmatrix} = \begin{Bmatrix} -\mathbf{f}_{\text{cL}} \\ +\mathbf{f}_{\text{cR}} \end{Bmatrix} \tag{12.24}$$

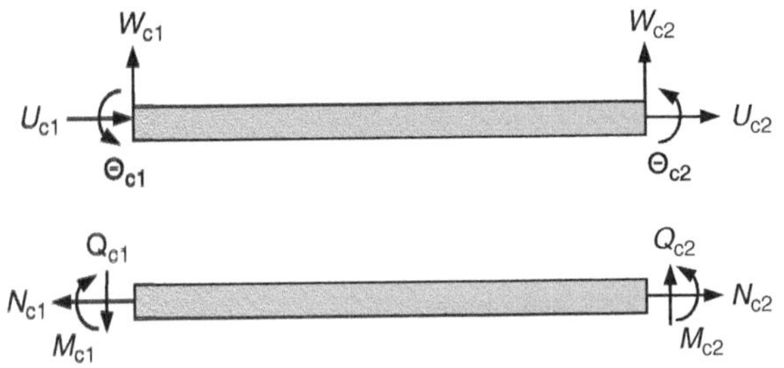

Figure 12.3 Sign convention for the transfer matrix

Equation 12.24 is the force-to-displacement relationship and it can be transformed into the input-to-output relationship as

$$\begin{Bmatrix} \boldsymbol{d}_{cR} \\ \boldsymbol{f}_{cR} \end{Bmatrix} = \boldsymbol{T}_{RLC}(\omega) \begin{Bmatrix} \boldsymbol{d}_{cL} \\ \boldsymbol{f}_{cL} \end{Bmatrix} \tag{12.25}$$

where

$$\boldsymbol{T}_{RLC}(\omega) = \begin{bmatrix} -\boldsymbol{S}_{cLR}^{-1}\boldsymbol{S}_{cLL} & -\boldsymbol{S}_{cLR}^{-1} \\ \boldsymbol{S}_{cRL} - \boldsymbol{S}_{cRR}\boldsymbol{S}_{cLR}^{-1}\boldsymbol{S}_{cLL} & -\boldsymbol{S}_{cRR}\boldsymbol{S}_{cLR}^{-1} \end{bmatrix} \tag{12.26}$$

$\boldsymbol{T}_{RLC}(\omega)$ is the transfer matrix for the RLC in the continuum nodal DOFs.

12.1.2 Transfer Matrix for an ET-Beam Element

The ET-beam model is a generalization of the classical Timoshenko-beam model to take account of the couplings between the axial, shear, and bending deformations. The dynamics of the ET-beam is represented by the equations of motion

$$\frac{\partial}{\partial x}\begin{Bmatrix} N \\ Q \\ M \end{Bmatrix} = \begin{bmatrix} \rho A & 0 & \rho R \\ 0 & \rho A & 0 \\ \rho R & 0 & \rho I \end{bmatrix} \begin{Bmatrix} \partial^2 u/\partial t^2 \\ \partial^2 w/\partial t^2 \\ \partial^2 \theta/\partial t^2 \end{Bmatrix} - \begin{Bmatrix} 0 \\ 0 \\ Q \end{Bmatrix} \tag{12.27}$$

in conjunction with the force–displacement relationships

$$\begin{Bmatrix} N \\ Q \\ M \end{Bmatrix} = \begin{bmatrix} EA & C_1 & C_2 \\ C_1 & \kappa GA & C_3 \\ C_2 & C_3 & EI \end{bmatrix} \begin{Bmatrix} \partial u/\partial x \\ \partial w/\partial x - \theta \\ \partial \theta/\partial x \end{Bmatrix} \tag{12.28}$$

where $u(x, t)$, $w(x, t)$, and $\theta(x, t)$ are the axial displacement, transverse displacement, and slope, respectively, and $N(x, t)$, $Q(x, t)$ and $M(x, t)$ are the axial tensile force, transverse shear force, and bending moment, respectively. The dot $(\cdot)$ and the prime $(')$ denote the partial derivatives with respect to the time t and the spatial coordinate x, respectively. EA is the extensional rigidity, κGA is the shear rigidity, EI is the bending (flexural) rigidity, C_k $(k = 1, 2, 3)$ are the coupling rigidities, and ρA, ρR, and ρI are the mass inertia properties defined by

$$(\rho A,\ \rho R,\ \rho I) = \int_z \rho(1,\ z,\ z^2) b \mathrm{d}z \tag{12.29}$$

where b is the width of the ET-beam and z is the coordinate in the thickness direction of the ET-beam.

Represent the dynamic responses in the spectral forms as

$$\{u(x,t),\ w(x,t),\ \theta(x,t)\} = \frac{1}{N}\sum_{n=0}^{N-1}\{U_n(x;\omega_n),\ W_n(x;\omega_n),\ \Theta_n(x;\omega_n)\}\mathrm{e}^{i\omega_n t} \tag{12.30}$$

Similarly, we represent the axial force, the transverse shear force, and the bending moment in the spectral forms as

$$\{N(x,t),\ Q(x,t),\ M(x,t)\} = \frac{1}{N}\sum_{n=0}^{N-1}\{N_n(x;\omega_n),\ Q_n(x;\omega_n),\ M_n(x;\omega_n)\}\mathrm{e}^{i\omega_n t} \tag{12.31}$$

Substitution of Equations 12.30 and 12.31 into Equations 12.27 and 12.28 gives

$$\frac{\mathrm{d}}{\mathrm{d}x}\begin{Bmatrix} N \\ Q \\ M \end{Bmatrix} = -\omega^2 \begin{bmatrix} \rho A & 0 & \rho R \\ 0 & \rho A & 0 \\ \rho R & 0 & \rho I \end{bmatrix} \begin{Bmatrix} U \\ W \\ \Theta \end{Bmatrix} - \begin{Bmatrix} 0 \\ 0 \\ Q \end{Bmatrix} \tag{12.32}$$

and

$$\begin{Bmatrix} N \\ Q \\ M \end{Bmatrix} = \begin{bmatrix} EA & C_1 & C_2 \\ C_1 & \kappa GA & C_3 \\ C_2 & C_3 & EI \end{bmatrix} \begin{Bmatrix} \mathrm{d}U/\mathrm{d}x \\ \mathrm{d}W/\mathrm{d}x - \Theta \\ \mathrm{d}\Theta/\mathrm{d}x \end{Bmatrix} \tag{12.33}$$

In Equations 12.32 and 12.33, the subscripts n are omitted for brevity.

We define a state-vector as

$$\boldsymbol{y}(x) = \{U(x) \quad W(x) \quad \Theta(x) \quad N(x) \quad Q(x) \quad M(x)\}^{\mathrm{T}} \tag{12.34}$$

By using the state-vector $\boldsymbol{y}(x)$, a state-vector equation can be derived from Equations 12.32 and 12.33 as

$$\frac{\mathrm{d}\boldsymbol{y}}{dx} = \boldsymbol{A}_{\mathrm{ETB}}(\omega)\boldsymbol{y} \tag{12.35}$$

where

$$\boldsymbol{A}_{\mathrm{ETB}}(\omega) = \begin{bmatrix} \begin{matrix} 0 & 0 & 0 \\ 0 & 0 & 1 \\ 0 & 0 & 0 \end{matrix} & \begin{bmatrix} EA & C_1 & C_2 \\ C_1 & \kappa GA & C_3 \\ C_2 & C_3 & EI \end{bmatrix}^{-1} \\ \begin{matrix} -\rho A\omega^2 & 0 & -\rho R\omega^2 \\ 0 & -\rho A\omega^2 & 0 \\ -\rho R\omega^2 & 0 & -\rho I\omega^2 \end{matrix} & \begin{matrix} 0 & 0 & 0 \\ 0 & 0 & 0 \\ 0 & -1 & 0 \end{matrix} \end{bmatrix} \tag{12.36}$$

Equation 12.36 can be solved for $\boldsymbol{y}(x)$ as

$$\boldsymbol{y}(x) = \mathrm{e}^{\boldsymbol{A}_{\mathrm{ETB}}(\omega)x}\boldsymbol{y}(0) \tag{12.37}$$

Thus we can relate the state-vector $\boldsymbol{y}(L)$ at $x = L$ to the state vector $\boldsymbol{y}(0)$ at $x = 0$ as follows:

$$\boldsymbol{y}(L) = \boldsymbol{T}_{\mathrm{ETB}}(\omega)\boldsymbol{y}(0) \tag{12.38}$$

where

$$\boldsymbol{T}_{\mathrm{ETB}}(\omega) = \mathrm{e}^{\boldsymbol{A}_{\mathrm{ETB}}(\omega)L} \tag{12.39}$$

The matrix $\boldsymbol{T}_{\mathrm{ETB}}(\omega)$ is the transfer matrix for the continuum ET-beam element of length L.

12.1.3 Determination of Equivalent Continuum Structural Properties

The equivalent structural properties of the continuum ET-beam model are determined by assuming the equivalence between the transfer matrix $\boldsymbol{T}_{\mathrm{RLC}}(\omega)$ for the RLC and transfer matrix

$\boldsymbol{T}_{\mathrm{ETB}}(\omega)$ for the equivalent continuum ET-beam element. Thus, we equate $\boldsymbol{T}_{\mathrm{ETB}}(\omega)$ with $\boldsymbol{T}_{\mathrm{RLC}}(\omega)$ as follows:

$$\boldsymbol{T}_{\mathrm{ETB}}(\omega) = \boldsymbol{T}_{\mathrm{RLC}}(\omega) \tag{12.40}$$

or

$$e^{\boldsymbol{A}_{\mathrm{ETB}}(\omega)L} = \boldsymbol{T}_{\mathrm{RLC}}(\omega) \tag{12.41}$$

From Equation 12.41, we obtain

$$\boldsymbol{A}_{\mathrm{ETB}}(\omega) = \frac{1}{L}\ln \boldsymbol{T}_{\mathrm{RLC}}(\omega) \tag{12.42}$$

The matrix $\boldsymbol{A}_{\mathrm{ETB}}(\omega)$ in the left-hand side of Equation 12.42 contains unknown equivalent structural properties of the equivalent continuum ET-beam model, while the right-hand side of Equation 12.42 is known because it can be numerically computed by using $\boldsymbol{T}_{\mathrm{RLC}}(\omega)$ given by Equation 12.26. Owing to the equality of Equation 12.42, we assume that the right-hand side of Equation 12.42 is numerically computed in the form of $\boldsymbol{A}_{\mathrm{ETB}}(\omega)$ as

$$\frac{1}{L}\ln \boldsymbol{T}_{\mathrm{RLC}}(\omega) = \begin{bmatrix} 0 & 0 & 0 & \begin{bmatrix} a_1 & a_2 & a_3 \\ a_2 & a_4 & a_5 \\ a_3 & a_5 & a_6 \end{bmatrix}^{-1} \\ 0 & 0 & 1 & \\ 0 & 0 & 0 & \\ b_1 & 0 & b_2 & \begin{matrix} 0 & 0 & 0 \end{matrix} \\ 0 & b_1 & 0 & \begin{matrix} 0 & 0 & 0 \end{matrix} \\ b_2 & 0 & b_2 & \begin{matrix} 0 & -1 & 0 \end{matrix} \end{bmatrix} \tag{12.43}$$

The logarithm of $\boldsymbol{T}_{\mathrm{RLC}}$ can be computed from [74]:

$$\ln \boldsymbol{T}_{\mathrm{RLC}} = \boldsymbol{U}(\ln\Lambda)\boldsymbol{U}^{-1} \tag{12.44}$$

where each column of $\boldsymbol{U}$ is an eigenvalue of $\boldsymbol{T}_{\mathrm{RLC}}$ and Λ is the diagonal matrix whose diagonal elements are the eigenvalues of $\boldsymbol{T}_{\mathrm{RLC}}$.

By substituting Equations 12.36 and 12.43 into the left- and right-hand sides of Equation 12.42, respectively, we can obtain

$$\begin{aligned} &EA = a_1 \qquad \kappa GA = a_4 \qquad EI = a_6 \\ &C_1 = a_2 \qquad C_2 = a_3 \qquad C_3 = a_5 \\ &\rho A = -\frac{b_1}{\omega^2} \quad \rho R = -\frac{b_2}{\omega^2} \quad \rho I = -\frac{b_3}{\omega^2} \end{aligned} \tag{12.45}$$

That is, the equivalent continuum structural properties can be determined from Equation 12.45 by using numerically computed values a_i ($i = 1, 2, \ldots, 6$) and b_i ($i = 1, 2, 3$).

Example 12.1

Consider a cantilevered plane lattice beam with 20 repeating lattice cells as shown in Figure 12.4(a). The RLC and its geometric and material properties are displayed in Figure 12.4(b). As the structure members within a repeating lattice cell are all clamped at the joints, they are represented by the beam models to derive the continuum ET-beam model.

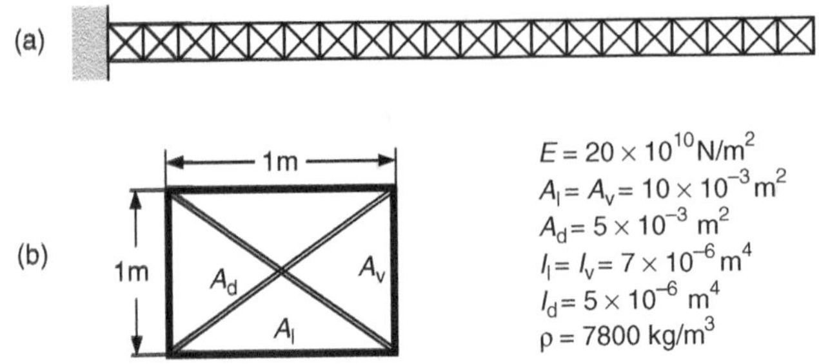

Figure 12.4 Geometric and material properties of an example plane lattice beam

Figure 12.5 shows the equivalent structural properties for the continuum ET-beam model of an RLC. As the RLC is geometrically symmetric with respect to its central axis, the coupling structural rigidities are all zero and thus they are not displayed in Figure 12.5. The equivalent structural properties seem to depend slightly on the frequency and, as shown in Figure 12.5, they have infinite values at the natural frequencies of each structure member within the RLC. This implies that the continuum modeling approach is valid only at the frequency below the lowest natural frequency of the structure members within a lattice structure.

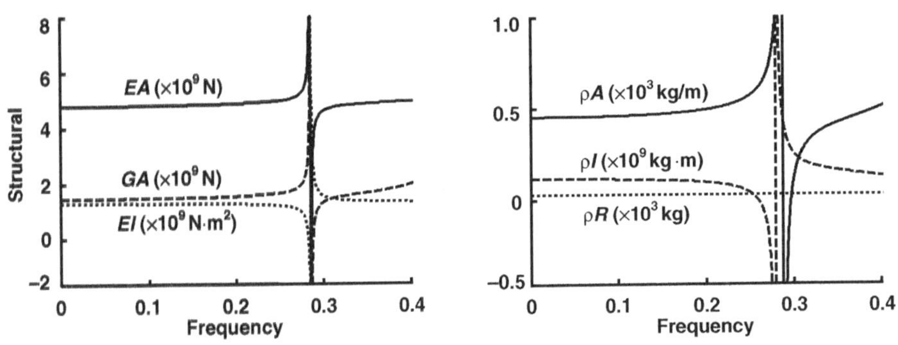

Figure 12.5 Equivalent structural properties of the continuum ETB model

The natural frequencies of the example plane lattice beam obtained by using the finite element method (FEM), the spectral element method (SEM) and the present continuum modeling method are compared in Table 12.1. For the FEM results, each structure member within a repeating lattice cell is represented by one finite element, two finite elements or four finite elements in order to improve the solution accuracy. For the exact results obtained by SEM, of course, one spectral element suffices to represent each structure member in an RLC. The equivalent structural properties at zero frequency are used to compute the natural frequencies of the continuum ET-beam model. It is obvious from Table 12.1 that the continuum ET-beam model provides reliable results, mainly within less than 10% error at low vibration modes.

Figure 12.6 compares the non-dimensional transverse displacement at the tip of the example plane lattice beam obtained by using the full-scale spectral element analysis with that obtained

Table 12.1 Natural frequencies of the example plane lattice beam

	FEM			SEM	
Natural frequency (Hz)	One-element model	Two-element model	Four-element model	One-element model	Continuum modeling method (present)
$f_1^{(b)}$	2.4940	2.4680	2.4680	2.4680	2.5059
$f_2^{(b)}$	14.998	14.977	14.957	14.949	14.975
$f_3^{(b)}$	37.235	36.378	36.362	36.345	39.219
$f_4^{(a)}$	39.175	38.925	38.915	38.905	41.920
$f_5^{(b)}$	70.428	69.908	69.738	69.398	70.778
$f_6^{(b)}$	100.95	98.40	98.190	97.950	107.16
$f_7^{(a)}$	133.35	126.73	125.49	125.35	125.76

Note: (b) = bending modes; (a) = axial modes.

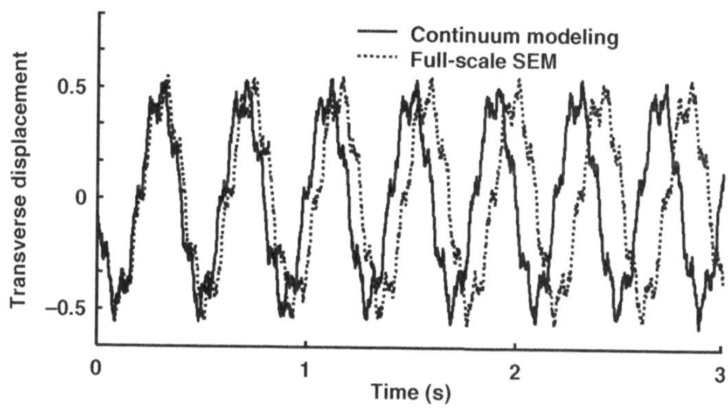

Figure 12.6 Dynamic responses of the example plane lattice beam

by using the continuum ETB model when an impulse force is applied at the tip of the lattice beam.

12.2 Spectral Transfer Matrix Method

As one of the important advantages of the SEM over the conventional FEM, the one spectral element modeling suffices for acquiring accurate solutions for a uniform structure element, regardless of its length. This will result in a drastic reduction in the total number of DOFs used in the SEM, reducing the computer capacity and computation time significantly. However, it is still inevitable that a huge number of DOFs have to be dealt with, even when we apply SEM to large beam-like (1-D) lattice structures that consist of many lattice cells. In the literature, the transfer matrix method (TMM) has been applied to 1-D structures as an efficient analysis method, with success [138]. Thus, by combining SEM and TMM, we develop a new dynamic analysis method called the "spectral transfer matrix method (STMM)" for the lattice structures that consists of many 1-D lattice substructures.

12.2.1 Transfer Matrix for a Lattice Cell

As an illustrative example, we will consider a cross-type plane lattice structure as shown in Figure 12.7(a). The cross-type plane lattice structure consists of four beam-like 1-D lattice substructures that are connected to a lattice cell placed at the junction. Each lattice substructure consists of many lattice cells. We can apply the SEM to such 1-D lattice structures in conjunction with the TMM in order to develop the STMM, which will be detailed in the following.

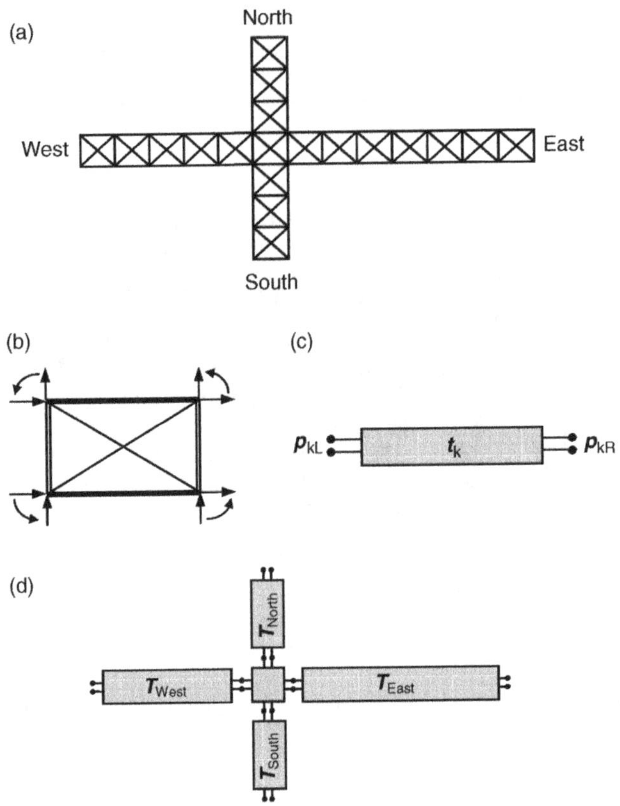

Figure 12.7 General procedure of the spectral transfer matrix method

Figure 12.7(b) shows a lattice cell isolated from a 1-D lattice substructure. The lattice cell consists of six lattice members: two longitudinal bars, two diagonal bars and two vertical bars. Each lattice member can be represented by the spectral beam element model or the spectral rod element model depending on whether the lattice members are welded or pin-connected at the joints. Both the spectral beam element model and the spectral rod element model are available in Chapter 3.

Following the general procedure introduced in Section 12.1.1, the spectral element modeling of the kth lattice cell can be accomplished by assembling the spectral element models of all lattice members in the global coordinates in the form of

$$\begin{bmatrix} S_{LL}(\omega) & S_{LR}(\omega) \\ S_{RL}(\omega) & S_{RR}(\omega) \end{bmatrix}_k \begin{Bmatrix} d_L \\ d_R \end{Bmatrix}_k = \begin{Bmatrix} f_L \\ f_R \end{Bmatrix}_k \tag{12.46}$$

where d_L and d_R denote the collection of all spectral nodal DOFs specified at the nodes in the left-section and the right-section of the lattice cell, respectively. Similarly, f_L and f_R denote the collection of all spectral nodal forces (and moments) specified at the nodes in the left-section and the right-section of the lattice cell, respectively. Equation 12.46 can be written in the spectral nodal DOFs and forces satisfying the sign convention used for the transfer matrix as follows:

$$\begin{bmatrix} S_{LL}(\omega) & S_{LR}(\omega) \\ S_{RL}(\omega) & S_{RR}(\omega) \end{bmatrix}_k \begin{Bmatrix} d_L \\ d_R \end{Bmatrix}_k = \begin{Bmatrix} -f_L \\ +f_R \end{Bmatrix}_k \tag{12.47}$$

Equation 12.47 can then be readily transformed into the input-to-output relationship form as

$$\begin{Bmatrix} d_R \\ f_R \end{Bmatrix}_k = \begin{bmatrix} -S_{LR}^{-1}S_{LL} & -S_{LR}^{-1} \\ S_{RL} - S_{RR}S_{LR}^{-1}S_{LL} & -S_{RR}S_{LR}^{-1} \end{bmatrix}_k \begin{Bmatrix} d_L \\ f_L \end{Bmatrix}_k \tag{12.48}$$

or, simply

$$p_{kR} = t_k(\omega) p_{kL} \tag{12.49}$$

where

$$p_{kR} = \begin{Bmatrix} d_R \\ f_R \end{Bmatrix}_k = \begin{Bmatrix} d_{kR} \\ f_{kR} \end{Bmatrix}, \quad p_{kL} = \begin{Bmatrix} d_L \\ f_L \end{Bmatrix}_k = \begin{Bmatrix} d_{kL} \\ f_{kL} \end{Bmatrix} \tag{12.50}$$

and

$$t_k(\omega) = \begin{bmatrix} -S_{LR}^{-1}S_{LL} & -S_{LR}^{-1} \\ S_{RL} - S_{RR}S_{LR}^{-1}S_{LL} & -S_{RR}S_{LR}^{-1} \end{bmatrix}_k \tag{12.51}$$

The matrix $t_k(\omega)$ is the transfer matrix that is derived from the spectral element model for the kth lattice cell and it relates the state-vector p_{kL} to the state-vector p_{kR}. The state-vector p_{kL} represents the spectral nodal DOFs and forces at the left cross-section (input) and the state-vector p_{kR} at right cross-section (output) of the kth lattice cell. They satisfy

$$p_{(k-1)R} = p_{kL} \quad (k = 1, 2, \ldots N) \tag{12.52}$$

Figure 12.7(c) shows the symbolic representation of the kth lattice cell in terms of transfer matrix $t_k(\omega)$ and associated state-vectors p_{kL} and p_{kR}.

If the 1-D lattice substructure is a periodic lattice structure with N identical lattice cells, each lattice cell will have the identical transfer matrix as

$$t_1(\omega) = t_2(\omega) = \ldots = t_N(\omega) = t(\omega) \tag{12.53}$$

12.2.2 *Transfer Matrix for a 1-D Lattice Substructure*

We first give each 1-D lattice substructure as shown in Figure 12.8(a) as a series of equivalent transfer matrices representation, which is shown in Figure 12.8(b). Then, by using the concept of TMM, we can transform it into an equivalent single lattice cell representation as shown in

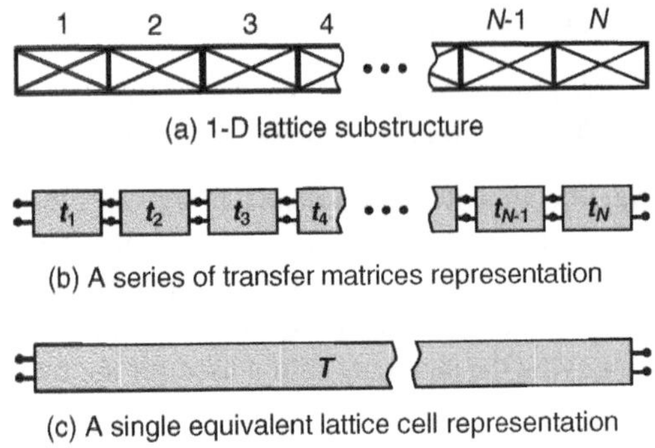

Figure 12.8 A single equivalent lattice cell representation of a 1-D lattice structure

Figure 12.8(c). To that end, we represent all lattice cells in the 1-D lattice substructure by the transfer matrix equations as

$$\begin{aligned} \boldsymbol{p}_{1\mathrm{R}} &= \boldsymbol{t}_1(\omega)\boldsymbol{p}_{1\mathrm{L}} \\ \boldsymbol{p}_{2\mathrm{R}} &= \boldsymbol{t}_2(\omega)\boldsymbol{p}_{2\mathrm{L}} \\ &\vdots \\ \boldsymbol{p}_{(N-1)\mathrm{R}} &= \boldsymbol{t}_{N-1}(\omega)\boldsymbol{p}_{(N-1)\mathrm{L}} \\ \boldsymbol{p}_{N\mathrm{R}} &= \boldsymbol{t}_N(\omega)\boldsymbol{p}_{N\mathrm{L}} \end{aligned} \tag{12.54}$$

Accordingly the 1-D lattice substructure can be symbolically represented as a series of the transfer matrices $t_k(\omega)$ ($k = 1, 2, \ldots, N$) as shown in Figure 12.8(b). By recursively using the transfer matrix equations (12.54), the state-vector $\boldsymbol{p}_{N\mathrm{R}}$ at the right-end of a 1-D lattice substructure can be related to the state-vector $\boldsymbol{p}_{1L}$ at its left-end as follows:

$$\boldsymbol{p}_{N\mathrm{R}} = \boldsymbol{T}(\omega)\boldsymbol{p}_{1\mathrm{L}} \tag{12.55}$$

where

$$\boldsymbol{T}(\omega) = \boldsymbol{t}_N(\omega)\boldsymbol{t}_{N-1}(\omega)\cdots\boldsymbol{t}_2(\omega)\boldsymbol{t}_1(\omega) \tag{12.56}$$

The matrix $\boldsymbol{T}(\omega)$ is the "global" transfer matrix for the whole 1-D lattice substructure. Thus, as shown in Figure 12.8(c), a 1-D lattice substructure can be simply represented as an equivalent single lattice cell whose transfer matrix is given by the global transfer matrix $\boldsymbol{T}(\omega)$ of Equation 12.56. If the 1-D lattice substructure is a periodic structure consisting of N identical lattice cells, the global transfer matrix $\boldsymbol{T}(\omega)$ can be further simplified as the Nth power of the transfer matrix $\boldsymbol{t}(\omega)$ of a representative lattice cell as follows:

$$\boldsymbol{T}(\omega) = \boldsymbol{t}^N(\omega) \tag{12.57}$$

12.2.3 Spectral Element Model for a 1-D Lattice Substructure

Equation 12.55 can be written in the partitioned form as

$$\begin{Bmatrix} \boldsymbol{d}_{N\mathrm{R}} \\ \boldsymbol{f}_{N\mathrm{R}} \end{Bmatrix} = \begin{bmatrix} \boldsymbol{T}_{11} & \boldsymbol{T}_{12} \\ \boldsymbol{T}_{21} & \boldsymbol{T}_{22} \end{bmatrix} \begin{Bmatrix} \boldsymbol{d}_{1\mathrm{L}} \\ \boldsymbol{f}_{1\mathrm{L}} \end{Bmatrix} \tag{12.58}$$

The spectral nodal DOFs and forces in Equation 12.58 satisfy the sign convention defined for the transfer matrix. Thus, by simply changing the sign of f_{1L} only, Equation 12.58 can be written in terms of the spectral nodal DOFs and forces defined to satisfy the sign convention for the spectral element matrix as

$$\begin{Bmatrix} d_{NR} \\ f_{NR} \end{Bmatrix} = \begin{bmatrix} T_{11} & T_{12} \\ T_{21} & T_{22} \end{bmatrix} \begin{Bmatrix} d_{1L} \\ -f_{1L} \end{Bmatrix} \tag{12.59}$$

From Equation 12.59, we can obtain the force–displacement relationship as

$$\begin{Bmatrix} f_{1L} \\ f_{NR} \end{Bmatrix} = S(\omega) \begin{Bmatrix} d_{1L} \\ d_{NR} \end{Bmatrix} \tag{12.60}$$

where

$$S(\omega) = \begin{bmatrix} -T_{12}^{-1} T_{11} & -T_{12}^{-1} \\ T_{21} - T_{22} T_{12}^{-1} T_{11} & -T_{22} T_{12}^{-1} \end{bmatrix} \tag{12.61}$$

The matrix $S(\omega)$ is the "global" spectral element matrix for the whole beam-like 1-D lattice substructure.

12.2.4 Spectral Element Model for the Whole Lattice Structure

So far, it has been shown that a beam-like 1-D lattice substructure consisting of many lattice cells can be simplified as an equivalent single lattice cell whose "global" transfer matrix and "global" spectral element matrix are given by Equations 12.56 and 12.61, respectively. This procedure will drastically reduce the total number of DOFs used in the spectral element analysis without, however, introducing any approximations or assumptions to degrade the accuracy of the solutions.

The next step is to apply the equivalent single lattice cell representation to all 1-D lattice substructures within the example cross-type plane lattice structure as shown in Figure 12.7(a). The cross-type plane lattice structure will be then transformed into an equivalent simple lattice structure representation as shown in Figure 12.7(d). The equivalent simple lattice structure representation now consists of only four equivalent single lattice cells, all connected to a lattice cell at the junction.

The original example cross-type plane lattice structure shown in Figure 12.7(a) has a total of 44 nodes (joints). If each node has three DOFs, the full-scaled spectral element model will have a total of 132 DOFs. However, as the equivalent simple lattice structure representation shown in Figure 12.7(d) will have a total of 12 nodes, the present STMM will require only total of 36 DOFs, without degrading the solution accuracy at all.

The remaining steps are: (i) to assemble all spectral elements within the equivalent simple lattice structure representation (i.e., four "global" spectral element matrices for four beam-like 1-D lattice substructures and a spectral element matrix for a lattice cell at the junction) and (ii) to conduct the spectral element analysis to compute the natural frequencies and dynamic responses in the frequency- and time-domains.

Example 12.2

As an illustrative example, we will consider a cross-type plane lattice structure as shown in Figure 12.9(a). It consists of a total of 51 identical lattice cells: 20 cells in the east wing, 10 cells

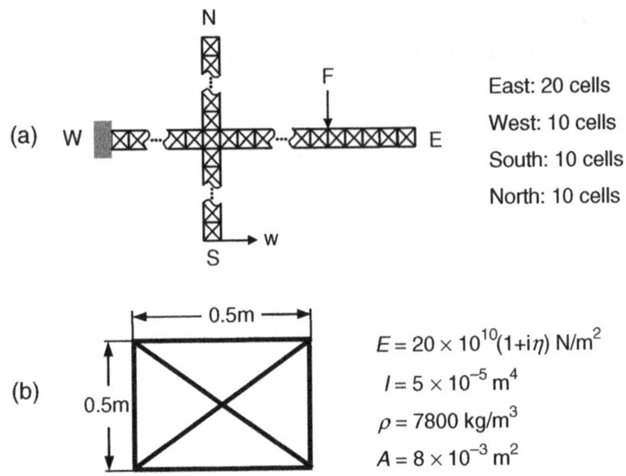

Figure 12.9 (a) A cross-type plane lattice structure and (b) its representative lattice cell

in the west wing, 10 cells in the south wing, 10 cells in the north wing, and a single cell at the junction. The end of the west wing is clamped and the ends of the other wings are free. The material and geometric properties for a representative lattice cell are displayed in Figure 12.9(b). As shown in Figure 12.9(a), an impulse force is applied at the joint between the fifth and sixth cells from the tip of the east wing, and the transverse vibration is calculated at the tip of the south wing.

Table 12.2 shows a comparison of the natural frequencies obtained by using the present STMM, SEM, and FEM. The number of total DOFs used in STMM is less than one tenth of those used in SEM and FEM. However, Table 12.2 certainly shows that the results obtained by STMM completely agree with the exact results obtained by SEM. On the other hand, although the FEM results are fairly accurate at low frequency modes for the example problem, they are shown to diverge from the exact results obtained by STMM or SEM at the high frequency modes.

Table 12.2 Natural frequencies (Hz) of the cross-type plane lattice structure

Mode	STMM (30)	SEM (306)	FEM (306)
1st	1.5882	1.5882	1.5882
2nd	7.1248	7.1248	7.1248
3rd	10.683	10.683	10.683
4th	14.619	14.619	14.619
5th	24.854	24.854	24.854
10th	89.585	89.585	89.585
20th	303.43	303.43	303.65
30th	559.66	559.66	561.36

Note: () indicates the total number of DOFs used in the analysis.

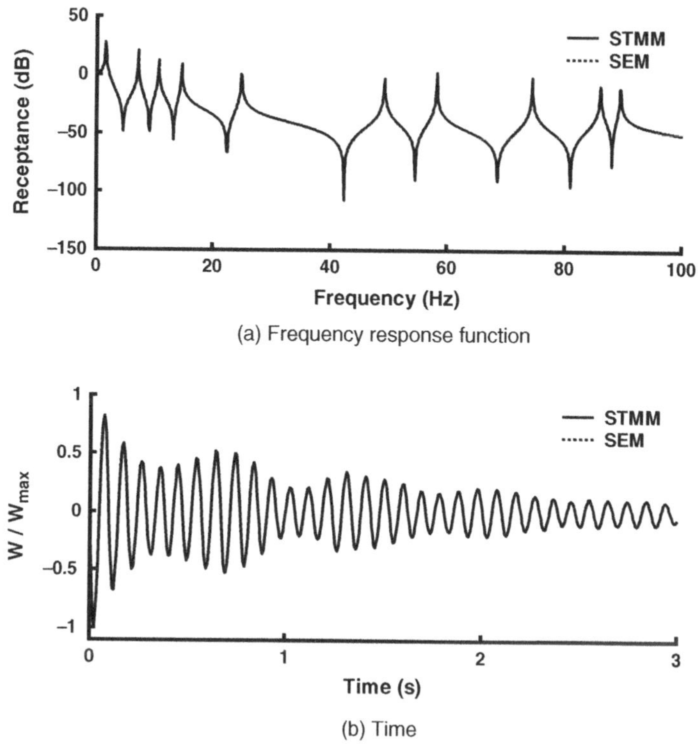

Figure 12.10 Dynamic responses of the cross-type plane lattice structure

Figure 12.10 shows the dynamic responses (transverse frequency response functions and time histories) at the tip of the south wing. A loss factor $\eta = 0.02$ is considered in the analysis. The dynamic responses obtained by the STMM are shown to be completely identical with those obtained by the full-scale spectral element analysis.

13

Biomechanics: Blood Flow Analysis

The arterial system transports blood from the heart to the tissues in the body. The blood flow characteristics (e.g., blood velocity and pressure) certainly influence the artery wall shear stress and tension, which regulate the lumen diameter and arterial wall thickness, respectively. As the cardiovascular diseases such as a stenosis and aneurysm are known to be strongly related to the artery wall shear stress and tension, it is very important to investigate the blood flow characteristics in an accurate and efficient way for the purposes of cardiovascular disease research, medical devise designs, and surgical planning. Accordingly, computational methods have emerged as powerful tools for the modeling and analysis of blood flow. The computational models of blood flow developed during the last half century can be classified into the lumped parameters models [140, 162], the one-dimensional (1-D) wave propagation models [57, 131, 149, 171], and the three-dimensional (3-D) fluid-structure interaction models [71, 153]. The 1-D models are very useful because they are able to provide clinically relevant information on local mean blood flow and pressure waves through the arterial system very efficiently in addiiotn to the suitable boundary conditions for the 3-D models. Thus, this chapter introduces a nonlinear 1-D spectral element model (SEM) for the blood flow through a blood vessel.

13.1 Governing Equations

13.1.1 One-Dimensional Blood Flow Theory

The 1-D theory of blood flow was derived in a general form by Hughes and Lubliner [57] and it consists of a continuity equation, an axial momentum balance equation, and a constitutive relationship for the flow of a Newtonian fluid in a permeable, deforming, elastic blood vessel. The continuity and momentum balance equations are given by

$$\frac{\partial A}{\partial t} + \frac{\partial Q}{\partial x} + \psi = 0 \tag{13.1}$$

Spectral Element Method in Structural Dynamics Usik Lee

$$\frac{\partial Q}{\partial t} + \frac{\partial}{\partial x}\left((1+\delta)\frac{Q^2}{A}\right) + \frac{A}{\rho}\frac{\partial P}{\partial x} = Af_{\mathrm{b}} - \nu H\frac{Q}{A} + \nu\frac{\partial^2 Q}{\partial x^2} \tag{13.2}$$

where $A(x, t)$ is the cross-sectional area of the blood vessel, $Q(x, t)$ the volumetric blood flow rate, $P(x, t)$ the blood pressure, ψ the outflow functionm which is zero for impermeable vessels, and f_{b} is the body force. The mass density and kinematic viscosity of the blood are given by ρ and ν, respectively. The parameters δ and H are determined by the blood flow velocity profile across the cross-section and they are given in Table 13.1. Notice that the last two terms in the right-hand side of Equation 13.2 are all related to the kinematic viscosity ν of the blood.

Table 13.1 The parameters δ and H for various velocity profiles

Flow type	Velocity profile	δ	H
Uniform flow	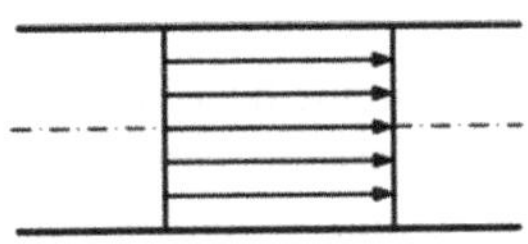	0	0
Parabolic flow	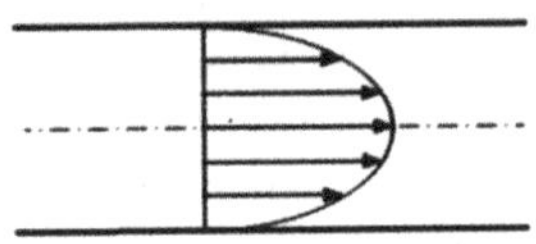	1/3	8π
Boundary layered flow	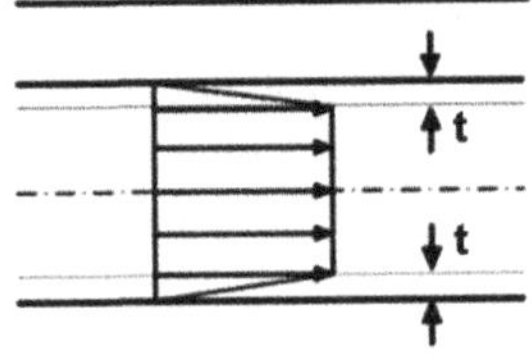	$\frac{2t}{3R}$	$\frac{2\pi R}{t}$

The constitutive relationship, which relates the blood pressure $P(x, t)$ to the cross-sectional area $A(x, t)$, is given by [131]

$$P(A(x,t),x,t) = P_{\mathrm{d}} + \frac{4}{3}\frac{Eh}{r_{\mathrm{d}}(x)}\left(1-\sqrt{\frac{A_{\mathrm{d}}(x)}{A(x,t)}}\right) \tag{13.3}$$

where E is the Young's modulus, h is the wall thickness, and $A_{\mathrm{d}}(x) = \pi r_{\mathrm{d}}(x)^2$ is the cross-sectional wetted area and $r_{\mathrm{d}}(x)$ is the inner radius of the blood vessel at the diastole pressure $P_{\mathrm{d}}(x)$. Olufsen [131] derived an empirical expression for Eh/r_{d} as being

$$\frac{Eh}{r_{\mathrm{d}}(x)} = \kappa_1\exp[\kappa_2 r_{\mathrm{d}}(x)] + \kappa_3 \cong \frac{Eh}{r_{\mathrm{a}}} = \kappa_1\exp(\kappa_2 r_{\mathrm{a}}) + \kappa_3 \tag{13.4}$$

where r_{a} is the mean radius of a tapered blood vessel, and κ_i $(i = 1, 2, 3)$ are given by

$$\begin{aligned} \kappa_1 &= 2\times 10^7\ \mathrm{g\,s^{-2}\,cm^{-1}} \\ \kappa_2 &= -22.53\ \mathrm{cm^{-1}} \\ \kappa_3 &= 8.65\times 10^5\ \mathrm{g\,s^{-2}\,cm^{-1}} \end{aligned} \tag{13.5}$$

13.1.2 Simplified Governing Equations

Ignore the body force (i.e., $f_b = 0$) and assume that the blood vessel is impermeable (i.e., $\psi = 0$) and its cross-sectional area varies linearly and slowly as follows:

$$A_d(x) = A_0(1 - \theta x) \qquad (\theta x < 1) \tag{13.6}$$

where $A_0 = \pi r_0^2$ is the cross-sectional area at the inlet of the blood vessel and θ is the parameter that represents the taper of the blood vessel (see Figure 13.1). In addition, we assume

$$\begin{aligned} Q(x,t) &= Q_d + q(x,t) \\ P(x,t) &= P_d + p(x,t) \\ A(x,t) &= A_d(x) + a(x,t) \end{aligned} \tag{13.7}$$

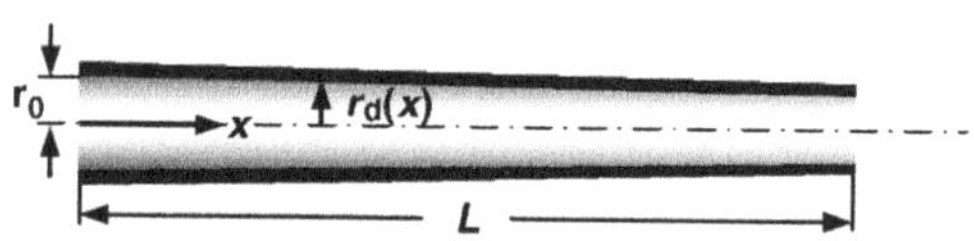

Figure 13.1 A blood vessel element

where $a(x, t)$ represents the small perturbation with respect to $A_d(x)$ so that $a(x, t) < A_d(x)$. Q_d and P_d are constant blood flow rate and blood pressure at the diastole phase, as shown in Figure 13.2.

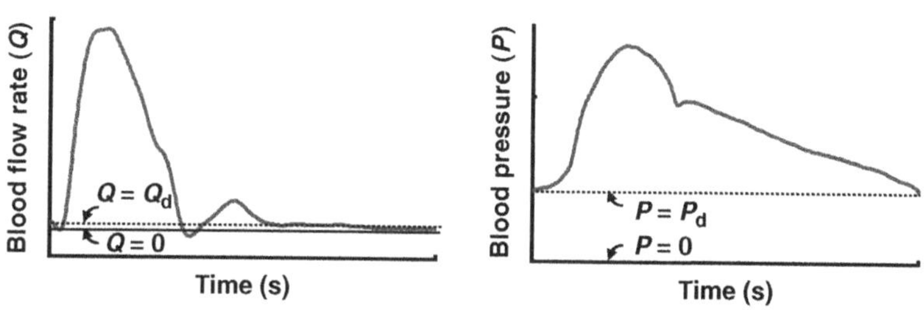

Figure 13.2 The blood flow rate and pressure as functions of time

By applying the last two lines of Equations 13.7 to Equation 13.3, we obtain the relationship

$$a(x,t) = \frac{3}{2}\frac{r_a}{Eh}A_d(x)p(x,t) \tag{13.8}$$

where r_a is the mean inner radius of a blood vessel. By substituting Equation 13.7 into Equations 13.1 and 13.2 and using Equation 13.8, we can simplify the governing equations as

$$c_1\dot{p} + q' = 0 \tag{13.9}$$

$$c_0p' + \dot{q} + c_2q + 2c_3q' - c_4q'' + g(p,q) = 0 \tag{13.10}$$

where

$$g(p,q) = c_5q^2 + c_6qq' + c_7 \tag{13.11}$$

and

$$
\begin{aligned}
c_0 &= \frac{A_0}{\rho}, & c_1 &= \frac{3}{2}\frac{r_a}{Eh}A_0 \\
c_2 &= \frac{\nu H}{A_0} + \frac{2(1+\delta)\theta}{A_0}Q_d, & c_3 &= \frac{1+\delta}{A_0}Q_d \\
c_4 &= \nu, & c_5 &= \frac{(1+\delta)\theta}{A_0} \\
c_6 &= \frac{2(1+\delta)}{A_0}, & c_7 &= \frac{(1+\delta)\theta}{A_0}Q_d^2 + \frac{\nu H}{A_0}Q_d
\end{aligned}
\tag{13.12}
$$

The derivative of Equation 13.10 with respect to x is given by

$$c_0 p'' + \dot{q}' + c_2 q' + 2c_3 q'' - c_4 q''' + f(p,q) = 0 \tag{13.13}$$

where

$$f(p,q) = g'(p,q) = 2c_5 q q' + c_6 q' q' + c_6 q q'' \tag{13.14}$$

From Equation 13.9, we get

$$q' = -c_1 \dot{p}, \quad \dot{q}' = -c_1 \ddot{p}, \quad q'' = -c_1 \dot{p}', \quad q''' = -c_1 \dot{p}'' \tag{13.15}$$

By using Equation 13.15, Equations 13.13 and 13.14 can be written as

$$c_0 p'' - c_1 \ddot{p} - c_1 c_2 \dot{p} - 2c_1 c_3 \dot{p}' + c_1 c_4 \dot{p}'' - f(x,t) = 0 \tag{13.16}$$

and

$$f(p,q) = 2c_1 c_5 q \dot{p} - c_1^2 c_6 \dot{p}\dot{p} + c_1 c_6 q \dot{p}' \tag{13.17}$$

Notice that all nonlinear terms are collected in the function $f(p, q; x, t)$. By considering $f(p, q; x, t)$ as a pseudo-force, we can use an iteration method to solve the coupled nonlinear equations (13.16) and (13.9) for $p(x, t)$ and $q(x, t)$ in the frequency domain, which will be discussed later in this chapter.

13.2 Spectral Element Modeling: I. Finite Element

The systemic arteries are organized in a bifurcating tree in which the larger vessels tapering along their length, while the smaller vessels do not taper significantly. The individual arteries between any two neighboring bifurcating points can be represented by the spectral element models while specifying physiologically correct boundary conditions at their inlets and outlets. We can apply the finite spectral element model to most short arteries, such as the aorta and iliac arteries (i.e., 4~6 cm). For long arteries, such as the femoral and brachial arteries (i.e., 38~45 cm), we can apply the finite spectral element model to the short upstream part (computational domain) that has the larger cross-sectional area and the semi-infinite spectral element model to the remaining, long downstream part with a smaller cross-sectional area. By applying the semi-infinite spectral element model to the long downstream parts of long arteries, the backward propagating blood waves reflected from the downstream end of the artery can be avoided. In the following, we present spectral elements for the finite and semi-infinite blood vessel elements.

13.2.1 Governing Equations in the Frequency Domain

By using the DFT theory, we represent $p(x, t)$, $q(x, t)$, and $g(x, t)$ in the spectral forms as

$$\begin{aligned} p(x,t) &= \frac{1}{N}\sum_{n=0}^{N-1} P_n(x)\mathrm{e}^{i\omega_n t} \\ q(x,t) &= \frac{1}{N}\sum_{n=0}^{N-1} Q_n(x)\mathrm{e}^{i\omega_n t} \\ f(p(x,t),q(x,t)) &= \frac{1}{N}\sum_{n=0}^{N-1} F_n(x;\boldsymbol{P},\boldsymbol{Q})\mathrm{e}^{i\omega_n t} \end{aligned} \tag{13.18}$$

where $P_n(x)$, $Q_n(x)$, and $F_n(x; \boldsymbol{P}, \boldsymbol{Q})$ are the spectral components of $p(x, t)$, $q(x, t)$, and $f(p(x, t), q(x, t))$, respectively, and $\boldsymbol{P}$ and $\boldsymbol{Q}$ are defined by

$$\begin{aligned} \boldsymbol{P}(x) &= \{P_n(x);\ n = 0,\ 1,\ 2,\ \ldots,\ N-1\} \\ \boldsymbol{Q}(x) &= \{Q_n(x);\ n = 0,\ 1,\ 2,\ \ldots,\ N-1\} \end{aligned} \tag{13.19}$$

Substituting Equation 13.18 into Equations 13.9 and 13.13, we obtain

$$i\omega_n c_0 P_n + c_1 Q_n' = 0 \tag{13.20}$$

$$P_n''(x) - 2b_{1n}P_n'(x) + b_{2n}P_n(x) - b_{3n}F_n(x) = 0 \tag{13.21}$$

where

$$\begin{aligned} b_{1n} &= \frac{c_1^2 c_3 c_4 \omega_n^2 + i c_0 c_1 c_3 \omega_n}{c_0^2 + c_1^2 c_4^2 \omega_n^2} \\ b_{2n} &= \omega_n^2 \frac{c_0 c_1 - c_1^2 c_2 c_4}{c_0^2 + c_1^2 c_4^2 \omega_n^2} - i\omega_n \frac{c_0 c_1 c_2 + c_1^2 c_4 \omega_n^2}{c_0^2 + c_1^2 c_4^2 \omega_n^2} \\ b_{3n} &= \frac{c_0 - i c_1 c_4 \omega_n}{c_0^2 + c_1^2 c_4^2 \omega_n^2} \end{aligned} \tag{13.22}$$

The Fourier components $F_n(x)$ can be represented in terms of the spectral components $\boldsymbol{P}$ and $\boldsymbol{Q}$ by applying the formula of Equation 2.59c to each of the nonlinear terms in the pseudo-force $f(p, q; x, t)$ as follows:

$$F_n(x;\boldsymbol{P},\boldsymbol{Q}) = \frac{1}{N}(2c_1 c_5 A_n + c_1^2 c_6 B_n + c_1 c_6 C_n) \tag{13.23}$$

where

$$\begin{aligned} A_n &= \sum_{j=0}^{n}(i\omega_{n-j})Q_j P_{n-j} + \sum_{j=n+1}^{N/2}(i\omega_{j-n})Q_j P_{j-n}^* + \sum_{j=1}^{(N/2)-1}(i\omega_{n+j})Q_j^* P_{n+j} \\ B_n &= \sum_{j=0}^{n}(\omega_j\omega_{n-j})P_j P_{n-j} + \sum_{j=n+1}^{N/2}(\omega_j\omega_{j-n})P_j P_{j-n}^* + \sum_{j=1}^{(N/2)-1}(\omega_j\omega_{n+j})P_j^* P_{n+j} \\ C_n &= \sum_{j=0}^{n}(i\omega_{n-j})Q_j P_{n-j}' + \sum_{j=n+1}^{N/2}(i\omega_{j-n})Q_j P_{j-n}'^* + \sum_{j=1}^{(N/2)-1}(i\omega_{n+j})Q_j^* P_{n+j}' \end{aligned} \tag{13.24}$$

13.2.2 Weak Form of Governing Equations

The weak form of Equation 13.21 can be derived from

$$\int_0^L [P_n''(x) - 2b_{1n}P_n'(x) + b_{2n}P_n(x) - b_{3n}F_n(x)]\,\delta P_n(x)\mathrm{d}x = 0 \qquad (13.25)$$

By applying the integration by parts, we can obtain

$$\begin{aligned} \int_0^L P_n''(x)\delta P_n(x)\mathrm{d}x &= P_n'(x)\delta P_n(x)\Big|_0^L - \int_0^L P_n'(x)\delta P_n'(x)\mathrm{d}x \\ \int_0^L P_n'(x)\delta P_n(x)\mathrm{d}x &= P_n(x)\delta P_n(x)\Big|_0^L - \int_0^L P_n(x)\delta P_n'(x)\mathrm{d}x \end{aligned} \qquad (13.26)$$

By using Equation 13.15, Equation 3.10 can be written as

$$p' = -\frac{1}{c_0}\dot{q} - \frac{c_2}{c_0}q + 2\frac{c_1c_3}{c_0}\dot{p} - \frac{c_1c_4}{c_0}\dot{p}' - \frac{c_5}{c_0}z_1 + \frac{c_1c_6}{c_0}z_2 - \frac{c_7}{c_0} \qquad (13.27)$$

where

$$z_1(x,t) = q^2, \qquad z_2(x,t) = \dot{p}q \qquad (13.28)$$

The spectral form of Equation 13.27 is given by

$$P_n' = -\frac{i\omega_n}{c_0}Q_n - \frac{c_2}{c_0}Q_n + 2\frac{i\omega_n c_1c_3}{c_0}P_n - \frac{i\omega_n c_1c_4}{c_0}P_n' - \frac{c_7}{c_0}C_{0n} - \frac{c_5}{c_0}Z_{1n} + \frac{c_1c_6}{c_0}Z_{2n} \qquad (13.29)$$

where

$$\begin{aligned} Z_{1n}(x;\boldsymbol{P},\boldsymbol{Q}) &= \frac{1}{N}\left(\sum_{j=0}^{n} Q_jQ_{n-j} + \sum_{j=n+1}^{N/2} Q_jQ_{j-n}^* + \sum_{j=1}^{N/2-1} Q_j^*Q_{n+j}\right) \\ Z_{2n}(x;\boldsymbol{P},\boldsymbol{Q}) &= \frac{1}{N}\left(\sum_{j=0}^{n}(i\omega_{n-j})Q_jP_{n-j} + \sum_{j=n+1}^{N/2}(i\omega_{j-n})Q_jP_{j-n}^* + \sum_{j=1}^{N/2-1}(i\omega_{n+j})Q_j^*P_{n+j}\right) \\ C_{0n} &= N\delta_{0n} = \begin{cases} N & \text{if} \quad n=0 \\ 0 & \text{otherwise} \end{cases} \end{aligned} \qquad (13.30)$$

Equation 13.29 can be rewritten in a simple form as

$$P_n' = -i\omega_n a_{1n}Q_n - a_{2n}Q_n + a_{3n}P_n - a_{4n}C_{0n} + Z_n(x;\boldsymbol{P},\boldsymbol{Q}) \qquad (13.31)$$

where

$$Z_n(x;\boldsymbol{P},\boldsymbol{Q}) = -a_{5n}Z_{1n}(x;\boldsymbol{P},\boldsymbol{Q}) + a_{6n}Z_{2n}(x;\boldsymbol{P},\boldsymbol{Q}) \qquad (13.32)$$

and

$$\begin{aligned} a_{1n} &= \frac{1}{c_0 + ic_1c_4\omega_n}, & a_{2n} &= \frac{c_2}{c_0 + ic_1c_4\omega_n} \\ a_{3n} &= \frac{2ic_1c_3\omega_n}{c_0 + ic_1c_4\omega_n} = 2b_{1n}, & a_{4n} &= \frac{c_7}{c_0 + ic_1c_4\omega_n} \\ a_{5n} &= \frac{c_5}{c_0 + ic_1c_4\omega_n}, & a_{6n} &= \frac{c_1c_6}{c_0 + ic_1c_4\omega_n} \end{aligned} \tag{13.33}$$

By substituting Equation 13.31 into the first term on the right-hand side of the first line of Equation 13.26, we obtain

$$\begin{aligned} \int_0^L P_n''(x)\delta P_n \mathrm{d}x =& (-i\omega_n a_{1n}Q_n - a_{2n}Q_n + a_{3n}P_n - a_{4n}C_{0n} + Z_n)\delta P_n(x)\Big|_0^L \\ & - \int_0^L P_n'(x)\delta P_n'(x)\mathrm{d}x \end{aligned} \tag{13.34}$$

Finally, by replacing the first term of Equation 13.25 with Equation 13.34, we obtain the weak form in the frequency domain as

$$\begin{aligned} &\alpha_{1n}\int_0^L P_n'(x)\delta P_n'(x)\mathrm{d}x + \alpha_{2n}\int_0^L [P_n'(x)\delta P_n(x) - P_n(x)\delta P_n'(x)]\mathrm{d}x \\ &-\alpha_{3n}\int_0^L P_n(x)\delta P_n(x)\mathrm{d}x = -\bar{\alpha}_{4n}Q_n(x)\delta P_n(x)\Big|_0^L \\ &-Q_n(x)\delta P_n(x)\Big|_0^L + \alpha_{2n}P_n(x)\delta P_n(x)\Big|_0^L - \alpha_{5n}\delta P_n(x)\Big|_0^L \\ &+\alpha_{1n}Z_n(x)\delta P_n(x)\Big|_0^L - \alpha_{6n}\int_0^L F_n(x)\delta P_n(x)\mathrm{d}x \end{aligned} \tag{13.35}$$

where

$$\begin{aligned} \alpha_{1n} &= \frac{1}{a_{2n}} = \frac{c_0 + i\omega_n c_1 c_4}{c_2}, \qquad \alpha_{2n} = \frac{b_{1n}}{a_{2n}} = \frac{i\omega_n c_1 c_3}{c_2} \\ \alpha_{3n} &= \frac{b_{2n}}{a_{2n}} = \omega_n^2 \frac{c_0c_1 - c_1^2c_2c_4}{c_2(c_0 - ic_1c_4\omega_n)} - i\omega_n \frac{c_0c_1c_2 + c_1^2c_4\omega_n^2}{c_2(c_0 - ic_1c_4\omega_n)} \\ \bar{\alpha}_{4n} &= \frac{i\omega_n a_{1n}}{a_{2n}} = \frac{i\omega_n}{c_2}, \qquad \alpha_{5n} = \frac{a_{4n}}{a_{2n}}C_{0n} = \frac{c_7}{c_2}C_{0n} \\ \alpha_{6n} &= \frac{b_{3n}}{a_{2n}} = \frac{1}{c_2} \end{aligned} \tag{13.36}$$

13.2.3 Spectral Nodal DOFs

Consider a finite blood vessel element of length L. As shown in Figure 13.3, the spectral nodal DOFs vectors for the blood flow rate and pressure are defined by

$$\boldsymbol{P}_n = \begin{Bmatrix} P_{n1} \\ P_{n2} \end{Bmatrix} = \begin{Bmatrix} +P_n(0) \\ +P_n(L) \end{Bmatrix}, \qquad \boldsymbol{Q}_n = \begin{Bmatrix} Q_{n1} \\ Q_{n2} \end{Bmatrix} = \begin{Bmatrix} +Q_n(0) \\ -Q_n(L) \end{Bmatrix} \tag{13.37}$$

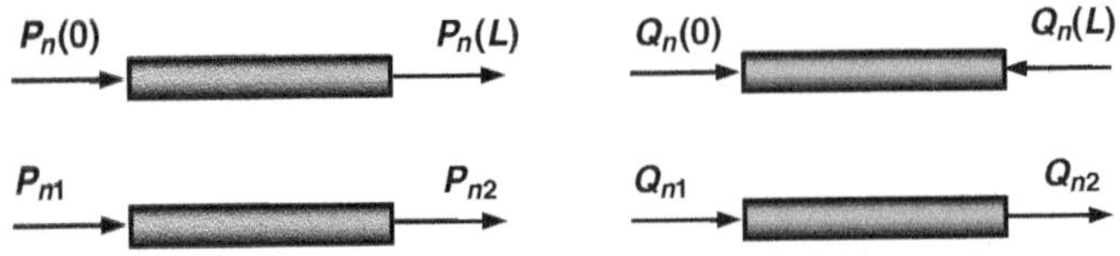

Figure 13.3 Sign convention for the blood vessel element

13.2.4 Dynamic Shape Functions

To derive the frequency dependent dynamic shape functions, we will consider the linear homogeneous equation reduced from Equation 13.21 by removing the pseudo-force term F_n as

$$P_n''(x) - 2b_{1n}P_n'(x) + b_{2n}P_n(x) = 0 \tag{13.38}$$

Assume the general solution of Equation 13.38 to be

$$P_n(x) = a e^{-ik_n x} \tag{13.39}$$

where k_n is the wavenumber. Substituting Equation 13.39 into Equation 13.38, we obtain a dispersion relation

$$k_n^2 - 2ib_{1n}k_n - b_{2n} = 0 \tag{13.40}$$

From Equation 13.40, we can obtain two wavenumbers as

$$k_{n1} = ib_{1n} + \sqrt{-b_{1n}^2 + b_{2n}}, \qquad k_{n2} = ib_{1n} - \sqrt{-b_{1n}^2 + b_{2n}} \tag{13.41}$$

The effects of nonlinear terms are neglected in the wavenumbers given by Equation 13.41.

Figure 13.4 shows the dispersion curves for a blood flow with a parabolic velocity profile as a dependence on the blood vessel taper θ and the kinematic viscosity ν of the blood. If the blood vessel is uniform ($\theta = 0$) and the blood is non-viscid ($\nu = 0$), k_{n1} represents a wavemode that

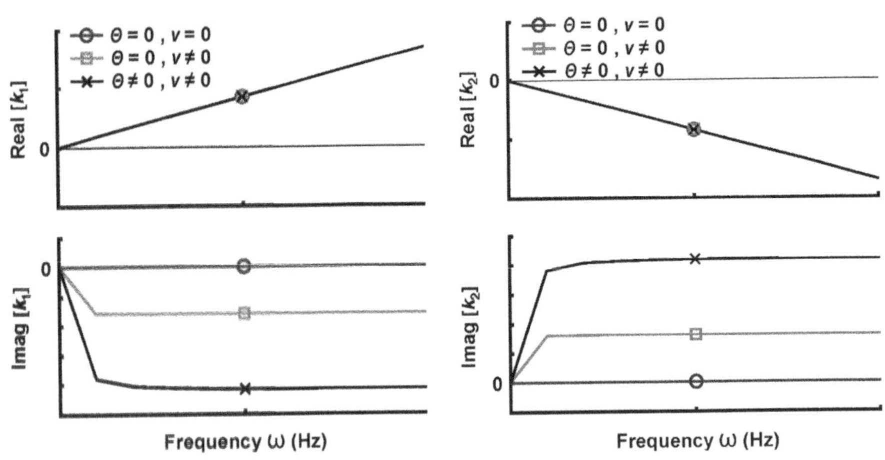

Figure 13.4 Dispersion curves

travels in the $+x$ direction without attenuation (forward propagating wave) while k_{n2} a wavemode that travels in the $-x$ direction without attenuation (backward propagating wave). However, as in real situations, if the blood vessel is not uniform ($\theta \neq 0$) and the blood is viscid ($\nu \neq 0$), both wavenumbers k_{n1} and k_{n2} will be complex with imaginary values. This means that k_{n1} and k_{n2} represent the wavemodes that attenuate as they travel in the $+x$ and $-x$ directions, respectively.

By using the wavenumbers, the general solution to Equation 13.38 can be written as

$$P_n(x) = \boldsymbol{e}_n(x)\boldsymbol{a} \tag{13.42}$$

where

$$\boldsymbol{e}_n(x) = \begin{bmatrix} \mathrm{e}^{-ik_{n1}x} & \mathrm{e}^{-ik_{n2}x} \end{bmatrix} \tag{13.43}$$

and

$$\boldsymbol{a} = \{a_1 \quad a_2\}^{\mathrm{T}} \tag{13.44}$$

By applying Equation 13.42 to the right-hand side of the first equation in Equation 13.37, we can obtain

$$\boldsymbol{P}_n = \boldsymbol{H}_n(\omega_n)\boldsymbol{a} \tag{13.45}$$

where

$$\boldsymbol{H}_n(\omega_n) = \begin{bmatrix} 1 & 1 \\ \mathrm{e}^{-ik_{n1}L} & \mathrm{e}^{-ik_{n2}L} \end{bmatrix} \tag{13.46}$$

By using Equation 13.45, the constant vector $\boldsymbol{a}$ can be removed from Equation 13.42 to obtain

$$P_n(x) = \boldsymbol{N}_n(x;\omega_n)\boldsymbol{P}_n \tag{13.47}$$

where $\boldsymbol{N}_n(x;\,\omega_n)$ is the dynamic shape function defined by

$$\begin{aligned} \boldsymbol{N}_n(x;\omega_n) &= \boldsymbol{e}_n(x)\boldsymbol{H}_n(\omega_n)^{-1} \\ &= \frac{1}{\Delta_n}\begin{bmatrix} \mathrm{e}^{-i(k_{n1}x+k_{n2}L)} - \mathrm{e}^{-i(k_{n2}x+k_{n1}L)} & \mathrm{e}^{-ik_{n2}x} - \mathrm{e}^{-ik_{n1}x} \end{bmatrix} \end{aligned} \tag{13.48}$$

where

$$\Delta_n = \mathrm{e}^{-ik_{n2}L} - \mathrm{e}^{-ik_{n1}L} \tag{13.49}$$

13.2.5 Spectral Element Equation

By using the spectral nodal DOFs vector defined by Equation 13.37, the weak form of Equation 13.35 can be rewritten as

$$\begin{aligned} &\alpha_{1n}\int_0^L P_n'(x)\delta P_n'(x)\mathrm{d}x + \alpha_{2n}\int_0^L [P_n'(x)\delta P_n(x) - P_n(x)\delta P_n'(x)]\mathrm{d}x \\ &-\alpha_{3n}\int_0^L P_n(x)\delta P_n(x)\mathrm{d}x = \bar{\alpha}_{4n}\delta\boldsymbol{P}_n^{\mathrm{T}}\boldsymbol{Q}_n + \delta\boldsymbol{P}_n^{\mathrm{T}}\boldsymbol{Q}_n + \alpha_{2n}\delta\boldsymbol{P}_n^{\mathrm{T}}\boldsymbol{T}\boldsymbol{P}_n \\ &+\alpha_{5n}\delta\boldsymbol{P}_n^{\mathrm{T}}\boldsymbol{U} + \alpha_{1n}\delta\boldsymbol{P}_n^{\mathrm{T}}\boldsymbol{Z}_n(\boldsymbol{P},\boldsymbol{Q}) - \alpha_{6n}\int_0^L F_n(x)\delta P_n(x)\mathrm{d}x \end{aligned} \tag{13.50}$$

where

$$T=\begin{bmatrix}-1 & 0\\ 0 & 1\end{bmatrix},\qquad U=\begin{Bmatrix}1\\ -1\end{Bmatrix} \tag{13.51}$$

and

$$Z_n(P,Q)=\begin{Bmatrix}-Z_n(0)\\ +Z_n(L)\end{Bmatrix}=\begin{Bmatrix}+a_{5n}Z_{1n}(0)-a_{6n}Z_{2n}(0)\\ -a_{5n}Z_{1n}(L)+a_{6n}Z_{2n}(L)\end{Bmatrix} \tag{13.52}$$

The weighed-integral statement of Equation 13.20 is given by

$$\int_0^L (ic_1\omega_n P_n + Q_n')\delta Q_n(x)\mathrm{d}x = 0 \tag{13.53}$$

From the linear homogenous wave equation for $Q_n(x)$, which is basically the same as Equation 13.38, we can derive

$$Q_n(x) = \boldsymbol{N}_n(x;\omega_n)\boldsymbol{Q}_n \tag{13.54}$$

By substituting Equations 13.47 and 13.54 into Equation 13.53, we can obtain a complementary equation as

$$\boldsymbol{Q}_n = -ic_1\omega_n \boldsymbol{R}_n \boldsymbol{P}_n \tag{13.55}$$

where

$$\boldsymbol{R}_n=\frac{1}{\Delta_n k_{n1}k_{n2}}\begin{bmatrix}-ik_{n1}\mathrm{e}_{n1}+ik_{n2}\mathrm{e}_{n2} & ik_{n1}-ik_{n2}\\ \mathrm{e}_{n1}\mathrm{e}_{n2}(-ik_{n1}+ik_{n2}) & ik_{n1}\mathrm{e}_{n2}-ik_{n2}\mathrm{e}_{n1}\end{bmatrix} \tag{13.56}$$

with

$$\Delta_n=\mathrm{e}_{n2}-\mathrm{e}_{n1},\qquad \mathrm{e}_{n1}=\mathrm{e}^{-ik_{n1}L},\qquad \mathrm{e}_{n2}=\mathrm{e}^{-ik_{n2}L} \tag{13.57}$$

Substituting Equation 13.47 into Equation 13.50 and applying Equation 13.55 to the first term on the right-hand side of Equation 13.50 yields

$$\begin{aligned}&\delta\boldsymbol{P}_n^{\mathrm{T}}\Bigg[\alpha_{1n}\int_0^L \boldsymbol{N}_n'^{\mathrm{T}}\boldsymbol{N}_n'\mathrm{d}x+\alpha_{2n}\left(\int_0^L \boldsymbol{N}_n^{\mathrm{T}}\boldsymbol{N}_n'\mathrm{d}x-\int_0^L \boldsymbol{N}_n'^{\mathrm{T}}\boldsymbol{N}_n\mathrm{d}x\right)\\ &\quad-\alpha_{3n}\int_0^L \boldsymbol{N}_n^{\mathrm{T}}\boldsymbol{N}_n\mathrm{d}x-\alpha_{4n}\boldsymbol{R}_n-\alpha_{2n}\boldsymbol{T}\Bigg]\boldsymbol{P}_n\\ &=\delta\boldsymbol{P}_n^{\mathrm{T}}\left[\boldsymbol{Q}_n+\alpha_{5n}\boldsymbol{U}+\alpha_{1n}\boldsymbol{Z}_n(\boldsymbol{P},\boldsymbol{Q})-\alpha_{6n}\int_0^L \boldsymbol{N}_n^{\mathrm{T}}F_n(x)\mathrm{d}x\right]\end{aligned} \tag{13.58}$$

where

$$\alpha_{4n}=\frac{c_1}{c_2}\omega_n^2 \tag{13.59}$$

As $\delta\boldsymbol{P}_n$ is arbitrary by definition, we can obtain the spectral element equation from Equation 13.58 as

$$\boldsymbol{S}(\omega_n)\boldsymbol{P}_n=\boldsymbol{Q}_n+\boldsymbol{f}_n \tag{13.60}$$

where

$$S(\omega_n) = \alpha_{1n}\int_0^L N_n'^{\mathrm{T}} N_n' \mathrm{d}x + \alpha_{2n}\left(\int_0^L N_n^{\mathrm{T}} N_n' \mathrm{d}x - \int_0^L N_n'^{\mathrm{T}} N_n \mathrm{d}x\right) - \alpha_{3n}\int_0^L N_n^{\mathrm{T}} N_n \mathrm{d}x - \alpha_{4n} R_n - \alpha_{2n} T \tag{13.61}$$

$$f_n = f_{Cn} + f_{NLn} \tag{13.62}$$

and

$$f_{Cn} = \alpha_{5n} U, \quad f_{NLn} = \alpha_{1n} Z_n(P, Q) - \alpha_{6n} F_n \tag{13.63}$$

By substituting the dynamic shape function Equation 13.48 into Equation 13.61 and taking an integration, we can obtain the spectral element matrix as

$$S(\omega_n) = \frac{1}{\Delta_n^2}\begin{bmatrix} S_{n11} & S_{n12} \\ S_{n21} & S_{n22} \end{bmatrix} - \alpha_{4n} R_n - \alpha_{2n} T \tag{13.64}$$

where

$$\begin{aligned}
S_{n11} &= i\beta_{n1}\lambda_{n1}\varepsilon_{n1}e_{n2}^2 + i\beta_{n2}\lambda_{n2}\varepsilon_{n2}e_{n1}^2 - \beta_{n3}(\lambda_{n3}+\lambda_{n4})\varepsilon_{n3}e_{n1}e_{n2} \\
S_{n12} &= -i\beta_{n1}\lambda_{n1}\varepsilon_{n1}e_{n2} - i\beta_{n2}\lambda_{n2}\varepsilon_{n2}e_{n1} + \beta_{n3}\varepsilon_{n3}(\lambda_{n3}e_{n2} + \lambda_{n4}e_{n1}) \\
S_{n21} &= -i\beta_{n1}\lambda_{n1}\varepsilon_{n1}e_{n2} - i\beta_{n2}\lambda_{n2}\varepsilon_{n2}e_{n1} + \beta_{n3}\varepsilon_{n3}(\lambda_{n3}e_{n1} + \lambda_{n4}e_{n2}) \\
S_{n22} &= i\beta_{n1}\lambda_{n1}\varepsilon_{n1} + i\beta_{n2}\lambda_{n2}\varepsilon_{n2} - \beta_{n3}(\lambda_{n3}+\lambda_{n4})\varepsilon_{n3}
\end{aligned} \tag{13.65}$$

with

$$\begin{aligned}
\beta_{n1} &= (1/2)k_{n1}^{-1}, \quad \beta_{n2} = (1/2)k_{n2}^{-1}, \quad \beta_{n3} = (k_{n1}+k_{n2})^{-1} \\
\varepsilon_{n1} &= 1 - e_{n1}^2, \quad \varepsilon_{n2} = 1 - e_{n2}^2, \quad \varepsilon_{n3} = 1 - e_{n1}e_{n2} \\
\lambda_{n1} &= \alpha_{1n}k_{n1}^2 + \alpha_{3n} \\
\lambda_{n2} &= \alpha_{1n}k_{n2}^2 + \alpha_{3n} \\
\lambda_{n3} &= i\alpha_{1n}k_{n1}k_{n2} + (k_{n1}-k_{n2})\alpha_{2n} + i\alpha_{3n} \\
\lambda_{n4} &= i\alpha_{1n}k_{n1}k_{n2} + (k_{n2}-k_{n1})\alpha_{2n} + i\alpha_{3n}
\end{aligned} \tag{13.66}$$

The nonlinear forcing term F_n in Equation 13.63 can be numerically computed as follows:

$$\begin{aligned}
F_n &\cong \sum_{j=1}^{M}\int_{x_j}^{x_{j+1}} F_n(x)N_n^{\mathrm{T}}(x)\mathrm{d}x = \sum_{j=1}^{M} F_n(x_{jc})\int_{x_j}^{x_{j+1}} N_n^{\mathrm{T}}(x)\mathrm{d}x \\
&= \frac{1}{\lambda_0}\sum_{j=1}^{M} F_n(x_{jc})\begin{Bmatrix} e_{n2}\lambda_1 - e_{n1}\lambda_2 \\ -\lambda_1 + \lambda_2 \end{Bmatrix}
\end{aligned} \tag{13.67}$$

where

$$\begin{aligned}
\lambda_0 &= k_{n1}k_{n2}\Delta_n \\
\lambda_1 &= ik_{n2}(e^{-ik_{n1}x_{j+1}} - e^{-ik_{n1}x_j}) \\
\lambda_2 &= ik_{n1}(e^{-ik_{n2}x_{j+1}} - e^{-ik_{n2}x_j})
\end{aligned} \tag{13.68}$$

For Equation 13.67, it is assumed that the blood vessel of length L is divided into M segments of equal length $\Delta = L/M$. Accordingly, the definitions of $x_j = (j-1)\Delta$, $x_{j+1} = j\Delta$, and $x_{jc} = (j-1/2)\Delta$ are used in Equations 13.67 and 13.68.

As Equation 13.60 is nonlinear in $\boldsymbol{P} = \{P_n; n = 0, 1, \ldots, N-1\}$ and $\boldsymbol{Q} = \{Q_n; n = 0, 1, \ldots, N-1\}$, we can use a proper iteration method to solve for $\boldsymbol{P}$ and $\boldsymbol{Q}$. Firstly, we ignore the nonlinear pseudo-force term $\boldsymbol{f}_{NLn}$ from Equation 13.60 to solve the linearized equation for $\boldsymbol{P}$ and $\boldsymbol{Q}$ by utilizing Equation 13.55. By using the first estimation of $\boldsymbol{P}$ and $\boldsymbol{Q}$, we compute F_n from Equation 13.23 and Z_n from Equation 13.32, followed by the computation of $\boldsymbol{F}_n$ from Equation 13.67 and $\boldsymbol{Z}_n$ from Equation 13.52. Next, by taking into account the nonlinear pseudo-force term $\boldsymbol{f}_{NLn}$ in Equation 13.60, we solve Equation 13.60 for $\boldsymbol{P}$ and $\boldsymbol{Q}$. The new results are an improved second estimation. We repeat the iteration until we obtain $\boldsymbol{P}$ and $\boldsymbol{Q}$ sufficiently converged to a certain required accuracy. Lastly, the time domain responses $p(x,t)$ and $q(x,t)$ are computed from the values for $\boldsymbol{P}$ and $\boldsymbol{Q}$ obtained at the final iteration step by using the FFT algorithm.

13.3 Spectral Element Modeling: II. Semi-Infinite Element

To formulate the spectral element for a semi-infinite blood vessel element, we start again from Equations 13.9 and 13.10. The governing equations are represented in the spectral forms as

$$ic_1\omega_n P_n(x) + Q_n'(x) = 0 \tag{13.69}$$

$$c_0 P_n'(x) + i\omega_n Q_n(x) + c_2 Q_n(x) + 2c_3 Q_n'(x) - c_4 Q_n''(x) + G_n(x) = 0 \tag{13.70}$$

where $G_n(x; \omega)$ are the spectral components of the function $g(x, t)$ defined by Equation 13.11. From Equations 13.69 and 13.70, we can obtain the wave equation for $Q_n(x; \omega)$ as

$$Q_n''(x) - 2b_{1n}Q_n'(x) + b_{2n}Q_n(x) - ic_1\omega_n b_{3n}F_n(x) = 0 \tag{13.71}$$

where b_{1n}, b_{2n} and b_{3n} are defined by Equation 13.22.

Consider the linear homogenous equation reduced from Equation 13.71 as

$$Q_n''(x) - 2b_{1n}Q_n'(x) + b_{2n}Q_n(x) = 0 \tag{13.72}$$

Assume the general solution to Equation 13.72 to be

$$Q_n(x) = ae^{-ik_n x} \tag{13.73}$$

Substituting Equation 13.73 into Equation 13.72 yields a dispersion relation as

$$k_n^2 - 2ib_{1n}k_n - b_{2n} = 0 \tag{13.74}$$

As Equation 13.72 for $Q_n(x)$ has the exactly same form as Equation 13.38 for $P_n(x)$, we obtain a dispersion relation (13.74) that is identical with Equation 13.40. Accordingly, we also obtain two identical wavenumbers k_{n1} and k_{n2} given by Equation 13.41. The wavenumber k_{n1} corresponds to the wave that is propagating in the $+x$ direction (the forward propagating wavemode), while the wavenumber k_{n1} corresponds to the wave that is propagating in the $-x$ direction (the backward propagating wavemode).

For the blood flow through a very long blood vessel, we may assume that the blood wave propagates forwards only so that the backwards propagating waves reflected from far downstream can be neglected. This approximation may lead to a non-resonant single-node semi-infinite spectral element. Thus, we represent the blood flow rate as

$$Q_n(x) = a e^{-ik_{n1}x} \tag{13.75}$$

The spectral nodal DOFs at the upstream end (inlet) of a semi-infinite blood vessel element are defined by

$$Q_n(0) = Q_{n1}, \qquad P_n(0) = P_{n1} \tag{13.76}$$

By using the first relationship of Equation 13.76, Equation 13.75 can be written as

$$Q_n(x) = e^{-ik_{n1}x} Q_{n1} \tag{13.77}$$

Substituting Equation 13.77 into the continuity equation (13.69), we obtain

$$P_n(x) = \frac{k_{n1}}{c_1 \omega_n} Q_{n1} e^{-ik_{n1}x} \tag{13.78}$$

Lastly, applying the second relationship in Equation 13.76 to Equation 13.78, we obtain the semi-infinite spectral element equation as

$$S_n^{SI} P_{n1} = Q_{n1} \tag{13.79}$$

where

$$S_n^{SI} = c_1 \frac{\omega_n}{k_{n1}} \tag{13.80}$$

13.4 Assembly of Spectral Elements

The spectral elements can be assembled in the same way as for the conventional finite element method. As the assembling is straightforward for straight blood vessels, the assembly process will be illustrated for a binary branch that consists of a parent vessel (denoted by "p") and two daughter vessels (denoted by "d_1" and "d_2") as shown in Figure 13.5.

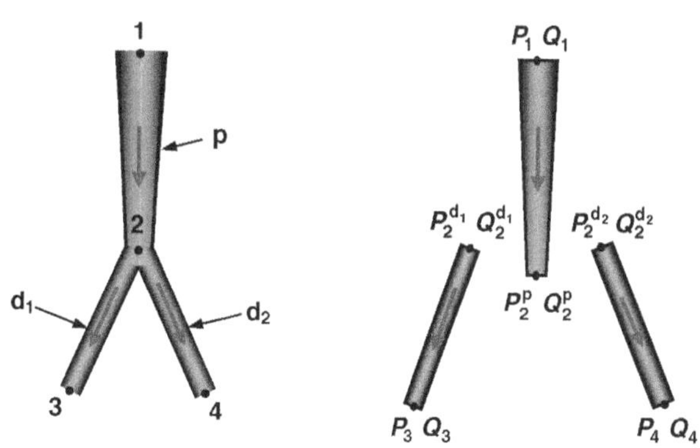

Figure 13.5 A binary branch element

At the bifurcation point 2, the following bifurcation conditions must be satisfied:

$$P_2^{\mathrm{p}} = P_2^{\mathrm{d}_1} = P_2^{\mathrm{d}_2} = P_2 \tag{13.81}$$

$$Q_2^{\mathrm{p}} = Q_2^{\mathrm{d}_1} + Q_2^{\mathrm{d}_2} \tag{13.82}$$

The spectral element equations for the parent vessel and two daughter vessels are given by

$$\begin{aligned}
&\begin{bmatrix} S_{n11}^{\mathrm{p}} & S_{n12}^{\mathrm{p}} \\ S_{n21}^{\mathrm{p}} & S_{n22}^{\mathrm{p}} \end{bmatrix}\begin{Bmatrix} P_1 \\ P_2^{\mathrm{p}} \end{Bmatrix} = \begin{Bmatrix} Q_1 \\ Q_2^{\mathrm{p}} \end{Bmatrix} + \begin{Bmatrix} f_{n1}^{\mathrm{p}} \\ f_{n2}^{\mathrm{p}} \end{Bmatrix} \quad \text{(parent)} \\
&\begin{bmatrix} S_{n11}^{\mathrm{d}_1} & S_{n12}^{\mathrm{d}_1} \\ S_{n21}^{\mathrm{d}_1} & S_{n22}^{\mathrm{d}_1} \end{bmatrix}\begin{Bmatrix} P_2^{\mathrm{d}_1} \\ P_3 \end{Bmatrix} = \begin{Bmatrix} Q_2^{\mathrm{d}_1} \\ Q_3 \end{Bmatrix} + \begin{Bmatrix} f_{n1}^{\mathrm{d}_1} \\ f_{n2}^{\mathrm{d}_1} \end{Bmatrix} \quad \text{(daughter 1)} \\
&\begin{bmatrix} S_{n11}^{\mathrm{d}_2} & S_{n12}^{\mathrm{d}_2} \\ S_{n21}^{\mathrm{d}_2} & S_{n22}^{\mathrm{d}_2} \end{bmatrix}\begin{Bmatrix} P_2^{\mathrm{d}_2} \\ P_4 \end{Bmatrix} = \begin{Bmatrix} Q_2^{\mathrm{d}_2} \\ Q_4 \end{Bmatrix} + \begin{Bmatrix} f_{n1}^{\mathrm{d}_2} \\ f_{n2}^{\mathrm{d}_2} \end{Bmatrix} \quad \text{(daughter 2)}
\end{aligned} \tag{13.83}$$

where the nonlinear pseudo-forces are omitted for brevity. By using the bifurcation conditions (13.81) and (13.82), three spectral element equations (13.83) can be readily assembled in the form of

$$\begin{bmatrix} S_{n11}^{\mathrm{p}} & S_{n12}^{\mathrm{p}} & 0 & 0 \\ S_{n21}^{\mathrm{p}} & S_{n22}^{\mathrm{p}} + S_{n11}^{\mathrm{d1}} + S_{n11}^{\mathrm{d2}} & S_{n12}^{\mathrm{d1}} & S_{n12}^{\mathrm{d2}} \\ 0 & S_{n21}^{\mathrm{d1}} & S_{n22}^{\mathrm{d1}} & 0 \\ 0 & S_{n21}^{\mathrm{d2}} & 0 & S_{n22}^{\mathrm{d2}} \end{bmatrix}\begin{Bmatrix} P_1 \\ P_2 \\ P_3 \\ P_4 \end{Bmatrix} = \begin{Bmatrix} Q_1 \\ 0 \\ Q_3 \\ Q_4 \end{Bmatrix} + \begin{Bmatrix} f_{n1}^{\mathrm{p}} \\ f_{n2}^{\mathrm{p}} + f_{n1}^{\mathrm{d}_1} + f_{n2}^{\mathrm{d}_2} \\ f_{n2}^{\mathrm{d}_1} \\ f_{n2}^{\mathrm{d}_2} \end{Bmatrix} \tag{13.84}$$

13.5 Finite Element Model

The weak form of the time-domain governing equation (13.16) is obtained from

$$\int_0^L [c_0 p'' - c_1 \ddot{p} - c_1 c_2 \dot{p} - 2c_1 c_3 \dot{p}' + c_1 c_4 \dot{p}'' - f(x,t)]\, \delta p\, \mathrm{d}x = 0 \tag{13.85}$$

By applying the integral by parts and using Equations 13.15 and 13.27, we obtain the weak form as

$$\begin{aligned}
&c_1 c_2^{-1} \int_0^L \ddot{p}\, \delta p\, \mathrm{d}x + c_1 \int_0^L \dot{p} \delta p\, \mathrm{d}x + c_1 c_2^{-1} c_3 \left(\int_0^L \dot{p}' \delta p\, \mathrm{d}x - \int_0^L \dot{p} \delta p'\, \mathrm{d}x \right) \\
&+ c_1 c_2^{-1} c_4 \int_0^L \dot{p}' \delta p'\, \mathrm{d}x + c_0 c_2^{-1} \int_0^L p' \delta p'\, \mathrm{d}x \\
&= (-c_2^{-1} \dot{q} - q + c_1 c_2^{-1} c_3 \dot{p} - c_2^{-1} c_7 + c_2^{-1} z)\, \delta p \Big|_0^L \\
&- 2c_1 c_2^{-1} c_5 \int_0^L q\dot{p}\, \delta p\, \mathrm{d}x + c_1^2 c_2^{-1} c_6 \int_0^L p\dot{p}\, \delta p\, \mathrm{d}x - c_1 c_2^{-1} c_6 \int_0^L q\dot{p}' \delta p\, \mathrm{d}x
\end{aligned} \tag{13.86}$$

where

$$z = -c_5 z_1 + c_1 c_6 z_2 \tag{13.87}$$

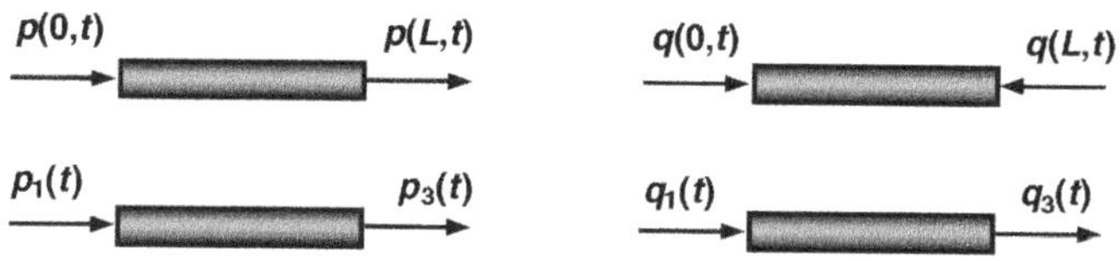

Figure 13.6 Three-node finite element model

For the three-node finite element shown in Figure 13.6, the blood flow rate and pressure are assumed from

$$p(x,t) = \boldsymbol{N}_p(x)\boldsymbol{p}(t), \qquad q(x,t) = \boldsymbol{N}_q(x)\boldsymbol{q}(t) \tag{13.88}$$

where

$$\begin{aligned} \boldsymbol{p}(t) &= \{p_1(t) \quad p_2(t) \quad p_3(t)\}^{\mathrm{T}} = \{p(0,t) \quad p(L/2,t) \quad p(L,t)\}^{\mathrm{T}} \\ \boldsymbol{q}(t) &= \{q_1(t) \quad q_3(t)\}^{\mathrm{T}} = \{q(0,t) \quad q(L,t)\}^{\mathrm{T}} \end{aligned} \tag{13.89}$$

and

$$\begin{aligned} \boldsymbol{N}_p(x) &= [1-3\xi+2\xi^2 \quad 4\xi-4\xi^2 \quad -\xi+2\xi^2] \\ \boldsymbol{N}_q(x) &= [1-3\xi+2\xi^2 \quad \xi-2\xi^2] \\ \xi &= x/L \end{aligned} \tag{13.90}$$

The weighted-integral statement of Equation 13.9 is given by

$$\int_0^L (c_1\dot{p}+q')\delta q \mathrm{d}x = 0 \tag{13.91}$$

Substitution of Equation 13.88 into Equation 13.91 yields a complementary relationship as

$$\boldsymbol{q} = -c_1\boldsymbol{R}_1\dot{\boldsymbol{p}} \tag{13.92}$$

where

$$\boldsymbol{R}_1 = -\frac{L}{40}\begin{bmatrix} 11 & 8 & 1 \\ 1 & 8 & 11 \end{bmatrix} \tag{13.93}$$

By substituting Equation 13.88 into Equation 13.86 and using Equation 13.92, we can obtain the finite element equation in the form of

$$\boldsymbol{M}\ddot{\boldsymbol{p}} + \boldsymbol{C}\dot{\boldsymbol{p}} + \boldsymbol{K}\boldsymbol{p} = \boldsymbol{q}_L + \boldsymbol{f}_{NL} + \boldsymbol{f}_C \tag{13.94}$$

where

$$\begin{aligned} \boldsymbol{M} &= c_1c_2^{-1}\int_0^L \boldsymbol{N}_p(x)^{\mathrm{T}}\boldsymbol{N}_p(x)\mathrm{d}x + c_1c_2^{-1}\boldsymbol{R}_2 \\ \boldsymbol{C} &= c_1\int_0^L \boldsymbol{N}_p(x)^{\mathrm{T}}\boldsymbol{N}_p(x)\mathrm{d}x + c_1c_2^{-1}c_3\int_0^L [\boldsymbol{N}_p(x)^{\mathrm{T}}\boldsymbol{N}'_p(x) - \boldsymbol{N}'_p(x)^{\mathrm{T}}\mathbf{N}_p(x)]\mathrm{d}x \\ &\quad + c_1c_2^{-1}c_4\int_0^L \boldsymbol{N}'_p(x)^{\mathrm{T}}\boldsymbol{N}'_p(x)\mathrm{d}x - c_1c_2^{-1}c_3\boldsymbol{N}_p^{\mathrm{T}}(x)\boldsymbol{N}_p(x)\Big|_0^L \\ \boldsymbol{K} &= c_0c_2^{-1}\int_0^L \boldsymbol{N}'_p(x)^{\mathrm{T}}\boldsymbol{N}'_p(x)\mathrm{d}x \end{aligned} \tag{13.95}$$

and

$$
\begin{aligned}
\boldsymbol{q}_L &= -\boldsymbol{N}_p^{\mathrm{T}}(x)\boldsymbol{N}_q(x)\Big|_0^L \boldsymbol{q} = \{q_1 \quad 0 \quad q_2\}^{\mathrm{T}} \\
\boldsymbol{f}_{NL} &= \Big[-2c_1c_2^{-1}c_5\int_0^L \boldsymbol{N}_p(x)^{\mathrm{T}}\boldsymbol{N}_q(x)\boldsymbol{q}(t)\boldsymbol{N}_p(x)\mathrm{d}x \\
&\quad + c_1^2c_2^{-1}c_6\int_0^L \boldsymbol{N}_p(x)^{\mathrm{T}}\boldsymbol{N}_p(x)\dot{\boldsymbol{p}}(t)\boldsymbol{N}_p(x)\mathrm{d}x \\
&\quad - c_1c_2^{-1}c_6\int_0^L \boldsymbol{N}_p(x)^{\mathrm{T}}\boldsymbol{N}_q(x)\boldsymbol{q}(t)\boldsymbol{N}_p'(x)dx\Big]\dot{\boldsymbol{p}}(t) \\
&\quad - c_5c_2^{-1}\boldsymbol{N}_p^{\mathrm{T}}(x)\boldsymbol{N}_q(x)\boldsymbol{q}\boldsymbol{N}_q(x)\Big|_0^L\boldsymbol{q} + c_1c_2^{-1}c_6\boldsymbol{N}_p^{\mathrm{T}}(x)\boldsymbol{N}_q(x)\boldsymbol{q}\boldsymbol{N}_p(x)\Big|_0^L\dot{\boldsymbol{p}} \\
\boldsymbol{f}_C &= -c_2^{-1}c_7\boldsymbol{N}_p(x)\Big|_0^L = c_2^{-1}c_7\{1 \quad 0 \quad -1\}^{\mathrm{T}}
\end{aligned}
\tag{13.96}
$$

with

$$
\boldsymbol{R}_2 = -\frac{L}{40}\begin{bmatrix} 11 & 8 & 1 \\ 0 & 0 & 0 \\ 1 & 8 & 11 \end{bmatrix} \tag{13.97}
$$

The finite element matrices $\boldsymbol{M}, \boldsymbol{C}, \boldsymbol{K}$, and the nodal force vector $\boldsymbol{f}_{NL}$ computed by using the shape functions (13.90) are displayed in Appendix 13.A.

13.6 Numerical Examples

Example 13.1

To demonstrate the performance of the spectral element model, consider a straight uniform blood vessel where the blood flow has a parabolic velocity profile, as an example problem. The blood vessel has the inlet radius $r_0 = 1.25$ cm and the taper $\theta = 0$. The properties of the blood are: density $\rho = 1.055$ g/cm^3 and kinematic viscosity $\nu = 0.046$ cm^2/s. For the parabolic velocity profile, the parameters $\delta = 1/3$ and $H = 8\pi$ are given from Table 13.1. Figure 13.7

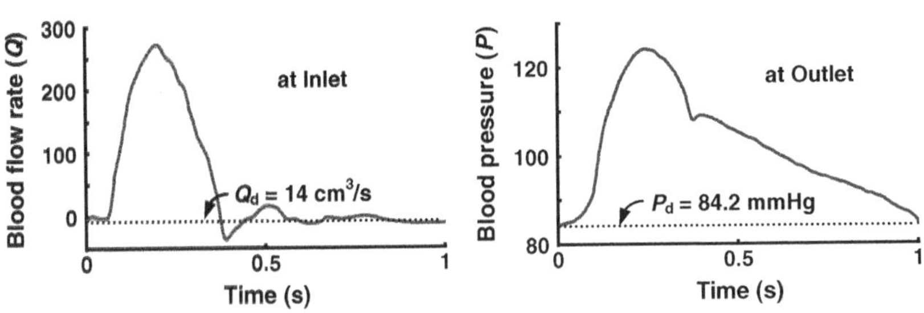

Figure 13.7 The inlet blood flow rate and the outlet blood pressure used as the boundary conditions

shows the blood flow rate at the inlet and the blood pressure at the outlet, which are used as the boundary conditions. The blood flow rates and pressures at an interim point of the blood vessel are predicted by using the one-element spectral model and they are compared with those predicted by using the FEM where the number of finite elements used in the analysis is varied. Figure 13.8 shows that FEM results converge towards the SEM results as the number of finite elements used in the finite element analysis is increased, which certainly shows the high accuracy of the present spectral element model.

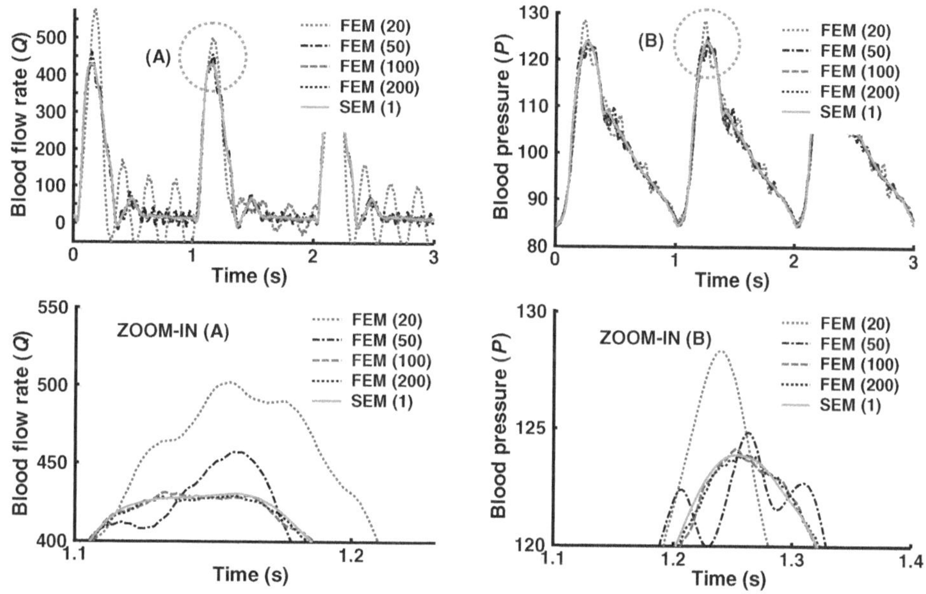

Figure 13.8 Comparison of the blood flow rates and pressures obtained by SEM and FEM(n), where n is the number of finite elements used in the analysis

Example 13.2

As the second illustrative example, two types of bifurcated blood vessels are considered: one is symmetric and the other one is asymmetric. As shown in Figure 13.9, each

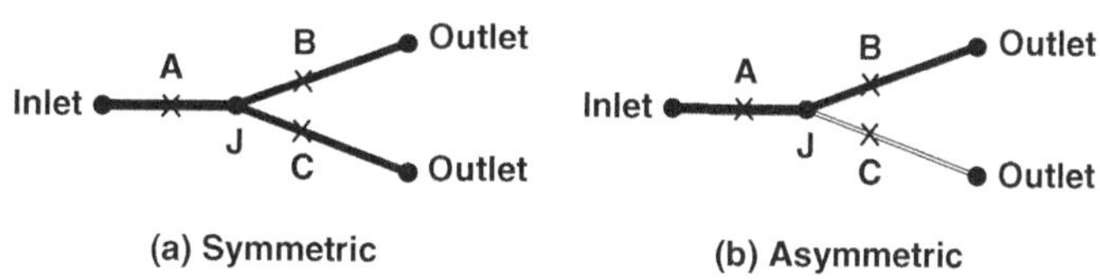

Figure 13.9 Two examples of bifurcated blood vessels

bifurcated blood vessel consists of a uniform parent vessel (from the inlet to junction J) and two uniform daughter vessels (from junction J to the two outlets). For the asymmetric bifurcated blood vessel, the inner radius of a daughter vessel is larger than that of the other daughter vessel. The blood flow rate at the inlet and the blood pressure at both outlets of each bifurcated blood vessel are shown in Figure 13.10 and they are used as the boundary conditions.

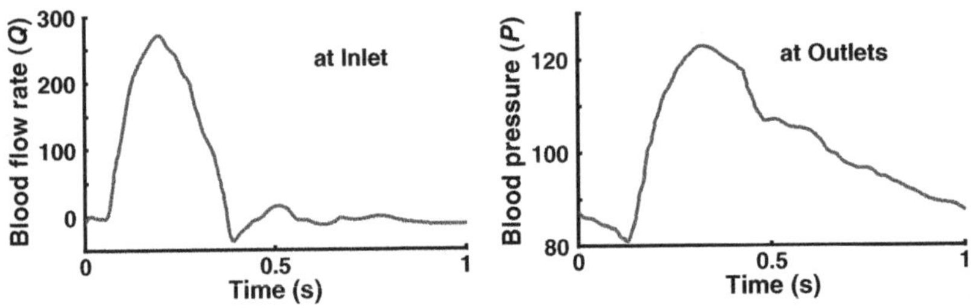

Figure 13.10 The blood flow rate pressure used as the boundary conditions

Figures 13.11 shows the blood flow rates and pressures predicted at A, B, and C of the symmetric bifurcated blood vessel. Similarly, Figure 13.12 shows the predicted values for the asymmetric bifurcated blood vessel. As would be expected, the predicted blood flow rate and pressure at B are same as those at C for the case of the symmetric bifurcated blood vessel, while it is not true for the case of the symmetric bifurcated blood vessel.

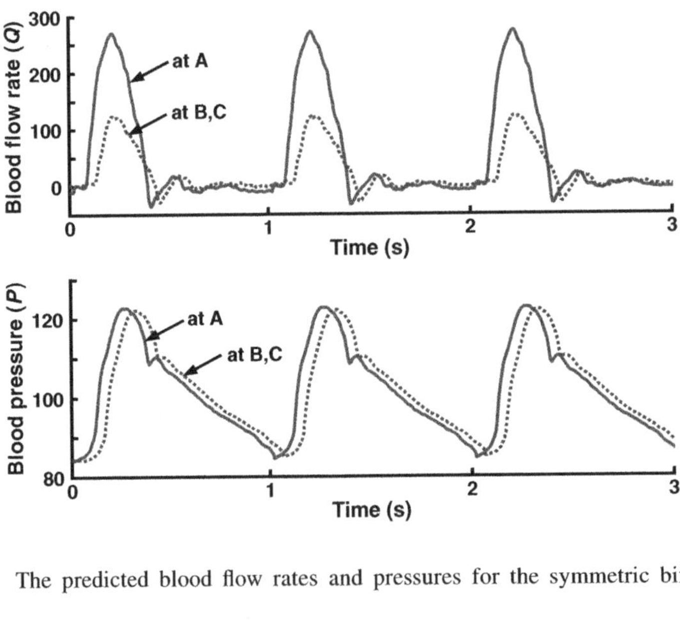

Figure 13.11 The predicted blood flow rates and pressures for the symmetric bifurcated blood vessel

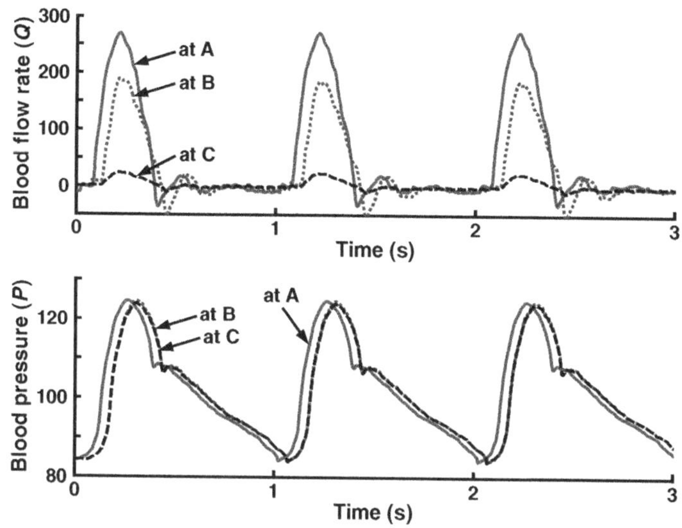

Figure 13.12 The predicted blood flow rates and pressures for the asymmetric bifurcated blood vessel

Appendix 13.A: Finite Element Model for the 1-D Blood Flow

The finite element matrices $\boldsymbol{M}$, $\boldsymbol{C}$, $\boldsymbol{K}$, and the nodal force vector $\boldsymbol{f}_{NL}$ defined in Equation 13.94 are given by

$$
\begin{aligned}
\boldsymbol{M} &= \frac{c_1}{c_2}\frac{L}{30}\begin{bmatrix} 4 & 2 & -1 \\ 2 & 16 & 2 \\ -1 & 2 & 4 \end{bmatrix} - \frac{c_1}{c_2}\frac{L}{40}\begin{bmatrix} 11 & 8 & 1 \\ 0 & 0 & 0 \\ 1 & 8 & 11 \end{bmatrix} \\
\mathbf{C} &= \frac{c_1 L}{30}\begin{bmatrix} 4 & 2 & -1 \\ 2 & 16 & 2 \\ -1 & 2 & 4 \end{bmatrix} + \frac{c_1 c_3}{3c_2}\begin{bmatrix} 0 & 4 & -1 \\ -4 & 0 & 4 \\ 1 & -4 & 0 \end{bmatrix} \\
&\quad + \frac{c_1 c_4}{3c_2 L}\begin{bmatrix} 7 & -8 & 1 \\ -8 & 16 & -8 \\ 1 & -8 & 7 \end{bmatrix} - \frac{c_1 c_3}{c_2}\begin{bmatrix} -1 & 0 & 0 \\ 0 & 0 & 0 \\ 0 & 0 & 1 \end{bmatrix} \\
\boldsymbol{K} &= \frac{c_0}{3c_2 L}\begin{bmatrix} 7 & -8 & 1 \\ -8 & 16 & -8 \\ 1 & -8 & 7 \end{bmatrix}
\end{aligned}
\tag{13.A.1}
$$

and

$$
\begin{aligned}
\boldsymbol{f}_{NL} &= \left(-\frac{c_1 c_5}{c_2}\frac{L}{210}\boldsymbol{F}_1 + \frac{c_1^2 c_6}{c_2}\frac{L}{420}\boldsymbol{F}_2 - \frac{1}{30}\frac{c_1 c_6}{c_2}\boldsymbol{F}_3\right)\dot{\boldsymbol{p}}(t) \\
&\quad - \frac{c_5}{c_2}\{-q_1^2 \quad 0 \quad q_3^2\}^{\mathrm{T}} + \frac{c_1 c_6}{c_2}\{-q_1\dot{p}_1 \quad 0 \quad -q_3\dot{p}_3\}^{\mathrm{T}}
\end{aligned}
\tag{13.A.2}
$$

where

$$
\begin{aligned}
\boldsymbol{F}_1 &= \begin{bmatrix} 3(13q_1+q_3) & 4(5q_1+2q_3) & 3(q_3-q_1) \\ 4(5q_1+2q_3) & 16(q_1-q_3) & -4(2q_1+5q_3) \\ 3(q_3-q_1) & -4(2q_1+5q_3) & -3(q_1+13q_3) \end{bmatrix} \\
\boldsymbol{F}_2 &= \begin{bmatrix} 39\dot{p}_1+20\dot{p}_2-3\dot{p}_3 & 4(5\dot{p}_1+4\dot{p}_2-2\dot{p}_3) & -(3\dot{p}_1+8\dot{p}_2+3\dot{p}_3) \\ 4(5\dot{p}_1+4\dot{p}_2-2\dot{p}_3) & 16(\dot{p}_1+12\dot{p}_2+\dot{p}_3) & 4(-2\dot{p}_1+4\dot{p}_2+5\dot{p}_3) \\ -(3\dot{p}_1+8\dot{p}_2+3\dot{p}_3) & 4(-2\dot{p}_1+4\dot{p}_2+5\dot{p}_3) & (-3\dot{p}_1+20\dot{p}_2+39\dot{p}_3) \end{bmatrix} \\
\boldsymbol{F}_3 &= \begin{bmatrix} -10q_1-q_3 & 12q_1 & q_3-2q_1 \\ -6q_1-2q_3 & 8q_1+8q_3 & -2q_1-6q_3 \\ q_1-2q_3 & 12q_3 & -q_1-10q_3 \end{bmatrix}
\end{aligned}
\tag{13.A.3}
$$

14

Identification of Structural Boundaries and Joints

The primary objective in the identification of structural parameters is to derive a successful mathematical model of a structure, which reproduces the observed dynamic behavior of the structure. However, experience has shown that a mathematical model often provides inaccurate dynamic behavior of a structure when compared with the measurements. This discrepancy is mostly due to the use of inaccurate structural parameters such as the mass, stiffness, and damping properties of a structure. The use of exact boundary conditions (BCs)is also important to predict the satisfactory dynamic behavior of a structure. However, as the exact boundary conditions are not always available in most practical situations, it is very common in most structural dynamic analyses to represent the real boundaries of a structure by the textbook-like ideal boundary conditions, such as the perfectly free, clamped, or simply-supported boundary conditions. Thus, the boundary condition should be considered as an additional structural parameter to be identified when providing more reliable structural dynamic analyses [86, 157]. This is also true for the structural joints in a structure [82, 128, 155, 173]. Thus, in this chapter, we introduce the spectral element model (SEM) based frequency-domain methods to identify the non-ideal boundary conditions and joints in a beam-type structure.

14.1 Identification of Non-Ideal Boundary Conditions

The real, non-ideal boundary conditions of a uniform beam can be represented by the use of two types of equivalent springs placed at the boundaries: a transverse spring K_w and a rotational spring K_θ. The stiffness of the equivalent boundary springs will in general be frequency-dependent, instead of being constant. The three well known types of ideal boundary conditions can be classified depending on the stiffness of the two springs at a boundary: the boundary is perfectly (ideally) "clamped" if all boundary springs have infinite stiffness, "free" if all boundary springs have zero stiffness, and "simply supported" if the transverse spring has infinite stiffness while the rotational spring has zero stiffness. As in most practical situations, if two equivalent springs at a boundary do not satisfy one of the above ideal boundary conditions, the boundary can not be considered as an ideal boundary condition. Thus, to conduct successful structural dynamics analysis for a structural system with such non-ideal boundaries, we must identify the dynamic characteristics of non-ideal boundaries in terms of the frequency-dependent (dynamic) stiffness of equivalent boundary springs. Figure 14.1 shows the general procedure for identifying the non-ideal boundary conditions.

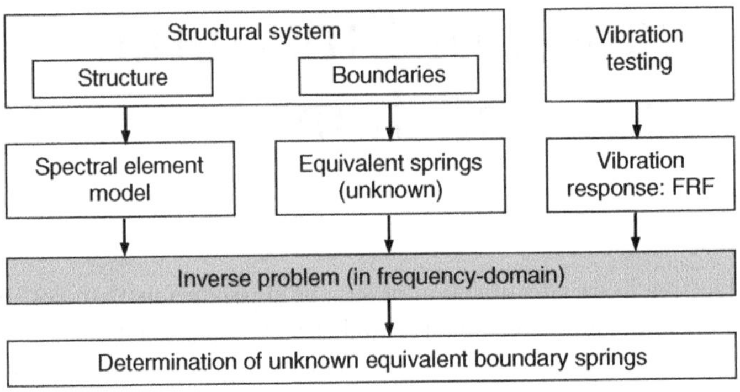

Figure 14.1 Procedure of the identification of the boundary conditions

In the following, we will discuss a method to identify the dynamic stiffness of equivalent boundary springs by using measured frequency response functions (FRFs) [86]. The identification method is then applied to a one-end supported (cantilever-type) beam and a two-ends supported (simply-supported type) beam as an illustrative problem.

14.1.1 One-End Supported Beam

As the first illustrative problem, consider a one-end supported beam as shown in Figure 14.2. The boundary is represented by two equivalent boundary springs: a transverse spring K_v and a rotational spring K_ϕ. We assume that an excitation force $p(t)$ is applied at an interim point to excite the beam.

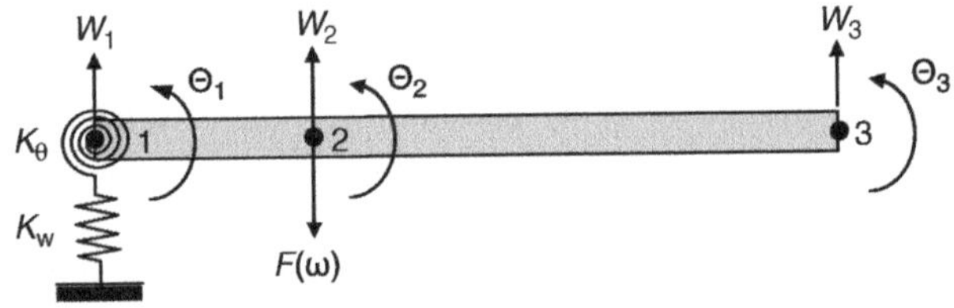

Figure 14.2 One-end supported beam

The beam is divided into two elements so that the interim point at which the excitation force is applied becomes a node connecting two elements. The spectral element models for various beam models are given in Chapter 3. Assembling two spectral Bernoulli–Euler beam elements, for instance, we can obtain the 6-by-6 global system equation as

$$\begin{bmatrix} s_{11}^1 & s_{12}^1 & s_{13}^1 & s_{14}^1 & 0 & 0 \\ s_{21}^1 & s_{22}^1 & s_{23}^1 & s_{24}^1 & 0 & 0 \\ s_{31}^1 & s_{32}^1 & s_{33}^1+s_{11}^2 & s_{34}^1+s_{12}^2 & s_{13}^2 & s_{14}^2 \\ s_{41}^1 & s_{42}^1 & s_{43}^1+s_{21}^2 & s_{44}^1+s_{22}^2 & s_{23}^2 & s_{24}^2 \\ 0 & 0 & s_{31}^2 & s_{32}^2 & s_{33}^2 & s_{34}^2 \\ 0 & 0 & s_{41}^2 & s_{42}^2 & s_{43}^2 & s_{44}^2 \end{bmatrix} \begin{Bmatrix} W_1 \\ \Theta_1 \\ W_2 \\ \Theta_2 \\ W_3 \\ \Theta_3 \end{Bmatrix} = \begin{Bmatrix} -K_w W_1 \\ -K_\theta \Theta_1 \\ F(\omega) \\ 0 \\ 0 \\ 0 \end{Bmatrix} \tag{14.1}$$

or, simply

$$\boldsymbol{S}(\omega)\boldsymbol{d} = \boldsymbol{f}(\omega) \tag{14.2}$$

where $F(\omega)$ are the spectral components of $f(t)$ and the superscripts 1 and 2 represent the quantities of the first and second spectral beam elements, respectively. The spectral nodal degrees-of-freedom (DOFs) appeared in Equation 14.1 can be classified into two groups: one group includes the DOFs that will be measured by experiments and the other group includes all remaining DOFs that will not or cannot be measured. In this typical example, as the transverse displacements W_2 and W_3 at nodes 2 and 3 can be readily measured by experiments, and they are considered to be the measured DOFs. It will not be as easy to access the boundary node 1 to measure W_1 and Θ_1. Thus W_1 andΘ_1 will be considered as the unmeasured DOFs. In general it is more difficult to measure the rotations than to measure the transverse displacements. Thus, Θ_2 and Θ_3 will be also considered as the unmeasured DOFs. Hence, there are a total of six unknowns: two springs K_w and K_θ and four unmeasured DOFs W_1, Θ_1, Θ_2, and Θ_3. These six unknowns will be determined from Equation 14.1 by using W_2 and W_3, which are measured by applying an excitation force $F(\omega)$ to the beam. In summary, we represent the measured DOFs by the vector $\boldsymbol{d}_m$ and the unmeasured DOFs by the vector $\boldsymbol{d}_u$.

Dividing the spectral nodal DOFs vector $\boldsymbol{d}$ into $\boldsymbol{d}_m$ (measured DOFs, known) and $\boldsymbol{d}_u$ (unmeasured DOFs, unknown) and the forcing vector $\boldsymbol{f}$ into $\boldsymbol{f}_b$ (boundary spring forces, unknown) and $\boldsymbol{f}_e$ (excitation force, known), we can write Equation 14.1 in a partitioned form as

$$\begin{bmatrix} \boldsymbol{S}_{bm} & \boldsymbol{S}_{bu} \\ \boldsymbol{S}_{em} & \boldsymbol{S}_{eu} \end{bmatrix} \begin{Bmatrix} \boldsymbol{d}_m \\ \boldsymbol{d}_u \end{Bmatrix} = \begin{Bmatrix} \boldsymbol{f}_b \\ \boldsymbol{f}_e \end{Bmatrix} \tag{14.3}$$

where

$$\begin{aligned} \boldsymbol{d}_m &= \{W_2 \quad W_3\}^T \\ \boldsymbol{d}_u &= \{W_1 \quad \Theta_1 \quad \Theta_2 \quad W_3\}^T \\ \boldsymbol{f}_b &= \{-K_w W_1 \quad -K_\theta \Theta_1\}^T \\ \boldsymbol{f}_e &= \{F \quad 0 \quad 0 \quad 0\}^T \end{aligned} \tag{14.4}$$

and

$$\boldsymbol{S}_{\mathrm{bm}} = \begin{bmatrix} s_{13}^1 & 0 \\ s_{23}^1 & 0 \end{bmatrix}, \qquad \boldsymbol{S}_{\mathrm{bu}} = \begin{bmatrix} s_{11}^1 & s_{12}^1 & s_{14}^1 & 0 \\ s_{21}^1 & s_{22}^1 & s_{24}^1 & 0 \end{bmatrix}$$

$$\boldsymbol{S}_{\mathrm{em}} = \begin{bmatrix} s_{33}^1 + s_{13}^2 & s_{13}^2 \\ s_{43}^1 + s_{21}^2 & s_{23}^2 \\ s_{31}^2 & s_{33}^2 \\ s_{41}^2 & s_{43}^2 \end{bmatrix}, \quad \boldsymbol{S}_{\mathrm{eu}} = \begin{bmatrix} s_{31}^1 & s_{32}^1 & s_{34}^1 + s_{12}^2 & s_{14}^2 \\ s_{41}^1 & s_{42}^1 & s_{44}^1 + s_{22}^2 & s_{24}^2 \\ 0 & 0 & s_{32}^2 & s_{34}^2 \\ 0 & 0 & s_{42}^2 & s_{44}^2 \end{bmatrix} \tag{14.5}$$

It is very convenient and customary for us to measure vibration responses by using accelerometers. Thus, it will be convenient to use the nodal transverse and rotational inertances in Equation 14.3 in place of the transverse and rotational nodal DOFs. The transverse and rotational inertances are the frequency response functions defined by the ratio of the transverse and rotational accelerations to the excitation force as follows:

$$a = -\frac{\omega^2}{F} W, \quad \alpha = -\frac{\omega^2}{F} \Theta \tag{14.6}$$

By using Equation 14.6, we can rewrite Equation 14.3 in terms of the nodal inertances as

$$\begin{bmatrix} \boldsymbol{S}_{\mathrm{bm}} & \boldsymbol{S}_{\mathrm{bu}} \\ \boldsymbol{S}_{\mathrm{em}} & \boldsymbol{S}_{\mathrm{eu}} \end{bmatrix} \begin{Bmatrix} \boldsymbol{D}_{\mathrm{m}} \\ \boldsymbol{D}_{\mathrm{u}} \end{Bmatrix} = \begin{Bmatrix} \boldsymbol{F}_{\mathrm{b}} \\ \boldsymbol{F}_{\mathrm{e}} \end{Bmatrix} \tag{14.7}$$

where

$$\begin{aligned} \boldsymbol{D}_{\mathrm{m}} &= -\frac{\omega^2}{F} \boldsymbol{d}_{\mathrm{m}} = \{a_2 \quad a_3\}^{\mathrm{T}} \\ \boldsymbol{D}_{\mathrm{u}} &= -\frac{\omega^2}{F} \boldsymbol{d}_{\mathrm{u}} = \{a_1 \quad \alpha_1 \quad \alpha_2 \quad \alpha_3\}^{\mathrm{T}} \\ \boldsymbol{F}_{\mathrm{b}} &= -\frac{\omega^2}{F} \boldsymbol{f}_{\mathrm{b}} = \{-K_w a_1 \quad -K_\theta \alpha_1\}^{\mathrm{T}} \\ \boldsymbol{F}_{\mathrm{e}} &= -\frac{\omega^2}{F} \boldsymbol{f}_{\mathrm{e}} = \{-\omega^2 \quad 0 \quad 0 \quad 0\}^{\mathrm{T}} \end{aligned} \tag{14.8}$$

where a_i ($i = 1, 2, 3$) are the nodal transverse inertances and α_i ($i = 1, 2, 3$) are the nodal rotational inertances.

The unmeasured nodal inertances vector $\boldsymbol{D}_{\mathrm{u}}$ and two boundary springs K_w and K_θ will be computed from Equation 14.7. To this end, we write the third row of Equation 14.8 in the form of

$$\begin{Bmatrix} K_w \\ K_\theta \end{Bmatrix} = -\begin{bmatrix} 1/a_1 & 0 \\ 0 & 1/\alpha_1 \end{bmatrix} \boldsymbol{F}_{\mathrm{b}} \tag{14.9}$$

From the second row of Equation 14.7, we obtain

$$\boldsymbol{D}_{\mathrm{u}} = \boldsymbol{S}_{\mathrm{eu}}^{-1} (\boldsymbol{F}_{\mathrm{e}} - \boldsymbol{S}_{\mathrm{em}} \boldsymbol{D}_{\mathrm{m}}) \tag{14.10}$$

By using Equation 14.10, we can eliminate $\boldsymbol{D}_{\mathrm{u}}$ from the first row of Equation 14.7 to obtain

$$\boldsymbol{F}_{\mathrm{b}} = (\boldsymbol{S}_{\mathrm{bm}} - \boldsymbol{S}_{\mathrm{bu}}\boldsymbol{S}_{\mathrm{eu}}^{-1}\boldsymbol{S}_{\mathrm{em}})\boldsymbol{D}_{\mathrm{m}} + \boldsymbol{S}_{\mathrm{bu}}\boldsymbol{S}_{\mathrm{eu}}^{-1}\boldsymbol{F}_{\mathrm{e}} \tag{14.11}$$

The two unknown springs K_w and K_θ can be then computed from Equation 14.9 by substituting Equation 14.11 into Equation 14.9.

14.1.2 Two-Ends Supported Beam

As the second illustrative problem, consider a two-ends supported beam as shown in Figure 14.3. The non-ideal boundary conditions at two end nodes (node 1 and node 6) are represented by four boundary springs: two transverse springs K_{w1} and K_{w2}, and two rotational springs $K_{\theta 1}$ and $K_{\theta 2}$.

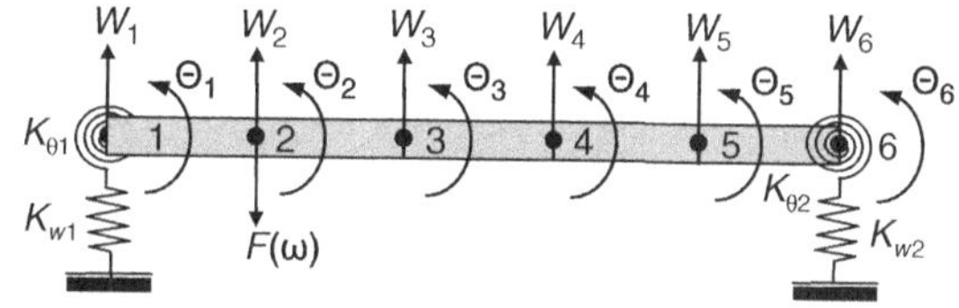

Figure 14.3 Two-ends supported beam

If the beam is represented by a two-element model as is the situation for the one end supported beam, the dimension of the assembled global system matrix equation (or the number of equations) will be six, while there will be more than six unknowns, including four boundary springs, four unmeasurable DOFs at both end nodes and one rotational DOF at the excitation point. As the number of unknowns is larger than the number of equations, we need more equations to make the inverse problem well-posed.

The number of equations can be increased by increasing the number of spectral elements. For the present problem, if we increase the number of spectral elements by one, the number of global nodal DOFs will be increased by two, while the number of unmeasurable DOFs (i.e., rotational DOFs) will be increased by one. Thus, as shown in Figure 14.3, the beam is represented by five spectral elements so that the total number of nodal DOFs and the dimensions of the global system matrix equation become 12. If W_2, W_3, W_4, and W_5 from among the 12 nodal DOFs are considered to be measured experimentally, the remaining eight nodal DOFs (W_1, W_6, Θ_1, Θ_2, Θ_3, Θ_4, Θ_5, and Θ_6) will be considered as the unmeasured DOFs. In this instance, the number of unknowns becomes 12 (eight unmeasured DOFs and four boundary springs), which is equal to the number of equations. The 12 unknowns can readily be solved from 12 equations by four measured DOFs, W_2, W_3, W_4, and W_5. If the number of measured DOFs happens to be smaller than the number of unknowns, more spectral elements will be required to increase the number of equations so that it becomes equal to the number of unknowns, and vice versa.

Going back to the illustrative problem of Figure 14.3, we assemble five spectral elements to obtain a 12-by-12 global system matrix equation as

$$\boldsymbol{S}(\omega)\boldsymbol{d} = \boldsymbol{f} \tag{14.12}$$

where

$$\begin{aligned} \boldsymbol{d} &= \{W_1 \quad \Theta_1 \quad W_2 \quad \Theta_2 \quad W_3 \quad \Theta_3 \quad W_4 \quad \Theta_4 \quad W_5 \quad \Theta_5 \quad W_6 \quad \Theta_6\}^{\mathrm{T}} \\ \boldsymbol{f} &= \{-K_{w1}W_1 \quad -K_{\theta 1}\Theta_1 \quad F \quad 0 \quad 0 \quad 0 \quad 0 \quad 0 \quad 0 \quad 0 \quad -K_{w2}W_6 \quad -K_{\theta 2}\Theta_6\}^{\mathrm{T}} \end{aligned} \tag{14.13}$$

Similarly, as done in the previous section, Equation 14.12 can be written in a partitioned form as

$$\begin{bmatrix} \boldsymbol{S}_{\mathrm{bm}} & \boldsymbol{S}_{\mathrm{bu}} \\ \boldsymbol{S}_{\mathrm{em}} & \boldsymbol{S}_{\mathrm{eu}} \end{bmatrix} \begin{Bmatrix} \boldsymbol{d}_{\mathrm{m}} \\ \boldsymbol{d}_{\mathrm{u}} \end{Bmatrix} = \begin{Bmatrix} \boldsymbol{f}_{\mathrm{b}} \\ \boldsymbol{f}_{\mathrm{e}} \end{Bmatrix} \tag{14.14}$$

where

$$\begin{aligned} \boldsymbol{d}_{\mathrm{m}} &= \{W_2 \quad W_3 \quad W_2 \quad W_3\}^{\mathrm{T}} \\ \boldsymbol{d}_{\mathrm{u}} &= \{W_1 \quad W_6 \quad \Theta_1 \quad \Theta_2 \quad \Theta_3 \quad \Theta_4 \quad \Theta_5 \quad \Theta_6\}^{\mathrm{T}} \\ \boldsymbol{f}_{\mathrm{b}} &= \{-K_{w1}W_1 \quad -K_{\theta 1}\Theta_1 \quad -K_{w2}W_6 \quad -K_{\theta 2}\Theta_6\}^{\mathrm{T}} \\ \boldsymbol{f}_{\mathrm{e}} &= \{F(\omega) \quad 0 \quad 0 \quad 0 \quad 0 \quad 0 \quad 0 \quad 0\}^{\mathrm{T}} \end{aligned} \tag{14.15}$$

and

$$\boldsymbol{S}_{\mathrm{bm}} = \begin{bmatrix} s_{13}^1 & 0 & 0 & 0 \\ s_{23}^1 & 0 & 0 & 0 \\ 0 & 0 & 0 & s_{31}^5 \\ 0 & 0 & 0 & s_{41}^5 \end{bmatrix}$$

$$\boldsymbol{S}_{\mathrm{bu}} = \begin{bmatrix} s_{11}^1 & 0 & s_{12}^1 & s_{14}^1 & 0 & 0 & 0 & 0 \\ s_{21}^1 & 0 & s_{22}^1 & s_{24}^1 & 0 & 0 & 0 & 0 \\ 0 & s_{33}^5 & 0 & 0 & 0 & 0 & s_{32}^5 & s_{34}^5 \\ 0 & s_{43}^5 & 0 & 0 & 0 & 0 & s_{42}^5 & s_{44}^5 \end{bmatrix}$$

$$\boldsymbol{S}_{\mathrm{em}} = \begin{bmatrix} s_{33}^1 + s_{11}^2 & s_{13}^2 & 0 & 0 \\ s_{43}^1 + s_{21}^2 & s_{23}^2 & 0 & 0 \\ s_{31}^2 & s_{33}^2 + s_{11}^3 & s_{13}^3 & 0 \\ s_{41}^2 & s_{43}^2 + s_{21}^3 & s_{23}^3 & 0 \\ 0 & s_{31}^3 & s_{33}^3 + s_{11}^4 & s_{13}^4 \\ 0 & s_{41}^3 & s_{43}^3 + s_{21}^4 & s_{23}^4 \\ 0 & 0 & s_{31}^4 & s_{33}^4 + s_{11}^5 \\ 0 & 0 & s_{41}^4 & s_{43}^4 + s_{21}^5 \end{bmatrix} \tag{14.16}$$

$$S_{eu} = \begin{bmatrix} s_{31}^1 & 0 & s_{32}^1 & s_{34}^1 + s_{12}^2 & s_{14}^2 & 0 & 0 & 0 \\ s_{41}^1 & 0 & s_{42}^1 & s_{44}^1 + s_{22}^2 & s_{24}^2 & 0 & 0 & 0 \\ 0 & 0 & 0 & s_{32}^2 & s_{34}^2 + s_{12}^3 & s_{14}^3 & 0 & 0 \\ 0 & 0 & 0 & s_{42}^2 & s_{44}^2 + s_{22}^3 & s_{24}^3 & 0 & 0 \\ 0 & 0 & 0 & 0 & s_{32}^3 & s_{34}^3 + s_{12}^4 & s_{14}^4 & 0 \\ 0 & 0 & 0 & 0 & s_{42}^3 & s_{44}^3 + s_{22}^4 & s_{24}^4 & 0 \\ 0 & s_{13}^5 & 0 & 0 & 0 & s_{32}^4 & s_{34}^4 + s_{12}^5 & s_{14}^5 \\ 0 & s_{23}^5 & 0 & 0 & 0 & s_{42}^4 & s_{44}^4 + s_{22}^5 & s_{24}^5 \end{bmatrix}$$

Equation 14.14 can be rewritten in terms of the nodal inertances defined by Equation 14.6 as

$$\begin{bmatrix} S_{bm} & S_{bu} \\ S_{em} & S_{eu} \end{bmatrix} \begin{Bmatrix} D_m \\ D_u \end{Bmatrix} = \begin{Bmatrix} F_b \\ F_e \end{Bmatrix} \tag{14.17}$$

where

$$\begin{aligned} D_m &= -\frac{\omega^2}{F} d_m = \{a_2 \quad a_3 \quad a_4 \quad a_5\}^T \\ D_u &= -\frac{\omega^2}{F} d_u = \{a_1 \quad a_6 \quad \alpha_1 \quad \alpha_2 \quad \alpha_3 \quad \alpha_4 \quad \alpha_5 \quad \alpha_6\}^T \\ F_b &= -\frac{\omega^2}{F} f_b = \{-K_{w1}a_1 \quad -K_{\theta 1}\alpha_1 \quad -K_{w2}a_6 \quad -K_{\theta 2}\alpha_6\}^T \\ F_e &= -\frac{\omega^2}{F} f_e = \{-\omega^2 \quad 0 \quad 0 \quad 0 \quad 0 \quad 0 \quad 0 \quad 0\}^T \end{aligned} \tag{14.18}$$

From the third equation of 14.18, the boundary springs can be written in terms of the boundary force F_b as

$$\begin{Bmatrix} K_{w1} \\ K_{\theta 1} \\ K_{w2} \\ K_{\theta 2} \end{Bmatrix} = -\begin{bmatrix} 1/a_1 & 0 & 0 & 0 \\ 0 & 1/\alpha_1 & 0 & 0 \\ 0 & 0 & 1/a_6 & 0 \\ 0 & 0 & 0 & 1/\alpha_6 \end{bmatrix} F_b \tag{14.19}$$

From the second row of Equation 14.17, we obtain

$$D_u = S_{eu}^{-1}(F_e - S_{em}D_m) \tag{14.20}$$

By substituting Equation 14.20 into the first row of Equation 14.17, the unknown boundary force F_b can be obtained in terms of the known quantities D_m and F_e as follows:

$$F_b = (S_{bm} - S_{bu}S_{eu}^{-1}S_{em})D_m + S_{bu}S_{eu}^{-1}F_e \tag{14.21}$$

Finally, the unknown stiffnesses of the boundary springs can be computed by substituting Equation 14.21 into Equation 14.19.

Example 14.1

An experiment is conducted for a one-end supported aluminum beam as shown in Figure 14.2. The exciting force at node 2 and the accelerations measured at nodes 2 and 3 are transformed into the frequency domain by using the fast Fourier transform (FFT) algorithm and they are used to compute the transverse inertance at node 1 and the rotational inertances at nodes 1,2, and 3 by using Equation 14.10. Figure 14.4 shows the measured transverse inertances (FRFs) at nodes 2 and 3, while Figure 14.5 shows the predicted rotational inertances (FRFs) at nodes 2 and 3. The equivalent boundary springs are identified by using Equation 14.9 and they are displayed in Figure 14.6. The identified equivalent boundary springs have frequency-dependent complex stiffnesses.

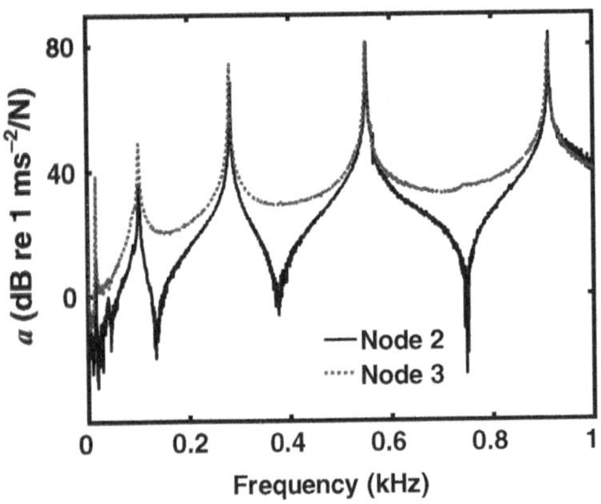

Figure 14.4 Measured transverse FRFs

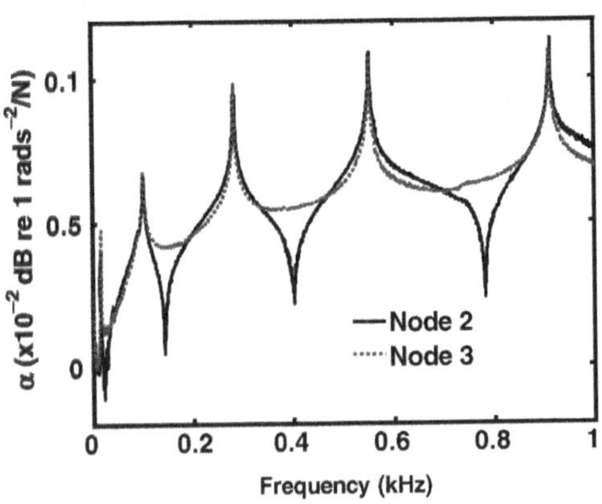

Figure 14.5 Predicted rotational FRFs

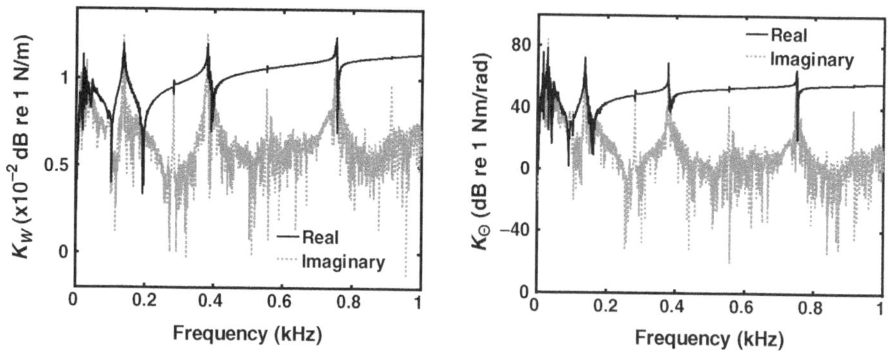

Figure 14.6 Predicted boundary spring stiffness

Figure 14.7 compares the transverse inertance measured at node 3 with the results predicted by using identified boundary conditions and ideal boundary conditions. When identified boundary conditions are used, the predicted transverse inertance is shown to be almost identical with the measured data. However, when ideal boundary conditions are used instead, the predicted transverse inertance deviates significantly from the measured data, especially at high frequency. Similarly, Figure 14.8 compares the measured time history of the acceleration with the results predicted by using identified boundary conditions and ideal boundary conditions. The time history predicted by using identified boundary conditions is shown to be very close to the measured data. A small discrepancy is observed with the lapse of time, which seems to be in part due to the inherent numerical errors or the lack of the higher frequency spectral components beyond the Nyquist frequency. In contrast to the prediction by using identified boundary conditions, Figure 14.8 certainly shows that a fairly large discrepancy exists between the measure data and the prediction when using ideal boundary conditions.

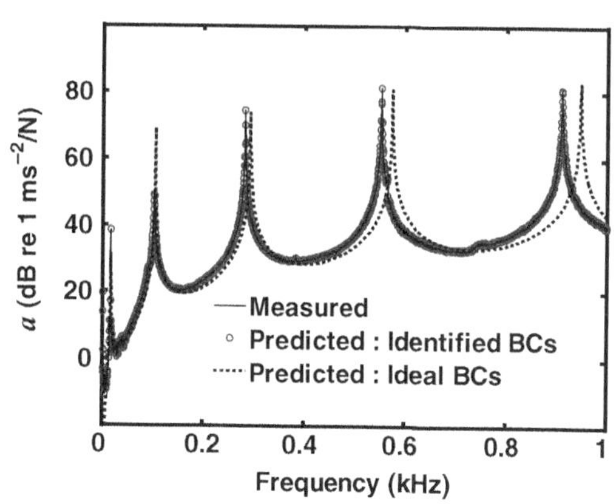

Figure 14.7 Measured and predicted transverse FRFs at node 3

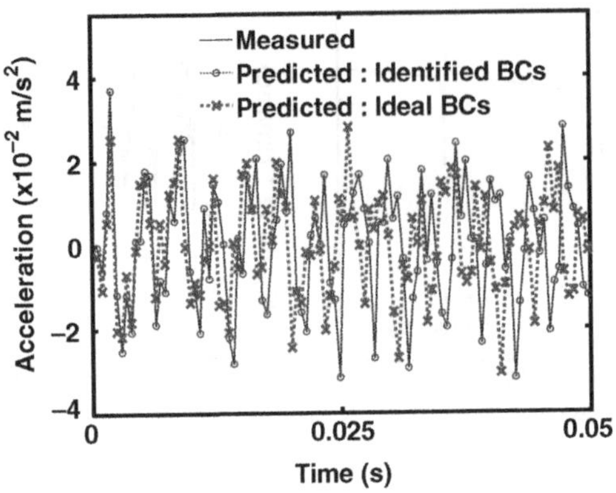

Figure 14.8 Measured and predicted time responses at node 3

Example 14.2

An experiment is also conducted for a two-ends supported aluminum beam as shown in Figure 14.3. The transverse inertances are measured at nodes 2, 3, 4, and 5 and only the transverse inertances at nodes 2 and 4 are displayed in Figure 14.9 for brevity. By using the measured transverse inertances, the transverse inertances at nodes 1 and 6 and the rotational inertances at nodes 1 through 6 are predicted from Equation 14.20 and only two rotational inertances at nodes 2 and 4 are displayed in Figure 14.10. Similarly, the transverse and rotational boundary spring stiffness predicted at node 1 are displayed in Figure 14.11. The transverse inertance and time history measured at node 4 are compared with those predicted by using identified boundary conditions and ideal boundary conditions in Figures 14.12 and 14.13, respectively. The dynamic responses predicted by using identified boundary conditions are shown to be almost identical with the measured data. This may imply the importance of the use of accurately identified non-ideal boundary conditions.

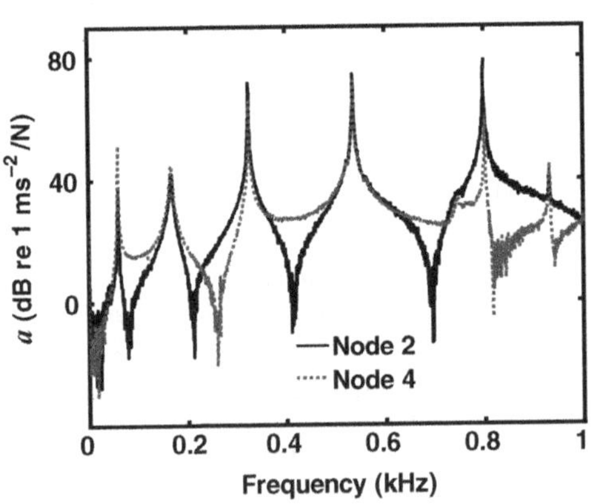

Figure 14.9 Measured transverse FRFs

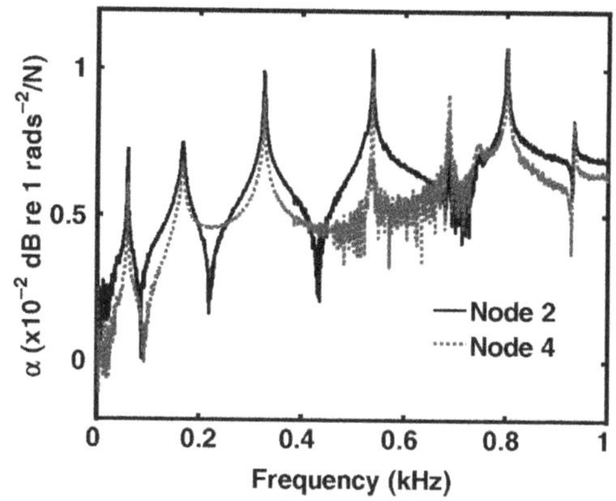

Figure 14.10 Predicted rotational FRFs

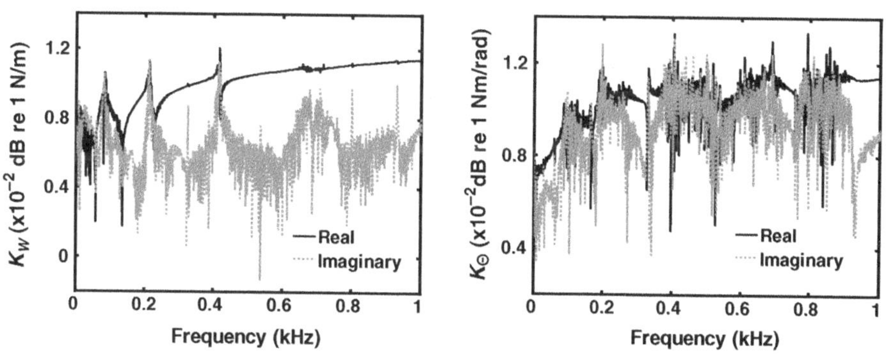

Figure 14.11 Predicted boundary spring stiffness at node 1

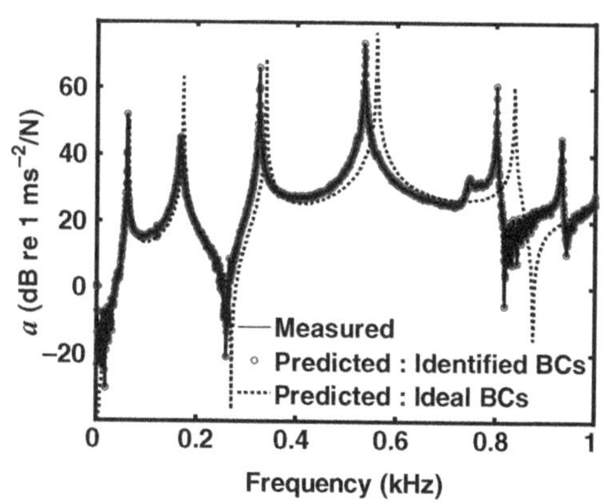

Figure 14.12 Measured and predicted transverse FRFs at node 4

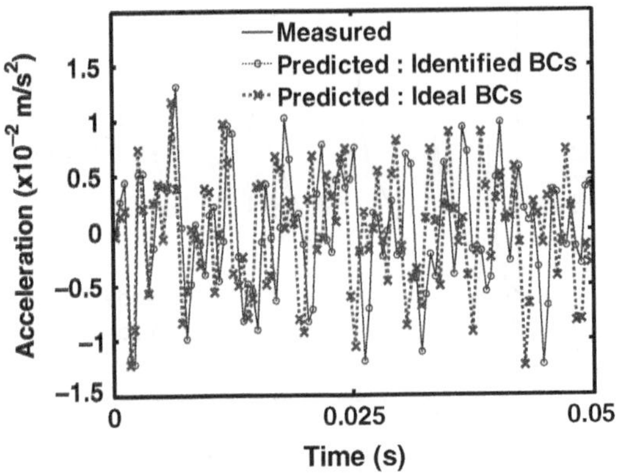

Figure 14.13 Measured and predicted time responses at node 4

14.2 Identification of Joints

A cantilevered beam structure shown in Figure 14.14 is considered as an example problem. The beam structure consists of two uniform beams that are connected to each other by using a bolt-joint at the midway point. To apply the spectral element analysis technique to the vibration analysis of the beam structure, two uniform beams are represented by the spectral Timoshenko beam (T-beam) element models that were introduced in Chapter 3. Accordingly, the joint part will be represented by the equivalent spectral element model characterized by four parameters to be determined by using experimentally measured FRFs [82].

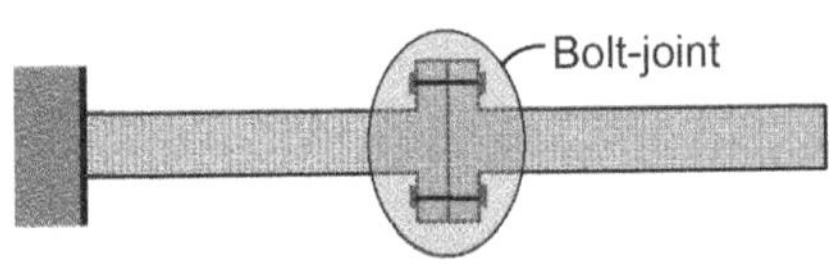

Figure 14.14 A cantilevered beam structure with a bolt-joint

14.2.1 Spectral T-Beam Element Model for Uniform Beam Parts

The uniform beam parts to the left and right of the joint part will be considered as the uniform T-beam elements and they will be represented by the spectral T-beam elements. For a uniform T-beam element, the spectral element equation is given by

$$\boldsymbol{S}_{\mathrm{T}}(\omega)\,\boldsymbol{d} = \boldsymbol{f} \tag{14.22}$$

where $\boldsymbol{S}_{\mathrm{T}}(\omega) = [s_{ij}(\omega)]$ is the 4-by-4 spectral element matrix given by Equation 3.86, $\boldsymbol{d}$ is the spectral nodal DOFs vector and $\boldsymbol{f}$ is the associated spectral nodal forces and moments vector.

Once the joint part has been represented by equivalent spectral element model, which will be detailed in the next section, it can be assembled with two spectral T-beam elements to form a global dynamic stiffness equation as an example beam structure.

14.2.2 Equivalent Spectral Element Model of the Joint Part

Consider an example joint part as shown in Figure 14.15(a). The joint part will be represented by an equivalent four-parameter model as shown in Figure 14.15(b). The four-parameter model consists of a lumped mass M, a lumped mass moment of inertia I, an effective extensional spring K_w, and an effective torsional spring K_θ. The inertia parameters M and I are connected to the spring parameters K_w and K_θ in series. The identification of the joint part will be completed by determining four parameters M, I, K_w, and K_θ by using experimentally measured vibration responses, which is detailed in the following.

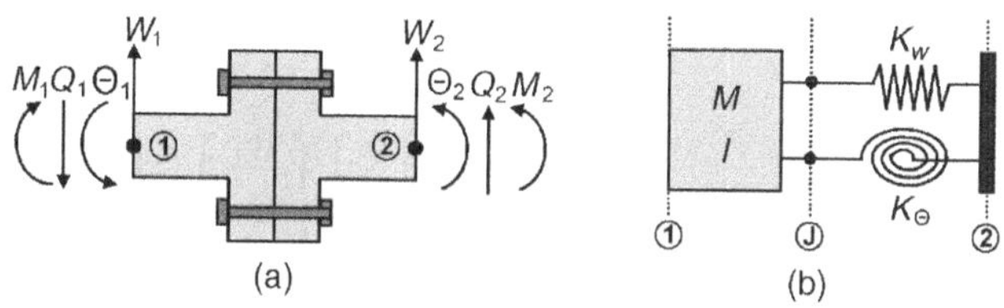

Figure 14.15 A joint part and its equivalent four-parameter model

14.2.2.1 Transfer Matrix for the Joint Model

To derive the transfer matrix of the joint model, we use the sign convention defined in Figure 14.15(a), which is commonly adopted in the transfer matrix method [138]. From the kinematics continuity and dynamic equilibrium of the mass-system, the transfer matrix equation for the mass-system can be derived as

$$\begin{Bmatrix} Q_1 \\ W_1 \\ M_1 \\ \Theta_1 \end{Bmatrix} = \begin{bmatrix} 1 & M\omega^2 & 0 & 0 \\ 0 & 1 & 0 & 0 \\ 0 & 0 & 1 & I\omega^2 \\ 0 & 0 & 0 & 1 \end{bmatrix} \begin{Bmatrix} Q_J \\ W_J \\ M_J \\ \Theta_J \end{Bmatrix} \tag{14.23}$$

where Q_J, W_J, M_J, and Θ_J are the quantities at the junction "J" between the inertia-system and the spring-system, as shown in Figure 14.15(b). Similarly, the transfer matrix equation for the spring-system can be derived as

$$\begin{Bmatrix} Q_J \\ W_J \\ M_J \\ \Theta_J \end{Bmatrix} = \begin{bmatrix} 1 & 0 & 0 & 0 \\ -1/K_w & 1 & 0 & 0 \\ 0 & 0 & 1 & 0 \\ 0 & 0 & -1/K_\theta & 1 \end{bmatrix} \begin{Bmatrix} Q_2 \\ W_2 \\ M_2 \\ \Theta_2 \end{Bmatrix} \tag{14.24}$$

By combining Equations 14.23 and 14.24, we can obtain a transfer matrix equation that relates the nodal values at the left of the joint part (node 1) to those at the right of the joint part (node 2) as follows:

$$\begin{Bmatrix} Q_2 \\ W_2 \\ M_2 \\ \Theta_2 \end{Bmatrix} = \begin{bmatrix} 1 & -M\omega^2 & 0 & 0 \\ 1/K_w & 1-M\omega^2/K_w & 0 & 0 \\ 0 & 0 & 1 & -I\omega^2 \\ 0 & 0 & 1/K_\theta & 1-I\omega^2/K_\theta \end{bmatrix} \begin{Bmatrix} Q_1 \\ W_1 \\ M_1 \\ \Theta_1 \end{Bmatrix} \tag{14.25}$$

or

$$\begin{Bmatrix} W_2 \\ \Theta_2 \\ Q_2 \\ M_2 \end{Bmatrix} = \begin{bmatrix} 1-M\omega^2/K_w & 0 & 1/K_w & 0 \\ 0 & 1-I\omega^2/K_\theta & 0 & 1/K_\theta \\ -M\omega^2 & 0 & 1 & 0 \\ 0 & -I\omega^2 & 0 & 1 \end{bmatrix} \begin{Bmatrix} W_1 \\ \Theta_1 \\ Q_1 \\ M_1 \end{Bmatrix} \tag{14.26}$$

14.2.2.2 Equivalent Spectral Element Model for the Joint Model

The input–output relationship of Equation 14.26 can be written in the force-displacement relationship as

$$\begin{bmatrix} -K_w+M\omega^2 & 0 & K_w & 0 \\ 0 & -K_\theta+I\omega^2 & 0 & K_\theta \\ -K_w & 0 & K_w & 0 \\ 0 & -K_\theta & 0 & K_\theta \end{bmatrix} \begin{Bmatrix} W_1 \\ \Theta_1 \\ W_2 \\ \Theta_2 \end{Bmatrix} = \begin{Bmatrix} Q_1 \\ M_1 \\ Q_2 \\ M_2 \end{Bmatrix} \tag{14.27}$$

In Equation 14.27, the sign convention defined for the transfer matrix is used for the nodal DOFs, forces, and moments. As shown in Figure 14.16, the sign conventions of Q_1 and M_1 defined for the transfer matrix are opposite to those defined for the spectral element matrix. Thus, by simply changing the signs of Q_1 and M_1 in Equation 14.27, we can obtain the spectral element equation for the joint part in the form of

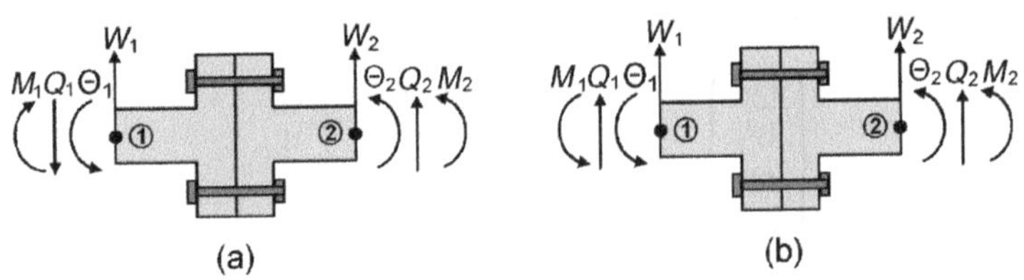

Figure 14.16 Sign convention defined for: (a) transfer matrix; (b) spectral element matrix

$$\boldsymbol{S}_J(\omega)\,\boldsymbol{d}_J = \boldsymbol{f}_J \tag{14.28}$$

where

$$\boldsymbol{S}_J(\omega) = \begin{bmatrix} K_w-M\omega^2 & 0 & -K_w & 0 \\ 0 & K_\theta-I\omega^2 & 0 & -K_\theta \\ -K_w & 0 & K_w & 0 \\ 0 & -K_\theta & 0 & K_\theta \end{bmatrix} \tag{14.29}$$

and

$$\begin{aligned} \boldsymbol{d}_J &= \{W_1\ \Theta_1\ W_2\ \Theta_2\}^{\mathrm{T}} \\ \boldsymbol{f}_J &= \{Q_1\ M_1\ Q_2\ M_2\}^{\mathrm{T}} \end{aligned} \tag{14.30}$$

The frequency-dependent symmetric matrix $S_J(\omega)$ is the spectral element matrix for the joint part and it contains four unknown parameters M, I, K_w, and K_θ. The four parameters will be determined from experimentally measured FRFs.

14.2.3 Determination of Joint Parameters

A uniform structure member can be represented by one spectral element, regardless of its length, without degrading the solution accuracy. However, if any discontinuities exist in the structural or material properties, the structure member needs to be discretized into multiple spectral elements.

The present example beam structure contains a bolt-joint and it is excited by a force applied at node 1 as shown in Figure 14.17. Thus, we perfom a first trial to discretize the beam structure into four segments: three T-beam elements and a joint element. In Figure 14.17, three T-beam

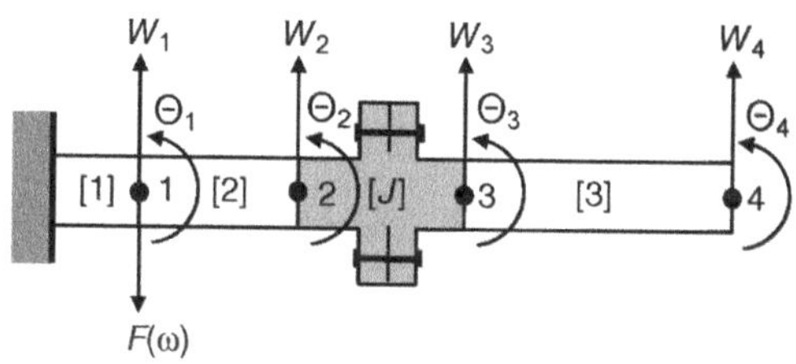

Figure 14.17 Spectral element modeling for a cantilevered beam structure with a bolt-joint

elements are denoted by [1], [2], [3] and the joint element by [J]. As each node of the spectral T-beam element has two nodal DOFs, there are a total of eight nodal DOFs: four transverse nodal DOFs (W_1, W_2, W_3, and W_4) and four rotational nodal DOFs (Θ_1,Θ_2,Θ_3, and Θ_4). Assembling three spectral T-beam elements and the spectral joint element and then imposing the clamped boundary conditions at the root of the beam structure, we can obtain an 8-by-8 global dynamic stiffness matrix equation in the form of

$$\boldsymbol{S}(\omega)\,\boldsymbol{d} = \boldsymbol{f} \tag{14.31}$$

where

$$\begin{aligned} \boldsymbol{d} &= \{W_1 \quad \Theta_1 \quad W_2 \quad \Theta_2 \quad W_3 \quad \Theta_3 \quad W_4 \quad \Theta_4\}^{\mathrm{T}} \\ \boldsymbol{f} &= \{F(\omega) \quad 0 \quad 0 \quad 0 \quad 0 \quad 0 \quad 0 \quad 0\}^{\mathrm{T}} \end{aligned} \tag{14.32}$$

where $F(\omega)$ represents the spectral components of the excitation force.

As it is somewhat easier to measure the transverse displacements than to measure the rotations, we may prefer to measure some or all of transverse DOFs experimentally. For the present example problem, we assume that all transverse DOFs (W_1, W_2, W_3, and W_4) are measured experimentally and grouped into the measured DOFs. Accordingly, the remaining rotational DOFs (Θ_1,Θ_2,Θ_3, and Θ_4) will be grouped into the unmeasured DOFs. In this case, there are a total of eight unknowns including four joint parameters (M, I, K_w, and K_θ) and four unmeasured DOFs (Θ_1,Θ_2,Θ_3, and Θ_4). As the number of unknowns is equal to the number of equations provided by Equation 14.31, the first trial of structural discretization is successful for the formulation of a well-posed inverse problem. However, if for any reason there are more unknowns, we may need another trial to discretize the beam structure into more segments so that more equations can be acquired to obtain a well-posed inverse problem. For the present example problem, assuming that all transverse DOFs are measured, the first trial to discretize the beam structure into four segments as shown in Figure 14.16 suffices to obtain a well-posed inverse problem.

For short, we represent the measured DOFs by $\boldsymbol{d}_{\mathrm{m}}$ and the unmeasured DOFs by $\boldsymbol{d}_{\mathrm{u}}$ as

$$\begin{aligned} \boldsymbol{d}_{\mathrm{m}} &= \{W_1 \quad W_2 \quad W_3 \quad W_4\}^{\mathrm{T}} \\ \boldsymbol{d}_{\mathrm{u}} &= \{\Theta_1 \quad \Theta_2 \quad \Theta_3 \quad \Theta_4\}^{\mathrm{T}} \end{aligned} \tag{14.33}$$

By rearranging the nodal DOFs and forces, Equation 14.31 can be written in a partitioned form as

$$\begin{bmatrix} \boldsymbol{S}_{0\mathrm{m}} & \boldsymbol{S}_{0\mathrm{u}} \\ \boldsymbol{S}_{\mathrm{em}} & \boldsymbol{S}_{\mathrm{eu}} \end{bmatrix} \begin{Bmatrix} \boldsymbol{d}_{\mathrm{m}} \\ \boldsymbol{d}_{\mathrm{u}} \end{Bmatrix} = \begin{Bmatrix} \boldsymbol{0} \\ \boldsymbol{f}_{\mathrm{e}} \end{Bmatrix} \tag{14.34}$$

where $\boldsymbol{0}$ is the zero vector and

$$\boldsymbol{f}_{\mathrm{e}} = \{F(\omega) \quad 0 \quad 0 \quad 0\}^{\mathrm{T}} \tag{14.35}$$

and

$$\begin{aligned} \boldsymbol{S}_{0\mathrm{m}} &= \begin{bmatrix} s_{31}^2 & s_{33}^2 + (K_w - M\omega^2) & -K_w & 0 \\ s_{41}^2 & s_{43}^2 & 0 & 0 \\ 0 & -K_w & s_{11}^3 + K_w & s_{13}^3 \\ 0 & 0 & s_{21}^3 & s_{23}^3 \end{bmatrix} \\ \boldsymbol{S}_{0\mathrm{u}} &= \begin{bmatrix} s_{32}^2 & s_{34}^2 & 0 & 0 \\ s_{42}^2 & s_{44}^2 + (K_\theta - I\omega^2) & -K_\theta & 0 \\ 0 & 0 & s_{12}^3 & s_{14}^3 \\ 0 & -K_\theta & s_{22}^3 + K_\theta & s_{24}^3 \end{bmatrix} \\ \boldsymbol{S}_{\mathrm{em}} &= \begin{bmatrix} s_{33}^1 + s_{11}^2 & s_{13}^2 & 0 & 0 \\ s_{43}^1 + s_{21}^2 & s_{23}^2 & 0 & 0 \\ 0 & 0 & s_{31}^3 & s_{33}^3 \\ 0 & 0 & s_{41}^3 & s_{43}^3 \end{bmatrix} \\ \boldsymbol{S}_{\mathrm{eu}} &= \begin{bmatrix} s_{34}^1 + s_{12}^2 & s_{14}^2 & 0 & 0 \\ s_{44}^1 + s_{22}^2 & s_{24}^2 & 0 & 0 \\ 0 & 0 & s_{32}^3 & s_{34}^3 \\ 0 & 0 & s_{42}^3 & s_{44}^3 \end{bmatrix} \end{aligned} \tag{14.36}$$

The superscripts 1, 2 and 3 in Equation 14.36 denote the finite element number defined in Figure 14.17.

By using the transverse and rotational inertances defined by Equation 14.6, we can write Equation 14.34 as

$$\begin{bmatrix} \boldsymbol{S}_{0\mathrm{m}} & \boldsymbol{S}_{0\mathrm{u}} \\ \boldsymbol{S}_{\mathrm{em}} & \boldsymbol{S}_{\mathrm{eu}} \end{bmatrix} \begin{Bmatrix} \boldsymbol{D}_{\mathrm{m}} \\ \boldsymbol{D}_{\mathrm{u}} \end{Bmatrix} = \begin{Bmatrix} \boldsymbol{0} \\ \boldsymbol{F}_{\mathrm{e}} \end{Bmatrix} \tag{14.37}$$

where

$$\begin{aligned} \boldsymbol{D}_{\mathrm{m}} &= -\frac{\omega^2}{F}\boldsymbol{d}_{\mathrm{m}} = \{a_1 \quad a_2 \quad a_3 \quad a_4\}^{\mathrm{T}} \\ \boldsymbol{D}_{\mathrm{u}} &= -\frac{\omega^2}{F}\boldsymbol{d}_{\mathrm{u}} = \{\alpha_1 \quad \alpha_2 \quad \alpha_3 \quad \alpha_4\}^{\mathrm{T}} \\ \boldsymbol{F}_{\mathrm{e}} &= -\frac{\omega^2}{F}\boldsymbol{f}_{\mathrm{e}} = \{-\omega^2 \quad 0 \quad 0 \quad 0\}^{\mathrm{T}} \end{aligned} \tag{14.38}$$

From the second row of Equation 14.37, the unmeasured nodal inertances $\boldsymbol{D}_{\mathrm{u}}$ can be obtained as

$$\boldsymbol{D}_{\mathrm{u}} = \boldsymbol{S}_{\mathrm{eu}}^{-1}\left(\boldsymbol{F}_{\mathrm{e}} - \boldsymbol{S}_{\mathrm{em}}\boldsymbol{D}_{\mathrm{m}}\right) \tag{14.39}$$

Equation 14.39 shows that unmeasured nodal inertances $\boldsymbol{D}_{\mathrm{u}}$ can be predicted by using the excitation force $\boldsymbol{F}_{\mathrm{e}}$ and measured nodal inertances $\boldsymbol{D}_{\mathrm{m}}$. From the first row of Equation 14.37, the joint parameters can be explicitly expressed in terms of the measured and unmeasured inertances as follows:

$$\begin{aligned} K_W &= \frac{1}{a_2 - a_3}\left(s_{11}^3 a_3 + s_{12}^3 \alpha_3 + s_{13}^3 a_3 + s_{14}^3 \alpha_4\right) \\ K_\Theta &= \frac{1}{\alpha_2 - \alpha_3}\left(s_{21}^3 a_3 + s_{22}^3 \alpha_3 + s_{23}^3 a_4 + s_{24}^3 \alpha_4\right) \\ M &= \frac{1}{a_2\omega^2}\left[K_w(a_2 - a_3) + s_{31}^2 a_1 + s_{32}^2 \alpha_1 + s_{33}^2 a_2 + s_{34}^2 \alpha_2\right] \\ I &= \frac{1}{\alpha_2\omega^2}\left[K_\theta(\alpha_2 - \alpha_3) + s_{41}^2 a_1 + s_{43}^2 a_2 + s_{42}^2 \alpha_1 + s_{44}^2 \alpha_2\right] \end{aligned} \tag{14.40}$$

The unmeasured inertances (a_1, a_2, a_3, a_4) are first predicted from Equation 14.39 and the results are then substituted into Equation 14.40 to obtain four unknown joint parameters. Once the joint part is identified in terms of four parameters, the vibration analysis can be readily accomplished for the beam structure by using the standard spectral element analysis procedure.

In above discussion, the ideal "clamped" boundary condition is applied to the root of the example beam structure. However, as in most practical situations, if the boundaries of a structural system cannot be represented by using such ideal boundary conditions, the present method of joint identification can be applied in conjunction with the method of boundary conditions identification discussed in the previous section, in order to identify the non-ideal boundary conditions simultaneously.

Example 14.3

Experiments are conducted for an aluminum beam structure with a bolt-joint as shown in Figure 14.18. Two uniform beams of equal length are connected at the mid-point by a bolt-joint.

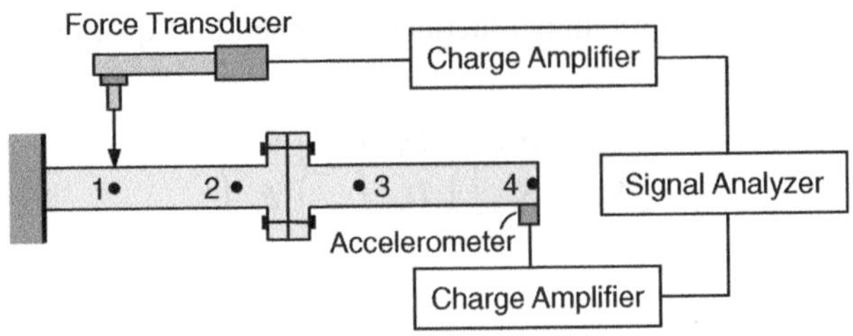

Figure 14.18 Schematic diagram of the experimental setup

The transverse inertances are measured at nodes 1–4 and, among them, the values at nodes 2 and 4 are shown in Figure 14.19. Figure 14.20 shows the rotational inertances at nodes 2 and 4 predicted by using measured inertances. Figures 14.21 and 14.22 show the predicted inertia properties and spring stiffness of the joint model, respectively. The transverse inertance at node

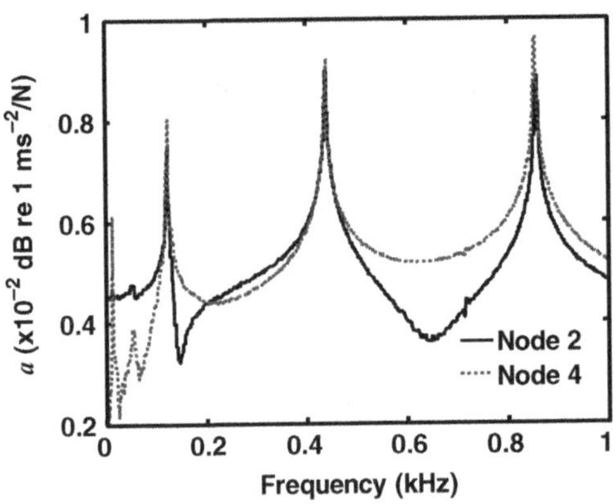

Figure 14.19 Measured transverse FRFs

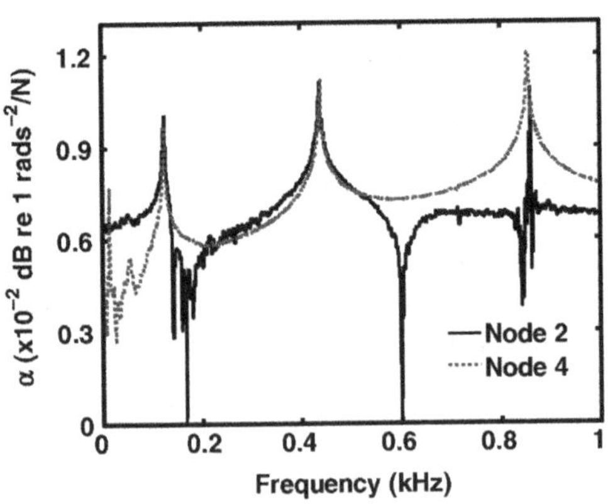

Figure 14.20 Predicted rotational FRFs

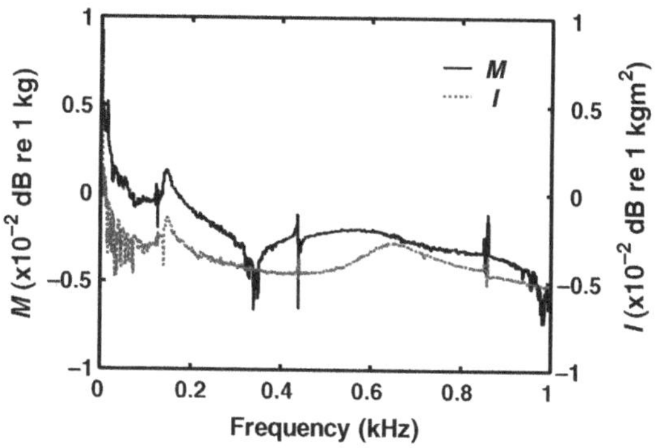

Figure 14.21 Predicted inertia properties of the joint

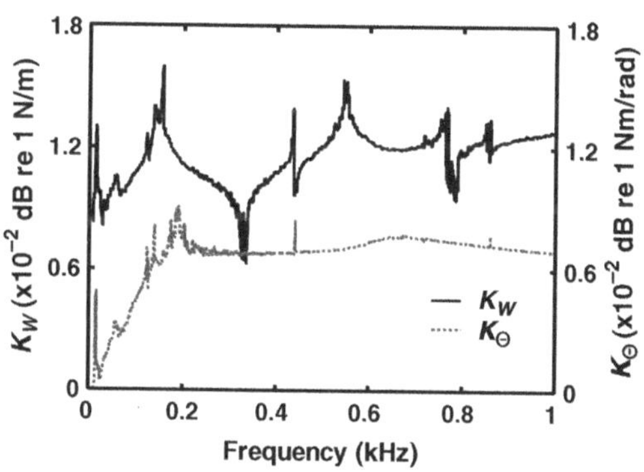

Figure 14.22 Predicted spring stiffness of the joint

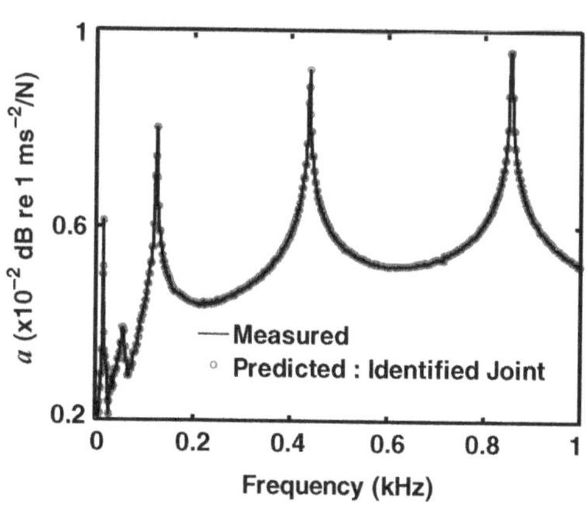

Figure 14.23 Measured and predicted transverse FRFs at node 4

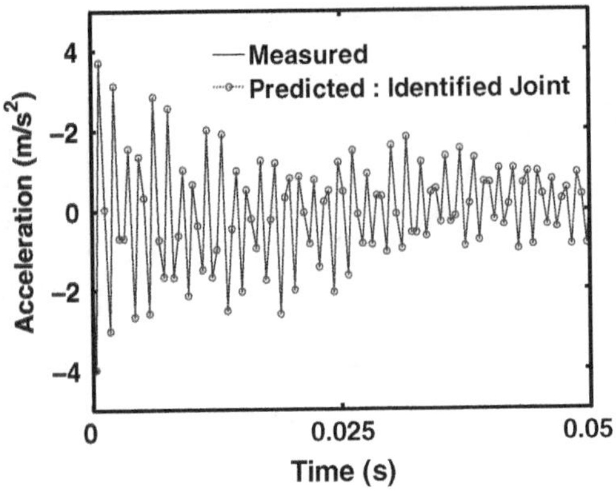

Figure 14.24 Measured and predicted time responses at node 4

4 (free-end) is predicted by using identified joint parameters and it is compared with the measured data in Figure 14.23. It is shown that the predicted and measured inertances are in good agreement with each other. Similarly, the time history of the transverse acceleration at node 4 is also predicted by using identified joint parameters and the result is compared with the measured data in Figure 14.24. The predicted and measured time histories are shown to be very close to each other, but with a small discrepancy. As the predicted time history is computed from the predicted FRF by using the inverse-FFT algorithm, the small discrepancy may, in part, be due to neglecting the high frequency spectral components beyond the Nyquist frequency and also in part due to the inherent computation errors.

15

Identification of Structural Damage

The structural damage generated within a structure may lead to changes in the dynamic responses and characteristics of the structure, such as the natural frequency, mode shape, and the modal damping, which, in turn, can be used to detect, locate, and quantify the damage. In the literature, there are a number of structural damage identification methods (SDIMs), which can be classified into several categories depending on the types of experimental data used in the methods. The experimental data used in the literature include the modal data, strain energy, transfer function parameters, flexibility matrix, residual forces, wave characteristics, mechanical impedances, frequency response function (FRF) data, and so forth [2, 35, 37, 46, 134, 145, 154, 170]. As the FRF data are measured directly from the structures, they will be free from any modal extraction errors. Furthermore, the FRF data can provide much more damage information in a desired frequency range when compared with the modal data, because the modal data are extracted mainly from a very limited number of FRF data around the resonance [160]. Motivated by the advantages of the FRF data over the modal data, this chapter introduces a spectral elment method (SEM) based SDIM in which the FRF data are used for structural damage identification. For the inverse problem, how to reduce the size of the problem has been an important research issue over a long period of time. Thus, a domain-reduction method is also introduced as a potential candidate approach.

15.1 Spectral Element Modeling of a Damaged Structure

15.1.1 Assembly of Spectral Elements

The spectral element equation for an intact structure is given in the form

$$\boldsymbol{s}_k(\omega)\boldsymbol{d}_k = \boldsymbol{f}_k(\omega) \tag{15.1}$$

where $\boldsymbol{s}_k$, $\boldsymbol{d}_k$, and $\boldsymbol{f}_k$ are the spectral element matrix, the spectral nodal degrees-of-freedom (DOFs) vector and the spectral nodal forces vector, respectively, for the kth element of the intact structure. When the structure has been damaged, but is still subjected to the same external forces, the spectral element equation is given in the form

$$\bar{\boldsymbol{s}}_k(\omega)\bar{\boldsymbol{d}}_k = \boldsymbol{f}_k(\omega) \tag{15.2}$$

Spectral Element Method in Structural Dynamics Usik Lee

where the over-bar indicates that the structure is damaged. The spectral elements represented by Equations 15.1 or 15.2 can be assembled in an analogous way to those used in the conventional finite element method (FEM) to obtain a global system equation for the intact structure as

$$\boldsymbol{S}_{\mathrm{g}}(\omega)\boldsymbol{d}_{\mathrm{g}} = \boldsymbol{f}_{\mathrm{g}}(\omega) \quad \text{(intact)} \tag{15.3}$$

where

$$\boldsymbol{S}_{\mathrm{g}}(\omega) = \sum_{k=1}^{N} \boldsymbol{L}_k^{\mathrm{T}} \boldsymbol{s}_k(\omega) \boldsymbol{L}_k \tag{15.4}$$

and for the damaged structure as

$$\bar{\boldsymbol{S}}_{\mathrm{g}}(\omega)\bar{\boldsymbol{d}}_{\mathrm{g}} = \boldsymbol{f}_{\mathrm{g}}(\omega) \quad \text{(damaged)} \tag{15.5}$$

where

$$\bar{\boldsymbol{S}}_{\mathrm{g}}(\omega) = \sum_{k=1}^{N} \boldsymbol{L}_k^{\mathrm{T}} \bar{\boldsymbol{s}}_k(\omega) \boldsymbol{L}_k \tag{15.6}$$

The matrix $\boldsymbol{L}_k$ ($k = 1, 2, \ldots, N$) is the locator matrix, which locates the kth spectral element matrices $\boldsymbol{s}_k$ and $\bar{\boldsymbol{s}}_k$ in the global dynamic stiffness matrices $\boldsymbol{S}_{\mathrm{g}}(\omega)$ and $\bar{\boldsymbol{S}}_{\mathrm{g}}(\omega)$, respectively, and N is the total number of finite elements used to represent a structure. The vector $\boldsymbol{f}_{\mathrm{g}}(\omega)$ is the global spectral nodal forces vector.

15.1.2 Imposition of Boundary Conditions

By applying the boundary conditions, the spectral nodal DOFs can be reduced to

$$\boldsymbol{d}_{\mathrm{g}} = \boldsymbol{T}_{\mathrm{B}}\boldsymbol{d}_{\mathrm{B}} \text{ (intact)}, \quad \bar{\boldsymbol{d}}_{\mathrm{g}} = \boldsymbol{T}_{\mathrm{B}}\bar{\boldsymbol{d}}_{\mathrm{B}} \text{ (damaged)} \tag{15.7}$$

where $\boldsymbol{d}_{\mathrm{B}}$ and $\bar{\boldsymbol{d}}_{\mathrm{B}}$ are the reduced spectral nodal DOFs vectors for the intact and damaged structures, respectively, and $\boldsymbol{T}_{\mathrm{B}}$ is the coordinates transformation matrix to impose the boundary conditions. By using the first part of Equation 15.7, Equation 15.3 can be transformed into the form

$$\boldsymbol{S}_{\mathrm{B}}(\omega)\boldsymbol{d}_{\mathrm{B}} = \boldsymbol{f}_{\mathrm{B}}(\omega) \quad \text{(intact)} \tag{15.8}$$

where

$$\boldsymbol{S}_{\mathrm{B}}(\omega) = \boldsymbol{T}_{\mathrm{B}}^{\mathrm{T}}\boldsymbol{S}_{\mathrm{g}}(\omega)\boldsymbol{T}_{\mathrm{B}}, \qquad \boldsymbol{f}_{\mathrm{B}}(\omega) = \boldsymbol{T}_{\mathrm{B}}^{\mathrm{T}}\boldsymbol{f}_{\mathrm{g}}(\omega) \tag{15.9}$$

Similarly, by using the second part of Equation 15.7, Equation 15.5 can be transformed into the form

$$\bar{\boldsymbol{S}}_{\mathrm{B}}(\omega)\bar{\boldsymbol{d}}_{\mathrm{B}} = \boldsymbol{f}_{\mathrm{B}}(\omega) \quad \text{(damaged)} \tag{15.10}$$

where

$$\bar{\boldsymbol{S}}_{\mathrm{B}}(\omega) = \boldsymbol{T}_{\mathrm{B}}^{\mathrm{T}}\bar{\boldsymbol{S}}_{\mathrm{g}}(\omega)\boldsymbol{T}_{\mathrm{B}}, \qquad \boldsymbol{f}_{\mathrm{B}}(\omega) = \boldsymbol{T}_{\mathrm{B}}^{\mathrm{T}}\boldsymbol{f}_{\mathrm{g}}(\omega) \tag{15.11}$$

15.1.3 Reordering of Spectral Nodal DOFs

The spectral nodal DOFs can be divided into two groups: the first group $\boldsymbol{d}_m$ contains the spectral nodal DOFs (called "master" DOFs) to be measured or to be excited by an external force and the second group $\boldsymbol{d}_s$ contains the remaining spectral nodal DOFs (called "slave" DOFs). The spectral nodal DOFs of $\boldsymbol{d}_\mathrm{B}$ and $\bar{\boldsymbol{d}}_\mathrm{B}$ can be reordered to form a new spectral nodal DOFs vectors as

$$\boldsymbol{d} = \begin{Bmatrix} \boldsymbol{d}_m \\ \boldsymbol{d}_s \end{Bmatrix} \quad \text{(intact)}, \qquad \bar{\boldsymbol{d}} = \begin{Bmatrix} \bar{\boldsymbol{d}}_m \\ \bar{\boldsymbol{d}}_s \end{Bmatrix} \quad \text{(damaged)} \tag{15.12}$$

The spectral nodal DOFs vectors $\boldsymbol{d}_\mathrm{B}$ and $\bar{\boldsymbol{d}}_\mathrm{B}$ can be related to the reordered spectral nodal DOFs vectors $\boldsymbol{d}$ and $\bar{\boldsymbol{d}}$, by introducing a transformation matrix $\boldsymbol{T}_\mathrm{O}$, as follows:

$$\boldsymbol{d}_\mathrm{B} = \boldsymbol{T}_\mathrm{O}\boldsymbol{d} \quad \text{(intact)}, \qquad \bar{\boldsymbol{d}}_\mathrm{B} = \boldsymbol{T}_\mathrm{O}\bar{\boldsymbol{d}} \quad \text{(damaged)} \tag{15.13}$$

By using the first part of Equation 15.13, Equation 15.8 can be transformed into the form

$$\boldsymbol{S}(\omega)\boldsymbol{d} = \boldsymbol{f}(\omega) \quad \text{(intact)} \tag{15.14}$$

where

$$\boldsymbol{S}(\omega) = \boldsymbol{T}_\mathrm{O}^\mathrm{T}\boldsymbol{S}_\mathrm{B}(\omega)\boldsymbol{T}_\mathrm{O} = \boldsymbol{T}_\mathrm{BO}^\mathrm{T}\boldsymbol{S}_\mathrm{g}(\omega)\boldsymbol{T}_\mathrm{BO} \tag{15.15}$$

and

$$\boldsymbol{f}(\omega) = \boldsymbol{T}_\mathrm{O}^\mathrm{T}\boldsymbol{f}_\mathrm{B}(\omega) = \boldsymbol{T}_\mathrm{BO}^\mathrm{T}\boldsymbol{f}_\mathrm{g}(\omega) \tag{15.16}$$

with the use of the definition

$$\boldsymbol{T}_\mathrm{BO} = \boldsymbol{T}_\mathrm{B}\boldsymbol{T}_\mathrm{O} \tag{15.17}$$

Similarly, by using the second part of Equation 15.13, Equation 15.10 can be transformed into the form

$$\bar{\boldsymbol{S}}(\omega)\bar{\boldsymbol{d}} = \boldsymbol{f}(\omega) \quad \text{(damaged)} \tag{15.18}$$

where

$$\bar{\boldsymbol{S}}(\omega) = \boldsymbol{T}_\mathrm{O}^\mathrm{T}\bar{\boldsymbol{S}}_\mathrm{B}(\omega)\boldsymbol{T}_\mathrm{O} = \boldsymbol{T}_\mathrm{BO}^\mathrm{T}\bar{\boldsymbol{S}}_\mathrm{g}(\omega)\boldsymbol{T}_\mathrm{BO} \tag{15.19}$$

Equations 15.14 and 15.18 can be written in the partitioned forms as

$$\begin{bmatrix} \boldsymbol{S}_{mm}(\omega) & \boldsymbol{S}_{ms}(\omega) \\ \boldsymbol{S}_{sm}(\omega) & \boldsymbol{S}_{ss}(\omega) \end{bmatrix} \begin{Bmatrix} \boldsymbol{d}_m \\ \boldsymbol{d}_s \end{Bmatrix} = \begin{Bmatrix} \boldsymbol{f}_m(\omega) \\ \boldsymbol{0} \end{Bmatrix} \quad \text{(intact)} \tag{15.20}$$

and

$$\begin{bmatrix} \bar{\boldsymbol{S}}_{mm}(\omega) & \bar{\boldsymbol{S}}_{ms}(\omega) \\ \bar{\boldsymbol{S}}_{sm}(\omega) & \bar{\boldsymbol{S}}_{ss}(\omega) \end{bmatrix} \begin{Bmatrix} \bar{\boldsymbol{d}}_m(\omega) \\ \bar{\boldsymbol{d}}_s(\omega) \end{Bmatrix} = \begin{Bmatrix} \boldsymbol{f}_m(\omega) \\ \boldsymbol{0} \end{Bmatrix} \quad \text{(damaged)} \tag{15.21}$$

Assuming that the structure is excited by an external force of spectral magnitude $F(\omega)$, the global spectral nodal forces vector $\boldsymbol{f}_\mathrm{g}$ can be written as

$$\boldsymbol{f}_\mathrm{g}(\omega) = \{0 \quad \cdots \quad F(\omega) \cdots 0\}^\mathrm{T} = F(\omega)\boldsymbol{\delta} \tag{15.22}$$

where

$$\boldsymbol{\delta} = \{0 \quad \cdots \quad 1 \cdots \quad 0\}^{\mathrm{T}} \tag{15.23}$$

By using Equation 15.22, Equation 15.16 can be written as

$$\boldsymbol{f}(\omega) = \begin{Bmatrix} \boldsymbol{f}_m(\omega) \\ \mathbf{0} \end{Bmatrix} = \boldsymbol{T}_{\mathrm{BO}}^{\mathrm{T}} \boldsymbol{f}_{\mathrm{g}}(\omega) = F(\omega) \boldsymbol{T}_{\mathrm{BO}}^{\mathrm{T}} \boldsymbol{\delta} \tag{15.24}$$

Introduce a transformation matrix $\boldsymbol{T}_f$ such that

$$\boldsymbol{f}_m(\omega) = \boldsymbol{T}_f \boldsymbol{f}(\omega) \tag{15.25}$$

Substitution of Equation 15.23 into Equation 15.24 gives

$$\boldsymbol{f}_m(\omega) = F(\omega) \boldsymbol{T}_f \boldsymbol{T}_{\mathrm{BO}}^{\mathrm{T}} \delta = F(\omega) \boldsymbol{\delta}_m \tag{15.26}$$

where

$$\boldsymbol{\delta}_m = \boldsymbol{T}_f \boldsymbol{T}_{\mathrm{BO}}^{\mathrm{T}} \boldsymbol{\delta} \tag{15.27}$$

15.2 Theory of Damage Identification

15.2.1 Uniform Damage Representation

As the location, geometry, and severity of the damage in a structure element are not known in advance, in most practical situations it is impossible to assign definitive representation for the stiffness at the damaged sites. Thus, as illustrated in Figure 15.1, a simple and rational approach might be to represent a damaged structure element by the elastic modulus degraded as follows: [175]

$$\bar{E} = E(1 - D) \tag{15.28}$$

where E is Young's modulus for an intact state and $\bar{E}$ is Young's modulus at a damaged state. D is the damage magnitude, which is assumed to be uniform over a structure element: $D = 0$ implies the intact state, while $D = 1$ implies complete material rupture due to damage.

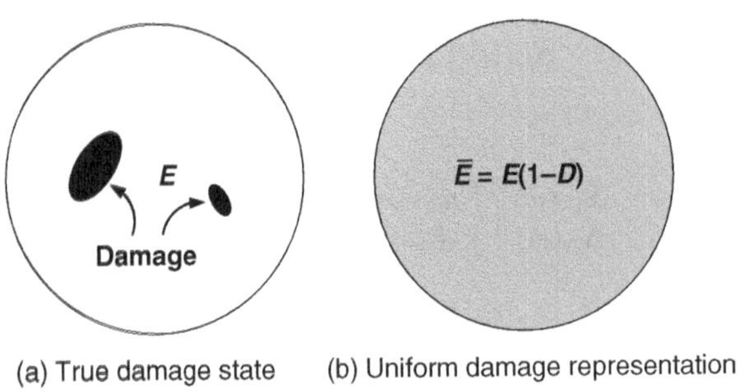

Figure 15.1 Uniform damage representation of the local damage in a structure element

Local damage may give rise to certain nonlinearities, for example, clearance or bilinear stiffness effects in systems with cracks. As early damage, long before structural failure, is mostly localized on very small spots rather than being spread wide over the structure, the nonlinearity due to such small local damage will be weak and confined to small areas surrounding the local damage. Thus, at the early stage of damage growth, the damage-induced nonlinearity may be negligible.

Assume that the spectral element matrix for an intact structure element is given by $\boldsymbol{s}(\omega)$. The spectral element matrix for a damaged element can be obtained by simply replacing E by $\bar{E} = E(1-D)$ as follows:

$$\bar{\boldsymbol{s}}(\omega) = \boldsymbol{s}(\omega)|_{E=\bar{E}} \tag{15.29}$$

The perturbed spectral element matrix $\Delta\boldsymbol{s}(\omega)$ can be computed from

$$\Delta\boldsymbol{s}(\omega) = \bar{\boldsymbol{s}}(\omega) - \boldsymbol{s}(\omega) \tag{15.30}$$

We can approximate $\Delta\boldsymbol{s}(\omega)$ as

$$\Delta\boldsymbol{s}_{\text{approx}}(\omega) \cong -D\boldsymbol{s}(\omega = 0) = -D\boldsymbol{k} \tag{15.31}$$

The accuracy of $\Delta\boldsymbol{s}_{\text{approx}}(\omega)$ may be improved by shortening the length of the spectral element and by lowering the frequencies chosen for damage identification [96].

15.2.2 Damage Identification Algorithms

Equation 15.21 can be written as

$$\bar{\boldsymbol{S}}_{mm}\bar{\boldsymbol{d}}_m + \bar{\boldsymbol{S}}_{ms}\bar{\boldsymbol{d}}_s = \boldsymbol{f}_m \tag{15.32}$$

$$\bar{\boldsymbol{S}}_{sm}\bar{\boldsymbol{d}}_m + \bar{\boldsymbol{S}}_{ss}\bar{\boldsymbol{d}}_s = 0 \tag{15.33}$$

From Equation 15.33, we obtain

$$\bar{\boldsymbol{d}}_s = \bar{\boldsymbol{t}}(\omega)\bar{\boldsymbol{d}}_m \tag{15.34}$$

where

$$\bar{\boldsymbol{t}}(\omega) = -\bar{\boldsymbol{S}}_{ss}^{-1}\bar{\boldsymbol{S}}_{sm} \tag{15.35}$$

Substitution of Equations 15.26 and 15.34 into Equation 15.32 gives

$$F(\omega)\boldsymbol{\delta}_m = [\bar{\boldsymbol{S}}_{mm} + \bar{\boldsymbol{S}}_{ms}\bar{\boldsymbol{t}}(\omega)]\bar{\boldsymbol{d}}_m \tag{15.36}$$

We define the perturbed quantities as follows:

$$\Delta\boldsymbol{S}(\omega) = \bar{\boldsymbol{S}}(\omega) - \boldsymbol{S}(\omega) \tag{15.37}$$

or

$$\begin{bmatrix} \Delta\boldsymbol{S}_{mm} & \Delta\boldsymbol{S}_{ms} \\ \Delta\boldsymbol{S}_{sm} & \Delta\boldsymbol{S}_{ss} \end{bmatrix} = \begin{bmatrix} \bar{\boldsymbol{S}}_{mm} & \bar{\boldsymbol{S}}_{ms} \\ \bar{\boldsymbol{S}}_{sm} & \bar{\boldsymbol{S}}_{ss} \end{bmatrix} - \begin{bmatrix} \boldsymbol{S}_{mm} & \boldsymbol{S}_{ms} \\ \boldsymbol{S}_{sm} & \boldsymbol{S}_{ss} \end{bmatrix} \tag{15.38}$$

and

$$\Delta \boldsymbol{t}(\omega) = \bar{\boldsymbol{t}}(\omega) - \boldsymbol{t}(\omega) \tag{15.39}$$

where

$$\boldsymbol{t}(\omega) = -\boldsymbol{S}_{ss}^{-1}\boldsymbol{S}_{sm} \tag{15.40}$$

Applying Equations 15.37 and 15.39 into Equation 15.36 yields

$$\boldsymbol{Y}(\boldsymbol{D}, \omega)\bar{\boldsymbol{d}}_m = F(\omega)\boldsymbol{\delta}_m - \boldsymbol{X}(\boldsymbol{\omega})\bar{\boldsymbol{d}}_m \tag{15.41}$$

where

$$\boldsymbol{X}(\omega) = \boldsymbol{S}_{mm} + \boldsymbol{S}_{ms}\boldsymbol{t}(\omega) \tag{15.42}$$

$$\boldsymbol{Y}(\boldsymbol{D}, \omega) = \Delta\boldsymbol{S}_{mm} + \Delta\boldsymbol{S}_{ms}\boldsymbol{t} + \boldsymbol{S}_{ms}\Delta\boldsymbol{t} + \Delta\boldsymbol{S}_{ms}\Delta\boldsymbol{t} \tag{15.43}$$

and $\boldsymbol{D}$ represents the damage magnitudes as

$$\boldsymbol{D} = \{D_k;\ k = 1, 2, \ldots, N\} \tag{15.44}$$

Define the inertance FRF as

$$\bar{\boldsymbol{a}}_m(\omega) = -\omega^2 \frac{\bar{\boldsymbol{d}}_m}{F(\omega)} \tag{15.45}$$

Equation 15.41 can be then be rewritten in terms of inertance FRF as

$$-\boldsymbol{Y}(\boldsymbol{D}, \omega)\bar{\boldsymbol{a}}_m(\omega) = \omega^2\boldsymbol{\delta}_m + \boldsymbol{X}(\omega)\bar{\boldsymbol{a}}_m(\omega) \tag{15.46}$$

The right-hand side of Equation 15.46 does not depend on the damage state because it is determined by the structural model (spectral element model) at the intact state, the inertance FRF measured at the damaged state, and the pre-specified excitation force. On the other hand, the left-hand side of Equation 15.46 does depend on the unknown damage magnitudes $\boldsymbol{D}$.

If $\boldsymbol{D} = \boldsymbol{0}$, Equation 15.46 is reduced to

$$\omega^2\boldsymbol{\delta}_m + \boldsymbol{X}(\omega)\boldsymbol{a}_m(\omega) = \boldsymbol{0} \tag{15.47}$$

Thus, applying Equation 15.47 to the right-hand side of Equation 15.46 gives

$$\boldsymbol{Z}(\boldsymbol{D}, \omega)\bar{\boldsymbol{a}}_m(\omega) = \Delta\boldsymbol{a}_m(\omega) \tag{15.48}$$

where

$$\boldsymbol{Z}(\boldsymbol{D}, \omega) = -\boldsymbol{X}(\omega)^{-1}\boldsymbol{Y}(\boldsymbol{D}, \omega) \tag{15.49}$$

and

$$\Delta\boldsymbol{a}_m(\omega) = \bar{\boldsymbol{a}}_m(\omega) - \boldsymbol{a}_m(\omega) \tag{15.50}$$

15.2.2.1 Linear Damage Identification Algorithm

Applying Equations 15.35 and 15.40 into Equation 15.39 gives

$$\Delta \boldsymbol{t}(\omega) = \bar{\boldsymbol{t}}(\omega) - \boldsymbol{t}(\omega) = -\bar{\boldsymbol{S}}_{ss}^{-1}\bar{\boldsymbol{S}}_{sm} + \boldsymbol{S}_{ss}^{-1}\boldsymbol{S}_{sm} \tag{15.51}$$

By using Equation 15.38, we can show that

$$\begin{aligned} \bar{\boldsymbol{S}}_{ss}^{-1} &= (\boldsymbol{S}_{ss} + \Delta\boldsymbol{S}_{ss})^{-1} = (\boldsymbol{I} + \boldsymbol{S}_{ss}^{-1}\Delta\boldsymbol{S}_{ss})^{-1}\boldsymbol{S}_{ss}^{-1} \\ &= \sum_{p=0}^{\infty}(-\boldsymbol{S}_{ss}^{-1}\Delta\boldsymbol{S}_{ss})^{p}\boldsymbol{S}_{ss}^{-1} \cong \boldsymbol{S}_{ss}^{-1} - \boldsymbol{S}_{ss}^{-1}\Delta\boldsymbol{S}_{ss}\boldsymbol{S}_{ss}^{-1} \end{aligned} \tag{15.52}$$

where $\boldsymbol{I}$ is the identity matrix. Substituting Equation 15.52 into Equation 15.51 gives

$$\begin{aligned} \Delta\boldsymbol{t}(\omega) &\cong -(\boldsymbol{S}_{ss}^{-1} - \boldsymbol{S}_{ss}^{-1}\Delta\boldsymbol{S}_{ss}\boldsymbol{S}_{ss}^{-1})(\boldsymbol{S}_{sm} + \Delta\boldsymbol{S}_{sm}) + \boldsymbol{S}_{ss}^{-1}\boldsymbol{S}_{sm} \\ &= \mathbf{S}_{ss}^{-1}\Delta\boldsymbol{S}_{ss}\boldsymbol{S}_{ss}^{-1}\boldsymbol{S}_{sm} - \boldsymbol{S}_{ss}^{-1}\Delta\boldsymbol{S}_{sm} + \boldsymbol{S}_{ss}^{-1}\Delta\boldsymbol{S}_{ss}\boldsymbol{S}_{ss}^{-1}\Delta\boldsymbol{S}_{sm} \\ &\cong \boldsymbol{S}_{ss}^{-1}\Delta\boldsymbol{S}_{ss}\boldsymbol{S}_{ss}^{-1}\boldsymbol{S}_{sm} - \boldsymbol{S}_{ss}^{-1}\Delta\boldsymbol{S}_{sm} \end{aligned} \tag{15.53}$$

Neglecting a small term $\Delta\boldsymbol{S}_{ms}\Delta\boldsymbol{t}$ from Equation 15.43 and then substituting Equation 15.53, $\boldsymbol{Y}(\boldsymbol{D}, \omega)$ can be approximated as

$$\begin{aligned} \boldsymbol{Y}(\boldsymbol{D}, \omega) &\cong \Delta\boldsymbol{S}_{mm} + \Delta\boldsymbol{S}_{ms}\boldsymbol{t} + \boldsymbol{S}_{ms}\Delta\boldsymbol{t} \\ &\cong \Delta\boldsymbol{S}_{mm} + \Delta\boldsymbol{S}_{ms}\boldsymbol{t} + \boldsymbol{t}^{\mathrm{T}}\Delta\boldsymbol{S}_{ss}\boldsymbol{t} + \boldsymbol{t}^{\mathrm{T}}\Delta\boldsymbol{S}_{sm} \\ &= \boldsymbol{T}_C^{\mathrm{T}}(\omega)\Delta\boldsymbol{S}(\omega)\boldsymbol{T}_C(\omega) \equiv \boldsymbol{Y}_{\mathrm{approx}}(\boldsymbol{D}, \omega) \end{aligned} \tag{15.54}$$

where

$$\boldsymbol{T}_C(\omega) = \begin{bmatrix} \boldsymbol{I} \\ \boldsymbol{t}(\omega) \end{bmatrix} = \begin{bmatrix} \boldsymbol{I} \\ -\boldsymbol{S}_{ss}^{-1}\boldsymbol{S}_{sm} \end{bmatrix} \tag{15.55}$$

Applying Equations 15.15 and 15.19 into Equation 15.37 gives

$$\Delta\boldsymbol{S}(\omega) = \bar{\boldsymbol{S}}(\omega) - \boldsymbol{S}(\omega) = \boldsymbol{T}_{\mathrm{BO}}^{\mathrm{T}}\Delta\boldsymbol{S}_{\mathrm{g}}(\omega)\boldsymbol{T}_{\mathrm{BO}} \tag{15.56}$$

Next, by substituting Equation 15.56 into Equation 15.54, we can obtain

$$\boldsymbol{Y}_{\mathrm{approx}}(\boldsymbol{D}, \omega) = \boldsymbol{T}^{\mathrm{T}}(\omega)\Delta\boldsymbol{S}_{\mathrm{g}}(\omega)\boldsymbol{T}(\omega) \tag{15.57}$$

where

$$\boldsymbol{T}(\omega) = \boldsymbol{T}_{\mathrm{BO}}\boldsymbol{T}_C(\omega) \tag{15.58}$$

By using Equations 15.4 and 15.6, we obtain

$$\begin{aligned} \Delta\boldsymbol{S}_{\mathrm{g}}(\omega) &= \bar{\boldsymbol{S}}_{\mathrm{g}}(\omega) - \boldsymbol{S}_{\mathrm{g}}(\omega) \\ &= \sum_{k=1}^{N}\boldsymbol{L}_k^{\mathrm{T}}(\bar{\boldsymbol{s}}_k - \boldsymbol{s}_k)\boldsymbol{L}_k = \sum_{k=1}^{N}\boldsymbol{L}_k^{\mathrm{T}}\Delta\boldsymbol{s}_k(\omega)\boldsymbol{L}_k \end{aligned} \tag{15.59}$$

Applying Equation 15.31 into Equation 15.59 gives

$$\Delta\boldsymbol{S}_{\mathrm{g}}(\omega) \cong -\sum_{k=1}^{N}\boldsymbol{L}_k^{\mathrm{T}}(D_k\boldsymbol{k}_k)\boldsymbol{L}_k = -\sum_{k=1}^{N}\boldsymbol{L}_k^{\mathrm{T}}\boldsymbol{k}_k\boldsymbol{L}_k D_k \tag{15.60}$$

By substituting Equation 15.60 into Equation 15.57, we can linearize $\boldsymbol{Y}_{\text{approx}}(\boldsymbol{D}, \omega)$ with respect to the damage magnitudes $\boldsymbol{D} = \{D_k;\ k = 1, 2, \ldots, N\}$ as

$$\begin{aligned}\boldsymbol{Y}_{\text{approx}}(\boldsymbol{D}, \omega) &\cong -\boldsymbol{T}^{\text{T}}\left(\sum_{k=1}^{N}\boldsymbol{L}_k^{\text{T}}\boldsymbol{k}_k\boldsymbol{L}_k D_k\right)\boldsymbol{T} \\ &= -\sum_{k=1}^{N}(\boldsymbol{T}^{\text{T}}\boldsymbol{L}_k^{\text{T}}\boldsymbol{k}_k\boldsymbol{L}_k\boldsymbol{T})D_k \equiv \boldsymbol{Y}_{\text{Linear}}(\boldsymbol{D}, \omega)\end{aligned} \tag{15.61}$$

We then replace $\boldsymbol{Y}(\boldsymbol{D}, \omega)$ by $\boldsymbol{Y}_{\text{Linear}}(\boldsymbol{D}, \omega)$ to write Equation 15.48 as

$$\boldsymbol{Z}_{\text{Linear}}(\omega)\bar{\boldsymbol{a}}_m(\omega) \cong \Delta\boldsymbol{a}_m(\omega) \tag{15.62}$$

where

$$\begin{aligned}\boldsymbol{Z}_{\text{Linear}}(\boldsymbol{D}, \omega) &= -\boldsymbol{X}(\omega)^{-1}\boldsymbol{Y}_{\text{Linear}}(\boldsymbol{D}, \omega) \\ &= \boldsymbol{X}(\omega)^{-1}\sum_{k=1}^{N}(\boldsymbol{T}^{\text{T}}\boldsymbol{L}_k^{\text{T}}\boldsymbol{k}_k\boldsymbol{L}_k\boldsymbol{T})D_k\end{aligned} \tag{15.63}$$

and

$$\boldsymbol{X}(\omega) = \boldsymbol{S}_{mm} - \boldsymbol{S}_{ms}\boldsymbol{S}_{ss}^{-1}\boldsymbol{S}_{sm} \tag{15.64}$$

The left-hand side of Equation 15.62 can be then be written as

$$\begin{aligned}\boldsymbol{Z}_{\text{Linear}}(\boldsymbol{D}, \omega)\bar{\boldsymbol{a}}_m(\omega) &= \boldsymbol{X}(\omega)^{-1}\sum_{k=1}^{N}(\boldsymbol{T}^{\text{T}}\boldsymbol{L}_k^{\text{T}}\boldsymbol{k}_k\boldsymbol{L}_k\boldsymbol{T}\bar{\boldsymbol{a}}_m)D_k \\ &= \boldsymbol{X}(\omega)^{-1}\sum_{k=1}^{N}\boldsymbol{h}_k(\omega)D_k \\ &\equiv \boldsymbol{A}(\omega)\boldsymbol{D}\end{aligned} \tag{15.65}$$

where

$$\boldsymbol{D} = \{D_1\ D_2\ D_3 \cdots D_N\}^{\text{T}} \tag{15.66}$$

and

$$\boldsymbol{A}(\omega) = \boldsymbol{X}(\omega)^{-1}\boldsymbol{H}(\omega) \tag{15.67}$$

with the use of the following definitions

$$\boldsymbol{H}(\omega) = [\boldsymbol{h}_1\ \boldsymbol{h}_2\ \boldsymbol{h}_3 \cdots \boldsymbol{h}_N] \tag{15.68}$$

$$\boldsymbol{h}_k(\omega) = \boldsymbol{T}^{\text{T}}\boldsymbol{L}_k^{\text{T}}\boldsymbol{k}_k\boldsymbol{L}_k\boldsymbol{T}\bar{\boldsymbol{a}}_m \tag{15.69}$$

Lastly, by using Equation 15.65, we can obtain the linear damage identification algorithm from Equation 15.62 as follows:

$$\boldsymbol{A}(\omega)_{(M\times N)}\boldsymbol{D}_{(N\times 1)} \cong \boldsymbol{b}(\omega)_{(M\times 1)} \tag{15.70}$$

where

$$b(\omega)_{(M\times1)} = \Delta a_m(\omega)_{(M\times1)} \tag{15.71}$$

The dimensions of the matrix $A(\omega)$ are M by N and the dimensions of the vector $b(\omega)$ are M by 1, where N is the number of unknown damage magnitudes, which is equal to the number of finite elements used to represent a complete structure under inspection, and M is the number of the spectral nodal DOFs to be measured for damage identification. Equations 15.70 and 15.71 show that $A(\omega)$ and $b(\omega)$ are determined by the spectral element model at the intact state and the damage-induced changes of inertance FRFs measured at M nodes.

As explained by Equation 15.21, the spectral nodal DOFs at which neither vibration responses are measured nor excitation force is applied are considered as the "slave" nodal DOFs to be removed, while all other spectral nodal DOFs are considered as the "master" nodal DOFs to be retained. Thus, in order to excite a damaged structure, an excitation force should be applied to one of the M nodes at which the vibration responses will be measured.

In general, the number of unknown damage magnitudes to be determined will be larger than the number of measured nodal DOFs (or the number of linear algebraic equations): that is, $N > M$. In such a situation, we may need to acquire more equations from Equation 15.70 to make the number of equations equal to or larger than the number of unknowns. Equation 15.70 is valid for any frequency. Thus, by varying the excitation frequency, we can obtain as many sets of linear algebraic equations as are needed to form a system of linear algebraic equations as follows:

$$\bar{A}(\omega)_{(QM\times N)} D_{(N\times1)} = \bar{b}(\omega)_{(QM\times1)} \tag{15.72}$$

where

$$\bar{A}(\omega)_{(QM\times N)} = \begin{bmatrix} A(\omega_1)_{(M\times N)} \\ A(\omega_2)_{(M\times N)} \\ \vdots \\ A(\omega_Q)_{(M\times N)} \end{bmatrix}, \quad \bar{b}(\omega)_{(QM\times1)} = \begin{Bmatrix} b(\omega_1)_{(M\times1)} \\ b(\omega_2)_{(M\times1)} \\ \vdots \\ b(\omega_Q)_{(M\times1)} \end{Bmatrix} \tag{15.73}$$

where Q is the number of chosen excitation frequencies and it is so determined that the number of equations is equal to or larger than the number of unknowns: that is, $(Q \times M) \geq N$. Solving Equation 15.72 for D is equal to estimating the damage distribution over N finite elements. This again implies identifying locations and magnitudes of multiple damages simultaneously.

The matrix $\bar{A}(\omega)$ might be very sensitive to the noises in the measured FRF data and to the chosen multiple frequencies (when $N > M$). Thus it is recommended to check the condition number of $\bar{A}(\omega)$ during the matrix inversion process. One may use the algorithms such as the singular value decomposition (MATHLAB function "svd") and the non-negative least squares (MATHLAB function "nnls") to solve Equation 15.72 for D [7].

15.2.2.2 Nonlinear Damage Identification Algorithm

For the derivation of the linear damage identification algorithm, Equation 15.70 or Equation 15.72, $Y(D, \omega)$ is linearized with respect to damage magnitudes D through the approximations of Equations 15.54 and 15.61. Owing to the approximations involved, the linear damage identification algorithm might be very sensitive to the chosen frequencies and unknown damage magnitudes and provide poor identification results. Thus, in this subsection, we will consider two iteration methods to solve the nonlinear damage identification algorithms: one is the direct iteration method and the other is the Newton–Raphson method.

Direct Iteration Method

We represent $\boldsymbol{Z}(\boldsymbol{D}, \omega)$ from Equation 15.49 as

$$\boldsymbol{Z}(\boldsymbol{D},\omega) = \boldsymbol{Z}_{\text{Linear}}(\boldsymbol{D},\omega) + \boldsymbol{Z}_{\text{Nonlinear}}(\boldsymbol{D},\omega) \tag{15.74}$$

where $\boldsymbol{Z}_{\text{Linear}}(\boldsymbol{D}, \omega)$ is defined by Equation 15.63. Thus, from Equation 15.74, we get

$$\begin{aligned}\boldsymbol{Z}_{\text{Nonlinear}}(\boldsymbol{D},\omega) &= \boldsymbol{Z}(\boldsymbol{D},\omega) - \boldsymbol{Z}_{\text{Linear}}(\boldsymbol{D},\omega)\\ &= -\boldsymbol{X}(\omega)^{-1}[\boldsymbol{Y}(\boldsymbol{D},\omega) - \boldsymbol{Y}_{\text{Linear}}(\boldsymbol{D},\omega)]\end{aligned} \tag{15.75}$$

Remember that exact definitions of $\boldsymbol{Y}(\boldsymbol{D}, \omega)$ and $\boldsymbol{Y}_{\text{Linear}}(\boldsymbol{D}, \omega)$ are given by Equations 15.43 and 15.61, respectively. Substituting Equation 15.74 into the exact equality, Equation 15.48, gives

$$\boldsymbol{Z}_{\text{Linear}}(\boldsymbol{D},\omega)\bar{\boldsymbol{a}}_m(\omega) = \Delta\boldsymbol{a}_m(\omega) - \boldsymbol{Z}_{\text{Nonlinear}}(\boldsymbol{D},\omega)\bar{\boldsymbol{a}}_m(\omega) \tag{15.76}$$

Applying Equations 15.65 and 15.75 to Equation 15.76 gives an iterative non-linear damage identification algorithm as

$$\boldsymbol{A}(\omega)\boldsymbol{D}^{r+1} = \boldsymbol{b}_{\text{NL}}(\boldsymbol{D}^r,\omega) \tag{15.77}$$

where

$$\boldsymbol{A}(\omega) = \boldsymbol{X}(\omega)^{-1}\boldsymbol{H}(\omega) \tag{15.78}$$

and

$$\begin{aligned}\boldsymbol{b}_{\text{NL}}(\boldsymbol{D}^r,\omega) &= \Delta\boldsymbol{a}_m(\omega) - \boldsymbol{Z}_{\text{Nonlinear}}(\boldsymbol{D}^r,\omega)\bar{\boldsymbol{a}}_m(\omega)\\ &= \Delta\boldsymbol{a}_m(\omega) + \boldsymbol{X}(\omega)^{-1}[\boldsymbol{Y}(\boldsymbol{D}^r,\omega) - \boldsymbol{Y}_{\text{Linear}}(\boldsymbol{D}^r,\omega)]\bar{\boldsymbol{a}}_m(\omega)\\ &= \Delta\boldsymbol{a}_m(\omega) + \boldsymbol{X}(\omega)^{-1}\boldsymbol{Y}(\boldsymbol{D}^r,\omega)\bar{\boldsymbol{a}}_m(\omega) + \boldsymbol{A}(\omega)\boldsymbol{D}^r\end{aligned} \tag{15.79}$$

In above equations, $\boldsymbol{D}^r$ and $\boldsymbol{D}^{r+1}$ represent the damage magnitudes computed at the rth and $(r+1)$th iterations, respectively.

If Q excitation frequencies are chosen, Equation 15.77 can be written as

$$\bar{\boldsymbol{A}}_{(QM\times N)}\boldsymbol{D}^{r+1}_{(N\times 1)} = \bar{\boldsymbol{b}}_{\text{NL}}(\boldsymbol{D}^r)_{(QM\times 1)} \tag{15.80}$$

where

$$\bar{\boldsymbol{A}}_{(QM\times N)} = \begin{bmatrix}\boldsymbol{A}(\omega_1)_{(M\times N)}\\ \boldsymbol{A}(\omega_2)_{(M\times N)}\\ \vdots\\ \boldsymbol{A}(\omega_Q)_{(M\times N)}\end{bmatrix}$$
$$\bar{\boldsymbol{b}}_{\text{NL}}(\boldsymbol{D}^r)_{(QM\times 1)} = \begin{Bmatrix}\boldsymbol{b}_{\text{NL}}(\boldsymbol{D}^r,\omega_1)_{(M\times 1)}\\ \boldsymbol{b}_{\text{NL}}(\boldsymbol{D}^r,\omega_2)_{(M\times 1)}\\ \vdots\\ \boldsymbol{b}_{\text{NL}}(\boldsymbol{D}^r,\omega_Q)_{(M\times 1)}\end{Bmatrix} \tag{15.81}$$

The iteration of Equation 15.80 is continued until a pre-specified convergence criterion is satisfied.

Newton–Raphson Method

Equation 15.48 can be rewritten in the form of

$$\boldsymbol{r}(\boldsymbol{D},\omega)=\boldsymbol{Z}(\boldsymbol{D},\omega)\bar{\boldsymbol{a}}_m(\omega)-\Delta\boldsymbol{a}_m(\omega)=0 \tag{15.82}$$

where $\boldsymbol{r}(\boldsymbol{D},\omega)$ denotes the residual. Let $\boldsymbol{D}^r$ and $\boldsymbol{D}^{r+1}$ be the damage magnitudes computed at the rth and $(r+1)$th iterations, respectively. Applying the Taylor series expansion to $\boldsymbol{r}(\boldsymbol{D}^{r+1},\omega)$ about $\boldsymbol{D}^r$ yields

$$\boldsymbol{r}(\boldsymbol{D}^{r+1},\omega)=\boldsymbol{r}(\boldsymbol{D}^r,\omega)+\boldsymbol{K}(\boldsymbol{D}^r,\omega)\Delta\boldsymbol{D}+\mathrm{O}(\Delta\boldsymbol{D}^2)=0 \tag{15.83}$$

where

$$\Delta\boldsymbol{D}=\boldsymbol{D}^{r+1}-\boldsymbol{D}^r \tag{15.84}$$

and $\boldsymbol{K}(\boldsymbol{D}^r,\omega)$ is the tangent stiffness matrix defined by

$$\boldsymbol{K}(\boldsymbol{D}^r,\omega)=\left.\frac{\partial\boldsymbol{r}(\boldsymbol{D},\omega)}{\partial\boldsymbol{D}}\right|_{\boldsymbol{D}=\boldsymbol{D}^r}=-\boldsymbol{X}(\omega)^{-1}\left.\frac{\partial\boldsymbol{y}(\boldsymbol{D},\omega)}{\partial\boldsymbol{D}}\right|_{\boldsymbol{D}=\boldsymbol{D}^r} \tag{15.85}$$

where

$$\boldsymbol{y}(\boldsymbol{D},\omega)=\boldsymbol{Y}(\boldsymbol{D},\omega)\bar{\boldsymbol{a}}_m(\omega) \tag{15.86}$$

Neglecting the higher order terms from Equation 15.83 gives

$$\boldsymbol{r}(\boldsymbol{D}^r,\omega)_{(M\times 1)}\cong-\boldsymbol{K}(\boldsymbol{D}^r,\omega)_{(M\times N)}\Delta\boldsymbol{D}_{(N\times 1)} \tag{15.87}$$

If Q excitation frequencies are chosen such that $Q\times M=N$, Equation 15.87 provides

$$\bar{\boldsymbol{r}}(\boldsymbol{D}^r)_{(QM\times 1)}\cong-\bar{\boldsymbol{K}}(\boldsymbol{D}^r)_{(QM\times N)}\Delta\boldsymbol{D}_{(N\times 1)} \tag{15.88}$$

where

$$\begin{gathered}\bar{\boldsymbol{r}}(\boldsymbol{D}^r)_{(QM\times 1)}=\begin{Bmatrix}\boldsymbol{r}(\boldsymbol{D}^r,\omega_1)\\ \boldsymbol{r}(\boldsymbol{D}^r,\omega_2)\\ \vdots\\ \boldsymbol{r}(\boldsymbol{D}^r,\omega_Q)\end{Bmatrix}=\begin{Bmatrix}\boldsymbol{Z}(\boldsymbol{D}^r,\omega_1)\bar{\boldsymbol{a}}_m(\omega_1)-\Delta\boldsymbol{a}_m(\omega_1)\\ \boldsymbol{Z}(\boldsymbol{D}^r,\omega_2)\bar{\boldsymbol{a}}_m(\omega_2)-\Delta\boldsymbol{a}_m(\omega_2)\\ \vdots\\ \boldsymbol{Z}(\boldsymbol{D}^r,\omega_Q)\bar{\boldsymbol{a}}_m(\omega_Q)-\Delta\boldsymbol{a}_m(\omega_Q)\end{Bmatrix}_{(QM\times 1)}\\ \bar{\boldsymbol{K}}(\boldsymbol{D}^r)_{(QM\times N)}=\begin{bmatrix}\boldsymbol{K}(\boldsymbol{D}^r,\omega_1)\\ \boldsymbol{K}(\boldsymbol{D}^r,\omega_2)\\ \vdots\\ \boldsymbol{K}(\boldsymbol{D}^r,\omega_Q)\end{bmatrix}_{(QM\times N)}\end{gathered} \tag{15.89}$$

By using Equation 15.84, we can derive the nonlinear damage identification algorithm from Equation 15.88 as

$$\boldsymbol{D}^{r+1}=\boldsymbol{D}^r-\bar{\boldsymbol{K}}(\boldsymbol{D}^r)^{-1}\bar{\boldsymbol{r}}(\boldsymbol{D}^r) \tag{15.90}$$

The iteration of Equation 15.90 is continued until a pre-specified convergence criteria is satisfied or the residual $\boldsymbol{r}(\boldsymbol{D},\omega)$ becomes less than a certain preselected value.

Example 15.1

Figure 15.2 shows a cantilevered uniform beam with three local areas damaged. The beam has the length $L = 0.4$ m, the bending stiffness $EI = 15\,\mathrm{N\,m^2}$, and the mass per length $\rho A = 0.3$ kg/m. Three local damaged areas of uniform magnitudes $D_1 = 0.2$, $D_2 = 0.3$ and $D_3 = 0.4$ are prespecified on the regions $7.0\,\mathrm{cm} \leq x \leq 7.5\,\mathrm{cm}$, $19.5\,\mathrm{cm} \leq x \leq 20\,\mathrm{cm}$, and $32.5\,\mathrm{cm} \leq x \leq 33\,\mathrm{cm}$, respectively. An impulse load is applied at $x = 28$ cm and the vibration responses are analytically calculated for damage identification.

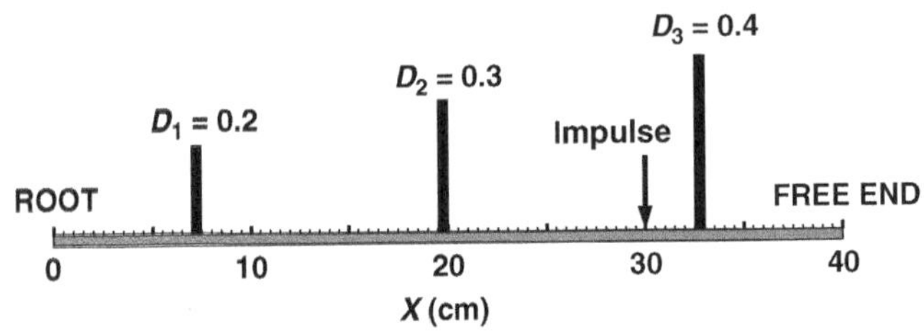

Figure 15.2 Cantilevered uniform beam with three local damaged areas

In Figure 15.3, the vibration response (inertance FRF) at $x = 15$ cm when the beam is at intact state is compared with that at a damaged state. It is shown that the FRF peaks are shifted to slightly lower frequencies due to the existence of the damage. That is, the natural frequencies are slightly reduced because the damage tends to degrade the Young's modulus and the structural stiffness.

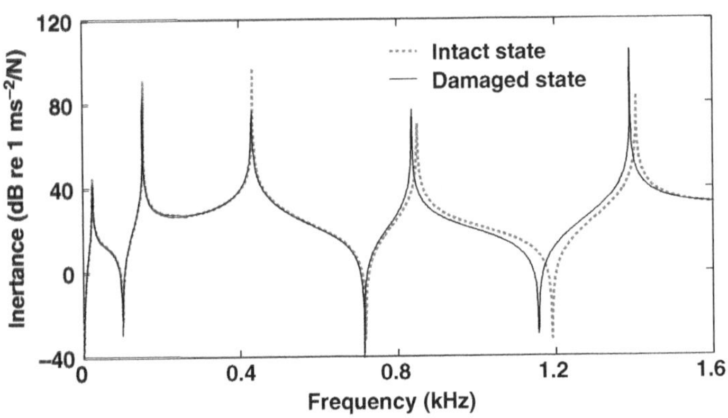

Figure 15.3 Inertance FRFs at the intact and damaged states

For a numerical test of the damage identification, the damaged beam shown in Figure 15.2 is represented by 80 finite elements of equal length. The inertance FRFs are then analytically computed at all nodes, in place of the experimental measurement. By using the analytically computed inertance FRF data at 150 Hz, the damage identification analyses are conducted by using both the linear and nonlinear damage identification algorithms. As expected, Figure 15.4 shows that the nonlinear algorithm certainly provides improved results when compared with the results obtained by using the linear algorithm.

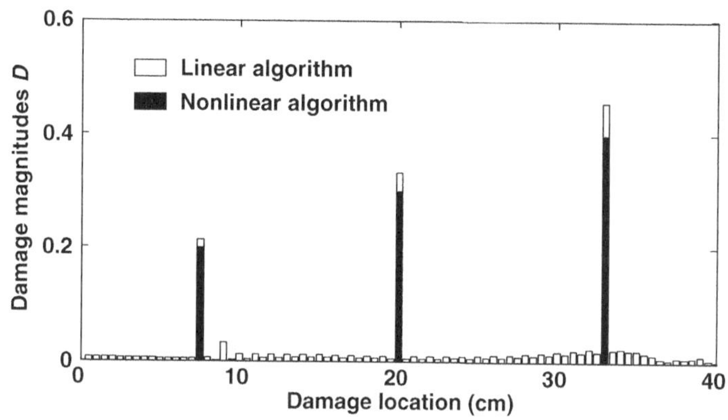

Figure 15.4 Damage identification results from the linear and nonlinear algorithm

15.3 Domain-Reduction Method

There have been continuous efforts to reduce the size of inverse problems such as the structural system identification and the structural damage identification. An approach commonly used in much of the existing literature is to reduce the number of DOFs used to formulate the inverse problems [62, 118]. As an alternative approach, we will introduce the domain-reduction method (DRM) of structural damage identification [83]. In the DRM, a three-step process is applied to reduce the initial spatial domain of damage inspection by iteratively searching out and removing damage-free zones during the damage identification analysis.

15.3.1 Domain-Reduction Method

Assume that, for the damage identification analysis, a damaged structure is divided into a total of N elements and only n elements among them are damaged: that is $n < N$. Thus, $D_{n+1} = D_{n+2} = \ldots = D_N = 0$. Accordingly, the summation in Equation 15.60 can be made only for n damaged elements as follows:

$$\Delta \boldsymbol{S}_{\rm g}(\omega) = -\sum_{k=1}^{N} \boldsymbol{L}_k^{\rm T} \boldsymbol{k}_k \boldsymbol{L}_k D_k = -\sum_{k=1}^{n} \boldsymbol{L}_k^{\rm T} \boldsymbol{k}_k \boldsymbol{L}_k D_k \quad (n < N) \tag{15.91}$$

Substitution of Equation 15.91 into Equation 15.57 gives

$$\begin{aligned} \boldsymbol{Y}_{\rm approx}(\boldsymbol{D}, \omega) &\cong \boldsymbol{Y}_{\rm Linear}(\boldsymbol{D}, \omega) \\ &= -\boldsymbol{T}^{\rm T} \left(\sum_{k=1}^{n} \boldsymbol{L}_k^{\rm T} \boldsymbol{k}_k \boldsymbol{L}_k D_k \right) \boldsymbol{T} = -\sum_{k=1}^{n} (\boldsymbol{T}^{\rm T} \boldsymbol{L}_k^{\rm T} \boldsymbol{k}_k \boldsymbol{L}_k \boldsymbol{T}) D_k \end{aligned} \tag{15.92}$$

Applying Equation 15.92 to Equation 15.49, the left-hand side of Equation 15.48 can be written as

$$\begin{aligned}\text{LHS of Equation 15.48} &= \boldsymbol{Z}(\boldsymbol{D},\omega)\bar{\boldsymbol{a}}_m(\omega)\\ &\cong \boldsymbol{X}(\omega)^{-1}\sum_{k=1}^{n}(\boldsymbol{T}^{\mathrm{T}}\boldsymbol{L}_k^{\mathrm{T}}\boldsymbol{k}_k\boldsymbol{L}_k\boldsymbol{T}\bar{\boldsymbol{a}}_m)\boldsymbol{D}_k\\ &= \boldsymbol{X}(\omega)^{-1}\sum_{k=1}^{n}\boldsymbol{h}_k(\omega)D_k = \hat{\boldsymbol{A}}(\omega)\hat{\boldsymbol{D}}\end{aligned} \tag{15.93}$$

where

$$\hat{\boldsymbol{D}} = \{D_1\, D_2\, D_3 \cdots D_n\}^{\mathrm{T}} \tag{15.94}$$

and

$$\hat{\boldsymbol{A}}(\omega) = \boldsymbol{X}(\omega)^{-1}\hat{\boldsymbol{H}}(\omega) \tag{15.95}$$

using the definitions

$$\hat{\boldsymbol{H}}(\omega) = [\boldsymbol{h}_1\boldsymbol{h}_2\boldsymbol{h}_3 \cdots \boldsymbol{h}_n] \tag{15.96}$$

$$\boldsymbol{h}_k(\omega) = \boldsymbol{T}^{\mathrm{T}}\boldsymbol{L}_k^{\mathrm{T}}\boldsymbol{k}_k\boldsymbol{L}_k\boldsymbol{T}\bar{\boldsymbol{a}}_m \qquad (k = 1, 2, \ldots, n) \tag{15.97}$$

Substituting Equation 15.93 into Equation 15.48 gives

$$\hat{\boldsymbol{A}}(\omega)_{(M\times n)}\hat{\boldsymbol{D}}_{(n\times 1)} = \boldsymbol{b}(\omega)_{(M\times 1)} \tag{15.98}$$

Equation 15.98 represents M linear algebraic equations in n unknown damage magnitudes $\hat{\boldsymbol{D}}$. If $M < n$, we may need to acquire more equations from Equation 15.98 by varying the excitation frequency ω to obtain a square matrix equation as

$$\bar{\hat{\boldsymbol{A}}}(\omega)_{(QM\times n)}\hat{\boldsymbol{D}}_{(n\times 1)} = \boldsymbol{b}(\omega)_{(QM\times 1)} \tag{15.99}$$

where

$$\bar{\hat{\boldsymbol{A}}}(\omega)_{(QM\times n)} = \begin{bmatrix}\hat{\boldsymbol{A}}(\omega_1)_{(M\times n)}\\ \hat{\boldsymbol{A}}(\omega_2)_{(M\times n)}\\ \vdots\\ \hat{\boldsymbol{A}}(\omega_Q)_{(M\times n)}\end{bmatrix} \qquad \bar{\boldsymbol{b}}(\omega)_{(QM\times 1)} = \begin{Bmatrix}\boldsymbol{b}(\omega_1)_{(M\times 1)}\\ \boldsymbol{b}(\omega_2)_{(M\times 1)}\\ \vdots\\ \boldsymbol{b}(\omega_Q)_{(M\times 1)}\end{Bmatrix} \tag{15.100}$$

and where Q is determined such that $(Q \times M) \geq n$. Equation 15.99 shows that, by removing all damage-free elements from the spatial domain under inspection, the size of the inverse problem can be reduced significantly. In theory, such domain reduction does not degrade the accuracy of damage identification. Thus, instead of inspecting the whole initial spatial domain under investigation all of the time to search out local damage (i.e., full-domain method), it will be more efficient to keep reducing the spatial domain, iteration after iteration, by removing definite damage-free zones from the spatial domain inspected at the previous iteration (i.e., domain-reduction method, simply DRM).

15.3.2 Three-Step Process

To begin the DRM, the locations and sizes of some damage-free zones should be known in advance. However, unfortunately this is impossible in most practical situations. Thus, we need a special technique to search out zones that are definitely damage-free during the iteration process of damage identification.

Assume that we start by dividing a damaged structure (the initial spatial domain of the problem) into a finite number of elements. Conducting damage identification analysis by using Equation 15.72, we obtain the first estimation of damage in each element. Next, we increase the number of elements, followed by the damage identification analysis to get an improved second damage estimation. By repeating such damage identification analysis, we may observe that the damage magnitudes estimated in the truly damaged elements become larger and converge to true damage magnitudes as the total number of elements is increased, while the damage magnitudes estimated in the truly intact elements become smaller and converge towards zero. This observation will be natural provided the structure model and the FRF data used for damage identification are sufficiently reliable.

From the above observation, the DRM can be realized by iteratively using a three-step process as illustrated in Figure 15.5. The details are as follows:

a. **The First Step:** (Figure 15.5(a)) Divide the initial spatial domain of damage inspection into p elements and make the first damage estimation for p elements; the results are represented by D_i $(i = 1, 2, \ldots, p)$.
b. **The Second Step:** (Figure 15.5(b)) Divide each element in the first step into q finer elements to produce a total of $(p \times q)$ elements and then make the second damage estimation for $(p \times q)$ elements; the results are represented by D_{ij} $(i = 1, 2 \ldots, p; j = 1, 2, \ldots, q)$.
c. **The Third Step:** (Figure 15.5(c)) Judge whether the jth fine element in the ith element for the first step is damage-free if $D_{ij} < D_i$, while it could possibly be damaged if $D_{ij} > D_i$.

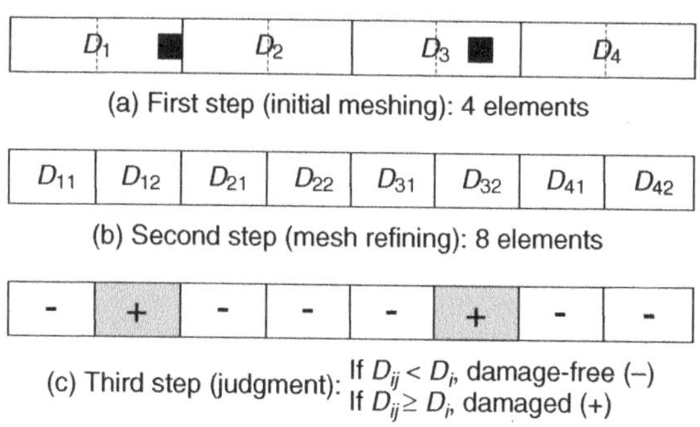

Figure 15.5 Three-step process of the domain-reduction method: in (a), ■ indicates local damage

Following the judgment in the third step, we remove all elements judged to be damage-free from the initial spatial domain of inspection to consider only the remaining elements in the next iteration, which will start again from the first step of the three-step process. By applying the three-step process repeatedly, all damage-free elements can be searched out and removed from the initial spatial domain to eventually leave only damaged elements: this means the identification of multiple local damage. The DRM will be valid when the above three-step (or equivalent) process can be successfully applied. The three-step process may work well when the noise in the FRF data are below a certain level, which can be met by well-prepared measurements.

16

Other Applications

As discussed in Chapter 1, the spectral element method (SEM) has many advantages because it can share the meshing and assembly features with the conventional finite element method (FEM) and the solution exactness feature with the dynamic stiffness method. However, in addition to such advantages, the SEM also has some difficulties in dealing with local discontinuities of geometry or material properties and also in avoiding the FFT errors (e.g., aliasing and leakage errors) to improve the accuracy of time-domain solutions. Because of such difficulties, application of the SEM has been limited mostly to simple one-dimensional (1-D) or 1-D-like problems. Although there are continuing efforts to apply the SEM to plates and shells, most studies have been confined to very specific boundary conditions or geometries. As application of the SEM to extended areas is still in itsinfancy, efforts to overcome associated difficulties are worth undertaking. In the previous chapters, the SEM has been applied to some specific problems: dynamics of beams and plates, pipelines conveying internal fluids, axially moving structures, rotor systems, multi-layered structures, smart structures, lattice structures, blood flow analysis, and the identification of structural boundaries, joints and damage. Thus, this chapter is devoted to a brief summary of other possible SEM applications, which could not be fully covered in this book.

16.1 SEM–FEM Hybrid Method

As the spectral elements and the conventional finite elements are all stiffness formulated, the same assembly technique can be applied in both SEM and FEM. The only the difference is that the conventional finite elements are formulated and assembled in the time domain, while the spectral elements are formulated in the frequency domain and assembled repeatedly at all discrete frequencies.

The spectral element model represented by an exact dynamic stiffness matrix can be formulated when exact frequency-domain wave solutions to the governing differential equations are available in analytical forms. As the spectral element model is formulated by using exact wave solutions, it can represent the mass distribution exactly. Thus it can capture the wave characteristics within a spatial domain of the problem exactly, of course within the accuracy limit of the governing differential equations adopted to formulate the spectral element model. Accordingly, a regular structure member (without discontinuities in geometry or material properties) of any length can be modeled by using a single spectral element. This will significantly reduce the total number of DOFs used in the spectral element analysis. Despite such advantages of the SEM, unfortunately it is not always possible to obtain exact wave solutions, especially for most complex multi-dimensional problems. Thus the development and application of spectral element models have been limited mostly to simple one-dimensional (1-D) or one-dimension-type problems. In contrast to the spectral element formulation, the conventional finite element formulation does not require exact time-domain solutions to the governing differential equations. Instead, simple polynomials are used as the shape (interpolation) functions to formulate finite element models. Thus, at present, it will be easier and more convenient to apply the FEM to more complicated geometries and multi-dimensional problems.

There are many practical problems that consist of multiple 1-D members and multiple complex 2-D or 3-D members. In such problems, it will be very effective to apply the spectral element modeling to 1-D members, and the finite element modeling to complex 2-D or 3-D members: this combination approach is thus called the "SEM–FEM hybrid method." A typical example is the cardiovascular system, which consists of a large number of blood vessels connected at many bifurcations as shown in Figure 16.1. The blood flow characteristics through the bifurcation regions will be more complicated than those through the blood vessels. Thus, for efficient blood flow analysis, one may apply the 1-D spectral element modeling (see Chapter 13) to most of the blood vessels, while the 2-D or 3-D finite element modeling to the bifurcation regions. The SEM–FEM hybrid method can be also applied to the multiple waveguides connected by arbitrarily shaped joints (e.g., pipeline networks and large space

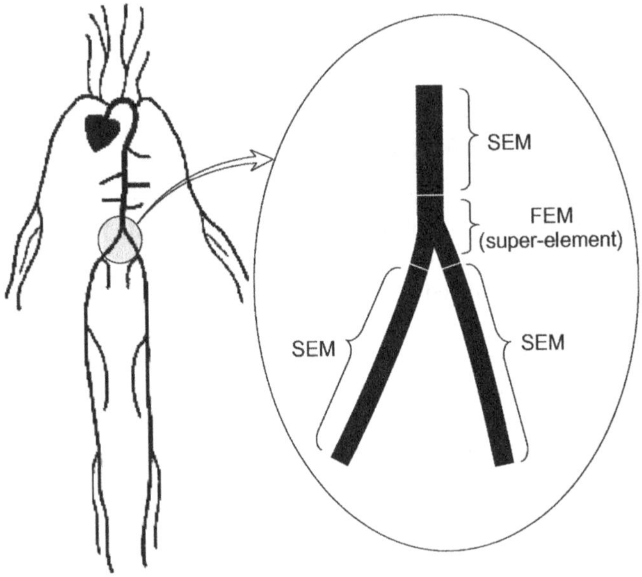

Figure 16.1 Application of the SEM–FEM hybrid method to the cardiovascular system

lattice structures) and to the structures containing local non-uniformities such as holes and cracks [40].

For the SEM–FEM hybrid method, the finite element models formulated in the time domain must be transformed into the frequency domain by using the discrete Fourier transform (DFT) theory. The finite element equation for a specific problem with viscous damping can be derived in a standard form as

$$\boldsymbol{M}\ddot{\bar{\boldsymbol{d}}}(t) + \boldsymbol{C}\dot{\bar{\boldsymbol{d}}}(t) + \boldsymbol{K}\bar{\boldsymbol{d}}(t) = \bar{\boldsymbol{f}}(t) \tag{16.1}$$

where $\bar{\boldsymbol{d}}(t)$ is the nodal DOFs vector, $\bar{\boldsymbol{f}}(t)$ is the nodal forces vector, and $\boldsymbol{M}$, $\boldsymbol{C}$, and $\boldsymbol{K}$ are the finite element mass, damping, and stiffness matrices, respectively. Throughout this section, the over bars (–) are used to denote the quantities derived from the finite element formulation.

Based on the DFT theory, $\bar{\boldsymbol{d}}(t)$ and $\bar{\boldsymbol{f}}(t)$ are expressed in the spectral forms as

$$\bar{\boldsymbol{d}}(t) = \frac{1}{N}\sum_{n=0}^{N-1} \bar{\boldsymbol{d}}_n e^{i\omega_n t}, \qquad \bar{\boldsymbol{f}}(t) = \frac{1}{N}\sum_{n=0}^{N-1} \bar{\boldsymbol{f}}_n e^{i\omega_n t} \tag{16.2}$$

By substituting Equation 16.2 into Equation 16.1 and omitting the subscripts n for brevity, we can obtain

$$\bar{\boldsymbol{S}}(\omega)\bar{\boldsymbol{d}}(\omega) = \bar{\boldsymbol{f}}(\omega) \tag{16.3}$$

where

$$\bar{\boldsymbol{S}}(\omega) = \boldsymbol{K} + i\omega\boldsymbol{C} - \omega^2\boldsymbol{M} \tag{16.4}$$

Thus, Equation 16.3 shows that the time-domain finite element Equation 16.1 can be transformed into the standard form of the frequency-domain spectral element equation

$$\boldsymbol{S}(\omega)\boldsymbol{d}(\omega) = \boldsymbol{f}(\omega) \tag{16.5}$$

where $\boldsymbol{d}(\omega)$ are the spectral nodal DOFs, $\boldsymbol{f}(\omega)$ are the spectral nodal forces, and $\boldsymbol{S}(\omega)$ is the spectral element matrix (exact dynamic stiffness matrix).

As an illustrative example, Figure 16.2(a) shows a 2-D structural joint connected to three 1-D beam members. The structural joint region is meshed with the 2-D finite elements and each 2-D finite element can be represented by the frequency-domain finite element in Equation 16.3. By assembling all 2-D finite elements in the structural joint region, we can obtain the global system equation in the form of

$$\bar{\boldsymbol{S}}_G(\omega)\bar{\boldsymbol{d}}_G(\omega) = \bar{\boldsymbol{f}}_G(\omega) \tag{16.6}$$

where the subscripts G denote the global quantities obtained by assembling all finite elements. Equation 16.6 can provide detailed local dynamic behaviors within the structural joint region. We can write Equation 16.6 in the partitioned form as

$$\begin{bmatrix} \bar{\boldsymbol{S}}_{Gmm} & \bar{\boldsymbol{S}}_{Gmi} \\ \bar{\boldsymbol{S}}_{Gim} & \bar{\boldsymbol{S}}_{Gii} \end{bmatrix} \begin{Bmatrix} \bar{\boldsymbol{d}}_{Gm} \\ \bar{\boldsymbol{d}}_{Gi} \end{Bmatrix} = \begin{Bmatrix} \bar{\boldsymbol{f}}_{Gm} \\ \boldsymbol{0} \end{Bmatrix} \tag{16.7}$$

where $\bar{\boldsymbol{d}}_{Gm}$ are the spectral nodal DOFs defined at the nodes located on the interfaces between the structural joint region and the 1-D beam members, $\bar{\boldsymbol{d}}_{Gi}$ are the spectral nodal DOFs defined

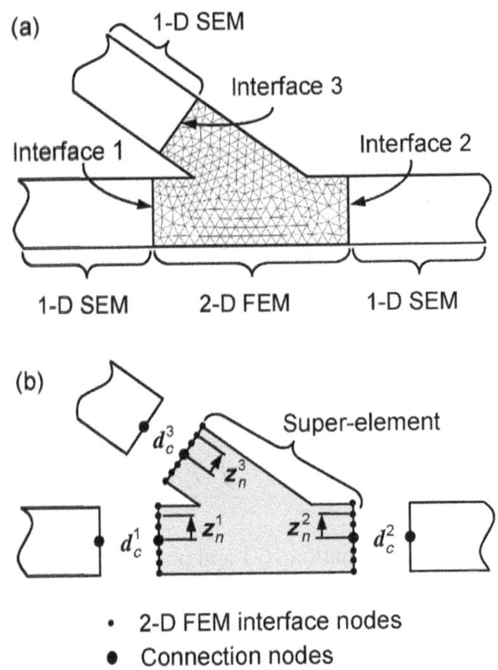

Figure 16.2 A structural joint region represented by a 2-D spectral super-element and three 1-D spectral beam elements

at the internal nodes, and $\bar{f}_{Gm}$ are the spectral nodal forces associated with $\bar{d}_{Gm}$: they are all defined in the global coordinates.

Applying the condensation procedure to Equation 16.7 without any approximation, $\bar{d}_{Gm}$ can be related to $\bar{f}_{Gm}$ as follows:

$$\bar{S}_{\text{super}}\,\bar{d}_{Gm} = \bar{f}_{Gm} \tag{16.8}$$

where

$$\bar{S}_{\text{super}} = \bar{S}_{Gmm} - \bar{S}_{Gmi}\bar{S}_{Gii}^{-1}\bar{S}_{Gim} \tag{16.9}$$

Equation 16.8 implies that the 2-D structural joint region can be regarded as a single equivalent spectral element represented in terms of the spectral nodal DOFs $\bar{d}_{Gm}$ and the spectral nodal forces $\bar{f}_{Gm}$. The equivalent spectral element is called the "spectral super-element" in the literature [40].

The remaining procedure involves assembling the spectral super-element along with its neighboring three-beam members, all represented by the 1-D spectral beam elements. As shown in Figure 16.2(b), the spectral super-element has multiple nodes across each interface, while the 1-D spectral beam element has a single node on the corresponding interface. Thus, for assembly, it is necessary to reduce the spectral nodal DOFs across each interface of the spectral super-element so as to make the spectral DOFs calculated at the center of each interface (termed the connection node) equal to the values at the same point on the interface of the corresponding spectral beam element side. One of the easy-to-use approximate approaches is as follows.

Assume that the kth interface of the 2-D spectral super-element has N_k nodes and each node has two (in-plane) nodal displacements $\hat{\bar{\boldsymbol{d}}}_n^k = \{\hat{\bar{U}}_n^k \ \ \hat{\bar{W}}_n^k\}^{\mathrm{T}}$ ($n = 1,2, \ldots, N_k$; $k = 1, 2, 3$), where the super hats ($\hat{}$) denote the quantities with respect to the local coordinates. The nodal displacements $\hat{\bar{\boldsymbol{d}}}_n^k$ in the local coordinates can be related to the values in the global coordinates by

$$\bar{\boldsymbol{d}}_n^k = \boldsymbol{T}^k \hat{\bar{\boldsymbol{d}}}_n^k \tag{16.10}$$

where $\bar{\boldsymbol{d}}_n^k = \{\bar{U}_n^k \ \ \bar{W}_n^k\}^{\mathrm{T}}$ are the nodal displacements in the global coordinates and $\boldsymbol{T}^k$ is the local-to-global coordinates transformation matrix.

It is worth mentioning that the present approach can be readily extended to more complicated situations. Thus, to simplify the following discussion, we assume that the 1-D beam members of the illustrative example are the axially deformable Bernoulli–Euler beams. Accordingly, we define four spectral DOFs at the connection node on the kth interface of the spectral super-element side and assume that they are identical with those defined at the connection node of the kth spectral beam element side as

$$\begin{aligned} \hat{\boldsymbol{d}}_C^k &= \{\hat{U}_C^k \ \ \hat{W}_C^k \ \ \hat{\Theta}_C^k\}^{\mathrm{T}} \quad \text{(in local coordinates)} \\ \boldsymbol{d}_C^k &= \{U_C^k \ \ W_C^k \ \ \Theta_C^k\}^{\mathrm{T}} \quad \text{(in global coordinates)} \end{aligned} \tag{16.11}$$

The spectral DOFs vector $\hat{\boldsymbol{d}}_C^k$ can be related to the spectral DOFs vector $\boldsymbol{d}_C^k$ by introducing a coordinates transformation matrix $\boldsymbol{T}^k$ as follows:

$$\hat{\boldsymbol{d}}_C^k = \boldsymbol{T}^k \boldsymbol{d}_C^k \tag{16.12}$$

The nodal displacements $(\hat{\bar{U}}_n^k, \hat{\bar{W}}_n^k)$ can be related to the spectral DOFs $(\hat{U}_C^k, \hat{W}_C^k, \hat{\Theta}_C^k)$ through the kinematic relationships as

$$\hat{\bar{U}}_n^k \cong \hat{U}_C^k - z_n^k \hat{\Theta}_C^k, \quad \hat{\bar{W}}_n^k \cong \hat{W}_C^k \tag{16.13}$$

or

$$\hat{\bar{\boldsymbol{d}}}_n^k = \boldsymbol{R}_n^k(z_n^k) \hat{\boldsymbol{d}}_C^k \tag{16.14}$$

where z_n^k is the distance of the nth node from the kth connection node. By using Equation 16.12, Equation 16.14 can be written as

$$\hat{\bar{\boldsymbol{d}}}_n^k = \boldsymbol{R}_n^k \boldsymbol{T}^k \boldsymbol{d}_C^k \tag{16.15}$$

Similarly, by using Equation 16.15, Equation 16.10 can be written as

$$\bar{\boldsymbol{d}}_n^k = \boldsymbol{Q}_n^k \boldsymbol{d}_C^k \tag{16.16}$$

where

$$\boldsymbol{Q}_n^k = \boldsymbol{T}^k \boldsymbol{R}_n^k(z_n^k) \boldsymbol{T}^k \tag{16.17}$$

The spectral nodal DOFs $\bar{\boldsymbol{d}}_{Gm}$ of Equation 16.8 can be written as

$$\bar{\boldsymbol{d}}_{Gm} = \begin{Bmatrix} \bar{\boldsymbol{d}}_{Gm,1} \\ \bar{\boldsymbol{d}}_{Gm,2} \\ \bar{\boldsymbol{d}}_{Gm,3} \end{Bmatrix} \tag{16.18}$$

where

$$\bar{\boldsymbol{d}}_{Gm,k} = \begin{Bmatrix} \bar{\boldsymbol{d}}_1^k \\ \bar{\boldsymbol{d}}_2^k \\ \vdots \\ \bar{\boldsymbol{d}}_{N_k}^k \end{Bmatrix} \quad (k = 1,\ 2,\ 3) \tag{16.19}$$

Substitution of Equation 16.16 into Equation 16.19 gives

$$\bar{\boldsymbol{d}}_{Gm,k} = \boldsymbol{Q}^k \boldsymbol{d}_C^k \quad (k = 1,\ 2,\ 3) \tag{16.20}$$

where

$$\boldsymbol{Q}^k = \begin{bmatrix} \boldsymbol{Q}_1^k \\ \boldsymbol{Q}_2^k \\ \vdots \\ \boldsymbol{Q}_{N_k}^k \end{bmatrix} \quad (k = 1,\ 2,\ 3) \tag{16.21}$$

By using Equation 16.20, Equation 16.18 can be then expressed as

$$\bar{\boldsymbol{d}}_{Gm} = \boldsymbol{Q}\boldsymbol{d}_C \tag{16.22}$$

where

$$\boldsymbol{Q} = \begin{bmatrix} \boldsymbol{Q}^1 & \boldsymbol{0} & \boldsymbol{0} \\ \boldsymbol{0} & \boldsymbol{Q}^2 & \boldsymbol{0} \\ \boldsymbol{0} & \boldsymbol{0} & \boldsymbol{Q}^3 \end{bmatrix}, \qquad \boldsymbol{d}_C = \begin{Bmatrix} \boldsymbol{d}_C^1 \\ \boldsymbol{d}_C^2 \\ \boldsymbol{d}_C^3 \end{Bmatrix} \tag{16.23}$$

By substituting Equation 16.22 into Equation 16.8 and then by multiplying both sides by $\boldsymbol{Q}^{\mathrm{T}}$, we obtain

$$\boldsymbol{S}_{\text{super}} \boldsymbol{d}_C = \boldsymbol{f}_C \tag{16.24}$$

where

$$\boldsymbol{S}_{\text{super}} = \boldsymbol{Q}^{\mathrm{T}} \bar{\boldsymbol{S}}_{\text{super}} \boldsymbol{Q}, \quad \boldsymbol{f}_C = \boldsymbol{Q}^{\mathrm{T}} \bar{\boldsymbol{f}}_{Gc} \tag{16.25}$$

As the final step, we assemble the spectral super-element represented by Equation 16.24 along with its neighboring three spectral beam elements. The remaining solution procedure is the same as has been discussed in previous chapters.

16.2 Identification of Impact Forces

Severe structural damage can be caused by various unidentified impact forces. Thus, identifying impact forces from structural response signals in an efficient and reliable way is necessary as part of structural health monitoring when it is impossible to install sensors on the correct force transfer paths in order to measure the impact forces directly.

The identification of an impact force can be divided into two major classes: the force-history identification and the force-location identification. For the force-history identification, the force-location is considered to be known in advance. A typical example is determining the impact of aircraft on landing. On the other hand, for the force-location identification, both the force-history and force-location are considered to be unknown. Thus, some form of iterative search to determine the force-location should be used in conjunction with identifying the force-history. A typical example is estimating the epicenter of an earthquake.

A lot of impact force identification methods in both the time- and frequency-domains have been proposed and intensive reviews of the related literature can be found in the papers by Inoue *et al.* [60] and Stevens [151]. In the following, we will consider an SEM-based frequency-domain method of impact force identification.

16.2.1 Force-History Identification

The dynamic response of a linear structural system measured at a point $\boldsymbol{x} = (x, y, z)$ can be related to an impact force applied at a point $\boldsymbol{x}_0 = (x_0, y_0, z_0)$ in the convolution integral form as

$$w(\boldsymbol{x}, t) = \int_0^t g(\boldsymbol{x}|\boldsymbol{x}_0, t-\tau) f(\boldsymbol{x}_0, \tau) \mathrm{d}\tau \tag{16.26}$$

where $w(\boldsymbol{x}, t)$ is the dynamic response measured at $\boldsymbol{x}$, $f(\boldsymbol{x}, t)$ and is the impact force applied at $\boldsymbol{x}_0$, and $g(\boldsymbol{x}|\boldsymbol{x}_0, t)$ is the structural transfer function (Green's function). The dynamic response $w(\boldsymbol{x}, t)$ can be in any form of displacement, velocity, acceleration, strain, and so on. By applying a discrete Fourier transform to both sides of Equation 16.26, we obtain the frequency-domain expression as

$$W(\boldsymbol{x}, \omega) = G(\boldsymbol{x}|\boldsymbol{x}_0, \omega) F(\boldsymbol{x}_0, \omega) \tag{16.27}$$

where $W(\boldsymbol{x}, \omega)$, $G(\boldsymbol{x}|\boldsymbol{x}_0, \omega)$ and $F(\boldsymbol{x}, \omega)$ are the spectral components of $w(\boldsymbol{x}, t)$, $g(\boldsymbol{x}|\boldsymbol{x}_0, t)$, and $f(\boldsymbol{x}, t)$, respectively. Equation 16.27 shows that the time-domain convolution of $g(\boldsymbol{x}|\boldsymbol{x}_0, t)$ with $f(\boldsymbol{x}, t)$ is completely deconvolved in the frequency domain, which makes the frequency-domain force identification method very effective.

If dynamic responses to n impact forces are measured at m locations, Equation 16.27 can be written in the matrix form as

$$\boldsymbol{d}_{m\times 1}(\omega) = \boldsymbol{G}_{m\times n}(\omega) \boldsymbol{f}_{n\times 1}(\omega) \tag{16.28}$$

where

$$\begin{aligned} \boldsymbol{d}_{m\times 1}(\omega) &= \{W(\boldsymbol{x}_1, \omega) \;\; W(\boldsymbol{x}_2, \omega) \cdots W(\boldsymbol{x}_m, \omega)\}^{\mathrm{T}} \\ \boldsymbol{f}_{n\times 1}(\omega) &= \{F(\boldsymbol{x}_{01}, \omega) \;\; F(\boldsymbol{x}_{02}, \omega) \cdots F(\boldsymbol{x}_{0n}, \omega)\}^{\mathrm{T}} \end{aligned} \tag{16.29}$$

and

$$\boldsymbol{G}_{m\times n}(\omega) = \begin{bmatrix} G_{11}(\omega) & G_{12}(\omega) & \cdots & G_{1n}(\omega) \\ G_{21}(\omega) & G_{22}(\omega) & \cdots & G_{2n}(\omega) \\ \vdots & \vdots & \ddots & \vdots \\ G_{m1}(\omega) & G_{m2}(\omega) & \cdots & G_{mn}(\omega) \end{bmatrix} \tag{16.30}$$

where $G_{ij}(\omega) = G(\boldsymbol{x}_i | \boldsymbol{x}_{0j}, \omega)$ are the frequency response functions (FRFs) and $\boldsymbol{G}(\omega)$ is the FRF matrix, which is also known as the mobility matrix.

If the dynamic responses $\boldsymbol{d}(\omega)$ and the FRF matrix $\boldsymbol{G}(\omega)$ are known, the desired impact forces $\boldsymbol{f}(\omega)$ can be computed from the following, when $m = n$,

$$\boldsymbol{f}(\omega) = \boldsymbol{G}(\omega)^{-1}\boldsymbol{d}(\omega) \tag{16.31}$$

and, when $m > n$,

$$\boldsymbol{f}(\omega) = \boldsymbol{G}(\omega)^{+}\boldsymbol{d}(\omega) \tag{16.32}$$

where $\boldsymbol{G}(\omega)^{+}$ is the pseudo-inverse of the non-square matrix $\boldsymbol{G}(\omega)$ and it is defined by

$$\boldsymbol{G}(\omega)^{+} = (\boldsymbol{G}^{\mathrm{H}}\boldsymbol{G})^{-1}\boldsymbol{G}^{\mathrm{H}} \tag{16.33}$$

where the superscript H denotes the Hermitian transpose. Once $\boldsymbol{f}(\omega)$ is computed from the deconvolution formulation in Equations 16.31 or 16.32 by using the measured dynamic responses $\boldsymbol{d}(\omega)$ and the FRF matrix $\boldsymbol{G}(\omega)$, the force-history of the impact force can be reconstructed in the time domain by using an inverse-FFT (fast Fourier transform).

The FRF matrix $\boldsymbol{G}(\omega)$ for an undamped or lightly damped structure tends to be ill-conditioned near the resonance frequencies, inducing a significant computation error during the matrix inversion process. In such instances, we can implement regulation filters, such as the truncated singular value decomposition filter and the Tikhonov filter in the above least-squares scheme, Equation 16.32, to overcome inversion instability. Further detailed discussion on this issue can be found in Ref. [114]. In addition, random noise can be added to $\boldsymbol{G}(\omega)$ in order to avoid the loss of information near the anti-resonance frequencies where $\boldsymbol{G}(\omega)$ is zero or near to zero [40].

The FRF matrix $\boldsymbol{G}(\omega)$ can be obtained either analytically or experimentally. The advantage of an analytically formulated FRF matrix is obvious: $\boldsymbol{G}(\omega)$ can be readily computed at any location on a structure. This may enhance the effectiveness of the frequency-domain force-history and force-location identification methods.

16.2.2 Force-Location Identification

For the SEM-model based force-location identification, consider two illustrative examples as shown in Figure 16.3. One is a one-dimensional beam element that has the length L and the other one is a two-dimensional plate element that has the lengths L_x and L_y in the x-direction and y-direction, respectively. Nodes are placed at all sensor locations where the dynamic responses will be measured (i.e., the sensor nodes 1 and 2 for the beam and the sensor nodes 1–8 for the plate) and at the impact location (i.e., the internal node i for both beam and plate).

By assembling two spectral beam elements (one has the length x and the other one has the length $L_x - x$) for the complete beam of length L and similarly four spectral plate elements for the complete plate of dimensions $(L_x \times L_y)$, we can obtain the global dynamic stiffness equation in a general form as

$$\widehat{\boldsymbol{S}}(\omega, x)\widehat{\boldsymbol{d}}(\omega) = \widehat{\boldsymbol{f}}(\omega) \tag{16.34}$$

where $\widehat{\boldsymbol{S}}(\omega, \boldsymbol{x})$ is the global dynamic stiffness matrix, $\widehat{\boldsymbol{d}}(\omega)$ is the global spectral nodal DOFs vector, and $\widehat{\boldsymbol{f}}(\omega)$ is the global spectral nodal forces vector. The global dynamic stiffness matrix

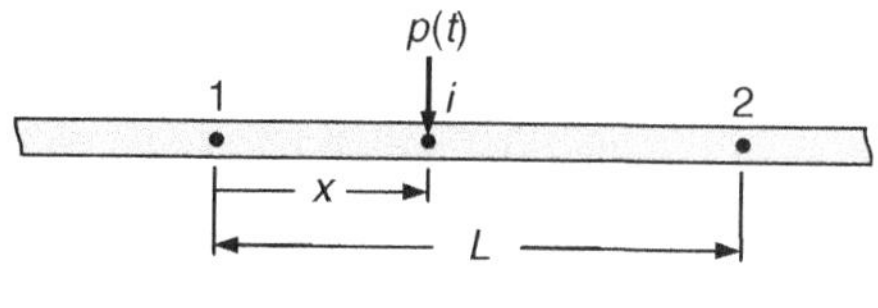

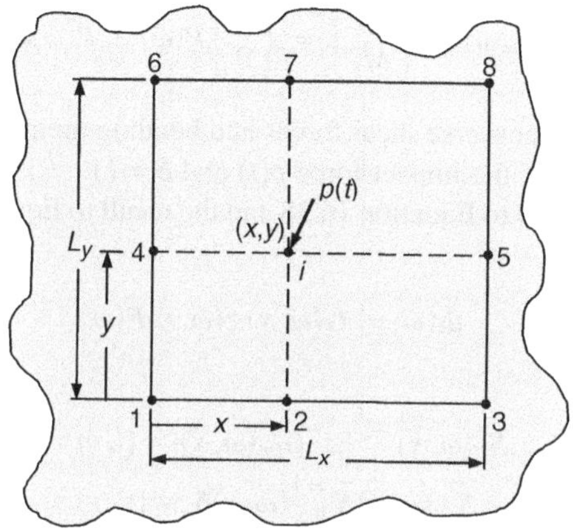

Figure 16.3 Force identifications for 1-D and 2-D structures

$\widehat{\boldsymbol{S}}(\omega, \boldsymbol{x})$ is the function of impact location: that is, $\boldsymbol{x} = x$ for the beam and $\boldsymbol{x} = (x, y)$ for the plate. The matrix in Equation 16.34 can be written in a partitioned form as

$$\begin{bmatrix} \widehat{\boldsymbol{S}}_{ss}(\omega, \boldsymbol{x}) & \widehat{\boldsymbol{S}}_{si}(\omega, \mathbf{x}) \\ \widehat{\boldsymbol{S}}_{is}(\omega, \boldsymbol{x}) & \widehat{\boldsymbol{S}}_{ii}(\omega, \mathbf{x}) \end{bmatrix} \begin{Bmatrix} \widehat{\boldsymbol{d}}_s \\ \widehat{\boldsymbol{d}}_i \end{Bmatrix} = \begin{Bmatrix} \widehat{\boldsymbol{f}}_s(\omega) \\ \widehat{\boldsymbol{f}}_i(\omega) \end{Bmatrix} \tag{16.35}$$

where $\widehat{\boldsymbol{d}}_s(\omega)$ are the spectral nodal DOFs specified on the sensor nodes and $\widehat{\boldsymbol{d}}_i(\omega)$ are the spectral nodal DOFs specified on the internal node i. The spectral nodal forces $\widehat{\boldsymbol{f}}_s(\omega)$ and $\widehat{\boldsymbol{f}}_i(\omega)$ are associated with $\widehat{\boldsymbol{d}}_s(\omega)$ and $\widehat{\boldsymbol{d}}_i(\omega)$, respectively.

We set $\widehat{\boldsymbol{f}}_s(\omega) = 0$ because no external forces will be applied at the sensor nodes. The internal spectral nodal DOFs $\widehat{\boldsymbol{d}}_i(\omega)$ can be then condensed out from Equation 16.35 to obtain an expression as

$$\widehat{\boldsymbol{S}}^*(\omega, \boldsymbol{x})\widehat{\boldsymbol{d}}_s(\omega) = \widehat{\boldsymbol{f}}^*(\omega, \boldsymbol{x}) \tag{16.36}$$

where

$$\widehat{\boldsymbol{S}}^*(\omega, \boldsymbol{x}) = \widehat{\boldsymbol{S}}_{ss}(\omega, \boldsymbol{x}) - \widehat{\boldsymbol{S}}_{si}(\omega, \boldsymbol{x})\widehat{\boldsymbol{S}}_{ii}^{-1}(\omega, \boldsymbol{x})\widehat{\boldsymbol{S}}_{is}(\omega, \boldsymbol{x}) \tag{16.37}$$

and

$$\widehat{\boldsymbol{f}}^*(\omega, \boldsymbol{x}) = -\widehat{\boldsymbol{S}}_{si}(\omega, \boldsymbol{x})\widehat{\boldsymbol{S}}_{ii}^{-1}(\omega, \boldsymbol{x})\widehat{\boldsymbol{f}}_i(\omega) \tag{16.38}$$

The matrix $\widehat{\boldsymbol{S}}^*(\omega, \boldsymbol{x})$ is the condensed dynamic stiffness matrix. Equation 16.38 shows that the impact force $\widehat{\boldsymbol{f}}_i(\omega)$ applied on the internal node i can be replaced by equivalent nodal forces $\widehat{\boldsymbol{f}}^*(\omega, \boldsymbol{x})$ placed on the nodes surrounding the internal node i.

To make the discussion simple, we will consider the beam shown in Figure 16.3(a) as an illustrative example. Assume that an impact force is applied on the internal node i in the transverse direction only. Then we can write $\widehat{\boldsymbol{f}}_i(\omega)$ as

$$\widehat{\boldsymbol{f}}_i(\omega) = \begin{Bmatrix} Q_i \\ M_i \end{Bmatrix} = \begin{Bmatrix} 1 \\ 0 \end{Bmatrix} P(\omega) \equiv P(\omega)\boldsymbol{\delta} \tag{16.39}$$

where Q_i and M_i are the transverse shear forces and bending moments, respectively. $P(\omega)$ are the spectral components of the impact force $p(t)$ and $\boldsymbol{\delta} = \{1 \quad 0\}^{\mathrm{T}}$.

Applying Equation 16.39 to Equation 16.38 and the result to Equation 16.36, we can obtain $\widehat{\boldsymbol{d}}_s(\omega)$ as

$$\widehat{\boldsymbol{d}}_s(\omega) = \boldsymbol{G}(\omega, x)\, \boldsymbol{z}(\omega, x) P(\omega) \tag{16.40}$$

where

$$\begin{aligned} \boldsymbol{G}(\omega, x) &= \widehat{\boldsymbol{S}}^*(\omega, x)^{-1} \equiv [G_{lm}(\omega, x)] \quad (l, m = 1,\ 2,\ 3,\ 4) \\ \boldsymbol{z}(\omega, x) &= -\widehat{\boldsymbol{S}}_{si}(\omega, x)\widehat{\boldsymbol{S}}_{ii}^{-1}(\omega, x)\boldsymbol{\delta} \equiv \{z_1\ z_2\ z_3\ z_4\}^{\mathrm{T}} \end{aligned} \tag{16.41}$$

Equation 16.40 can be written as

$$\widehat{d}_{sl}(\omega) = \sum_{m=1}^{4} G_{lm}(\omega, x)\, z_m(\omega, x) P(\omega) \tag{16.42}$$

Thus, from Equation 16.42, we can obtain

$$P(\omega) = \frac{\widehat{d}_{sl}(\omega)}{\sum_{m=1}^{4} G_{lm}(\omega, x)\, z_m(\omega, x)} \tag{16.43}$$

Once the transfer functions $G_{lm}(\omega, x)$ ($m = 1, 2, 3, 4$) are provided in analytical forms and the dynamic response $\widehat{d}_{sl}(\omega)$ is measured from the lth sensor node, the impact force $P(\omega)$ at a guessed impact location x can be computed from Equation 16.43. The force-history can be then reconstructed from $P(\omega)$ by using the inverse-FFT. A good guess of the force-location will provide matching impact forces from each dynamic response $\widehat{d}_{sl}(\omega)$ measured at the sensor nodes, while a poor guess will provide dissimilar impact forces. Thus the amount of agreement between the estimated impact forces in the time- or frequency-domain can be used as a measure of how accurate a guessed force-location is. Based on this concept, the correlation function between the estimated impact forces from each sensor pair can be used to iteratively search for an accurate force-location [116].

Although the theory from Equations 16.39–16.43 is described for the case of the example beam shown in Figure 16.3(a), we can readily generalize it for the application to other more complex structures, including the example plate as shown in Figure 16.3(b).

16.3 Other Applications

Some other spectral element modeling and applications, which could not be discussed fully in the previous chapters, are listed in the following. Although not all of the related literature can be cited herein due to a lack of information, the following brief summary will be useful for readers to understand the areas where the SEM has been, or can be, applied.

- **Curved beams:** Issa [61]
- **Twisted beams:** Banerjee [15]
- **Helical springs:** Lee and Thompson [78], Leung [107], Pearson and Wittrick [136]
- **Beams of non-uniform cross-section:** Banerjee and Williams [17], Eisenberger [43], Gopalakrishinan and Doyle [51], Huang and Liu [56]
- **Layered solids:** Al-Khoury *et al.* [5], Doyle [39]
- **Sandwich structures:** Benerjee [14], Howson and Zare [55], Kim and Lee [66], Lee and Kim [87, 88], Wang and Wereley [158]
- **Plates (isotropic, orthotropic, composite, honeycomb):** Berçin [22], Berçin and Langley [23], Casimir *et al.* [27], Kim *et al.* [68], Lee and Lee [92], Leung and Zhou [110], Massimo and Panos [119]
- **Thin-walled structures:** Friberg [48], Kim *et al.* [69], Leung and Zeng [109]
- **Aircraft wings (flutter analysis):** Banerjee *et al.* [21]
- **Structures on elastic foundation:** Issa [61], Williams and Kennedy [167]
- **Functionally graded material (FGM) structures:** Chakraborty and Gopalakrishinan [28]
- **Acoustic radiation:** Doyle [41]
- **Power flow analysis:** Ahmida and Arruda [3], Langley [75]
- **Structural intensity analysis:** Oya *et al.* [132]
- **Continuum modeling of periodic lattice structures:** Lee and Lee [90]

References

[1] Abrate, S. (1991) Simple models for periodic lattices. *Proceedings of 1991 ASME Design Engineering Technical Conference, DE-38*, pp. 59–65.

[2] Adams, R.D., Cawley, P., Pye, C.J. and Stone, B.J. (1978) A vibration technique for non-destructively assessing the integrity of structures. *Journal of Mechanical Engineering Science*, **20** (2), 93–100.

[3] Ahmida, K.M. and Arruda, J.R.F. (2001) Spectral element-based prediction of active power flow in timoshenko beams. *International Journal of Solids and Structures*, **38** (10–13), 1669–1679.

[4] Åkesson, B. (1976) PFVIBAT – a computer program for plane frame vibration analysis by an exact method. *International Journal for Numerical Methods in Engineering*, **10** (6), 1221–1231.

[5] Al-Khoury, R., Scarpas, A., Kasbergen, C. and Blaauwendraad, J. (2001) Spectral element technique for efficient parameter identification of layered media. *International Journal of Solids and Structures*, **38** (9), 1605–1623.

[6] Alford, R.M., Kelly, K.R. and Boore, D.M. (1974) Accuracy of finite difference modeling of the acoustic wave equation. *Geophysics*, **39** (6), 834–842.

[7] Araujo dos Santos, J.V., Mota Soares, C.M., Mota Soares, C.A. and Maia, N.M.M. (2005) Structural damage identification in laminate structures using FRF data. *Composite Structures*, **67** (2), 239–249.

[8] Ashley, H. and Haviland, G. (1950) Bending vibrations of a pipe line containing glowing fluids. *Journal of Applied Mechanics*, **72**, 229–232.

[9] Bai, J.M. and Sun, C.T. (1993) A refined theory of flexural vibration for viscoelastic damped sandwich beams. *Proceedings Damping '93, San Francisco, CA*, pp. 319–329.

[10] Banerjee, J.R. (1989) Coupled bending-torsional dynamic stiffness matrix for beam elements. *International Journal for Numerical Methods in Engineering*, **28** (6), 1283–1289.

[11] Banerjee, J.R. (1997) Dynamic stiffness formulation for structural elements: a general approach. *Computers & Structures*, **63** (1), 101–103.

[12] Banerjee, J.R. (1998) Free vibration of axially loaded composite timoshenko beams using the dynamic stiffness matrix method. *Computers & Structures*, **69** (2), 197–208.

[13] Banerjee, J.R. (1999) An Exact Analytical Method of Flutter Analysis Using Symbolic Computation, *AIAA Paper 99-1354*, AIAA.

[14] Banerjee, J.R. (2003) Free vibration of sandwich beams using the dynamic stiffness method. *Computers & Structures*, **81** (18–19), 1915–1922.

[15] Banerjee, J.R. (2004) Development of an exact dynamic stiffness matrix for free vibration analysis of a twisted timoshenko beam. *Journal of Sound and Vibration*, **270** (1–2), 379–401.

[16] Banerjee, J.R. and Fisher, S.A. (1992) Coupled bending-torsional dynamic stiffness matrix for axially loaded beam elements. *International Journal for Numerical Methods in Engineering*, **33** (4), 739–751.

[17] Banerjee, J.R. and Williams, F.W. (1985) Exact Bernoulli-Euler dynamic stiffness matrix for a range of tapered beams. *International Journal for Numerical Methods in Engineering*, **21** (12), 2289–2302.

[18] Banerjee, J.R. and Williams, F.W. (1992) Coupled bending-torsional dynamic stiffness matrix for timoshenko beam elements. *Computers & Structures*, **42** (3), 301–310.

Spectral Element Method in Structural Dynamics Usik Lee

[19] Banerjee, J.R. and Williams, F.W. (1994) An exact dynamic stiffness matrix for coupled extensional-torsional vibration of structural members. *Computers & Structures*, **50** (2), 161–166.

[20] Banerjee, J.R. and Williams, F.W. (1995) Vibration of composite beams–an exact method using symbolic computation. *Journal of Aircraft*, **32** (3), 636–642.

[21] Banerjee, J.R., Su, H. and Jayatunga, C. (2008) A dynamic stiffness element for free vibration analysis of composite beams and its application to aircraft wings. *Computers & Structures*, **86** (6), 573–579.

[22] Berçin, A.N. (1995) Analysis of orthotropic plate structures by the direct-dynamic stiffness method. *Mechanics Research Communications*, **22** (5), 461–466.

[23] Berçin, A.N. and Langley, R.S. (1996) Application of the dynamic stiffness technique to the in-plane vibrations of plate structures. *Computers & Structures*, **59** (5), 869–875.

[24] Blevins, R.D. (1979) *Formulas for Natural Frequency and Mode Shape*, Van Nostrand Reinhold Co, New York.

[25] Cameron, T.M. and Griffin, J.H. (1989) An alternating frequency/time domain method for calculating the steady-state response of nonlinear dynamic systems. *Journal of Applied Mechanics*, **56**, 149–154.

[26] Capron, M.D. and Williams, F.W. (1988) Exact dynamic stiffness for an axially loaded uniform timoshenko member embedded in an elastic medium. *Journal of Sound and Vibration*, **124** (3), 453–466.

[27] Casimir, J.B., Kevorkian, S. and Vinh, T. (2005) The dynamic stiffness matrix of two-dimensional elements: application to Kirchhoff's plate continuous elements. *Journal of Sound and Vibration*, **287** (3), 571–589.

[28] Chakraborty, A. and Gopalakrishinan, S. (2003) A spectrally formulated finite element for wave propagation analysis in functionally graded beams. *International Journal of Solids and Structures*, **40** (10), 2421–2448.

[29] Chandrashekhara, K., Krishnamurthy, K. and Soy, S. (1990) Free vibration of composite beams including rotary inertia and shear deformation. *Composite Structures*, **14** (4), 269–279.

[30] Cheng, F.Y. (1970) Vibration of timoshenko beams and frameworks. *Journal of the Structural Division*, **96** (3), 551–571.

[31] Cheng, F.Y. and Tseng, W.H. (1973) Dynamic matrix of timoshenko beam column. *Journal of the Structural Division*, **99** (3), 527–549.

[32] Cheung, Y.K. (1976) *Finite Strip Method in Structural Analysis*, Pergamon Press, Oxford.

[33] Cho, J. and Lee, U. (2006) An FFT-based spectral analysis method for linear discrete dynamic systems with non-proportional damping. *Shock and Vibration*, **13** (6), 595–606.

[34] Christensen, R.M. (1982) *Theory of Viscoelasticity: An Introduction*, 2nd edn, Academic Press, New York.

[35] Cornwell, J. and Lew, S. (1995) Using transfer function parameter changes for damage detection of structures. *AIAA Journal*, **33** (11), 2189–2193.

[36] Crawley, E.F. and de Luis, J. (1987) Use of piezoelectric actuators as elements of intelligent structures. *AIAA Journal*, **25** (10), 1373–1385.

[37] Doebling, P.S.W. and Farrar, C.R. (1999) Application of the strain energy damage detection method to plate-like structures. *Journal of Sound and Vibration*, **224** (2), 359–374.

[38] Doyle, J.F. (1988) A spectrally formulated finite elements for longitudinal wave propagation. *International Journal of Analytical and Experimental Modal Analysis*, **3**, 1–5.

[39] Doyle, J.F. (1992) A spectral element approach to wave motion in layered solids. *Journal of Vibration and Acoustics*, **114** (4), 569–577.

[40] Doyle, J.F. (1997) *Wave Propagation in Structures: Spectral Analysis Using Fast Discrete Fourier Transforms*, 2nd edn, Springer-Verlag, New York.

[41] Doyle, J.F. (2000) Application of the Spectral Element Method to Acoustic Radiation, NASA/CR-2000-210642.

[42] Ehrich, F.F. (1992) *Handbook of Rotordynamics*, McGraw-Hill, New York.

[43] Eisenberger, M. (1991) Exact solution for general variable cross-section members. *Computers & Structures*, **41** (4), 765–772.

[44] Eisenberger, M., Abramovich, H. and Shulepov, O. (1995) Dynamic stiffness analysis of laminated beams using a first order shear deformation theory. *Composite Structures*, **31** (4), 265–271.

[45] Eshleman, R.L. and Eubanks, R.A. (1969) On the critical speeds of a continuous rotor. *Journal of Engineering for Industry*, **91**, 1180–1188.

[46] Feroz, K.T. and Oyadiji, S.O. (1999) Damage detection in rods using wave propagation and regression analysis. *Proceedings of 1999 ASME Design Engineering Technical Conferences, Las Vegas, NV.*

[47] Friberg, P.O. (1983) Coupled vibration of beams–an exact dynamic element stiffness matrix. *International Journal for Numerical Methods in Engineering*, **19** (4), 479–493.

[48] Friberg, P.O. (1985) Beam element matrices derived from Vlasov's theory of thin-walled elastic beams. *International Journal for Numerical Methods in Engineering*, **21** (7), 1205–1228.

[49] Ginsberg, J.H. (2001) *An Introduction to Random Vibrations, Spectral and Wavelet Analysis*, Addison Wesley Longman Ltd, Essex.

[50] Ginsberg, J.H. (2001) *Mechanical and Structural Vibrations: Theory and Applications*, John Wiley & Sons, Inc., New York.

[51] Gopalakrishinan, S. and Doyle, J.F. (1994) Wave propagation in connected waveguides of varying cross-section. *Journal of Sound and Vibration*, **175** (3), 347–363.

[52] Hallauer, W.L. and Liu, R.Y.L. (1982) Beam bending-torsion dynamic stiffness method for calculation of exact vibration modes. *Journal of Sound and Vibration*, **85** (1), 105–113.

[53] Hess, M.S. (1969) The end problem for a laminated elastic strip – I. The general solution. *Journal of Composite Materials*, **3** (2), 262–280.

[54] Howson, W.P. and Williams, F.W. (1973) Natural frequencies of frames with axially loaded timoshenko members. *Journal of Sound and Vibration*, **26** (4), 503–515.

[55] Howson, W.P. and Zare, A. (2005) Exact dynamic stiffness matrix for flexural vibration of three-layered sandwich beams. I. Forward calculation. *Journal of Sound and Vibration*, **282** (3–5), 753–767.

[56] Huang, K.J. and Liu, T.S. (2001) Dynamic analysis of rotating beams with nonuniform cross sections using the dynamic stiffness method. *Journal of Vibration and Acoustics*, **123** (4), 536–539.

[57] Hughes, T.J.R. and Lubliner, J. (1973) On the one-dimensional theory of blood flow in the larger vessels. *Mathematical Biosciences*, **18** (1–2), 161–170.

[58] Humar, J.L. (2001) *Dynamics of Structures*, A.A. Balkema Publishers, Lisse.

[59] Humar, J.L. and Xai, H. (1993) Dynamic response analysis in the frequency domain. *Earthquake Engineering and Structural Dynamics*, **22**, 1–22.

[60] Inoue, H., Harrigan, J.J. and Reid, S.R. (2001) Review of inverse analysis for indirect measurement of impact force. *Applied Mechanics Review*, **54** (6), 503–524.

[61] Issa, M.S. (1988) Natural frequencies of continuous curved beams on winkler-type foundation. *Journal of Sound and Vibration*, **127** (2), 291–310.

[62] James, G.H. and Zimmerman, D.C. (1997) Reduction/expansion studies for damage identification of continuous aerospace structures. *Proceedings of International Modal Analysis Conference*, pp. 1772–1778.

[63] Kammler, D.W. (2000) *A First Course in Fourier Analysis*, Prentice Hall, New Jersey.

[64] Kantorovich, L.V. and Krylov, V.I. (1958) *Approximate Methods of Higher Analysis*, Interscience Publishers, New York.

[65] Kerwin, E.D. Jr. (1959) Damping of flexural waves by a constrained viscoelastic layer. *Journal of Acoustical Society of America*, **31** (7), 952–962.

[66] Kim, J. and Lee, U. (2001) Modal Analysis of the Passive Constrained Layer Beams Based on Spectrally Formulated Exact Normal Modes, *AIAA Paper 2001-1564*, AIAA.

[67] Kim, J., Cho, J., Lee, U. and Park, S. (2003) Modal spectral element formulation for axially moving plates subjected to in-plane axial tension. *Computers & Structures*, **81** (20), 2011–2020.

[68] Kim, J., Cho, J., Lee, U. and Park, S. (2003) Modal spectral element formulation for axially moving plates subjected to in-plane axial tension. *Computers & Structures*, **81** (20), 2011–2020.

[69] Kim, M.Y., Kim, N.I. and Yoon, H.T. (2003) Exact dynamic and static stiffness matrices of shear deformable thin-walled beam-columns. *Journal of Sound and Vibration*, **267** (1), 29–55.

[70] Kolousek, V. (1941) Anwendung des gesetzes der virtuellen verschiebungen und des reziprozitatssatzes in der stabwerksdynamic. *Ingenieur Archiv*, **12**, 363–370.

[71] Ku, J.P., Draney, M.T., Arko, F.R., Lee, W.A., Chan, F.P., Pelc, N.J., Zarins, C.K. and Taylor, C.A. (2002) *In vivo* validation of numerical prediction of blood flow in arterial bypass grafts. *Annals of Biomedical Engineering*, **30** (6), 743–752.

[72] Kwon, K. and Lee, U. (2006) Spectral element modeling and analysis of an axially moving thermoelastic beam-plate. *Journal of Mechanics of Materials and Structures*, **1** (4), 605–632.

[73] Lam, M.J., Saunders, W.R. and Inman, D.J. (1995) Modeling active constrained layer damping using Golla-Hughes-McTavish approach. *Proceedings of the Conference on Smart Structures and Materials: Passive Damping, San Diego, CA*, pp. 86–97.

[74] Lancaster, P. and Tismenetsky, M. (1985) *The Theory of Matrices*, 2nd edn, Academic Press, London.

[75] Langley, R.S. (1990) Analysis of power flow in beams and frameworks using the direct-dynamic stiffness method. *Journal of Sound and Vibration*, **136** (3), 439–452.

[76] Le-Ngoc, L. and McCallion, H. (1995) Wide bandsaw blade under cutting conditions, part I: vibration of a plate moving in its plane while subjected to tangential edge loading. *Journal of Sound and Vibration*, **186** (1), 125–142.

[77] Le-Ngoc, L. and McCallion, H. (1999) Dynamic stiffness of an axially moving string. *Journal of Sound and Vibration*, **220** (4), 749–756.

[78] Lee, J. and Thompson, D.J. (2001) Dynamic stiffness formulation, free vibration and wave motion of helical springs. *Journal of Sound and Vibration*, **239** (2), 297–320.

[79] Lee, U. (1990) Dynamic continuum modeling of beamlike space structures using finite-element matrices. *AIAA Journal*, **28** (4), 725–731.

[80] Lee, U. (1998) Equivalent continuum representation of lattice beams: spectral element approach. *Engineering Structures*, **20** (7), 587–592.

[81] Lee, U. (2000) Vibration analysis of one-dimensional structures using the spectral transfer matrix method. *Engineering Structures*, **22** (6), 681–690.

[82] Lee, U. (2001) Dynamic characterization of the joints in a beam structure by using spectral element method. *Shock and Vibration*, **8** (6), 357–366.

[83] Lee, U. (2003) A reduced-domain method of structural damage identification: application to a spectral element beam model. *Shock and Vibration*, **10** (5–6), 313–324.

[84] Lee, U. and Cho, J. (2008) FFT-based spectral element analysis for the linear continuum dynamic systems subjected to arbitrary initial conditions by using the pseudo-force method. *International Journal for Numerical Methods in Engineering*, **74** (1), 159–174.

[85] Lee, U. and Jang, I. (2007) On the boundary conditions for axially moving beams. *Journal of Sound and Vibration*, **306** (3–5), 675–690.

[86] Lee, U. and Kim, J. (2000) Determination of non-ideal beam boundary conditions: a spectral element approach. *AIAA Journal*, **38** (2), 309–316.

[87] Lee, U. and Kim, J. (2000) Dynamics of elastic-piezoelectric two-layer beams using spectral element method. *International Journal of Solids and Structures*, **37** (32), 4403–4417.

[88] Lee, U. and Kim, J. (2001) Spectral element modeling for the beams treated with active constraining layer damping. *International Journal of Solids and Structures*, **38** (32–33), 5679–5702.

[89] Lee, U. and Lee, C. (2008) Spectral element modeling for extended timoshenko beams. *Journal of Sound and Vibration*, **319** (3–5), 993–1002.

[90] Lee, U. and Lee, J. (1996) Dynamic continuum modeling of truss-type space structures using spectral elements. *Journal of Spacecraft and Rockets*, **33** (3), 404–409.

[91] Lee, U. and Lee, J. (1998) Vibration analysis of the plates subject to distributed dynamic loads by using spectral element method. *KSME International Journal*, **12** (4), 565–571.

[92] Lee, U. and Lee, J. (1999) Spectral-element method for levy-type plates subject to dynamic loads. *Journal of Engineering Mechanics*, **125** (2), 243–247.

[93] Lee, U. and Oh, H. (2003) The spectral element model for pipelines conveying internal steady fluid. *Engineering Structures*, **25** (8), 1045–1055.

[94] Lee, U. and Oh, H. (2005) Dynamics of an axially moving viscoelastic beam subject to axial tension. *International Journal of Solids and Structures*, **42** (8), 2381–2398.

[95] Lee, U. and Park, J. (2006) Spectral element modeling and analysis of the pipeline conveying internal unsteady fluid. *Journal of Fluids and Structures*, **22** (2), 273–292.

[96] Lee, U. and Shin, J. (2002) A frequency-domain method of structural damage identification formulated from dynamic stiffness equation of motion. *Journal of Sound and Vibration*, **257** (4), 615–634.

[97] Lee, U., Kim, J. and Leung, A.Y.T. (2000) The spectral element method in structural dynamics. *The Shock and Vibration Digest*, **32** (6), 451–465.

[98] Lee, U., Kim, J. and Leung, A.Y.T. (2001) Vibration analysis of the active multi-layer beams by using spectrally formulated exact natural modes. *KSME International Journal*, **15** (2), 199–209.

[99] Lee, U., Kim, J. and Oh, H. (2004) Spectral analysis for the transverse vibration of an axially moving timoshenko beam. *Journal of Sound and Vibration*, **271** (3–5), 685–703.

[100] Lee, U., Kim, J., Shin, J. and Leung, A.Y.T. (2002) Development of a Wittrick-Williams algorithm for the spectral element model of elastic-piezoelectric two-layer active beams. *International Journal of Mechanical Sciences*, **44** (2), 305–318.

[101] Lee, U., Kim, S. and Cho, J. (2005) Dynamic analysis of the linear discrete dynamic systems subjected to the initial conditions by using an FFT-based spectral analysis method. *Journal of Sound and Vibration*, **288** (1–2), 293–306.

[102] Lee, U., Pak, C.H. and Hong, S.C. (1995) The dynamics of a piping system with internal unsteady flow. *Journal of Sound and Vibration*, **180** (2), 297–311.

[103] Lee, U., Seo, B. and Jang, I. (2007) Blood Flow Analysis by Using Spectral Element Method, Dynamics of Continuous. *Discrete and Impulsive Systems-Series B-Applications & Algorithms*, **14** (S4), 180–184.

[104] Lee, U., Seo, B. and Jang, I. (2008) Spectral element modeling and analysis of the blood flow through a human blood vessel. *Journal of Mechanical Science and Technology*, **22** (8), 1612–1619.

[105] Leibowitz, M.M. and Vinson, J.R. (1993) On active (piezoelectric) constrained layer damping in composite sandwich structures. *Proceedings of the 4th International Conference on Adaptive Structures, Cologne, Germany*, pp. 530–541.

[106] Leung, A.Y.T. (1988) Dynamic stiffness analysis of follower force. *Journal of Sound and Vibration*, **126** (3), 533–543.

[107] Leung, A.Y.T. (1991) Exact stiffness matrix for twisted helix beam. *Finite Element in Analysis and Design*, **9** (1), 23–32.

[108] Leung, A.Y.T. (1993) *Dynamic Stiffness and Substructures*, Springer-Verlag, London.

[109] Leung, A.Y.T. and Zeng, S.P. (1994) Analytical formulation of dynamic stiffness. *Journal of Sound and Vibration*, **177** (4), 555–564.

[110] Leung, A.Y.T. and Zhou, W.E. (1996) Dynamic stiffness analysis of laminated composite plates. *Thin-Walled Structures*, **25** (2), 109–133.

[111] Li, J., Shen, R., Hua, H. and Jin, X. (2004) Bending-torsional coupled dynamic response of axially loaded composite timoshenko thin-walled beam with closed cross-section. *Composite Structures*, **64** (1), 23–35.

[112] Liao, W.H. (1997) Active-Passive Hybrid Structural Control: An Enhanced Active Constrained Layer Damping Treatment with Edge Elements, Ph.D. Thesis. Pennsylvania State University.

[113] Lin, C.C. (1997) Stability and vibration characteristics of axially moving plates. *International Journal of Solids and Structures*, **34** (24), 3179–3190.

[114] Liu, Y. and Shepard, W.S. Jr. (2005) Dynamic force identification based on enhanced least squares and total least-squares schemes in the frequency domain. *Journal of Sound and Vibration*, **282** (1–2), 37–60.

[115] Lundén, R. and Åkesson, B. (1983) Damped second-order Rayleigh-Timoshenko beam vibration in space – an exact complex dynamic member stiffness matrix. *International Journal for Numerical Methods in Engineering*, **19** (3), 431–449.

[116] Martin, M.T. and Doyle, J.F. (1996) Impact force identification in frame structures. *International Journal of Impact Engineering*, **18** (1), 79–97.

[117] Martins, C.J., Carrer, J.A.M., Mansur, W.J. and Araújo, F.C. (2006) On the use of pseudo-forces to consider initial conditions in 3D time- and frequency-domain acoustics analysis. *Computer Methods in Applied Mechanics and Engineering*, **195**, 4371–4382.

[118] Marwala, T. and Heyns, P.S. (1998) Multiple-criterion method for determining structural damage. *AIAA Journal*, **36** (8), 1494–1501.

[119] Massimo, R. and Panos, T. (2003) Control of wave propagation in sandwich plate rows with periodic honeycomb core. *Journal of Engineering Mechanics*, **129** (9), 975–986.

[120] McIver, D.B. (1972) Hamilton's principle for systems of changing mass. *Journal of Engineering Mathematics*, **7**, 249–261.

[121] McTavish, D.J. and Hughes, P.C. (1993) Modeling of linear viscoelastic space structures. *Journal of Vibration and Acoustics*, **115** (11), 103–110.

[122] Mead, D.J. (1982) A comparison of some equations for the flexural vibration of damped sandwich beams. *Journal of Sound and Vibration*, **83** (3), 363–377.

[123] Mead, D.J. and Markus, S. (1969) The forced vibration of a three-layer, damped sandwich beam with arbitrary boundary conditions. *Journal of Sound and Vibration*, **10** (2), 163–175.

[124] Narayanan, G.V. and Beskos, D.E. (1978) Use of dynamic influence coefficients in forced vibration problems with the aid of fast fourier transform. *Computers & Structures*, **9** (2), 145–150.

[125] Nayfeh, A.M. and Hefzy, M.S. (1978) Continuum modeling of the mechanical and thermal behavior of discrete large structures. *AIAA Journal*, **19** (6), 766–773.

[126] Nelson, F.C. (2003) A brief history of early rotor dynamics. *Sound and Vibration*, **37** (6), 8–11.

[127] Newland, D.E. (1993) *An Introduction to Random Vibrations, Spectral and Wavelet Analysis*, 3rd edn, Addison Wesley Longman Ltd, Essex.
[128] Nobari, A.S., Robb, D.A. and Ewins, D.J. (1933) Model updating and joint identification methods, restrictions and overlap. *International Journal of Analytical and Experimental Modal Analysis*, **8** (2), 95–103.
[129] Noor, A.K., Anderson, M.S. and Green, W.H. (1978) Continuum models for beam-and platelike lattice structures. *AIAA Journal*, **16** (12), 1219–1228.
[130] Oh, H., Lee, U. and Park, D.H. (2004) Dynamics of an axially moving Bernoulli-Euler beam: spectral element modeling and analysis. *KSME International Journal*, **18** (3), 382–393.
[131] Olufsen, M.S. (1999) Structured tree outflow condition for blood flow in larger systemic arteries. *American Journal of Physiology*, **276** (1), H257–H268.
[132] Oya, T., Suzuki, Y. and Ito, K. (2005) A method for structural intensity measurement on a plate by using spectral element method. *Acoustical Science and Technology*, **26** (3), 301–304.
[133] Païdoussis, M.P. and Issid, N.T. (1974) Dynamic stability of pipes conveying fluid. *Journal of Sound and Vibration*, **33** (3), 267–294.
[134] Pandey, A.K. and Biswas, M. (1995) Damage diagnosis of truss structures by estimation of flexibility change. *International Journal of Analytical and Experimental Modal Analysis*, **10** (2), 104–117.
[135] Patera, A.T. (1984) A spectral element method for fluid dynamics – laminar flow in a channel expansion. *Journal of Computational Physics*, **54**, 468–488.
[136] Pearson, D. and Wittrick, W.H. (1986) An exact solution for the vibration of helical springs using a Bernoulli-Euler model. *International Journal of Mechanical Science*, **28** (2), 83–96.
[137] Pellicano, F. and Vestroni, F. (2001) Non-linear dynamics and bifurcations of an axially moving beam. *Journal of Vibration and Acoustics*, **22**, 21–30.
[138] Pestel, E.C. and Leckie, F.A. (1993) *Matrix Methods in Elastomechanics*, McGraw-Hill, New York.
[139] Petyt, M. (1990) *Introduction to Finite Element Vibration Analysis*, Cambridge University Press, Cambridge.
[140] Pietrabissa, R., Mantero, S., Marotta, T. and Menicanti, S. (1996) A lumped parameter model to evaluate the fluid dynamics of different coronary bypasses. *Medical Engineering & Physics*, **18** (6), 477–484.
[141] Plump, J.M. and Hubbard, J.E. Jr. (1986) Modeling of an active constrained layer damper. *Proceedings of 12th International Congress on Acoustics, Toronto, Canada.*
[142] Przemieniecki, J.S. (1968) *Theory of Matrix Structural Analysis*, McGraw-Hill Inc., New York.
[143] Rao, J.S. (1996) *Rotor Dynamics*, New Age International Ltd, New Delhi.
[144] Reddy, J.N. (1997) *Mechanics of Laminated Composite Plates: Theory and Analysis*, CRC Press, New York.
[145] Ricles, J.M. and Kosmatka, J.B. (1992) Damage detection in elastic structures using vibratory residual forces and weighted sensitivity. *AIAA Journal*, **30** (9), 2310–2316.
[146] Rizzi, S.A. and Doyle, J.F. (1992) A spectral element approach to wave motion in layered solids. *Journal of Vibration and Acoustics*, **114** (4), 569–577.
[147] Shen, I.Y. (1994) Hybrid damping through intelligent constrained layer treatments. *Journal of Vibration and Acoustics*, **116** (3), 341–349.
[148] Spyrakso, C.C. and Beskos, D.E. (1982) Dynamic response of frameworks by fast fourier transform. *Computers & Structures*, **15** (5), 495–505.
[149] Steele, B.N., Wan, J., Ku, J.P., Hughes, T.J.R. and Taylor, C.A. (2003) *In vivo* validation of a one-dimensional finite-element method of predicting blood flow in cardiovascular bypass grafts. *IEEE Transactions on Biomedical Engineering*, **50** (6), 649–656.
[150] Stephenson, R.W. and Rouch, K.E. (1993) Modeling rotating shafts using axisymmetric solid finite elements with matrix reduction. *Journal of Vibration and Acoustics*, **115**, 484–489.
[151] Stevens, K.K. (1987) Force identification problems: an overview. *Proceeding of the SEM Spring Conference on Experimental Mechanics, Houston*, pp. 838–844.
[152] Sun, C.T., Kim, B.J. and Bogdanoff, T.L. (1981) On the derivation of equivalent simple models of beam-like and plate-like structures in dynamic analysis. *AIAA Paper*, **81** (624), 523–532.
[153] Taylor, C.A., Hughes, T.J.R. and Zarins, C.K. (1998) Finite element modeling of three-dimensional pulsatile flow in the abdominal aorta: relevance to atherosclerosis. *Annals of Biomedical Engineering*, **26** (6), 1–14.
[154] Thyagarajan, S.K., Schulz, M.J. and Pai, P.F. (1998) Detecting structural damage using frequency response functions. *Journal of Sound and Vibration*, **210** (1), 162–170.
[155] Tsai, J.S. and Chou, Y.F. (1988) The identification of dynamic characteristics of a single bolt joint. *Journal of Sound and Vibration*, **125** (3), 487–502.
[156] The MathWorks, Inc. (2000) *Using Matlab Version 6*, MathWorks, Inc., Natick, MA.

[157] Wang, F. and Chen, S. (1996) A method to determine the boundary condition of the finite element model of a slender beam using measured modal parameters. *Journal of Vibration and Acoustics*, **118** (3), 474–478.

[158] Wang, G. and Wereley, N.M. (2002) Spectral finite element analysis of sandwich beams with passive constrained layer damping. *Journal of Vibration and Acoustics*, **124** (3), 376–386.

[159] Wang, T.M. and Kinsman, T.A. (1971) Vibration of frame structures according to the timoshenko theory. *Journal of Sound and Vibration*, **14** (2), 215–227.

[160] Wang, Z.R., Lin, M. and Lim, M.K. (1997) Structural damage detection using measured FRF data. *Computer Methods in Applied Mechanics and Engineering*, **147** (1), 187–197.

[161] Weisshaar, T.A. and Foist, B.L. (1985) Vibration tailoring of advanced composite lifting surfaces. *Journal of Aircraft*, **22** (2), 141–147.

[162] Westerhof, N., Bosman, F., De Vries, C.J. and Noordergraaf, A. (1969) Analog studies of the human systemic arterial tree. *Journal of Biomechanics*, **2** (2), 121–143.

[163] White, F.M. (1999) *Fluid Mechanics*, McGraw-Hill, Singapore.

[164] Wickert, J.A. and Mote, C.D. (1988) Current research on the vibration and stability of axially moving materials. *Shock and Vibration Digest*, **20**, 3–13.

[165] Wickert, J.A. and Mote, C.D. (1990) Classical vibration analysis of axially moving continua. *Journal of Applied Mechanics*, **57**, 738–744.

[166] Wiggert, D.C., Hatfield, F.J. and Stuckenbruck, S. (1987) Analysis of liquid and structural transients by the method of characteristics. *Journal of Fluid Engineering*, **109**, 161–165.

[167] Williams, F.W. and Kennedy, D. (1987) Exact dynamic member stiffness for a beam on an elastic foundation. *Earthquake Engineering and Structural Dynamics*, **15** (1), 133–136.

[168] Williams, F.W., Yuan, S., Ye, K., Kennedy, D. and Djoudi, M.S. (2002) Towards deep and simple understanding of the transcendental eigenproblem of structural vibrations. *Journal of Sound and Vibration*, **356** (4), 681–693.

[169] Wittrick, W.H. and Williams, F.W. (1971) A general algorithm for computing natural frequencies of elastic structures. *Quarterly Journal of Mechanics and Applied Mathematics*, **24** (3), 263–284.

[170] Wong, B.S., Guan, T.C. and King, L.M. (1993) Mechanical impedance inspection of composite structures. *British Journal of NDT*, **35** (1), 3–9.

[171] Womersley, J.R. (1957) Oscillatory flow in arteries: the constrained elastic tube as a model of arterial flow and pulse transmission. *Physics in Medicine & Biology*, **2** (2), 178–187.

[172] Yan, M.J. and Dowell, E.H. (1972) Governing equations for vibrating constrained-layer damping sandwich plates and beams. *Journal of Applied Mechanics*, **39**, 1041–1046.

[173] Yuan, J.X. and Wu, X.M. (1985) Identification of joint structural parameters of machine tool by DDS and FEM. *Journal of Engineering for Industry*, **107**, 64–69.

[174] Yuan, S., Ye, K. and Williams, F.W. (2004) Second order mode-finding method in dynamic stiffness matrix methods. *Journal of Sound and Vibration*, **269** (3–5), 689–708.

[175] Yuen, M.M.F. (1985) A numerical study of the eigenparameters of a damaged cantilever. *Journal of Sound and Vibration*, **103** (3), 301–310.

[176] Zu, J.W.Z. and Han, R.P.S. (1992) Natural frequencies and normal modes of a spinning timoshenko beam with general boundary conditions. *Journal of Applied Mechanics*, **59**, S197–S204.

Index

Printed and bound by CPI Group (UK) Ltd, Croydon, CR0 4YY
16/09/2026
14963405-0001